Sustainable Management of Arthropod Pests of Tomato

Sustainable Management of Arthropod Pests of Tomato

Edited by

Waqas Wakil
Department of Entomology
University of Agriculture
Faisalabad, Pakistan

Gerald E. Brust
CMREC-UMF
University of Maryland
Upper Marlboro, Maryland, USA

Thomas M. Perring
Department of Entomology
University of California
Riverside, California, USA

ACADEMIC PRESS
An imprint of Elsevier

Academic Press is an imprint of Elsevier
125 London Wall, London EC2Y 5AS, United Kingdom
525 B Street, Suite 1800, San Diego, CA 92101-4495, United States
50 Hampshire Street, 5th Floor, Cambridge, MA 02139, United States
The Boulevard, Langford Lane, Kidlington, Oxford OX5 1GB, United Kingdom

Notices
Knowledge and best practice in this field are constantly changing. As new research and experience broaden our understanding, changes in research methods, professional practices, or medical treatment may become necessary.

Practitioners and researchers must always rely on their own experience and knowledge in evaluating and using any information, methods, compounds, or experiments described herein. In using such information or methods they should be mindful of their own safety and the safety of others, including parties for whom they have a professional responsibility.

To the fullest extent of the law, neither the Publisher nor the authors, contributors, or editors, assume any liability for any injury and/or damage to persons or property as a matter of products liability, negligence or otherwise, or from any use or operation of any methods, products, instructions, or ideas contained in the material herein.

Library of Congress Cataloging-in-Publication Data
A catalog record for this book is available from the Library of Congress

British Library Cataloguing-in-Publication Data
A catalogue record for this book is available from the British Library

ISBN: 978-0-12-802441-6

For information on all Academic Press publications visit our website at
https://www.elsevier.com/books-and-journals

Working together
to grow libraries in
developing countries

www.elsevier.com • www.bookaid.org

Publisher: Andre Gerhard Wolff
Acquisition Editor: Nancy Maragioglio
Editorial Project Manager: Billie Jean Fernandez
Production Project Manager: Punithavathy Govindaradjane
Designer: Matthew Limbert

Typeset by TNQ Books and Journals

This narrative was written by Dr. Marshall W. Johnson, one of Dr. Oatman's Ph.D. students.

We dedicate this work to the memory and contributions of Earl R. Oatman (1920–2015), Emeritus Professor of Entomology, University of California, Riverside (UCR), United States. Professor Oatman grew up in the agricultural setting of rural, eastern Oklahoma. He joined the United States Army prior to America's involvement in WWII. Following the Japanese invasion of the Philippines, he was taken prisoner in April 1942 at the age of 22. He experienced the infamous Bataan Death March from which he escaped and evaded the Japanese for about 1 year. He was later retaken prisoner and held for 3 years, eventually being sent to Japan where he did slave labor in a zinc mine. He was freed following the destruction of Hiroshima and Nagasaki.

Following the war's end, Professor Oatman earned B.S. and M.S. degrees from the University of Missouri at Columbia. He continued his education at the University of California, Berkeley, where he obtained a Ph.D. in Entomology in 1956, studying under Professor A. E. Michelbacher, an early proponent of the "integrated control" concept. Afterward, he joined the Entomology faculty of the University of Wisconsin, Madison, where his research focused on integrated control of arthropod pests of deciduous fruit trees. In 1962, he was hired into the Division of Biological Control of UCR's Department of Entomology, where his primary responsibilities were research and instruction (undergraduate and graduate).

Professor Oatman's research focused on the biological control and integrated pest management (IPM) of arthropods attacking vegetable crops (e.g., tomatoes, broccoli, and potatoes) and small fruit (strawberries). These cropping systems historically received frequent applications of pesticides due to low cosmetic thresholds. In the 1960s and 1970s, he was one of a handful of researchers who pioneered the use of alternative controls and IPM approaches in tomatoes and other vegetable crops. Major pests that he and his team of colleagues and students targeted included the tomato fruitworm, *Heliocoverpa zea* (Boddie), beet armyworm, *Spodoptera exigua* (Hübner), tomato pinworm, *Keiferia lycopersicella* (Walsingham), cabbage looper, *Trichoplusia ni* (Hübner), tobacco hornworm, *Manduca sexta* (Johannson), and vegetable leafminer, *Liriomyza sativae* Blanchard. Studies were conducted on population monitoring, biological control (conservation, augmentation, and classical), timing of chemical applications based on pest densities, selective pesticides, natural enemy biology and ecology, non-target impacts of broad spectrum pesticide use, economic analysis of management approaches, and mating disruption using pheromones.

Results from the studies of Professor Oatman and his associates demonstrated that calendar-based, unilateral chemical control for the complete array of tomato pests was unnecessary and sometimes resulted in upsets of secondary pests (e.g., *Liriomyza* leafminers) within the crop system. By integrating selective pesticides, such as formulations of *Bacillus thuringiensis* spores, and augmentative releases of the egg parasitoid *Trichogramma pretiosum* Riley, it was possible to control the direct fruit pests (e.g., *H. zea, S. exigua, K. lycopersicella*) without decimating the natural enemies that held leafminers in check.

The desire to use the most effective egg parasitoids in augmentative releases stimulated Professor Oatman to conduct foreign exploration for effective *Trichogramma* species and become a world expert on the biology and systematics of this genus in collaboration with Professor John Pinto. These efforts contributed to the development and success of today's commercial insectaries that produce *Trichogramma* for commercial farming operations.

Contents

SECTION I
Introduction

1. Tomato and Management of Associated Arthropod Pests: Past, Present, and Future

Waqas Wakil, Gerald E. Brust and Thomas M. Perring

SECTION II
Global Pests of Tomato

2. Aphids: Biology, Ecology, and Management

Thomas M. Perring, Donatella Battaglia, Linda L. Walling, Irene Toma and Paolo Fanti

3. Thrips: Biology, Ecology, and Management

David Riley, Alton Sparks Jr., Rajagopalbab Srinivasan, George Kennedy, Greg Fonsah, John Scott and Steve Olson

4. Whiteflies: Biology, Ecology, and Management

Thomas M. Perring, Philip A. Stansly, T.X. Liu, Hugh A. Smith and Sharon A. Andreason

SECTION III
Integrated Pest Management of Tomato Pests

9. Host-Plant Resistance in Tomato

Michael J. Stout, Henok Kurabchew and
Germano Leão Demolin Leite

10. Engineering Insect Resistance in Tomato by Transgenic Approaches

Manchikatla V. Rajam and Sneha Yogindran

11. Biological Control in Tomato Production Systems: Theory and Practice

Sriyanka Lahiri and David Orr

12. Entomopathogenic Nematodes as Biological Control Agents of Tomato Pests

Fernando Garcia-del-Pino, Ana Morton and David Shapiro-Ilan

13. Applications and Trends in Commercial Biological Control for Arthropod Pests of Tomato

Norman C. Leppla, Marshall W. Johnson, Joyce L. Merritt and Frank G. Zalom

14. Protection of Tomatoes Using Bagging Technology and Its Role in IPM of Arthropod Pests

Germano Leão Demolin Leite and Amanda Fialho

15. Integrated Pest Management Strategies for Tomato Under Protected Structures

Srinivasan Ramasamy and Manickam Ravishankar

List of Contributors

Sharon A. Andreason, University of California, Riverside, CA, United States

Donatella Battaglia, Universita delgli Studi della Basilicata, Potenza, Italy

Michael Braverman, Rutgers University, Princeton, NJ, United States

Gerald E. Brust, CMREC-UMF, University of Maryland, Upper Marlboro, MD, United States

Keith Dorschner, Rutgers University, Princeton, NJ, United States

Paolo Fanti, Universita delgli Studi della Basilicata, Potenza, Italy

Amanda Fialho, Universidade Federal de Minas Gerais, Montes Claros, Brazil

Greg Fonsah, UGA, Tifton, GA, United States

Fernando Garcia-del-Pino, Universitat Autònoma de Barcelona, Barcelona, Spain

Tetsuo Gotoh, Ibaraki University, Ami, Japan

Marshall W. Johnson, University of California, Riverside, CA, United States

George Kennedy, NCSU, Raleigh, NC, United States

Thomas P. Kuhar, Virginia Tech, Blacksburg, VA, United States

Daniel Kunkel, Rutgers University, Princeton, NJ, United States

Henok Kurabchew, Hawassa University, Hawassa, Ethiopia

Sriyanka Lahiri, North Carolina State University, Raleigh, NC, United States

Germano Leão Demolin Leite, Universidade Federal de Minas Gerais, Montes Claros, Brazil

Norman C. Leppla, University of Florida, Gainesville, FL, United States

T.X. Liu, Northwest A&F University, Yangling, China

Joyce L. Merritt, University of Florida, Gainesville, FL, United States

Ana Morton, Universitat Autònoma de Barcelona, Barcelona, Spain

Steve Olson, UF North Florida Research and Education Center, Institute of Food and Agricultural Sciences, Quincy, FL, United States

David Orr, North Carolina State University, Raleigh, NC, United States

Thomas M. Perring, University of California, Riverside, CA, United States

Christopher R. Philips, University of Minnesota, Grand Rapids, MN, United States

Sean M. Prager, University of Saskatchewan, Saskatoon, SK, Canada

Mirza A. Qayyum, University of Agriculture, Faisalabad, Pakistan; Muhamamd Nawaz Sharif University of Agriculture, Multan, Pakistan

Manchikatla V. Rajam, University of Delhi South Campus, New Delhi, India

Srinivasan Ramasamy, AVRDC – The World Vegetable Center, Tainan, Taiwan

Manickam Ravishankar, World Vegetable Center, Ranchi, India

David Riley, UGA, Tifton, GA, United States

John Scott, UF Gulf Coast REC, Wimauma, FL, United States

David Shapiro-Ilan, USDA-ARS, Byron, GA, United States

Alvin M. Simmons, USDA, ARS, Charleston, SC, United States

Hugh A. Smith, University of Florida – IFAS, Wimauma, FL, United States

Alton Sparks Jr., UGA, Tifton, GA, United States

Rajagopalbab Srinivasan, UGA, Tifton, GA, United States

Philip A. Stansly, University of Florida – IFAS, Immokalee, FL, United States

Michael J. Stout, Louisiana State University Agricultural Center, Baton Rouge, LA, United States; Hawassa University, Hawassa, Ethiopia

Irene Toma, Comprehensive School "Jannuzzi-Di Donna", Andria, Italy

John T. Trumble, University of California, Riverside, CA, United States

Waqas Wakil, University of Agriculture, Faisalabad, Pakistan

James F. Walgenbach, North Carolina State University, Mills River, NC, United States

Linda L. Walling, University of California, Riverside, CA, United States

Sneha Yogindran, University of Delhi South Campus, New Delhi, India

Frank G. Zalom, University of California, Davis, CA, United States

Foreword

Tomato (*Solanum lycopersicum* L.) is the second most important vegetable crop in the world after potato. World production and consumption of tomato has grown quickly over the past 25 years. Current world production is about 170.75 million tons of fresh fruit produced on 5.02 million hectares in over 150 countries. The tomato plant has been bred to improve productivity and fruit quality. Because of its popularity and use in cooking and processing, tomatoes are one of the most profitable vegetable crops. However, tomato production is also labor-intensive and prone to production problems that can reduce both yield and quality of fruit which in turn reduces grower's income. Tomatoes can be subjected to attack by a number of insect pests from the time plants first emerge until harvest. The damage can result from feeding on roots, foliage, and fruit or by spreading certain diseases, such as viruses.

Much has been published over the years about the insect pests of tomato and their control. What is missing is a comprehensive synthesis of all the information in one place. This book *Sustainable Management of Arthropod Pests of Tomato* integrates and evaluates all this information into one volume. This is accomplished by 46 authors who have substantial knowledge and experience in the field of pest management in tomato production systems across the world where tomatoes are grown. The first chapters of the book detail the most important pests of tomato throughout the world. Each of these pest chapters is arranged in the same general way discussing identification, biology, distribution, hosts, damage, crop losses, economic thresholds, and management practices.

With most insects, outbreaks are difficult to predict, and choosing management tactics and timing of those tactics can be challenging. Scheduled sprays are frequently considered the most practical management program due to the variety of insect pests on this crop. However, dependence on chemical pesticides is not likely to provide a sustained solution to many of the pest problems. This book provides the knowledge of insect behaviors, pest monitoring, and effective ecological and economic strategies that will enable producers to either avoid or greatly reduce the damage they incur in their tomato crop. Additional chapters discuss the principles behind alternative controls to chemicals such as host-plant resistance, transgenic approaches for pest control, and biological control using predators, parasitoids, and entomopathogenic nematodes and other pathogens. The ideas and principles behind integrated pest management are discussed in two chapters dealing with IPM in protected environments and in field production systems.

Sustainable Management of Arthropod Pests of Tomato is an important book for anyone interested in tomato arthropod pest management.

Prof. Dr. Iqrar Ahmad Khan
Vice Chancellor
University of Agriculture
Faisalabad, Pakistan
April 2017

Preface

Tomatoes are one of the most commonly grown vegetable crops in the world. In 2014, 5.02 million hectares of land were devoted to tomato cultivation with a total production of almost 170.75 million tons. It is a major money-making crop and few other agricultural commodities can match the income potential of fresh market and processed tomatoes. While tomatoes are grown throughout the world, it is not the easiest crop to grow profitably. The level of training or education in tomato production has often been found to be a major factor determining whether or not a grower produces a profitable crop. Extension educators and other forms of education and information have been found to be essential in providing valuable information about tomato production and pest management to small and large growers.

Our goal in writing *Sustainable Management of Arthropod Pests of Tomato* was to create a resource for tomato producers, field workers, extension educators, university personnel, and others about tomato insect pests. In addition to describing the pests' biology, life history, and damage, a great deal of this book is dedicated to management. Many publications have descriptions of the pests, but oftentimes offer only generic information as to how the pest can be controlled with chemicals, cultural management practices, or biological control agents. The chapters in this book are written by authors who have decades of experience in pest management in vegetables and more specifically in tomatoes. Rather than just list some general approaches for the pests' management, they discuss management programs that have been proven under various field and tomato production systems. Much time and expertise has gone into describing different management practices and their advantages, disadvantages, and limitations. Management of some of these pests requires wide-ranging combinations of control strategies and the authors of these chapters have the experience and knowledge to provide this type of information.

The first part of this book provides information about specific insects and mite pests of tomato throughout the world. These chapters are arranged in the same general way discussing identification and biology, distribution and hosts, damage, crop losses, economic thresholds, and management practices, which is the largest section of each chapter. The management section deals with how best to monitor the pest, what chemical, biological, or cultural controls are most practical to use, and other IPM tactics such as host-plant resistance, and pheromone-based strategies including mass trapping and mating disruption. Also included for each pest are the economic considerations that examine the most cost-effective methods to manage the pests in tomato production systems. Following these chapters are broader treatments of the principles behind management alternatives to chemical control, including host-plant resistance, transgenic approaches for pest control, biological control using parasitoids, predators, pathogens, and nematodes, and using the cultural control of bagging technology. The ideas and principles behind integrated pest management are discussed in two chapters dealing with IPM in protected environments and in field production systems. Finally, because of the continued importance they play in insect management, there is a discussion of the registration process for insecticides.

The authors of this book work in universities and government agencies located in North and South America, Europe, Africa, and Asia, across the world where tomatoes are grown. This gives not only a global perspective on the different pests and their management, but also the local production practices and concerns of different areas throughout the world. The combined experience and knowledge of the authors in the field of pest management in tomato production systems is substantial, and we appreciate the dedication of each author in bringing their expertise to the pages of this book. We are also thankful to Nancy Maragioglio and Billie Jean Fernandez for their continuous support and endurance throughout this project.

Waqas Wakil
Faisalabad, Pakistan

Gerald E. Brust
Maryland, USA

Thomas M. Perring
California, USA

Section I

Introduction

Chapter 1

Tomato and Management of Associated Arthropod Pests: Past, Present, and Future

Waqas Wakil[1], Gerald E. Brust[2], Thomas M. Perring[3]

[1]*University of Agriculture, Faisalabad, Pakistan;* [2]*CMREC-UMF, University of Maryland, Upper Marlboro, MD, United States;* [3]*University of California, Riverside, CA, United States*

1. INTRODUCTION

Tomato, *Solanum lycopersicum* L., is the most widely grown vegetable and leading non-grain commodity in the global production system (Bai and Lindhout, 2007; Srinivasan, 2010; Testa et al., 2014). It belongs to the family Solanaceae, which has over 3000 plant species of economic importance, including potato, eggplant, petunia, tobacco, pepper (*Capsicum*), and *Physalis. Solanum* is the largest genus in the Solanaceae family encompassing 1250 to 1700 plant species which are widespread in distribution, remarkable in morphological and ecological diversity, and present on almost all temperate and tropical continents (Weese and Bohs, 2007). It was in the 16th century that tomato was assessed as a close relative of the genus *Solanum* and declared as *Solanum pomiferum* Cav. (Bergougnoux, 2014). In 1753 Linnaeus first classified tomato as *S. lycopersicum*; however, many revisions were suggested later on by different researchers (Foolad, 2007; Peralta and Spooner, 2007). It took about 200 years to confirm the contribution of Linnaeus for recognition of tomato in the genus *Solanum* when phylogenetic classification of the Solanaceae and the genus *Lycopersicon* were revised through molecular data (Bergougnoux, 2014).

Botanically, tomato is a fruit berry, and not a vegetable, and this fact was featured in a historical debate in 19th century United States in a special legal hearing of Nix versus Hedden-149 United States 304 (1893) (Bergougnoux, 2014). In the spring of 1886, Nix challenged the tax collection at the port of New York on tomatoes imported from the West Indies categorizing tomatoes under vegetable. There was a long debate based on literal and scientific claims about tomato but in the end the court opined that "Though botanically tomatoes are fruit of a vine like cucumbers, squashes, beans and peas but these have common usage as vegetable for gardens and kitchens and commonly used as vegetables" (https://supreme.justia.com/cases/federal/us/149/304/case.html). Initially, tomatoes were flattened in shape, segmented, and golden in color (Matthiolus, 1544). In 1554, Matthiolus (1554) reported another variety comparable in shape, but red in color. The first cultivated tomatoes were yellow and cherry-sized and hence this fruit was named golden apples. Tomatoes were regarded as poisonous for a time, but their beauty was still appreciated for ornamental purposes. After its arrival in Europe, tomatoes were known as the Peruvian apples. Patrick Bellow of Castletown is known as the first British tomato grower who grew plants from seeds in 1554 (Bauchet and Causse, 2012). Different species of tomatoes are now available in different shapes (Tanksley, 2004).

Tomatoes were distributed globally after their import in the 16th century from the Andean region to Europe (Bergougnoux, 2014). The word "tomato" in English has its origin from the Aztec word *tomatl*, clearly depicting its domestication history. The introduction of Europeans to tomatoes probably took place during a voyage of Cortez in 1519, when he picked up some tomato plants from Mexico. McCue (1952) provided bibliographic investigations about the domestication history of tomato. He found that the Spanish conquistador Cortes introduced the small yellow tomato to Spain when Tenochtitlan, the Aztec city known as Mexico City today, was captured in 1521. Following this tomatoes were brought to Italy through Naples (a Spanish possession at that time). The first known European name tomato as "pome dei Moro" (Moor's apple) and the French as pommes d'amour, or love apples, because tomatoes were thought to possess aphrodisiacal properties. "Tomate" was introduced in the 17th century, which later was modified to "tomato," most likely due to inspiration from the more familiar potato.

Native to western South America, wild tomatoes were spread to a wide variety of habitats ranging from sea level on the Pacific coast up to 3300 m above sea level in the Andean Highlands, and from arid to rainy climates. Wild tomatoes are believed to be restricted to a narrow range of isolated valleys which possess a particular climate and soil type (Nakazato and

Housworth, 2011). This hypothesis is supported by studies using two close tomato relatives *S. lycopersicum* and *Solanum pimpinellifolium* L. This diversity is further expressed through morphological, physiological, and sexual characteristics (Peralta and Spooner, 2005; Peralta et al., 2005). Modern molecular techniques have helped to determine that tomatoes from Europe and North America share similar isozymes and molecular markers with those from Mexico and Central America (Brazil, Guatemala, El Salvador, Honduras, Nicaragua, Costa Rica, and Panama), which clearly shows that both regions transferred tomatoes to Europe and back to North America (Peralta and Spooner, 2007; Bauchet and Causse, 2012).

Because tomatoes have an enriched nutritional package, are easily cultivated, and are highly adaptable, there has been a dramatic increase in tomato production around the world. The world's drastic fluctuations and uncertainties in food supply have placed this crop in the upper echelon to fight food security issues. Tomatoes are not only used as fresh produce but also in a broad range of processed products such as juice, paste, powder, soup, sauce, and concentrate. They are enriched with nutrients such as β-carotene, lycopene, and vitamin C, all known for their positive impacts on human health (Bergougnoux, 2014).

2. PRODUCTION, AREA, AND YIELD

At the global level, tomato consumption surpasses all other vegetables after potato (FAO, 2017), making it one of the most popular garden crops. Tomato is an important vegetable of Asia and Africa with a global production of ~70% (Srinivasan, 2010). Recently, Europe surpassed Africa in production and the combined share of Asia and Europe is 72.83% of the world's total production (FAO, 2017) (Fig. 1.1A). The Republic of China is the world leader in tomato

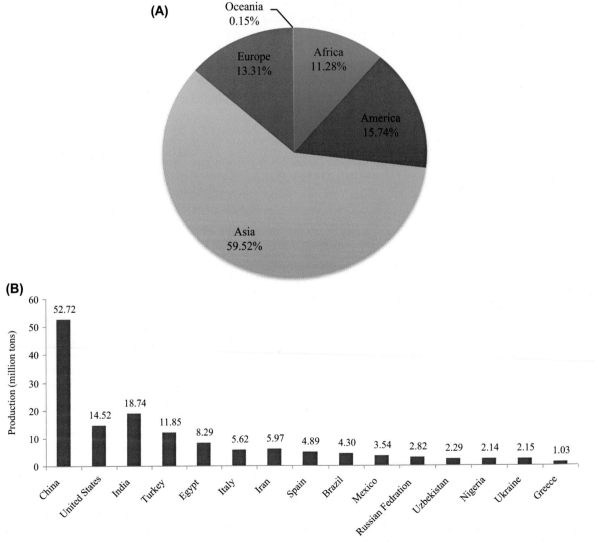

FIGURE 1.1 The average share of continents for tomato production (A) and summary of global tomato production trend represented by 15 leading producers of the world (B) during 2014 (FAO, 2017).

production, providing more than 50% of the world's tomato acreage (Fig. 1.1B). After China, the United States and India add more than one-third of the world's production; Turkey and Egypt also have a notable contribution. During the last 20 years, tomato production and area under its cultivation is continuously increasing. It is very interesting to note that ~20 years ago, the United States and Europe were the leading tomato producers, but now the scenario has changed. The area under tomato cultivation has gradually increased globally, reaching 5.02 million ha in 2014 which was 3.27 million ha in 1995 (Fig. 1.2A). Correspondingly, the total annual production also enhanced from 87.44 to 170.75 million tons since 1995 to 2014, respectively (Fig. 1.2B). The recent trend for increased production also corresponds to the increase in public consumption reaching an average of about 20.5 kg/capita/year in 2009. Libya, Egypt, and Greece represent the nations with the largest tomato consumption of 100 kg/capita/year. A general perception is that in the Mediterranean and Arabian region, the consumption of tomato is highest in the world averaging between 40 and 100 kg/capita/year (Bergougnoux, 2014).

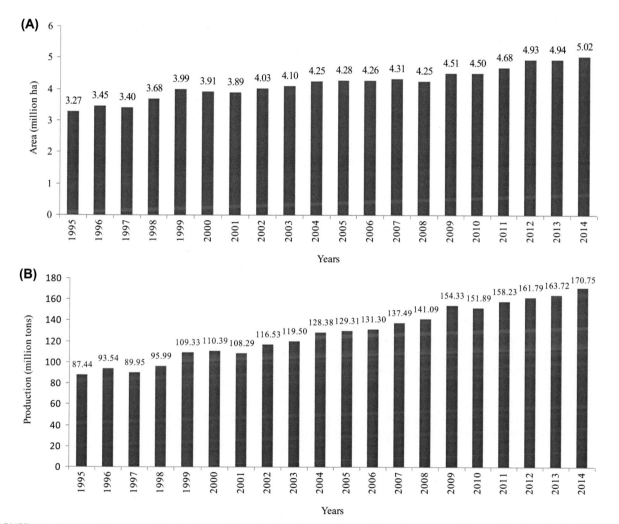

FIGURE 1.2 Metrics of tomato production in the world showing increasing trend; area dedicated to tomato cultivation (A) and production (B) during last 20 years (1995–2014) (FAO, 2017).

3. TOMATO AND GLOBAL FOOD SECURITY

If one considers only lipids, proteins, and sugars, tomatoes undoubtedly have little nutrition. However, in the true sense, tomato encompasses an important nutritional package for human health, including antioxidants such as lycopene, vitamin A (β-carotene), and ascorbic acid (vitamin C) (Bergougnoux, 2014). The antioxidant properties of lycopene are considered to protect against cancer and cardiovascular diseases (Rao and Agarwal, 2000). Wild tomatoes, in contrast to cultivated cultivars, possess about 5 times more ascorbic acid (Stevens, 1986). Nutritional improvements have been bred into some cultivars, but these can result in lower yields that restrict the commercial success of these cultivars (Causse et al., 2007). The development of improved varieties of tomatoes through modern techniques is a key factor that will promote tomato in the area of food insecurity.

4. PESTS OF TOMATO: CURRENT STATUS, CHALLENGES, AND FUTURE PRIORITIES

Damage caused by arthropod feeding presents one of the greatest economic challenges faced by tomato growers in greenhouse and open-field situations. From the time the seed germinates until the fruit are ready to harvest, tomato is under constant threat by a diversity of insect and mite pests. These can be categorized broadly as those that feed on the vegetative plant causing indirect yield loss and those that feed directly on the fruit. In the early vegetative stage, the germinating seedling is exposed to mole crickets and flea beetles. Mole crickets (Orthoptera: Gryllotalpidae) feed on the roots and stems of emerging and newly transplanted tomatoes (Silcox, 2011; Bailey, 2012), which can kill the plant. Often the damage due to mole crickets is not apparent until the grower notices dead plants, and by then it is too late to prevent the yield loss. In extreme situations, the grower may need to replant the field. Once plants are established, the next pest of concern is flea beetles (species *Epitrix* and *Phyllotreta*). Adult flea beetles feed on cotyledons and early true leaves, leaving small shot holes in the leaves (Cranshaw, 2013), which can kill the plant outright, or at minimum reduce photosynthesis and water balance of the leaves.

As the tomato plant continues to grow, it faces with additional pests. Mites in the family Tetranychidae (*Tetranychus urticae* Koch, *Tetranychus evansi* Baker and Pritchard), Eriophyidae (*Aculops lycopersici* (Massee), *Aceria lycopersici* (Wolffenstein)), and Tarsonemidae (*Polyphagotarsonemus latus* (Banks)) feed on the leaves and stems of the plant; under severe infestations mite feeding can cause plant death. More commonly, mites rasp the epidermal cells of leaves and stems that reduce photosynthesis which can reduce the yield (Jeppson et al., 1975; Alagarmalai et al., 2009; Meck, 2010; Navajas et al., 2013).

Aphids, specifically *Macrosiphum euphorbiae* Thomas, *Myzus persicae* (Sulzer) and whiteflies, *Trialeurodes vaporariorum* Westwood, and *Bemisia tabaci* (Gennadius) feed on the foliage, extracting phloem sap which weakens the plant reducing yield (Hussey et al., 1969; Dedryver et al., 2010). Aphids and whiteflies can kill small seedlings under very high densities. In addition to sap feeding, they also produce copious amounts of honeydew that covers fruit and foliage. This sugary material serves as a substrate for sooty mold fungi which turns the leaves and fruit black (Johnson et al., 1992). Sooty mold can reduce photosynthesis and must be washed from fruit which adds expenses incurred by the grower. In addition, feeding by immature stages of *B. tabaci* (MEAM1) on tomato foliage causes a fruit abnormality known as tomato irregular ripening (Schuster et al., 1996; Dinsdale et al., 2010). This abnormality is characterized by sections of the fruit that remain green, while other sections ripen normally. Another hemipteran that feeds on foliage is the potato psyllid, *Bactericera cockerelli* (Sulc). This insect also feeds on plant phloem, injecting salivary toxins that result in "psyllid yellows." Feeding by nymphal stages is most often associated with psyllid yellows (Cranshaw, 1994).

Leafminers in the Diptera order can be severe pests, mining the leaves of tomato foliage. There are four species, i.e., *Liriomyza sativae* (Blanchard), *Liriomyza trifolii* (Burgess), *Liriomyza bryoniae* (Kaltenbach), and *Liriomyza huidobrensis* (Blanchard) to be considered, and under high densities they can severely damage leaves causing reduced yields (Walker, 2012). Leafminers tend to be secondary pests that normally are under control of various parasitoids. However they can be severe when pesticides are overused reducing the natural enemies.

Another foliage feeder is the Colorado potato beetle, *Leptinotarsa decemlineata* Say; adults and larvae feed on the tender stems of the plant as well as the young foliage. Under heavy infestations, these beetles can completely defoliate tomato plants, causing heavy crop loss (Hare and Moore, 1988).

Thrips (Thysanoptera: Thripidae) can be pests of young seedlings where they rasp epidermal tissue causing a reduction in photosynthesis. It is rare in established plants, but high densities can stunt the plant. Western flower thrips, *Frankliniella occidentalis* Pergande, feed on pollen and flowers, which lead to bud damage and blossom drop (Kirk, 1997). They also feed on young fruit, causing discoloration and scarring making them unacceptable for fresh fruit marketing (Salguero Navas et al., 1991).

Stink bugs (Hemiptera: Pentatomidae) can be a pest problem in tomato. Early instar stink bugs prefer to feed on the foliage, which generally does not affect tomato unless the bug density is high. Later instar and adult bugs prefer to feed on young, green fruit (Lye and Story, 1988). Stink bug feeding leaves puncture marks on the fruit, often surrounded by discoloration; these marks downgrade the quality of fresh market tomatoes. Similar to stink bug, leaf-footed bugs, *Leptoglossus phyllopus* (L.), *Phthia picta* (Drury) also cause damage due to feeding. With their long proboscis, leaf-footed bugs probe deep within the plant. When this occurs to fruit, the fruit may abort or become discolored as they mature, making them unmarketable (Ingels and Haviland, 2014).

Finally, insects of the order Lepidoptera feed on tomato foliage, stems, and fruit. While some species such as the tomato leafminer, *Tuta absoluta* (Meyrick) (Desneux et al., 2010) feed predominantly on foliage, other species feed on foliage in the early larval instars and move on to the fruit during later instars. These include *Helicoverpa armigera* (Hübner) (Venette et al., 2003), *Helicoverpa zea* (Boddie) (Capinera, 2001), *Spodoptera exigua* (Hübner) (Liburd et al., 2000), *Spodoptera litura* (F.) (Ahmad et al., 2013), and *Keiferia lycopersicella* (Walsingham) (Capinera, 2001). There are several lepidopterans that feed mainly on the fruit, including *Manduca quinquemaculata* (Haworth), *Manduca sexta* (L.) (Foster and Flood, 2005), and *Phthorimaea operculella* (Zeller) (Kroschel, 1995), attacking tomato by boring into the fruit or by chewing damage on the foliage and/or fruit.

Compounding the diversity of pests in tomato is the fact that global trade and ease of travel across the globe has resulted in the movement of pests from one region to another. These invasive pest introductions upset the existing integrated pest management (IPM) programs, complicating the decisions that growers must make to preserve their economic viability. A current example of an introduction of a non-native pest of tomato is that of the South American tomato leafminer *T. absoluta*. This small moth has invaded 40% of the world's tomato crop where it almost causes total crop loss (Anonymous, 2005). It is apparent that more effort is needed to prevent invasive pests from entering new areas of tomato production, and such exclusion is one of the main techniques of IPM. Increasing efforts to inhibit international pest movement in the future will greatly reduce crop loss and pest management costs.

Direct-feeding damage on tomato by many insects and mites challenges growers in the development of holistic IPM programs. One simply cannot manage a single pest; all potential pests must be considered. This is compounded by the fact that many of the insects that feed on tomato also serve as vectors of plant pathogens. The most significant of these are the aphids and whiteflies. Their unique way of feeding in the phloem causes minimal damage to the sieve elements so that inoculation of virions results in successful inoculation and the formation of disease. There are more than 150 viruses known to be transmitted by whiteflies (Polston and Anderson, 1997; Jones, 2003) and over 100 viruses that are transmitted by aphids (Kennedy et al., 1962; Blackman and Eastop, 2007). Thrips also transmit viruses in the genus *Tospovirus*, the most widespread of which is tomato spotted wilt virus (Riley et al., 2011). Management of insect-borne viruses presents particular challenges for tomato growers. Due to a very short inoculation time, insecticides and biological control agents are not effective in reducing spread. In these situations, areawide control to reduce vector numbers outside the target tomato field or exclusion strategies for greenhouses or the open field must be used, often at a high expense to the growers. In these cases, the future depends on varieties that are resistant to the virus or to insect feeding. Other insects transmit bacterial pathogens. The tomato psyllid, *B. cockerelli* is the vector of the bacterial pathogen, *Candidatus* Liberibacter Jagoueix et al. solanacearum (CLso). This bacterium causes a disease in tomato known as vein greening, which is known to reduce fruit set and can cause plant death (Butler and Trumble, 2012b; Sengoda et al., 2013). Stink bugs puncture green fruit and can infect them with bacterial and yeast pathogens that lead to fruit rot (Zalom and Zalom, 1992).

Fortunately, there is a wealth of information from research conducted on the various arthropod pests and vectors of pathogens, which is used to form management strategies. Foremost in the development of IPM programs is determining cost-effective methods for sampling pests in tomato; many of these are outlined in this book for specific insects and mites. Sampling strategies include using light traps, colored sticky traps, and pheromone traps for monitoring pest population levels. Information from traps and field sampling has been used to develop a variety of sampling plans. With good estimates of pest density, growers then can select appropriate management tools for their pest situation.

The most utilized control method is to treat pest populations with insecticides and/or acaricides. The reason behind this is the ease of application, inexpensive cost of pesticides, and rapid reduction of pest numbers. Pesticides also have a long history of use for pest control. Sulfur, herbs, oils, and soaps were used as controls in the beginning (Brown, 1951; Jones, 1973). In the 1600s nicotine mixtures, herbs, and arsenic were the most important constituents used for insect pest control. In the early 1800s entomologists began to understand the importance of temperature in the development and distribution of insects, a concept we use today to know when a particular pest species is active (Martin, 1940; Ware and Whitacre, 2004). In the 1860s the Colorado potato beetle, *L. decemlineata*, became one of the first invasive insect pests when it was introduced from the United States to Europe resulting in major losses of solanaceous crops, including tomato. This led governments to begin inspecting agricultural products entering their countries, which at times resulted in quarantines to be implemented (Ware and Whitacre, 2004). The agricultural inspections and quarantines caused an increase in the importance of pest control procedures and products. One of these products was Paris green, a mixture of arsenic and copper sulfate, for the control of Colorado potato beetle (O'Kane, 1915). Over the next 50 years, there was expansive development of equipment that could more effectively apply insecticides. In the early 1900s overreliance on pesticides for the control of arthropod pests in vegetables resulted in field failures; they were dangerous to farm workers and had negative impacts on the environment (Brown, 1951; Mrak, 1969).

It was not until the early 1940s that chemicals effective in killing insects with moderate mammalian toxicity were synthesized and became commercially available (Shepard, 1951). In the 1950s and 1960s there was a rapid increase of synthesized pesticides most notably the chlorinated hydrocarbons and organophosphates (Shepard, 1951; Ware and Whitacre, 2004). These products were very successful because they were highly effective, relatively inexpensive, fairly easy to apply, and relatively safe to humans.

However, in 1962 Rachel Carson's seminal book, *Silent Spring*, brought public awareness to the possible deleterious effects of unrestrained pesticide use (Carson, 1962). Issues included pesticides and their metabolites found throughout the environment and in humans, adverse effects on wildlife, resistance development, and secondary pest outbreaks. In response, government established agencies to manage pesticides such as the Environmental Protection Agency (EPA) in the United States and The European Environment Agency. Since the 1970s, IPM has been emphasized in academia and government agencies and promoted in the agricultural community. IPM can be defined most simply as utilizing multiple tactics of pest control (i.e., resistant plant varieties, pesticides, natural enemies, cultural control, etc.) to keep pests below economic thresholds in order to conserve environmental quality (Ordish, 1976; EPA, 1998).

To accomplish better management of arthropod pests, in the early 1990s reduced-risk pesticides were given expedited review by the Environmental Protection Agency. For a pesticide to be considered reduced risk, it must have one or more of the following qualities: a reduced impact on human health and very low mammalian relative to alternative materials; a reduced impact on non-target organisms; a lower potential for contaminating groundwater; and a lower pest-resistance potential (EPA, 1998). The reduced-risk pesticides can be integrated into IPM programs easily as they tend not to cause some of the negative drawbacks of older and more toxic pesticides, such as secondary pest outbreaks and reduction of natural enemies.

Second to pesticides, biological control has become a prominent component of IPM programs. There are specific and generalist natural enemies that have been identified for each of the insect and mite pests attacking tomato. A few examples are tachinid flies that parasitize adults and older nymphs of stink bugs, entomopathogenic nematodes that control flea beetle larvae overwintering in the soil (Miles et al., 2012), parasitoids and predators that attack aphids and whiteflies, predatory mites from the family Phytoseiidae that feed on mites and small insects, and generalist predators for the Colorado potato beetle (Brust, 1994), leaf-footed bugs (Ingels and Haviland, 2014), and potato psyllid (Butler and Trumble, 2012a).

Additional tools used in IPM for tomato pests fall under the broad categories of cultural control. Cultural control strategies are those that are applied to the crop in such a manner that they create a suboptimal environment for herbivores. Examples include planting date, manipulating fertilizer and irrigation regimes, crop-free periods, and natural and synthetic mulches (Weintraub and Berlinger, 2004). For greenhouses, cultural controls involve using insect exclusion screens and adding UV-blocking materials to plastic houses (Vatsanidou et al., 2011).

Perhaps the most exciting area of IPM for tomato pests is plant resistance. Tomato has been the topic of considerable research which has identified a variety of genes in wild tomato relatives that have been bred into commercial tomato cultivars. In addition, tomato has proven to be a model plant for identifying novel genes and defensive pathways that are helping us understand how herbivores interact with the tomato plant. This research holds tremendous promise for the future in the development of tomatoes that not only are resistant against the pests, but against the pathogens that many of them vector. Additional successes in biotechnology, such as RNA interference (RNAi) hold promise for incorporation into tomato plants that selectively targets pests (Zhang et al., 2015).

5. CLIMATE CHANGE

One of the main challenges to face agriculture in the next 50 years and beyond is climate change. While climate change is sometimes presented in disastrous hyperbolic terms, it may be more pragmatic to look at the coming changes as challenges, much as we would with any invasive new pest. Agricultural crop production will be affected under future climate change, but the magnitude and direction of impacts on crops will vary locally and are difficult to predict because of the complexity of the interacting variables involved, such as CO_2 levels, temperatures, precipitation, and near-surface winds. To date, research has focused on single factors usually in controlled environments, creating uncertainty about climate-change effects on crop production. Changes in average climate conditions are important, as are changes in the timing and incidence of extreme climate events. For instance, if we look only at elevated CO_2 levels, which now are 30% higher than during pre-industrial times, we might expect an increase in crop yields (Kimball, 1983; Lawlor and Mitchell, 1991). However, if we include the impact of higher temperatures on pest biology into the mix, the resultant scenario may counter balance the beneficial effect of the increase in CO_2 (Cushman et al., 1988; Adams et al., 1990; Lawlor and Mitchell, 1991).

An increase in temperatures could lead to an increase in crop water demand, increasing water use (Rosenzweig and Parry, 1994). On the other hand, the trend of diminished near-surface winds over the last 20 years and projections for continued declines may decrease evapotranspiration of cropping regions. The anticipated greater spring–summer air temperatures would be beneficial to crop production at northern sites, where the length of the growing season is presently a limiting factor for growth. For vegetables, a critical period of exposure to temperatures is the pollination stage, when pollen is released to fertilize the plant and trigger the development of fruit. Temperature thresholds are normally cooler for each crop during pollination than for optimum vegetative growth. Exposure to high temperatures during this period can greatly reduce crop yields and/or quality, and increase the risk of crop failure. Tomato plants exposed to nighttime temperatures above 20°C during flowering and fruit set can experience reduced yield and quality (Brust, 2016). And these are only a few of the climate-change effects when considering just the crop. We have to take into account the effects on the pests of these crops and their natural enemies.

Generally, an increase in air temperature will benefit insect pests, as long as upper limits are not exceeded. Greater temperatures accelerate an insect's life cycle while warmer winters reduce cold-stress mortality. The overall positive effect of increasing temperatures on expansion of insect geographical ranges in natural systems is well known (Parmesan, 2006; Gregory et al., 2009; Bale and Hayward, 2010). With increasing temperatures, there is early migration and maturation resulting in successful establishment in habitats that were previously beyond the range of insect population (Bale and Hayward, 2010). However, as is the case for crops, insects have optimal temperatures under which they thrive; therefore not all insect populations will increase with increasing temperature.

Extension of the growing season will likely have a profound effect on crop injury from some insect pests (Bradshaw and Holzapfel, 2010). The anticipated expansion of the geographic range of the corn earworm, *H. zea* and European corn borer *Ostrinia nubilalis* (Hübner) will result in increased losses from these pests (Diffenbaugh et al., 2008). Climate change will alter the environmental thresholds currently keeping some pests in check, making pest outbreaks more common as a result. For example, because of global warming over the last century, the northern extension of some crop ranges may have altered aphid community composition. In Europe, autumn sowing of winter wheat, barley, and rape provides a substrate for non-diapausing aphids to survive the winter (Roos et al., 2011). In Scotland the genetic variability of green peach aphid populations is increasing in association with warmer winters and earlier dispersal (Malloch et al., 2006).

Management costs may increase in the future due to shifting geographic ranges and decreasing generation times requiring more frequent pesticide applications. If we examine Lepidopteran pests on sweet corn in the United States, we see that pesticide applications decrease with increasing latitude, from 15 to 32 times per year in Florida, four to eight times per year in Delaware, and zero to five times per year in New York (Hatfield et al., 2011). Because insects develop more rapidly at higher temperatures, their populations will increase more quickly than under current climate conditions and crop damage will occur more frequently. Therefore, action thresholds based on insect density may need to be reduced to prevent undesirable losses (Trumble and Butler, 2009). Climate change is also likely to affect virus diseases of plants indirectly by altering the geographic range of both vectors and non-crop reservoirs, and the feeding habits of vectors although these effects are likely to vary by geographic region (Canto et al., 2009; Navas-Castillo et al., 2011).

Results from climate-change studies have indicated an increased number of extreme weather events over the next 50 years (Adams et al., 1990; Rosenzweig and Parry, 1994). From periods of drought followed by torrential downpours to warm late winter temperatures that suddenly develop into one or two nights of a devastating late frost, weather patterns in many places of the world will be subject to wild swings, but with a consistent steady rise in CO_2 levels and temperature. Increases in extreme precipitation events may result in similar fluctuations in pest populations as pest outbreaks are often associated with dry years, although extreme drought and extremely wet years also are unfavorable to insects (Hawkins and Holyoak, 1998; Fuhrer, 2003). The effect of increased atmospheric CO_2 on insect pests is much more complex than that of increasing temperature because insect fitness is greatly dependent upon the response of the host plant to increased CO_2 concentrations. This indirect action of CO_2 makes for more variable interactions between plants and insect pests.

The extensive amount of research on climate change and agriculture is beyond the scope of these few pages. This brief section is intended to show the possibilities that may come in the next few decades, and the complexity of predicting what the effects of climate change will be on the agricultural community. The economic consequences of climate change will be contingent upon the reactions by growers, industry, and governments to specific regional changes in climate parameters. Responses could vary from individual farmers adjusting their horticultural practices and pest management programs in response to more variable weather patterns, to the agricultural industry developing drought and heat-tolerant cultivars, to increased government investment in climate-change research and more federal risk-management programs. The complexity of the agricultural system will demand substantial efforts and cooperation between all stakeholders to develop effective strategies to adapt to changing climate patterns.

REFERENCES

Adams, R.M., Rosenzweig, C., Peart, R.M., Ritchie, J.T., McCarl, B.A., Glyer, J.D., Curry, R.B., Jones, J.W., Boote, K.J., Allen, L.H., 1990. Global climate change and US agriculture. Nature 345, 219–224.

Ahmad, M., Ghaffar, A., Rafiq, M., 2013. Host plants of leaf worm, *Spodoptera litura* (Fabricius), Lepidoptera: Noctuidae) in Pakistan. Asian Journal of Agricultural Biology 1, 23–28.

Alagarmalai, J., Grinberg, M., Soroker, V., 2009. Host selection by the herbivorous mite *Polyphagotarsonemus latus* (Acari: Tarsonemidae). Journal of Insect Behavior 22, 375–387.

Anonymous, 2005. *Tuta absoluta* – data sheets on quarantine pests. OEPP/EPPO Bulletin 35, 434–435.

Bai, Y., Lindhout, P., 2007. Domestication and breeding of tomatoes: what have we gained and what can we gain in the future? Annals of Botany 100, 1085–1094.

Bailey, D.L., 2012. Characterization of Biopores Resulting from Mole Crickets (*Scapteriscus* spp.) (M.Sc. thesis). Graduate Faculty of Auburn University, Auburn, AL, USA, p. 57.

Bale, J.S., Hayward, S.A.L., 2010. Insect overwintering in a changing climate. Journal of Experimental Biology 213, 980–994.

Bauchet, G., Causse, M., 2012. In: Caliskan, M. (Ed.), Genetic Diversity in Tomato (*Solanum lycopersicum*) and its Wild Relatives. InTech, Rijeka, Croatia, pp. 133–162.

Bergougnoux, V., 2014. The history of tomato: from domestication to biopharming. Biotechnology Advances 32, 170–189.

Blackman, R.L., Eastop, V.F., 2007. Aphids on the World's Herbaceous Plants and Shrubs. The Aphids, vol. 2. John Wiley and Sons with the Natural History Museum, London, UK, pp. 1025–1439.

Bradshaw, W.E., Holzapfel, C.M., 2010. Insects at not so low temperature: climate change in the temperate zone and its biotic consequences. In: Denlinger, D.L., Lee, R.E. (Eds.), Low Temperature Biology of Insects. Cambridge University Press, Cambridge, UK, pp. 242–275.

Brown, A.W.A., 1951. Insect Control by Chemicals. John Wiley and Sons Inc., New York, USA, p. 817.

Brust, G., 2016. Tomato Pollination and How to Increase it in High Tunnels. University of Maryland Extension, Maryland, USA. http://extension.umd.edu/sites/extension.umd.edu/files/_docs/programs/mdvegetables/TomatoPollination-WebArticle_2016_03.pdf.

Brust, G.E., 1994. Natural enemies in straw-mulch reduce Colorado potato beetle populations and damage in potato. Biological Control 4, 163–169.

Butler, C.D., Trumble, J.T., 2012a. Identification and impact of natural enemies of *Bactericera cockerelli* (Hemiptera: Triozidae) in Southern California. Journal of Economic Entomology 105 (5), 1509–1519.

Butler, C.D., Trumble, J.T., 2012b. The potato psyllid, *Bactericera cockerelli* (Sulc) (Hemiptera: Triozidae): life history, relationship to plant diseases, and management strategies. Terrestrial Arthropod Reviews 5, 87–111.

Canto, T., Aranda, M.A., Fereres, A., 2009. Climate change effects on physiology and population processes of hosts and vectors that influence the spread of hemipteran-borne plant viruses. Global Change Biology 15, 1884–1894.

Capinera, J.L., 2001. Handbook of Vegetable Pests. Academic Press, San Diego, CA, USA, p. 729.

Carson, R., 1962. Silent Spring. A Mariner Book Houghton Mifflin Company, New York, USA, p. 375.

Causse, M., Chaïb, J., Lecomte, L., Buret, M., Hospital, F., 2007. Both additivity and epistasis control the genetic variations for fruit quality traits in tomato. Theoretical and Applied Genetics 115, 429–442.

Cranshaw, W.S., 1994. The potato (tomato) psyllid *Paratrioza cockerelli* (Sulc), as a pest of potatoes. In: Zehnder, G.W., Powelson, R.K., Jansson, R.K., Raman, K.V. (Eds.), Advances in Potato Biology and Management. APS Press, Saint Paul, MN, USA, p. 655.

Cranshaw, W.S., 2013. Flea Beetles, Fact Sheet No. 5.592, Insect Series, Home and Garden. Colorado State University Extension, Fort Collins, CO, USA. http://extension.colostate.edu/topic-areas/insects/flea-beetles-5-592/.

Cushman, R.M., Farrel, M.P., Koomanoff, F.A., 1988. Climate and regional resource analysis: the effect of scale on resource homogeneity. Climate Change 13, 129–147.

Dedryver, C.A., Le Ralec, A., Fabre, F., 2010. The conflicting relationships between aphids and men: a review of aphid damage and control strategies. Comptes Rendus Biologies 333, 539–553.

Desneux, N., Wajnberg, E., Wyckhuys, K.A.G., Burgio, G., Arpaia, S., Narváez-Vasquez, C.A., González-Cabrera, J., Catalán Ruescas, D., Tabone, E., Frandon, J., Pizzol, J., Poncet, C., Cabello, T., Urbaneja, A., 2010. Biological invasion of European tomato crops by *Tuta absoluta*: ecology, geographic expansion and prospects for biological control. Journal of Pest Science 83, 197–215.

Diffenbaugh, N.S., Giorgi, F., Pal, J.S., 2008. Climate change hotspots in the United States. Geophysical Research Letters 35, L16709.

Dinsdale, A., Cook, L., Riginos, C., Buckley, Y.M., De Barro, P., 2010. Refined global analysis of *Bemisia tabaci* (Hemiptera: Sternorrhyncha: Aleyrodoidea: Aleyrodidae) mitochondrial cytochrome oxidase 1 to identify species level genetic boundaries. Annals of the Entomological Society of America 103, 196–208.

EPA (Environmental Protection Agency), 1998. Status of Pesticides in Registration, Re-registration, and Special Review. Office of Pesticide Programs, Washington, DC, USA.

FAO, 2017. FAOSTAT: Data-crops. Food and Agriculture Organization of the United Nations, Rome, Italy. http://www.fao.org/faostat/en/#data/QC.

Foolad, M.R., 2007. Genome mapping and molecular breeding of tomato. International Journal Plant Genomics 2007, 1–52.

Foster, R., Flood, B., 2005. Vegetable Insect Management. Meister Media Worldwide Press, Willoughby, OH, USA.

Fuhrer, J., 2003. Agroecosystem responses to combinations of elevated CO_2, ozone, and global climate change. Agriculture, Ecosystems and Environment 97, 1–20.

Gregory, P.J., Johnson, S.N., Newton, A.C., Ingram, J.S.I., 2009. Integrating pests and pathogens into the climate change/food security debate. Journal of Experimental Botany 60, 2827–2838.

Hare, J.D., Moore, R.E., 1988. Impact and management of late-season populations of the Colorado potato beetle (Coleoptera: Chrysomelidae) on potato in Connecticut. Journal of Economic Entomology 81, 914–921.

Hatfield, J.L., Boote, K.J., Kimball, B.A., Ziska, L.H., Izaurralde, R.C., Ort, D., Thomson, A.M., Wolfe, D., 2011. Climate Impacts on Agriculture: Implications for Crop Production. .

Hawkins, B.A., Holyoak, M., 1998. Transcontinental crashes of insect populations? American Naturalist 152, 480–484.

Hussey, N.W., Read, W.H., Hesling, J.J., 1969. The Pests of Protected Cultivation: The Biology and Control of Glasshouse and Mushroom Pests. Edward Arnold Publishers Limited, London, UK, p. 404.

Ingels, C., Haviland, D., 2014. Leafooted Bug, Integrated Pest Management for Landscape Professionals and Home Gardeners. Pest Notes, Publication No. 74168. UC Statewide Integrated Pest Management Program. University of California, Davis, USA.

Jeppson, L.R., Keifer, H.H., Baker, E.W., 1975. Mites Injurious to Economic Plants. University of California Press, Berkeley and Los Angles, CA, USA, p. 615.

Johnson, M.W., Caprio, L.C., Coughlin, J.A., Tabashnik, B.E., Rosenheim, J.A., Welter, S.C., 1992. Effect of *Trialeurodes vaporarorum* (Homoptera: Aleyrodidae) on yield of fresh market tomatoes. Journal of Economic Entomology 85, 2370–2376.

Jones, D.P., 1973. Agricultural Entomology. In: Smith, R.F., Mittler, T.E., Smith, C.N. (Eds.), A History of Entomology. Annual Reviews Palo Alto, CA, USA.

Jones, D.R., 2003. Plant viruses transmitted by whiteflies. European Journal of Plant Pathology 109, 195–219.

Kennedy, J.S., Day, M.F., Eastop, V.F., 1962. A Conspectus of Aphids as Vectors of Plant Viruses. Commonwealth Institute of Entomology, London, UK.

Kimball, B.A., 1983. Carbon dioxide and agricultural yield: an assemblage and analysis of 430 prior observations. Agronomy Journal 75, 779–786.

Kirk, W.D.J., 1997. Distribution, abundance and population dynamics. In: Lewis, T. (Ed.), Thrips as Crop Pests. CABI, Oxon, UK, pp. 217–257.

Kroschel, J., 1995. Integrated Pest Management in Potato Production in the Republic of Yemen with Special Reference to the Integrated Biological Control of the Potato Tuber Moth (*Phthorimaea Operculella* Zeller). Tropical Agriculture 8. Margraf Verlag, Weikersheim, Germany, p. 227.

Lawlor, D.W., Mitchell, R.A.C., 1991. The effects of increased CO_2 on crop photosynthesis and productivity: a review of field studies. Plant, Cell & Environment 14, 729–739.

Liburd, O.E., Funderburk, J.E., Olson, S.M., 2000. Effect of biological and chemical insecticides on *Spodoptera* species (Lepidoptera, Noctuidae) and marketable yields of tomatoes. Journal of Applied Entomology 124, 19–25.

Lye, B.H., Story, R.N., 1988. Feeding preference of the southern green stink bug (Hemiptera: Pentatomidae) on tomato fruit. Journal of Economic Entomology 81, 522–526.

Malloch, G., Highet, F., Kasprowicz, L., Pickup, J., Neilson, R., Fenton, B., 2006. Microsatellite marker analysis of peach – potato aphids (*Myzus persicae*, Homoptera: Aphididae) from Scottish suction traps. Bulletin of Entomological Research 96, 573–582.

Martin, H., 1940. The Scientific Principles of Plant Protection with Special Reference to Chemical Control. Longmans, Green Company, New York, USA.

Matthiolus, P.A., 1544. Di Pedacio Dioscoride Anazerbeo libri cinque della historia et materia medicinale trodottie in lingua vulgare Italiana. Venetia, Italia.

Matthiolus, P.A., 1554. Commentarii in libros sex Pedaeii Dioscoridis Anazerbeii, de medica materia. Venetiis, Italia.

McCue, G.A., 1952. The history of the use of the tomato: an annotated bibliography. Annals of the Missouri Botanical Garden 39 (4), 289–348.

Meck, E.D., 2010. Management of the Twospotted Spider Mite *Tetranychus Urticae* (Acari: Tetranychidae) in North Carolina Tomato Systems (Ph.D. thesis). Graduate Faculty North Carolina State University, Raleigh, NC, USA, p. 86.

Miles, C., Blethen, C., Gaugler, R., Shapiro-Illan, D., Murray, T., 2012. Using Entomopathogenic Nematodes for Crop Insect Pest Control. A Pacific Northwest Extension Publication No. PNW544 Washington State University Extension, USA. http://cru.cahe.wsu.edu/CEPublications/PNW544/PNW544.pdf.

Mrak, E.M., 1969. Report of the Second Commission on Pesticides and Their Relationship to Environmental Health. U.S. Department of Health, Education and Welfare, Washington, DC, USA.

Nakazato, T., Housworth, E.A., 2011. Spatial genetics of wild tomato species reveals roles of the Andean geography on demographic history. American Journal of Botany 98, 88–98.

Navajas, M., Moraes, G.J., Auger, P., Migeon, A., 2013. Review of the invasion of *Tetranychus evansi*: biology, colonization pathways, potential expansion and prospects for biological control. Experimental & Applied Acarology 59, 43–65.

Navas-Castillo, J., Fiallo-Olivé, E., Sánchez-Campos, S., 2001. Emerging virus diseases transmitted by whiteflies. Annual Review of Phytopathology 49, 219–248.

O'Kane, W.C., 1915. Injurious insects. How to Recognize and Control Them. The Macmillan Company, New York, USA.

Ordish, G., 1976. The Constant Pest: A Short History of Pests and Their Control. The Macmillan Company, New York, USA.

Parmesan, C., 2006. Ecological and evolutionary responses to recent climate change. Annual Review of Ecology, Evolution and Systematics 37, 637–669.

Peralta, I.E., Knapp, S., Spooner, D.M., 2005. New species of wild tomatoes (*Solanum* section Lycopersicon: Solanaceae) from northern Peru. Systematic Botany 30, 424–434.

Peralta, I.E., Spooner, D.M., 2007. History, origin and early cultivation of tomato (Solanaceae). In: Razdan, M.K., Mattoo, A.K. (Eds.), Genetic Improvement of Solanaceous Crops, vol. 2, Tomato. Science Publishers, Enfield, NH, USA, pp. 1–27.

Peralta, I.E., Spooner, D.M., 2005. Morphological characterization and relationships of wild tomatoes (*Solanum L.* Section *Lycopersicon*). In: Croat, T.B., Hollowell, V.C., Keating, R.C. (Eds.), A Festschrift for William G. D'Arcy. vol. 104. Missouri Botanical Garden Press, Saint Louis Missouri, USA, pp. 227–257.

Polston, J.E., Anderson, P.K., 1997. The emergence of whitefly transmitted geminiviruses in tomato in the Western Hemisphere. Plant Disease 81, 1358–1369.

Rao, A.V.R., Agarwal, S., 2000. Role of antioxidant lycopene in cancer and heart disease. Journal of the American College of Nutrition 19, 563–569.

Riley, D.G., Joseph, S.V., Srinivasan, R., Diffie, S., 2011. Thrips vectors of tospoviruses. Journal of Integrated Pest Management 1, 1–10.

Roos, J., Hopkins, R., Kvarnheden, A., Dixelius, C., 2011. The impact of global warming on plant diseases and insect vectors in Sweden. European Journal of Plant Pathology 129, 9–19.

Rosenzweig, C., Parry, M.L., 1994. Potential impact of climate change on world food supply. Nature 367, 133–138.

Salguero Navas, V.E., Funderburk, J.E., Olson, S.M., Beshear, R.J., 1991. Damage to tomato fruit by the western flower thrips (Thysanoptera: Thripidae). Journal of Entomological Science 26, 436–442.

Schuster, D.J., Stansly, P.A., Polston, J.E., 1996. Expressions of plant damage from *Bemisia*. In: Gerling, G., Mayer, R.T. (Eds.), *Bemisia* 1995: Taxonomy, Biology, Damage Control and Management. Intercept Limited, Andover, UK, pp. 153–165.

Sengoda, V.G., Buchman, J.L., Henne, D.C., Pappu, H.R., Munyaneza, J.E., 2013. "*Candidatus Liberibacter solanacearum*" titer over time in *Bactericera cockerelli* (Hemiptera : Triozidae) after acquisition from infected potato and tomato plants. Journal of Economic Entomology 106 (5), 1964–1972.

Shepard, H.H., 1951. The Chemistry and Action of Insecticides. McGrawHill, New York, USA, p. 504.

Silcox, D.E., 2011. Response of the Tawny Mole Cricket (Orthoptera: Gryllotalpidae) to Synthetic Insecticides and Their Residues (M.Sc. thesis). Department of Entomology, North Carolina State University, Raleigh, NC, USA, pp. 1–164.

Srinivasan, R., 2010. Safer Tomato Production Methods: A Field Guide for Soil Fertility and Pest Management. AVRDC – The World Vegetable Center Publication No. 10-710. , p. 97 Shanhua, Taiwan.

Stevens, M.A., 1986. Inheritance of tomato fruit quality components. Plant Breeding Review 4, 274–311.

Tanksley, S.D., 2004. The genetic, developmental, and molecular bases of fruit size and shape variation in tomato. The Plant Cell 16, S181–S189.

Testa, R., Trapani, A.M.D., Sgroi, F., Tudisca, S., 2014. Economic sustainability of Italian greenhouse cherry tomato. Sustainability 6, 7967–7981.

Trumble, J.T., Butler, C.D., 2009. Climate change will exacerbate California's insect pest problems. California Agriculture 63, 73–78.

Vatsanidou, A., Bartzanas, T., Papaioannou, C., Kittas, C., 2011. Efficiency of physical means of IPM on insect population control in greenhouse crops. Acta Horticulturae 893, 1247–1254.

Venette, R.C., Davis, E.E., Zaspel, J., Heisler, H., Larson, M., 2003. Mini Risk Assessment: Old World Bollworm, *Helicoverpa Armigera* Hübner (Lepidoptera: Noctuidae). University of Minnesota, Saint Paul, MN, USA. https://www.aphis.usda.gov/plant_health/plant_pest_info/owb/downloads/mini-risk-assessment-harmigerapra.pdf.

Walker, C.S., 2012. The Application of Sterile Insect Technique against the Tomato Leafminer *Liriomyza bryoniae* (Ph.D. thesis). Department of Life Sciences, Imperial College, London, UK, pp. 1–122.

Ware, G.W., Whitacre, D.M., 2004. History of pesticides. In: The Pesticide Book. sixth ed. Meister Pro Information Resources, Willoughby, OH, USA.

Weese, T.L., Bohs, L., 2007. A three gene phylogeny of the genus *Solanum* (Solanaceae). Systematic Botany 32, 445–463.

Weintraub, P.G., Berlinger, M.J., 2004. Physical control in greenhouses and field crops. In: Horowitz, A.R., Ishaaya, I. (Eds.), Insect Pest Management, Field and Protected Crops. Springer-Verlag Berlin Heidelberg, Germany, pp. 301–318.

Zalom, F.G., Zalom, J.S., 1992. Stink bugs in California tomatoes. California Tomato Grower 35, 8–11.

Zhang, J., Khan, S.A., Hasse, C., Ruf, C., Heckel, D.G., Bock, R., 2015. Full crop protection from an insect pest by expression of long double-stranded RNAs in plastids. Science 347 (6225), 991–994.

Section II

Global Pests of Tomato

Chapter 2

Aphids: Biology, Ecology, and Management

Thomas M. Perring[1], Donatella Battaglia[2], Linda L. Walling[1], Irene Toma[3], Paolo Fanti[2]

[1]University of California, Riverside, CA, United States; [2]Universita delgli Studi della Basilicata, Potenza, Italy; [3]Comprehensive School "Jannuzzi-Di Donna", Andria, Italy

1. INTRODUCTION

Aphids, as a group, represent some of the most damaging insects to agriculture in the world. They were the subject of numerous research studies, and the topic of many books, including Dixon (1973, 1985), Harris and Maramorosh (1977), Minks and Harrewijn (1987), Blackman and Eastop (2000), and van Emden and Harrington (2007). While it is difficult to quantify the worldwide losses and control costs related to aphid feeding on tomato, it is clearly substantial. Aphids damage plants in three major ways. First, they remove plant sap from phloem sieve elements, which weaken the plants resulting in lower quality and quantity of fruit. Heavy infestations can result in plant death. Second, due to the fact that phloem is an amino acid poor substrate, aphids must process a large quantity of phloem to gain the products necessary for protein synthesis which they use to produce offspring. With a modification of their gut into a filter chamber, aphids are able to shunt large quantities of phloem which is excreted in a carbohydrate-rich exudate that is termed "honeydew." Aphids produce large quantities of honeydew which can cover the leaves and fruit. The sugary substrate provides a suitable medium for the growth of black sooty mold fungi (from the genera *Capnodium*, *Fumago*, *Scorias*, *Antennariella*, *Aureobasidium*, and *Limacinula*). On the foliage this sooty mold can become so thick that it reduces photosynthetic activity on the leaves resulting in poor quality and quantity of fruit. On the fruit, the honeydew and sooty mold must be removed prior to processing the tomatoes or packaging them for fresh market. This is an added cost for tomato producers. The third, and perhaps most costly type of aphid damage, is as efficient vectors of a number of plant viruses. This chapter discusses the biology, ecology, and management of aphids that attack tomato.

2. IDENTIFICATION, BIOLOGY, DISTRIBUTION, HOST RANGE, AND SEASONAL OCCURRENCE OF MAJOR APHID PESTS ON TOMATO

Blackman and Eastop (2000) noted 12 "mostly very polyphagous aphid species" found on tomato. These included *Aphis craccivora* Koch, *Aphis fabae* Scopoli, *Aphis gossypii* Glover, *Aphis spiraecola* Patch, *Aulacorthum solani* (Kaltenbach), *Brachycaudus helichrysi* Kaltenbach, *Macrosiphum euphorbiae* Thomas, *Myzus ornatus* Laing, *Myzus persicae* (Sulzer), *Rhopalosiphum rufiabdominale* (Sasaki), *Smynthurodes betae* Westwood, and *Toxoptera aurantii* (Boyer de Fonscolombe). Holman (2009) listed 15 species on tomato, which included many of those mentioned by Blackman and Eastop (2000) plus *Aphis nasturtii* Kaltenbach, *Brachycaudus (Acaudus) cardui* L., *Brachyunguis harmalae* B. Das, *Brachyunguis plotnikovi* Nevsky, *Macrosiphum rosae* L., *Myzakkaia verbasci* Chowdhuri, Basu, Chakrabarti & Raychaudhuri, bringing the total number of aphid species that feed on tomato to 18. Many of these species invade tomatoes when they are grown in close proximity to the aphid's primary host plant. This was the case for *R. rufiabdominale*, a grain aphid which moved from outside into glasshouse tomatoes and overwintered on tomato and pepper (Zilahi-Balogh et al., 2005). While any of the species mentioned may be economic pests of tomatoes in the local context, such as *A. gossypii* (Kobari and Noori, 2012; Helmi and Mohamed, 2016), two species are considered to be global pests, *M. euphorbiae* and *M. persicae*. These aphids will be the focus of this chapter.

The potato aphid, *M. euphorbiae*, "undoubtedly originated in North America" (Blackman and Eastop, 2007). It was first identified in the UK in 1917 and quickly spread throughout Europe on potato (Eastop, 1958). In addition to North America and Europe, it has long been established in Central and South America and Africa, and recently has spread in eastern Asia (Blackman and Eastop, 2007). It is a highly polyphagous species feeding on more than 200 plant species in 20 different families (Blackman and Eastop, 2007). During the winter the eggs are laid on many different primary hosts (for example *Rosa*, *Solanum*, *Euphorbia*, and *Lycium* spp.), while migrants spend the summer on a variety of secondary

Sustainable Management of Arthropod Pests of Tomato. http://dx.doi.org/10.1016/B978-0-12-802441-6.00002-4

hosts. In particular, they colonize the Solanaceae, especially potato and tomato (Blackman and Eastop, 2000), and plants in the Cucurbitaceae (melon), Chenopodiaceae (beet), Compositae (lettuce), Leguminosae (bean), Rosaceae (citrus). They are commonly found on numerous ornamental plant species. There are two color morphs, pink and green, and the younger tomato leaves are preferred by the pink form, while the younger and middle leaves are preferred by the green form (Walker et al., 1984a). The pink form tends to predominate older plants (Walker et al., 1984b).

Blackman and Eastop (2000) and Stoetzel and Miller (1998) provide general descriptions of *M. euphorbiae*. It is a relatively large aphid with two color biotypes; yellowish-green or pinkish-red (Fig. 2.1). Wingless adult females (apterae) are 1.7–3.6 mm long, and pear-shaped in appearance. The body is often shiny and the eyes are distinctly reddish. The antennae have six segments, with two to six secondary sensoria on the basal half of segment three. Antennae are generally dark apically, but sometimes are entirely dark. The legs, siphunculi (cornicles) and cauda (tail) are usually the same color as the body, but the siphunculi can be darker toward the apices. The siphunculi have a slight apical constriction with several rows of polygonal reticulations in the constricted areas. They are approximately 6–11 times as long as wide. The cauda is also relatively long, more than twice as long as wide. It is elongate, finger-shaped and with 8–10 lateral setae and 2–3 dorsal pre-apical setae. The legs are also noticeably long. Wingless immature forms are rather long-bodied and paler than adults, with a dark spinal stripe and a light dusting of whitish-gray wax. The dark stripe on the back is sometimes seen in adults.

Winged adult females (alatae) are 1.7–3.4 mm long, although sometimes they can be noticeably larger than apterae of the same population. Different shades from green to pink characterize the color of the body. They have pale-green to yellow-brown thoracic lobes, with the antennae and siphunculi being darker than in the apterae. The hind wings have two characteristic oblique veins. The antennae are 6-segmented, with 10–18 secondary sensoria on the basal third of the third segment. The central stripe on the back is much less distinct in winged forms (Stoetzel and Miller, 1998; Blackman and Eastop, 2000).

The main characteristics distinguishing *M. euphorbiae* from other aphid species are the length of the reticulated area on the siphunculi, the number of sensoria on the third antennal segment, and the tapering shape and presence of lateral hairs on the cauda (Palmer, 1952). *M. euphorbiae* has a diploid chromosome number of $2n = 10$ (Blackman and Eastop, 2000; Monti et al., 2011). Keys to distinguish *M. euphorbiae* from other aphid species were presented by Denmark (1990), Stoetzel et al. (1996), and Stoetzel and Miller (2001).

The biology of *M. euphorbiae* can be heteroecious holocyclic (=alternating host, sexual reproduction) in certain temperate climates, and autoecious anholocyclic (=single host, parthenogenetic reproduction) in warmer tropical climates. For both types, aphid females can produce apterae (wingless) and alate (winged) forms. The alates are produced when the population needs to migrate to another host, which can be triggered by crowding or queues that the female receives from a declining host plant.

For holocylic populations, day length, temperature, host quality, parent type, and genetic factors play a role in the production of different offspring (MacGillivray and Anderson, 1964; Lamb and MacKay, 1997). In the spring, when day length is expanding, alate females migrate from overwintering hosts into summer hosts and produce a large proportion of apterous individuals. Multiple generations of alates and apterae will utilize the secondary summer hosts until the day length and temperature signal the upcoming winter months. When conditions are right, the females will give rise to a generation of males

FIGURE 2.1 Pink color morph of the potato aphid, *Macrosiphum euphorbiae*.

and gynoparae (females) who migrate to the primary overwintering host where they mate (Shands et al., 1972). Males are attracted to females by a sex pheromone that is released from the waving hind leg of "calling" virgin females (Goldansaz and McNeil, 2003). The sex pheromone was identified as a mixture of (1R, 4aS, 7S, 7aR)-nepetalactol and (4aS, 7S, 7aR)-nepetalactol in a ratio that varied with age from 4:1 to 2:1 (Goldansaz et al., 2004). In laboratory tests, males walked toward synthetic blends of the two components in the ratios released by females, but did not respond to the individual components alone (Goldansaz et al., 2004). The female offspring (oviparae) then lay eggs on the bark of *Rosa* spp. and the aphids overwinter in the egg stage. In the spring, eggs hatch into a new generation of apterous females (fundatrix). The offspring of these fundatrices feed on the new growth in rose buds. By the second generation, some alatae are produced and most individuals of the third generation are alatae. These individuals migrate toward secondary host plants, including tomato, in the late spring and summer. This migration is influenced by the time of egg hatching and the rate of aphid development.

In Europe, and most other areas where *M. euphorbiae* is an exotic species, the life cycle is mainly anholocyclic with continuous asexual reproduction on secondary hosts. Sexual morphs are sometimes produced in small numbers (Möller, 1971). However, overwintering as eggs on *Rosa* spp. is rare in Europe (Meier, 1961); most aphids remain mobile through the winter months on weeds, potato sprouts in storage or chitting houses, or on lettuce and other crops in greenhouses. In early May or June winged forms are produced that migrate to potatoes or other field crops. A second winged dispersal occurs in July if populations are high, while a smaller migration occurs in the autumn.

The distinction between different life cycles in *M. euphorbiae* is less clear-cut than for most aphid species. A degree of "life cycle plasticity" and "genotype plasticity" has been described. *M. euphorbiae* can remain on its primary host *Rosa* spp. throughout the year, eliminating the summer secondary host plants altogether (MacGillivray and Anderson, 1964). *M. euphorbiae* also deviates from the normal pattern seen in host-alternating aphids as apterous females on secondary hosts are capable of producing winged females (gynoparae) and males. This enables *M. euphorbiae* to reproduce sexually on both primary and secondary hosts, although sexual reproduction on secondary hosts is much less common (Lamb and MacKay, 1997). Aphid populations can survive year-round in warm climates and in greenhouses, infesting foliage, stems, and fruit late in the season.

The green peach aphid or peach potato aphid, *M. persicae* (Fig. 2.2) is extremely polyphagous and has a worldwide distribution (Jansson, 2003) except where there are temperature and humidity extremes. It is probably of Asian origin where it's primary host, *Prunus persica* (L.) Batsch, is native. Within its range, it is relatively cold resistant. Howling et al. (1994) described the mortality of aphids at various cold temperatures and their results suggested that an acclimatized overwintering population of *M. persicae* would persist without significant mortality after a period of 7–10 days with −5°C temperatures each night.

M. persicae is a highly variable species; strains, races, and biotypes have been distinguished based on morphology, color, biology, host-plant preference, ability to transmit viruses and insecticide resistance (van Emden et al., 1969). The eggs measure about 0.6 mm long and 0.3 mm wide, and are elliptical in shape. Eggs initially are yellow or green, but soon turn black. Mortality in the egg stage might be quite high. Nymphs are initially greenish, but soon turn yellowish, greatly resembling viviparous (parthenogenetic, nymph-producing) adults. Adult wingless parthenogenetic females are oval-bodied, 1.2–2.1 mm in body length (Blackman and Eastop, 2000), and can vary in color from whitish green, pale yellow-green, gray green, mid-green, dark green, to pink or red (Blackman and Eastop, 2007). Apart from genetically determined color

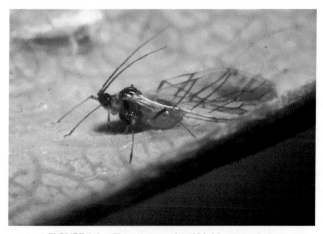

FIGURE 2.2   The green peach aphid, *Myzus persicae*.

variations, they are more deeply pigmented green or magenta in cold conditions. Immature stages are quite shiny, but adults are less so. Winged morphs have a black central dorsal patch on the abdomen. Immatures of the winged females are often pink or red, especially in autumn populations, and immature males are yellowish (Blackman and Eastop, 1984). The distinguishing characters of the *M. persicae* group that can be seen with a hand lens or under the microscope are the convergent inner faces of the antennal tubercles in dorsal view, and the very slightly clavate (club-like) siphunculi, which are usually dark-tipped and more or less as long as the terminal processes of the antennae. *M. persicae* alate virginoparae derived from overwintering eggs on *Prunus* have cylindrical cornicles, whereas those from populations derived from overwintering virginoparae have clavate cornicles.

van Emden et al. (1969) reviewed the life cycle of *M. persicae*. This aphid can complete a generation in 10–12 days, with over 20 generations per year in mild climates. Blackman (1974) discussed the life cycle variability of *M. persicae* on a worldwide basis. It is heteroecious holocyclic (host alternating, with sexual reproduction during part of life cycle) between *Prunus* and summer host plants, but anholocyclic on secondary (summer) hosts in many parts of the world where peach is absent, and where a mild climate permits active stages to survive throughout the winter. It is usually anholocyclic in the tropics and subtropics, with exceptions. For example, Ghosh and Verma (1990) reported apterous oviparous females of *M. persicae* for the first time from India, collected on *P. persica*.

For the holocyclic populations, alate virginoparae created from the fundatrices migrate from the overwintering hosts to summer hosts in the spring. From these females, a number of apterous female generations occur, with successive generations of apterous and alate virginoparae born viviparously by thelytokous (all female) parthenogenesis. These females develop on summer hosts until the day length shortens in the fall (critical photoperiod between 12.5 and 14 h in Europe) and the temperature falls below a certain threshold. These environmental conditions induce autumn migrants (gynoparae) to migrate back to peach. Gynoparae will attempt to colonize a range of trees and shrubs, but the sexual part of the cycle is only completed on *P. persica* and close relatives. Gynoparae produce oviparae (mating females) that feed on peach leaves. Males are produced after gynoparae (1 month later in a study from Italy) on the summer hosts, and migrate independently to peach, where they mate with the oviparae, which by then have become adults. Males appear to be attracted by a sex pheromone released by sexual females, and are also attracted to the smell of the winter host (Tamaki et al., 1970). Oviparae lay 4–13 eggs, usually in crevices around and in axillary buds. Up to 20,000 eggs may occur per *P. persica* tree, although they average around 4000. Clearly, there is a large variation in egg numbers between trees (van Emden et al., 1969). The eggs overwinter in diapause, requiring a period of chilling to develop, and are extremely cold resistant (surviving temperatures as low as −46°C). Hatching coincides with the swelling of flower buds, which provide food for fundatrices (van Emden et al., 1969). In the second generation, both apterous and alate females are produced. Both of these morphs continue producing offspring, and as the nutritional suitability of the peach tree declines, the production becomes primarily alates, which migrate to summer hosts. Although the winter host is limited to *Prunus* and close relatives, the array of summer hosts is vast with plants from over 40 families (Blackman and Eastop, 2000). These include Brassicaceae, Solanaceae, Poaceae, Leguminosae, Cyperaceae, Convolvulaceae, Chenopodiaceae, Compositae, Cucurbitaceae, and Umbelliferae. These summer hosts involve many economically important plants, including tomato.

Genetically, *M. persicae* normally has $2n=12$ chromosomes (Blackman et al., 1978), but chromosomal variation exists (Blackman, 1971). In this aphid, $2n$ has been reported to be 12, 13, and 14. This variation can be related to local populations. For example Cognetti (1961) reported $2n=14$ from aphids collected in Italy and the same diploid number was also reported from India (Sethi and Nagaich, 1972). According to Blackman (1971), all of the dark green biotypes have a chromosome complement of either $2n=13$ or 14. The aphids with $2n=13$ differed from normal diploids ($2n=12$) in having one unpaired element and two shorter autosomes. The aphids with $2n=14$ seemed to have two unpaired elements and four shorter fragments (Gautam et al., 1993). Morphological differences were recorded in the $2n=13$ and $2n=14$ aphids. Blackman (1980) stated that the aphids with $2n=13$ are evolutionarily older with the $2n=14$ aphids evolving later.

3. DAMAGE, ECONOMIC THRESHOLDS, AND LOSSES

Aphids damage tomato directly by inserting their stylets into phloem sieve elements and withdrawing sap from the plant's phloem, and indirectly by excreting honeydew on which sooty mold grows and by vectoring plant viruses. As phloem-feeders, they withdraw plant resources for their growth and reproduction (Dedryver et al., 2010). Common visible symptoms may be chlorosis, necrosis, wilting, growth stunting, and malformations (Goggin, 2007). Among the aphid species that attack tomato, *M. euphorbiae* causes the most significant direct damage (Strand, 1998), particularly during the period from 6 to 8 weeks before harvest (University of California IPM program, http://ipm.ucanr.edu/PMG/r783301711.html). Symptoms of *M. euphorbiae* attack are foliar necrosis at the feeding site and chlorosis that extends beyond the point of infestation (Strand, 1998; Goggin, 2007). Continued feeding by large colonies results in distortion of leaves and stems, and the plant can be stunted; severe infestations can result in plant death. In comparison, severe *M. persicae* infestation results in moderate wilting, which is usually not considered a problem unless plants are water-stressed (University of California IPM Program, http://ipm.ucanr.edu/PMG/r783300711.html). Excretion of large quantities of honeydew promotes the development of black sooty mold fungi which covers foliage and fruit. This sooty mold reduces the photosynthetic capacity of foliage and must be washed from tomato fruit before it can be marketed, adding cost to the tomato grower.

The extent of damage caused by *M. euphorbiae* to tomato depends upon both aphid and plant genotype. There is considerable difference in tomato variety susceptibility to potato aphid feeding. For example, in an experimental trial where plants were artificially infested with the same initial number of *M. euphorbiae* females, plant height and dry mass reduction caused by aphid infestation was significantly greater in the cultivar Rio Grande than the other two cultivars tested, Scintilla and Beefmaster (Rivelli et al., 2012). In the case of the last two cultivars, aphids didn't cause any significant growth reduction compared to the noninfested control. Tomato cultivars carrying the *Mi* gene are known to be resistant to the potato aphid (Rossi et al., 1998; Kaloshian et al., 2000; De Ilarduya et al., 2003). However, some aphid clones are able to overcome the *Mi* gene plant resistance (Goggin et al., 2001). Indeed there is tremendous genetic variability in aphid populations as demonstrated by Raboudi et al. (2011) and this variability is likely important in aphid/tomato variety interactions. They showed three distinct genetic groups among just 15 *M. euphorbiae* populations in Tunisia. Virulent potato aphid genotypes can establish comparable population densities on resistant (*Mi+*) and susceptible (*Mi–*) tomato cultivars (Rossi et al., 1998; Goggin et al., 2001). In addition, avirulent aphid clones, whose population growth is significantly lower on resistant versus susceptible tomato plants, show quantitative and qualitative differences in their virulence (Cooper et al., 2004). The mode of action of *Mi*-mediated resistance seems to differ from one aphid population to another, and includes feeding deterrence and antibiosis (Hebert et al., 2007).

The virulence variability of different aphid populations within a single aphid species, together with variations due to weather, timing of infestation, and diversity of cultural conditions, makes it difficult to precisely assess the economic losses caused by aphids. In general, economic loss assessment requires information about the relationship between pest abundance and yield loss. This relationship involves a complex set of interactions between the insect, the host plant, and the environment. Two approaches are possible: to build a model for yield losses based on plant growth analysis or, to empirically assess yield loss through observations and experimental trials in field and semi-field conditions (Dent, 1991). West (1946) and Walgenbach (1997) attempted the empirical approach to assess the economic damage caused by the potato aphid to tomato crops and to establish an economic threshold level for this insect. In the earlier study, tomato plants grown in a very controlled greenhouse environment were used to show that 40 *Macrosiphum solanifolii* Ashmead (=*M. euphorbiae*) aphids per gram of plant tissue was a critical infestation level, resulting in 33.7–42.2% plant weight reductions (West, 1946). Walgenbach's (1997) observations were based on aphid infestations in the field. Infestations were maintained below fixed threshold levels by applying an aphicide. Threshold levels of 10%, 25%, 50%, 65%, and 100% infested leaves were compared. Walgenbach (1997) found that yields were consistently lower in the non-treated control, 100%, and 65% infested leaf treatments compared with the other treatments, although differences were not significant. He also observed a reduction in fruit quality that appeared to be the major cause of profit–loss in the high aphid density treatments. Reductions in fruit quality related to aphids were caused by sunscald.

Ideally, the number of insects feeding on a plant and the duration of the infestation will influence the extent of yield loss (Dent, 1991), so the level of attack should be expressed as "insect-days." In addition, the effect of each developmental stage should be experimentally quantified and aphid counts should be adjusted to "adult equivalents" (Wratten et al., 1979). Walgenbach's (1997) approach to the assessment of the economic damage caused by potato aphid on fresh market tomato, which used the percentage of infested leaves as an attack index, may not be precise, but is practical. To assess infestation percentage is much less time consuming than to count the number of aphids on plants. In this way the cost of sampling is reduced and the practice can be adopted more readily by growers. The main limit of the results obtained by Walgenbach (1997) lies in the fact that it describes a single situation. The economic threshold proposed may not be valid, for instance, in the case of other tomato cultivars, other areas of cultivation, or when adopting different cultural practices. However, Walgenbach (1997) provided the first reference point for making decisions on insecticide applications against potato aphid in tomato. Threshold levels for *M. euphorbiae* recommended by Walgenbach (1997) were 50% infested leaves when using broad-spectrum insecticides, and 25% when using narrow-spectrum aphicides. Subsequent work by Wittenborn and Olkowski (2000) determined a treatment threshold on processing tomatoes to be 37% aphid positive leaflets or two aphids/leaflets. Both of these studies contributed to the current action threshold of 50% infested leaves in the United States (University of California IPM Program, http://www.ipm.ucanr.edu/PMG/selectnewpest.tomatoes.html) (Kuhar et al., 2009; Bessin, 2010). Lower action thresholds are sometimes suggested in other countries, such as Italy (10% of infested plants or less) (Viggiani, 2002).

Aphids also damage tomato by vectoring viruses. This type of damage often is much more serious than that caused by feeding activity. Blackman and Eastop (2000) note that *M. persicae* is the most important aphid virus vector, known to transmit over 100 plant viruses (Kennedy et al., 1962; Blackman and Eastop, 2007). *M. euphorbiae* is also an efficient vector, transmitting over 40 non-persistently transmitted viruses and 5 persistently transmitted viruses (Blackman and Eastop, 2007). The economic threshold for aphids as virus vectors is much lower than it is for them as direct pests (Hooks and Fereres, 2006), and it is sometimes assumed to be zero (Viggiani, 2002).

Based on the persistence of the pathogen in the vector and the way in which the vector carries the pathogen, Nault (1997) classified viruses into four categories, namely (1) non-persistently transmitted, stylet borne; (2) semi-persistently transmitted, foregut borne; (3) persistently transmitted, circulative; and (4) persistently transmitted, propagative. Katis et al. (2007) shortened these to three mechanisms. In their description, viruses that require brief (<1 min) stylet probes are non-persistent. There is no latent period between acquisition and transmission and these viruses are rapidly lost after acquisition. The second type is semi-persistent transmission in which aphids require a longer feeding probe (15 min) to acquire and transmit the pathogen. In this type of transmission also, there is no latent period and the virus generally is retained in the aphid for up to 2 days. Persistent transmission is characterized by long probes required for acquisition and inoculation. There is a latent period required for the virus to circulate through the aphid gut, through the hemolymph, and to the salivary glands where they can be transmitted. In persistent transmission, the pathogen is retained for the life of the aphid.

Most viruses transmitted by aphids to tomato are non-persistently transmitted, stylet-borne viruses. These include alfalfa mosaic virus (AMV), cucumber mosaic virus (CMV), potato virus Y (PVY), potato leafroll virus (PLV), tomato aspermy virus (TAV), and tobacco etch virus (TEV). General symptoms of viral diseases are stunting, chlorosis and necrosis, leaf malformations, poor fruit set, fruit malformation, and uneven ripening (Crescenzi and D'Agrosa, 2001). Aphid behavior during host selection is particularly important in virus epidemiology and management. Perring et al. (1999) provided a conceptual model of the sequences required for acquisition and transmission of a plant virus. The first step is that the aphid receives stimuli to land and feed on the infected plant in such a way as to acquire virions from the infected plant. When searching for a host, migrant aphids first respond to visual stimuli, landing on both host and non-host plants (Hooks and Fereres, 2006; Katis et al., 2007). After landing, aphids make brief (<1 min) stylet insertions as a conditioned reflex to tarsal contact with any solid surface (Powell et al., 1999). This probing behavior is not inhibited by the presence of repellents or deterrent cues (Katis et al., 2007). During these initial exploratory probes limited to the epidermis, aphids acquire non-persistently and semi-persistently transmitted viruses. The second step is that the aphid must receive stimuli to fly from the infected plant and stimuli to land on another (non-infected) plant. Since aphids land and probe on a series of plants while searching for their appropriate host, stylet-borne virus can be spread very rapidly, even if aphids tend to lose them quite readily when the probe is extended for a long time after virus acquisition (Goggin, 2007). During this host searching process, the virus must be retained in the aphid, the third step in the successful transmission process (Perring et al., 1999). Fourth, after landing on the non-infected plant, the aphid must receive stimuli to feed on the new plant in such a way as to inoculate it with virions. The fifth and final step in the process is the plant must be susceptible to the virions, which then cause disease.

Because of the host searching behavior of aphids and the fact that most aphid-borne tomato viruses are non-persistently transmitted, there is a high risk for the transmission of viral pathogens. Thus, the best management strategies aim to prevent the incoming attack of aphid vectors rather than to control the aphid population in response to a particular action threshold. Insecticide application to the crop is not effective in preventing the spread of stylet-borne viruses (Goggin, 2007), perhaps with the exception of applying oils (Perring et al., 1999; Katis et al., 2007). Oils are distinguished from other insecticides by the fact that they inhibit virus transmission without altering aphid-probing behavior.

4. MANAGEMENT

The management of *M. persicae* and *M. euphorbiae* on tomato has been the focus of a great deal of research. Much of the early work was on the use of insecticides because of their low cost, ease of use, and effectiveness. This single-strategy approach gave way to the concept of integrated control introduced by Stern et al. (1959) who were working on the spotted alfalfa aphid, *Therioaphis maculata* (Buckton) on alfalfa. Stern et al. (1959) suggested using a lower rate of insecticide which would allow natural enemies to survive, thus providing two mechanisms of control. From this early beginning, strategies that bring added layers of control have been studied to the point that now we think in terms of integrated pest management (IPM). This can be defined as "a comprehensive pest technology that uses combined means to reduce the status of a pest to tolerable levels while maintaining a quality environment" (Pedigo and Rice, 2006). First and foremost in any IPM program is the development of methods to quantitatively assess pest densities (Pedigo and Rice, 2006).

4.1 Monitoring and Sampling

In tomato, aphids can be monitored in a variety of ways depending on the intended purpose of the information. Harrington et al. (2007) provide an excellent review of aphid monitoring and many of the techniques they present are used for aphids on tomato. In their chapter, they divide the sampling methods into (1) "Crop Sampling," which includes in situ plant counts, destructive plant counts, vacuum sampling, sweeping, and beating, and (2) "Aerial Sampling" which includes colored water pan traps, clear or plant-colored water pan traps, sticky traps, sex pheromones (which are rare for aphids), filter nets, and suction traps (Harrington et al., 2007). To evaluate alate aphid immigration into the field, growers can use pan traps, sticky traps, filter nets, or suction traps. The yellow sticky trap (Fig. 2.3) is quick and easy to employ, and aphids are highly attracted to these traps because of the yellow color. In addition, aphids are easy to count on sticky traps. The second type of trap is the yellow water pan trap (Fig. 2.4), which is highly attractive to aphids because of the yellow color (Eastop, 1955). Lykoyressis et al. (1993) determined the relationship between aphids caught in yellow pan traps and numbers of alate and apterous aphids on the plants over 2 years. They found no relationship between trap catches and plant counts for *M. persicae* throughout the study. However, they found significant relationships for the total numbers of *M. euphorbiae* caught in traps and the total number of alates on plants ($r^2 = 0.83$) and the total aphid population on plants ($r^2 = 0.69$) for 1 year of the study. When they observed the immigration period data, they found even stronger relationships with alate numbers on plants ($r^2 = 0.93$) and total aphid numbers on plants ($r^2 = 0.80$). These data suggest that the yellow trap is not that reliable for estimating field densities, but if they are used, they will be most effective when aphids are immigrating into a field such as during the early growth cycle. Wise and Lamb (1955) came to the same conclusion when sampling *M. euphorbiae* on oilseed flax (*Linum usitatissimum* L.) finding that neither pan traps nor sweep samples were as reliable as plant counts.

Early work on sampling *M. euphorbia* in processing tomato showed that *M. euphorbiae* were found more often on upper leaves than inner canopy leaves (Walker et al., 1984a). Further studies on a large number of fields (50 fields equaling 2500 acres) supported a presence/absence sampling program, but better accuracy was determined when leaflets were sampled from upper and lower parts of the plant canopy (Wittenborn and Olkowski, 2000). Walgenbach (1994) sampling *M. euphorbiae* throughout the season determined that the distribution changed as the plant grew. He found that the highest

FIGURE 2.3   Yellow sticky traps used to trap aphids. Flat trap (left) and circular trap (right). Sometimes sticky traps catch non-target insects, such as the alfalfa butterflies in the circular trap.

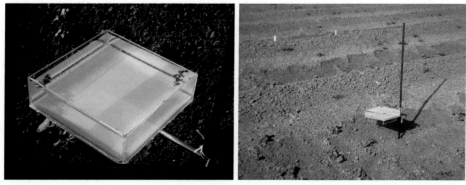

FIGURE 2.4   Yellow water pan trap used to trap aphids. Close up of trap (left) and deployed in the field (right).

percentage of aphids was on the middle plant stratum through the first 60 days of plant growth, followed by a majority of aphids on the upper stratum after 60 days. In addition, he noted that the third most recently expanded leaf from the uppermost part of the plant gave the highest level of precision among all single-leaf samples. He presented a binomial (presence/absence) sampling plan based on this third leaf, which he labeled "T3." Similar findings were presented by Hummel et al. (2004), who determined that aphid numbers on the upper canopy leaves were predictive of the inner canopy aphid densities. They also showed that the proportion of infested leaves were predictive of the average number of aphids per leaf in the upper canopy, thereby confirming that a presence/absence sampling plan would predict the aphid density on the plants. Field distributions were shown to be highly aggregated (Walgenbach, 1994) and Hummel et al. (2004) suggested that this aggregated distribution resulted from alate aphids infesting fields and females parthenogenetically reproducing where they alight. Given this clumped distribution, it is important to adequately sample a field to determine the field distribution. The University of California IPM guidelines (http://ipm.ucanr.edu/PMG/r783301711.html) note that potato aphids should be monitored from flower bloom to early fruit set by determining the presence/absence of potato aphids on "the leaf below the highest open flower on 30 plants selected at random throughout the field." An interesting variation of this was reported by Boll and Lapchin (2002), who used "project pursuit non-parametric regression" to estimate *M. euphorbiae* densities in greenhouse grown tomatoes. Citing the need to sample many plants in the greenhouse to determine the level of infestation, they suggest counting aphids in an area for 1 min, and assigning the counts to abundance classes, which then could be used to determine treatment thresholds. Boll and Lapchin (2002) noted that this strategy saved sampling time by 10-fold, was non-destructive, and best suited for large scale surveys and optimizing IPM strategies.

4.2 Cultural Control

In addition to using insecticides, biological control agents, and host-plant resistance, agricultural producers can employ a variety of agronomic techniques that influence pest densities. Wratten et al. (2007) lists 18 techniques that can be used to manipulate the crop environment. By using these methods, growers can alter how aphids settle upon and leave plants, can impact aphid reproduction and mortality and the amount of damage caused by each aphid. It is important to note that agronomic practices and environmental factors may interact in a complex way and these practices can either help to reduce pest populations, or create favorable conditions for infestations. In most cases, agronomic techniques directly affect yield, independent of pest presence and control (Clark et al., 1999).

Manipulating the physical environment is very different between the open field and under greenhouse conditions. When tomato is grown in greenhouses, the walls act as physical barriers that deny aphid access to plants. Greenhouse walls are made of solid plastic or glass in cold climate regions, and they require heating in the winter months. In the Mediterranean climatic zones, which do not require heating in the winter, greenhouses may be entirely covered with insect exclusion screens (IES) (Weintraub, 2009). The advantage of mesh screening is that it permits movement of air and helps reduce humidity, which is important for the control of plant pathogens. IESs have proven to be very effective in preventing pest infestations (Vatsanidou et al., 2011), and this is important in the prevention of diseases caused by viral pathogens transmitted by aphids and other phloem-feeding insects. Of course, greenhouses must be properly constructed with double screened entrances, and be properly maintained. Technical features of IES have been reviewed by Weintraub and Berlinger (2004). Polyethylene films and nets may contain color additives that alter the transmitted light spectrum, interfering with the vision of insects (Edser, 2002; Diaz and Fereres, 2007; Ben Yakir et al., 2012). UV-blocking materials reduce the attraction of winged aphids to the outside walls of the greenhouse and prevent ingress of aphids when gates are opened for ventilation (Costa et al., 2002; Kumar and Poehling, 2006). However, the effects of these plastic films on biological control agents and pollinator species must also be considered. Results concerning predator and parasitoid foraging behavior in altered conditions of light are few and discordant (Weintraub, 2009). Performances of individual species are expected to be more or less negatively affected by UV-blocking materials according to the relative importance of chemical cues compared to visual ones in host location and acceptance (Diaz and Fereres, 2007; Vatsanidou et al., 2011). In the case of the aphid parasitoids *Aphidius colemani* Viereck and *Aphidius matricariae* Haliday, host finding and fecundity are not adversely affected by UV-absorbing plastics (Chyzik et al., 2003; Chiel et al., 2006). Effects of UV-light blocking materials on pollinator's behavior may be important as ultraviolet-sensitive photoreceptors play a pivotal role in orientation and in the perception of color and shape of flowers (Diaz and Fereres, 2007). All these must be considered carefully as the incorporation of the bumblebee (*Bombus terrestris* L.) hives into greenhouses is a common practice to increase tomato yield. So if IES have an important role in preventing aphid infestations in greenhouses, the pros and cons of using a UV-absorbent material should be considered.

In the case of open-field tomatoes, mulches may reduce aphid settling to some extent; the amount of impact varies according to type of mulch (Weintraub and Berlinger, 2004). Mulches may be natural or synthetic. Natural mulches are made of straw, compost, peat moss, bark chips, sawdust, or other natural material. Synthetic materials include aluminum and various types of plastics. Mulches used to control weeds, to retain soil water, and to promote crop maturation may also repel insect pests. Aphid landing preference is mainly determined by the degree of contrast between the yellow-green color of plants and the color of the soil (background color) (A'Brook, 1968). The best results have been obtained by using plastic UV reflective mulches (aluminum-painted or metallized mulch) as a crop cover (Wolfenbarger and Moore, 1968; Schalk and Robbins, 1987; Kring and Schuster, 1992; Csizinszky et al., 1995; Yoltas et al., 2003; Doring et al., 2004; Summers et al., 2004). Another strategy that has been tested is to use floating row covers. These have been shown to reduce *M. persicae* on tomato (Atta-Aly et al., 1998), but the plant morphology of tomato makes them useful only in the early growth stages. With the high cost of row covers, it is unlikely they will be used for such a short time period in the early vegetative growth.

Once aphids become establish in the crop, cultural controls may be used to counteract population growth and spread. For example, tomato planting date has been shown to impact the number of aphids and their predators. Perring et al. (1988) found more aphids on young tomato compared to older tomatoes; thus they recommended planting early so that plants were older when aphid abundance was greatest. Furthermore, they found that early infestation by *M. persicae*, which did little damage to the tomatoes, provided hosts for the predator *Aphidoletes aphidimyza* (Rondani), which established and provided biological control against subsequent infestations of *M. euphorbiae* (Perring et al., 1988). Another study on planting date conducted in Bangladesh found that aphids were more abundant in tomatoes planted early (Ahsan et al., 2005). These two studies illustrate the importance of aphid seasonality in local regions when considering a planting date strategy. Another practice that has been evaluated is the use of intercropping other plants with tomato. Afifi et al. (1990) found that intercropping with onion or garlic significantly reduced *M. persicae* densities in the tomato. Among other agronomic practices, fertilization and irrigation have the highest impact on aphid populations as they significantly alter the food quality of host plants. Trends in the amino acid concentration of the phloem sap of host plant are highly correlated with changes in the aphid fecundity (Awmack and Leather, 2002), and fertilization with high nitrogen levels has long been known to be associated with aphid infestations (Kennedy, 1958). In addition to the total amount of nitrogen, whether it is supplied in organic or inorganic form is important. For example, the application of inorganic fertilizers may result in nutrient pulses. Nitrogen pulses cause nutrient imbalances and lead to high concentrations of foliar nitrogen. In contrast, organic farming practices promote a gradual release of plant nutrients which does not lead to spikes in nitrogen accumulation in plant tissues (Altieri and Nicholls, 2003). As a consequence, organic crops would be expected to be less susceptible to insect pests. As evidence of that, Morales et al. (2001) reported a significant reduction of *Rhopalosiphum maidis* (Fitch) infestations in corn after the application of organic fertilizers compared with synthetic fertilizers. A similar study also showed lower *M. euphorbiae* and *M. persicae* densities when organic fertilizers were used compared to synthetic fertilizers (Yardim and Edwards, 2003). Some field observations seem to contradict the link between high nitrogen concentrations in plant tissues and high level of insect damage in tomato (Letourneau et al., 1996; Letourneau and Goldstein, 2001), although aphid infestations tend to be higher in conventional farms (Letourneau and Goldstein, 2001). In agreement with field observations, Inbar et al. (2001), in a laboratory experiment with potted tomato plants, found no significant relationships between C/N ratio and insect performance. Nevertheless, they found that fertilized tomato plants were more vigorous and more suitable for insect pests. Thus, the relationship between fertilization and performance of phytophages in tomato seems to be more complex and not immediately inferable from the nitrogen content in plant tissues. Inbar et al. (2001) concluded that, "regardless of insect species or bioassay method, the results in the tomato system support the plant vigor hypothesis (Price, 1991) that predicts positive association between insect performance and plant growth."

With specific regard to aphids in tomato, Leite et al. (1999) tried to link aphid longevity to plant fertilization. They found that high nitrogen and potassium fertilization level in tomato led to an increase in leaf area, which in turn caused a reduction in the trichome density. *Solanum lycopersicum L.* trichomes, despite being mainly nonglandular, play a role in protection against insect pests (see above). Therefore, the reduction of trichome density subsequent to high levels of NK fertilization is likely to favor aphid infestations. However, Leite et al. (1999) did not observe any significant effect of plant fertilization on survival and longevity of *M. persicae* in tomato. These authors focused only on aphid longevity and did not report any observation on aphid fecundity, even though fecundity is the life history trait that we would expect to be most affected by nitrogen fertilization (Awmack and Leather, 2002). Latigui and Dellal (2009) found that percentage of NH_4^- had a direct impact on aphid fertility; the higher the rate of NH_4^+ the higher the *M. euphorbiae* fertility.

Management of soil fertility should also consider the soil microflora since it may play an important role in plant nutrition and health (Hyakumachi and Kubota, 2004). Among plant growth–promoting microorganisms, fungi of the genus *Trichoderma* and *Glomus* are known to have a symbiotic relationship with tomato. Arbuscular mycorrhizal fungi (AMF), to which fungi in the genus *Glomus* belong, are particularly important for phosphorus uptake (Conversa et al., 2007) and mycorrhizal inoculation in the soil enhances tomato yield (Conversa et al., 2007; Colella et al., 2014). However, *Glomus* (mycorrhizal fungi) and *Trichoderma* (non-mycorrhizal fungi) species seem to significantly differ in their direct effect on aphid populations, at least in the case of potted tomato plants in controlled conditions (Guerrieri et al., 2004; Battaglia et al., 2013). Tomato plants colonized by *Trichoderma longibrachiatum* Rifai MK1 were shown to be more suitable hosts for *M. euphorbiae* (Battaglia et al., 2013). Feeding on colonized plants, aphids displayed faster developmental time and higher reproduction. However, aphid performance was reduced when tomato roots were colonized by *Glomus mosseae* (T. H. Nicolson & Gerd) Gerd & Trappe (Guerrieri et al., 2004). Indirectly, both plant growth–promoting fungi enhance the attractiveness of the aphid parasitoid *Aphidius ervi* Haliday and the generalist predator *Macrolophus pygmaeus* (Rambur) (Guerrieri et al., 2004; Battaglia et al., 2013). However, the levels of parasitoid attractiveness to plants colonized by *T. longibrachiatum* MK1 were lower than the attractiveness to aphid-infested plants (Sasso et al., 2007). *M. pygmaeus* develops faster on *Trichoderma*-colonized plants (Battaglia et al., 2013). Furthermore, *Rhizophagus irregularis* (Syn. *Glomus intraradices* Schenk and Smith) positively affects the foraging behavior and the life history traits of *M. pygmaeus* (Duran Prieto et al., 2016). This mirid is a zoo-phytophagous predator (Alomar et al., 2002) that uses the plant mainly as a source of water but keeps a phytophagous behavior in the youngest stages when there is a scarcity of prey. These fungi somehow promote a faster development of this predator, possibly due to an increased nutritional value of plant tissues. Thus, while colonization by *Glomus* reduces aphid infestations, the effect of colonization by *Trichoderma* is difficult to predict. The final effect depends on whether the enhanced performance of natural enemies is sufficient to counteract the positive nutritional impact of *Trichoderma*-colonized plants on aphid performance. It should also be noted that there is no information on how these fungi affect damage thresholds. It is possible that plant tolerance to pests might increase as a consequence of better plant growth and enhanced vigor.

In addition to plant growth–promoting fungi, a number of plant growth–promoting rhizobacteria (PGPR) have also been identified. Among them, the rhizobacterium *Bacillus subtilis* Ehrenberg is promising for protection of tomato from diseases and insect pests. *B. subtilis* isolates, obtained from the rhizosphere soil of healthy tomato plants, negatively affect several growth parameters of *A. gossypii*, resulting in a reduction of aphid numbers and CMV incidence in the field (Sudhakar et al., 2011).

Field application of growth-promoting bacteria and fungi in tomato has not always produced the expected effects in terms of control of aphids and viral pathogens transmitted by them (Zehnder et al., 2001; Colella et al., 2014). Variability in the response of plants to growth-promoting organisms may be the result of specific fungal or bacterial strains used (Klironomos, 2000; Hart and Reader, 2002). Furthermore, tomato is characterized by a high degree of genetic variability and plant responses to environmental stress and cultivation techniques can change with the cultivar (Srinivasa Rao et al., 2001; Rivelli et al., 2013). As a consequence, suitability for aphids in tomato may be very different among genotypes (Goggin et al., 2001; Cooper and Goggin, 2005; Rivelli et al., 2013).

Volatile organic compounds (VOCs), which are products of defensive pathways (Colby et al., 1998), are directly involved in plant–aphid interactions by playing a role in the attraction of winged morphs (Visser et al., 1996; Digilio et al., 2012a), in host acceptance, and eventually in aphid reproduction (Hildebrand et al., 1993). Moreover, VOCs can act as synomones by attracting aphid parasitoids (Sasso et al., 2007, 2009). Cultural practices that increase VOCs attractiveness to natural enemies tend to increase the efficiency of biological control.

4.3 Resistance of Tomato Plant to Aphids

Plants offer a wide variety of resistance mechanisms to aphids. In his review, van Emden (2007) categorized 12 distinct mechanisms of resistance to aphids using the 3 classifications originally proposed by Painter (1951): antixenosis, antibiosis, and tolerance. A number of studies have evaluated the performance of commercial tomato varieties against *M. euphorbiae*, *M. persicae*, and *A. gossypii* (Kobari and Noori, 2012; Rivelli et al., 2012, 2013; Helmi and Mohamed, 2016). These studies provide valuable information when selecting certain varieties to grow in specific geographical locations.

Tomatoes have been used extensively in resistance research due to unique genes that have been introduced from wild tomato into commercial varieties. Several wild species have been studied extensively, including *Solanum* (=*Lycopersicon*) *cheesmanii* Riley (Simmons et al., 2003, 2005; Simmons and Gurr, 2004, 2006), *Solanum* (=*Lycopersicon*) *hirsutum* (=*Solanum habrochaites* S. Knapp & D.M. Spooner) (Quiros et al., 1977; Kok-Yokomi, 1978; Musetti and Neal, 1997a,b; Leite et al., 1999; Simmons et al., 2003; Simmons and Gurr, 2004; Kohler and St. Clair, 2005; Antonious and Snyder, 2008; Antonious et al., 2014), *Solanum* (=*Lycopersicon*) *pennellii* (Correll) D'Arcy (Goffreda et al., 1988, 1989; Goffreda and Mutschler, 1989; Rodriguez et al., 1993; Simmons and Gurr, 2004, 2006; Simmons et al., 2003, 2005; Kohler and St. Clair, 2005; Antonious and Snyder, 2008; Leckie et al., 2016), and *Solanum* (=*Lycopersicon*) *peruvianum* L. (Kok-Yokomi, 1978; Kaloshian et al., 1995). From these wild species have come a number of traits that confer resistance to aphids, including leaf volatile and non-volatile metabolites, many of which are produced and stored in trichomes.

4.3.1 Leaf Trichomes

The constitutive defense of the tomato plant is associated with the trichomes, plant hairs that are unicellular or multicellular appendages arising from epidermal cells. "Pubescence" refers to the set of trichomes that cover the surface of a plant; it confers resistance to a broad spectrum of pests. Trichomes are of various shapes and sizes (see review by Glas et al., 2012). In many species, the intraspecific variability of trichome type and density is often related to environmental parameters (Levin, 1973). Although trichomes have been used by many authors for taxonomic purposes, their real significance was ignored for a long time. It is now clear that trichomes play an important role in plant defense, especially with regard to insects (Glas et al., 2012).

The defensive role of trichomes has been studied in detail in Solanaceae. *Solanum* spp. have seven trichomes types with two primary roles in plant defense (Luckwill, 1943; Reeves, 1977; Seithe, 1979; Glas et al., 2012). First, the presence of a dense mat of hairs on plant surfaces constitutes a physical barrier and confers resistance, since it reduces the access of insects to the surface and limits the possibility of feeding. For example, *M. euphorbiae* avoid excessively hairy plants, including *S. hirsutum* (Quiros et al., 1977). Trichomes may also hinder the ovipositional activity of insects. In some cases, however, females have better adhesion due to pubescence and therefore they prefer leaves with the presence of trichomes (Robinson et al., 1980). The importance of trichome density for aphid resistance is also exemplified in the *Woolly* mutant of *S. lycopersicum*, which appears gray due the dramatic increases in the number of non-glandular trichomes (types II, III, and V) (Shilling, 1959; Tian et al., 2012). This increase in trichome density reduces the amount of aphid and leaf miner feeding on tomatoes (Chai et al., 2002).

Second, trichomes can possess exudates that are toxic to aphids (Levin, 1973). These exudates are released when broken by insects (Kennedy, 2003), and they can poison aphids by contact, ingestion, and/or inhalation. They can prevent insects from moving and feeding, thereby impacting survival. In their review, Glas et al. (2012) described the various compounds found in glandular trichomes and they identified the biochemical pathways of these trichomes. They noted the value of these structures for breeding resistant plants and touched on methods that might be employed for engineering resistant plants based on glandular trichomes.

The glandular type IV trichomes are abundant in wild tomatoes and are correlated with broad-spectrum arthropod resistance (Glas et al., 2012). Luckwill (1943) observed that the type IV trichomes were completely absent in cultivated *S. lycopersicum*. As early as 1938, McKinney (1938) suggested that *M. persicae* migrants were unable to colonize tomato because the insects became entangled in a "gum-like secretion or exudate from the tomato foliage." While the main mechanism of resistance conferred by type IV glandular trichomes is entrapment (Johnson, 1956; Goffreda et al., 1988), aphid mortality is often noted (Simmons et al., 2003, 2005).

The chemical constituents of type IV trichomes differ in the wild tomato species ranging from sesquiterpenoids to methyl ketones to acyl sugars. In relation to resistance to aphids, the trichomes of *S. pennellii* are best characterized. *S. pennellii* has a high density of type IV glandular trichomes on leaves, stems, sepals, and fruit epidermal surfaces. The exudates of these type IV trichomes irritate, entrap, and kill insects, including aphids (Lemke and Mutschler, 1984; Rodriguez et al., 1993). In addition, Goffreda et al. (1988) noted that *M. euphorbiae* settled less frequently on *S. pennellii* with type IV trichomes, had a delay in time to first probe, had fewer probes, and spent less time feeding on the plant than they did on *S. lycopersicum*, which lacked type IV trichomes. Goffreda et al. (1989) demonstrated that the active compound in the *S. pennellii* trichomes was 2,3,4-tri-*O*-acylglucose, as plants treated with this acyl sugar reduced aphid settling relative to untreated plants (Goffreda et al., 1989; Rodriguez et al., 1993). While early attempts to cross *S. pennellii* with *Solanum esculentum* found only moderate level of heritability in the crosses (Goffreda and Mutschler, 1989), it is now known that production of acyl sugars is controlled by multiple genes (Leckie et al., 2012, 2013, 2014). Furthermore, Da Silva et al. (2013) determined that tomato lines with high levels of acyl sugars exhibited antibiosis-type resistance to *M. persicae*. The differences in the genes from *S. pennellii* and cultivated tomatoes that control acyl sugar production have been identified demonstrating the possibility of metabolically engineering cultivated tomato to express a spectrum of these toxic metabolites (Fan et al., 2016).

The volatiles of *S. habrochaites* var *glabratum* type IV trichomes are acutely toxic for *A. gossypii*, *M. euphorbiae*, and *M. persicae* (Williams et al., 1980; Musetti and Neal, 1997a; Da Silva et al., 2013). The key toxicity factor is the methyl ketone, 2-tridecanone (Eigenbrode and Trumble, 1993), which alone caused 100% mortality in *M. euphorbiae* and *A. gossypii*. The type IV trichomes of *S. habrochaites* accession LA1777 contain volatile terpenoids that provide resistance to insects (Muigai et al., 2002, 2003; Bleeker et al., 2009, 2011). The genes underlying these important resistance traits have been identified and metabolic engineering of cultivated tomatoes for these compounds is possible (Besser et al., 2009; Bleeker et al., 2009, 2011). While LA1777's monoterpenes and sesquiterpenes provide a strong resistance to lepidopteran larvae and whiteflies and have been intensively studied, their potential for aphid control was considered less compelling than *S. pennellii's* acyl sugars (Simmons et al., 2003). Entrapment and mortality of *M. persicae* was observed in the first 24 h after contacting LA1777 plants.

Finally, while not stored in trichomes, tomato's volatile six-carbon (C6) aldehydes directly reduce aphid fecundity and the C6 aldehydes and alcohols indirectly induce defenses in the leaves, which interfere with aphid population expansion (Hildebrand et al., 1993).

4.3.2 Genetics of Resistance

When host plants are attacked by aphids, there is a massive transcriptional, physiological, and metabolic reprogramming in the host plant. Despite the activation of defense-signaling networks, in many interactions, aphids have adapted to a host plant and are able to establish residency and proliferate. To deter adapted aphids, selected host-plant genotypes have evolved resistance (*R*) genes that serve as a molecular surveillance system to recognize a specific aphid species/biotype, and rapidly induce effective defenses that interfere with insect growth, development, viability, fecundity, and/or repellence. Numerous *R* genes that confer resistance to microbial pathogens have also been isolated and characterized. The majority of *R* genes encode nucleotide-binding leucine-rich receptor (NLR) proteins that recognize specific pathogen/pest effectors and enable rapid deployment of defenses (van der Hoorn and Kamoun, 2008; Jacob et al., 2013; Cesari et al., 2014; Wu et al., 2015a). While numerous genes conferring resistance to aphids have been identified genetically, only two *R* genes that control resistance to aphids have been cloned to date. Tomato's *Mi1.2* and the melon *Vat* both encode NLRs with a coiled domain (Milligan et al., 1998; Boissot et al., 2016).

The tomato *Mi* gene was the first aphid *R* gene characterized at the molecular level. The *Mi* locus contains two NLR-encoding genes, namely *Mi-1.1* and *Mi-1.2* (Milligan et al., 1998). The *Mi* locus originally was identified as conferring resistance to several root-knot nematode species (*Meloidogyne* spp.) (Roberts and Thomason, 1986) and later demonstrated to confer resistance to *M. euphorbiae* (Vos et al., 1998). *Mi1.2* was independently identified as *Meu-1* in a screen for *M. euphorbiae* resistance in tomato (Kaloshian et al., 1995, 1997; Rossi et al., 1998). Finally, *Mi-1.2* also confers resistance to three additional hemipteran pests: two whitefly species (*Bemisia tabaci* (Gennadius) B and Q), and one tomato psyllid (*Bactericerca cockerelli* Sulc) (Nombela et al., 2003; Casteel et al., 2006). Feeding on *Mi1.2* plants increased *M. euphorbiae* mortality, decreased female longevity, and altered feeding behaviors (Kaloshian et al., 1997; Kielkiewicz et al., 2010). However, it does not confer resistance to *M. persicae* (Civolani et al., 2010).

Mi-1.2-mediated resistance to *M. euphorbiae* is developmentally regulated. Aphid resistance is first displayed in fully expanded leaves of tomato plants that are 5 weeks old (Kaloshian et al., 1995, 1997); surprisingly, expanding leaves remain susceptible to *M. euphorbiae* throughout the life of a *Mi-1.2* plant. *Mi-1.2*-induced resistance are factors in the phloem and the mesophyll and/or epidermis (Kaloshian et al., 1997; Pallipparambil et al., 2010). De Ilarduya et al. (2001, 2004) identified another gene, *Rme1* ("resistance to *Meloidogyne* spp.") required for *Mi-1.2*-mediated resistance. *Rme1* is functional at the same physiological step as *Mi-1.2* or is upstream of *Mi-1.2*.

4.3.3 Tomato Response to Aphid Feeding

Plants express a complex defense network to deploy the chemical defenses that control pathogens and pests. At the core of this network are the defense hormones salicylic acid (SA) and jasmonic acid (JA). These hormones interact with ethylene, gibberellic acid, abscisic acid, auxin, and cytokinins to fine tune responses to insect and pathogen attack (Pieterse et al., 2012). Treatments with SA and JA and the use of mutants that interfere with the production or perception of SA and JA have demonstrated the key roles of these hormones in resistance to insects. The relative importance of SA- and JA-mediated responses on resistance to aphids and other phloem-feeding insects varies in different host plants (Walling, 2009). In tomato, JA treatments decrease *M. euphorbiae* longevity and net population growth in susceptible plants (Cooper et al., 2004; Cooper and Goggin, 2005). Congruent with these data, aphids prefer and perform better on tomato plants that are unable to perceive JA (the *jai1* mutant) and, therefore, cannot express JA-regulated defense traits (Bhattarai et al., 2007). While JA-regulated defenses are critical in aphid-susceptible genotypes, JA treatments did not enhance, nor did the *jai1* impair the aphid resistance conferred by the *Mi1.2* gene.

Unlike JA, SA-regulated defenses contribute to aphid resistance in both aphid-susceptible (*mi-1.2*) and resistant (*Mi-1.2*) tomato genotypes. Treatments of *mi-1.2* or *Mi-1.2* tomato plants with the SA analog BTH (benzo (1,2,3)-thiaiazole-7-carbothioic acid *S*-methyl ester) decreased *M. euphorbiae* population growth (Cooper et al., 2004; Cooper and Goggin, 2005; Li et al., 2006). Furthermore, aphid performance on the transgenic tomato line (*NahG*), which has low SA levels, showed that SA-regulated defenses are important in resistance in both *mi-1.2* and *Mi-1.2* tomato plants (Li et al., 2006; Thaler et al., 2010). The importance of both SA- and JA-regulated defenses in basal resistance to *M. persicae* has also been documented (Boughton et al., 2006).

Insects cause varying amounts of cellular and tissue damage depending on their feeding behaviors (Walling, 2000). It is widely thought that the amount of tissue damage that occurs during aphid feeding is less than that which occurs with caterpillars or beetles, and more than that which occurs with stealthy phloem-feeders such as whiteflies (Tjallingii, 2006; Walling, 2008). Therefore the activation of the signaling pathways activated by tissue damage varies in response to aphids and is dependent on the amount of tissue damage and introduction of effectors that influence damage-induced defense signaling. With tissue damage, there is a rise in JA, the volatile methyl-JA, and the bioactive JA conjugate (JA-Ile), which leads to activation of JA-regulated defenses (Wasternack et al., 2006). JA induces the expression of a wide variety of JA biosynthetic enzymes, antinutritive proteins (e.g., proteinase inhibitors, polyphenol oxidases, and arginase) and chemicals (e.g., chlorogenic acid and rutin), and changes in volatile blend to activate defense gene expression and attract predators and parasitoids to insect-damaged plants. JA not only increases trichome densities, it also alters trichome chemical composition to enhance resistance to insects, including aphids. However, while cellular damage occurs in response to aphid feeding, activation of the JA-regulated genes appears to occur early and transiently and is subsequently followed by activation of SA-response genes (Bhattarai et al., 2007). Recently the changes in the tomato transcriptome in response to 24, 48, and 96 h of *M. euphorbiae* feeding was determined (Coppola et al., 2013). This study showed that in an aphid-susceptible tomato plant, SA-regulated responses dominate.

In addition to the importance of SA and JA in defense responses to aphid feeding, ethylene (ET) has an important role in aphid–tomato interactions. There is an ET burst and increased expression of ET biosynthesis and response genes in response to *M. euphorbiae* feeding on both aphid-resistant and -susceptible tomato plants (Mantelin et al., 2009). In contrast, Anstead et al. (2010) found that some ethylene-biosynthetic gene RNAs were upregulated in resistant tomato compared to the susceptible control plants after *M. euphorbiae* feeding, while ET-perception genes were not regulated in resistant or susceptible tomato plants. It should be noted that relative to changes in ET-response genes in aphid-resistant melon plants after *A. gossypii* feeding, the tomato ET-response to *M. euphorbiae* was relatively weak.

Using the tomato plants with impaired ET biosynthesis or perception, Mantelin et al. (2009) showed that when ET signaling is reduced in an aphid-susceptible tomato plant (*mi1.2*), the plants have smaller *M. euphorbiae* populations. In contrast, ET does not appear to contribute to *Mi-1.2*-mediated resistance to *M. euphorbiae*. To further examine the role of ET-response factors in aphid-susceptible and resistant plants, Wu et al. (2015b) silenced the ET-response factor Pti5, which is known to have a role in resistance to the microbial pathogen *Pseudomonas syringae* Van Hall pv. *tomato*. Wu et al. (2015b) showed that *Pti5* RNAs increase in response to aphid feeding but this increase was independent of ET. Silencing of *Pti5* levels caused increases in aphid populations on both aphid-resistant and susceptible tomato plants. This raises an important feature in the tomato resistance to aphids. Multiple pathways are important in aphid resistance, particularly when combined with R-gene-mediated resistance (Cooper et al., 2004; Cooper and Goggin, 2005).

The importance of another signaling pathway in aphid resistance was recently revealed. The study of *M. persicae*-Arabidopsis interactions indicated that the disaccharide trehalose and trehalose-6-phosphate constitute a novel defense-signaling branch for curtailing aphid success. Both molecules are present in trace amounts and using *Trehalose Phosphate Synthase11* (*TPS11*) mutants, Singh et al. (2011) showed *TPS11* controls starch accumulation and the levels of *PAD4* in Arabidopsis. *PAD4* is important in regulating senescence-associated genes and in reducing aphid success. In 2012, Singh and Shah demonstrated that there are strong parallels in tomato and Arbidopsis responses to *M. persicae*. Like Arabidopsis, tomato *TPS11* and *PAD4* RNAs are induced after aphid attack, the levels of trehalose and trehalose-6-phosphate increase, and starch accumulates. Tomato's *PAD4* is induced by exogenous trehalose, and aphid success is antagonized by trehalose. While not yet supported genetically in tomato, these results suggest that trehalose signaling may be an important new defense-signaling branch in tomato's resistance against aphids.

Finally, volatiles are potent molecules that attract natural enemies to insect-infested plants and are also important in plant-to-plant communication and induction of defenses. For example, when tomato plants are in close proximity to tobacco plants engineered to emit high levels of the monoterpene β-ocimene, tomato plants had enhanced resistance to aphids (Cascone et al., 2015). The β-ocimene-exposed tomato plants had a lower number of *M. euphorbiae* settle on the plants, fewer nymphs born, and aphid body masses were lower. In addition, there were higher numbers of the parasitoid *A. ervi*. In addition, β-ocimene-exposed tomato plants had higher levels of methyl-SA and *cis*-3-hexen-1-ol (a C6 volatile), both of which are known to reduce aphid development.

4.3.4 Aphid Strategies to Overcome Resistance and "Effectors"

As aphids have coevolved with their host plants, they have developed a number of strategies that allow them to overcome plant resistance. These strategies include avoidance of plant surveillance, manipulation of signal transduction by insect effectors, and resistance to plant defense chemicals (Bansal and Michel, 2015). Aphid effectors have been identified in aphid–Arabidopsis and aphid–tomato interactions (Kaloshian and Walling, 2016). These effectors target unknown steps in plant defense signaling, but they result in enhanced aphid success. For example, three *M. euphorbiae* effectors, Me10, Me23, and Me47, have been characterized to date (Atamian et al., 2013; Kettles and Kaloshian, 2016). When these effectors were introduced into *Nicotiana benthamiana* Domin, aphid fecundity was increased, suggesting that these actively suppress *N. benthamiana* defenses against aphids. Similarly, when expressed in tomato, Me10 and Me47 increase aphid fecundity.

Aphid effectors may also be generated by aphid secondary endosymbionts. Chaudhary et al. (2014) showed that *Buchnera's* GroEL was present in the saliva of *M. euphorbiae* and it was able to induce pattern-triggered immunity in the model plant *Arabidopsis*. Aphids feeding on plants treated with GroEL had reduced fecundity. Furthermore plants engineered with GroEL had reduced fecundity in *M. persicae* and *M. euphorbiae*.

The effectors that are recognized by resistance genes are not known for any aphid–plant interaction (Kaloshian and Walling, 2016). However, there is evidence that there is diversity in *M. euphorbiae* isolates from the United States and Europe with regard to their ability to grown on tomato plants containing the *Mi-1.2* resistance gene (Goggin et al., 2001; Cooper et al., 2004; Hebert et al., 2007). While *M. euphorbiae* from France and the Netherlands were killed on the *Mi-1.2* plants, *M. euphorbiae* isolates from California, North Carolina, and New Jersey were able to overcome the *Mi1.2*-mediated resistance. These data suggest that the European *M. euphorbiae* isolates have effectors that are directly or indirectly recognized by *Mi-1.2* and thereby activate rapid and effective defenses. The *M. euphorbiae* isolates from the United States either do not express an effector recognized by *Mi-1.2* or have found an additional route to circumvent this potent aphid resistance mechanism.

4.3.5 Impact of Abiotic Factors on Resistance

Agronomic factors such as fertilization and water stress can play an important role in the development of the population of pests. Indirectly they may affect the adaptation phase of the pests to the host plant or change the level of plant resistance to the insect. For example Leite et al. (1999) found that nitrogen and potassium levels were inversely related to trichome density and idioblasts in *S. esculentum* and to idioblasts in *S. hirsutum*.

Water stress is also a common abiotic environmental factor that causes physical and chemical changes in plants, consequently influencing the relationship with insect pests (Holtzer et al., 1988). It is evident that a plant characterized by conditions of water deficit is most susceptible to infestations by phytophagous arthropods, as well as more easily infected by diseases of bacterial and fungal origin (Mattson and Haack, 1987). The literature on the response of arthropods to water-stressed plants is vast. A number of variables including the host, the aphid species, and the level and duration of water stress impacts aphid survival, growth, development, and reproduction (Huberty and Denno, 2004; Agele, 2006). Intermittent stress can have a completely different effect than water stress that is constant. Rivelli et al. (2012, 2013) subjected tomato plants infested with *M. euphorbiae* to water stress and found that *M. euphorbiae* response to water stress depended on the tomato cultivar, the intensity of the stress applied, and the speed with which the stress was developed. Interestingly, the density of *M. euphorbiae* increased during the early days of the experiment, perhaps as a function of increased fecundity initiated by elevated soluble amino acids in the early phases of the water stress. Another explanation is the deactivation of plant resistance genes.

4.4 Biological Control

The aphidophagous guild is particularly rich in species, and includes predators, parasitoids and pathogens. Nevertheless, biological control of aphids often is not satisfactory (Dixon, 2000; Snyder and Ives, 2003). This can be explained by the fact that natural enemies of aphids usually are deficient in one or more traits that an effective biological control agent should have. According to Debach and Rosen (1991), an effective biological control agent should have a high prey specificity, as well as a reproductive rate and a tolerance of environmental conditions similar to that of the pest. An important feature of aphids is that they have a very high reproductive potential and an unstable relationship with the host plant over time (Dixon, 1992; Dixon et al., 1993). Adult females that settle on a suitable host plant maximize their investment in reproduction (Powell et al., 2006). In this way, aphid colonies which are largely apterous expand rapidly as long as favorable plant and environmental conditions persist. Once these conditions change, alates appear and the aphid density on the infested plant quickly decreases. Aphid parasitoids are sufficiently specialized and reproduce rapidly in such a manner to show density-dependent responses to increasing host populations. Yet, often this occurs with a time lag which is generally equal to one generation of the host (Snyder and Ives, 2009). During this time the aphid population grows and can reach very high levels. Parasitoid densities become large enough to suppress the pest later on, but eventually too late. The economic threshold may be passed even if the parasitization rate reaches levels close to 100%.

The scientific literature regarding the use of predators and parasitoids for biological control of aphids most often refers to crops other than tomatoes (Powell and Pell, 2007). These include pepper (Rabasse et al., 1983; van Schelt, 1999), cucumber (Benninson, 1992; Benninson and Corless, 1993; Hágvar and Hofsvang, 1995; Burgio et al., 1997; Mulder et al., 1999; El Habi et al., 1999, 2000; Kuroda and Miura, 2003), strawberry (Sterk and Meesters, 1997; Tommasini and Mosti, 2001), and ornamental plants (Scopes, 1969; Hämäläien, 1980; Snyder et al., 2004). Knowledge on delivery methods, including natural enemy/prey ratio, and the effectiveness of control is widely available. But this information must be applied with caution to crops other than those referred to in the experimental trials; the host plant may play a key role (Beglyarov and Smetnik, 1977).

4.4.1 Parasitoids

For biological control of aphids, parasitoids have been used more often than predators and pathogens (Powell and Pell, 2007). One of the key advantages of aphid parasitoids is that they prey exclusively on aphids, although some parasitoids will feed on multiple aphid species. Aphid parasitoids belong to the families Braconidae and Aphelinidae, with the majority of species in Branconidae. In their review, Powell and Pell (2007) listed no successful incidences of using parasitoids for biological control of aphids on tomato. However, in the glasshouse, for *M. persicae* control they did cite augmentative releases of *A. matricariae* and *A. colemani* in sweet pepper, *Lysiphlebus testaceipes* (Cresson) in beans, and *Aphelinus abdominalis* (Dalman) on cut roses. Cota and Isufi (2009) noted decreases in potato aphid populations when *A. colemani* was released. For *M. euphorbiae*, *A. abdominalis*, and *Aphelinus asychis* Walker have been used on roses. These parasitoids species should be evaluated for aphid control in greenhouse-grown tomato. One such study showed some impact of *A. abdominalis* for control of *M. euphorbiae* particularly when other predators (*A. aphidimyza*, *Macrolophus caliginosus* Wanger, and *Dicyphus tamaninii* Wanger) were present (Alomar et al., 1997).

Other studies have revealed some interesting biological interactions with aphid parasitoids, their host and the host plant. For example, Sasso et al. (2007) and Digilio et al. (2012a) found that, as a consequence of *M. euphorbiae* attack, tomato plants released methyl salicylate and *cis*-hex-3-en-1-ol, two compounds which negatively impact the aphid but serve to attract the aphid parasitoid *A. ervi*. In another study, Digilio et al. (2012b) determined that transgenic tomato plants expressing Cry3Bb *Bacillus thuringiensis* Berliner toxin had no impact on *M. euphorbiae* that fed on the plants, nor was there any impact on the natural enemies *M. caliginosus* or *A. ervi*.

4.4.2 Predators

On tomatoes, Syrphidae (Gilbert, 1986, 2005; Verheggen et al. 2009a), Coccinellidae (Hodek and Honěk, 1996; Sarmento et al., 2007), Chrysopidae (Principi and Canard, 1984), Cecidomyiidae (Meadow et al., 1984; Nijveldt, 1988; Cota and Isufi, 2009), and spiders (Sunderland et al., 1986) are major components of the aphidophagous guild. Aphid predators are generalist in the sense that they feed on several prey species, both aphids and non-aphids. However, it is necessary to make some distinction. Coccinellids, syrphids, and cecidomyiids such as *A. aphidimyza* feed primarily on aphids (although multiple species of aphids) and only opportunistically upon other prey (Ankersmit et al., 1986; Chambers and Adams, 1986; Hodek and Honěk, 1996; Snyder and Ives, 2003). Other predator species are broad generalists. Among this group are Hemiptera (Alvarado et al., 1997), Carabidae, and spiders. In addition to polyphagy, predators have relatively long generation times (Dixon and Hemptinne, 2003; Snyder and Ives, 2009), which is not a favorable attribute for an effective biological control agent. For instance, ladybird developmental time is comparable to the average duration of an aphid colony. Therefore, coccinellid larvae have difficulty completing development in a single aphid patch and they must be able to explore multiple patches, often on different plants, to complete development (Dixon and Hemptinne, 2003). This situation is exacerbated when there is an early decline in aphid abundance. Additionally coccinellid larvae have to cope with the fact that the prey becomes scarce just when the food requirements are greatest. As a reproductive strategy, coccinellids lay just a few eggs in patches with aphid colonies at an early stage, well before aphid populations peak in abundance. The number of predator larvae is regulated further by cannibalism that may dramatically reduce survival (Dixon, 2000). This helps explain why aphidophagous predators have limited effect on aphid abundance peaks (Kindlmann and Dixon, 1999; Dixon and Hemptinne, 2003; Snyder and Ives, 2003). Moreover, predators prey on each other and they often prey on the juvenile stages of parasitoids inside aphids (Polis et al., 1989; Rosenheim et al., 1993, 1995; Ferguson and Stiling, 1996; Lucas et al., 1998; Rosenheim, 1998). This intraguild predation may have a disruptive effect on biological control (Rosenheim et al., 1993, 1995).

In certain systems, generalist predators have favorable characteristics for the biological control of aphids. These predators feed on several different prey species and sometimes they have phytophagous habits as well. For these reasons their density is independent of the densities of individual prey species (Harmon and Andow, 2004). In addition, generalist predators may already be present in the crop at the beginning of aphid infestations. Conversely, generalist predator predation on aphids may be influenced by the presence of other more favored prey. Parasitoids and generalist predators can exert a complementary effect, despite the unidirectional intraguild predation (Snyder and Ives, 2003; Weisser, 2003). Parasitoids are capable of mounting a strong numerical response to host increase, but with a time lag equal to the time required for the development of juveniles into adults. Feeding activity of generalist predators can prevent aphid populations from reaching a high peak density during this time.

Host-plant preferences must be considered when exploiting aphid natural enemies as biological control agents. High trichome density can provide major resistance against aphids (Farrar and Kennedy, 1991; Simmons et al., 2003), but also may reduce or disrupt biological control. In tomato, trichomes may interact directly or indirectly with aphid natural enemies (Kennedy, 2003; Economou et al., 2006). For instance, the ability of the lacewing *Mallada signatus* (Schneider) to prey on tomato aphids is significantly lower in the presence of densely arranged trichomes (Simmons and Gurr, 2004, 2006). Similarly, *Episyrphus balteatus* (DeGeer) is reported to fail in controlling aphids on tomato plants (Verheggen et al., 2009a), even though it is attracted to tomato volatiles released in response to *M. persicae* feeding (Verheggen et al., 2009b). These hoverfly females have difficulty landing, presumably hampered by the long type I and III trichomes (Verheggen et al., 2009a). The larvae are strongly affected by trichomes because they do not possess legs (apodal) (Rotheray, 1987). They adhere to plant surfaces by greasy smeared saliva and move by means of body contractions. Thus their attachment and mobility is hindered on hairy tomato leaves (Verheggen et al., 2009a). This negative effect depends on the combination of larval morphology, size, and mode of movement. For example, larvae of the predatory midge *A. aphidimyza* are apodal and their locomotion is similar to that of hoverflies. However their survival increases in the presence of hairy surfaces (Lucas and Brodeur, 1999). They have a smaller body size compared to syrphid fly larvae, and they move easily between trichomes. Negative effects of trichomes may also be less apparent on leg-bearing predators, especially if they have long legs allowing them to maintain some distance from the plant's surface. Some predators actually prefer plants with trichomes that provide their eggs a refuge from cannibalism and predation. Both *A. aphidimyza* and the coccinellid *Coleomegilla maculata* DeGeer lay more eggs on plants with high density of trichomes (Lucas and Brodeur, 1999; Seagraves and Yeargan, 2006). In addition, the mirid bug *Dicyphus errans* (Wolff) prefers foraging on pubescent plants (Voigt et al., 2007).

Recently, the use of predatory mirid species for the control of tomato pests in greenhouse has spread in Europe. Three mirid species, native to the Mediterranean region, have been selected for this purpose: *D. tamaninii*, *M. pygmaeus*, and *Nesidiocoris tenuis* (Reuter), all belonging to the subfamily Dicyphinae (Lykouressis et al., 2000; Alomar et al., 2002). Another species, *Macrolophus basicornis* (Stal), has also been shown to have population parameters that suggest it would be a good predator on *M. euphorbiae* and *M. persicae* (Banos-Diaz et al., 2014). A fifth species, *Dicyphus hesperus* Knight, was selected for greenhouse pest management in Canada after screening several native mirid species (McGregor et al., 1999; Shipp and Wang, 2006; Gillespie et al., 2007; Buitenhuis et al., 2013). These generalist predators are known to readily colonize tomato crops, having an impact on whiteflies, aphids, lepidopterans, and mites (Gabarra et al., 1988; Perdikis and Lykouressis, 1996; Alomar et al., 2002; Castañé et al., 2004; Gabarra et al., 2004). They may also be able to establish on crops early in the growing season, when pests are colonizing, and they remain on the target crop when prey are scarce. Unfortunately, as omnivores, their feeding on the host plant carries the risk of causing economically significant damage to the crop. The risk of crop damage in tomato varies with predator species and it is dependent on the predator-to-prey ratio. The differences in the damage caused by three of the predatory species mentioned above are mainly due to the preference for various parts of the plant (Castañé et al., 2011). For example, *D. tamaninii* prefers to feed on fruit when prey are scarce. This behavioral characteristic discourages commercial release of this predator. *N. tenuis* has a strong preference for the three uppermost leaves and the apical bud. Damage includes the abortion of flowers and small fruits, as well as stunted plant growth (El-Dessouki et al., 1976; Vacante and Tropea-Garzia, 1994; Sanchez and Lacasa, 2008; Calvo et al., 2009; Arnó et al., 2010). Yet Sanchez and Lacasa (2008) showed an increase in the final average weight of fruits that compensated for the reduction in fruit numbers due to the abortion of flowers. Nevertheless, the damage caused by *N. tenuis* may be severe in the presence of continued feeding on the plant due to lack of prey (Arnó et al., 2010). In spite of the risk of crop damage, *N. tenuis* is being broadly released in tomato greenhouses (van der Blom et al., 2009). Finally, the use of *M. pygmaeus* is generally considered safe as injury caused by this predator has not been observed on commercial crops (Malausa and Trottin-Caudal, 1996; van Schelt et al., 1996; Castañé et al., 2003). This species has been released throughout European tomato greenhouses since 1994 (Malausa and Trottin-Caudal, 1996; van Lenteren, 2003; Arnó et al., 2009).

The predatory behavior of *M. pygmaeus* against *M. persicae* and *M. euphorbiae* has been widely studied (Perdikis et al., 1999; Perdikis and Lykouressis, 2002; Lykouressis et al., 2007; Fantinou et al., 2008, 2009; Moerkens et al., 2014). This predator shows a higher predation rate and preference for smaller aphid instars. In the presence of aphids as the sole prey available, *M. pygmaeus* exhibits a type II functional response (Holling, 1959, 1966), which relies on a constant rate of attack on each prey throughout prey densities. As a result, *M. pygmaeus* might remain effective at low prey densities. *N. tenuis* also feeds on *M. persicae* under laboratory conditions (Valderrama et al., 2007). In practice, these two predatory species are not used for the control of aphids although they could be effective. They both are used against whiteflies and *Tuta absoluta* (Meyrick). Prospects for predatory mirid bugs as biocontrol agents of aphids have been recently considered for sweet pepper (De Backer, 2012; Messelink and Janssen, 2014; Pérez-Hedo and Urbaneja, 2015). A release of four *M. pygmaeus* per plant has been evaluated and shown to control small densities of aphids (De Backer, 2012). A drawback of this strategy is that *M. pygmaeus* numbers increase slowly, so that it takes a long time to establish in the crop (Lenfant et al., 2000; Castañé et al., 2006). Thus a better strategy might be to release *M. pygmaeus* into the crop at a very early, preventative stage. In addition, it would be best to follow the common practice of providing predators with eggs of *Ephestia kuehniella* Zeller as supplementary food (Lenfant et al., 2000).

4.4.3 Pathogens (Entomopathogenic Fungi)

While few aphids are impacted by bacteria and viruses, they are infected by entomopathogenic fungi. Currently, about 700 species of fungi that are pathogenic to insects and mites have been described (Vega et al., 2009). Most belong to the division Ascomycota, order Hypocreales, and the division Zygomycota, order Entomophthorales (Powell and Pell, 2007). Entomopathogenic fungal spores attach to the host surface and germinate, producing hyphae that penetrate the exoskeleton (Butt et al., 1990; Hajek and Leger, 1994). The ability to penetrate through host cuticle is a unique feature that distinguishes fungi from other pathogenic organisms, such as bacteria, viruses, and microsporidia. Because they do not need to be ingested, this feature increases the spectrum of hosts that can be used as a food source. The trophic relationship with the host can be of various types (Vega et al., 2009). Some fungal species are biotrophic, exclusively feeding on living cells. Other species are necrotrophic, feeding on dead tissues. Still other species have a mixed behavior, biotrophic in the first phase and subsequently necrotrophic. The level of specificity of entomopathogenic fungi varies from extreme specialization (Jensen et al., 2001) to extreme polyphagia (Meyling et al., 2009). Some fungal species have an intermediate level of specialization, infecting hosts of a specific systematic group. One example is *Pandora neoaphidis* (Remaudière and Hennebert) which attacks aphid species (Ekesi et al., 2005).

Symptoms associated with infection by entomopathogenic fungi are variable and often nonspecific (Deseö Kovács and Rovesti, 1992). Sometimes it is possible to observe the presence of dark spots on the cuticle at the points of hyphae penetration. These spots are caused by integument melanization following the host defensive reaction. In some cases a change of color of the entire body can be observed. In adult insects, the fungus redirects host physiological activities in its favor, effectively causing sterility in the host insect. This effect is evident within 24 h after *P. neoaphidis* or *Beauveria bassiana* (Balsamo-Crivelli) Vuillemin infection (Baverstock et al., 2006). Finally, entomopathogenic fungi can induce behavioral changes in the host. For example *P. neoaphidis* alters the response to aphid alarm pheromone ((E)-β-farnesene). Aphids, the second and third day after infection, become less sensitive to the pheromone but continue to produce it (Roy et al., 1998). The percentage of aphids that does not respond to the pheromone increases as the disease progresses.

Fungi in the order Entomophthorales have been the subject of several studies on the environmental distribution and interaction with the host plant, in order to develop techniques for biological control using fungi. *P. neoaphidis* is an example, which is widespread in temperate regions, but not presently useful for inoculation or augmentation. This species, unlike other Entomophthorales and most entomopathogenic fungi, does not produce dormant spores. However, in the presence of unfavorable environmental conditions, it can produce loricoconidia with thickened walls (Nielsen et al., 2003). Several studies have attempted to clarify the mechanism of environmental persistence of this entomopathogenic species in the absence of aphid hosts. *P. neoaphidis* can remain active during the winter, infecting small anholocyclic (=single host, parthenogenetic reproduction) populations of aphids and it can remain as inoculum deposited on plant or soil substrates (Remaudière et al., 1981; Nielsen et al., 2003). The optimum temperature for infection is around 20°C, but *P. neoaphidis* can also infect aphids at lower temperatures (5°C), compatible with winter temperatures in some temperate regions. The time required to kill the host increases with decreasing temperature (Wilding, 1973). The propagules of the fungus that do not reach the host can have a different survival rate depending on the substrate on which they fall. Conidia on vegetation germinate producing secondary conidia that can persist for up to 14 days. In the soil, however, the conidia may persist for several months, but their infectivity decreases over time (Baverstock et al., 2008). Conidia need light to germinate and infect aphids (Baverstock et al., 2008). It has also been speculated that they are sensitive to photoperiod, but this is not completely understood.

Particularly interesting are the interactions between Entomophthorales fungi and other organisms from a community ecology point of view. For example, larvae and adults of *Coccinella septempunctata* (L.) are not infected by *P. neoaphidis*

(Roy et al., 2001); therefore they are able to prey on aphids infected by the fungus (Rosenheim et al., 1995; Brodeur and Rosenheim, 2000). In addition, adults of *A. ervi* are not infected by the fungus, but the larvae of the parasitoid directly compete with the fungus inside the aphid host (Powell et al., 1986). The parasitoid is able to outcompete the fungus and complete larval development only if the deposition of the egg takes place before the fungal infection. In addition to the direct interactions described above, indirect interactions between these organisms exist. It was shown that the presence of entomophagous insects, competing with the fungus for the same resource, may facilitate the infection by *P. neoaphidis* (Pell et al., 1997; Roy et al., 1998; Roy and Pell, 2000; Baverstock et al., 2007). Although the aphidophagous species, such as *C. septempunctata*, can carry fungal conidia and accidentally contaminate the host, it is thought that the increase in the percentage of infection is mainly linked to the escape behavior of aphids, which offers greater possibilities of contact with the pathogen (Roy and Pell, 2000; Baverstock et al., 2007). In aphids, winged migrants also play an important role in the spread of the fungus (Chen et al., 2008). The percentage of infected females landing on a new host plant can exceed 30%. Fungal species that spread in this way are mainly *P. neoaphidis*, *Entomophthora planchoniana* Cornu, and *B. bassiana*. Another aspect to be considered is the ecological interaction between entomopathogenic fungi and host plants. Of particular interest is the discovery of endophyte behavior of *B. bassiana* in tomato (Ownley et al., 2008). In this capacity, it also plays a role in protecting the plant from diseases caused by plant fungal pathogens.

Fungal entomopathogens, as well as other biological control agents, are implemented with various strategies (Fuxa, 1987). In the context of augmentative biological control of aphids, entomopathogenic fungi are used to inundate the crop in much the same manner as a chemical pesticide. Insect control is achieved directly by the applied organisms without the need for epizootic spread. Several species of entomopathogenic fungi can be reared and stored easily (Vega et al., 2003). However, factors related to the production and storage of large quantities have resulted in only a dozen species of entomopathogenic fungi available in about 170 products (Faria and Wraight, 2007). These species belong mainly to the genera *Beauveria*, *Metarhizium*, *Lecanicillium*, and *Isaria*. *B. bassiana* is effectively used against whiteflies, thrips, and aphids (Gindin et al., 1996; Shipp et al., 2003; Quesada-Moraga et al., 2006). Under laboratory conditions, *Lecanicillium* isolates have proven to be more virulent against aphids species compared to *B. bassiana* and other genera belonging to the order Hypocreales (Loureiro and Moino, 2006; Åsman, 2007). *Lecanicillium longisporum* R. Zare & W. Gams (=*Verticillium lecanii*) has been used successfully against aphids in glasshouses in North European countries, where temperature and humidity requirements of this fungus are satisfied (Dedryver et al., 2010).

It should be noted that different isolates of the same fungus can have different virulence against individual insect species, and different capacity of adaptation to the environmental conditions (Yeo et al., 2003; Quesada-Moraga et al., 2006; van Hanh et al., 2007). Moreover, in the context of IPM, to consider the effectiveness of entomopathogenic fungi against the target species is not enough. We must consider possible conflicts with plant disease control and with the use of predators, parasitoids, and pollinators. Observations on pollinators visiting crops treated with *B. bassiana* seem to indicate that pollinator susceptibility varies widely depending on the pollinator species. *Apis mellifera* L. is not susceptible to this fungus (Goettel and Jaronski, 1997), while epidemics have developed in the brood of *Bombus impatiens* Cresson (Al-mazra'awi et al., 2006). Thus the effect of bioinsecticides on pollinators must be quantified depending on the fungal isolate, doses of use, and pollinating species. *B. bassiana* and *Metarhizium anisopliae* (Metchnikoff) Sorokin can infect both pests and predators (Roy and Cottrell, 2008). Some predators, such as *Anthocoris nemorum* L. and *C. septempunctata*, avoid parts of the plant contaminated with *B. bassiana* (Meyling and Pell, 2006; Meyling and Hajek, 2010). This behavior reduces predation rate and can have significant negative effects, particularly when the prey is not very sensitive to the entomopathogenic fungus.

4.4.4 Application of Biological Control for Aphids in Tomato

Aphid control by use of augmentation, both with fungi or entomophagous insects, is mainly limited to greenhouses (Dedryver et al., 2010). Several species of aphid natural enemies, including entomopathogenic fungi, are commercially available and can be used for augmentative biological control in tomato (Powell and Pell, 2007). The aim of this method is to control pests by mass-rearing and releasing natural enemies which do not occur naturally in sufficient number or at the optimal time. Four parasitoid species, all belonging to family Braconidae, are available in Europe and in the United States. These include *A. colemani*, *A. ervi*, *A. matricariae*, and *L. testaceipes* (Powell and Pell, 2007). Among them, *A. colemani* has become the most popular for management of aphids in greenhouses. Parasitoids can be introduced into the greenhouse by direct release or by using "banker plants" (Frank, 2010). These plants are infested by an aphid species that is not harmful to the crop but suitable for the parasitoid. This strategy allows the buildup of parasitoid numbers at an early stage of crop infestation (Powell and Pell, 2007). This approach has also been used to establish populations of *A. aphidimiza* in the greenhouse. Alternatively, *A. aphidimyza* can be distributed throughout the tomatoes as pupae. One interesting study determined that by rearing *A. aphidimyza* in cages containing spider webs, there was more mating, and females laid more eggs than in cages without spider webbing (van Schelt and Mulder, 2000).

There are very few reports of successful augmentative releases of natural enemies against aphids in field situations. One such report concerns augmentative releases of *Crysoperla carnea* (Stephens) that successfully controlled aphids in field tomato in the former USSR (Beglyarov and Smetnik, 1997). This result was obtained with releases of one larva per five aphids, a very high rate. For another predator, *H. convergens*, resistant tomato varieties had a negative impact on consumptive rates and led to higher densities and dispersal of *M. euphorbiae* (Kersch-Becker and Thaler, 2015) and *Orius insidious* (Say) (Pallipparambil et al., 2015). Thus there are some general constraints of augmentative biological control of aphids in tomato fields, some of which are technical and some economic. For this reason research efforts have focused on developing a different strategy known as conservation biological control. Conservation aims at enhancing the efficacy and local abundance of the existing community of natural enemies by means of habitat management, modification of existing practices, and/or manipulation of natural enemy behavior (Zehnder et al., 2007; Dedryver et al., 2010). We mentioned that in natural conditions, individual species of aphidophagous predators and parasitoids are not very effective in aphid biological control due to the long developmental time in comparison to that of the prey. However, many different types of natural enemies attack aphids, and they could be combined to maintain the pest population under the economic threshold (Gillespie et al., 2006; Yano, 2006). In this manner, conservation biological control may be a promising way for limiting aphid damage and reducing pesticide use in the field. In order for this type of biological control to be successful, it may be necessary to manage insects that negatively impact natural enemies. For example, the presence of red imported fire ants, *Solenopsis invicta* Buren, has been shown to increase aphid abundance in field tomatoes, presumably by interfering with biological control (Coppler et al., 2007). One of the major factors in conservation biological control of aphids seems to be the plant biodiversity. Letourneau and Goldstein (2001) found that the relative amount of natural vegetation surrounding tomato fields is the best parameter for explaining parasitoid and mirid predator abundance. These authors observed a greater proportion of parasitized aphids in tomato fields near natural areas. In addition, the predatory mirids were much more abundant, on average, on organic farms than on conventional farms, logically because of the minimal use of pesticides. Balzan and Moonen (2014) conducted studies proving that field margin vegetation promoted biological control and reduced damage in tomato fields. Togni et al. (2016) found that, when intercropping tomato and coriander, the coccinellid predator *Cycloneda sanguinea* (L.) used the coriander for oviposition, and fed on pollen and nectar before aphids became available in the tomato. The intercropping supported their conservation biological control agenda.

In general, the limitation of pesticide use can be considered a prerequisite for the application of conservation biological control. If fungicides are avoided as well, natural activity of Entomophthorales fungi can be conserved (Fritz et al., 1981). Fungi in the Entomophthorales are not as easily to rear as those in the Hypocreales, but they are particularly suitable for conservation biological control because they can cause natural epizootics that reduce aphid densities within a short time (Dedryver et al., 2010).

4.5 Chemical Control

Despite clear evidence for the benefits of a holistic-IPM program for aphids, insecticidal control remains the first and most popular strategy used by conventional tomato growers. The reason for this is twofold. First, there is a continued tradition of using insecticides that date back to the earliest days of tomato production (e.g., Miles, 1927; Morgan, 1937; Sloan, 1945; Taschenberg, 1949; Schuder, 1950; Wolfenbarger, 1954; Taschenberg and Avens, 1955; Harding, 1959). These papers report control with a wide variety of materials including oils, botanical insecticides like pyrethrum, rotenone, and nicotine, early compounds including nicotine sulfate, calcium cyanide, and lead arsenate, through the "synthetic chemical era" which saw organochlorines, organophosphates, carbamates, formamadines, and pyrethroids being used for insect (including aphid) control. Second, the materials used had the advantages of being fast acting, very effective, inexpensive, easy for the grower to apply, long lasting, and were broad spectrum, killing many pest insects with single applications. While these materials provided outstanding results for killing insects, they also had negative effects on the environment and non-target species. These effects became apparent through the 1960s and 1970s giving support to the integrated control concepts proposed by Stern et al. (1959). Insecticide development moved toward materials that were effective, expensive, short-lived, narrow spectrum, selective for the pest insect, and therefore, environmentally friendly. These attributes remain today for the majority of insecticides in use and a substantial amount of research on controlling aphids using insecticides is ongoing. In addition, a great deal of information is available on the modes of actions for the various materials (see Insecticide Resistance Action Committee at http://www.irac-online.org/modes-of-action/). By rotating among these modes of action, resistance to insecticides can be reduced or eliminated.

The University of California Pest Management website lists the currently available materials for aphid control on tomatoes in California (http://ipm.ucanr.edu/PMG/r783301711.html). This list includes synthetic materials in various insecticide classes as well as some natural products and products that are registered for organic use. In addition, a variety of other materials have been evaluated, including acetone fumigation in greenhouses (Tunc et al., 2004), cyantraniliprole in the field (Misra and Mukherjee, 2012), endosulfan and monocrotophos (Ao, 2015), translaminar action of cyantraniliprole (Berry et al., 2015), and seed treatment with thiamethoxam (Maurya et al., 2015). In this last study, the insecticidal effect was strong early in the plant growth and declined as the plant matured. In addition, there was minimal impact on natural enemies.

Neem-based products from the neem tree, *Azadirachta indica* Jussieu, also have been evaluated for aphid control on tomatoes. Soliman and Tarasco (2008) and El-Shafie and Abdelraheem (2012) found that Neem Azal-T/S provided good control of aphids on greenhouse tomato and cucumber. Neem soap was also found to provide 92% mortality of *M. persicae* 48 h after treatment in the laboratory (Satyavati et al., 2014). On the other hand, neem products were found to be poor aphid controls by Ao (2015). Nzanza and Mashela (2012) found that mixing neem extract and wild garlic (*Tulbaghia violaceae* Harvey) was more effective for aphid and whitefly control than either plant extract alone, suggesting a synergistic effect.

Other plant-based products also have been tested. Antonious and Snyder (2008) made crude extracts from tomato leaves, separated the products, and screened them against insects. The methyl ketones provided the best insecticidal properties against *M. persicae*. Dubey et al. (2012) made a soap-based insecticide containing a natural pyrethroid made from Chrysanthemum plant extracts, which significantly reduced aphids on tomato and the incidence of tomato yellow leaf, an aphid-borne virus. Bamboo vinegar has also been shown to reduce *M. persicae* densities in tomato, although there may be other side effects to using this material on tomato (Yao et al., 2012). In laboratory and greenhouse studies, *A. gossypii* was reduced using extracts of geranium and chinaberry, while basil, onion, and garlic gave moderate to low control (Ghanim, 2014).

A final interesting study evaluated vermicomposts for *M. persicae* control on tomato. Vermicomposts are produced when microorganism break down the waste products of earthworms (Arancon et al., 2007). Edwards et al. (2010) prepared an aqueous solution of vemicompost and compared dilutions of 20%, 10%, and 5% in soil drenches of tomato seedlings, after which the plants were infested with aphids. The authors reported that all of the treatments significantly reduced pest establishment on the plants, reduced their reproductive rates, and in the higher rates of solutions, caused the insects to die 14 days after treatment. Edwards et al. (2010) attributed the response to the uptake of phenolic materials that came from the vermicompost.

4.6 Management of Viruses Transmitted by Aphids

Previously, we mentioned that insecticides generally are not effective to reduce the incidence of tomato plants infected by aphid-borne viruses. The exception to this general rule is in the case of persistently transmitted viruses (Perring et al., 1999). The best long-term strategy is the development of tomato varieties resistant to the pathogen and/or the aphid vector. Other strategies for reducing disease spread of aphid-borne viruses focus on preventative measures. Katis et al. (2007) reviewed these measures and categorized them under strategies that reduce sources of the virus and those that prevent or reduce virus spread. To reduce virus sources, they suggest seed certification to insure non-infected seed, controlling weeds that harbor the virus and vectors, removing diseased plants in the field, eliminating volunteer crop plants that germinate prior to planting the crop, and isolating the field from virus and aphid sources. To prevent or reduce virus spread, Katis et al. (2007) lists the biological control of aphids, which may be effective in persistently transmitted virus systems. However biological control can result in an increase in aphid movement which actually promotes virus spread (Weber et al., 1996). The use of colored and/or reflective mulches, ground sprays, row covers, barrier crops, and intercrops also have shown promise for reducing aphid-borne virus incidence (Csizinszky et al., 1995; Katis et al., 2007; Wratten et al., 2007).

Considering biological control in the reduction of aphid-borne virus spread, it generally is not effective for reducing non-persistently transmitted plant viruses spread. This is because the natural enemy does not act quickly enough to prevent inoculation. However, biological control may have more relevance when secondary spread is accomplished by aphids that develop in the field. These systems are characterized by persistently transmitted pathogens and by reducing the overall number of vectors in a field, the virus spread can be reduced. Care should be taken when trying to use natural enemies to reduce virus spread because aphids may emit an alarm pheromone when attacked by predators and parasitoids, thus eliciting escape behavior that may accelerate virus spread within the field (Hooks and Fereres, 2006; Kaplan and Thaler, 2012).

4.7 Integrated Pest Management

From the discussion above (Sections 4.1–4.6), it is clear that the management of aphids on tomato, both in the field and in greenhouses, has been the focus of many researchers. There are a few other ideas that have been evaluated which may be useful in certain situations. For field tomatoes, we mentioned the concept of creating more stable plant communities by planting a variety of permanent plants at the field margins. This practice holds natural enemies and serves to reduce aphid pests (Balzan and Moonen, 2014). Earlier in this chapter, we discussed the JA-mediated induced resistance pathway, which has been shown to have a negative impact on aphids. A novel use of this system was reported by Worrall et al. (2012), who treated tomato seeds with JA. Plants grown from these seeds were more resistant to a number of arthropods, including aphids, than non-treated plants. They also found that this seed priming did not reduce plant growth, suggesting it to be explored for commercial use. A final novel technique involves aphid pathogens. Recently a virus has been isolated from *M. euphorbiae* (Teixeira et al., 2016). This virus, named MeV-1, is thought to be in the genus *Flavivirus*, and it is found only in nymphal and adult aphids, but not in the tomato host plants. Interestingly, the virus was found in aphid populations from Europe, but not North America. If this virus causes disease in the aphids, it may present a novel management strategy.

For greenhouses, there are additional possibilities. For example, Latimer and Oetting (1994) found that brushing tomato plants (a practice used to control plant height in the greenhouse), prior to infestation with *M. persicae*, reduced the number of aphids that colonized the plants. The authors suggested that this may have been due to changes in plant biochemistry elicited by the brushing. It is possible that brushing the leaves disturbed the trichomes, providing volatiles that repelled the aphids. Other ideas have focused on managing CO_2 levels. Sun et al. (2011) reviewed the literature on studies that have evaluated CO_2 levels on plant growth and insect pests. They found that aphids, in general, responded positively to higher CO_2 levels. It would be interesting to determine if this relationship is true for aphids on tomato and if so, does reducing CO_2 levels in greenhouses provide an aphid management tool. Another greenhouse strategy that provides a more airy environment compared to screening, used an electric field to prevent insects. With insulated conductor wires at 5 mm intervals Kakutani et al. (2012) created an electric field that captured insects and airborne conidia. This electric field provided stable pest exclusion and abundant air circulation. Along these same lines, Takikawa et al. (2016) described an electrostatic nursery shelter that successfully caught *M. persicae*. Another study applied low-frequency acoustic signals to tomato and found that this increased yield, improved crop quality, and had a reduced incidence of infestation by aphids (Hou et al., 2009). Further research with sound wave technology has been shown to stimulate plant growth of a variety of vegetable crops, including tomato (Hassanien et al., 2014). These authors suggest that sound waves strengthen the plant immune system, reducing infestation of mites and aphids. Another paper reports the genetic modification of wheat to express and release (E)-beta-farnesene, the aphid alarm pheromone (Bruce et al., 2015). The plants produced pure (E)-beta-farnesene and repelled three aphid species in laboratory assays. However field trials showed no reduction in aphids. In these field studies, the authors noted that insect abundance was low and the climatic conditions were erratic, suggesting the need for further research. This is an interesting idea that could have great applicability for aphid management on tomato, particularly in the greenhouse.

5. ECONOMICS OF MANAGEMENT STRATEGIES

From the discussion above, it is clear that substantial research has been done in the various management areas that can lead to aphid reduction in tomatoes. While it is challenging to assign direct cost–benefit to these strategies, the most effective, quick acting, and frequently used strategy is insecticides. Systemic materials that target phloem-feeding insects, such as aphids, provide long-term control, and the application to the roots through various irrigation practices that are already being done for the crop, make insecticides attractive to tomato growers in field and greenhouse settings. These types of applications also enable the use of biological control in many situations promoting a trend toward more integrated management systems. Continued application of management options that integrate cultural control practices provides additional reduction in aphid numbers. Recent advances in identifying plant resistance genes are moving toward the development of commercial varieties that have high quality traits and aphid resistance. This is particularly important when we consider aphid-borne pathogens that are difficult to manage through conventional insecticide, biological, and cultural control methods. IPM must also take into consideration the multiple pests of tomato. For example, Cameron et al. (2009) noted that, after removing sources of aphid and thrips-borne viruses, New Zealand processing tomato growers were able to reduce overall insecticide use for *Helicoverpa armigera* (Hübner) by 60%. Similarly, densities of aphids, whiteflies, and lepidopteran larvae were lowest when tomato growers followed IPM modules in Gujarat, India (Chavan et al., 2013). Put another way, Zalom et al. (2001) noted that established IPM programs fail when there is too much emphasis on one strategy versus another. In addition, pest management strategies are often most effective when implemented on an areawide basis. Bauske et al. (1998) reported on the success of IPM teams from seven states in southeastern United States. These teams, comprising growers, pest consultants, cooperative extension agents, and university faculty, developed best IPM practices for tomato growers in the region. Through their work, they found that growers in many of the states adopted the IPM strategies on more than 75% of the fresh-market tomato acreage. Programs like these that integrate the scientific evidence obtained in the studies above will result in economically and environmentally sustainable management of aphids in tomato.

6. FUTURE PROSPECTS AND CONCLUSION

Depending on the location and seasonality of tomatoes, aphids can be primary, secondary, or occasional pests of this valuable commodity. The literature is rich with studies that help us understand the biology of the two primary aphid pests, *M. euphorbiae* and *M. persicae*, and how they interact with their biotic community and within their abiotic environment. These studies give a historical perspective on where we have been, and the "other IPM section" above gives insight into where we might be headed in the future. Greater specificity and efficiency of insecticides, recent success with biological control, and our expanding knowledge of tomato genetics all point to a positive future for aphid management.

REFERENCES

A'Brook, J., 1968. The effect of plant spacing on the numbers of aphids trapped over the groundnut crop. Annals of Applied Biology 61, 289–294.

Afifi, F.M.L., Haydar, M.F., Oma, H.I.H., 1990. Effect of different intercropping systems on tomato infestation with major insect pests; *Bemisia tabaci* (Genn.) (Hemiptera: Aleyrodidae), *Myzus persicae* Sulzer (Homoptera: Aphididae) and *Phthorimaea operculella* Zeller (Lepidoptera: Gelechiidae). Cairo University, Faculty of Agriculture, Bulletin 41, 885–900.

Agele, S.O., 2006. Effects of watering regimes on aphid infestation and performance of selected varieties of cowpea (*Vigna unguiculata L. Walp*) in a humid rainforest zone of Nigeria. Crop Protection 25, 73–78.

Ahsan, M.I., Hossain, M.S., Parvin, S., Karim, Z., 2005. Effect of varieties and planting dates on the incidence of aphid and white fly attack on tomato. International Journal of Sustainable Agricultural Technology 1, 26–30.

Al-mazra'awi, M.S., Shipp, L., Broadbent, B., Kevan, P., 2006. Biological control of *Lygus lineolaris* (*Hemiptera: Miridae*) and *Frankliniella occidentalis* (*Thysanoptera: Thripidae*) by *Bombus impatiens* (Hymenoptera: Apidae) vectored *Beauveria bassiana* in greenhouse sweet pepper. Biological Control 37, 89–97.

Alomar, O., Gabarra, R., Castana, C., 1997. The aphid parasitoid *Aphelinus abdominalis* (Hym.: Aphelinidae) for biological control of *Macrosiphum euphorbiae* on tomatoes grown in unheated plastic greenhouses. IOBC/WPRS Bulletin 20, 203–205.

Alomar, O., Goula, M., Albajes, R., 2002. Colonization of tomato fields by predatory mirid bugs (Hemiptera: Heteroptera) in Northern Spain. Agriculture, Ecosystems and Environment 89, 105–115.

Altieri, M.A., Nicholls, C.I., 2003. Soil fertility management and insect pests: harmonizing soil and plant health in agroecosystems. Soil and Tillage Research 72, 203–211.

Alvarado, P., Balta, O., Alomar, O., 1997. Efficiency of four heteroptera as predators of *Aphis gossypii* and *Macrosiphum euphorbiae* (Homptera: Aphididae). Entomophaga 42, 215–226.

Ankersmit, G.W., Dijkman, H., Keuning, N.J., Mertens, H., Sins, A., Tacoma, H., 1986. *Episyrphus balteatus* as a predator of the aphid *Sitobion avenae* on winter wheat. Entomologia Experimentalis et Applicata 42, 271–277.

Anstead, J., Samuel, P., Song, N., Wu, C., Thompson, G.A., Goggin, F., 2010. Activation of ethylene-related genes in response to aphid feeding on resistant and susceptible melon and tomato plants. Entomologia Experimentalis et Applicata 134, 170–181.

Antonious, G., Snyder, J., 2008. Tomato leaf crude extracts for insects and spider mite control. In: Preedy, V.R., Watson, R.R. (Eds.), Tomatoes and Tomato Products: Nutritional, Medicinal and Therapeutic Properties. Science Publishers, USA, pp. 269–297.

Antonious, G.F., Kamminga, K., Snyder, J.C., 2014. Wild tomato leaf extracts for spider mite and cowpea aphid control. Journal of Environmental Science and Health, Part B. Pesticides, Food Contaminants, and Agricultural Wastes 49, 527–531.

Ao, W.M.A., 2015. Efficacy of some insecticides against aphids (*Aphis gossypii*) and serpentine leaf miner (*Liriomyza trifolii*) in tomato under Nagaland condition. Environment and Ecology 33, 10–13.

Arancon, N.Q., Edwards, C.A., Oliver, T.J., Byrne, R.J., 2007. Suppression of two-spotted spider mite (*Tetranychus urticae*), mealy bugs (*Pseudococcus* sp.) and aphid (*Myzus persicae*) populations and damage by vermicomposts. Crop Protection 26, 26–39.

Arnó, J., Gabarra, R., Estopà, M., Gorman, K., Peterschmitt, M., Bonato, O., Vosman, B., Hommes, M., Albajes, R., 2009. Implementation of IPM Programs in European Greenhouse Tomato Production Areas. Tools and Constraints. Edicions de la Universitat de Lleida, Lleida, Spain.

Arnó, J., Castañé, C., Riudavets, J., Gabarra, R., 2010. Risk of damage to tomato crops by the generalist zoophytophagous predator *Nesidiocoris tenuis* (Reuter) (Hemiptera: Miridae). Bulletin of Entomological Research 100, 105–115.

Åsman, K., 2007. Aphid infestation in field grown lettuce and biological control with entomopathogenic fungi (Deuteromycotina: Hyphomycetes). Biological Agriculture and Horticulture 25, 153–173.

Atamian, H.S., Chaudhary, R., Dal Cin, V., Bao, E., Girke, T., Cin, V.D., Kaloshian, I., 2013. In planta expression or delivery of potato aphid *Macrosiphum euphorbiae* effectors Me10 and Me23 enhances aphid fecundity. Molecular Plant-Microbe Interactions 26, 67–74.

Atta-Aly, M.A., Abdel-Megeed, M.I., Hegab, M.F., Kamel, M.H., 1998. Increasing tomato plant growth and yield with improving fruit quality by controlling sap-sucking insects (whitefly and aphid) without insecticides. Annals of Agricultural Science 3, 845–863.

Awmack, C.S., Leather, S.R., 2002. Host plant quality and fecundity in herbivorous insects. Annual Review of Entomology 47, 817–844.

Balzan, M.V., Moonen, A., 2014. Field margin vegetation enhances biological control and crop damage suppression from multiple pests in organic tomato fields. Entomologia Experimentalis et Applicata 150, 45–65.

Banos-Diaz, H.L., Louzada, E., Moura, N., Paez-Bueno, V.E., 2014. Life table of *Macrolophus basicornis* (Hemiptera: Miridae) preying on *Myzus persicae* (Sulzer) and *Macrosiphum euphorbiae* (Thomas) (Hemiptera: Aphididae). Revista de Protección Vegetal 29, 94–98.

Bansal, R., Michel, A., 2015. Molecular adaptations of aphid biotypes in overcoming host-plant resistance. In: Raman, C., Goldsmith, M.R., Agunbiade, T.A. (Eds.), Short Views on Insect Genomics and Proteomics. Insect Genomics, vol. 1. Springer International Publishing, Switzerland, pp. 75–93.

Battaglia, D., Bossi, S., Cascone, P., Digilio, M.C., Duran Prieto, J., Fanti, P., Guerrieri, E., Iodice, L., Lingua, G., Lorito, M., Maffei, M.E., Massa, N., Ruocco, M., Sasso, R., Trotta, V., 2013. Tomato belowground-aboveground interactions: *Trichoderma longibrachiatum* affects the performance of *Macrosiphum euphorbiae* and its natural antagonists. Molecular Plant-Microbe Interactions 26, 1249–1256.

Bauske, E.M., Zehnder, G.M., Sikora, E.J., Kemble, J., 1998. Southeastern tomato growers adopt integrated pest management. HortTechnology 8, 40–44.

Baverstock, J., Roy, H.E., Clark, S.J., Alderson, P.G., Pell, J.K., 2006. Effect of fungal infection on the reproductive potential of aphids and their progeny. Journal of Invertebrate Pathology 91, 136–139.

Baverstock, J., Alderson, P.G., Pell, J.K., 2007. Intraguild interactions involving the entomopathogenic fungus *Pandora neoaphidis*. In: Papierok, B. (Ed.), Insect Pathogens and Insect Parasitic NematodesProceedings of the 10th European Meeting IOBC Working Group at Locorotondo, Bari (Italy), 23–29 June 2005, vol. 30, pp. 95–98.

Baverstock, J., Clark, S.J., Pell, J.K., 2008. Effect of seasonal abiotic conditions and field margin habitat on the activity of *Pandora neoaphidis* inoculum on soil. Journal of Invertebrate Pathology 97, 282–290.

Beglyarov, G.A., Smetnik, A.I., 1997. Seasonal colonization of entomophages in the USSR. In: Ridgeway, R.L., Vinson, S.B. (Eds.), Biological Control by Augmentation of Natural Enemies. Plenum, New York, USA, pp. 283–328.

Ben Yakir, D., Antignus, Y., Offir, Y., Shahak, Y., 2012. Colored shading nets impede insect invasion and decrease the incidences of insect-transmitted viral diseases in vegetable crops. Entomologia Experimentalis et Applicata 144, 249–257.

Benninson, J.A., Corless, S.P., 1993. Biological control of aphids on cucumbers: further development of open rearing units or 'banker plants' to aid establishment of aphid natural enemies. IOBC/WPRS Bulletin 16, 5–8.

Benninson, J.A., 1992. Biological control of aphids on cucumbers, use of open rearing systems or 'banker plants' to aid stablishment of *Aphidius matricariae* and *Aphidoletes aphidimyza*. Mededelingen van de Faculteit Landbouwwetenschappen Universiteit Gent 57, 457–466.

Berry, J.D., Portillo, H.E., Annan, I.B., Cameron, R.A., Clagg, D.G., Dietrich, R.F., Watson, L.J., Leighty, R.M., Ryan, D.L., McMillan, J.A., Swain, R.S., Kaczmarczyk, R.A., 2015. Movement of cyantraniliprole in plants after foliar applications and its impact on the control of sucking and chewing insects. Pest Management Science 71, 395–403.

Besser, K., Harper, A., Welsby, N., Schauvinhold, I., Slocombe, S., Li, Y., Dixon, R.A., Broun, P., 2009. Divergent regulation of terpenoid metabolism in the trichomes of wild and cultivated tomato species. Plant Physiology 149, 499–514.

Bessin, R., 2010. Tomato Insect IPM Guidelines. University of Kentucky Cooperative Extension Services, ENTFACT-313. http://www2.ca.uky.edu/entomology/entfacts/ef313.asp.

Bhattarai, K.K., Xie, Q.G., Pourshalimi, D., Younglove, T., Kaloshian, I., 2007. Coi1-dependent signaling pathway is not required for Mi-1-mediated potato aphid resistance. Molecular Plant-Microbe Interactions 20, 276–282.

Blackman, R.L., Eastop, V.F., 1984. Aphids on the World's Crops. An Identification and Information Guide. John Wiley, Chichester, UK.

Blackman, R.L., Eastop, V.F., 2000. Aphids on the World's Crops, second ed. John Wiley and Sons, Chichester, UK.

Blackman, R.L., Eastop, V.F., 2007. Aphids on the World's Herbaceous Plants and Shrubs. The Aphids, vol. 2. John Wiley and Sons with the Natural History Museum, London, pp. 1025–1439.

Blackman, R.L., Takada, H., Kawakami, K., 1978. Chromosomal rearrangement involved in insecticide resistance of *Myzus persicae*. Nature UK 271, 450–452.

Blackman, R.L., 1971. Chromosomal abnormalities in an anholocyclic biotype of *Myzus persicae* (Sulzer). Experimentia 27, 704–706.

Blackman, R.L., 1974. Life-cycle variation of *Myzus persicae* (Sulz.) (Hom. Aphididae) in different parts of the world, in relation to genotype and environment. Bulletin of Entomological Research 63, 595–607.

Blackman, R.L., 1980. Chromosomes numbers in the Aphididae and their taxonomic significance. Systematic Entomology 5, 7–25.

Bleeker, P.M., Diergaarde, P.J., Ament, K., Guerra, J., Weidner, M., Schutz, S., de Both, M.T.J., Haring, M.A., Schuurink, R.C., 2009. The role of specific tomato volatiles in tomato-whitefly interaction. Plant Physiology 151, 925–935.

Bleeker, P.M., Diergaarde, P.J., Ament, K., Schutz, S., Johne, B., Dijkink, J., Hiemstra, H., de Gelder, R., de Both, M.T.J., Sabelis, M.W., Haring, M.A., Schuurink, R.C., 2011. Tomato-produced 7-epizingiberene and R-curcumene act as repellents to whiteflies. Phytochemistry 72, 68–73.

Boissot, N., Schoeny, A., Vanlerberghe-Masutti, F., 2016. Vat, an amazing gene conferring resistance to aphids and viruses they carry: from molecular structure to field effects. Frontiers in Plant Science 7, 1420.

Boll, R., Lapchin, L., 2002. Projection pursuit nonparametric regression applied to field counts of the aphid *Macrosiphum euphorbiae* (Homoptera: Aphididae) on tomato crops in greenhouses. Journal of Economic Entomology 95, 493–498.

Boughton, A.J., Hoover, K., Felton, G.W., 2006. Impact of chemical elicitor applications on greenhouse tomato plants and population growth of the green peach aphid, *Myzus persicae*. Entomologia Experimentalis et Applicata 120, 175–188.

Brodeur, J., Rosenheim, J.A., 2000. Intraguild interactions in aphid parasitoids. Entomologia Experimentalis et Applicata 97, 93–108.

Bruce, T.J.A., Aradottir, G.I., Smart, L.E., Martin, J.L., Caulfield, J.C., Doherty, A., Sparks, C.A., Woodcock, C.M., Birkett, M.A., Napier, J.A., Jones, H.D., Pickett, J.A., 2015. The first crop genetically engineered to release an insect pheromone for defense. Scientific Reports 5, 11183. http://dx.doi.org/10.1038/srep11183.

Buitenhuis, R., Murphy, G., Shipp, L., 2013. *Aphis gossypii* glover, melon/cotton aphid, *Aulacorthum solani* (Kaltenbach), foxglove aphid, and other arthropod pests in greenhouse crops. In: Mason, P.G., Gillespie, D.R. (Eds.), Biological Control Programmes in Canada 2001–2012, pp. 98–107.

Burgio, G., Ferrari, R., Nicoli, G., 1997. Biological and integrated control of *Aphis gossypii* Glover (Hom., Aphididae) in protected cucumber and melon. Bollettino dell'Istituto di Entomologia 'Guido Grandi' della Università degli Studi di Bologna 51, 171–178.

Butt, T.M., Beckett, A., Wilding, N., 1990. A histological study of the invasive and developmental processes of the aphid pathogen *Erynia neoaphidis* (*Zygomycotina, Entomophthorales*) in the pea aphid *Acyrthosiphon pisum*. Canadian Journal of Botany 68, 2153–2163.

Calvo, J., Bolckmans, K., Stansly, P., Urbaneja, A., 2009. Predation by *Nesidiocoris tenuis* on *Bemisia tabaci* and injury to tomato. BioControl 54, 237–246.

Cameron, P.J., Walker, G.P., Hodson, A.J., Kale, A.J., Herman, T.J.B., 2009. Trends in IPM and insecticide use in processing tomatoes in New Zealand. Crop Protection 28, 421–427.

Cascone, P., Iodice, L., Maffei, M.E., Bossi, S., Arimura, G., Guerrieri, E., 2015. Tobacco overexpressing beta-ocimene induces direct and indirect responses against aphids in receiver tomato plants. Journal of Plant Physiology 173, 28–32.

Castañé, C., Alomar, O., Riudavets, J., 2003. Potential risk of damage to zucchinis caused by mirid bugs. IOBC/WPRS Bulletin 26, 135–138.

Castañé, C., Alomar, O., Goula, M., Gabarra, R., 2004. Colonization of tomato greenhouses by the predatory mirid bugs *Macrolophus caliginosus* and *Dicyphus tamaninii*. Biological Control 30, 591–597.

Castañé, C., Alomar, O., Riudavest, J., Gemeno, C., 2006. Reproductive traits of the generalist predator *Macrolophus caliginosus*. IOBC/WPRS Bulletin 29, 229–234.

Castañé, C., Arnó, J., Gabarra, R., Alomar, O., 2011. Plant damage to vegetable crops by zoophytophagous mirid predators. Biological Control 59, 22–29.

Casteel, C.L., Walling, L.L., Paine, T.D., 2006. Behavior and biology of the tomato psyllid, *Bactericerca cockerelli*, in response to the *Mi-1.2* gene. Entomologia Experimentalis et Applicata 121, 67–72.

Cesari, S., Bernoux, M., Moncuquet, P., Kroj, T., Dodds, P.N., 2014. A novel conserved mechanism for plant NLR protein pairs: the "integrated decoy" hypothesis. Frontiers in Plant Science 5, 606. http://dx.doi.org/10.3389/fpls.2014.00606.

Chai, M., YunHua, D., Ding, Y.H., 2002. Inheritance of tomato *Womz* gene and the value of the gene in tomato breeding. Acta Horticulturae Sinica 29, 133–136.

Chambers, R.J., Adams, T.H.L., 1986. Quantification of the impact of hoverflies (Diptera: Syrphidae) on cereal aphids in winter wheat: an analysis of field populations. Journal of Applied Ecology 23, 895–904.

Chaudhary, R., Atamian, H.S., Shen, Z., Brigg, S.P., Kaloshian, I., Briggs, S.P., 2014. GroEL from the endosymbiont *Buchnera aphidicola* betrays the aphid by triggering plant defense. Proceedings of the National Academy of Sciences of the United States of America 111, 8919–8924.

Chavan, S.M., Kumar, S., Arve, S.S., 2013. Population dynamics and development of suitable pest management module against major insect pests of tomato (*Solanum lycopersicum*). Journal of Applied Horticulture 15, 150–155.

Chen, B., Li, Z.Y., Feng, M.G., 2008. Occurrence of entomopathogenic fungi in migratory alate aphids in Yunnan Province of China. BioControl 53, 317–326.

Chiel, E., Messika, Y., Steinberg, S., Antignus, Y., 2006. The effect of UV-absorbing plastic sheet on the attraction and host location ability of three parasitoids: *Aphidius colemani*, *Diglyphus isaea* and *Eretmocerus mundus*. BioControl 51, 65–78.

Chyzik, R., Dobrinin, S., Antignus, Y., 2003. Effect of a UV-deficient environment on the biology and flight activity of *Myzus persicae* and its hymenopterous parasite *Aphidius matricariae*. Phytoparasitica 31, 467–477.

Civolani, S., Marchetti, E., Chicca, M., Castaldelli, G., Rossi, R., Pasqualini, E., Dindo, M.L., Baronio, P., Leis, M., 2010. Probing behavior of *Myzus persicae* on tomato plants containing *Mi* gene or BTH-treated evaluated by electrical penetration graph. Bulletin of Insectology 63, 265–271.

Clark, M.S., Horwath, W.R., Shennan, C., Scow, K.M., Lantni, W.T., Ferris, H., 1999. Nitrogen, weeds and water as yield-limiting factors in conventional, low-input, and organic tomato systems. Agriculture, Ecosystems and Environment 73, 257–270.

Cognetti, G., 1961. Citogenetica della partenogenesi negli afidi. Archivio Zoologico Italiano 46, 89–122.

Colby, S., Crock, J., Dowdle-Rizzo, B., Lemaux, P., Croteau, R., 1998. Germacrene C synthase from *Lycopersicon esculentum* cv. VFNT Cherry tomato: cDNA isolation, characterization, and bacterial expression of the multiple product sesquiterpene cyclase. Proceedings of the National Academy of Sciences of the United States of America 95, 2216–2221.

Colella, T., Candido, V., Campanelli, G., Camele, I., Battaglia, D., 2014. Effect of irrigation regimes and artificial mycorrhization on insect pest infestations and yield in tomato crop. Phytoparasitica 42, 235–246.

Conversa, G., Elia, A., La Rotonda, P., 2007. Mycorrhizal inoculation and phosphorus fertilization effect on growth and yield of processing tomato. Acta Horticuturae 758, 333–338.

Cooper, W.R., Goggin, F.L., 2005. Effects of jasmonate-induced defenses in tomato on the potato aphid, *Macrosiphum euphorbiae*. Entomologia experimentalis et applicata 115, 107–115.

Cooper, W., Jia, L.C., Goggin, F.L., 2004. Acquired and R-gene-mediated resistance against the potato aphid in tomato. Journal of Chemical Ecology 30, 2527–2542.

Coppler, L.B., Murphy, J.F., Eubanks, M.D., 2007. Red imported fire ants (Hymenoptera: Formicidae) increase the abundance of aphids in tomato. The Florida Entomologist 90, 419–425.

Coppola, V., Coppola, M., Rocco, M., Digilio, M.C., D'Ambrosio, C., 2013. Transcriptomic and proteomic analysis of a compatible tomato-aphid interaction reveals a predominant salicylic acid-dependent plant response. BioMed Central Genomics 14, 1–18.

Costa, H.S., Robb, K.L., Wilen, C.A., 2002. Field trials measuring the effects of ultraviolet-absorbing greenhouse plastic films on insect populations. Journal of Economic Entomology 95, 113–120.

Cota, E., Isufi, A., 2009. Non-chemical control of aphids in protected crops in Albania. IOBC/WPRS Bulletin 49, 265–266.

Crescenzi, A., D'Agrosa, G., 2001. Patologia e Difesa Integata del Pomodoro. Calderini Edagricole, Bologna, Italy.

Csizinszky, A.A., Schuster, D.J., Kring, J.B., 1995. Color mulches influence yield and insect pest populations in tomatoes. Journal of the American Society for Horticultural Science 120, 778–784.

Da Silva, A.A., Maluf, W.R., Moraes, J.C., Alvarenga, R., Rodrigues Costa, E.M., 2013. Resistance to *Myzus persicae* in tomato genotypes with high levels of foliar allelochemicals. Bragantia 72, 173–179.

De Backer, L., 2012. Evaluation of *Macrolophus pygmaeus* (Heteroptera: Miridae) as biocontrol agent against aphids. In: Travail De Fin D'études Présenté En Vue De L'obtention Du Diplôme De Master Bioingénieur En Sciences Agronomiques. Année Académique 2011–2012. Université de Liege, pp. 1–85. http://orbi.ulg.ac.be/bitstream/2268/159459/1/De%20Backer,%20Evaluation%20of%20Macrolophus%20pygmaeus%20as%20biocontrol%20agent%20against%20aphids.pdf.

De Ilarduya, O.M., Moore, A.E., Kaloshian, I., 2001. The tomato *Rme1* locus is required for *Mi1*-mediated resistance to root-knot nematodes and the potato aphid. The Plant Journal 27, 417–425.

De Ilarduya, O.M., Xie, Q., Kaloshian, I., 2003. Aphid-induced defense responses in *Mi-1*-mediated compatible and incompatible tomato interactions. Molecular Plant-Microbe Interaction 16, 699–708.

De Ilarduya, O.M., Nombela, G., Hwang, C., Williamson, V.M., Muniz, M., Kaloshian, I., 2004. *Rme1* is necessary for *Mi-1*-mediated resistance and acts early in the resistance pathway. Molecular Plant-Microbe Interactions 17, 55–61.

Debach, P., Rosen, D., 1991. Biological Control by Natural Enemies. Cambridge University Press, New York, USA.

Dedryver, C.A., Le Ralec, A., Fabre, F., 2010. The conflicting relationships between aphids and men: a review of aphid damage and control strategies. Comptes Rendus Biologies 333, 539–553.

Denmark, H.A., 1990. A Field Key to the Citrus Aphids in Florida (Homoptera: Aphididae). Entomology Circular No. 335. Florida Department of Agriculture and Consumer Services, Division of Plant Industry, Gainesville, Florida, USA.

Dent, D., 1991. Insect Pest Management. CAB International, Wallingford, UK.

Deseö Kovács, K.V., Rovesti, L., 1992. Lotta microbiologica contro i fitofagi: teoria e pratica. Edagricole, Bolonga, p. 296.

Diaz, B.M., Fereres, A., 2007. Ultraviolet-blocking materials as a physical barrier to control insect pests and plant pathogens in protected crops. Pest Technology 1, 85–95.

Digilio, M.C., Cascone, P., Iodice, L., Guerrieri, E., 2012a. Interactions between tomato volatile organic compounds and aphid behavior. Journal of Plant Interactions 7, 322–325.

Digilio, M.C., Sasso, R., Di Leo, M.G., Iodice, L., Monti, M.M., Santeramo, R., Arpaia, S., Guerrieri, E., 2012b. Interactions between Bt-expressing tomato and non-target insects: the aphid *Macrosiphum euphorbiae* and its natural enemies. Journal of Plant Interactions 7, 71–77.

Dixon, A.F.G., Hemptinne, J.L., 2003. Ladybirds and the biological control of aphid populations. In: Soares, A.O., Ventura, M.A., Garcia, V., Hemptinne, J.L. (Eds.), Proceedings of the 8th International Symposium on Ecology of Aphidophaga: Biology, Ecology and Behavior of Aphidophagous InsectsArquipélago: Life and Marine Sciences – Supplement 5, , pp. 1–10.

Dixon, A., Wellings, P.W., Carter, C., Nichols, J., 1993. The role of food quality and competition in shaping the seasonal cycle in the reproductive activity of the sycamore aphid. Oecologia 95, 89–92.

Dixon, A.F.G., 1973. Biology of Aphids. Edward Arnold, London, p. 55.

Dixon, A.F.G., 1985. Aphid Ecology. Springer Science and Business Media, New York, USA, p. 157.

Dixon, A.F.G., 1992. Constraints on the rate of parthenogenetic reproduction and pest status of aphids. Invertebrate Reproduction and Development 22, 159–163.

Dixon, A.F.G., 2000. Insect Predator-Prey Dynamics. Ladybird Beetles & Biological Control. Cambridge University Press.

Doring, T.F., Kirchner, S.M., Kuhne, S., Saucke, H., 2004. Response of alate aphids to green targets on colored backgrounds. Entomologia Experimentalis et Applicata 113, 53–61.

Dubey, S., Verghese, S.P., Jain, D., Khemani, L.D., Srivastava, M.M., 2012. Studies on efficacy of eco-friendly insecticide obtained from plant products against aphids found on tomato plant. In: Khemani, L.D., Srivastava, M.M., Srivastava, S. (Eds.), Chemistry of Phytopotentials: Health, Energy and Environmental Perspectives. Springer, Berlin, pp. 265–267.

Duran Prieto, J., Castañé, C., Calvet, C., Camprubi, A., Battaglia, D., Trotta, V., Fanti, P., 2016. Tomato belowground–aboveground interactions: *Rhizophagus irregularis* affects foraging behavior and life history traits of the predator *Macrolophus pygmaeus* (Hemiptera: Miridae). Arthropod-Plant Interactions. http://dx.doi.org/10.1007/s11829-016-9465-5.

Eastop, V.F., 1955. Selection of aphid species by different kinds of insect traps. Nature 176, 936.

Eastop, V.F., 1958. The history of *Macrosiphum euphorbiae* (Thomas) in Europe. Entomologist London 91, 198–201.

Economou, L.P., Lykouressis, D.P., Barbetaki, A.E., 2006. Time allocation of activities of two heteropteran predators on the leaves of three tomato cultivars with variable glandular trichome density. Environmental Entomology 35, 387–393.

Edser, C., 2002. Light manipulating additives extend opportunities for agricultural plastic films. Plastics, Additives and Compounding 4, 20–24.

Edwards, C.A., Arancon, N.Q., Vasko Bennett, M., Askar, A., Keeney, G., Little, B., 2010. Suppression of green peach aphid (*Myzus persicae*) (Sulz.), citrus mealybug (*Planococcus citri*) (Risso), and two spotted spider mite (*Tetranychus urticae*) (Koch.) attacks on tomatoes and cucumbers by aqueous extracts from vermicomposts. Crop Protection 29, 80–93.

Eigenbrode, S.D., Trumble, J.T., 1993. Antibiosis to beet armyworm (*Spodoptera exigua*) in *Lycopersicon* accessions. Hortscience 28, 932–934.

Ekesi, S., Shah, P.A., Clark, S.A.J., Pell, J.K., 2005. Conservation biological control with the fungal pathogen *Pandora neoaphidis*: implications of aphid species, host plant and predator foraging. Agricultural and Forest Entomology 7, 21–30.

El Habi, M., El Jadd, L., Sekkat, A., Boumezzough, A., 1999. Lutte contre *Aphis gossypii* Glover (Homoptera: Aphididae) sur concombre sous serre par *Coccinella septempunctata* Linnaeus (Coleoptera: Coccinellidae). Insect Science and Its Application 19, 57–63.

El Habi, M., Sekkat, A., El Jadd, L., Boumezzough, A., 2000. Biology of *Hippodamia variegate* Goeze (Col. Coccinellidae) and its suitability against *Aphis gossypii* Glov. (Hom., Aphididae) on cucumber under greenhouse conditions. Journal of Applied Entomology 124, 365–374.

El-Dessouki, S.A., El-Kifl, A.H., Helal, H.A., 1976. Life cycle, hosts plants and symptoms of damage of the tomato bug, *Nesidiocoris tenuis* Reut. (Heteroptera: Miridae), in Egypt. Journal of Plant Diseases and Protection 83, 204–220.

El-Shafie, H.A.F., Abdelraheem, B.A., 2012. Field evaluation of three biopesticides for integrated management of major pests of tomato, *Solanum lycopersicum* L. in Sudan. Agriculture and Biology Journal of North America 3, 340–344.

Fan, P.X., Miller, A.M., Schilmiller, A.L., Liu, X.X., Ofner, I., Jones, A.D., Zamir, D., Last, R.L., 2016. In vitro reconstruction and analysis of evolutionary variation of the tomato acylsucrose metabolic network. Proceedings of the National Academy of Sciences 113, E239–E248.

Fantinou, A.A., Perdikis, D.Ch., Maselou, D.A., Lambropoulos, P.D., 2008. Prey killing without consumption: does *Macrolophus pygmaeus* show adaptive foraging behavior? Biological Control 47, 187–193.

Fantinou, A.A., Perdikis, D.Ch., Lambropoulos, P.D., Maselou, D.A., 2009. Preference and consumption of *Macrolophus pygmaeus* preying on mixed instar assemblages of *Myzus persicae*. Biological Control 51, 76–80.

Faria, M.R., Wraight, S.P., 2007. Mycoinsecticides and mycoacaricides: a comprehensive list with worldwide coverage and international classification of formulation types. Biological Control 43, 237–256.

Farrar, R.R., Kennedy, G.G., 1991. Insect and mite resistance in tomato. In: Kalloo, G. (Ed.), Genetic Improvement of Tomato. Monographs on Theoretical and Applied Genetics, vol. 14. Springer-Verlag, Berlin, pp. 121–142.

Ferguson, K.I., Stiling, P., 1996. Nonadditive effects of multiple natural enemies on aphid populations. Oecologia 108, 375–379.

Frank, S.D., 2010. Biological control of arthropod pests using banker plant systems: past progress and future directions. Biological Control 52, 8–16.

Fritz, R., Delorme, R., Dedryver, C.A., 1981. Effets de traitements fongicides sur les entomophthorales. In: Société Française de Phytiatrie Phytopharmacie (Ed.), Troisième Colloque Sur Les Effets Non Intentionnels Des Fongicides, Versailles, pp. 91–100.

Fuxa, J.R., 1987. Ecological considerations for the use of entomopathogens in IPM. Annual Review of Entomology 32, 225–251.

Gabarra, R., Castañé, C., Bordas, E., Albajes, R., 1988. *Dicyphus tamaninii* Wagner as a beneficial insect and pest in tomato crops in Catalonia, Spain. Entomophaga 33, 219–228.

Gabarra, R., Alomar, O., Castañé, C., Goula, M., Albajes, R., 2004. Movement of the greenhouse whitefly and its predators between in- and outside of Mediterranean greenhouses. Agriculture, Ecosystems and Environment 102, 341–348.

Gautam, D.C., Crema, R., Bonvicini Pagliai, A.M., 1993. Cytogenetic mechanisms in aphids. Bollettino di Zoologia 60, 233–244.

Ghanim, N.M., 2014. Control of *Tuta absoluta* (Lepidoptera: Gelechildae) and *Aphis gossypii* (Hemiptera: Aphididae) by some aqueous plant extracts. Egyptian Journal of Biological Pest Control 24, 45–52.

Ghosh, L.K., Verma, K.D., 1990. Discovery of sexual female of *Myzus persicae* (Sulzer) (Homoptera: Aphididae) with redescription of its alate male in India. Journal of Aphidology 4, 30–35.

Gilbert, F.S., 1986. Hoverflies. Cambridge University Press, Cambridge, UK.

Gilbert, F.S., 2005. Syrphid aphidophagous predators in a food-web context. European Journal of Entomology 102, 325–333.

Gillespie, D., Brodeur, J., Cloutier, C., Goettel, M., Jaramillo, P., Labbe, R., Roitberg, B., Thompson, C., Van Laerhoven, S., 2006. Combining pathogens and predators of insects in biological control. In: Castane, C., Sanchez, J.A. (Eds.), International Organization of Biological Control, West Palearctic Regional Section Bulletin, vol. 29, pp. 3–8.

Gillespie, D.R., McGregor, R., Sanchez, J.A., Van Laerhoven, S.L., Quiring, D.M.J., Roitberg, B.D., Foottit, R.G., Schwartz, M.D., Shipp, J.L., 2007. An endemic omnivorous predator for control of greenhouse pests. In: Vincent, C., Goettel, M., Lazarovits, G. (Eds.), Biological Control: A Global Perspective. CABI Publishing, UK, pp. 128–135.

Gindin, G., Barash, I., Raccah, B., Singer, S., Ben-Zeev, I.S., Klein, M., 1996. The potential of some entomopathogenic fungi as biocontrol agents against the onion thrips, *Thrips tabaci* and the western flower thrips, *Frankliniella occidentalis*. Folia Entomologica Hungarica 57 (Suppl.), 37–42.

Glas, J.J., Schimmel, B.C.J., Alba, J.M., Escobar-Bravo, R., Schuurink, R.C., Kant, M.R., 2012. Plant glandular trichomes as targets for breeding or engineering of resistance to herbivores. International Journal of Molecular Sciences 13, 17077–17103.

Goettel, M.S., Jaronski, S.T., 1997. Safety and registration of microbial agents for control of grasshoppers and locusts. Memoirs of the Entomological Society of Canada 171, 83–99.

Goffreda, J.C., Mutschler, M.A., 1989. Inheritance of potato aphid resistance in hybrids between *Lycopersicon esculentum* and *L. pennellii*. Theoretical and Applied Genetics 78, 210–216.

Goffreda, J.C., Mutschler, M.A., Tingey, W.M., 1988. Feeding behavior of potato aphid affected by glandular trichomes of wild tomato. Entomologia Experimentalis et Applicata 48, 101–107.

Goffreda, J.C., Mutschler, M.A., Ave, D.A., Tingey, W.M., Steffens, J.C., 1989. Aphid deterrence by glucose esters in glandular trichome exudate of the wild tomato, *Lycopersicon pennellii*. Journal of Chemical Ecology 15, 2135–2147.

Goggin, F.L., Williamson, V.M., Ullman, D.E., 2001. Variability in the response of *Macrosiphum euphorbiae* and *Myzus persicae* (Hemiptera: Aphididae) to the tomato resistance gene *Mi*. Environmental Entomology 30, 101–106.

Goggin, F.L., 2007. Plant–aphid interactions: molecular and ecological perspectives. Current Opinion in Plant Biology 10, 399–408.

Goldansaz, S.H., McNeil, J.N., 2003. Calling behaviour of the potato aphid *Macrosiphum euphorbiae* oviparae under laboratory and field conditions. Ecological Entomology 28 (3), 291–298.

Goldansaz, S.H., Dewhirst, S., Birkett, M.A., Hooper, A.M., Smiley, D.W.M., Pickett, J.A., Wadhams, L., McNeil, J.N., 2004. Identification of two sex pheromone components of the potato aphid, *Macrosiphum euphorbiae* (Thomas). Journal of Chemical Ecology 30, 819–834.

Guerrieri, E., Lingua, G., Digilio, M.C., Massa, N., Berta, G., 2004. Do interactions between plant roots and the rhizosphere affect parasitoid behavior? Ecological Entomology 29, 753–756.

Hágvar, E.B., Hofsvang, T., 1995. Colonization behavior and parasitization by *Ephedrus cerasicola* (Hymenoptera: Aphidiidae) in choice studies with two species of plants and aphids. Journal of Applied Entomology 118, 23–30.

Hajek, A.E., Leger, R.J., 1994. Interactions between fungal pathogens and insect hosts. Annual Review of Entomology 39, 293–322.

Hämäläien, M., 1980. Introduction of *Aphidoletes aphidimyza* (Rond.) (Diptera: Cecidomyiidae) from an open rearing unit for the control of aphids in glasshouses. IOBC/WPRS Bulletin 3, 59–64.

Harding, J., 1959. Potato aphid control on tomatoes. Journal of Economic Entomology 52, 355–356.

Harmon, J., Andow, D.A., 2004. Indirect effects between shared prey: predictions for biological control. BioControl 49, 605–626.

Harrington, R., Hulle, M., Plantegenest, M., 2007. Monitoring and forecasting. In: van Emden, H.F., Harrington, R. (Eds.), Aphids as Crop Pests. CAB International, Wallingford, Oxfordshire, UK, pp. 515–536.

Harris, K.F., Maramorosh, K., 1977. Aphids as Virus Vectors. Academic Press, Inc., New York, USA.

Hart, M.M., Reader, R.J., 2002. Taxonomic basis for variation in the colonization strategy of arbuscular mycorrhizal fungi. New Phytologist 153, 335–344.

Hassanien, R.H.E., Tian zhen, H., Yu feng, L., Bao ming, L., 2014. Advances in effects of sound waves on plants. Journal of Integrative Agriculture 13, 335–348.

Hebert, S.L., Jia, L.L., Goggin, F.L., 2007. Quantitative differences in aphid virulence and foliar symptom development on tomato plants carrying the *Mi* resistance gene. Environmental Entomology 36, 458–467.

Helmi, A., Mohamed, H.I., 2016. Biochemical and ultrastructural changes of some tomato cultivars after infestation with *Aphis gossypii* Glover (Hemiptera: Aphididae) at Qalyubiyah, Egypt. Gesunde Pflanzen 68, 41–50.

Hildebrand, D.F., Brown, G.C., Jackson, D.M., Hamilton Kemp, T.R., 1993. Effects of some leaf-emitted volatile compounds on aphid population increase. Journal of Chemical Ecology 19, 1875–1887.

Hodek, I., Honěk, A., 1996. Ecology of the Coccinellidae. Kluwer, Academic Publishers, Dordrecht, Netherlands.

Holling, C.S., 1959. Some characteristics of simple types of predation and parasitism. Canadian Entomologist 91, 385–398.

Holling, C.S., 1966. The functional response of invertebrate predators to prey density. Memoirs of the Entomological Society of Canada 48, 86.

Holman, J., 2009. Host Plant Catalog of Aphids, Palaearctic Region. Springer, Netherlands.

Holtzer, T.O., Archer, T.L., Norman, J.M., 1988. Host plant suitability in relation to water stress. In: Heinrichs, E.A. (Ed.), Plant Stress-Insect Interactions. John Wiley and Sons, Chichester, UK, pp. 111–137.

Hooks, C.R.R., Fereres, A., 2006. Protecting crops from non-persistently aphid-transmitted viruses: a review on the use of barrier plants as a management tool. Virus Research 120, 1–16.

Hou, T., BaoMing, Li., GuangHui, T., Qing, Z., YingPing, X., Qi, L., 2009. Application of acoustic frequency technology to protected vegetable production. Transactions of the Chinese Society of Agricultural Engineering 25, 156–160.

Howling, G.G., Bale, J.S., Harrington, R., 1994. Effects of extended and repeated exposures to low temperature on mortality of the peach-potato aphid *Myzus persicae*. Ecological Entomology 19, 361–366.

Huberty, A.E., Denno, R.E., 2004. Plant water stress and its consequences for herbivorous insects: a new synthesis. Ecology 85, 1383–1398.

Hummel, N.A., Zalom, F.G., Miyao, G.M., Underwood, N.C., Villalobos, A., 2004. Potato aphid, *Macrosiphum euphorbiae* (Thomas), in tomatoes: plant canopy distribution and binomial sampling on processing tomatoes in California. Journal of Economic Entomology 97, 490–495.

Hyakumachi, M., Kubota, M., 2004. Fungi as plant growth promoter and disease suppressor. In: Arora, D.K. (Ed.), Fungal Biotechnology in Agricultural, Food, and Environmental Applications. Marcel Dekker, New York, USA, pp. 101–110.

Inbar, M., Doostdar, H., Mayer, R.T., 2001. Suitability of stressed and vigorous plants to various insect herbivores. Oikos 94, 228–235.

Jacob, F., Vernaldi, S., Maekawa, T., 2013. Evolution and conservation of plant NLR functions. Frontiers in Immunology 4, 297.

Jansson, J., 2003. The Influence of Plant Fertilization Regime on Plant-Aphid-Parasitoid Interactions (Ph.D. thesis). Swedish University of Agricultural Sciences, Uppsala, Sweden.

Jensen, A.B., Thomsen, L., Eilenberg, J., 2001. Intraspecific variation and host specificity of *Entomophthora muscae sensu strict* isolates revealed by random amplified polymorphic DNA, universal primed PCR, PCR-restriction fragment length polymorphism, and conidial morphology. Journal of Invertebrate Pathology 42, 267–278.

Johnson, B., 1956. The influence on aphids of the glandular hairs on tomato plants. Plant Pathology 5, 131–132.

Kakutani, K., Matsuda, Y., Nonomura, T., Toyoda, H., Kimbara, J., Osamura, K., Kusakari, S., 2012. Practical application of an electric field screen to an exclusion of flying insect pests and airborne fungal conidia from greenhouses with a good air penetration. Journal of Agricultural Science 4, 51–60.

Kaloshian, I., Walling, L.L., 2016. Hemipteran and dipteran pests: effectors and plant host immune regulators. Journal of Integrative Plant Biology 58, 350–361.

Kaloshian, I., Lange, W.H., Williamson, V.M., 1995. An aphid-resistance locus is tightly linked to the nematode-resistant gene *Mi* in tomato. Proceedings of the National Academy of Sciences 92, 622–625.

Kaloshian, I., Kinsey, M.G., Ullman, D.E., Williamson, V.M., 1997. The impact of *Meu1*-mediated resistance in tomato on longevity, fecundity and behavior of the potato aphid, *Macrosiphum euphorbiae*. Entomologia Experimentalis et Applicata 83, 181–187.

Kaloshian, I., Kinsey, M.G., Williamson, V.M., Ullman, D.E., 2000. *Mi*-mediated resistance against the potato aphid *Macrosiphum euphorbiae* (Hemiptera: Aphididae) limits sieve element ingestion. Environmental Entomology 29, 690–695.

Kaplan, I., Thaler, J.S., 2012. Phytohormone-mediated plant resistance and predation risk act independently on the population growth and wing formation of potato aphids, *Macrosiphum euphorbiae*. Arthropod-Plant Interactions 6, 181–186.

Katis, N.I., Tsitsipis, J.A., Stevens, M., Powell, G., 2007. Transmission of plant viruses. In: van Emden, H.F., Harrington, R. (Eds.), Aphids as Crop Pests. CAB International, Wallingford, Oxfordshire, UK, pp. 353–389.

Kennedy, J.S., Day, M.F., Eastop, V.F., 1962. A Conspectus of Aphids as Vectors of Plant Viruses. Commonwealth Institute of Entomology, London, UK.

Kennedy, J.S., 1958. Physiological condition of the host-plant and susceptibility to aphid attack. Entomologia Experimentalis et Applicata 1, 50–65.

Kennedy, G., 2003. Tomato, pests, parasitoids, and predators: tritrophic interactions involving the genus *Lycopersicon*. Annual Review of Entomology 48, 51–72.

Kersch-Becker, M.F., Thaler, J.S., 2015. Plant resistance reduces the strength of consumptive and non-consumptive effects of predators on aphids. Journal of Animal Ecology 84, 1222–1232.

Kettles, G.J., Kaloshian, I., 2016. The potato aphid salivary effector Me47 is a glutathione-S-transferase involved in modifying plant responses to aphid infestation. Frontiers in Plant Science 7, 1142.

Kielkiewicz, M., Godzina, M., Staniaszek, M., 2010. Relevance of the *Mi23* marker and the potato aphid biology as indicators of tomato plant (*Solanum lycopersicum* L.) resistance to some pests. Vegetable Crops Research Bulletin 72, 25–33.

Kindlmann, P., Dixon, A., 1999. Strategies of aphidophagous predators: lessons for modelling insect predator-prey dynamics. Journal of Applied Entomology 123, 397–399.

Klironomos, J.N., 2000. Host specificity and functional diversity among arbuscular mycorrhizal fungi. In: Bell, C.R., Brylinsky, M., Johnson-Green, P. (Eds.), Microbial Biosystems: New Frontiers. Proceedings of the 8th International Symposium of Microbial Ecology, Atlantic Canada Society for Microbial Ecology, Halifax, UK, pp. 845–851.

Kobari, E., Noori, G., 2012. A comparison of the relative resistance of some tomato cultivars to the cotton-melon aphid, *Aphis gossypii* (Hom.: Aphidiae), under greenhouse conditions. Iranian Journal of Plant Protection Science 43, 133–141.

Kohler, G.R., St. Clair, D.A., 2005. Variation for resistance to aphids (Homoptera: Aphididae) among tomato inbred backcross lines derived from wild *Lycopersicon* species. Journal of Economic Entomology 98, 988–995.

Kok-Yokomi, M.L., 1978. Mechanisms of host plant resistance in tomato to the potato aphid *Macrosiphum euphorbiae* (Thomas). Dissertation Abstracts International B 39, 1121.

Kring, J.B., Schuster, D.J., 1992. Management of insects on pepper and tomato with UV-reflective mulches. Florida Entomologist 75, 119–129.

Kuhar, T., Reiter, S., Doughty, H., 2009. Potato Aphid on Tomatoes – Homoptera: Aphididae, *Macrosiphum euphorbiae*. Virginia Cooperative Extension, Virginia Tech, USA. https://pubs.ext.vt.edu/2901/2901-1031/2901-1031_pdf.pdf.

Kumar, P., Poehling, H.M., 2006. UV-blocking plastic films and nets influence vectors and virus transmission on greenhouse tomatoes in the humid tropics. Environmental Entomology 35, 1069–1082.

Kuroda, T., Miura, K., 2003. Comparison of the effectiveness of two methods for releasing *Harmonia axyridis* (Pallas) (Coleoptera: Coccinellidae) against *Aphis gossypii* Glover (Homoptera: Aphididae) on cucumbers in a greenhouse. Applied Entomology and Zoology 38, 271–274.

Lamb, R.J., MacKay, P.A., 1997. Photoperiodism and life cycle plasticity of an aphid, *Macrosiphum euphorbiae* (Thomas), from central North America. Canadian Entomologist 129, 1035–1048.

Latigui, A., Dellal, A., 2009. Effect of different variation of NH_4^+ compared to $N \left(NH_4^+ + NO_3^- \right)$ fertilization of tomato (*Lycopersicum esculentum*) cultivated in inert media on the fecundity of the aphids *Macrosiphum euphorbiae* (Homptera-Aphididae). American Journal of Plant Physiology 4, 80–88.

Latimer, J.G., Oetting, R.D., 1994. Brushing reduces thrips and aphid populations on some greenhouse-grown vegetable transplants. HortScience 29, 1279–1281.

Leckie, B.M., De Jong, D.M., Mutschler, M.A., 2012. Quantitative trait loci increasing acylsugars in tomato breeding lines and their impacts on silverleaf whiteflies. Molecular Breeding 30, 1621–1634.

Leckie, B.M., De Jong, D.M., Mutschler, M.A., 2013. Quantitative trait loci regulating sugar moiety of acylsugars in tomato. Molecular Breeding 31, 957–970.

Leckie, B.M., Halitschke, R., De Jong, D.M., Smeda, J.R., Kessler, A., Mutschler, M.A., 2014. Quantitative trait loci regulating the fatty acid profile of acylsugars in tomato. Molecular Breeding 34, 1201–1213.

Leckie, B.M., D'Ambrosio, D.A., Chappell, T.M., Halitschke, R., De Jong, D.M., Kessler, A., Kennedy, G.G., Mutschler, M.A., 2016. Differential and synergistic functionality of acylsugars in suppressing oviposition by insect herbivores. PLoS One 11, e0153345.

Leite, G., Picanco, M., Guedes, R., Skowronski, L., 1999. Effect of fertilization levels, age and canopy height of *Lycopersicon hirsutum* on the resistance to *Myzus persicae*. Entomologia Experimentalis et Applicata 91, 267–273.

Lemke, C.A., Mutschler, M.A., 1984. Inheritance of glandular trichomes in crosses between *Lycopersicon esculentum and L. pennellii*. Journal of American Society for Horticultural Science 109, 592–596.

Lenfant, C., Ridray, G., Schoen, L., 2000. Biopropagation of *Macrolophus caliginosus* Wagner for a quicker establishment in Southern tomato greenhouses. OILB SROP Bulletin 23, 247–251.

Letourneau, D.K., Goldstein, B., 2001. Pest damage and arthropod community structure in organic vs conventional tomato production in California. Journal of Applied Ecology 38, 557–570.

Letourneau, D.K., Drinkwater, L.E., Shennon, C., 1996. Effects of soil management on crop nitrogen and insect damage in organic versus conventional tomato fields. Agriculture, Ecosystems and Environment 57, 174–187.

Levin, D.A., 1973. The role of trichomes in plant defense. The Quartely Review of Biology 48, 3–15.

Li, Qi., Xie, Q., Smith Becker, J., Navarre, D.A., Kaloshian, I., 2006. *Mi-1*-mediated aphid resistance involves salicylic acid and mitogen-activated protein kinase signaling cascades. Molecular Plant-Microbe Interactions 19, 655–664.

Loureiro, E.S., Moino Jr., A., 2006. Pathogenicity of hyphomycete fungi to aphids *Aphis gosyypii* Glover and *Myzus persicae* (Sultzer) (Hemiptera: Aphididae). Neotropical Entomology 35, 660–665.

Lucas, E., Brodeur, J., 1999. Oviposition site selection by the predatory midge *Aphidoletes aphidimyza* (Diptera: Cecidomyiidae). Environmental Entomology 28, 622–627.

Lucas, E., Coderre, D., Brodeur, J., 1998. Intraguild predation among aphid predators: characterization and influence of extraguild prey density. Ecology 79, 1084–1092.

Luckwill, L.C., 1943. The Genus *Lycopersicon*: Historical, Biological, and Taxonomic Survey of the Wild and Cultivated Tomatoes. Aberdeen University Press, Scotland.

Lykouressis, D.P., Perdikis, D.C., Tsagarakis, A., 2000. Polyphagous mirids in Greece: host plants and abundance in traps placed in some crops. Bollettino del Laboratorio di Entomologia Agraria "Filippo Silvestri" 56, 57–68.

Lykouressis, D.P., Perdikis, D.C., Gaspari, M.D., 2007. Prey preference and biomass consumption of *Macrolophus pygmaeus* (Hemiptera: Miridae) fed *Myzus persicae* and *Macrosiphum euphorbiae* (Hemiptera: Aphididae). European Journal of Entomology 104, 199–204.

Lykoyressis, D.P., Perdikis, D.C., Chalkia, C.A., Vardaki, S.C., 1993. Comparisons between alate aphids caught in yellow water traps and aphid populations on tomato plants. Entomologia Hellenica 11, 29–34.

MacGillivray, M.E., Anderson, G.B., 1964. The effect of photoperiod and temperature on the production of gamic and agamic forms in *Macrosiphum euphorbiae* (Thomas). Canadian Journal of Zoology 42, 491–510.

Malausa, J.C., Trottin-Caudal, Y., 1996. Advances in the strategy of use of the predaceous bug *Macrolophus caliginosus* (Heteroptera: Miridae) in glasshouse crops. In: Alomar, O., Wiedenmann, R.N. (Eds.), Zoophytophagous Heteroptera: Implications for Life History and Integrated Pest Management. Entomological Society of America, Lanham, MD, USA, pp. 178–189.

Mantelin, S., Bhattarai, K., Kaloshian, I., 2009. Ethylene contributes to potato aphid susceptibility in a compatible tomato host. New Phytologist 183, 444–456.

Mattson, W.J., Haack, R.A., 1987. The role of drought in outbreaks of plant-eating insects. Bioscience 37, 110–118.

Maurya, S.K., Maurya, R.P., Singh, D., 2015. Evaluation of thiamethoxam 70% WS as seed treatment against early sucking pests of tomato. Journal of Applied and Natural Science 7, 763–767.

McGregor, R.R., Gillespie, D.R., Quiring, D.M.J., Mitch, R.J.F., 1999. Potential use of *Dicyphus hesperus* Knight (Heteroptera: Miridae) for biological control of pests of greenhouse tomatoes. Biological Control 16, 104–110.

McKinney, K.B., 1938. Physical characteristics on the foliage of beans and tomatoes that tend to control some small insect pests. Journal of Economic Entomology 31, 630–631.

Meadow, R.H., Kelly, W.C., Shelton, A.M., 1984. Evaluation of *Aphidoletes aphidimyza* (Diptera: Cecidomyiidae) for control of *Myzus persicae* (Homoptera: Aphididae) in greenhouse and field experiments in the United States. Entomophaga 30, 385–392.

Meier, W., 1961. Beiträge zur Kenntnis der grünstreifigen Kartoffelblattlaus, *Macrosiphum euphorbiae* Thomas 1870, und verwandter Arten (Hemipt., Aphid.). Mitteilungen der Schweizerische Entomologischen Gesellschaft 34, 127–186.

Messelink, G.J., Janssen, A., 2014. Increased control of thrips and aphids in greenhouses with two species of generalist predatory bugs involved in intraguild predation. Biological Control 79, 1–7.

Meyling, N., Hajek, A.E., 2010. Principles from community and metapopulation ecology: application to fungal entomopathogens. BioControl 55, 39–54.

Meyling, N., Pell, J.K., 2006. Detection and avoidance of an entomopathogenic fungus by a generalist insect predator. Ecological Entomology 31, 162–171.

Meyling, N., Lubeck, M., Buckley, E.P., Eilenberg, J., Rehner, S.A., 2009. Community composition, host range and genetic structure of the fungal entomopathogen *Beauveria* in agricultural and seminatural habitats. Molecular Ecology 18, 1282–1293.

Miles, H.W., 1927. On the control of glasshouse insects with calcium cyanide. Annals of Applied Biology 14, 240–246.

Milligan, S.B., Bodeau, J., Yaghoobi, J., Kaloshian, I., Zabel, P., Williamson, V.M., 1998. The root-knot nematode gene *Mi* from tomato is a member of the leucine zipper-nucleotide binding leucine-rich repeat family of plant genes. Plant Cell 10, 1307–1319.

Minks, A.K., Harrewijn, P., 1987. Aphids: Their Biology, Natural Enemies and Control. Elsevier, Oxford, New York, Tokyo.

Misra, H.P., Mukherjee, S.K., 2012. Management of aphid, *Aphis gossypii* Glov. in Tomato (*Solanum lycopersicum* L.) by newer insecticides and their effect on the aphid predator *Coccinella septempunctata* L. Journal of Plant Protection and Environment 9, 11–15.

Moerkens, R., Berckmoes, E., van Damme, V., van den Broecke, S., Wittemans, L., 2014. Optimization of pest control by the predatory bug *Macrolophus pygmaeus* in greenhouse tomato production. IOBC/WPRS Bulletin 102, 157–162.

Möller, F.W., 1971. Hybridisation experiments within the complex of species around the green-striped potato aphid *Macrosiphum euphorbiae* (Thomas). Beitrage zur Entomologie 21, 531–537.

Monti, V., Manicardi, G.C., Mandrioli, M., 2011. Cytogenetic and molecular analysis of the holocentric chromosomes of the potato aphid *Macrosiphum euphorbiae* (Thomas, 1878). Comparative Cytogenetics 5, 163–172.

Morales, H., Perfecto, I., Ferguson, B., 2001. Traditional fertilization and its effect on corn insect populations in the Guatemalan highlands. Agriculture, Ecosystems and Environment 84, 145–155.

Morgan, W.L., 1937. Green aphids on tomatoes. Add white oil emulsion to the nicotine sulphate – bordeaux spray. Agricultural Gazette of New South Wales 48, 616.

Muigai, S.G., Schuster, D.J., Snyder, J.C., Scott, J.W., Bassett, M.J., McAuslane, H.J., 2002. Mechanisms of resistance in *Lycopersicon* germplasm to the whitefly *Bemisia argentifolii*. Phytoparasitica 30, 347–360.

Muigai, S.G., Bassett, M.J., Schuster, D.J., Scott, J.W., 2003. Greenhouse and field screening of wild *Lycopersicon* germplasm for resistance to the whitefly *Bemisia argentifolii*. Phytoparasitica 31, 27–38.

Mulder, S., Hoogerbrugge, H., Altena, K., Bolckmans, K., 1999. Biological pest control in cucumbers in the Netherlands. IOBC/WPRS Bulletin 22, 177–180.

Musetti, L., Neal, J., 1997a. Toxicological effects of *Lycopersicon hirsutum* f. *glabratum* and behavioral response of *Macrosiphum euphorbiae*. Journal of Chemical Ecology 23, 1321–1332.

Musetti, L., Neal, J., 1997b. Resistance to the pink potato aphid, *Macrosiphum euphorbiae*, in two accessions of *Lycopersicon hirsutum* f. *glabratum*. Entomologia Experimentalis et Applicata 84, 137–146.

Nault, L.R., 1997. Arthropod transmission of plant viruses: a new synthesis. Annals of the Entomological Society of America 90, 521–541.

Nielsen, C., Hajek, A.E., Humber, R.A., Bresciani, J., Eilenberg, J., 2003. Soil as an environment for winter survival of aphid-pathogenic *Entomopthorales*. Biological Control 28, 92–100.

Nijveldt, W., 1988. Cecidomyiidae. In: Minks, A.K., Harrewijn, P., Helle, W. (Eds.), World Crop Pests, Aphids. Elsevier Science, New York, USA, pp. 271–277.

Nombela, G., Williamson, V.M., Muniz, M., 2003. The root-knot nematode resistance gene *Mi-1.2* of tomato is responsible for resistance against the whitefly *Bemisia tabaci*. Molecular Plant-Microbe Interactions 16, 645–649.

Nzanza, B., Mashela, P.W., 2012. Control of whiteflies and aphids in tomato (*Solanum lycopersicum* L.) by fermented plant extracts of neem leaf and wild garlic. African Journal of Biotechnology 11, 16077–16082.

Ownley, B.H., Griffin, M.R., Klingeman, W.E., Gwinn, K.D., Moulton, J.K., Pereira, R.M., 2008. *Beauveria bassiana*: endophytic colonization and plant disease control. Journal of Invertebrate Pathology 98, 267–270.

Painter, R.H., 1951. Insect Resistance in Crop Plants. University of Kansas Press, Lawrence, KS, USA.

Pallipparambil, G.R., Reese, J.C., Avila, C.A., Louis, J.M., Goggin, F.L., 2010. *Mi*-mediated aphid resistance in tomato: tissue localization and impact on the feeding behavior of two potato aphid clones with differing levels of virulence. Entomologia Experimentalis et Applicata 135, 295–307.

Pallipparambil, G.R., Sayler, R.J., Shapiro, J.P., Thomas, J.M.G., Kring, T.J., Goggin, F.L., 2015. *Mi*-1.2, an R gene for aphid resistance in tomato, has direct negative effects on a zoophytophagous biocontrol agent, Orius insidiosus. Journal of Experimental Biology 66, 549–557.

Palmer, M.A., 1952. Aphids of the Rocky Mountain Region. The Thomas Say Foundation, Colorado, USA.

Pedigo, L.P., Rice, M.E., 2006. Entomology and Pest Management, fifth ed. Pearson Education Inc., Upper Saddle River, New Jersey, USA.

Pell, J.K., Pluke, R., Clark, S.J., Kenward, M.G., Alderson, P.G., 1997. Interactions between two aphid natural enemies, the entomopathogenic fungus *Erynia neoaphidis*, Remaudiere and Hennebert (*Zygomycetes:Entomopthorales*) and the predatory beetle *Coccinella septempunctata* L. *Coleoptera: Coccinellidae*). Journal of Invertebrate Pathology 69, 261–268.

Perdikis, D.C., Lykouressis, D.P., 1996. Aphid populations and their natural enemies on fresh market tomatoes in Central Greece. OILB/SROP Bulletin 19, 33–37.

Perdikis, D.C., Lykouressis, D.P., 2002. Life table and biological characteristics of *Macrolophus pygmaeus* when feeding on *Myzus persicae* and *Trialeurodes vaporariorum*. Entomologia Experimentalis et Applicata 102, 261–272.

Perdikis, D.C., Lykouressis, D.P., Economou, L.P., 1999. The influence of temperature, photoperiod and plant type on the predation rate of *Macrolophus pygmaeus* Rambur on *Myzus persicae* (Sulzer). BioControl 44, 281–289.

Pérez-Hedo, M., Urbaneja, A., 2015. Prospects for predatory mirid bugs as biocontrol agents of aphids in sweet peppers. Journal of Pest Science 88, 65–73.

Perring, T.M., Farrar, C.A., Toscano, N.C., 1988. Relationships among tomato planting date, potato aphids (Homoptera: Aphididae), and natural enemies. Journal of Economic Entomology 81, 1107–1112.

Perring, T.M., Gruenhagen, N.M., Farrar, C.A., 1999. Management of plant viral diseases through chemical control of insect vectors. Annals of the Entomological Society of America 44, 457–481.

Pieterse, C.M.J., Van der Does, D., Zamioudis, C., Leon-Reyes, A., Van Wees, S.C.M., 2012. Hormonal modulation of plant immunity. Annual Review of Cell and Developmental Biology 28, 489–521.

Polis, G.A., Myers, C., Holt, R., 1989. The ecology and evolution of intraguild predation: potential competitors that eat each other. Annual Review of Ecology and Systematics 20, 297–330.

Powell, W., Pell, J.K., 2007. Biological control. In: van Emden, H.F., Harrington, R. (Eds.), Aphids as Crop Pests. CAB International, Wallingford, Oxfordshire, UK, pp. 469–513.

Powell, W., Wilding, N., Brobyn, P.J., Clark, S.J., 1986. Interference between parasitoids (Hym. Aphidiidae) and fungi (*Entomophthorales*) attacking cereal aphids. Entomophaga 31, 293–302.

Powell, G., Maniar, S.P., Pickett, J.A., Hardie, J., 1999. Aphid responses to non-host epicuticular lipids. Entomologia Experimentalis et Applicata 91, 115–123.

Powell, G., Tosh, C.R., Hardie, J., 2006. Host plant selection by aphids: behavioral, evolutionary and applied perspectives. Annual Review of Entomology 51, 309–330.

Price, P.W., 1991. The plant vigor hypothesis and herbivore attack. Oikos 62, 244–251.

Principi, M.M., Canard, M., 1984. Feeding habits. In: Canard, M., Semeria, Y., New, T.T. (Eds.), Biology of Chrysopidae. Junk Publishers, The Hague, The Netherlands, pp. 76–92.

Quesada-Moraga, E., Maranhao, E.A.A., Valverde-Garcia, P., Santiago-Alvarez, C., 2006. Selection of *Beauveria bassiana* isolates for the control of the whiteflies *Bemisia tabaci* and *Trialeurodes vaporariorum* on the basis of their virulence, thermal requirements, and toxicogenic activity. Biological Control 36, 274–287.

Quiros, C.F., Stevens, M.A., Rick, C.M., Kokyokomi, M.L., 1977. Resistance in tomato to pink form of potato aphid (*Macrosiphum euphorbiae* Thomas) – role of anatomy, epidermal hairs, and foliage composition. Journal of the American Society for Horticultural Science 102, 166–171.

Rabasse, J.M., Lafont, J.P., Delpuech, I., Silvie, P., 1983. Progress in aphid control in protected crops. IOBC/WPRS Bulletin 7, 151–162.

Raboudi, F., Makni, H., Makni, M., 2011. Genetic diversity of potato aphid, *Macrosiphum euphorbiae*, populations in Tunisia detected by RAPD. African Entomology 19, 133–140.

Reeves II, A.F., 1977. Tomato trichomes and mutations affecting their development. American Journal of Botany 64, 186–189.

Remaudière, G., Latgè, J.P., Michel, M.F., 1981. Ecologie comparée des entomophthoracées pathogènes de pucerons en France littorale et continentale. Entomophaga 26, 157–178.

Rivelli, S.R., Toma, I., Trotta, V., Fanti, P., De Maria, S., Battaglia, D., 2012. Combined effect of water stress and *Macrosiphum euphorbiae* infestation on plant growth in tomato. In: Stoddard, F., Mäkelä, P. (Eds.), 12th Congress of the European Society for Agronomy, Helsinki, 20–24 August 2012, vol. 1, pp. 334–335 Helsinki, Finland.

Rivelli, A.R., Trotta, V., Toma, I., Fanti, P., Battaglia, D., 2013. Relation between plant water status and *Macrosiphum euphorbiae* (Hemiptera: Aphididae) population dynamics on three cultivars of tomato. European Journal of Entomology 110, 617–625.

Roberts, P.A., Thomason, I.J., 1986. Variability in reproduction of isolates of *Meloidogyne incognita* and *M. javanica* on resistant tomato genotypes. Plant Disease 70, 547–551.

Robinson, S.H., Wolfenbarger, D.A., Dilday, R.H., 1980. Antixenosis of smooth leaf cotton to the ovipositional response of tobacco budworm. Crop Science 20, 646–649.

Rodriguez, A.E., Tingey, W.M., Mutschler, M.A., 1993. Acylsugars of *Lycopersicon pennellii* deter settling and feeding of the green peach aphid (Homoptera, Aphididae). Journal of Economic Entomology 86, 34–39.

Rosenheim, J.A., Wilhoit, L.R., Armer, C.A., 1993. Influence of intraguild predation among generalist insect predators on the suppression of an herbivore population. Oecologia 96, 439–449.

Rosenheim, J.A., Kaya, H.K., Ehler, L.E., Marois, J.J., Jaffee, B.A., 1995. Intraguild predation among biological control agents: theory and practice. Biological Control 5, 303–335.

Rosenheim, J.A., 1998. Higher-order predators and the regulation of insect herbivore populations. Annual Review of Entomology 43, 421–447.

Rossi, M., Goggin, F.L., Milligan, S.B., Kaloshian, I., Ullman, D.E., Williamson, V.M., 1998. The nematode resistance gene *Mi* of tomato confers resistance against the potato aphid. Proceedings of the National Academy of Sciences 95, 9750–9754.

Rotheray, G.E., 1987. Larval morphology and searching efficiency in aphidophagous syrphid larvae. Entomologia Experimentalis et Applicata 43, 49–54.

Roy, H.E., Cottrell, T.E., 2008. Forgotten natural enemies: interactions between coccinellids and insect-parasitic fungi. European Journal of Entomology 105, 391–398.

Roy, H.E., Pell, J.K., 2000. Interactions between entomopathogenic fungi and other natural enemies: implications for biological control. Biocontrol Science and Technology 10, 737–752.

Roy, H.E., Pell, J.K., Clark, S.J., Alderson, P.G., 1998. Implications of predator foraging on aphid pathogen dynamics. Journal of Invertebrate Pathology 71, 236–247.

Roy, H.E., Pell, J.K., Alderson, P.G., 2001. Targeted dispersal of the aphid pathogenic fungus *Erynia neoaphidis* by the aphid predator *Coccinella septempunctata*. Biocontrol Science and Technology 11, 99–110.

Sanchez, J.A., Lacasa, A., 2008. Impact of the zoophytophagous plant bug *Nesidiocoris tenuis* (Heteroptera: Miridae) on tomato yield. Journal of Economic Entomology 101, 1864–1870.

Sarmento, R.A., Pallini, A., Venzon, M., de Souza, O.F.F., Molina Rugama, A.J., de Oliveira, C.L., 2007. Functional response of the predator *Eriopis connexa* (Coleoptera: Coccinellidae) to different prey types. Brazilian Archives of Biology and Technology 50, 121–126.

Sasso, R., Iodice, L., Digilio, M.C., Carretta, A., Ariati, L., Guerrieri, E., 2007. Host-locating response by the aphid parasitoid *Aphidius ervi* to tomato plant volatiles. Journal of Plant Interactions 2, 175–183.

Sasso, R., Iodice, L., Woodcock, C.M., Pickett, J.A., Guerrieri, E., 2009. Electrophysiological and behavioral responses of *Aphidius ervi* (Hymenoptera: Braconidae) to tomato plant volatiles. Chemoecology 19, 195–201.

Satyavati, Nagaraju, N., Padmaja, A.S., 2014. Bio-management of vectors and viral diseases in tomato (*Solanum esculentum* Mill.) and chili (*Capsicum annuum* Linn.). Mysore Journal of Agricultural Sciences 48, 29–36.

Schalk, J.M., Robbins, M.L., 1987. Reflective mulches influence plant-survival, production, and insect control in fall tomatoes. HortScience 22, 30–32.

Schuder, D.L., 1950. The effect of some of the newer insecticides on tomatoes and tomato insects. Proceedings of the Indiana Academy of Science 60, 211–221.

Scopes, N.E.A., 1969. The potential of *Crysopa carnea* as a biological control agent of *Myzus persicae* on glasshouse chrysanthemums. Annals of Applied Biology 64, 433–439.

Seagraves, M.P., Yeargan, K.V., 2006. Selection and evaluation of a companion plant to indirectly augment densities of *Coleomegilla maculata* (Coleoptera: Coccinellidae) in sweet corn. Environmental Entomology 35, 1334–1341.

Seithe, A., 1979. Hairs as taxonomic characters in *Solanum*. In: Hawks, J.G., Lester, R.N., Skelding, A.D. (Eds.), The Biology and Taxonomic of Solanaceae. Linnaean Society Symposium Series No. 7. Academic Press, London, pp. 307–320.

Sethi, J., Nagaich, B.B., 1972. Chromosome number of different clones of *Myzus persicae* with varying virus transmission efficiency. Indian Journal of Experimental Biology 10, 154–155.

Shands, W.A., Simpson, G.W., Wave, H.E., 1972. Seasonal Population Trends and Productiveness of the Potato Aphid on Swamp Rose in Northeastern Maine. Technical Bulletin, Life Sciences and Agriculture Experiment Station. University of Maine, pp. 35–52.

Shilling, P.R., 1959. An investigation of the hereditary character, woolly, in the tomato. Ohio Journal of Science 59, 289–302.

Shipp, J.L., Wang, K., 2006. Evaluation of *Dicyphus hesperus* (Heteroptera: Miridae) for biological control of *Frankliniella occidentalis* (Thysanoptera: Thripidae) on greenhouse tomato. Journal of Economic Entomology 99, 414–420.

Shipp, J.L., Zhang, Y., Hunt, D.W.A., Ferguson, G., 2003. Influence of humidity and greenhouse microclimate on the efficacy of *Beauveria bassiana* (Balsamo) for control of greenhouse arthropod pests. Environmental Entomology 32, 1154–1163.

Simmons, A.T., Gurr, G.M., 2004. Trichome-based host plant resistance of *Lycopersicon species* and the biocontrol agent *Mallada signata*: are they compatible? Entomologia Experimentalis et Applicata 113, 95–101.

Simmons, A.T., Gurr, G.M., 2006. The effect on the biological control agent *Mallada signata* of trichomes of F1 *Lycopersicon esculentum × L. cheesmanii* f. minor and *L. esculentum × L. pennellii* hybrids. Biological Control 38, 174–178.

Simmons, A.T., Gurr, G.M., McGrath, D., Nicol, H.I., Martin, P.M., 2003. Trichomes of *Lycopersicon* spp. and their effect on *Myzus persicae* (Sulzer) (Hemiptera: Aphididae). Australian Journal of Entomology 42, 373–378.

Simmons, A.T., McGrath, D., Gurr, G.M., 2005. Trichome characteristics of F1 *Lycopersicon esculentum × L. cheesmanii* f. *minor* and *L. esculentum × L. pennellii* hybrids and effects on *Myzus persicae*. Euphytica 144, 313–320.

Singh, V., Shah, J., 2012. Tomato responds to green peach aphid infestation with the activation of trehalose metabolism and starch accumulation. Plant Signaling and Behavior 7, 605–607.

Singh, V., Louis, J., Ayre, B.G., Reese, J.C., Shah, J., 2011. Trehalose Phosphate Synthase 11-dependent trehalose metabolism promotes *Arabidopsis thaliana* defense against the phloem-feeding insect *Myzus persicae*. Plant Journal 67, 94–104.

Sloan, W.J.S., 1945. The control of tomato pests. Queensland Agricultural Journal 61, 17–41.

Snyder, W.E., Ives, A.R., 2003. Interactions between specialist and generalist natural enemies: parasitoids, predators, and pea aphid biocontrol. Ecology 84, 91–107.

Snyder, W.E., Ives, A.R., 2009. Population dynamics and species interactions. In: Radcliffe, E.B., Hutchison, W.D., Cancelado, R.E. (Eds.), Integrated Pest Management. Cambridge University Press, Cambridge, Massachusetts, USA, pp. 62–74.

Snyder, W.E., Ballard, S.N., Yang, S., Clevenger, G.M., Miller, T.D., Ahn, J.J., Hatten, T.D., Berryman, A.A., 2004. Complementary biocontrol of aphids by the ladybird beetle *Harmonia axyridis* and the parasitoid *Aphelinus asychis* on greenhouse roses. Biological Control 30, 229–235.

Soliman, M.M.M., Tarasco, E., 2008. Toxic effects of four biopesticides (Mycotal, Vertalec, Vertemic and Neem Azal-T/S) on *Bemisia tabaci* (Gennadius) and *Aphis gossypii* (Glover) on cucumber and tomato plants in greenhouses in Egypt. Entomologica 41, 195–217.

Srinivasa Rao, N.K., Bhatt, R.M., Sadashiva, A.T., 2001. Tolerance to water stress in tomato cultivars. Photosynthetica 38, 465–467.

Sterk, G., Meesters, P., 1997. IPM in strawberries in glasshouses and plastic tunnels in Belgium, new possibilities. Acta Horticulture 439, 905–911.

Stern, V.M., Smith, R.F., van den Bosch, R., Hagen, K.S., 1959. The integrated control concept. Hilgardia 29, 81–101.

Stoetzel, M.B., Miller, G.L., 1998. Aphids (Homoptera: Aphididae) colonizing peach in the United States or with potential for introduction. Florida Entomologist 81, 325–345.

Stoetzel, M.B., Miller, G.L., 2001. Aerial feeding aphids of corn in the United States with reference to the root-feeding Aphis maidiradicis (Homoptera: Aphididae). Florida Entomologist 84, 83–98.

Stoetzel, M.B., Miller, G.L., O'Brien, P.J., Graves, J.B., 1996. Aphids (Homoptera: Aphididae) colonizing cotton in the United States. Florida Entomologist 79, 193–205.

Strand, L.R., 1998. Integrated Pest Management for Tomatoes, fourth ed. University of California Division of Agriculture and Natural Resources Publication 3274, University of California, Oakland, USA, p. 118.

Sudhakar, N., Thajuddin, N., Murugesan, K., 2011. Plant growth-promoting rhizobacterial mediated protection of tomato in the field against cucumber mosaic virus and its vector *Aphis gossypii*. Biocontrol Science and Technology 21, 367–386.

Summers, C.G., Mitchell, J.P., Stapleton, J.J., 2004. Management of aphid-borne viruses and *Bemisia argentifolii* (Homoptera: Aleyrodidae) in zucchini squash by using UV reflective plastic and wheat straw mulches. Environmental Entomology 33, 1447–1457.

Sun, Y., Yin, J., Chen, F., Wu, G., Ge, F., 2011. How does atmospheric elevated CO_2 affect crop pests and their natural enemies? Case histories from China. Insect Science 18, 393–400.

Sunderland, K.D., Fraser, A.M., Dixon, A.F.G., 1986. Field and laboratory studies on money spiders (Linyphilidae) as predators of cereal aphids. Journal of Applied Ecology 23, 433–447.

Takikawa, Y., Matsuda, Y., Nonomura, T., Kakutani, K., Okada, K., Morikawa, S., Shiboa, M., Kusakari, S., Toyoda, H., 2016. An electrostatic nursery shelter for raising pest and pathogen free tomato seedlings in an open-window greenhouse environment. Journal of Agricultural Science (Toronto) 8, 13–25.

Tamaki, G., Butt, B.A., Landis, B.J., 1970. Arrest and aggregation of male *Myzus persicae* (Hemiptera: Aphididae). Annals of the Entomological Society of America 63, 955–960.

Taschenberg, E.F., Avens, A.W., 1955. Further studies on control of potato aphid on tomatoes. Journal of Economic Entomology 48, 685–688.

Taschenberg, E.F., 1949. Control of Potato Aphid on Tomatoes. New York State Agricultural Experiment Station Bulletin 736. Cornell, University, Geneva, New York, USA, p. 21.

Teixeira, M., Sela, N., Ng, J., Casteel, C.L., Peng, H.C., Bekal, S., Girke, T., Ghanim, M., Kaloshian, I., 2016. A novel virus from *Macrosiphum euphorbiae* with similarities to members of the family Flaviviridae. Journal of General Virology 97, 1261–1271.

Thaler, J.S., Agrawal, A.A., Halitschke, R., 2010. Salicylate-mediated interactions between pathogens and herbivores. Ecology 91, 1075–1082.

Tian, D.L., Tooker, J., Peiffer, M., Chung, S.H., Felton, G.W., 2012. Role of trichomes in defense against herbivores: comparison of herbivore response to woolly and hairless trichome mutants in tomato (*Solanum lycopersicum*). Planta (Berlin) 236, 1053–1066.

Tjallingii, W.F., 2006. Salivary secretions by aphids interacting with proteins of phloem wound responses. Journal of Experimental Botany 57, 739–745.

Togni, P.H.B., Venzon, M., Muniz, C.A., Martins, E.F., Pallini, A., Sujii, E.R., 2016. Mechanisms underlying the innate attraction of an aphidophagous coccinellid to coriander plants: implications for conservation biological control. Biological Control 92, 77–84.

Tommasini, M.G., Mosti, M., 2001. Control of aphids by *Chrysoperla carnea* on strawberry in Italy. In: McEwen, P.K., New, T.R., Whittington, A.E. (Eds.), Lacewings in the Crop Environment. Cambridge University Press, Cambridge, Massachusetts, USA, pp. 481–486.

Tunc, I., Unlu, M., Dagli, F., 2004. Bioactivity of acetone vapors against greenhouse pests, *Tetranychus cinnabarinus*, *Aphis gossypii* and *Franklinielia occidentalis*. Zeitschrift für Pflanzenkrankheiten und Pflanzenschutz 111, 225–230.

Vacante, V., Tropea-Garzia, G., 1994. Nesidiocoris tenuis: antagonista naturale di aleurodidi. Informatore Fitopatologico 4, 23–28.

Valderrama, K., Granobles, J., Valencia, E., Sanchez, M., 2007. *Nesidiocoris tenuis* (Hemiptera: Miridae) predator in tobacco crops (*Nicotiana tabacum*). Revista Colombiana de Entomología 33, 141–145.

van der Blom, J., Robledo, A., Torres, S., Sánchez, J.A., 2009. Consequences of the wide scale implementation of biological control in greenhouse horticulture in Almería, Spain. IOBC/WPRS Bulletin 49, 9–13.

van der Hoorn, R.A.L., Kamoun, S., 2008. From guard to decoy: a new model for perception of plant pathogen effectors. Plant Cell 20, 2009–2017.

van Emden, H.F., Harrington, R. (Eds.), 2007. Aphids as Crop Pests. CAB International, Wallingford, Oxfordshire, UK.

van Emden, H.F., Eastop, V.F., Hughes, H.D., Way, M.J., 1969. The ecology of *Myzus persicae*. Annual Review of Entomology 14, 197–270.

van Emden, H.F., 2007. Host plant resistance. In: van Emden, H.F., Harrington, R. (Eds.), Aphids as Crop Pests. CAB International, Wallingford, Oxfordshire, UK, pp. 447–468.

van Hanh, V., Hong, S.I.I., Kim, K., 2007. Selection of entomopathogenic fungi for aphid control. Journal of Bioscience and Bioengineering 104, 498–505.

van Lenteren, J.C., 2003. Commercial availability of biological control agents. In: van Lenteren, J.C. (Ed.), Quality Control and Production of Biological Control Agents: Theory and Testing Procedures. CABI Publishing, UK, pp. 167–179.

van Schelt, J., Mulder, S., 2000. Improved methods of testing and release of *Aphidoletes aphidimyza* (Diptera: Cecidomyiidae) for aphid control in glasshouses. European Journal of Entomology 97, 511–515.

van Schelt, J., Klapwijk, J., Letard, M.Y., Aucouturier, C., 1996. The use of *Macrolophus caliginosus* as a whitefly predator in protected crops. In: Gerling, D., Mayer, R.T. (Eds.), Bemisia 1995: Taxonomy, Biology, Damages, Control and Management. Intercept, Andover, UK, pp. 515–521.

van Schelt, J., 1999. Biological control of sweet pepper pests in the Netherlands. IOBC/WPRS Bulletin 22, 217–220.

Vatsanidou, A., Bartzanas, T., Papaioannou, C., Kittas, C., 2011. Efficiency of physical means of IPM on insect population control in greenhouse crops. Acta Horticulturae 893, 1247–1254.

Vega, F.E., Jackson, M.A., Mercadier, G., Poprawski, T.J., 2003. The impact of nutrition on spore yields for various fungal entomopathogens in liquid culture. World Journal of Microbiology and Biotechnology 19, 363–368.

Vega, F.E., Goettel, M.S., Blackwell, M., Chandler, D., Jackson, M.A., Keller, S., Koike, M., Maniania, N.K., Monzón, A., Ownley, B.H., Pell, J.K., Rangel, D.E.N., Roy, H.E., 2009. Fungal entomopathogens: new insights on their ecology. Fungal Ecology 2, 149–159.

Verheggen, F.J., Capella, Q., Schwartzberg, E.G., Voigt, D., Haubruge, E., 2009a. Tomato-aphid-hoverfly: a tritrophic interaction incompatible for pest management. Arthropod-Plant Interactions 3, 141.

Verheggen, F.J., Capella, Q., Wathelet, J.P., Haubruge, E., 2009b. What makes *Episyrphus balteatus* (Diptera: Syrphidae) oviposit on aphid infested tomato plants? Communications in Agricultural and Applied Biological Sciences 73, 371–381.

Viggiani, G., 2002. Lotta Biologica e Integrate Nella Difesa Fitosanitaria, vol. 2. Liguori Editore, Italy, p. 445.

Visser, J.H., Piron, P.G.M., Hardie, J., 1996. The aphid's peripheral perception of plant volatiles. Entomologia Experimentalis et Applicata 80, 35–38.

Voigt, D., Gorb, E., Gorb, S., 2007. Plant surface-bug interactions: *Dicyphus errans* stalking along trichomes. APIS 1, 221–243.

Vos, P., Simons, G., Jesse, T., Wijbrandi, J., Heinen, L., Hogers, R., Frijters, A., Groenendijk, J., Diergaarde, P., Reijans, M., FierensOnstenk, J., deBoth, M., Peleman, J., Liharska, T., Hontelez, J., Zabeau, M., 1998. The tomato *Mi-1* gene confers resistance to both root-knot nematodes and potato aphids. Nature Biotechnology 16, 1365–1369.

Walgenbach, J.F., 1994. Distribution of parasitized and nonparasitized potato aphid (Homoptera, Aphididae) on stake tomato. Environmental Entomology 23, 795–804.

Walgenbach, J.F., 1997. Effect of potato aphid (Homoptera: Aphididae) on yield, quality, and economics of staked-tomato production. Journal of Economic Entomology 90, 996–1004.

Walker, G.P., Madden, L.V., Simonet, D.E., 1984a. Spatial dispersion and sequential sampling of the potato aphid, *Macrosiphum euphorbiae* (Homoptera: Aphididae) on processing tomatoes in Ohio. Canadian Entomologist 116, 1069–1075.

Walker, G.P., Nault, L.R., Simonet, D.E., 1984b. Natural mortality factors acting on potato aphid (*Macrosiphum euphorbiae*) populations in processing-tomato fields in Ohio. Environmental Entomology 13, 724–732.

Walling, L.L., 2000. The myriad plant responses to herbivores. Journal of Plant Growth Regulation 19, 195–216.

Walling, L.L., 2008. Avoiding effective defenses: strategies employed by phloem-feeding insects. Plant Physiology 146, 859–866.

Walling, L.L., 2009. Adaptive defense responses to pathogens and pests. Advances in Botanical Research 51, 551–612.

Wasternack, C., Stenzel, I., Hause, B., Hause, G., Kutter, C., Maucher, H., Neumerkel, J., Feussner, I., Miersch, O., 2006. The wound response in tomato – role of jasmonic acid. Journal of Plant Physiology 163, 297–306.

Weber, C.A., Godfrey, L.D., Mauk, P.A., 1996. Effects of parasitism by *Lysiphlebus testaceipes* (Hymenoptera: Aphididae) on transmission of beet yellows closterovirus by bean aphid (Homoptera: Aphididae). Journal of Economic Entomology 89, 1431–1437.

Weintraub, P.G., Berlinger, M.J., 2004. Physical control in greenhouses and field crops. In: Horowitz, A.R., Ishaaya, I. (Eds.), Novel Approaches to Insect Pest Management. Springer, Dordrecht, Germany, pp. 301–318.

Weintraub, P.G., 2009. Physical control: an important tool in pest management programs. In: Ishaaya, I., Horowitz, A.R. (Eds.), Biorational Control of Arthropod Pests. Application and Resistance Management. Springer, Dordrecht, Germany, pp. 317–324.

Weisser, W.W., 2003. Additive effects of pea aphid natural enemies despite intraguild predation. In: Soares, A.O., Ventura, V., Garcia, V., Hemptinne, J.L. (Eds.), Proceedings of the 8th International Symposium on Ecology of Aphidophaga: Biology, Ecology and Behavior of Aphidophagous InsectsArquipélago Life and Marine Sciences – Supplement 5, , pp. 11–15.

West, F.T., 1946. Ecological effects of an aphid population upon weight gains of tomato plants. Journal of Economic Entomology 38, 338–343.

Wilding, N., 1973. The survival of *Entomopthora* spp. in mummified aphids at different temperatures and humidities. Journal of Invertebrate Pathology 21, 309–311.

Williams, W.G., Kennedy, G.G., Yamamoto, R.T., Thacker, J.D., Bordner, J., 1980. 2-tridecanone – naturally occurring insecticide from wild tomato *Lycopersicon hirsutum* f. *glabratum*. Science 207, 888–889.

Wise, I.L., Lamb, R.J., 1995. Spatial distribution and sequential sampling methods for the potato aphid, *Macrosiphum euphorbiae* (Thomas) (Homoptera: Aphididae), in oilseed flax. Canadian Entomologist 127, 967–976.

Wittenborn, G., Olkowski, W., 2000. Potato aphid monitoring and biocontrol in processing tomatoes. The IPM Practitioner 22, 1–7.

Wolfenbarger, D.O., Moore, W.D., 1968. Insect abundances on tomatoes and squash mulched with aluminum and plastic sheetings. Journal of Economic Entomology 61, 34–37.

Wolfenbarger, D.O., 1954. A comparison of dilute and concentrate sprays for control of insect of potato and tomato. Journal of Economic Entomology 47, 537–539.

Worrall, D., Holroyd, G.H., Moore, J.P., Glowacz, M., Croft, P., Taylor, J.E., Paul, N.D., Roberts, M.R., 2012. Treating seeds with activators of plant defense generates long-lasting priming of resistance to pests and pathogens. New Phytologist 193, 770–778.

Wratten, S.D., Lee, G., Stevens, D.J., 1979. Duration of cereal aphid populations and the effects on wheat yield and quality. In: British Crop Protection Council, Proceedings of the 1979 British Crop Protection Conference – Pests and Diseases, pp. 1–8.

Wratten, S.D., Gurr, G.M., Tylianakis, J.M., Robinson, K.A., 2007. Cultural control. In: van Emden, H.F., Harrington, R. (Eds.), Aphids as Crop Pests. CAB International, Wallingford, Oxfordshire, UK, pp. 423–445.

Wu, C.H., Krasileva, K.V., Banfield, M.J., Terauchi, R., Kamoun, S., 2015a. The "sensor domains" of plant NLR proteins: more than decoys? Frontiers in Plant Science 6, 134.

Wu, C.J., Avila, C.A., Goggin, F.L., 2015b. The ethylene response factor Pti5 contributes to potato aphid resistance in tomato independent of ethylene signaling. Journal of Experimental Biology 66, 559–570.

Yano, E., 2006. Ecological considerations for biological control of aphids in protected culture. Population Ecology 48, 333–339.

Yao, Y., ZhengZhou, C., Ke, H., HaoBing, C., Jia, L., Ting, H., HouZhang, W., 2012. Effects of bamboo vinegar with different concentrations on the growth and development of tomato and *Myzus persicae*. Journal of Anhui Agricultural University 39, 646–650.

Yardim, E.N., Edwards, C.A., 2003. Effects of organic and synthetic fertilizer sources on pest and predatory insects associated with tomatoes. Phytoparasitica 31, 324–329.

Yeo, H., Pell, J.K., Alderson, P.G., Clark, S.J., Pye, B.J., 2003. Laboratory evaluation of temperature effects on the germination and growth of entomo-pathogenic fungi and on their pathogenicity to two species. Pest Management Science 59, 156–165.

Yoltas, T., Baspinar, H., Aydin, A.C., Yildirim, E.M., 2003. The effect of reflective and black mulches on yield, quality and aphid populations on processing tomato. Acta Horticulturae 613, 267–270.

Zalom, F.G., Lanini, W.T., Miyao, G., Davis, R.M., 2001. A continuum of integrated pest management practices in processing tomatoes. Acta Horticulturae 542, 55–62.

Zehnder, G.W., Murphy, I.F., Sikora, E.J., Kloepper, J.W., 2001. Application of rhizobacteria for induced resistance. European Journal of Plant Pathology 107, 39–50.

Zehnder, G., Gurr, G.M., K¨uhne, S., Wade, M.R., Wratten, S.D., Wyss, E., 2007. Arthropod pest management in organic crops. Annual Review of Entomology 52, 57–80.

Zilahi-Balogh, G.M.G., Foottit, R.G., Ferguson, G., Shipp, J.L., 2005. New records for *Rhopalosiphum rufiabdominale* (Sasaki) (Hemiptera: Aphididae) on greenhouse tomatoes and peppers. Journal of the Entomological Society of Ontario 136, 85–87.

Chapter 3

Thrips: Biology, Ecology, and Management

David Riley[1], Alton Sparks Jr.[1], Rajagopalbab Srinivasan[1], George Kennedy[2], Greg Fonsah[1], John Scott[3], Steve Olson[4]

[1]UGA, Tifton, GA, United States; [2]NCSU, Raleigh, NC, United States; [3]UF Gulf Coast REC, Wimauma, FL, United States; [4]UF North Florida Research and Education Center, Institute of Food and Agricultural Sciences, Quincy, FL, United States

1. INTRODUCTION

Tomato spotted wilt (TSW) was reported as a disease of tomato for the first time in Australia (Brittlebank, 1919) which was attributed to tomato spotted wilt virus (TSWV) transmission by *Thrips tabaci* Lindeman (Pittman, 1927). Later, "*Frankliniella insularis* (Franklin)" (Smith, 1931), a misidentification of *Frankliniella lycopersici* Steele (Andrewartha, 1937) which was synonymized with *Frankliniella schultzei* (Trybom) (Moulton, 1948), was also implicated. The incidence of TSW was not significantly reduced by the available inorganic and nicotine insecticides at that time (Samuel and Pittman, 1928). By 1933, studies regarding host-plant resistance to "Kromnek disease" caused by thrips (*Frankliniella* sp.)—transmitted virus—were conducted in the Cape Province of South Africa, but with little success (Moore, 1933). "Kromnek disease" was later described as identical to the Australian TSW (Moore and Anderssen, 1939). Smith (1933) reported that spotted wilt of tomato had been observed at that time in Australia, the British Isles (in glasshouses), and in the United States. In the United States, McWhorter and Milbrath (1938) summarized a 4-year study on thrips-transmitted "tip blight" in southern Oregon cannery tomato production areas (the vectors *Frankliniella occidentalis* Pergande and *T. tabaci* were reported as present) and was the first to document that weeds could serve as a virus reservoir. By 1942, insecticide treatment and greenhouse sanitation proved effective for managing this problem in Oregon (Leach and Berg, 1944). In 1939, Bonnemaison (1939) reported TSW in several locations in France. By the 1940s, tomato was being bred for TSW resistance in Hawaii (Kikuta and Frazier, 1946) and spraying programs for thrips control to reduce TSW were being developed in Brazil (Costa et al., 1950). Sakimura (1961) stated that neither *F. occidentalis* nor *T. tabaci* were observed to reproduce readily in California (United States) tomato, whereas *T. tabaci* did reproduce readily in lettuce. This supported his observation that lettuce was the major source of the virus at that time in California. Through the 1970s, TSW in tomato remained a sporadic pest problem in the United States. In a 1980 California study of arthropod pests of tomatoes, thrips did not make the list of primary or secondary pests of tomato, and were not mentioned in standard field survey report forms (Lange and Bronson, 1981).

Thrips, as a major pest of tomatoes causing widespread economic losses in the United States, emerged about the same time as TSW showed up in major peanut production areas in the southcentral and southeastern United States. The first detection of TSW in peanut was in 1973 in Texas (Halliwel and Philley, 1974) and by the 1980s severe TSW problems were reported in peanut in Louisiana (Bond et al., 1983), Alabama (Gudauskas et al., 1988), Georgia (Culbreath et al., 1991), and North Carolina (Brandenburg, 1986). At the same time, TSW problems in tomato were also reported in Arkansas (Gergerich, 1985), Louisiana (Greenough et al., 1985), and northern Florida (Olson and Funderburk, 1986). Early reports in Louisiana suggested that a wide array of weeds were the hosts of the virus (Bond et al., 1983) and that there was a possible link between the increased incidence of western flower thrips and TSW (Greenough et al., 1985). The southeastern United States was in a regional epidemic by the early 1990s. Thrips-transmitted TSWV was confirmed to be a severe and widespread disease in tomato (Fig. 3.1) in Georgia (Gitaitis et al., 1998), which lead to intensive pesticide use and other tactics to control thrips and reduce economic losses in this high value crop (Greenough et al., 1990; Gitaitis et al., 1998; Riley and Pappu, 2000).

The two main thrips vector species associated with the initial TSW epidemic in the southeastern United States were the tobacco thrips, *Frankliniella fusca* (Hinds), and the western flower thrips, *F. occidentalis* (Greenough et al., 1985; Culbreath et al., 1991). Riley et al. (2011a) estimated that from 1996 to 2006 there was an annual average loss of $12.3 million in peanut, $11.3 million in tobacco, and a combined $9 million in tomato and pepper for a total of $326 million over the 10 years in the state of Georgia, alone. Losses as high as 95% in spring-grown susceptible

FIGURE 3.1   Hybrid Florida 47 susceptible tomato plant damaged early by TSW. This early infection results in total loss of fruit from this plant. *www. tswvramp.org, with permission.*

tomatoes were observed at individual sites during the mentioned time period. Early attempts at control consisted of one of the following: the systemic insecticide imidacloprid applied as a seedling drench to reduce thrips feeding; foliar insecticides to suppress thrips populations; and reflective mulch to deter thrips from alighting on the plants (Riley and Pappu, 2000). A TSW-resistant tomato cultivar, Stevens (Stevens et al., 1991), was available but was not as productive as standard hybrids (Riley and Pappu, 2000). Effective integrated management was beginning to be developed by 2004 (Momol et al., 2004; Riley and Pappu, 2004). However, by 2008 the use of high quaility TSW-resistant commercial cultivars of tomato had become widespread in the southeastern United States (Riley et al., 2011b), which greatly mitigated the crisis. At the time of writing of this chapter, the status of TSW in tomato in the southeastern United States is considered adequately managed with host-plant resistance, reflective mulch, and various chemical treatments. How this management strategy and the resulting recommendations were developed is summarized from the body of work presented by the authors at http://www.tswvramp.org/ and discussed in the following sections. A critical first step to thrips management in tomato begins with proper identification and understanding the biology of the individual thrips species relative to their role in the epidemiology of tospoviruses. The following section focuses on the two main vector species, *F. occidentalis* and *F. fusca*, in the southeastern United States.

2. IDENTIFICATION

Thrips fall under the insect order Thysanoptera (fringe/hair-winged insects) and all known TSWV vector species are from the family Thripidae (Riley et al., 2011a). Vector-species adults are around 1–2 mm long (Figs. 3.2–3.4). Therefore, these small insects can be difficult to recognize unless the crop scout is trained to identify them (Hoddle et al., 2012). The importance of thrips as agricultural pests is mostly a result of their ability to transmit tospoviruses. Worldwide, TSW is the most serious disease caused by a tospovirus (Moyer, 2000). TSWV can be vectored by at least nine species of thrips (Ullman et al., 1997; Pappu et al., 2009; Riley et al., 2011a). It appears from regional studies that thrips responsible for TSW outbreaks in the southeastern United States are western flower thrips, *F. occidentalis*, tobacco thrips, *F. fusca* (Navas et al., 1991b; Riley and Pappu, 2000, 2004), and possibly *Frankliniella bispinosa* (Morgan) (Webb et al., 1997). Other thrips species that have been reported to transmit TSWV include: *F. schultzei*, *Frankliniella intonsa* (Trybom), *Frankliniella gemini* (Bagnall), *Frankliniella cephalica* (Crawford), *T. tabaci*, and *Thrips setosus* (Moulton) (Riley et al., 2011a), but currently *F. occidentalis*, and *F. fusca* are likely responsible for the majority of TSWV epidemics in the southeastern United States (Groves et al., 2003).

2.1 *Frankliniella occidentalis* Pergande

Western flower thrips are 1.2–1.9 mm long, yellow to brown with eight antennal segments. Setae on the head and thorax, a comb on the abdomen, and antennal features help to distinguish it from other closely related species (Oetting et al., 1993). Briefly, features that can alert the scout that they are looking at a western flower thrips include (1) the postocular

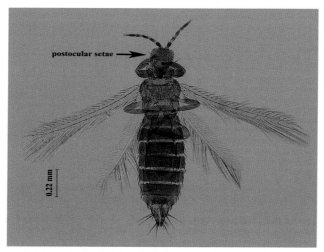

FIGURE 3.2 Slide mount of *Frankliniella occidentalis* Pergande showing antennal segments and pronounced postocular setae. *Riley, D.G., Joseph, S.V., Srinivasan, R., Diffie, S., 2011a. Thrips vectors of tospoviruses. Journal of Integrated Pest Management 1. http://dx.doi.org/10.1603/IPM10020, with permission.*

FIGURE 3.3 Slide mount of *Frankliniella fusca* (Hinds), common name Tobacco thrips. *Riley, D.G., Joseph, S.V., Srinivasan, R., Diffie, S., 2011a. Thrips vectors of tospoviruses. Journal of Integrated Pest Management 1. http://dx.doi.org/10.1603/IPM10020, with permission.*

FIGURE 3.4 *Frankliniella occidentalis* Pergande feeding on pine pollen. *www.tswvramp.org, with permission.*

seta IV is distinct under 40× magnification and (2) the length of the pronotal antero-marginal setae are equal to that of the antero-angular setae (Fig. 3.2). Also, the row of setae or comb on tergite VIII is complete with long and irregular setae (Moritz et al., 2001; Mound et al., 2009); however, this is only visible using greater than 40× magnification on a properly slide-mounted specimen. The interactive keys by Moritz et al. (2001) clearly distinguish this species from other pestiferous thrips species.

2.2 *Frankliniella fusca* (Hinds)

Briefly, tobacco thrips are 1.0–1.3 mm in length, generally dark brown to black, and also possess an antenna with eight segments (Fig. 3.3). Features that can alert the scout that they are looking at a tobacco thrips are: (1) the postocular seta IV is mostly absent under 40× magnification and (2) the length of the pronotal antero-marginal setae is distinctly shorter to that of the antero-angular setae. Using higher magnification, tergite VIII does not exhibit a distinct row of setae or a "comb" (Moritz et al., 2001; Mound et al., 2009). A reduced-wing (micropterous) form of adult tobacco thrips can occur with the typical winged (macropterous) form (Oetting et al., 1993). The reduced-wing form of tobacco thrips is common in Georgia peanuts (Todd et al., 1995). Again, refer to Moritz et al. (2001) or the online key provided by Hoddle et al. (2012).

3. BIOLOGY

The population dynamics of *F. occidentalis* and *F. fusca* have been documented in many landscape systems in the Southeast including crops (e.g., tomato [Navas et al., 1991b; Riley and Pappu, 2000, 2004], cover crops [Toapanta et al., 1996], peanut [Barbour and Brandenburg, 1994; Todd et al., 1995], and tobacco [McPherson et al., 1999]), and weeds (Chamberlin et al., 1992; Groves et al., 2001b, 2002; Srinivasan et al., 2014). In peanuts, all thrips species tend to peak early in the season, e.g., less than 30 days post-April planting (Todd et al., 1995). In fact, most of the population dynamics studies in the southeastern United States cited above show a distinct late April to early June peak, which is also reflected in regional trapping studies for thrips (Groves et al., 2003). Temperature largely controls when thrips generation turnovers occur in the spring (Morsello et al., 2010), but a possible explanation for the intensity of late spring thrips populations is the regional availability of pine pollen (Riley et al., 2010). Pine pollen is an excellent food source for thrips (Fig. 3.4). Pine pollen deposited on thrips host plants increases western flower thrips and tobacco thrips net reproduction by severalfold (Hulshof et al., 2003; Angelella and Riley, 2010; respectively).

Seasonal factors that can alter thrips population dynamics include temperature (Lowry et al., 1992; Morsello and Kennedy, 2009), photo-period (Whittaker and Kirk, 2004; Chaisuekul and Riley, 2005), ambient humidity (Kirk, 1997), precipitation (Morsello et al., 2010), and the availability of suitable host plants for oviposition and development (Lewis, 1973). Temperature is likely the major factor affecting the rate of development. The number of degree days needed for *F. fusca* and *F. occidentalis* to complete one generation (egg to adult) are reported to be 234 (lower threshold 10.5°C) and 253 (6.5°C), respectively (Lowry et al., 1992). These thrips require approximately 2 weeks to develop from eggs to adults at 25°C (Lowry et al., 1992). Adults begin the process by inserting bean-shaped eggs into plant tissues, which can be either leaves or floral parts (Hansen et al., 2003). The eggs are usually visible only after plant-tissue staining (Fig. 3.5). After oviposition, the eggs hatch in 3–4 days. The first instar nymph is off-white to yellowish in color and adult western flower thrips is yellow to orange in color. Tobacco thrips adults are dark brown to black as previously mentioned. The heavily feeding, first to second instar stages take 10–14 days to complete, while the pre-pupa and pupa stages can last approximately 1 week (Lowry et al., 1992). As thrips develop into the pre-pupal stage, they typically fall from the plant into the soil and pupate; however, Broadbent et al. (2003) reported that some pupae can remain on the plant. The non-feeding pupal stage, distinguished by well-developed wing pads, has limited mobility (Lewis, 1973). Adults emerge within 3 days at 30°C (Lowry et al., 1992). Adult *F. occidentalis* females can survive for 4–5 weeks at 30°C and lay an average of 50 eggs (Reitz, 2008). Thrips are typically haplodiploid with virgin females producing only haploid male offspring, and mated females producing diploid female and haploid male offspring in a female-biased sex ratio (Lewis, 1973; Moritz, 1997). Western flower thrips adults have fringed wings (Figs. 3.2 and 3.4) like most adult Thripidae, but some adult tobacco thrips can be wingless (brachypterous) which can affect overwintering and dispersal (Wells et al., 2002). In addition to understanding this general thrips biology, it is helpful to understand the biological relationship between thrips and tospoviruses.

Tospoviruses, in the virus genus *Tospovirus* and family Bunyaviridae, are transmitted only by thrips in the family Thripidae, subfamily Thripinae (Moyer, 2000; Riley et al., 2011a). The main tospoviruses in tomato in the United States are tomato spotted wilt virus (TSWV) and, to a lesser extent, tomato chlorotic spot virus (TCSV), impatiens necrotic spot virus, and groundnut ringspot virus (GRSV); (German et al., 1992; Moyer, 2000; Adkins et al., 2013). For a review of the global status of tospoviruses, see Pappu et al. (2009). As previously stated, the main thrips vector species of these viruses in

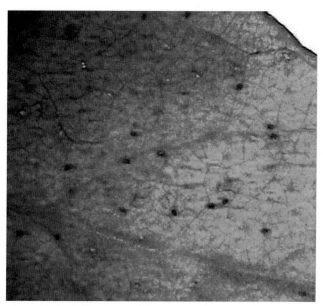

FIGURE 3.5 Stained thrips eggs in young leaf tissue. *www.tswvramp.org, with permission.*

the United States are *F. occidentalis* and *F. fusca*, and to a lesser extent, *F. bispinosa*, *T. tabaci*, *Scirtothrips dorsalis* Hood, and *F. schultzei* (Riley et al., 2011a). *T. setosus* is also reported to transmit TSWV to tomato in Japan (Nakao et al., 2011).

The well-documented relationship between TSWV and western flower thrips (Wijkamp et al., 1995; Ullman et al., 1997; Whitfield et al., 2005) is our best guide to the potential relationship between TSWV and other thrips vectors. The first and second instars of western flower thrips acquire TSWV by feeding on infected plant tissue. The virus enters the thrips through infection sites in the midgut, replicates, and migrates to the salivary gland, and after a latent period, viruliferous adults inoculate the virus when they settle and feed on other plants (Ullman et al., 1997). Virus replication takes place in the thrips vector (Ullman et al., 1997) and the highly polyphagous nature of thrips (Kirk, 1997) makes it difficult to manage TSWV epidemics. This is because, once infected adult thrips can disperse, sometimes kilometers depending on weather conditions (Lewis, 1973), to a wide range of host plants and transmit the virus. TSWV acquisition by first instar thrips can be as short as 1–2 h (Ullman et al., 1992) and median latent period at 27°C is 84 h (Wijkamp and Peters, 1993). However, it is not until the thrips complete their development and the adults move to other host plants that the spread of virus across the landscape occurs. Transmission can occur in as little as 5 min of feeding (Wijkamp, 1995), which is often the time in the field before thrips can be killed by insecticides. It is important to note that each cohort of first to second instar thrips must acquire the virus since there is no transovarial passage of the virus (Wijkamp et al., 1996). Due to this requirement of re-acquisition for subsequent thrips generations, the frequency of viruliferous adults fluctuates greatly over time. However, even low percentages of viruliferous thrips can cause significant yield loss. The percentage of viruliferous thrips typically increases over time in the spring, but even 3% infected thrips in field populations can be associated with a severe TSW year (McPherson et al., 2005). Several generations occur per year in the southeastern United States (Chappell et al., 2013), but only one or two generations occur in a single tomato season. Both *F. fusca* and *F. occidentalis* were reported in multiple native, non-crop or "weed" plant species over the winter (Groves et al., 2002; Beaudoin and Kennedy, 2012; Srinivasan et al., 2014). The growth stage of the most frequently encountered thrips sampled from the foliage of weeds during the winter months are nymphs, not adults (Chamberlin et al., 1992; Srinivasan et al., 2014). *F. fusca* can reproduce throughout the winter months in southern Georgia in Vidalia onions (Fig. 3.6, Sparks et al., 2010) and in weed hosts (Srinivasan et al., 2014). Of course, the lower winter temperatures greatly lengthens the life cycle of thrips (Chappell et al., 2013) and can alter the relative proportions of thrips species (Riley and Sparks, 2014).

4. DISTRIBUTION AND HOST RANGE

The distribution of thrips and its associated host range is complex and extensive (Mound, 2005). Even the worldwide distribution of the single species, *F. occidentalis*, is partly due to its invasive biology (Morse and Hoddle, 2006). It is better researched than most pest thrips species because of its wide distribution and economic impact (Kirk, 2002; Kirk and Terry, 2003; Reitz, 2009). International shipments of ornamental plants are thought to be linked with *F. occidentalis* movement

FIGURE 3.6 Thrips adults and immatures and associated leaf damages in onions. *www.tswvramp.org, with permission.*

around the world (Perrings et al., 2005). Western flower thrips are highly polyphagous, feeding on >60 host botanical families (Yudin et al., 1986). *F. occidentalis* settle and feed mostly on flowers and pollen (Navas et al., 1991b), but can also settle and feed on plant leaves (Todd et al., 1995; Joost and Riley, 2004). The action of feeding on fruit can lead to a "dimpling" and discoloration of the tomato fruit surface (Navas et al., 1991a). *F. fusca* is somewhat different from *F. occidentalis* in that it settles and feeds more on plant leaves (preferably young leaf tissue) than on flowering and fruiting structures (Joost and Riley, 2007). *F. fusca* is reported to be the dominant thrips species on pre-flowering tomato foliage in Georgia (Joost and Riley, 2004). Leaf damage and associated feeding behavior has been documented for *F. fusca* (Joost and Riley, 2005). As with all thrips feeding, a very high frequency of feeding can disfigure and stunt the plant. *F. fusca* is prevalent in the southeastern United States (Diffie et al., 2008), but found throughout the United States (Hoddle et al., 2012). It now occurs in the Netherlands (Vierbergen, 1995), southeastern Asia (Wang et al., 2010), and Japan (Nakao et al., 2011), but is thought to be native to the new world (Mound et al., 2009). The plant host range for this thrips is very broad. It is one of the main thrips pest of cotton (DuRant et al., 1994; Groves et al., 2003), onion (Sparks et al., 2010; Riley and Sparks, 2014), peanut (Lowry et al., 1992; Todd et al., 1995), ornamentals (Oetting et al., 1993), peppers (Hansen et al., 2003), and tomato (Navas et al., 1991b; Riley and Pappu, 2000, 2004; Joost and Riley, 2004, 2007).

5. SEASONAL OCCURRENCE

The timings of thrips movement before and during the tomato growing season are important from the standpoint of TSWV epidemiology. Relative to this epidemiology, it is also important to understand the complex relationship between TSWV, its thrips vectors, and plant hosts at the landscape level (Groves et al., 2003; Morsello and Kennedy, 2009; Morsello et al., 2010; Chappell et al., 2013). At the level of individual thrips vectors, factors affecting virus acquisition and latency periods regulate the flow of TSWV through vectors from one host plant to another (Whitfield et al., 2005). As previously mentioned, viruliferous adults occur only when first instar nymphs acquire TSWV from an infected host plant (Ullman et al., 1997). Following ingestion, the virus infects cells in the midgut and migrates in the haemolymph to the salivary glands (Wijkamp, 1995). TSWV continues to replicate in the salivary glands and is inoculated to healthy plants when the thrips salivate during feeding. Due to the time required for the virus to migrate to the salivary glands and replicate, there is a temperature-dependent latent period of 3.5–7 days following acquisition of the virus before the thrips is able to inoculate another plant. Although some thrips that acquire TSWV as a first instar are able to inoculate the virus in the second instar, most of them are not able to transmit until they become adults. Moreover, because second instars do not disperse readily and subsequent preadult stages do not feed, it is the adults that disperse causing TSWV epidemics. The virus is unable to be acquired by adult thrips via ingestion of infected plant sap. Therefore, non-infected adults that feed on TSWV-infected host plants are not able to transmit TSWV to new plant tissues (Ullman et al., 1997). An important consequence of this relationship is that only infected host plants which support the reproduction of vector thrips species can act as a source for spread to other plants.

After an infected adult thrips inoculates a plant with TSWV by feeding, it takes one to three or more weeks, depending on plant age and environmental conditions, before the virus spreads through the plant and is available to be acquired by thrips larvae feeding on the plant. This delay, combined with the fact that only adults who acquired the virus during the first instar are able to spread TSWV, means that a plant infected in the field cannot become a source for spread of the virus to other plants until a minimum of 2 to 3 weeks have passed (Riley et al., 2012a). The length of this delay is influenced by numerous factors including plant age and temperature, and is important in determining the potential for TSWV to be spread from plant to plant within a susceptible crop field. In addition, as plants age they become more resistant to infection (Beaudoin et al., 2009) and less preferred by thrips for settling and feeding (Joost and Riley, 2007). In the case of pepper, susceptibility to infection under a given level of inoculation pressure (number of infectious thrips feeding on the plants) decreases by an average of approximately 50% in the first 7–9 days after transplanting (Beaudoin et al., 2009). This increase in resistance with plant age explains why tobacco, pepper, and tomato crops transplanted during periods of peak virus spread suffer higher infection than those transplanted several weeks prior to the period of peak virus spread. It also explains why within-field spread of TSWV tends to be limited unless large densities of reproducing vectors develop within the crop. Because age-related resistance is not absolute, within-field spread of TSWV can be significant when large densities of vectors are allowed to develop.

Although both tobacco thrips and western flower thrips spread TSWV in tomato, the tobacco thrips, *F. fusca*, is primarily responsible for early season spread, which most often causes the greatest yield losses (Chaisuekul et al., 2003; Joost and Riley, 2004). In the southeastern United States, tobacco thrips and the virus overwinter mostly on non-crop plant hosts, primarily seasonal, annual weeds. Many of these winter weeds tend to senesce in the spring when temperatures rise, causing a migration of viruliferous thrips from these plants in the landscape to freshly planted crops, such as tomato (Groves et al., 2002; Beaudoin and Kennedy, 2012). The prior infection of winter weeds resulted from fall migration of viruliferous *F. fusca* and *F. occidentalis* coming out of senescing summer host plants that were infected with TSWV (Kahn et al., 2005). It appears that *F. fusca* as a TSWV vector plays a critical role in maintaining TSWV-infected plants in the fall, winter, and spring, but that *F. occidentalis* serves a critical role as TSWV vector during the warmer months in the annual cycle (Chappell et al., 2013). The annual peak in thrips densities occurs in the spring in the southeastern United States (Riley et al., 2010), specifically in North Carolina between Julian dates 281 and 309 (Fig. 3.7).

There is good evidence that for the southeastern United States, *F. fusca* initiates the primary virus inoculation in spring in the tomato crop in North Carolina, South Carolina, and Georgia from overwintering weed sources (Chamberlin et al., 1992; Cho et al., 1995; Groves et al., 2002, 2003) as an early crop invader (Joost and Riley, 2004). Tomato plants infected early in the season will likely produce few marketable fruits, if any (Moriones et al., 1998; Chaisuekul et al., 2003). However, in situations where large densities of *F. occidentalis* develop within tomato crops following bloom, they can continue to cause later-season spread of TSWV that results in fruit damage and suppresses marketable yield (Riley et al., 2012a).

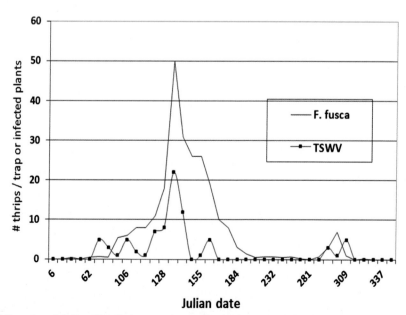

FIGURE 3.7 Numbers of dispersing adult *Frankliniella fusca* caught on yellow sticky traps and sentinel plants infected with TSWV throughout the year at one location in North Carolina. Note the large spring peaks in numbers of dispersing thrips and spread of TSWV in spring and summer. *www.tswvramp. org, with permission; adapted from Groves, R.L., Walgenbach, J.F., Moyer, J.W., Kennedy, G.G., 2003. Seasonal dispersal patterns of* Frankliniella fusca *(Thysanoptera: Thripidae) and tomato spotted wilt virus occurrence in central and eastern North Carolina. Journal of Economic Entomology 96, 1–11.*

Since *F. occidentalis* feeds directly on the fruit (Navas et al., 1991a), mature plant resistance may not be a significant factor. Mature plant resistance is an observed phenomenon in pepper, as when the plant leaf tissue matures it becomes less attractive for thrips feeding and TSWV transmission (Beaudoin et al., 2009). Each of these situations present different pest management challenges.

The population levels of the primary vector in the southeastern United States, *F. fusca* are dependent on the availability of host plants suitable for reproduction, favorable temperatures, and the lack of heavy rainfall, a significant regional factor in *F. fusca* mortality (Morsello and Kennedy, 2009; Morsello et al., 2010). Regional modeling of *F. fusca* population dynamics based on empirical validation using these factors has been mostly successful in the southeastern United States (Chappell et al., 2013). Even then, we know that other biotic and abiotic thrips mortality factors can greatly influence within-crop population dynamics, such as chemical and physical control tactics (Riley and Pappu, 2000, 2004; Riley et al., 2012b), naturally occurring biological control by predaceous minute pirate bugs, *Orius* spp. (Baez et al., 2004), and a parasitic nematode, *Thripinema fuscum* Tipping and Nguyen (Sims et al., 2009).

6. DAMAGE, LOSSES, AND ECONOMIC THRESHOLDS

Regional, multi-crop economic losses in the southeastern United States due to thrips-transmitted TSWV were presented in the introduction to this chapter. Since tomato production typically occurs near other major crops infested with thrips, e.g., peanut (Todd et al., 1995), cotton (DuRant et al., 1994; Groves et al., 2003), tobacco (McPherson et al., 1992, 1999), onion (Sparks et al., 2010), and tomato can serve as a sink or a source of thrips vectors. It is challenging to relate thrips numbers, and even more difficult to link numbers of viruliferous thrips to regional crop loss (McPherson et al., 2005). The most common in-season crop metric that has the strongest relationship to tomato crop yield is TSW foliar symptoms (Fig. 3.1). The type of relationship reported in tomato is that the earlier the TSW symptoms appear, the lower the yield (Moriones et al., 1998; Chaisuekul et al., 2003; Riley et al., 2012a). This is also the case in other solanaceous crops such as tobacco (McPherson et al., 2005). Therefore, preventative control, not a threshold-based on thrips numbers, is critical to the management of this pest as a vector of tospoviruses. On the other hand, thrips and virus dispersal in the landscape can be predicted with weather-based models (Chappell et al., 2013). Therefore, although there is no economic threshold in the traditional sense, there are triggers for control action similar to those used in regional plant-disease management based on an assessment of risk (see http://climate.ncsu.edu/thrips/).

Relative to the direct damage that thrips cause to the plant, tobacco thrips prefer young leaves to older tissue (Joost and Riley, 2007) and they have been associated with early season TSW development in the crop (Riley et al., 2012a). Plant-tissue scarring caused by thrips feeding can be severe enough to disfigure leaves or even stunt the plant. The same crop also can have feeding-associated injury by western flower thrips (Yudin et al., 1986). Plant pollens and flower tissues are mainly targeted by the flower thrips for feeding which can lead to flower bud damage and blossom dehiscence (Kirk, 1997). As previously stated, although flower thrips are prevalent on flowers (Navas et al., 1991b), western flower thrips have been reported in relatively low numbers on the leaves of tomato seedlings (Joost and Riley, 2004). One of the more serious concerns about direct damage caused by flower thrips is the dimpling and discoloration of fruit (Fig. 3.8) rendering them unacceptable for the fresh tomato market (Navas et al., 1991a).

Thrips use piercing-sucking mouthparts to feed on plant tissues (Kirk, 1997). They use a single mandible as a sharp probe and a rocking motion of the head to penetrate the leaf cuticle. After the outer cell wall is penetrated, a pair of maxillary stylets extend deeper into cellular structure of the leaf. The cellular contents are digested by saliva that has been injected into the cells and then sucked up through the maxillary stylets (Kindt et al., 2003). The adults inoculate plant tissue with TSWV present in their saliva. The reduction of tomato fruit quality caused by this inoculation can be severe, making some plants non-harvestable, directly reducing overall weight of tomato produced per plant, and causing severe post-harvest tomato quality losses due to the irregular ripening. This symptom appears on fruit perceived to be marketable at the time of harvest that develops after the commercial ripening process in the packing sheds (Riley and Pappu, 2004).

A typical example of fruit damage in tomato associated with thrips virus transmission to the tomato plant is shown in Fig. 3.9. The concentric ring pattern is commonly thought to be indicative of TSWV infection (Moyer, 2000), but is similar to the infections of TCSV and GRSV in tomato (Adkins et al., 2013). This damage can show up many weeks after virus transmission occurred (Riley et al., 2012a). Direct scarring or dimpling of the fruit surface can show within a week of feeding and oviposition. The halo spotting of mature tomato fruit caused by thrips reported by Navas et al. (1991a) can sometimes be confused with stink bug damage. Stink bug damage often causes larger areas of discoloration around the feeding site due to much larger size of the stink bug and deeper probing into the fruit. Thrips damage to fruit is typically associated with western flower thrips or other flower thrips species, and not tobacco thrips, which is more of a foliage feeder. Late-season control of thrips to prevent fruit damage must take into consideration how effective the insecticide is against flower thrips species. An average 1 adult per plant can cause no damage but growers should make some decision at this threshold level (Funderburk et al., 2014).

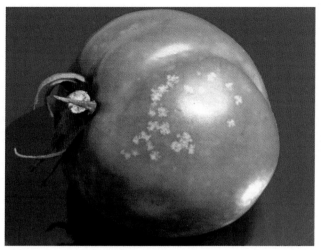

FIGURE 3.8   Halo spots around thrips feeding sites. *www.tswvramp.org, with permission.*

FIGURE 3.9   TSWV symptoms/damage to FL 47 susceptible tomato fruit. *Riley, D.G., Sparks Jr., A., Langston, D., 2011c. Managing Tomato Spotted Wilt in Georgia. The University of Georgia Cooperative Extension Service, Circular 1002, with permission.*

7. MANAGEMENT

The primary solution to TSWV problems in tomato in the southeastern United States has been the development and use of TSWV-resistant tomato cultivars through traditional breeding programs and regional field evaluations (Riley et al., 2011b). However, with the potential threat of host-plant resistance-breaking strains of TSWV (Roselló et al., 1997, 1998), the use of multiple tactics seems to be a prudent approach to long-term TSWV management in tomato (Momol et al., 2004; Riley and Pappu, 2004). Most of the management practices used for TSWV mitigation in the tomato crop, at least at the time of writing this chapter, occur within the tomato field during the growing season. The epidemiological conditions outside of the field are considered only when trying to estimate potential risk of TSWV infection for that season (Chappell et al., 2013). One view of thrips management in fruiting vegetables can be found in the study by Demirozer et al. (2012) which emphasizes the role of natural enemies.

7.1 Cultural and Physical Control

Silver agricultural film (Fig. 3.10) was first used as a physical/cultural control for TSW in the panhandle area of North Florida in 1997 based on the study of Greenough et al. (1990). Metalized reflective mulch (Repelgro-Full Reflective, ReflecTek Foils Inc., Lake Zurich, IL, similar to Clarke Ag Plastics—silver on black, Greenwood, Virginia, United States) was field-tested in 1998 (Stavisky et al., 2002), and in 1999 commercial tomato growers used the metalized mulch to produce improved tomato yields. Grower trials expanded in 2000 and the results of the largest trial comparing 25 acres of

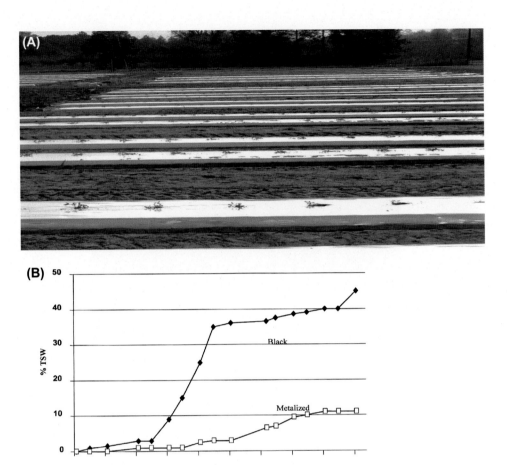

FIGURE 3.10 (A) Tomato seedlings transplanted into metalized reflective-mulch beds, and (B) commercial percentage of tomato spotted wilt disease symptoms in plots in black versus metallic plastic mulch treatments. *www.tswvramp.org, modified with permission.*

metalized mulch to 100 acres of black mulch, with data collected by Glades Crop Care, Inc (Fig. 3.10). There have been consistently documented benefits of using reflective mulch for managing thrips and TSW over the years (Momol et al., 2004; Riley and Pappu, 2004; Andersen et al., 2012).

Reflective mulches presumably disorient thrips, thereby reducing the level of initial settling of the thrips on the tomato crop (Greenough et al., 1990; Stavisky et al., 2002; Riley and Pappu, 2004). In large-scale, reflective-mulch field plots represented in Fig. 3.10, TSW incidence was mostly observed around the field edges. The metalized reflective mulch consists of a very thin (100–300 Å) layer of aluminum deposited on a polyethylene mulch base (Clarke and Scruby, 2001). It is important to note that soil temperature can be reduced by highly reflective plastic mulch films (Diaz-Perez and Batal, 2002), and this can lead to slightly delayed tomato harvests in the spring (Riley and Pappu, 2004). On the other hand, during the summer months when soil temperatures can be excessive, higher plant stem weights may occur with the metalized mulch (Andersen et al., 2012). Also, the use of reflective mulch can reduce the incidence of tomato pests other than thrips, such as whiteflies (Summers et al., 2010).

7.1.1 Weed Management

Although a great deal of effort has focused on implicating weeds sources for TSWV epidemics (Johnson et al., 1995; Groves et al., 2001b, 2002, 2003; Srinivasan et al., 2014) and the management of weeds can affect *F. fusca* dispersal (Beaudoin and Kennedy, 2012), a validated weed management practice has not been shown to be effective in reducing overall TSWV incidence in large-scale commercial tomato fields to date. Glyphosate (herbicide) treatment of winter weeds in conservation tillage can lead to earlier and higher levels of thrips dispersal from treated weeds (Beaudoin and Kennedy, 2012). When this occurs in a field adjacent to a tomato field that is being transplanted, it can lead to increased spread of TSWV into the field. Therefore, keeping weed sources of thrips and TSWV adjacent to the tomato crop green and actively growing could delay thrips movement from weeds sources into tomato. We have observed that if infected weeds adjacent to tomato are killed just prior to transplanting tomatoes, the incidence of TSWV was significantly higher in the tomato relative to plots where weeds remained intact (Riley, unpublished data). Therefore, it might be best to view weeds as both a source (as illustrated in Fig. 3.11) and a sink for thrips vectors of TSWV. Perhaps the best use of weeds is simply a mechanism for monitoring and assessing risk levels associated with thrips vectors of TSWV (Groves et al., 2003; Morsello and Kennedy, 2009).

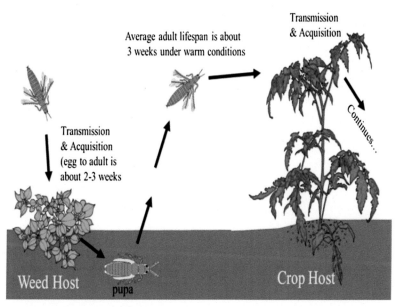

FIGURE 3.11 Movement of thrips in the weedy landscape during the spring. *David and Angelika Riley, www.tswvramp.org, with permission.*

7.2 Host-Plant Resistance

Since the early 1990s when annual TSW epidemics broke out in the southeastern United States, host-plant resistance was considered a top priority to both growers and pest managers dealing with the problem. This was largely because of the ongoing success of transgenic virus resistant vegetables, particularly mosaic resistant squash, during that time (Gaba et al., 2004). In fact, traditional (non-transgenic) TSWV resistance had been reported in tomato during that period (Stevens et al., 1991, 1992; Kumar et al., 1993, 1995), but very few commercial cultivars were widely available in the 1990s. The tomato variety "Stevens" (Stevens et al., 1991), was observed to be mostly asymptomatic in field tests in Georgia, but "Stevens" produced less marketable fruit than the tomato "FL 47," a common TSWV-susceptible tomato hybrid (Riley and Pappu, 2000). Even so, there were active commercial breeding efforts producing new releases in the late 1990s. BHN444, a high yielding TSW-resistant tomato, was released in 1998 and successfully field-tested in the southeast (Momol et al., 2004; Riley and Pappu, 2004). By 2008, multiple TSW-resistant commercial tomato lines became available to grower in the southeastern United States (Table 3.1). Currently, host-plant resistance is the main tool for mitigating the impact of TSWV in regions where this virus has become established. Therefore, it is important to know the origin and potential stability of resistance in the future.

The earliest TSWV resistance genes originated from *Solanum pimpinellifolium* (L.) in the 1950s (Stevens et al., 1992). Two resistance alleles at the *Sw-1* locus (*Sw-1ᵃ*, *Sw-1ᵇ*) and three recessive resistance genes; sw-2, sw-3, and sw-4 were identified by Finlay (1953). In 1959, Clayberg et al. (1960) reported that the cultivar "Pearl Harbor" had *Sw-1ᵃ* and "Rey de los Tempranos" had *Sw-1ᵇ* and the three recessive genes which were TSWV strain specific in their ability to reduce disease symptoms. Paterson et al. (1989) field-tested Finlay's resistant lines and observed that they were susceptible to isolates of TSWV in Arkansas. In 1986, the *Solanum esculentum* Mill. cultivar "Stevens" bred in South Africa from an introgression from *Lycopersicon peruvianum* (L.) Mill. was reported to be strongly resistant (van Zijl et al., 1986). The exact source of resistance in "Stevens" is likely from PI 128654 (Stevens et al., 1992) which was characterized by a single dominant gene, *Sw-5*, observed to provide a broader spectrum of protection from a range of TSWV isolates. The *Sw-5* gene also provided protection from TCSV and GRSV (Boiteux and Giordano, 1993). The *Sw-5* gene is reported to be located on the long arm of chromosome 9 between CT71 and CT220 (Stevens et al., 1995; Brommonschenkel and Tanksley, 1997; Folkertsma et al., 1999). Stevens et al. (1996) developed a genetic marker tightly linked to the *Sw-5* locus using RAPD primer UBC#421

TABLE 3.1 Tomato Cultivars, Their Source, Resistance Designation, and TSW-Resistance Expression for the Years Evaluated From 2006 to 2010

Cultivar[a]	Plant Source	TSW-Resistant Gene[b]	TSW-Resistance Expression[c]
Quincy	Seminis	+	*****(5)
BHN 640	BHN/Siegers	+	*****(5)
Crista	Harris Moran	+	*****(5)
Bella Rosa	Sakata/Siegers	+	*****(5)
Amelia	Harris Moran	+	*****(5)
Talladega	Syngenta	+	****(4)
Red Defender	Harris Moran	+	****(4)
BHN 444	BHN/Siegers	+	****(4)
Redline	Syngenta	+	****(4)
BHN 602	BHN/Siegers	+	****(4)
Top Gun	Twilley	+	****(4)
Inbar	Hazera	+	****(4)
Tycoon	Hazera	+	**(2)
Carson	Hazera	+	**(2)
Picus	Seminis	+	**(2)
BHN 685	Siegers	+	**(2)
Shanty (Roma)	Hazera	+	**(2)
Tribeca	Vilmorin	+	**(2)
Mountain Glory	Siegers	+	**(2)
Fletcher	Reimer Seeds	+	**(2)
Muriel	Sakata/Siegers	+	**(2)
Finishline	Syngenta	+	**(2)
8746	UF	+	*(1)
Tous 91	Hazera	+	*(1)
8751	UF	^	*(1)
Nico	Harris Moran	+	*(1)
8612	UF	+	*(1)
8749	UF	+^	*(1)
Tribute	Sakata	+	*(1)
HAZ 3089	Hazera	+	*(1)
HAZ 3084	Hazera	+	*(1)
Rubia (Roma)	Sakata	+	*(1)
8768	UF	+	*(1)
8688	UF	+^	*(1)
8793	UF	^	*(1)
8685	UF	+	*(1)
HAZ 3088	Hazera	+	*(1)

Continued

TABLE 3.1 Tomato Cultivars, Their Source, Resistance Designation, and TSW-Resistance Expression for the Years Evaluated From 2006 to 2010—cont'd

Cultivar[a]	Plant Source	TSW-Resistant Gene[b]	TSW-Resistance Expression[c]
FTM 2305	Sakata	+	*(1)
8684	UF	+	*(1)
8687	UF	∧	*(1)
8686	UF	∧	*(1)
Hedvig	Hazera	+	*(1)
Galilea	Hazera	+	*(1)
Cupid	Seminis	−	***(3)
SecuriTY 28	Harris Moran	−	(3)
Big Boy	Ferry Morse	−	(1)
Tygress	Seminis	−	(3)
Mariana (Roma)	Sakata/Rupp	−	(3)
FL 47	Seminis/Siegers	−	(5)
Caporal	Vilmorin/Clifton	−	(3)
Pony Express	Harris Moran	−	(1)
Marglobe	USDA	−	(2)

[a]Cultivars ranked (from highest to lowest) based on the marketable fruit yield (kg) within number of years they were tested.
[b]The abbreviations "+" = Sw-5; "∧" = Sw-7 present.
[c]Number of year(s) when significant TSWV-resistance was detected ("*") on a selected cultivar compared with susceptible-cultivars per total years evaluated (in parenthesis). Cultivars are arranged within expression group from highest to lowest yield.
Reproduced with permission from Riley, D.G., Joseph, S.V., Kelley, W.T., Olson, S., Scott, J., 2011b. Host-plant resistance to tomato spotted wilt virus (Bunyaviridae: tospovirus) in tomato. HortScience 46, 1626–1633.

(which has greatly aided breeding efforts). An improved sequence characterized amplified regions (SCAR) marker was developed, which was easier to read on the gel and subsequently became widely used (Garland et al., 2005). More recently, Shi et al. (2011) developed a PCR-based marker and two single nucleotide polymorphism (SNP) markers for marker-assisted selection (MAS). These advances have facilitated the deployment of *Sw-5* in a number of cultivars (Table 3.1).

Unfortunately, some reports showed that certain strains of TSWV had overcome the *Sw-5* resistance (Roselló et al., 1997, 1998). TSWV strains that cause disease symptoms in *Sw-5* cultivars were documented in South Africa (Thompson and van Zijl, 1996), Hawaii (Cho et al., 1996), and Australia (Latham and Jones, 1998). The threat of resistance-breaking strains (Lopez et al., 2011) and direct transmission to *Sw-5* tomato fruit that subsequently expresses damage (Aramburu et al., 2000) are causes for concern about the long-term viability of this tactic. Potential sources of resistance were found in *L. peruvianum* accessions PI 126935, PI 126944, CIAPAN 16, PE-18, and CIAPAN 17, and Roselló et al. (1998) described the resistance gene, *Sw-6*, from PE-18 that provided partial resistance (Roselló et al., 1999). The level of resistance of *Sw-6* was not as high as that provided by *Sw-5*. Roselló et al. (2001) found that a heterozygous tomato line with both *Sw-5* and *Sw-6* genes had a higher level of resistance to TSWV. Resistance derived from *Lycopersicon chilense* Dun. in a line called Ck-12 has a broad spectrum of resistance (Canady et al., 2001). A gene derived from Ck-12 (*Sw-7*) was mapped to chromosome-12 (Price et al., 2007). Preliminary studies indicate that hybrids heterozygous for *Sw-7* have intermediate resistance to TSWV (Scott, unpublished data) and do not confer resistance to GRSV (Adkins, personnel communication). Furthermore, using lines with *Sw-7* (e.g., Fla. 8516) does not seem to confer resistance to *Sw-5* resistance-breaking strains in Hawaii (Scott, unpublished data). It may be that Ck-12 has another resistance gene lost in *Sw-7* lines developed from it, and therefore further selection may reveal a modification of *Sw-7* that can be used in *Sw-5* resistance management. Transformed tomato plants with the nucleocapsid protein (*N*) gene from TSWV were achieved fairly early during the time of TSWV outbreaks in the southeastern United States (Kim et al., 1994). Gubba et al. (2002) was able to develop broad-spectrum resistance to TSWV by combining transgenic resistance with *Sw-5* resistance. Recent research continues to enhance tospovirus resistance with NSs (nonstructural protein encoded by the "s" segment) fragments increasing the range of resistance across serologically unrelated tospoviruses (Yazhisai et al., 2015). Whitfield et al. (2008) reported that a soluble form of the TSWV envelope glycoprotein could inhibit virus transmission in thrips. Small interfering RNAs or RNAi were commercially deployed in squash and tomato as a powerful tool to reduce viral disease along with an assessment of potential ecological risks (Auer and Frederick, 2009). Additionally, tomato hybrids engineered with TSWV sequences for virus resistance have proven effective for mitigating the virus' ability to develop resistance-breaking strains (Accotto et al., 2005). In the southeastern United States, the traditionally bred *Sw-5* resistance has proven stable and effective on commercial farms (Scott et al., 2005, 2011), but it is reassuring to know that transgenic options are available in the event that resistance-breaking strains of TSWV increase.

Currently, commercially field-grown tomato cultivars with *Sw-5* (Riley et al., 2011b) in the southeastern United States have not shown significant damage symptoms at any period during plant development. However, late infection by *F. occidentalis* directly to fruit of *Sw-5* plants can result in fruit blotch (Aramburu et al., 2000). Though fruit blotch occurs in crops that are harvested mature-green, it is not evident until the fruits ripen during shipping, causing serious price reductions for the tomato packer and affect their reputation. This is a particularly insidious problem because it occurs after processing and shipping costs have been incurred. There was preliminary evidence that lines with *Sw-7* did not have the fruit blotch problem of *Sw-5*, which still needs to be verified. *Sw-5* or *Sw-7* lines continued to be tested for TSWV resistance at the University of Florida through 2014. Private companies were also actively developing resistant cultivars, and by 2011 had quite a large selection available (Table 3.1), including Roma or plum-type tomato lines (Riley et al., 2011c).

7.3 Biological Control

Biological control of thrips has been reported to be quite effective in pepper (Funderburk et al., 2000, 2002, 2013) and is recommended for fruiting vegetable crops in general (Funderburk, 2009). The most effective natural enemies are the minute pirate bugs, *Orius* spp. (Hemiptera: Anthocoridae) (Funderburk et al., 2000) and nematode parasites of thrips *Thripinema* spp. (Tylenchida: Allantonematidae) (Lim et al., 2001; Funderburk et al., 2002). Insecticides such as pyrethroids and neonicotinoids have been reported to significantly reduce *Orius* sp. in pepper, while other insecticides, such as spinetoram and cyantraniliprole, are less disruptive to the natural enemies (Funderburk et al., 2013). In tomato, the effectiveness of natural enemies on the tomato crop to prevent primary spread of TSWV is somewhat questionable, given the rapidity of virus transmission once the vector lands on the plant (Wijkamp, 1995). However, biological control agents do help to reduce overall thrips densities and thereby reduce secondary spread in the crop (Funderburk, 2009). In addition, since thrips persists in the landscape in weeds (Groves et al., 2001b; Srinivasan et al., 2014), naturally occurring biological control by natural enemies in non-crop areas may be important in generally suppressing thrips vector densities, presumably leading to reduced TSWV inoculum in the landscape.

7.4 Chemical Control

Thrips, and in particular western flower thrips, are highly polyphagous (Mound, 2005), which implies that they can adapt to a wide range of plant compounds. Similarly thrips can adapt to a wide range of insecticidal compounds (Gao et al., 2012). Therefore, it is important to have a range of insecticidal modes of action for thrips control and insecticide resistance management (Espinosa et al., 2002; Broughton and Herron, 2009; Willmot et al., 2013). Resistance-management strategies for thrips include a resistance monitoring program, rotation of insecticides for each generation of thrips, no more than two consecutive treatments with the same insecticide in a season, and rotating insecticide chemistries that do not have known cross-resistance tendencies in species-specific thrips populations (Gao et al., 2012). The diversity of thrips (*F. fusca, F. occidentalis, F. bispinosa,* and *Frankliniella tritici* (Fitch)) and differences in individual species' response to insecticides (Willmot et al., 2013) add to the difficulty in thrips control. In order to assess insecticide efficacy, proper identification of thrips by a trained individual is required both pre- and post-insecticide application.

Probably the largest impediment to effective use of insecticides to mitigate TSWV spread is the fact that transmission occurs relatively quickly, within 5 min of thrips feeding time (Wijkamp, 1995). Thus treatments need to be preventative rather than reactive. In addition, TSWV symptom development in the field occurs 2–5 weeks after inoculation; therefore disease symptoms cannot be used to time insecticide application (Riley et al., 2012a). Sticky traps (Riley et al., 2012a) and beat cup samples from tomato seedlings (Joost and Riley, 2004) can be useful in detecting early invasion of thrips into a tomato field. Early preventative, systemically applied insecticide treatments have been recommended for various solanaceous cropping systems (Riley and Pappu, 2000, 2004; McPherson et al., 2005; Jacobson and Kennedy, 2013). However, where thrips vectors are not a concern during early season in tomato, preventative treatments are not warranted (Cameron et al., 2009).

Perhaps the most effective use of insecticides for TSWV disease management is as a chemical deterrent to thrips feeding (Groves et al., 2001a; Joost and Riley, 2005; Jacobson and Kennedy, 2013). Unfortunately, an insecticide's ability to influence thrips feeding behavior can be species-specific (Joost and Riley, 2005) and even insecticide-rate-specific (Jacobson and Kennedy, 2013). In one documented case, thrips settling on peanut and tomato foliage systemically treated with imidacloprid varied between species, with *F. fusca* being deterred by the insecticide in both crops, but *F. occidentalis* actually settling more on peanuts treated with imidacloprid (Riley, 2007). Therefore, if *F. occidentalis* is the main TSWV vector present, imidacloprid is not recommended. The new systemic diamide insecticide, cyantraniliprole, has inhibitory activity on thrips vectors of TSWV in pepper although not quite as strong an effect on reducing thrips feeding as imidacloprid (Jacobson and Kennedy, 2013).

Insecticides with primarily contact activity against thrips often are used when thrips densities reach direct damaging levels, i.e., in the absence of TSWV. There are many insecticide groups with distinct modes of action (Insecticide Resistance Action Committee Groups, http://www.irac-online.org/modes-of-action/; accessed July 26, 2016) that control thrips including organophosphates (1B), carbamates (1A), pyrethroids (3A), neonicotinoids (4A), spinosyns (5), avermectins (6), and others (Shan et al., 2012). Spinosyns can be very effective against thrips while preserving the thrips predator, minute pirate bugs (Funderburk et al., 2000). But they can negatively affect beneficial Hymenoptera (Biondi et al., 2012). It is important to note that thrips immature stages do not tend to be more susceptible to insecticide than the adult stage, but also that selection for resistance in the same thrips populations occurs at much different rates between adult and immature thrips stages (Contreras et al., 2010). In addition, later-season sprays have been recommended to reduce secondary spread of TSWV (Andersen et al., 2012). Once a plant expresses severe TSW, insecticide sprays do not affect tomato production on that plant. The use of insecticides can affect the spread of TSWV in tomato, often enhancing other control tactics (Riley and Pappu, 2000, 2004; Coutts and Jones, 2005; Andersen et al., 2012). As a general recommendation, growers should use preventative insecticide treatments only where thrips and TSWV reoccur annually for a given tomato production season. In addition, thrips vectors need to be identified in a given location to be able to select a more appropriate insecticide (Table 3.2). Finally, if prevention is warranted, use effective systemic treatments, such as cyantraniliprole, that not only control thrips, but also reduce thrips feeding on the plant.

7.4.1 Plant Activators

Plant activators are compounds that can illicit an immune response in the plant when applied as a spray or drench. In the case of TSW disease suppression there are systemic acquired resistance (SAR) compounds that have been shown to

TABLE 3.2 Insecticide Efficacy Against Thrips in Tomato and Pepper in the southeastern United States, 2012

Insecticide	Efficacy Rating[a]		Registration		Mode of Action (MOA)
	Tobacco Thrips	Flower Thrips	Tomato	Pepper	
Acephate (e.g., Orthene 97)	G	G	No[b]	Yes	1B
Dinotefuran (e.g., Venom 70SG)	G	G	Yes	Yes	4A
Endosulfan[c] (e.g., Endosulfan 3EC)	F	F	Yes	Yes	2A
Imidacloprid (e.g., Admire 2F)	G-AF	P	Yes	Yes	4A
Methomyl (e.g., Lannate 2.4LV)	G	G	Yes	Yes	1A
Oxamyl (e.g., Vydate 2L)	G	G	Yes	Yes	1A
Spinetoram (e.g., Radiant 1SC)	G-E	G-E	Yes	Yes	5
Pyrethroids[d]	G	F-P	Yes	Yes	3
Cyantraniliprole (e.g., Verimark)	G	G	Yes	Yes	28

[a]Efficacy rating: AF, anti-feeding effect; E, excellent; F, fair; G, good; P, poor.
[b]Individual products may not be labeled for the specific pest but should provide indicated level of control when applied according to label instructions for the crop.
[c]Endosulfan is labeled through 2015 on pepper and tomato.
[d]Labeled pyrethroid insecticides include bifenthrin (e.g., Capture 2EC), lambda-cyhalothrin (e.g., Warrior 1CS), permethrin (e.g., Pounce 3.2EC), zeta-cypermethrin (e.g., MustangMax 0.8EC), esfenvalerate (e.g., Asana XL), and cyfluthrin (e.g., Tombstone 2EC, Baythroid XL).
Modified with permission from www.tswvramp.org.

suppress TSWV disease development if applied at, or just prior to, virus inoculation by the thrips (Louws et al., 2001; Obradovic et al., 2004; Vavrina et al., 2004). Specifically, the compound acibenzolar-S-methyl (Actigard 50WG, Syngenta Crop Protection, Greensboro, NC), is an analog of salicylic acid that can cause tomato plants to exhibit less disease symptom development early in the season under field conditions (Riley et al., 2012b). Prevention of early disease symptoms in tomato can lead to significantly increased yields in tomato (Chaisuekul et al., 2003). The use of SAR compounds can enhance other TSW mitigation practices such as reflective mulch (Momol et al., 2004; Awondo et al., 2012; Riley et al., 2012b). Typically these compounds provide similar levels of control as preventative insecticide treatments (Riley et al., 2012b). Foliar sprays of acibenzolar-S-methyl have been shown to reduce TWV disease symptoms when applied before transplanting followed by foliar treatments on a biweekly basis (Momol et al., 2004). Typical rates of acibenzolar-S-methyl are low to avoid phytotoxicity in young plants but can be increased as the plant matures (Mandal et al., 2008). As an additional benefit, other tomato plant diseases can be reduced with the use of SAR compounds (Louws et al., 2001).

7.5 Integrated Pest Management

In crops other than tomato, there has been some indication that tillage practices can affect TSWV incidence, specifically that conservation/minimum tillage reduces TSWV incidence in peanut (Hurt et al., 2006). Since tomato production in the southeastern United States is primarily for fresh market and is grown on plastic mulch beds, minimum tillage practices will not be applicable for most commercial fields. However, rye cover crops between plastic mulch beds are occasionally used, and results from peanut trials indicate that rye covers can reduce thrips and TSWV (Olson et al., 2006). Altering tomato crop fertility can affect thrips populations, and greater nitrogen has been associated with higher thrips numbers (Stavisky et al., 2002); excess nitrogen should be avoided. Intercropping, living mulches, and other practices that reduce thrips ability to efficiently find and feed on the tomato crop from overwintering sources should provide some degree of reduction of TSW, but the question comes down to the cost–benefit and cultural acceptability of the tactic.

8. ECONOMICS OF MANAGEMENT STRATEGIES

Based on surveys in 2008–09 in Georgia, Florida, South Carolina, and North Carolina, 75% of tomato growers and 58% of pepper growers used at least some improved method (e.g., resistant cultivar, metallic reflective mulch, Actigard, imidacloprid) in managing TSW. Over 67% of tomato growers used some TSW-resistant tomato cultivar and imidacloprid, while 60% reported using Actigard. Fewer tomato (19%) and pepper (16%) growers used metallic reflective mulch. Most of the respondents (71%) indicated that TSW-resistant cultivars reduced TSW and produced fruits of acceptable yield and quality. Over half of the respondents (62%) indicated having problems with more than one virus in tomato production, while only 36% face a similar problem with pepper. Statistical analysis revealed that the odds of growers satisfied with the use of Actigard alone in controlling TSW were 7 to 1, while the odds of growers satisfied with the use of reflective mulch alone were 9 to 1 when compared to not using one of the tactics. The odds those growers were satisfied using a combination of Actigard and reflective mulch in the control of TSW were 59 to 1 as opposed to not using both (Awondo et al., 2012).

A cost–benefit analysis comparing improved management practices (IMPs) with conventional management practices (CMPs) using results from on-station experiments and cost (fixed, variable) data revealed a nominal total production cost of $11,013/ac for the CMP and $13,314/ac for IMP. A breakeven (BE) variable cost of $2.88 and $2.33 was estimated for CMP and IMP, respectively, and a BE yield of 1366 cartons/ac (25 lb) for CMP and 1664 cartons/ac (25 lb) for IMP was realized (Fonsah et al., 2010). The IMP consisted of using one of the Sw-5 tomato hybrids (see Table 3.1). Second, the IMP used a systemic insecticide effective against the thrips vector species at the time of transplanting. Third, IMP used metallic reflective mulch if slight soil cooling had no negative economical consequence. Considering the margin of error in the estimates (using a risk-rated sensitivity analysis), the results of Table 3.3 indicate that the increase in yield and net return from using IMP are large enough to compensate tomato growers for the additional production cost incurred in using IMP (Fonsah et al., 2010). Most commercial tomato farmers in regions of the southeast United States, where TSW is a known problem, currently use tomato cultivars with host-plant resistance to TSWV. Marketing and trade will continue to be a major influence on choices made by the tomato growers in this area and economics ultimately will determine if the industry can adapt to this and other threats to production. At present, it appears that economical solutions to TSW will prevail in the southeastern United States.

9. FUTURE PROSPECTS

Ideally, management of thrips and tospovirus in tomato should be based on key economic factors that control the use of tactics to provide the most profitable production of tomato. Decision tools need to be developed to control the use of tactics beginning with regional scale factors such as weather (e.g., Chappell et al., 2013). Then a risk-assessment

TABLE 3.3 Risk-Rated Sensitivity Analysis of Producing Tomatoes Using a Conventional Management Strategy and an Improved Management Strategy Developed for the southeastern United States, 2009

Returns by Strategy	Best	Optimistic	Median	Pessimistic	Worst
IMP returns[a]	$6052	$3882	$2798	$1713	$−456
CMS returns[b]	$3645	$1926	$1067	$207	$−1512
Risk-rated % chances	7%	31%	50%	31%	7%

[a]Based on CMS risk-rated sensitivity costs.
[b]Based IMP risk-rated sensitivity costs.
Reproduced with permission from www.tswvramp.org.

model should be developed for each region, as has been done for TSWV management in peanut (Brown et al., 2005). The durability of *Sw-5* resistance in the southeast United States and elsewhere may be enhanced by the addition of *Sw-7* to resistant cultivars. If new virulent strains of TSWV occur or new tospoviruses emerge that are not controlled by existing genes, then new sources of resistance must be discovered. This will result in excessive work especially if the resistance is found in species that do not cross easily with tomato such as *S. peruvianum* or *S. chilense*. Evolving genomic advances should facilitate the introgression of such resistance genes, but there will still be considerable time between resistance discovery and acceptable resistant cultivars. Ideally, early generation lines from any introgression projects should be made available to gene banks so that they can be tested for resistance to tospovirus and other problems as they emerge. This would save considerable breeding time in attaining the first crosses in distantly related species. Otherwise, transgenic approaches could make future impacts if an economically justified need spurs its commercial development. New technologies involving RNA silencing could accelerate the development of transgenic resistant cultivars (Auer and Frederick, 2009).

10. SUMMARY

Thrips have become a major worldwide pest of tomato in the last few decades mainly because of their ability to transmit tospoviruses to the crop. The main *Tospovirus* in tomato in the United States is TSWV and the main thrips vector species in the United States are *F. occidentalis* and *F. fusca*. Current management of thrips relies heavily on preventative tactics to reduce thrips settling and/or feeding (and thus virus inoculation) on the tomato plant, and host-plant resistance to the *Tospovirus*. The main tactics for reducing the impact of TSW on tomato, ranked from most to least effective, are: (1) tomato plant resistance to TSW, (2) cultural/physical controls such as reflective mulch, (3) properly chosen, timed, and applied insecticides for the thrips species present on the crop, (4) plant activators to reduce the development of TSW symptoms, and (5) biological and other control tactics. This ranking may change as new techniques and strategies are developed. However, the thrips/*Tospovirus* crisis that challenged the southeastern United States has much abated in the last decade with the aforementioned management tactics.

REFERENCES

Accotto, G.P., Nervo, G., Acciarri, N., Tavella, L., Vecchiati, M., Schiavi, M., Mason, G., Vaira, A.M., 2005. Field evaluation of tomato hybrids engineered with tomato spotted wilt virus sequences for virus resistance, agronomic performance, and pollen-mediated transgene flow. Phytopathology 95, 800–807.

Adkins, S., Zitter, T., Momol, T., 2013. Tospoviruses (Family Bunyaviridae, Genus Tospovirus). Fact Sheet PP-212, Florida Cooperative Extension Services. Institute of Food and Agricultural Sciences, University of Florida.

Andersen, P.C., Olson, S.M., Momol, M.T., Freeman, J.H., 2012. Effect of plastic mulch type and insecticide on incidence of tomato spotted wilt, plant growth, and yield of tomato. HortScience 47, 861–865.

Andrewartha, H.V., 1937. A species of Thysanoptera of economic importance in South Australia. Transactions of the Royal Society of South Australia 61, 163–165.

Angelella, G., Riley, D.G., 2010. Life table bioassay for pine pollen effects on *Frankliniella fusca* reproduction on onion. Environmental Entomology 39, 505–512.

Aramburu, J., Rodriguez, M., Ariño, J., 2000. Effect of tomato spotted wilt tospovirus (tswv) infection on the fruits of tomato (*Lycopersicon esculentum*) plants of cultivars carrying the *SW-5* gene. Journal of Phytopathology 148, 569–574.

Auer, C., Frederick, R., 2009. Crop improvement using small RNAs: applications and predictive ecological risk assessments. Trends in Biotechnology 27, 644–651.

Awondo, S.N., Fonsah, E.G., Riley, D., Abney, M., 2012. Effectiveness of tomato spotted wilt virus management tactics. Journal of Economic Entomology 105, 943–948.

Baez, I., Reitz, S.E., Funderburk, J.E., 2004. Predation by *Orius insidiosus* (Heteroptera: Anthocoridae) on life stages and species of *Frankliniella* flower thrips (Thysanoptera: Thripidae) in pepper flowers. Environmental Entomology 33, 662–670.

Barbour, J.D., Brandenburg, R.L., 1994. Vernal infusion of thrips into North Carolina peanut fields. Journal of Economic Entomology 87, 446–451.

Beaudoin, A.L., Kahn, N.D., Kennedy, G.G., 2009. Bell and banana pepper exhibit mature-plant resistance to tomato spotted wilt Tospovirus transmitted by *Frankliniella fusca* (Thysanoptera: Thripidae). Journal of Economic Entomology 102, 30–35.

Beaudoin, A.L.P., Kennedy, G.G., 2012. Management of winter weeds affects *Frankliniella fusca* (Thysanoptera: Thripidae) dispersal. Environmental Entomology 41, 362–369.

Biondi, A., Mommaerts, V., Smagghe, G., Viñuela, E., Zappalà, L., Desneux, N., 2012. The non-target impact of spinosyns on beneficial arthropods. Pest Management Science 68, 1523–1536.

Boiteux, L.S., Giordano, L.D., 1993. Genetic-basis of resistance against 2 tospovirus species in tomato (*Lycopersicon esculentum*). Euphytica 71, 151–154.

Bond, W.P., Whitman, H.K., Black, L.L., 1983. Indigenous weeds as reservoirs of tomato spotted wilt virus in Louisiana. Phytopathology 73, 499.

Bonnemaison, L., 1939. La maladie bronzee de la tomate. Annales Epiphyties et Phytogenetique 5, 267–308.

Brandenburg, R.L., 1986. Peanut insect management. In: Peanut 1986. North Carolina Agriculture Extension Service Raleigh, North Carolina, pp. 49–56.

Brittlebank, C.C., 1919. Tomato diseases. Journal of Department of Agriculture Victoria 17, 231–235.

Broadbent, A.B., Rhainds, M., Shipp, L., Murphy, G., Wainman, L., 2003. Pupation behavior of western flower thrips (Thysanoptera: Thripidae) on potted chrysanthemum. Canadian Entomologist 135, 741–744.

Brommonschenkel, S.H., Tanksley, S.D., 1997. Map-based cloning of the tomato genomic region that spans the *Sw-5* tospovirus resistance gene in tomato. Molecular and General Genetics 256, 121–126.

Broughton, S., Herron, G.A., 2009. Potential new insecticides for the control of western flower thrips (Thysanoptera: Thripidae) on sweet pepper, tomato, and lettuce. Journal of Economic Entomology 102, 646–651.

Brown, S.L., Culbreath, A.K., Todd, J.W., Gorbet, D.W., Baldwin, J.A., Beasley Jr., J.P., 2005. Development of a method of risk assessment to facilitate integrated management of spotted wilt of peanut. Plant Disease 89, 348–356.

Cameron, P.J., Walker, G.P., Hodson, A.J., Kale, A.J., Herman, T.J.B., 2009. Trends in IPM and insecticide use in processing tomatoes in New Zealand. Crop Protection 28, 421–427.

Canady, M.A., Stevens, M.R., Barineau, M.S., Scott, J.W., 2001. Tomato spotted wilt virus (TSWV) resistance in tomato derived from *Lycopersicon chilense* Dun. LA 1938. Euphytica 117, 19–25.

Chaisuekul, C., Riley, D., Pappu, H., 2003. Transmission of tomato spotted wilt virus to tomato plants of different ages. Journal of Entomological Science 38, 126–135.

Chaisuekul, C., Riley, D.G., 2005. Host plant, temperature, and photoperiod effects on ovipositional preference of *Frankliniella occidentalis* and *F. fusca* (Thysanoptera: Thripidae). Journal of Economic Entomology 98, 2107–2113.

Chamberlin, J.R., Todd, J.W., Beshear, R.J., Culbreath, A.K., Demski, J.W., 1992. Overwintering hosts and wingform of thrips, *Frankliniella* spp., in Georgia (Thysanoptera: Thripidae): implications for management of spotted wilt disease. Environmental Entomology 21, 121–128.

Chappell, T.M., Beaudoin, A.L.P., Kennedy, G.G., 2013. Interacting virus abundance and transmission intensity underlie tomato spotted wilt virus incidence: an example weather-based model for cultivated tobacco. PLoS One 8, e73321.

Cho, J.J., Custer, D.M., Brommonschenkel, S.H., Tanksley, S.D., 1996. Conventional breeding: host-plant resistance and the use of molecular markers to develop resistance to tomato spot wilt virus in vegetables. Acta Horticulturae 431, 367–378.

Cho, K.J., Eckel, C.S., Walgenbach, J.F., Kennedy, G.G., 1995. Overwintering of thrips (Thysanoptera, Thripidae) in North Carolina. Environmental Entomology 24, 58–67.

Clarke, P.W.W., Scruby, T.M., 2001. Visual Characteristics Alteration for Synthetic Agricultural Mulch Sheet, Involving Attaching Isolated Synthetic Mulch Layer of Reset Ref on Existing Synthetic Mulch Layer. US Patent 6202344-B1.

Clayberg, C.D., Butler, L., Rick, C.M., Young, P.A., 1960. Second list of known genes in the tomato. Journal of Heredity 51, 167–174.

Contreras, J., Espinosa, P.J., Quinto, V., Abellàn, J., Gràvalos, C., Fernàndez, E., Bielza, P., 2010. Life-stage variation in insecticide resistance of the western flower thrips (Thysanoptera: Thripidae). Journal of Economic Entomology 103, 2164–2168.

Costa, A.S., Forster, R., Fraga, C., 1950. Controle de vira-cabeca do tomate pela destruicao do vetor. Bragantia 10, 1–9.

Coutts, B.A., Jones, R.A.C., 2005. Suppressing spread of tomato spotted wilt virus by drenching infected source or healthy recipient plants with neonicotinoid insecticides to control thrips vectors. Annals of Applied Biology 146, 95–103.

Culbreath, A.K., Csinos, A.S., Bertrand, P.F., Demski, J.W., 1991. Tomato spotted wilt virus epidemic in flue cured tobacco in Georgia. Plant Disease 75, 483–485.

Demirozer, O., Tyler-Julian, K., Funderburk, J., Leppla, N., Reitz, S., 2012. *Frankliniella occidentalis* (Pergande) integrated pest management programs for fruiting vegetables in Florida. Pest Management Science 69, 1537–1545.

Diaz-Perez, J.C., Batal, K.D., 2002. Color plastic film mulches affect tomato growth and yield via changes in root-zone temperature. Journal of the American Society of Horticultural Science 127, 127–136.

Diffie, S., Edwards, G.B., Mound, L.A., 2008. Thysanoptera of southeastern USA: a checklist for Florida and Georgia. Zootaxa 1787, 45–62.

DuRant, J.A., Roof, M.E., Cole, C.L., 1994. Early season incidence of thrips (Thysanoptera) on wheat, cotton, and three wild host plant species in South Carolina. Journal of Agricultural Entomology 11, 61–71.

Espinosa, P.J., Bielza, P., Contreras, J., Lacasa, A., 2002. Field and laboratory selection of *Frankliniella occidentalis* (Pergande) for resistance to insecticides. Pest Management Science 58, 920–927.

Finlay, K.W., 1953. Inheritance of spotted wilt resistance in tomatoes; five genes controlling spotted wilt resistance in four tomatoes. Australian Journal of Biological Science 6, 153–163.

Folkertsma, R.T., Spassova, M.I., Prins, M., Stevens, M.R., Hille, J., Goldbach, R.W., 1999. Construction of a bacterial artificial chromosome (BAC) library of *Lycopersicon esculentum* cv. Stevens and its application to physically map the *Sw-5* locus. Molecular Breeding 5, 197–207.

Fonsah,E.G.,Ferrer,C.M.,Riley,D.G.,Sparks,S.,Langston,D.,2010.CostandBenefitAnalysisofTomatoSpottedWiltVirus(TSWV)ManagementTechnology in Georgia. The Southern Agricultural Economic Association Conference (SAEA). http://ageconsearch.umn.edu/bitstream/56386/2/51028.pdf.

Funderburk, F., 2009. Management of the western flower thrips (Thysanoptera: Thripidae) in fruiting vegetables. Florida Entomologist 92, 1–6.

Funderburk, J., Adkins, S., Freeman, J., Stansly, P., Smith, H., McAvoy, G., Demirozer, O., Snodgrass, C., Paret, M., Leppla, N., 2014. Managing thrips and tospoviruses in tomato. Institute of Food and Agriculture Science Extension Publication No. ENY859, University of Florida, Gainesville, Florida, USA. pp. 1–10.

Funderburk, F., Srivastava, M., Funderburk, C., McManus, S., 2013. Evaluation of imidacloprid and cyantraniliprole for suitability in conservation biological control program for *Orius insidiosus* (Hemiptera: Anthocoridae) in field pepper. Florida Entomologist 96, 229–231.

Funderburk, F., Stavisky, J., Olson, S., 2000. Predation of *Frankliniella occidentalis* (Thysanoptera: Thripidae) in field peppers by *Orius insidiosus* (Hemiptera: Anthocoridae). Environmental Entomology 29, 376–382.

Funderburk, J., Ripa, R., Espinoza, F., Rodriguez, F., 2002. Parasitism of *Frankliniella australis* (Thysanoptera: Thripidae) by *Thripinema khrustalevi* (Tylenchida: Allantonematidae) isolate Chile. Florida Entomologist 85, 645–649.

Gaba, V., Zelcer, A., Gal-On, A., 2004. Cucurbit biotechnology – the importance of virus resistance. In Vitro Cellular and Development Biology – Plant 40, 346–358.

Gao, Y., Lei, Z., Reitz, S.R., 2012. Western flower thrips resistance to insecticides: detection, mechanisms and management strategies. Pest Management Science 68, 1111–1121.

Garland, S., Sharmon, M., Persley, A.D., McGrath, D., 2005. The development of an improved PCR-based marker system for *Sw-5*, an important TSWV resistance gene of tomato. Australian Journal of Agricultural Research 56, 285–289.

Gergerich, R.C., 1985. Tomato spotted wilt virus–a continuing problem. In: Proceedings of Arkansas State Horticultural Society, pp. 45–47.

German, T.L., Ullman, D.E., Moyer, J.W., 1992. Tospoviruses: diagnosis, molecular biology, phylogeny, and vector relationships. Annual Review of Phytopathology 30, 315–348.

Gitaitis, R.D., Dowler, C.C., Chalfant, R.B., 1998. Epidemiology of tomato spotted wilt in pepper and tomato in Southern Georgia. Plant Disease 82, 752–756.

Greenough, D.R., Black, L.L., Bond, W.P., 1990. Aluminum-surfaced mulch: an approach to the control of tomato spotted wilt virus in solanaceous crops. Plant Disease 74, 805–808.

Greenough, D.R., Black, L.L., Story, R.N., Newsom, L.D., Bond, W.P., 1985. Occurrence of *Frankliniella occidentalis* in Louisiana – a possible cause for the increased incidence of tomato spotted wilt virus. Phytopathology 75, 1362.

Groves, R.L., Sorenson, C.E., Walgenbach, J.F., Kennedy, G.G., 2001a. Effects of imidacloprid on transmission of tomato spotted wilt tospovirus to pepper, tomato, and tobacco by *Frankliniella fusca* Hinds (Thysanoptera: Thripidae). Crop Protection 20, 439–445.

Groves, R.L., Walgenbach, J.F., Moyer, J.W., Kennedy, G.G., 2002. The role of weed hosts and tobacco thrips, *Frankliniella fusca*, in the epidemiology of tomato spotted wilt virus. Plant Disease 86, 573–582.

Groves, R.L., Walgenbach, J.F., Moyer, J.W., Kennedy, G.G., 2003. Seasonal dispersal patterns of *Frankliniella fusca* (Thysanoptera: Thripidae) and tomato spotted wilt virus occurrence in central and eastern North Carolina. Journal of Economic Entomology 96, 1–11.

Groves, R.L., Walgenbach, J.F., Moyer, J.W., Kennedy, J.J., 2001b. Overwintering of *Frankliniella fusca* (Thysanoptera: Thripidae) on winter annual weeds infested with tomato spotted wilt virus and patterns of virus movement between susceptible hosts. Phytopathology 91, 891–899.

Gubba, A., Gonsalves, C., Stevens, M.R., Tricoli, D.M., Gonsalves, D., 2002. Combining transgenic and natural resistance to obtain broad resistance to tospovirus infection in tomato (*Lycopersicon esculentum* Mill.). Molecular Breeding 9, 13–23.

Gudauskas, R.T., Hagan, A.K., Weeks, J.R., Shelby, R.A., Gazaway, W.S., French, J.C., 1988. Incidence of tomato spotted wilt in peanuts in Alabama USA. Phytopathology 78, 1564.

Halliwel, R.S., Philley, G., 1974. Spotted wilt of peanut in Texas. Plant Disease Reporter 58, 23–25.

Hansen, E.A., Funderburk, J.E., Reitz, S.R., Ramachandran, S., Eger, J.E., McAuslane, H., 2003. Within-plant distribution of *Frankliniella* species (Thysanoptera: Thripidae) and *Orius insidiosus* (Heteroptera: Anthocoridae) in field pepper. Environmental Entomology 32, 1035–1044.

Hoddle, M.S., Mound, L.A., Paris, D.L., 2012. Thrips of California. Centre for Biological Information Technology Publishing, Queensland, Australia. http://keys.lucidcentral.org/keys/v3/thrips_of_california/Thrips_of_California.html.

Hulshof, J., Ketoja, E., Vaenninen, I., 2003. Life history characteristics of *Frankliniella occidentalis* on cucumber leaves with and without supplemental food. Entomologia Experimentalis et Applicata 108, 19–32.

Hurt, C.A., Brandenburg, R.L., Jordan, D.L., Royals, B.M., Johnson, P.D., 2006. Interactions of tillage with management practices designed to minimize tomato spotted wilt of peanut (*Arachis hypogaea* L.). Peanut Science 33, 83–89.

Jacobson, A.L., Kennedy, G.G., 2013. Effect of cyantraniliprole on feeding behavior and virus transmission of *Frankliniella fusca* and *Frankliniella occidentalis* (Thysanoptera: Thripidae) on *Capsicum annuum*. Crop Protection 54, 251–258.

Johnson, R.R., Black, L.L., Hobbs, H.A., Valverde, R.A., Story, R.N., Bond, W.P., 1995. Association of *Frankliniella fusca* and 3 winter weeds with tomato spotted wilt virus in Louisiana. Plant Disease 79, 572–576.

Joost, H., Riley, D.G., 2005. Imidacloprid differentially affects probing and settling behavior of *Frankliniella fusca* (Hinds) and *Frankliniella occidentalis* (Pergrande) (Thysanoptera: Thripidae). Journal of Economic Entomology 98, 1622–1629.

Joost, H., Riley, D.G., 2007. Tomato plant and leaf age effects on the probing and settling behavior of *Frankliniella fusca* and *Frankliniella occidentalis* (Thysanoptera: Thripidae). Environmental Entomology 37, 213–223.

Joost, P.H., Riley, D.G., 2004. Sampling techniques for thrips (Thysanoptera: Thripidae) in pre-flowering tomato. Journal of Economic Entomology 97, 1450–1454.

Kahn, N.D., Walgenbach, J.F., Kennedy, G.G., 2005. Summer weeds as hosts for *Frankliniella occidentalis* and *Frankliniella fusca* (Thysanoptera: Thripidae) and as reservoirs for Tomato spotted wilt *Tospovirus* in North Carolina. Journal of Economic Entomology 98, 1810–1815.

Kikuta, K., Frazier, W.A., 1946. Breeding tomatoes for resistance to spotted wilt in Hawaii. Proceedings American Society of Horticultural Science 47, 271–276.

Kim, J.W., Sun, S.S.M., German, T.L., 1994. Diseases resistance in tobacco and tomato plants transformed with the tomato spotted wilt virus nuclesocapsid gene. Plant Disease 6, 615–621.

Kindt, F., Joosten, N.N., Peters, D., Tjallingii, W.F., 2003. Characterisation of the feeding behaviour of western flower thrips in terms of electrical penetration graph (EPG) waveforms. Journal of Insect Physiology 49, 183–191.

Kirk, W.D.J., 1997. Distribution, abundance and population dynamics. In: Lewis, T. (Ed.), Thrips as Crop Pests. CABI, Oxon, UK, pp. 217–257.

Kirk, W.D.J., 2002. The pest and vector from the West: *Frankliniella occidentalis*. Thrips and Tospoviruses. In: Marullo, R., Mound, L.A. (Eds.), Proceedings of the Seventh International Symposium on Thysanoptera Australian National Insect Collection, Canberra, Australia, pp. 33–44.

Kirk, W.J., Terry, L.I., 2003. The spread of the western flower thrips, *Frankliniella occidentalis* (Pergande). Agricultural Forest Entomolology 5, 301–310.

Kumar, N.K.K., Ullman, D.E., Cho, J.J., 1993. Evaluation of *Lycopersicon* germ plasm for tomato spotted wilt tospovirus resistance by mechanical and thrips transmission. Plant Disease 77, 938–941.

Kumar, N.K.K., Ullman, D.E., Cho, J.J., 1995. Resistance among *Lycopersicon* species to *Frankliniella occidentalis* (Thysanoptera: Thripidae). Journal of Economic Entomology 88, 1057–1065.

Lange, W.H., Bronson, L., 1981. Insect pests of tomatoes. Annual Review of Entomology 26, 345–371.

Latham, L.J., Jones, R.A.C., 1998. Selection of resistance breaking strains of tomato spotted wilt tospovirus. Annals of Applied Biology 133, 385–402.

Leach, J.G., Berg, A., 1944. Successful control of tip blight of tomato. Plant Disease Reporter 27, 590.

Lewis, T., 1973. Thrips: Their Biology, Ecology and Economic Importance. Academic Press, London, UK, p. 349.

Lim, U.T., Van Driesche, R.G., Heinz, K.M., 2001. Biological attributes of the nematode, *Thripinema nichlewoodi*, a potential biological control agent of western flower thrips. Biological Control 22, 300–306.

Lopez, C., Aramburu, J., Galipienso, L., Soler, S., Nuez, F., Rubio, F., 2011. Evolutionary analysis of tomato *Sw-5* resistance breaking isolates of tomato spotted wilt virus. Journal of General Virology 92, 210–215.

Louws, F.J., Wilson, M., Campbell, H.L., Cuppels, D.A., Jones, J.B., Shoemaker, P.B., Sahin, F., Miller, S.A., 2001. Field control of bacterial spot and bacterial speck of tomato using plant activator. Plant Disease 85, 481–488.

Lowry, V.K., Smith, J.W., Mitchell, F.L., 1992. Life fertility tables for *Frankliniella fusca* (Hinds) and *F. Occidentalis* (Pergande) (Thysanoptera: Thripidae) on peanut. Annals of the Entomological Society of America 85, 744–754.

Mandal, B., Mandal, S., Csinos, A.S., Martinez, N., Culbreath, A.K., Pappu, H.R., 2008. Biological and molecular analyses of the acibenzolar-S-methyl-induced systemic acquired resistance in flue-cured tobacco against tomato spotted wilt virus. Phytopathology 98, 196–204.

McPherson, R.M., Beshear, R.J., Culbreath, A.K., 1992. Seasonal abundance of thrips (Thysanoptera, suborders Terebrantia and Tubulifera) in Georgia flue-cured tobacco and impact of management-practices on the incidence of Tomato spotted wilt virus. Journal of Entomological Science 27, 257–268.

McPherson, R.M., Pappu, H.R., Jones, D.C., 1999. Occurrence of five thrips species on flue-cured tobacco and impact on spotted wilt disease incidence in Georgia. Plant Disease 83, 765–767.

McPherson, R.M., Stephenson, M.G., Lahue, S.S., Mullis, S.W., 2005. Impact of early-season thrips management on reducing the risks of spotted wilt virus and suppressing aphid populations in flue-cured tobacco. Journal of Economic Entomology 98, 129–134.

McWhorter, F.P., Milbrath, J.A., 1938. The tipblight disease of tomato. Oregon Agricultural Experiment Station Circular 128, 1–14.

Momol, M.T., Olson, S.M., Funderburk, J.E., Stavisky, J., Marois, J.J., 2004. Integrated management of tomato spotted wilt on field-grown tomatoes. Plant Disease 88, 882–890.

Moore, E.S., 1933. The Kromnek disease of tobacco and tomato in the eastern Cape Province. Farming in South Africa 8, 379–380.

Moore, E.S., Anderssen, E.E., 1939. Notes on plant virus diseases in South Africa. In: the Kromnek disease of tobacco and tomato. Union South Africa Department Agriculture and Forestry Science Bulletin 182, 1–23.

Moriones, E., Aramburu, J., Ruidavets, J., Arno, J., Lavina, A., 1998. Effect of plant age at time of infection by Tomato spotted wilt virus on the yield of field-grown tomato. European Journal of Plant Pathology 104, 295–300.

Moritz, G., 1997. Structure, growth and development. In: Lewis, T. (Ed.), Thrips as Crop Pests. CAB International, New York, USA, pp. 15–63.

Moritz, G., Morris, D.C., Mound, L.A., 2001. Thrips ID–Pest Thrips of the World. An Interactive Identification and Information System. CD-ROM Published by ACIAR. CSIRO Publishing, Melbourne, Australia.

Morse, J.G., Hoddle, M.S., 2006. Invasion biology of thrips. Annual Review of Entomology 51, 67–89.

Morsello, S., Kennedy, G.G., 2009. Spring temperature and precipitation affect tobacco thrips, *Frankliniella fusca* (Thysanoptera: Thripidae) population growth and tomato spotted wilt virus within patches of the winter weed *Stellaria media*. Entomologia Experimentalis et Applicata 130, 138–148.

Morsello, S.C., Beaudoin, A.L.P., Groves, R.L., Nault, B.L., Kennedy, G.G., 2010. The influence of temperature and precipitation on spring dispersal of *Frankliniella fusca* changes as the season progresses. Entomologia Experimentalis et Applicata 134, 260–271.

Moulton, D., 1948. The genus *Frankliniella* Karny, with keys for the determination of species (Thysanoptera). Revista de Entomologia 19, 55–114.

Mound, L.A., 2005. Thysanoptera: diversity and interactions. Annual Review of Entomology 50, 247–269.

Mound, L.A., Paris, D., Fisher, N., 2009. World Thysanoptera. CSIRO, Black Mountain, Australia. http://anic.ento.csiro.au/thrips/.

Moyer, J.W., 2000. Tospoviruses (Bunyaviridae). In: Granoff, A., Webster, R.G. (Eds.), Encyclopedia of Virology. Academic Press, Elsevier Inc., California, USA, pp. 1803–1807.

Nakao, S., Chikamori, C., Okajima, S., Narai, Y., Murai, T., 2011. A new record of the tobacco thrips *Frankliniella fusca* (Hinds) (Thysanoptera: Thripidae) from Japan. Applied Entomology and Zoology 46, 131–134.

Navas, S.V.E., Funderburk, J.E., Beshear, R.J., Olson, S.M., Mack, T.P., 1991b. Seasonal patterns of *Frankliniella* spp. (Thysanoptera: Thripidae) in tomato flowers. Journal of Economic Entomology 84, 1818–1822.

Navas, S.V.E., Funderburk, J.E., Olson, S.M., Beshear, R.J., 1991a. Damage to tomato fruit by the western flower thrips (Thysanoptera: Thripidae). Journal of Entomological Science 26, 436–442.

Obradovic, A., Jones, J.B., Momol, M.T., Balogh, B., Olson, S.M., 2004. Management of tomato bacterial spot in the field by foliar applications of bacteriophages and SAR inducers. Plant Disease 88, 736–740.

Oetting, R.D., Beshear, R.J., Liu, T.X., Braman, S.K., Baker, J.R., 1993. Biology and identification of thrips on greenhouse ornamentals. Georgia Agricultural Experiment Station. Research Bulletin 414, 20.

Olson, D.M., Davis, R.F., Brown, S.L., Roberts, P., Phatak, S.C., 2006. Cover crop, rye residue and in-furrow treatment. Journal of Applied Entomology 130, 302–308.

Olson, S.M., Funderburk, J.E., 1986. A new threatening pest in Florida – western flower thrips. In: Stall, W.M. (Ed.). Stall, W.M. (Ed.), Proceedings of the Florida Tomato Institute, vol. VEC 86-1. University of Florida Extension Report, Gainesville, Florida, USA.

Pappu, H.R., Jones, R.A.C., Jain, R.K., 2009. Global status of tospovirus epidemics in diverse cropping systems: successes achieved and challenges ahead. Virus Research 141, 219–236.

Paterson, R.G., Scott, S.J., Gergerich, R.C., 1989. Resistance in two *Lycopersicon* species to an Arkansas isolate of tomato spotted wilt virus. Euphytica 43, 173–178.

Perrings, C., Dehnen-Schmutz, K., Touza, J., Williamson, M., 2005. How to manage biological invasions under globalization. Trends in Ecology & Evolution 20, 212–215.

Pittman, H.A., 1927. Spotted wilt of tomatoes. Journal of the Council on Science and Industry Research Australia 1, 74–77.

Price, D.L., Memmott, F.D., Scott, J.W., Olson, S.M., Stevens, M.R., 2007. Identification of molecular markers linked to a new tomato spotted wilt virus resistance source in tomato. Report of the Tomato Genetics Cooperative 57, 35–36.

Reitz, S.R., 2008. Comparative bionomics of *Frankliniella occidentalis* and *Frankliniella tritici*. Florida Entomologist 91, 474–476.

Reitz, S.R., 2009. Biology and ecology of the western flower thrips (Thysanoptera: Thripidae): the making of a pest. Florida Entomologist 92, 7–13.

Riley, D.G., 2007. Effect of imidacloprid on the settling behavior of *Frankliniella occidentalis* and *Frankliniella fusca* (Thysanoptera: Thripidae) on tomato and peanut. Journal of Entomological Science 42, 74–83.

Riley, D.G., Angelella, G.M., McPherson, R.M., 2010. Pine pollen dehiscence relative to thrips (Thysanoptera: Thripidae) population dynamics. Entomologia Experimentalis et Applicata 138, 223–233.

Riley, D.G., Joseph, S.V., Kelley, W.T., Olson, S., Scott, J., 2011b. Host plant resistance to tomato spotted wilt virus (Bunyaviridae: tospovirus) in tomato. HortScience 46, 1626–1633.

Riley, D.G., Joseph, S.V., Srinivasan, R., 2012a. Temporal relationship of thrips populations to tomato spotted wilt incidence in tomato in the field. Journal of Entomological Science 47, 65–75.

Riley, D.G., Joseph, S.V., Srinivasan, R., 2012b. Reflective Mulch and Acibenzolar-S-methyl treatments relative to thrips (Thysanoptera: Thripidae) and tomato spotted wilt virus incidence in tomato. Journal of Economic Entomology 105, 1302–1310.

Riley, D.G., Joseph, S.V., Srinivasan, R., Diffie, S., 2011a. Thrips vectors of tospoviruses. Journal of Integrated Pest Management 1. http://dx.doi.org/10.1603/IPM10020.

Riley, D.G., Pappu, H., 2000. Evaluation of tactics for management of thrips-vectored tomato spotted wilt *Tospovirus* in tomato. Plant Disease 84, 847–852.

Riley, D.G., Pappu, H., 2004. Tactics for management of thrips (Thysanoptera: Thripidae) and tomato spotted wilt *Tospovirus* in tomato. Journal of Economic Entomology 97, 1648–1658.

Riley, D.G., Sparks Jr., A., Langston, D., 2011c. Managing Tomato Spotted Wilt in Georgia. The University of Georgia Cooperative Extension Service. Circular 1002.

Riley, D.G., Sparks Jr., A.N., 2014. Current status of thrips (Thysanoptera: Thripidae) in Vidalia onions in Georgia. Florida Entomologist 97, 355–361.

Roselló, S., Diez, J., Nuez, F., 1998. Genetics of tomato spotted wilt virus resistance coming from *Lycopersicon peruvianum*. European Journal of Plant Pathology 104, 499–509.

Roselló, S., Diez, M.J., Lacasa, A., Jordá, C., Nuez, F., 1997. Testing resistance to TSWV introgressed from *Lycopersicon peruvianum* by artificial transmission techniques. Euphytica 98, 93–98.

Roselló, S., Ricarte, B., Jose, M., Diez, J., Nuez, F., 2001. Resistance to tomato spotted wilt virus introgressed from *Lycopersicon peruvianum* in line UPV 1 may be allelic to *Sw-5* and can be used to enhance the resistance of hybrids cultivars. Euphytica 119, 357–367.

Roselló, S., Soler, S., Díez, M.J., Rambla, J.L., Richarte, C., Nuez, F., 1999. New sources for high resistance of tomato to the Tomato spotted wilt virus from *Lycopersicon peruvianum*. Plant Breeding 118, 425–429.

Sakimura, K., 1961. Field observations on thrips vector species of tomato spotted wilt virus in San Pablo area, California. Plant Disease Reporter 45, 772.

Samuel, G., Pittman, H.A., 1928. Dusting and spraying experiments for the control of spotted wilt of tomatoes. Report Australasian Association for the Advancement of Science Hobart Tasmania 19, 588–590.

Scott, J.W., Hutton, S.F., Olson, S.M., Stevens, M.R., 2011. Spotty Results in Our Sw-7 Tomato Spotted Wilt Virus Research. http://vegetablemdonline.ppath.cornell.edu/TDW/Abstracts/11%20Jay%20ScottTDW_abstract_2011.pdf.

Scott, J.W., Stevens, M.R., Olson, S.M., 2005. An alternative source of resistance to tomato spotted wilt virus. Report of the Tomato Genetics Cooperative 55, 40–42.

Shan, C., Ma, S., Want, M., Gao, G., 2012. Evaluation of insecticides against the western flower thrips, *Frankliniella occidentalis* (Thysanoptera: Thripidae), in the laboratory. Florida Entomologist 95, 454–460.

Shi, A., Vierling, R.G., Chen, P., Caton, H., Panthee, D., 2011. Identification of molecular markers for *Sw-5* gene of tomato spotted wilt virus resistance. American Journal of Biotechnology and Molecular Science 1, 8–16.

Sims, K.R., Funderburk, J.E., Reitz, S.R., Boucias, D.G., 2009. The impact of a parasitic nematode, *Thripinema fuscum*, on the feeding behavior and vector competence of *Frankliniella fusca*. Entomologia Experimentalis et Applicata 132, 200–208.

Smith, K.M., 1931. *Thrips tabaci* Lind. As a vector of plant virus disease. Nature 127, 852–853.

Smith, K.M., 1933. Spotted wilt: an important virus disease of the tomato. Journal of the Ministry of Agriculture 39, 1097–1103.

Sparks, A.N., Diffie, S., Riley, D.G., 2010. Thrips species composition on onions in the Vidalia production region of Georgia. Journal of Entomological Science 46, 1–6.

Srinivasan, R., Riley, D., Diffie, S., Shrestha, A., Culbreath, A., 2014. Winter weeds as inoculum sources of tomato spotted wilt virus and as reservoirs for its vector, *Frankliniella fusca* (Thysanoptera: Thripidae) in farmscapes of Georgia. Environmental Entomology 43, 410–420.

Stavisky, J., Funderburk, J., Brodbeck, B.V., Olson, S.M., Andersen, P.C., 2002. Population dynamics of *Frankliniella* spp. and tomato spotted wilt incidence as influenced by cultural management tactics in tomato. Journal of Economic Entomology 95, 1216–1221.

Stevens, M.R., Heiny, D.K., Griffiths, P.D., Scott, J.W., Rhoads, D.D., 1996. Identification of co-dominant RAPD markers tightly linked to the tomato spotted wilt virus (TSWV) resistance gene *Sw-5*. Report of the Tomato Genetics Cooperative 46, 27.

Stevens, M.R., Lamb, E.M., Rhoads, D.D., 1995. Mapping the *Sw-5* locus for tomato spotted wilt virus resistance in tomatoes using RAPD and RFLP analyses. Theoretical and Applied Genetics 90, 451–456.

Stevens, M.R., Scott, S.J., Gergerich, R.C., 1991. Inheritance of a gene for resistance to tomato spotted wilt virus (TSWV) from *Lycopersicon*-PE. Euphytica 59, 9–17.

Stevens, M.R., Scott, S.J., Gergerich, R.C., 1992. Inheritance of a gene for resistance to tomato spotted wilt virus (TSWV) from *Lycopersicon peruvianum* Mill. Euphytica 59, 9–17.

Summers, C.G., Newton, A.S., Mitchell, J.P., Stapleton, J.P., 2010. Population dynamics of arthropods associated with early-season tomato plants as influenced by soil surface microenvironment. Crop Protection 29, 249–254.

Thompson, G.J., van Zijl, J.J.B., 1996. Control of tomato spotted wilt virus in tomatoes in South Africa. Acta Horticulturae 431, 379–384.

Toapanta, M., Funderburk, J., Webb, S., Chellemi, D., Tsai, J., 1996. Abundance of *Frankliniella* spp (Thysanoptera: Thripidae) on winter and spring host plants. Environmental Entomology 25, 793–800.

Todd, J.W., Culbreath, A.K., Chamberlin, J.R., Beshear, R.J., Mullinix, B.G., 1995. Colonization and population dynamics of thrips in peanuts in the southern United States. In: Parker, B.L., Skinner, M., Lewis, T. (Eds.), Thrips Biology and Management. Plenum Press, New York, USA, pp. 453–460.

Ullman, D.E., Sherwood, J.L., German, T.L., 1997. Thrips as vectors of plant pathogens. In: Lewis, T.L. (Ed.), Thrips as Crop Pests. CAB International, NY, pp. 539–566.

Ullman, D.E., Cho, J.J., Mau, R.F.L., Westcot, D.M., Custer, D.M., 1992. A midgut barrier to tomato spotted wilt virus acquisition by adult western flower thrips. Phytopathology 82, 1333–1342.

van Zijl, J.J.B., Bosch, S.E., Coetzee, C.P.J., 1986. Breeding tomatoes for processing in South Africa. Acta Horticulturae 194, 69–75.

Vavrina, C.S., Roberts, P.D., Kokalis-Burelle, N., Ontermaa, E.O., 2004. Systemic resistance in tomato: greenhouse screening of commercial products and application programs. HortScience 39, 433–437.

Vierbergen, G., 1995. Entomologische, Berichten Amsterdam, Netherlands. The Genus *Frankliniella* in The Netherlands, with a Key to the Species (Thysanoptera: Thripidae), vol. 55, pp. 185–192.

Wang, C.L., Lin, F.C., Chiu, Y.C., Shih, H.T., 2010. Species of *Frankliniella* Trybom (Thysanoptera: Thripidae) from the Asian-Pacific area. Zoological Studies 49, 824–838.

Webb, S.E., Kok-Yokomi, M.L., Tsai, J.H., 1997. Evaluation of *Frankliniella bispinosa* as a potential vector of *Tomato spotted wilt virus* (Abstract). Phytopathology 87 (Suppl. 6), S102.

Wells, M.L., Culbreath, A.K., Todd, J.W., Csinos, A.S., Mandal, B., McPherson, R.M., 2002. Dynamics of spring tobacco thrips (Thysanoptera: Thripidae) populations: implications for tomato spotted wilt virus management. Environmental Entomology 31, 1282–1290.

Whitfield, A.E., Kumar, N.K.K., Rotenberg, D., Ullman, D.E., Wyman, E.A., Zietlow, C., Willis, D.K., German, T.L., 2008. A soluble form of the tomato spotted wilt virus (TSWV) glycoprotein G_N-S) inhibits transmission of TSWV by *Frankliniella occidentalis*. Phytopathology 98, 45–50.

Whitfield, A.E., Ullman, D.E., German, T.L., 2005. Tospovirus-thrips interactions. Annual Review of Phytopathology 43, 459–489.

Whittaker, M.S., Kirk, W.D.J., 2004. The effect of photoperiod on walking, feeding, and oviposition in the western flower thrips. Entomologia Experimentalis et Applicata 111, 209–214.

Wijkamp, I., 1995. Virus Vector Relationships in the Transmission of Tospoviruses (Ph.D. thesis). Agricultural University Wageningen, Netherlands, p. 143.

Wijkamp, I., Goldbach, R., Peters, D., 1996. Propagation of tomato spotted wilt virus in *Frankliniella occidentalis* does neither result in pathological effects nor in transovarial passage of the virus. Entomologia Experimentalis et Applicata 81, 285–292.

Wijkamp, I., Peters, D., 1993. Determination of the median latent period of 2 tospoviruses in *Frankliniella-occidentalis*, using a novel leaf disk assay. Phytopathology 83, 986–991.

Willmot, A.L., Cloyd, R.A., Zhu, K.Y., 2013. Efficacy of pesticide mixtures against the western flower thrips (Thysanoptera: Thripidae) under laboratory and greenhouse conditions. Journal of Economic Entomology 106, 247–256.

Wijkamp, I., Almarza, N., Goldbach, R., Peters, D., 1995. Distinct levels of specificity in thrips transmission of tospoviruses. Phytopathology 85, 1069–1074.

Yazhisai, U., Rajagopalan, P.A., Raja, J.A.J., Chen, T.C., Yeh, S.D., 2015. Untranslatable tospoviral NSs fragment coupled with L conserved region enhances transgenic resistance against the homologous virus and a serologically unrelated tospovirus. Transgenic Research 24, 635–649.

Yudin, L.S., Cho, J.J., Mitchell, W.C., 1986. Host range of western flower thrips, *Frankliniella occidentalis* (Thysanoptera: Thripidae), with special references to *Leucaena glauca*. Environmental Entomology 15, 1292–1295.

Chapter 4

Whiteflies: Biology, Ecology, and Management

Thomas M. Perring[1], Philip A. Stansly[2], T.X. Liu[3], Hugh A. Smith[4], Sharon A. Andreason[1]

[1]*University of California, Riverside, CA, United States;* [2]*University of Florida – IFAS, Immokalee, FL, United States;* [3]*Northwest A&F University, Yangling, China;* [4]*University of Florida – IFAS, Wimauma, FL, United States*

1. INTRODUCTION

Whiteflies have long been considered as an important insect pest on a global scale, attacking a wide variety of agricultural commodities (Mound and Halsey, 1978; Gerling, 1990b). There are 1556 accepted species names in 161 genera that exist among the known extant whiteflies in the world (Martin and Mound, 2007). However, two species, the sweetpotato whitefly, *Bemisia tabaci* (Gennadius), and the greenhouse whitefly, *Trialeurodes vaporariorum* Westwood cause most of the damage (Nauen et al., 2014). This is the case for tomato, and both species are pests in both the open field and greenhouse tomato agroecosystems. Whiteflies, using mouthparts modified for sucking, damage plants by removing sap, excreting honeydew, vectoring plant-limiting viruses, and for one member of *B. tabaci* species complex, causing tomato irregular ripening. Because of these four types of damages, research on whiteflies has varied widely. Sometimes studies have focused on solving crop loss due to one or several of these damage types and the results are most applicable on a local scale. Other research has broader application for tomato growers throughout the world. In this chapter, we focus on the basic biology, ecology, and management of *B. tabaci* and *T. vaporariorum* on tomato.

2. IDENTIFICATION, BIOLOGY, DISTRIBUTION, HOST RANGE, AND SEASONAL OCCURRENCE OF WHITEFLY PESTS ON TOMATO

Both *B. tabaci* and *T. vaporariorum* are pests of crops in open field and greenhouse situations, but their importance in each of these types of agriculture have varied over time. *T. vaporariorum* has been a problem in greenhouses for a very long time; in fact its common name of greenhouse whitefly or glasshouse whitefly is symbolic of this long-term relationship with greenhouses. *B. tabaci*, while undergoing a number of changes regarding its systematics (see below), originally was named the tobacco whitefly because of its association with that open field crop. Both of these whiteflies have a large host range, of which tomato is a key crop. A search of the Web of Science Core Collection search engine found 251/855 (29.3%) and 938/3796 (25.9%) of the citations were related to tomato for *T. vaporariorum* and *B. tabaci*, respectively. Historically, *T. vaporariorum* has been a problem of greenhouse tomatoes and *B. tabaci* has been a problem in field tomatoes, mostly as a virus vector. However, within the last 25 years, we have experienced these two whitefly species as pests, sometimes severe, in both greenhouse and field tomato crops.

The interactions of these two whiteflies on tomato have been studied with some interesting results. In Japan, both species can occur on tomatoes in the greenhouse. When this occurred, *T. vaporariorum* selected upper leaflets, while *B. tabaci* selected middle leaflets (Tsueda et al., 2014). These choices were reflected in the number of eggs laid by each species on their selected leaflets. The authors further determined that whitefly selection was related to specific volatiles emitted by the tomato leaflets in their respective positions. Additional studies conducted in China determined that *B. tabaci* produced up to 4.9 times the progeny as *T. vaporariorum* (Zhang et al., 2011). In addition, *B. tabaci* developed faster, lived longer, and had a higher female:male sex ratio than *T. vaporariorum*. All of these parameters lead to displacement of *T. vaporariorum* by *B. tabaci* in as few as four generations. Chun-Li et al. (2015) found that these two species influence each other through their impact on tomato. Prior infestation of tomato with *B. tabaci* resulted in improved development, longevity, and fecundity of *T. vaporariorum*. However, prior infestation with *T. vaporariorum* did not have an impact on the development of *B. tabaci*. This study suggests a negative impact of *B. tabaci* feeding on plant defenses that does not occur with *T. vaporariorum* feeding.

Sustainable Management of Arthropod Pests of Tomato. http://dx.doi.org/10.1016/B978-0-12-802441-6.00004-8

2.1 *Bemisia tabaci* (Gennadius)

B. tabaci is a global pest causing significant loss to a wide variety of agricultural commodities. Arguably one of the most damaging insects in the world, three books have been written specifically on *B. tabaci* in the last 2 decades (Gerling and Mayer, 1996; Stansly and Naranjo, 2010; Thomson, 2011). The literature on this insect is expansive and was thoroughly reviewed by Cock (1986). *B. tabaci* was first described by Gennadius (1889) as *Aleurodes tabaci*, a whitefly collected in Greece on tobacco. It has undergone a number of synonymization events with other whiteflies (Perring, 2001), and after several decades of research into the nomenclature of *B. tabaci*, it is becoming clear that this entity is a complex of morphologically indistinguishable species (Dinsdale et al., 2010; Tay et al., 2012). Whitefly researchers have utilized a variety of naming conventions including race, biotype, genetic groups, and species to identify the variants of whiteflies in the complex. The major variants on tomato are New World whitefly (New World), previously known as Biotype A, Middle East Asia Minor 1 whitefly (MEAM1), previously known as Biotype B or *Bemisia argentifolii* Bellows and Perring (Dinsdale et al., 2010; Boykin, 2014), and the Mediterranean (MED) whitefly, previously known as Biotype Q and now known to be the original *B. tabaci* (Cuthbertson and Vänninen, 2015). Although we recognize the distinction between the *B. tabaci* species, we will refer to these simply as *B. tabaci* unless there is information specific to the New World, MEAM1, and MED whiteflies.

Whiteflies have six life stages: the egg, four nymphal instars, and the adult stage (Walker et al., 2010). The egg stage of all whiteflies has a stalk at their larger end, called the "pedicel." Quaintance and Baker (1913) stated that this stalk serves as an attachment for the egg, and it also "seems to direct the spermatozoa at the time of fertilization." During this time the stalk is filled with protoplasm. After fertilization the protoplasm dries up and the stalk becomes a hollow tube. Most whiteflies insert the pedicel into leaf stomata (Paulson and Beardsley, 1985). But *B. tabaci* pedicles are inserted directly into epidermal tissue. The apex of the pedicel has a porous fibrous structure which absorbs water and possibly nutrients from the leaf (Walker et al., 2010). The eggs of *B. tabaci* are typically scattered about the leaf surface although sometimes they are laid in partial egg circles. A "fecundity of well over 500 progeny per female has been demonstrated" (Naranjo et al., 2010).

All whitefly nymphal stages are oval in shape and dorsoventrally flattened. The first nymphal instar is called the "crawler" stage, because it is highly mobile and probes the leaf surface for a suitable location to insert its stylets into the phloem sieve elements. The lateral margins of the crawlers have many setae that are absent in later instars. These are thought to be mechanosensory in nature enabling the crawler to sense leaf hairs (Walker et al., 2010). Once the crawler locates a suitable feeding site, it begins feeding and molts into the subsequent instars, which are sessile on the plant, not moving from the original location selected by the crawler (Gill, 1990). The insect goes through three additional nymphal instars (second–fourth). The fourth instar is sometimes referred to as the "pupal" stage (Bemis, 1904; Sharaf and Batta, 1985; López-Avila, 1986; Gill, 1990), but Gill (1990) noted that the early fourth instar feeds, and thus is not a pupa in the normal sense of holometabolous insects. Later in the instar, the insect enters a transitional substage during which apolysis takes place and the adult cuticle is formed. This stage has red eyes and yellow body pigment. Byrne and Bellows (1991) suggested that the term pupa, if used, be reserved for the last nymphal stadium found after apolysis.

Walker et al. (2010) provided a detailed review of the adult morphology of *B. tabaci*, noting that they have the typical anatomy of adult insects in the Sternorrhyncha. The wings of *B. tabaci* are covered in white powder wax that is secreted soon after the adult ecloses from the last nymphal instar (Gill, 1990). The wings are held "roof-like" over the abdomen (Fig. 4.1), which gives them a long, narrow appearance on the leaf. Adults typically are found on the lower leaf surfaces of tomato. There is a sexual dimorphism in *B. tabaci*. Females are larger than the males and have a rounded abdomen where the male abdomen is more pointed (Fig. 4.1). Perring and Symmes (2006) described 4 phases in the elaborate courtship behavior of MEAM1 whiteflies leading to copulation. MEAM1 males also have been shown to interrupt courtship and impact mating of other *B. tabaci* sibling species (Perring, 1996, 2001; Liu et al., 2007; Walker et al., 2010). Whiteflies have arrhenotokous reproduction systems in which non-fertilized eggs are haploid males and fertilized eggs are diploid females.

Naranjo et al. (2010) reviewed the life history of *B. tabaci* and summarized the research that had been conducted up to that time. Typical of all poikilothermic insects, development, survival, and reproduction of *B. tabaci* is linked closely to temperature (Wagner, 1995; Wang and Tsai, 1996; Nava-Camberos et al., 2001; Xie et al., 2011; Guo et al., 2013; Hai Lin et al., 2014; Zeshan et al., 2015). These studies present a range of egg to adult developmental times from 105 days at 15°C to 14 days at 30°C. Fecundity also varied from a mean of 324 eggs per female at 20°C to 22 eggs per female at 30°C. Yang and Chi (2006) constructed life tables calculating an intrinsic rate of natural increase of −0.0176, 0.0667, 0.1469, 0.1611, 0.1745, and 0.0989 d^{-1} at 15, 20, 25, 28, 30, and 35°C, respectively. The mean generation time was determined to be 81.9, 48.6, 28.4, 25.3, 22.1, and 18.2 days, respectively. In addition to temperature, the host plant plays a major role in the biology of *B. tabaci* as shown in studies by Butler et al. (1983), Coudriet et al. (1985), Bethke et al. (1991), Perring et al. (1992, 1993), Nava-Camberos et al. (2001), Kakimoto et al. (2007), and Lorenzo et al. (2016). In fact, Hai Lin et al. (2014) found

FIGURE 4.1 *Bemisa tabaci*, one female (center) and two males displaying courtship behavior (top). Male on right is attempting to disrupt male on left by using his wings. Adults and eggs (Bottom). Notice wings held roof-like over the abdomen.

that host plant had a greater impact on egg development than temperature, humidity, and photoperiod. Another interesting study conducted in China found that MEAM1 whiteflies had an increase in number of eggs laid after feeding on tomato plants infected with tomato yellow leaf curl China virus (Guo et al., 2010). Whiteflies carrying this virus have been shown to feed more often from phloem sieve elements, make more phloem contacts, and spend more time in the salivation phase than whiteflies that were not carrying the virus (Moreno-Delafuente et al., 2013). These behaviors are known to enhance inoculation of tomato yellow leaf curl virus (TYLCV) by *B. tabaci* and result in more efficient feeding for the whitefly. Thus, the virus–whitefly relationship is mutually beneficial to the whitefly and the virus (Moreno-Delafuente et al., 2013).

Yet even on the same host plant, developmental rate may vary across different *B. tabaci* sibling species, locations, and experimental conditions. Two studies, both conducted at 25°C and 65% RH illustrate this point. Salas and Mendoza (1995) studied Biotype A (New World group) in a lab setting in Venezuela on detached tomato leaflets. Jamuna et al. (2016) studied Biotype B (MEAM1 group) in a greenhouse setting in India on attached tomato leaves. These studies showed some marked differences (Table 4.1). The total pre-adult duration was 4 days difference between the New World whiteflies (22.3 days) and the MEAM1 whiteflies (18.1 days). The ovipositional period also was different with 16.7 days for New World whiteflies compared to 2.25 days for MEAM1, resulting in vastly different average numbers of eggs per female (194.9 for New World and 94.5 for MEAM1). This variability makes the interpretation of biological data on *B. tabaci* in the literature challenging; one must be aware of the specific whitefly variant and the host plant on which the studies are performed.

B. tabaci has a worldwide distribution, inhabiting every continent (Ghahari et al., 2013; CABI, 2017a). Originally it was known as a pest in subtropical and arid regions of the world, but has now been moved to temperate environments where it exists in greenhouse and open-field tomatoes. It is absent only in the most northern regions of Europe and Asia. This whitefly also is polyphagous, estimated to feed on over 900 host plants (McKenzie et al., 2014). This host range has been

TABLE 4.1 Life Stage Parameters for *B. tabaci* Reported in 2 Studies

Life Stage Parameter	Salas and Mendoza (1995)	Jamuna et al. (2016)
Egg stage duration (days)	7.3	6.3
First instar duration (days)	4.0	3.65
Second instar duration (days)	2.7	2.75
Third instar duration (days)	2.5	2.9
Fourth instar duration (days)	5.8	3.5
Total pre-adult duration (days)	22.3	18.1
Pre-oviposition period (days)	1.4	1.45
Oviposition period (days)	16.7	2.25
Fecundity (No. of eggs/female)	194.9	94.5
Longevity of female (days)	19.0	5.65
Longevity of male (days)	19.4	3.3

compiled from all of the known variants that have been assigned to the single species *B. tabaci*. Most of the species in the *B. tabaci* complex have limited host ranges, some with only a single known host-plant species. The MEAM1 and MED whiteflies are anomalies in the complex by feeding on many different host plants.

2.2 *Trialeurodes vaporariorum* (Westwood)

T. vaporariorum, known as the greenhouse whitefly or glasshouse whitefly, is the most economically important species in the genus (Russell, 1948). It was first described as *Aleurodes vaporariorum* by Westwood in 1856, from whiteflies collected on tomato in glasshouses throughout Europe (Quaintance, 1900; Cockerell, 1902). However, Cockerell (1902) noted that this whitefly "is supposed to have originated in Brazil." Russell (1948) listed 7 synonymies for *T. vaporariorum* and Ghahari et al. (2013) listed 11 whitefly species that have been synonymized.

Like other whiteflies, *T. vaporariorum* has six life stages; the egg, four nymphal instars, and the adult. Various authors have described these life stages in detail (Morrill, 1905; Hargreaves, 1915; Russell, 1948; Burnett, 1949; Gill, 1990). On tomato, eggs are 0.23 mm long and 0.10 mm wide (Gerk et al., 1995). The egg pedicel is inserted in the leaf, similar to *B. tabaci*. However, the placement of eggs is quite different than *B. tabaci*. Hargreaves (1915) states that the eggs "generally are laid in circles on the undersides of leaves." The insect inserts its stylets into the leaf tissue and, using that point as center and its body as radius, revolves as each egg is deposited. *T. vaporariorum* is known to lay large numbers of eggs reaching as high as 534 per female on *Lamium purpureum* L. (Lloyd, 1922). In this same study, the number laid on tomato was 110 per female.

The nymphal stages are oval in shape and dorsoventrally flattened as in *B. tabaci*, but they are distinct in having a marginal fringe and long wax rays on the dorsum (Gill, 1990) (Fig. 4.2). Gerk et al. (1995) reported the length and width of first–fourth instars on tomato to be 0.30:0.16 (length:width), 0.41:0.23, 0.54:0.34, and 0.75:0.47 mm, respectively, similar to the sizes reported by other authors (Quaintance, 1900; Hargreaves, 1915). Adult whiteflies have wax covered wings which are held flat over the abdomen (Fig. 4.3); a trait that can be used to distinguish from *B. tabaci*. Males (length = 0.75 mm, width = 0.26 mm) are slightly smaller than females (length = 0.90 mm, width = 0.30 mm) (Gerk et al., 1995). This species has an arrhenotokous reproduction system in which non-fertilized eggs are haploid males and fertilized eggs are diploid females. Similar to *B. tabaci*, the courtship and mating behavior of *T. vaporariorum* is elaborate (Las, 1979; Ahman and Ekbom, 1981; Li and Maschwitz, 1985).

The development of *T. vaporariorum* is optimal at lower temperatures than *B. tabaci*. Xie et al. (2011) evaluated both species at 15, 18, 21, and 24°C, and found that *T. vaporariorum* produced more eggs and survived better at 15 and 18°C while *B. tabaci* produced more eggs and had higher survival at 24°C. This information is consistent with previous studies on *T. vaporariorum*, which showed higher fitness at cooler temperatures. For example, Burnett (1949) reported the fecundity and longevity of females at temperatures ranging from 9 to 33°C. He reported maximum longevity at 15°C with very short survival at 9°C and temperatures over 27°C. Similarly, females laid more eggs at 18 and 21°C than lower and higher temperatures. Hussey and Gurney (1958) found maximum development of immatures and highest fecundity at 27°C compared

FIGURE 4.2 *Trialeurodes vaporariorum* adult female and immatures. Two of the individuals are fourth instar nymphs (yellow in color) and two of them have enclosed to the adult stage leaving the exoskeleton of the last nymphal instar. Notice the long marginal wax filaments.

FIGURE 4.3 *Trialeurodes vaporariorum*, three females, eggs, and crawlers (Top). Adults and eggs (Bottom). Notice wings held flat over the abdomen.

to 23 and 21°C. However, the maximum female longevity was at 23°C. Specific studies on tomato found an average total life cycle of 25 days, with pre-adult development lasting 16.2 days (Gerk et al., 1995).

T. vaporariorum is distributed throughout the world, found on every continent (Ghahari et al., 2013; CABI, 2017b). Its pest status in greenhouses has resulted in its range expansion into far temperate regions of the globe. The host range of *T. vaporariorum* is very large, which was recognized very early, as Russell (1948) listed 138 plant species as host for this whitefly. Currently, CABI (2017b) estimates 859 species of plants from 469 genera and 121 families on which *T. vaporariorum* is known to exist. The biology and life history parameters of *T. vaporariorum* vary with host plant. For example, they were found to have a higher intrinsic rate of natural increase on zucchini ($r_m = 0.12$) than on tomato ($r_m = 0.10$) (Calvitti and Buttarazzi, 1995).

3. DAMAGE, ECONOMIC THRESHOLDS, AND LOSSES

Whiteflies, using mouthparts modified for sucking plant sap, damage plants in four ways, three of which are similar to aphids (see Chapter 2). They remove plant sap which can reduce yields and weaken plants to the point of killing them under high whitefly densities. Hussey et al. (1969) presented results for yield loss due to assimilate removal by *T. vaporariorum* and noted that these losses varied by season, being greater in the July to October crop compared to the March to July crop. The authors speculated that the reduced loss from feeding in summer was due to production of excess photosynthate under long day conditions. For *B. tabaci* an economic injury level (EIL) of four nymphs per leaf and one adult per beat tray (30 cm width × 45 cm length × 5 cm depth) were obtained in a study of open field tomato in Brazil (Gusmao et al., 2006). This study was conducted in processing tomatoes valued at \$218.2/ton, which is somewhat lower than a 5-year average of \$269.7/ton reported for the United States from 2011 to 2015 (ERS, 2017). This higher value in the United States would reduce the EIL to 3.25 nymphs per leaf, all other factors being equal.

Second, a large amount of the phloem sap is not utilized for whitefly metabolism or egg production, and this sap is excreted in a high carbohydrate-rich form known as honeydew. Whitefly honeydew varies over the different life stages and adults produce more than nymphs (Henneberry et al., 1999). This sugary substance covers tomato foliage and fruit and provides a substrate for a variety of saprophytic fungal species, turning the surface sticky and black. These include *Penicillium* sp., *Cladosporium herbarum* (Pers.) Link, *Fumago vagans* Pers. (Lloyd, 1922), and *Cladosporium sphaerospermum* Link (Eggenkamp-Rotteveel Mansfeld et al., 1982). This "sooty mold" reduces photosynthesis on covered leaves and fruit, and must be washed from tomatoes before the fruit can be marketed. Johnson et al. (1992) conducted a 2-year field study to quantify loss of yield and quality due to feeding of *T. vaporariorum* and subsequent sooty mold. They reported a 5% loss caused by sooty mold deposition caused by an infestation of 298 whitefly days per cm² of leaf surface for both years. This corresponding to a maximum density of about 8.3 nymphs per cm² which agreed with a threshold of 7 nymphs per cm² suggested earlier by Hussey et al. (1969) to avoid losses from sooty mold-contaminated fruit.

Third, both *T. vaporariorum* and *B. tabaci* transmit plant viruses which can be very debilitating to tomato crops. More than 150 plant viruses are known to be transmitted by whiteflies and the number continues to grow (Polston and Anderson, 1997; Jones, 2003). Although the majority of these are vectored by *B. tabaci* (Polston et al., 2014), the criniviruses are vectored by *Trialeurodes* species. McKenzie et al. (2014) lists 100 viruses vectored by *B. tabaci* and CABI (2017a) lists 121 virus species. Thus *B. tabaci* has been labeled as a "supervector" (Gilbertson et al., 2015). They noted that the high reproductive rate, dispersal ability, polyphagy, and development of insecticide resistance contributed to disease outbreaks throughout the world.

Finally, whiteflies in the MEAM1 group of *B. tabaci* possess salivary components that are injected into tomato during feeding which causes a fruit abnormality known as tomato irregular ripening (TIR) (Fig. 4.4). This abnormality has been linked to nymphal feeding (Maynard and Cantliffe, 1989; Schuster et al., 1990; Schuster, 2001; McCollum et al., 2004; Dinsdale et al., 2010). Schuster et al. (1996) determined the relationship between nymphal densities and TIR symptom severity. Whiteflies in most other *B. tabaci* groups do not cause TIR. An action threshold of one whitefly nymph per two leaflets prevented TIR of tomato (Schuster, 2002). More effective control measures would presumably increase this threshold, whereas less effective control would have the opposite effect.

These four types of damage can be severe in their own right, but taken together, the hierarchy of whitefly damage potential is: plant viruses > irregular ripening > sap removal, resulting in the reverse relationship for (EILs) (Stansly and Natwick, 2010).

4. MANAGEMENT

Severe crop losses due to the damage and control efforts of *B. tabaci* and *T. vaporariorum* have resulted in over a century of research being conducted on the management of these whiteflies. In this chapter, we cover the individual topics of chemical, biological, and cultural control, as well as host-plant resistance as it relates to whiteflies. While these are presented as independent topics, the best management schemes integrate strategies and consider not just whiteflies but other pests on tomato (Zalom et al., 2001). Chavan et al. (2013) found that an integrated pest management (IPM) module was equally

FIGURE 4.4 Tomato irregular ripening caused by salivary components of *Bemisia tabaci* MEAM1 group whiteflies. Fruit showing external symptom of red and green coloring (Top). Fruit showing internal red and green sectoring (Bottom).

effective at controlling whiteflies, aphids, and lepidopteran pests as was the insecticide standard with the added benefit that natural enemies were conserved in the IPM module. IPM programs can be particularly effective in the production of organic tomatoes (Shojai et al., 2003). In this study, they employed intercropping with cucumber and tomato (both *T. vaporariorum* hosts), organic and inorganic fertilizers, release of predators and entomopathogenic fungi, and trapping to manage *B. tabaci*, leaf miners, and lepidopteran pests. In addition to integrated management programs for direct whitefly damage, Gilbertson et al. (2011) and Pratissoli et al. (2015) present IPM programs to reduce whitefly-vectored plant viruses.

4.1 Monitoring and Sampling

Prior to the implementation of any IPM program, suitable methods for pest density assessment must be developed (Pedigo and Rice, 2006). Based on similar life histories of *B. tabaci* and *T. vaporariorum* on tomato, the physical aspects of sampling are similar. However, there may be distinctions between the two whiteflies regarding when and where to sample, particularly in open field and greenhouse environments. In their review of whitefly sampling, Ekbom and Rumei (1996) noted that specific sampling plans were determined by the distributions of insects, which was a function of three mechanisms; (1) plant physiology, biochemistry, and architecture; (2) the degree of heterogeneity in the environment; and (3) the specific insect behavior and movement.

A more recent review revealed the breadth of sampling studies that have been conducted on a wide variety of crops (Naranjo et al., 2010). Specific studies on tomato have been conducted by numerous authors (Martin et al., 1991; Kim et al., 1999, 2001; Gusmao et al., 2005, 2006; Arnó et al., 2006; Park et al., 2011; Song et al., 2014; Böckmann et al., 2015). These studies have determined that a presence-absence sampling program should be restricted to the 3–5 leaf zone on infested plants (Martin et al., 1991). Arnó et al. (2006) also evaluated whitefly distribution in tomato and found that *T. vaporariorum* eggs were located mainly in the top plant stratum, while *B. tabaci* eggs were more in the middle stratum; nymphs of both species were in the bottom stratum. They suggested sampling a 1.15 cm leaf disc from the top stratum as the most efficient sampling unit for simultaneously sampling the eggs of both species. Gusmao et al. (2005), after evaluating 12 sampling methods, determined that the best sampling method for *B. tabaci* in open field tomatoes was to beat one apical leaf into a tray to assess adults and to count nymphs on a basal leaf. Furthermore, they noted that 45 samples per field needed to be taken for nymphs and 24 samples were needed for adults. Additional research used these sampling protocols and established EILs and sequential sampling plans which vastly reduced the number of samples required to reach a decision compared to their conventional sampling protocol (Gusmao et al., 2006). Wang et al. (2014) developed a novel counting algorithm for counting whiteflies in computer images. This method utilized two features of the whiteflies, color and shape, to provide low error rates across various commodities, including tomato.

One of the most widely used methods for sampling whiteflies is the yellow sticky trap (see Fig. 2.3 in Chapter 2). A number of variables have been evaluated with regard to yellow sticky traps, including the color or shade of the yellow used (Okadome and Amano, 2014), the trap shape, vertical/horizontal orientation, and placement with respect to the crop being sampled. Studies evaluating these variables were reviewed by Ekbom and Rumei (1996) who noted that sometimes the trap counts provide a good estimate of whiteflies on plants and other times they do not. Song et al. (2014) related trap counts to treatment threshold, stating that for a treatment threshold of 10 adults *B. tabaci* per trap, the required number of traps in the greenhouse was 15 with a fixed precision level of 0.25. They also determined that the mean whitefly density per trap was efficiently estimated when the trap had two or more whiteflies. Böckmann et al. (2015) determined that yellow trap catches provided reliable information on nymphal density in greenhouse tomatoes. Further evaluation of yellow sticky traps was made in the review of the topic by Pinto-Zevallos and Vanninen (2013). These authors cover various aspects of using yellow sticky traps including their reliability, representativeness (i.e., the difference between the expectation of the sample estimate, and the population parameter being estimated), the relevance of the sticky trap estimate to crop loss, and the practicality of the sticky trap program. Based on the considerations of these parameters, Pinto-Zevallos and Vanninen (2013) offered suggestions for improving the use of yellow sticky cards using automated counting tools and with geostatistics, as in Park et al. (2011).

4.2 Cultural Control

Cultural control involves the deliberate manipulation of environmental factors and production practices to limit pest damage. Cultural controls can involve both temporal and spatial manipulations.

4.2.1 Crop Management: Fertilizer and Irrigation

Whitefly fitness is a function of the quality of its host plant. As such, different fertilizers can impact the density of whiteflies developing on the leaves. This has been demonstrated in *T. vaporariorum*, which has been shown to have a higher intrinsic rate of increase on tomatoes with higher nitrogen levels (Jauset et al., 2000). Similarly, higher nitrogen-containing plants positively influenced the production of greater number of eggs by *T. vaporariorum*, which obtained sufficient nutrients for metabolism and oviposition with less feeding (Park et al., 2009). The positive relationship between host nitrogen content and whitefly fitness is also apparent in tomatoes grown hydroponically (Hosseini et al., 2015). Irrigation methods have also been shown to impact *B. tabaci* infestations and virus incidence. Abd Rabou and Simmons (2012a) compared drip, furrow, and sprinkler irrigation, and found that daily drip irrigation resulted in lower whitefly densities and lower incidence of virus.

4.2.2 Host-Free Period

One of the more drastic forms of cultural control is the imposition of region-wide host-free periods to reduce whitefly densities and viral inoculum. While a host-free period in the greenhouse can be created using lethal high temperatures (Tsueda et al., 2007), the implementation is much more difficult in the open field on an area-wide basis. However, there are examples that have proven effective. One strategy is to alter planting dates to provide as much time as possible between successive crops to reduce whitefly density (Ahsan et al., 2005). A host-free period was imposed in the Dominican Republic in response to devastating epidemics of TYLCV that threatened their tomato industry (Hilje et al., 2001; Gilbertson et al., 2007). The host-free period was imposed for June, July, and August when commercial production of tomato ceased and prohibited cultivation of a range of whitefly hosts, including eggplant, peppers, beans, cotton, cucurbits, and tomato. Salati et al. (2002) recorded a significant decrease in viruliferous whiteflies in areas where compliance with the host-free period was high, and an increase in the viruliferous whitefly population as tomato production resumed. In Cyprus in the early

1980s, a program to destroy overwintering tomato as a source of viral inoculum for spring planting greatly reduced incidence TYLCV (Ioannou, 1987). In the Arava desert region of Israel, a 1 month crop-free period was established during peak summer heat (June–July) to reduce vector habitat and eliminated losses due to TYLCV and other viruses (Ucko et al., 1998). Examples of successful host-free periods in Cyprus and Israel were carried out on a limited scale compared to the Dominican Republic. A voluntary host-free period for managing *B. tabaci* in tomato and other vegetable crops was successfully implemented in Florida (Hilje et al., 2001). The campaign was particularly effective in the southwestern production area where incidence of whiteflies and associated viruses were reduced by maintaining fields free of host crops for at least 2 months during the rainy summer period.

The success of a host-free period depends, in part, on the presence of weeds and other non-economically important plants in the surrounding area serving as hosts of whiteflies and viruses. The contribution of weeds acting as reservoir hosts of TYLCV on viral dissemination remains ambiguous. In Israel, where TYLCV first was identified, the virus has been detected in the weeds *Cynanchum acuta* L. and *Malva parviflora* L. (Cohen et al., 1988). The virus has also been detected in other regions in the weed species *Cleome viscosa* L., *Croton lobatus* L, *Datura stramonium* L., *Mercurialis ambigua* L.f., *Physalis* spp., and *Solanum nigrum* L. (Mansour and Al-Musa, 1992; Sanchez-Campos et al., 2000; Salati et al., 2002). Surveying weeds in Cyprus, Papayiannis et al. (2011) detected TYLCV in over 460 samples of almost 50 different weed species representing 15 plant families. Polston et al. (2009) tested 1920 plants from 45 known species of wild plants representing 15 plant families in Florida, but did not detect wild plant reservoirs of TYLCV. Smith et al. (2015) confirmed that *Amaranthus retroflexus* L. is a host of TYLCV in Florida and that *Bidens alba* (L.) DC., *Emilia fosbergii* Nicolson, and *Raphanus raphanistrum* L. are not hosts of the virus.

Tomato is considered the primary source of TYLCV inoculum in Florida. For this reason, prompt and complete destruction of the crop after harvest is considered the foundation of TYLCV management (Schuster et al., 2007). These authors recommend that plants be treated with a contact desiccant herbicide to destroy plant tissue combined with a heavy crop oil to kill adult *B. tabaci* and prevent their dispersal from the field. Time of planting can also be manipulated to avoid damage from whiteflies and whitefly-vectored viruses. In Cyprus, tomatoes planted early in spring or late in fall usually escaped infection with TYLCV (Ioannou, 1987). Planting date for tomatoes in Egypt has been effectively shifted to reduce whitefly damage (El-Gendi et al., 1997).

4.2.3 Trap Crops

Orientation toward the yellow-green spectrum of vegetation is one of the first steps in host finding for *B. tabaci* and *T. vaporariorum*. Economic and wild hosts of *B. tabaci* vary in terms of their suitability as substrates for oviposition and development (Smith et al., 2014). *B. tabaci* will respond to volatiles produced by some host plants (Cao et al., 2008; Li et al., 2014), however, evidence indicates that whiteflies require gustatory information to ultimately accept a host (van Lenteren and Noldus, 1990; Hunter et al., 1996; Lei et al., 1998). The need for *B. tabaci* and *T. vaporariorum* to taste a host in order to accept it may help explain the ambiguity surrounding the use of trap crops to manage whiteflies. A trap crop is a plant that is more attractive to the pest than the main crop which is being protected (Hokkanen, 1991). The assumption with trap cropping is that the trap crop will influence host selection before the pest has landed on the main crop, drawing the pest away from the main crop. Schuster (2003) consistently found more *B. tabaci* on squash than tomato in both field and greenhouse choice experiments. In a follow-up study, Schuster (2004) found that tomato surrounded by squash had lower numbers of whiteflies and lower incidence of TYLCV than tomato surrounded by tomato. In contrast, trap cropping with eggplant or beans did not reduce whitefly numbers on tomato (Al-Musa, 1982; Arias and Hilje, 1993; Peralta and Hilje, 1993), although Choi et al. (2015) found that eggplant attracted *B. tabaci* better than tomato in the greenhouse. Furthermore, these authors found that treating the eggplant with dinotefuran decreased the density of whiteflies in tomato greenhouses (Choi et al., 2016). Maize was shown to be a good barrier crop, and eggplant was shown to be a good trap crop for *B. tabaci* (Rajasri et al., 2009). Both crops resulted in low incidence of *tomato leaf curl virus* (ToLCV). Lee et al. (2010) found that *T. vaporariorum* tended to settle and accumulate more on eggplant than poinsettia (*Euphorbia pulcherrima* Willd. ex. Klotzsch) in a greenhouse setting, with the eggplant essentially functioning as a trap crop. Previous host experience can also influence whitefly acceptance of new crops (van Lenteren and Noldus, 1990).

4.2.4 Intercropping

Intercropping to prevent *B. tabaci* from finding host plants has shown promise for whitefly management. The presence of multiple hosts may in itself affect the behavior of *B. tabaci*. Bernays (1999) found that when *B. tabaci* was presented with a variety of acceptable hosts, it tended to move more and feed for shorter periods of time than when only one host was available. Studies on intercropping tomato with squash (Youssef et al., 2001; Schuster, 2004), corn (Abd Rabou and

Simmons, 2012b), and *Capiscum* (El-Serwiy et al., 1987) resulted in lower *B. tabaci* numbers and lower incidence of TYLCV. Al-Musa (1982) observed reduced TYLCV in tomato intercropped with cucumber. Similar studies have lowered the incidence of ToLCV and whiteflies in tomato. Verma et al. (2013) used French bean as a tomato intercrop and showed reduced *B. tabaci* numbers and ToLCV incidence. Celery has also been used for intercropping with tomato in the greenhouse, resulting in a 98% reduction (Pei-Xiang et al., 2011) and 89% reduction (Zhou et al., 2014) in *T. vaporariorum* densities. Intercropping can also impact the natural enemy complex. Smith et al. (2001) observed increased whitefly (both *B. tabaci* and *T. vaporariorum*) parasitoid diversity, but no increase in overall parasitism on tomato intercropped with roselle (*Hibiscus sabdariffa* L.) and corn (*Zea mays* L.) compared to tomato grown in monoculture.

4.2.5 Mulches

Whitefly flight behavior is known to be influenced by light. As whiteflies leave unfavorable host conditions (poor resources and/or crowding), they orient toward UV light (320–400 nm), the spectrum associated with the sky. Once airborne, whiteflies become attracted to the yellow-green spectrum of light (510–590 nm) reflected by vegetation (Mound, 1962; Coombe, 1982). This orientation behavior is physiologically predisposed to successful location of new host plants (van Lenteren and Noldus, 1990; Blackmer and Byrne, 1993a,b) and can manipulated for whitefly management.

One method is to use living mulches; surrounding the protected crop with other plants that "mask" the target plant from the whiteflies. Research with living mulches in Costa Rica demonstrated that planting tomato into an established mulch of either perennial peanut (*Arachis pintoi* Krapov. & W.C. Gregory), the weed cinquillo (*Drymaria cordata* (L.) Willd. ex J.A. Schultes), or coriander (*Coriandrum sativum* L.) reduced the number of *B. tabaci* adults entering plots and the incidence of tomato yellow mottle virus. The presence of living mulches also delayed the onset of viral symptoms compared to crops grown on bare ground (Hilje and Stansly, 2008). Living mulches are best suited to control whiteflies during the first 5 weeks after planting, when protection from whitefly-vectored viruses is crucial, and before the mulch plants must be suppressed to prevent competition with the main crop (Hilje et al., 2001).

Another method is to use plastic mulches on which the protected crop is grown. These plastic mulches have been evaluated extensively and are known to impact whitefly alighting behavior and virus disease spread in tomato (Csizinszky et al., 1995). Many different colors have been evaluated, and these studies were reviewed by Greer and Dole (2003) who noted that aluminum foil and aluminum-painted mulches were the most effective at repelling whiteflies. This is logical given the relationship between whitefly flight and UV light; the creation of UV reflectance close to plants repels whiteflies from "seeing" the yellow-green spectrum of target host plants (Antignus, 2000). One specific study on tomato has shown that plants grown over aluminum mulch had lower densities of *B. tabaci* and lower incidence of tomato mottle virus (TMoV) (Csizinszky et al., 1999). The repellant effect of reflective mulches diminishes over time as the mulch becomes soiled and shaded by growing crops (Smith et al., 2000).

4.2.6 Habitat Manipulation

Another strategy that has been used to manage pests involves the provisioning of habitat, floral resources, or alternate hosts and prey to enhance biological control of the pest, though the effectiveness of this strategy on whitefly control in tomato is unclear. Habitat plantings that provide floral resources in the form of nectar and pollen have been used to enhance biological control for many pests (Naranjo and Hagler, 1998; Jervis and Heimpel, 2005). Providing floral resources to enhance parasitism of whiteflies might be of value in tomato if it does not increase attraction of flower-loving, phytophagous pests such as *Frankliniella* thrips, leafminers, and hemipterans. In general, Hymenoptera parasitoids rarely feed on pollen, although they may take advantage of floral nectar (Sivinski et al., 2011). Host feeding is a common approach for accessing protein among parasitic Hymenoptera, including *Encarsia* and *Eretmocerus*, the genera most used for whitefly suppression (Gerling, 1990a). Whitefly parasitoids also take advantage of honeydew produced by their hosts as a sugar-rich energy source (van Lenteren et al., 1987; Hirose et al., 2009). Given that key whitefly parasitoids rely on host feeding and honeydew rather than nectar and pollen for food as adults, provisioning floral resources might not enhance parasitism of whiteflies.

4.2.7 Protected Agriculture

In protective structures such as greenhouses and screen houses, whiteflies can be physically excluded from the crop by the use of plastic film or fine mesh materials. There are logistical challenges to hermetically sealing a structure because of the need for ventilation to manage temperature and relative humidity (Stansly et al., 2004). Mesh sizes of 230×900 μm or less are needed to exclude whiteflies (Bethke and Paine, 1991). Tanaka et al. (2008), Kakutani et al. (2012), Nonomura et al. (2012, 2014), and Takikawa et al. (2016) described the use of electrically charged screens in greenhouses to repel and/or kill *B. tabaci* while allowing proper ventilation for tomato production. Once inside a structure, whiteflies are protected from natural enemies and may find conditions for establishment more favorable than in the field. To address this issue, Takikawa et al. (2015) developed a battery-operated electrostatic insect sweeper for eliminating whitefly adults. A combination of resistant varieties, and chemical and biological control may be needed to manage whiteflies in protected structures (Stansly et al., 2004).

Based on the principles of exclusion in protective houses, several other strategies have been adopted in open field tomatoes. For many years, growers in Israel and Egypt have used plastic (Berlinger et al., 1983) and muslin (Haydar and Aly, 1990) coverings to protect young tomato plants from infestation by *B. tabaci*. A recent study determined that the optimal covering period for row covers to reduce the incidence of whitefly-vectored viruses and maintain high yields was 6–7 weeks (Al-Shihi et al., 2016). Floating row covers can be used in the field to exclude whitefly and other pests from crops grown in small plots (Qureshi et al., 2007; Ben-Yakir et al., 2012; Gogo et al., 2014). It has also been shown that row covers combined with basil intercropping can improve tomato yields (Mutisya et al., 2016).

4.2.8 Ultraviolet Light

While UV and other intensely reflected lights repel whiteflies and are useful as aluminum mulches (see above), there is evidence that whiteflies rely on near-UV light to navigate during flight (Antignus, 2000). Polyethylene films used in greenhouses are typically made with UV-absorbing material to maintain the longevity of the film while transmitting sufficient visible light for crop production. UV-absorbing plastic has been used in high tunnels and other forms of protected culture to reduce crop colonization by whitefly and other pests. Antignus et al. (1996) found that UV-absorbing plastic sheeting could reduce incidence of TYLCV to as low as 20% in tomato grown in high tunnels compared to 93% under non-UV-absorbing plastic. This is apparently because *B. tabaci* is significantly less likely to enter a tunnel constructed with UV-absorbing material than one constructed with conventional plastic (Costa and Robb, 1999; Kumar and Poehling, 2006b). Follow-up studies showed a 30% disease reduction after 50 days in tunnels with the UV-absorbing plastics (Antignus et al., 1998). In addition to whiteflies not entering the tunnel, Antignus et al. (2001) suggested that the lack of UV light in the tunnels interfered with whitefly flight orientation, resulting in low dispersal capabilities. The color of greenhouse or tunnel netting may also influence *B. tabaci* activity. Ben-Yakir et al. (2012) found that the incidence of TYLCV was 2–4 times lower in high tunnels covered with yellow or pearl-colored netting than in tunnels covered with black or red netting.

4.3 Tomato Plant Resistance to Whiteflies

Host-plant resistance is a key strategy in successful IPM programs and often a last line of plant defense when focused preventative biological and cultural controls are in place along with insecticide management. In tomatoes, efforts for breeding resistance to both the whitefly pests and their transmitted viruses have focused on wild relatives of *Solanum lycopersicum* L. As with resistance to aphids and aphid-borne viruses (see Section 4.3 in Chapter 2), these wild tomatoes offer resistance traits encoded by unique genes, which have been introduced into domesticated tomato. Resistance to both *B. tabaci* and/or *T. vaporariorum* has been studied with some traits offering resistance or tolerance to both species.

Early development of whitefly (as well as aphid) resistance in tomatoes focused on screening existing tomato varieties and related species for reduced pest survival and fitness (Curry and Pimentel, 1971; Clayberg and Kring, 1974; De Ponti et al., 1975). These studies indicated that domesticated *S. lycopersicum* has lower levels of resistance than some wild relatives and that one of the biggest determinants of increased resistance was the presence and density of trichomes. Trichomes are hair-like structures that are either glandular or non-glandular, providing a physical defense against pest feeding or oviposition, and, as with glandular trichomes, a chemical deterrent, including the secretion of acyl sugars. Manipulation of the types of trichomes and secondary metabolites secreted by glandular trichomes has been a major focus of pest-resistant tomato breeding.

S. lycopersicum has several types of trichomes that confer various levels of constitutive defense against insect pests. For more information on the types of trichomes present in tomatoes and their roles in plant defense against hemipteran pests, see Chapter 2 . The means of resistance conferred by trichomes against whiteflies are similar to those in aphids. In *S. lycopersicum*, types II, III, and V trichomes are non-glandular and vary in size (Channarayappa et al., 1992; Glas et al., 2012). As purely a physical deterrent, a higher density of branched non-glandular trichomes on a wooly mutant tomato line demonstrated significant resistance against *T. vaporariorum* (Clayberg and Kring, 1974). This resistance, however, does not seem to be the case with *B. tabaci*. Studies have shown that higher densities of non-glandular trichomes are correlated with increased oviposition rates by *B. tabaci* (Heinz and Zalom, 1995; Toscano et al., 2002). The presence of non-glandular trichomes alone is not a reliable character on which to base whitefly resistance, particularly as mixed infestations of *B. tabaci* and *T. vaporariorum* do occur on tomato (Arnó et al., 2006).

Most efforts in tomato-resistance discovery and breeding have focused on the effects and manipulation of glandular trichomes. Glandular trichomes (Types I, IV, VI, and VII) on *S. lycopersicum* are involved in the production, storage, and release of defensive compounds, including different classes of tomato secondary metabolites (Bleeker et al., 2012). In *Solanum* spp. glandular trichome types IV and VI are associated with resistance to arthropods including whiteflies (Antonious and Snyder, 2008; Rodriguez-Lopez et al., 2012). Secreted substances can entrap whiteflies killing them or may have repellent or toxic properties (Kisha, 1981; Freitas et al., 2002; Muigai et al., 2002; Fancelli et al., 2003; Antonious and Snyder, 2008). Secondary metabolites secreted from the trichomes of wild tomatoes that have demonstrated resistant properties against whiteflies include methyl ketones, sesquiterpene hydrocarbons, and glucolipids (Antonious and Snyder, 2008).

Different species and varieties of undomesticated tomato have contributed to whitefly-resistant *S. lycopersicum* breeding efforts. Resistance to *T. vaporariorum* has been demonstrated in *Solanum pennellii* Correll D'Arcy due to its production of sugar esters (glucolipids) from type IV trichomes (Rodriguez et al., 1993). Acyl sugars are also implicated in resistance against *B. tabaci*. *S. pennellii*-derived acyl sugars have been shown to reduce oviposition by *B. tabaci* with several varieties showing resistance; LA716 (containing the highest acyl sugar content) showed the highest (Nombela et al., 2000; Freitas et al., 2002; Muigai et al., 2003; De Resende et al., 2009; Oriani and Vendramim, 2010). Studies have also demonstrated the potential for foliar application of acyl sugars as a pesticide against whiteflies (Leckie et al., 2016). Crosses of *S. lycopersicum* with *S. pennellii* that have superior fruit production along with high whitefly resistance have been challenging but potentially achievable (Dias et al., 2016), particularly with the acyl sugar quantitative trait loci (QTL) identified (Leckie et al., 2012). This cross confers resistance linked to repellence by sesquiterpenes, including zingiberene and curcumene, and monoterpenes, including p-cymene, alpha-terpiniene, and alpha-phellandrene (Bleeker et al., 2009).

Other wild tomato species including *Solanum galapagense* S.C. Darwin & Peralta (Firdaus et al., 2012) and *Solanum hirsutum* (=*habrochaites*) Dunal (Heinz and Zalom, 1995; Muigai et al., 2003) have also demonstrated resistance properties. Compounds such as zingiberene (a sesquiterpene) produced by *S. hirsutum* var. *hirsutum* (not produced by commercial tomato cultivars (Bleeker et al., 2012)) show reduced oviposition and nymphs of *B. tabaci* (Lima et al., 2016). QTL has been identified in *S. hirsutum* that reduce oviposition in both *B. tabaci* and *T. vaporariorum* (Maliepaard et al., 1995; Lucatti et al., 2014). Some methyl ketones, derived from *Solanum glabratum* Dunal, are toxic to *B. tabaci*, showing significant mortality rates at certain concentrations (Antonious and Snyder, 2008). The discovery, manipulation, and testing of trichome exudates and their contributions to tomato resistance against whiteflies has been extensive because of its promise and continuation in advancing host-plant resistance research.

Plant defenses against attack by pests and pathogens follow two courses of action, which are triggered differentially based on the location of first detection by plant cells. When pests or their products are detected by cellular plasma membrane receptors, or pattern recognition receptors (PRRs), pattern-triggered immunity (PTI) is activated (Kaloshian and Walling, 2015). When plant intracellular receptors recognize pest virulence products (effectors), effector-triggered immunity (ETI) is activated (Dodds and Rathjen, 2010). These plant immunity pathways have been studied and continue to be elucidated primarily via research with plant pathogens; however, these pathways have also been implicated in plant defense response to hemipteran (piercing-sucking) insects, including whiteflies.

In tomato, a broad-spectrum resistance (R) gene, *Mi*, which encodes a coiled-coil, nucleotide-binding, leucine-rich repeat receptor, has been discovered (Roberts and Thomason, 1986) and cloned (Milligan et al., 1998). First discovered for its resistance against nematodes, *Mi-1.2* has since been shown to confer resistance to *B. tabaci* (Nombela et al., 2001, 2003b) in the form of reduced whitefly infestation, oviposition, and subsequent number of fourth instar nymphs developing on plants carrying the gene. Further study on this R gene has identified a specific tomato locus, *Rme1* that is required for *Mi-1*-mediated resistance (Nombela et al., 2003a; Martinez de Ilarduya et al., 2004). Questions remain about *Mi-1* resistance against *T. vaporariorum*. Lucatti et al. (2010) found that greenhouse whitefly resistance in tomato is independent of the REX-1 marker (Williamson et al., 1994), which is used for detection of the *Mi-1* gene in tomato lines. Evidences for a post-whitefly stylet penetration, pre-phloem ingestion defense mechanism, as well as a phloem-located resistance factor in wild tomato species against *T. vaporariorum* has been demonstrated (McDaniel et al., 2016), but *Mi-1*-mediated resistance has not yet been concluded for *T. vaporariorum*.

Plant innate immunity against pests and pathogens involves complex signaling networks controlled by salicylic acid (SA), jasmonic acid (JA), ethylene (ET), and interactions with other pathways (Kaloshian and Walling, 2005). Whitefly secreted compounds have been shown to interact with these pathways as a means of manipulating or thwarting plant defense response. As with pathogen response following plant-triggered defenses, hemipteran salivary products act as effectors that alter plant immune responses. Whitefly effectors delivered into tomato and other host plants, including one originating from an endosymbiont of *B. tabaci*, *Hamiltonella defensa* Moran et al. (Su et al., 2015), have been reviewed recently by Kaloshian and Walling (2015). Small RNAs that may act to manipulate plant defense have also been detected recently in the saliva of *B. tabaci* feeding on tomato and in the phloem of whitefly-infested leaves (van Kleeff et al., 2016). Our understanding of tomato defense responses and whitefly–tomato interactions deepens as research in this area continues.

4.4 Biological Control

Natural enemies of whiteflies include a wide variety of predators, parasitoids, and entomopathogens that have been the subjects of various reviews. For example, Vet et al. (1980) reviewed the biological control of *T. vaporariorum* noting that 25% of the glasshouses producers in the Netherlands were using *Encarsia formosa* Gahan for biological control as early as 1978. Ekbom (1981) listed nine predators that had been evaluated for control of *T. vaporariorum*. Gerling (1990a) reviewed the predators and parasitoids of whiteflies, noting 13 predators and 9 parasitoid species of *T. vaporariorum*, and 34 predators and 7 parasitoid species attacking *B. tabaci*. More recently, Gerling et al. (2001) reviewed the biological control of *B. tabaci* using predators and parasitoids, whereas Naranjo (2001) focused more specifically on conservation and evaluation of these natural enemies in IPM systems. Arnó et al. (2010) covered the biology and ecology, interactions, utilization, monitoring, and impact of natural enemies of *B. tabaci*. Most recently, Liu et al. (2015) reviewed the distribution, life history, bionomics, and utilization of whitefly parasitoids, noting that at least 115 species in 23 genera have been reported on *B. tabaci*. Faria and Wraight (2001) discussed fungal pathogens as biological control agents of *B. tabaci*. CABI (2017a) lists 24 predator species, 21 parasitoid species, and 10 pathogen species that have been reported on *T. vaporariorum* and 48 predators, 33 parasitoids, and 7 pathogens species that are known from *B. tabaci* (CABI, 2017b). In this section, we summarize current studies of natural enemies of *B. tabaci* and *T. vaporariorum* for biological control as a component of IPM.

Although there are many predators, parasitoids, and microbial agents that attack whiteflies, there are "few definitive examples of successful biological control" of whiteflies (Naranjo, 2001). Difficulties managing whiteflies with biological control in general may be exacerbated by difficulties achieving biological control of whiteflies on tomato in particular. Trichomes, glandular and non-glandular, can negatively impact the behavior and life-history parameters of predatory insects and mites, including survival, foraging, oviposition, and movement (Riddick and Simmons, 2014; Schmidt, 2014).

4.4.1 Predators

Of the more than 150 arthropod species currently described as predators of whiteflies, only a handful have been studied in detail, leaving many with limited laboratory observations or qualitative field records (Gerling et al., 2001; Arnó et al., 2010). Most of the known predators of whiteflies are ladybird beetles, predaceous bugs, lacewings, phytoseiid mites, and spiders.

4.4.1.1 Coccinellidae

Ladybird beetles are important natural enemies of whiteflies that may exhibit various degrees of oligophagy. About 40 species have been reported to prey on *B. tabaci* (Gerling et al., 2001; Arnó et al., 2010) and 10 species are known to prey on *T. vaporariorum* (Gerling, 1990a). Deligeorgidis et al. (2005) determined that *Coccinella septempunctata* (L.) provided good control of *T. vaporariorum* on tomato. They suggested a 1:30 predator:prey ratio to achieve success. They also found that this predator preferred *B. tabaci* over *T. vaporariorum*. Most other studies have focused on three beetle species: *Serangium parcesetosum* Sicard, *Delphastus catalinae* (Horn) (=*Delphastus pusillus* (LeConte)), and *Nephaspis oculatus* (Blatchley).

Larvae and adults of *S. parcesetosum* can prey on all whitefly immature stages but they prefer fourth instar nymphs (Al-Zyoud and Sengonca, 2004). Larvae developed normally within a temperature range of 18 and 30°C (Sengonca et al., 2004). Throughout the developmental periods they consumed approximately 1440 first–third instar nymphs and 250 fourth instar nymphs of *B. tabaci* at 18°C, or 970 first–third instar nymphs and 170 fourth instar nymphs at 30°C (Sengonca et al., 2005). Adults survived an average of 5–6 months at 18°C and 2–3 months at 30°C on *B. tabaci*. Al-Zyoud et al. (2005) reported that mean total fecundity of *S. parcesetosum* was greater on cucumber than on cotton (means of 97.7 and 31.0 eggs/female, respectively), although there were no significant host-plant effects on developmental time from egg to adult. However, longevity varied by host-plant species and sex, with a mean of 63.4 day for females and 50.3 day for males on cucumber compared to 92.4 day for females and 52.5 day for males on cotton. *S. parcesetosum* avoided *B. tabaci* parasitized by *En. formosa*, preferring to feed on non-parasitized prey (Al-Zyoud and Sengonca, 2004; Al-Zyoud, 2007). Al-Zyoud et al. (2007) found that a single pair of *S. parcesetosum* per plant released 1 or 2 weeks in an initial infestation of 50 *B. tabaci* adults per plant resulted in maximum reduction of 90% after 5 weeks.

D. catalinae is the most commonly used predacious natural enemy being commercially reared for controlling whiteflies on various ornamental and vegetable crops under greenhouse conditions. *D. catalinae* consumed immature stages of *B. tabaci*, especially eggs, and completed larval development between 22 and 30°C (Legaspi et al., 2008). Liu (2005) reported that the developmental time of *D. catalinae* from oviposition to adult emergence was 19.0 days for females and 18.8 days for males; females laid a mean of 5.6 eggs per day over a 97.0-day period. Simmons and Legaspi (2004, 2007) studied the life history of this predator. They determined the longevity was reduced from 174 days at 25°C to 16 days at either 5°C or 35°C. No eggs hatched when maintained at 5°C, while 48% hatched at 15°C, although none of the resulting larvae reached the pupal stage. In a field study, a few individuals were still able to survive during winter when temperatures dropped to –8°C. *D. catalinae* females consumed more than 150 whitefly eggs or nymphs per day, and each larva ate nearly 1000 eggs during a 2 week developmental period. They were also found to discriminate between parasitized and nonparasitized whitefly nymphs and avoided feeding on the parasitized nymphs. Simmons et al. (2008) determined a negative effect of low relative humidity on oviposition, egg hatching, and immature survival. Liu and Stansly (2004) found that pyriproxyfen and buprofezin sprayed on *B tabaci* eggs reduced *D. catalinae* egg fertility although the process was largely reversible following removal of treated eggs.

N. oculatus has shown good potential for biological control of *B. tabaci*, especially in greenhouses (Liu et al., 1997). *N. oculatus* consumed fewer whiteflies but had a more efficient searching behavior than *D. catalinae* (Liu and Stansly, 1999; Ren et al., 2002; Huang et al., 2006; Legaspi et al., 2006). *N. oculatus* consumed immature stages of *B. tabaci*, especially eggs, and completed development between 20 and 33°C (Ren et al., 2002). At 26.7°C, 55% RH, and photoperiod of 14:10 (L:D), *N. oculatus* adult longevity averaged 56.1 days for males and 67.5 days for females. After an average 11.3 day pre-oviposition period, females laid a mean of 3.03 eggs per day. Developmental time was 19.4 days from oviposition to adult emergence for females and 18.3 days for males. Male–female pairs of adults consumed 184.1 eggs per day over a period of 16 weeks. Net reproductive rate was 54.27, generation time was 51.27 days, and doubling time was 8.89 days. The intrinsic rate of population increase was estimated at 0.078 (Ren et al., 2002).

4.4.1.2 Heteroptera

12 Anthocoridae, 2 Berytidae, 3 Lygaeidae, 18 Miridae, 1 Nabidae, and 2 Reduviidae species have been recorded as whitefly predators (Gerling et al., 2001; Arnó et al., 2010). *Macrolophus pygmaeus* (Rambur) and *Nesidiocoris tenuis* (Reuter) are widely used as biological control agents of *B. tabaci*, and *Dicyphus hesperus* Knight is being used for control of *T. vaporariorum* (Lambert et al., 2005) and considered for control of *B. tabaci* (Calvo et al., 2016). In their study, Calvo et al. (2016) showed no plant damage by this predator and suggested augmentative releases to help improve biological control. *Tupiocoris cucurbitaceus* (Spinola) also has been shown to survive, develop, and reproduce on *T. vaporariorum* (Noemi Lopez et al., 2012), and it also shows promise for control of *B. tabaci* (Orozco Munoz et al., 2012).

M. pygmaeus attacks all pre-imaginal stages of *T. vaporariorum* (Mohd Rasdi et al., 2009) but prefers older nymphs of *B. tabaci* (Bonato et al., 2006). Prey consumption of *B. tabaci* averaged 5.94 per day regardless of stage, and adult longevity averaged 33.7 days. For *T. vaporariorum*, this predator can consume 94 eggs and 56.5, 2.4, and 11.8 nymphs in the second, third, and fourth nymphal instars, respectively (Lykouressis et al., 2009). This predator can cause damage to gerbera flowers and may reduce tomato fruit set on some cultivars when prey is limited. Another negative is that this species can interfere with biological control by consuming whitefly parasitoids (Moreno Ripoll et al., 2012b). *M. pygmaeus* can be successfully reared on a meat diet supplemented with potato sprouts for nymphs and as oviposition substrate for adults (Castane and Zapata, 2005). *N. tenuis* is another mirid species, endemic in Mediterranean regions, which has been shown to make a significant natural contribution to the control of pests such as whiteflies, leafminers, thrips, and spider mites within that region (Ryckewaert and Alauzet, 2002; Urbaneja et al., 2005). *N. tenuis* is also zoophytophagous and can feed on certain host plants (e.g., tomato), causing feeding punctures leading to possible infection and flower abortion when prey is scarce or absent. Intraspecific evaluations using both *M. pygmaeus* and *N. tenuis* found no intra-guild predation of these two predators (Moreno-Ripoll et al., 2014).

In vegetable greenhouses, success in using mirids for biological control of *B. tabaci* includes the inoculative and augmentative release of *M. pygmaeus* and *N. tenuis*. Adults of *M. pygmaeus* were released at two rates (2 and 6 individuals/ plant) in an initial infestation of 10 *B. tabaci* adults per plant, and the results showed that the high release rate could control whitefly populations (Alomar et al., 2006). Calvo et al. (2009b) compared two different release rates of *N. tenuis* (1 and 4 individuals/plant) in large exclusion cages and found significant reduction of *B. tabaci* populations (>90%) with both release rates. However, bug feeding can weaken the apex and arrest plant growth. Calvo et al. (2012a,b) evaluated different release rates (0.5, 1, and 2 individuals/plant in winter as well as 0.5 and 1 individuals/plant) for tomato greenhouse application of *N. tenuis* and found significant reduction of the *B. tabaci* populations with all release rates. However, only the 0.5 *N. tenuis*/plant rate did not result in appreciable crop damage.

Predation on *B. tabaci* was somewhat compromised where *M. pygmaeus* and *N. tenuis* coexisted with each other on tomato plants but was improved when the parasitoid *Eretmocerus mundus* Mercet was included in the mix (Moreno-Ripoll et al., 2012a). This result supported previous reports that the combined use of the parasitoid and predator provided better results than the use of any single natural enemy, especially in spring when whitefly populations were very high in an experimental tomato greenhouse (Gabarra et al., 2006; Trottin-Caudal et al., 2006). Figuls et al. (1999) found that chlorpyrifos, formetanate, and methamidophos were toxic to third and fourth instar mirids for 30 days, whereas foliar applications of imidacloprid and endosulfan became harmless by 21 days after treatment. There were no effects on the reproductive capacity of females that were exposed as nymphs to these insecticides.

In addition to these hemipterans, several species of *Orius* have been evaluated for whitefly control. Arnó et al. (2008) reported that *Orius laevigatus* (Fieber) and *Orius majusculus* (Reuter), both known as thrips predators, were able to feed on whitefly eggs, nymphs, and adults, and that they completed their respective preimaginal development with *B. tabaci* nymphs as their only available prey.

4.4.1.3 Neuroptera

Twenty one Chrysopidae and two Coniopterygidae species have been recorded as *B. tabaci* predators (Gerling et al., 2001; Arnó et al., 2010). However, only *Chrysoperla carnea* (Stephens) and *Chrysoperla rufilabris* (Burmeister) have wide commercial availability and use as biological control agents of *B. tabaci*. Syed et al. (2005) determined that *B. tabaci* was better prey for *C. carnea* than the cotton leafhopper *Amrasca devastans* (Dist.). Breene et al. (1992) evaluated first and second instar *C. rufilabris* larvae against *B. tabaci* on *Hibiscus rosa-sinensia* L. in the greenhouse and found that 25 or 50 larvae per plant at 2 week intervals maintained marketability. Legaspi et al. (1996) found that *C. carnea* and *C. rufilabris* fed on *B. tabaci* had faster development, increased survival, and greater weight on cucumbers and cantaloupes compared to poinsettia and lima bean.

4.4.1.4 Phytoseiidae Predatory Mites

Seventeen Phytoseidae and one Stigmaeidae species have been recorded as whitefly predators (Gerling et al., 2001; Arnó et al., 2010). *Amblyseius swirskii* Athias-Henriot and *Euseius scutalis* (Athias-Henriot) appear to be promising biological control agents against *B. tabaci* based on their high intrinsic rates of increase (Nomikou et al., 2001). They feed mainly on whitefly eggs and nymphs. Immature mites needed 15–20 whitefly eggs or nymphs for development, and adults consumed up to three eggs or two nymphs daily. Enhanced survival, development, and reproduction was seen when they also fed on pollen and honeydew (Nomikou et al., 2002, 2003). Both mite species were able to suppress *B. tabaci* populations on cucumber (Nomikou et al., 2001, 2002), and thus may have utility in tomato. Polyphagy and the ability to feed on alternative foods were found to promote persistence in the crop, even if *B. tabaci* was scarce, enabling the inoculative release of mites before pest colonization (Nomikou et al., 2002, 2004; Messelink et al., 2006). Other phytoseiids, *Amblyseius barkeri* Hughes and *Amblyseius cucumeris* (Oudemans) also prey on whiteflies. Cheng et al. (2014) showed release of both predators resulted in good control of *B. tabaci* on greenhouse tomatoes. Control lasted for 50 days after initial release which was better than pesticide-treated plots (Cheng et al., 2014).

4.4.2 Parasitoids

To date more than 500 species of parasitoids in 23 genera within six families have been described for whiteflies (Evans, 2007; Noyes, 2012; Liu et al., 2015). While whitefly parasitoids represent six families of Hymenoptera, the majority are from the genera *Encarsia* and *Eretmocerus* (Aphylinidae). *Encarsia* species are endoparasitic, whereas eggs of *Eretmocerus* are laid under the host. However, the first instar larva penetrates the host, later emerging from the fourth instar nymph. *En. formosa*, *Er. mundus* and *Eretmocerus eremicus* Rose, and Zolnerowich have been widely used for biological control of *B. tabaci*, and *En. formosa* has been the parasitoid most used for *T. vaporariorum*.

En. formosa is a thelytokous parthenogenetic and synovigenic species that has been used and studied around the world for control of *T. vaporariorum* under greenhouse conditions (Hoddle et al., 1998). Studies relating to *B. tabaci* were inspired by outbreaks in European glasshouses (Drost et al., 1996). *En. formosa* oviposits in any of the four whitefly nymphal instars but develops mainly in the fourth from which they emerge as adults. The first instar nymph is the least suitable stage, resulting in highest parasitoid mortality and longest immature development. Larvae are elongate and caudate in the first instar with stout, prominent tails (Gerling, 1966), whereas second and third instars of both sexes are hymenopteriform. *En. formosa* host feeds on all nymphal stages of whitefly, although more frequently on second and late fourth instars than on first or third instars.

En. formosa is the most widely used insect for biological control of greenhouse whitefly in vegetable crops including tomatoes. Both inundative and seasonal inoculative release methods have been utilized to control *T. vaporariorum* with *En. formosa* for a long time (Wilson, 1931; Scopes and Biggerstaff, 1971; Liu et al., 2015). Van Lenteren et al. (1996) described the searching behavior of this parasitoid, noting that on some plants a single parasitoid and her offspring are able to kill more whiteflies per unit of time than a single whitefly female can produce. However, on certain plants the whitefly reproduces so quickly that the parasitoid cannot keep up. Hoddle et al. (1997a,b) achieved better control of *B. tabaci* on poinsettia with a low inundative release rate of *En. formosa* (one compared to three wasps/plant/week). Studies on demographic parameters with *B. tabaci* as host have shown lower reproductive rates for *En. formosa* than the more proovigenic *Eretmocerus* spp., especially at high temperatures favored by *B. tabaci* (Qiu et al., 2004; Arnó et al., 2010). *Eretmocerus* spp. also are able to locate patches of the whitefly more quickly (Hoddle et al., 1997b). Therefore, interest has turned to *Eretmocerus* species, in particular *Er. eremicus* and *Er. mundus*, for control of *B. tabaci* (Hoddle, 2004; Stansly et al., 2004, 2005a,b).

Er. mundus and *Er. eremicus* parasitize all immature instars of *B. tabaci* but especially the second and third. *Er. eremicus* also attacks *T. vaporaiorum*. The first instar larva penetrates into the host from underneath during the early fourth nymphal stadium (Gelman et al., 2005). The second and third instar nymphs are preferred for oviposition, and development is delayed when the first instar nymph is attacked. First instars are pear-shaped, becoming globular after penetrating the host. The globular shape is retained by second and third instars with an indentation in the oral area (Gerling and Blackburn, 2013; Gerling et al., 1991). *Eretmocerus* females host-feed by penetrating the host's vasiform orifice with their spatulate ovipositor (Gerling et al., 1998). Demographic parameters generally favor *Er. mundus* over *Er. eremicus* and *En. formosa* for control of *B. tabaci* (Arnó et al., 2010).

The release rates are also important for *Er. mundus* success in greenhouse tomatoes. Rates of six parasitoids per m² resulted in 92% reduction in *B. tabaci* (Stansly et al., 2005b). The ovipositing *Er. mundus* female deposits methyl-branched cuticular hydrocarbons (C31 and C33 dimethyl alkanes) on *B. tabaci* nymphs (Buckner and Jones, 2005). The host marking is relatively nonvolatile and enables conspecific females to discriminate parasitized from nonparasitized hosts through antennation. *Er. eremicus* also avoided these marked hosts, although *Er. mundus* did not respect hosts marked by *Er. eremicus*.

Er. eremicus attacks both *B. tabaci* and *T. vaporariorum* with apparently equal facility (Stansly and Natwick, 2010). Thus, it is especially useful for controlling mixed infestations of the two whiteflies. It has been used to control pure infestations of *B. tabaci*, albeit with limited success. *Er. mundus* became the dominant species in greenhouses in Spain where it was released along with *Er. eremicus*. Although immigration from outside the greenhouse partly explained this dominance, behavioral traits such as inclination to parasitize hosts parasitized by the other species may have assisted *Er. mundus* in competition with *Er. eremicus*. Similarly, introduced Old World species of *Eretmocerus* gained over native *Eretmocerus* spp. following invasion of exotic Old World *Bemisia* spp. in the southern United States and Australia.

As the combination of *A. swirskii* and *Er. mundus* can provide a significantly better suppression of *B. tabaci* than *Er. mundus* alone, this combination should be the strategy advised for commercial sweet pepper greenhouses (Calvo et al., 2009a). Control in tomato was shown to be possible, although higher release rates were required to obtain the same level of control as in pepper (Stansly et al., 2005b). While *Er. eremicus* seemed less well adapted to controlling *B. tabaci* in Spanish greenhouses (Stansly et al., 2004, 2005a), *Er. mundus* has limited commercial availability due to sterility of unknown etiology affecting both males and females in mass rearing facilities (Chiel et al., 2012).

For decades, workers have utilized one of the three methods for introducing parasitoids into greenhouses for control of greenhouse whiteflies. Parr et al. (1976) introduced the "pest in first" method, which called for establishing a low, evenly distributed whitefly population on which introduced parasitoids could establish. While this makes sense to the biological control specialist who wants to ensure establishment of the parasitoid, it is not favored by growers who are reluctant to have whiteflies established in their crop. This method also requires repeated introductions of the parasitoids. A second method, the "banker plant method," involves placing potted plants with parasitized whiteflies throughout the greenhouse (explained in more detail in Section 4.4.4) (Stacey, 1977). For this method, growers should not use fewer than 90 infested plants per hectare. This method has the advantage of easy implementation and a single introduction is all that is needed for season-long control. A third method "introduction after the pest is seen" (also known as the "multiple introduction," "dribble method," or "Dutch method") was developed by Woets (1978). This method relies on scouting to detect whiteflies and once they are found, releases need to be made. Multiple releases are typically required to keep the whitefly population in check. The rate of introduction is approximately two parasitoids per plant (Woets, 1978). Eggenkamp-Rotteveel Mansfeld et al. (1982) evaluated the Dutch methods with the parasitoid *En. formosa* and determined that an even distribution of *T. vaporariorum* was not essential for success of this method, which provided satisfactory control.

4.4.3 Pathogens

Most reports of entomopathogenic fungi of whiteflies refer to species of *Isaria* (*Paecilomyces*), *Lecanicillium* (formally *Verticillium*), *Beauveria*, and *Aschersonia*. *Beauveria bassiana* (Balsamo-Crivelli) Vuillemin, *Isaria fumosorosea* Wize (=*Paecilomyces fumosoroseus* (Wize) A.H.S. Br. & G. Sm.), and *Lecanicillium lecanii* Zare and Gams are generalist entomopathogenic fungi that infect *B. tabaci*. Studies on *T. vaporariorum* have shown *Aschersonia* sp. (Spasova et al., 1980),

Aschersonia aleyrodis Webber (Fransen et al., 1987), and *L. lecanii* (Lee et al., 2002) to be effective. These fungi require high humidity for infection and development typical of greenhouse environments (Faria and Wraight, 2001).

Second and third instar nymphs are highly susceptible to *B. bassiana* and *I. fumosorosea*. Conidia of *B. bassiana* germinate most readily on the cuticle of second instars (54% germinated), and *I. fumosorosea* germination is highest on third instar cuticle (45%). Fourth instars have low susceptibility to these pathogens, and spore germination on the cuticle of fourth instars is very low for *B. bassiana* (7%) and intermediate for *I. fumosorosea* (33%) (James et al., 2003). More recent studies have shown the highest production in cherry tomatoes when plants were treated with *I. fumosorosea* and *L. lecanii* (Espinel et al., 2008). *I. fumosorosea* also has epizootic potential against *B. tabaci* adults under favorable conditions. Different strains of these pathogens have different impacts on whitefly mortality (Junaid et al., 2016).

L. lecanii was shown to have a high pathogenicity to all developmental stages of *B. tabaci*; however, the second instar nymphs were shown to be most susceptible to infection. Pathogenicity of *L. lecanii* involves adhesion of spores to the insect cuticle, germination, penetration, and internal colonization culminating in host death. *L. lecanii* produces secondary metabolites with insecticidal properties during colonization of the host tissue, which may play an important role in host mortality (Wang et al., 2007). Cuthbertson et al. (2005b) reported that *L. lecanii* can significantly reduce *B. tabaci* survival on tomato and verbena foliage held at high humidity and that second instars were generally most susceptible to infection. Certain strains of this pathogen were also shown to be effective for controlling *T. vaporariorum* (Lee et al., 2002; Koike et al., 2004).

The entomopathogenic fungi *B. bassiana*, *I. fumosorosea*, and *L. lecanii* have been commercially developed as mycopesticides to control *B. tabaci* (Faria and Wraight, 2001). *I. fumosorosea* significantly reduced *B. tabaci* populations on ornamentals when applied weekly. Under normal greenhouse conditions, infection was detectable 7–10 days after application. However, many of the whiteflies killed by the fungus did not show typical coloration unless placed in humid conditions to allow sporulation. *I. fumosorosea* was compatible with *Encarsia*, *Eretmocerus*, and *D. pusillus*, and was tolerant of some fungicides. *B. tabaci* can be suppressed for several months following a single application of *L. lecanii* if the temperature remains between 15 and 25°C and the humidity is greater than 90% for at least 10 h/day. *L. lecanii* can kill 80–97% of nymphs, and subsequent infection kills many adults emerging from surviving nymphs. However, it is also pathogenic to *Encarsia* when applied directly, but whiteflies are even more susceptible, so the fungus and wasp can coexist.

Entomopathogenic fungi are easy to apply, although good coverage is required on the abaxial (lower) foliar surfaces where whiteflies reside. Electrostatic sprayers have been evaluated to improve coverage, and this application was shown not to have a negative impact on the fungus *P. fumosorosea* (Saito, 2005). Some fungi also have been shown to provide additive control when applied in combination with insecticides (Cuthbertson et al., 2005a; Borisade, 2015) or the photoactive dye Phloxine B (Kim et al., 2010). Entomopathogenic fungi present essentially no risk to human health, and most studies show that they are relatively innocuous to other natural enemies (Goettel et al., 2001; Vestergaard et al., 2003; Zimmermann, 2008). However, fungi are slow acting compared to chemical insecticides, exhibit poor adulticidal activity, and are incompatible with many commonly used fungicides. In addition, they are relatively expensive, have limited shelf life, and are dependent on favorable environmental conditions (Inglis et al., 2001; Faria and Wraight, 2001; Vidal et al., 2003; Fargues et al., 2005; Saito, 2005; Rivas et al., 2014).

A final pathogen to discuss is the entomopathogenic nematode *Steinernema feltiae* (Filipjev) which has shown promise for *B. tabaci* control in the greenhouse (Cuthbertson et al., 2007). These authors evaluated the nematodes in the formulation Nemasys on five host plants, including tomato. The infested plants were sprayed with the nematodes, and after 72 h the mortality of whiteflies was 80%. Similar to entomopathogenic fungi, nematodes require specific environmental conditions, particularly high humidity (Cuthbertson et al., 2007) to thrive and provide control.

4.4.4 Application of Biological Control for Whiteflies on Tomato

The trend in whitefly management is toward greater use of non-insecticide based control methods involving deployment of natural enemies. In fact, progress on the control of whiteflies with natural enemies has been outstanding and will continue to be used on a more regular basis. However, although large numbers of arthropod predators, parasitoids, and fungal species are known to attack whiteflies in a variety of agricultural systems worldwide, few have been studied and evaluated for use in IPM programs. Thus, much work will be required before biological control assumes a more dominant role in pest management systems for *B. tabaci* and *T. vaporariorum* in affected crops. More comparative studies on pathogens (Soliman and Tarasco, 2008), parasitoids (Pang et al., 2011), and predators (Lambert et al., 2005) as well as intra-guild impacts (Gillespie et al., 2006; Bao Fundora et al., 2016) need to be conducted.

In general, natural enemies are most effective when introduced early and whitefly populations are low. Otherwise, non-residual sprays such as insecticidal soap, horticultural oil, natural pyrethrins, or insect growth regulators (IGRs) can be used before releasing natural enemies. Biological control is becoming a cost-effective standard for greenhouse vegetable production, especially in Europe but also in Canada, the United States, and elsewhere. This is in response to increasing public sentiment against pesticides, which in turn is driving development of improved pest exclusion techniques, virus-resistant cultivars, more affordable and increased natural enemy options, and better quality control. There remains vast potential for expansion in biological control of whiteflies in open field and greenhouse-grown tomatoes.

An interesting strategy for applying entomopathogens was reported by Kapongo et al. (2008a,b). They used bumblebees to transmit *B. bassiana* and *Clonostachys rosea* (Link) Schroers, Samuels, Seifert, and W. Gams to whiteflies on tomatoes in a greenhouse. In their studies, the pathogens resulted in 49% control of *T. vaporariorum*. While this is not adequate control on its own, it may contribute to other practices in an IPM program. Another novel idea combined an application of *Pseudomonas* sp. with chitosan and reduced disease severity of ToLCV (Shefali et al., 2014).

Another strategy gaining momentum that not only is compatible with biological control but also enhances success is the use of banker plants. Broadly defined, banker plants are plants added to a crop system to provide a population of prey species that supports and maintains a constant population of natural enemies as the target crop continues to grow. There are several examples of banker plants used for management of whiteflies in tomato. Stacey (1977) used tomato plants infested with parasitized *T. vaporariorum* to initiate and maintain a population of *En. formosa* in greenhouse tomatoes. This strategy was has also been to control *B. tabaci* in greenhouse tomatoes by releasing the predator *N. tenuis* onto banker plants (Nakano et al., 2016). Xiao et al. (2011) used papaya infested with another whitefly, *Trialeurodes variabilis* (Quaintance), to raise and maintain the parasitoid *Encarsia sophia* (Girault & Dodd) for control of MEAM1 whiteflies on greenhouse tomatoes. Mullein (*Verbascum thapsus* L.) is a banker plant used to establish the whitefly predator *D. hesperus* in greenhouses to enhance predation of *T. vaporariorum* on tomato (Sanchez et al., 2003). *D. hesperus*, similar to other predatory mirids used for biological control of whiteflies, will feed on plants such as tomato when there is a paucity of prey. Sanchez et al. (2003) and Nguyen-Dang et al. (2016) found that *D. hesperus* was more abundant in greenhouses with mullein than in houses only containing tomato and that mullein sustained significant numbers of *D. hesperus* when whitefly densities on tomato were low. While there have been some failures using banker plants to enhance biological control of whiteflies (Parolin et al., 2013, 2015), the identified successes suggest that banker plants could improve biological control.

4.5 Chemical Control

Chemical insecticides provide a major control strategy in the management of whiteflies. Historically, there was more reliance on insecticides, as the discovery of other IPM components such as natural enemies and plant resistance were rare. Very early work for greenhouse whiteflies used fumigation materials such as hydrocyanic acid gas (Lloyd, 1922; Speyer and Owen, 1926; Miles, 1927), calcium cyanide (Miles, 1927), and tetrachloroethane (Parker, 1928) with recommendations to treat at night and to open the greenhouse at dawn to minimize the risk to workers. During the 1940s and 1950s, additional research on pesticides was conducted including nicotine fumigation (Richardson et al., 1943), DDT, and hexaethyl tetraphosphate aerosols (Smith et al., 1947). Additional research was conducted through the 1960s and 1970s with organophosphate materials such as Dibrom (Condron et al., 1962; Adamson et al., 1972). Azodrin, Guthion, and Parathion were sprayed weekly for 3 weeks, which controlled *T. vaporariorum* (Smith, 1970). These materials were evaluated as plant sprays and as aerosols and fumigants in the greenhouse. In addition, carbamates such as aldicarb (Lindquist et al., 1972b) and methomyl (Krueger et al., 1973) were introduced into greenhouse whitefly management (Lindquist et al., 1972b).

Heavy reliance on insecticides is still present in tomato, especially in open field crops. With *B. tabaci*, the older broad-spectrum insecticides (organophosphates and pyrethroids) have not proved very useful except perhaps in combination. In contrast, the systemic neonicotinoids have been key players since their advent in the mid-1990s, particularly those that are relatively stable in the soil and effectively absorbed through the root system. Other modes of action widely used as foliar sprays primarily against nymphs include the IGRs pyriproxyfen and buprofezin, Insecticide Resistance Action Committee (IRAC) mode of action (MoA) 23 products (spiromesifen, spirotetramat) and the anthranilic diamides, cyantraniliprole, and chlorantraniliprole (IRAC MoA28). Oils, soaps, and detergents also have been used widely for whitefly control.

4.5.1 Pyrethroids and Organophosphates (IRAC MoA 3A and 1B)

These older chemistries have long histories of use against whiteflies, especially *B. tabaci* on cotton. Not surprisingly, resistance against both has been equally widespread, with resistance ratios sometimes well above 1000 (Denholm et al., 1996). One way of dealing with this issue has been to use mixtures, some but not all of which are synergistic (Ishaaya et al., 1987).

For instance, acephate and methamidophos were potentiating with bifenthrin and fenpropathrin although acephate was antagonistic with cypermethrin (Ahmad, 2007). However, the acephate-fenpropatherine mixture can be extremely active against *B. tabaci*, even when the individual products are almost inert (Denholm et al., 1996). These authors hypothesized that the synergism was most likely based on inhibition of pyrethroid-hydrolyzing esterases by OPs.

4.5.2 Nicotinic Receptor Agonists (IRAC MoA 4)

Nicotine (MoA 4B) derived from tobacco was the first insecticide used in this class (Ujváry, 1999), but this material was accompanied with high mammalian toxicity ($LD_{50}=3$ mg/kg in mice) (Okamoto et al., 1994). The neonicotinoids (MoA 4A) are more toxic to insects because they bind more tightly to insect neuroreceptors (Tomizawa and Casida, 1999). Imidacloprid was the first synthetic member of this group, eventually becoming the most widely used insecticide worldwide (Yamamoto, 1999). Its advent revolutionized whitefly control in many crops, including tomato, where dramatic declines in densities of *B. tabaci* were documented (Stansly, 1996). Through reducing whiteflies, imidacloprid has also been shown to reduce TIR (Powell and Stoffella, 1998). It has been shown that the effectiveness of imidacloprid can be enhanced with fertilizers (Sun and Liu, 2016). Other neonicotinoids, including thiamethoxam, dinotefuron, and acetamiprid, soon followed and have been shown to be effective toward *B. tabaci* (Esashika et al., 2016) and *T. vaporariorum* (Sachin et al., 2016). Those recommended for soil application (all but acetamiprid) are xylem mobile and translocated efficiently from roots to emerging leaf tissue where whitefly reproduction mostly occurs (Nauen et al., 1999). At the same time, residue concentrations are low in fruit which mainly utilize phloem as their primary source of water and in which transpiration rates are negligible compared to leaves (Juraske et al., 2009). Soil application is the most efficient method of delivery and also minimizes exposure to beneficial fauna (Smith and Krischik, 1999; Prabhaker et al., 2011), workers, and consumers. Newer nicotinic receptor agonists with different chemistries include sulfoxaflor (MoA 4C) applied as a foliar spray and flupyradifurone (MoA 4D) applied both to foliage and soil.

4.5.3 Insect Growth Regulators

Pyriproxyfen, a juvenile hormone mimic (IRAC MoA 7C), and buprofezin, an inhibitor of chitin biosynthesis (IRAC MoA 16) are the two IGRs used most for whitefly control. Activity is primarily against nymphs, although they are also active against the adult female through suppression of embryogensis (Ishaaya and Horowitz, 1992). These products have been useful in many crops, but are most intensively used in cotton where resistance has been notable, especially to pyriproxyfen in the MED group of *B. tabaci* in Israel (Horowitz et al., 2003).

4.5.4 IRAC MoA 23

Spiromesifen and spirotetramat are classed as tetronic acid derivatives that cause reduction in lipid biosynthesis by targeting the acetyl-coenzyme A carboxylase. Both are effective against mites and whiteflies as well as a wider range of sucking insects in the case of spirotetramat, thanks to both xylem and phloem mobility (Nauen et al., 2008). Both products are used effectively to control whitefly nymphs in tomato and other vegetable pests, although resistance to spiromesifen has already been described in *T. vaporariorum* (Karatolos et al., 2012). Both are applied as foliar sprays rather than soil treatments, despite the fact that spirotetramat has systemic activity. The additional effectiveness against mites is an advantage over the neonicotinoids which are known to exacerbate spider mite infestations (James and Price, 2002).

4.5.5 Cyantraniliprole (IRAC MoA 28)

This anthranilic diamide belongs to a class of insecticides that function by activating insect ryanodine receptors causing mortality from uncontrolled release of calcium ion stores in muscle cells (Selby et al., 2013). In contrast to other insecticides of this class, cyantraniliprole is effective against sucking insects due to greater xylem mobility and its inherent activity (Barry et al., 2015). It has shown excellent activity against *B. tabaci* on tomato in the field, both as foliar and soil application (Smith and Giurcanu, 2013; Stansly and Kostyk, 2013, 2016a,b; Barry et al., 2015) and against *T. vaporariorum* in the open field (Sachin et al., 2016). It is also effective against Lepidoptera and dipteran leafminers. Another apparent advantage is the ability to apply an entire seasonal rate to seedlings prior to transplanting (Stansly, unpublished data). Resistance to cyantraniliprole has not yet been reported, but might be expected soon because the closely related chlorantraniliprole has been used widely for control of Lepidoptera and dipteran leafminers in vegetable production during the last decade.

4.5.6 Soaps and Oils

Soaps and oils have been used as insecticides at least since the 18th century (Metcalf and Metcalf, 1993). Oils are thought to kill by suffocation, whereas soaps, detergents, and other surfactants are believed to kill by breaking the surface tension of water, which then fills the trachea and results in drowning. Thus both materials kill by blocking the spiracles, in one case with a thin film of oil and in the other by filling the spiracles with super-wetted water (Stansly et al., 1996). Oils are also known to reduce whitefly settling and may impact transmission of plant viruses (Schuster et al., 2009). While generally less effective than many synthetic products, these physical modes of action are unlikely to be subject to selection for resistance. In addition, they provide a degree of selectivity, targeting whiteflies with limited impact on their natural enemies (Butler et al., 1993; Stansly et al., 1996, 2002). These materials may also be more effective in combinations (Liu and Stansly, 2000). It is essential to get good coverage, especially in the case of oils. Highly refined horticultural mineral oils are characterized by narrow boiling point ranges and minimal concentrations of unsaturated ring compounds to limit phytotoxicity (Agnello, 2002). Soaps and household detergents can also be phytotoxic at high concentrations but are generally safe for frequent use on tomato at concentrations of 1% or less (Vavrina et al., 1995).

4.5.7 Botanicals and Other Materials

A variety of plant-based products have been evaluated for whitefly control. For example, compounds identified from crude extracts of tomato relatives *Solanum* (=*Lycopersicum*) *hirsutum* (=*Solanum habrochaites* S. Knapp & D.M. Spooner) f. *glabratum*, *S. habrochaites* f. *typicum*, and *S.* (=*Lycopersicum*) *pennellii* were evaluated for *B. tabaci* control (Antonious and Snyder, 2008). They found that the compound 2-tridecanone caused 72% mortality of whiteflies. Oil from the seeds of sugar apple, *Annona squamosa* L. were shown to cause MEAM1 whitefly nymphs to shrink and detach from the leaf surface (Singh et al., 2000; Lin et al., 2009). Another natural product, polygodial has been shown to repel *B. tabaci* adults (Prota et al., 2014). The authors caution that this material can cause moderate phytotoxicity in tomato. Extracts from milkweed, *Calotropis procera* (Aiton) W.T. Aiton and garlic, *Allium sativum* L., reduced oviposition by MEAM1 whiteflies by 56.6% and 56.5%, respectively, compared to non-treated controls (Barati et al., 2014). Neem-based products from *Azadirachta indica* A. Jussie have also been shown to negatively impact whiteflies (El-Shafie and Abdelraheem, 2012; Diabate et al., 2014; Chavan et al., 2015; Asmita and Ukey, 2016). Kumar and Poehling (2006a) found that soil drenching with NeemAzal, a commercial neem-based product, resulted in the best control and efficiency for *B. tabaci* control in tropical open field and greenhouse grown tomatoes. Similarly, Karanja et al. (2015) showed excellent control of *T. vaporariorum* nymphs on tomato with soil applications of neem. There is some evidence for synergism with fermented extracts of neem and wild garlic, *Tulbaghia violacea* Harvey (Nzanza and Mashela, 2012). They found a mixture of these two plant extracts were more effective at reducing *B. tabaci* densities than either extract alone. Fumigation with extracts from the fruit of chinaberry, *Melia azedarach* L., were shown by Palacios et al. (2008) to reduce *T. vaporariorum* in the greenhouse on cucumber which can also be considered for tomato. Extracts from *Jatropha curcas* L. were shown to reduce *B. tabaci* on open-field tomato equal to the reduction by using synthetic pesticides (Diabate et al., 2014) and extracts from *Ageratum conyzoides* L., *Plectranthus neochilus* Schltr., and *Tagetes erecta* L. were shown to be repellant to MEAM1 adults, resulting in reduced oviposition on tomato (Baldin et al., 2013). Other studies by Emilie et al. (2015) that showed *B. tabaci* were repelled by extracts from *Aframomum pruinosum* Gagnepain, *Cinnamomum zeylanicum* Presl, *Pelargonium graveolens* L'Héritier, *Anethum graveolens* L., *Cymbopogon winterianus* Jowitt ex Bor, *Litsea cubeba* (Lour.) Pers., and *Satureja montana* L. *P. graveolens* were also found to be effective by Baldin et al. (2015), particularly the vapor toxicity of the main chemical constituents geraniol, linalool, and citronellol. Additional essential oils of *Piper callosum* Ruiz & Pav., *Adenocalymma alliaceum* (Lam.) Miers, *P. graveolens*, and *P. neochilus* inhibited the settlement and oviposition of MEAM1 adults on tomato plants (Fanela et al., 2016).

In summary, synthetic and natural insecticides have served as effective tools for controlling whiteflies on tomato. A current list of materials currently approved for use in California (United States) is provided by UC-IPM (2016). Novel modes of action providing evermore safe, effective, and selective control continue to appear in the market place. Recent activity in the area of plant-based natural products is showing a variety of materials that have efficacy against whiteflies. Yet with these advances, costs of whitefly control with chemicals continue to rise, and the grace period free of resistance is typically short. These economic and biological considerations, as well as increasing public concern, continue to stimulate the search for pesticide alternatives to manage whiteflies and other pests of tomato and food crops in general.

4.6 Insecticide Resistance

Horowitz et al. (2007) reviewed the topic of insecticide resistance in *B. tabaci*. In this review, they discuss various resistance mechanisms that exist in whiteflies, show a number of bioassays used to evaluate resistance, and review the various studies that have been done to monitor for resistance in *B. tabaci* throughout the world. Furthermore, they present the known cases of resistance to various insecticidal classes and give strategies for delaying or reducing resistance. Other studies have also shown resistance of *T. vaporariorum* to neonicotinoid compounds (Pappas et al., 2013). McKenzie et al. (2014) emphasize the importance of rotating chemicals among different modes of action to minimize resistance in whiteflies.

4.7 Management of Viruses Transmitted by Whiteflies

B. tabaci transmits over 150 virus species, the majority of which are Begomoviruses (family Geminiviridae) (Lapidot and Polston, 2010). More than 35 of the described Begomoviruses infect tomato (Ji et al., 2007). Several Begomovirus species cause tomato yellow leaf curl disease (TYLCD), one of the most important diseases affecting tomato production (Moriones and Navas-Castillo, 2000). Viruses in the TYLCD group have a monopartite genome. Others, including TMoV, have bipartite genomes (Abhary et al., 2007).

The management of viruses transmitted by whiteflies is a broad topic which has been the subject of several reviews (Czosnek et al., 2010; Lapidot et al., 2014). Most whitefly-vectored viruses belong to the genus Begomovirus in the family Geminiviridae, characterized by a twin isohedral structure containing a monopartite or bipartite genome consisting of circular single-stranded DNA. All Begomoviruses are persistently transmitted by *B. tabaci* and many infect tomato. The introduction of different members of the *B. tabaci* species complex into new geographic areas can result in expanding disease problems (Pan et al., 2015). Perhaps the most devastating Begomovirus is the monopartite TYLCV group, first described in 1939 from Israel (Picó et al., 1996). Outbreaks of viruses in this group have been encountered worldwide.

Symptoms of TYLCV consist of stunting, small chlorotic leaves that roll inward, and flower abortion. Symptom expression begins in emerging buds above which few flowers set fruit. Consequently, yield loss is greatest early in the crop cycle. Given the potential severity of TYLCV, protection from whitefly attack may be the most important pest management task facing the grower or consultant. Because the virus is persistent in the whitefly vector, insecticides can provide effective management (Smith and Giurcanu, 2014). However, no thresholds have been proposed for this or any other plant virus vector. Such a threshold would have to be based not only on number of whiteflies per sample unit, but also the proportion carrying the virus and capable of transmitting the pathogen. Clearly, it is difficult to assess the number or proportion of viruliferous whiteflies coming into a tomato crop, although some idea can be gained by knowing the likely origin of immigrating whiteflies.

Mohamed (2010) reported a 67.5% decrease in tomato yield from plants inoculated at a young age with TYLCV from Egypt. Saikia and Muniyappa (1989) reported losses of up to 91% from TYLCV in southern India. This level of loss corresponded to that observed for tomato plants showing first symptoms of TYLCV 30 days after transplanting in the field (Stansly unpublished data). For TYLCV, there is a direct relationship between the number of days until first symptom expression and the yield loss (Fig. 4.5A). This linear relationship is the same for TMoV, but the slope of the relationship is much lower reflecting the lower impact of the TMoV disease on yield (Fig. 4.5B) (Stansly and Schuster, 1990).

Several sources of resistance against TYLCV are available in commercial cultivars that greatly reduce the potential damage from *B. tabaci*, thus increasing tolerance to whitefly attack (Ji et al., 2007; Pereira Carvalho et al., 2015; Elbaz et al., 2016). In Israel *Solanum peruvianum* L. has been used as source of resistance in commercially available varieties; the resistance is partially dominant and involves at least three genes (Friedmann et al., 1998). Resistance genes from *Solanum chilense* (Dunal) Reiche have been used to produce varieties resistant to both TYLCV and ToMoV (Scott et al., 1996). The gene *Ty-1*, which confers partial dominant resistance, has been identified from *S. chilense*. The *Ty-2* gene, which confers tolerance to several stains of TYLCV, has been identified from *S. hirsutum* and incorporated into several breeding lines.

Studies on "pyramiding" multiple genes have also shown improved performance (Prasanna et al., 2015), as have the use of RNAi technologies (Leibman et al., 2015; Chen et al., 2016). Nevertheless, growers often opt for susceptible varieties due to preference for particular horticultural characteristics (Ozores-Hampton et al., 2013). Begomoviruses are diverse and impact tomato production in environmentally diverse production areas globally. Environment influences the degree of tolerance exhibited by resistant varieties, which are more effective against some strains of Begomovirus than others. The result is that resistant varieties may be highly effective in some regions but susceptible to virus in other regions (Ji et al., 2007). In addition, like other begamoviruses, TYLCV has a relatively narrow host range, and cultural controls such as field sanitation, crop rotation, and especially host-free periods are effective in reducing primary spread (Hilje et al., 2001).

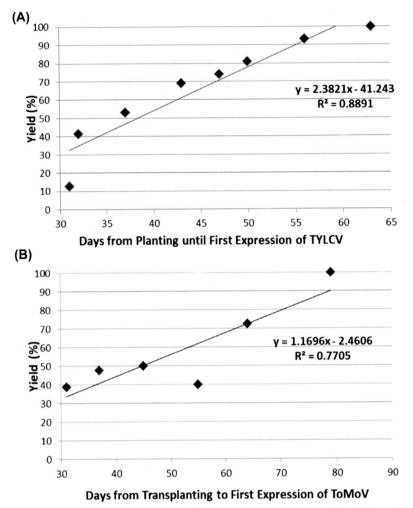

FIGURE 4.5 Production of virus-infected tomato plants expressed as percentage of yield loss from noninfected plants; (A) TYLCV, Florida 1999, (B) TMoV, Florida 1990. *Adapted from Stansly, P.A., Schuster, D.J., 1990. Update on sweetpotato whitefly. In: Stall, W.M. (Ed.), Proceedings of Florida Tomato Institute, Vegetable Crops Special Series SS-VEC-001. IFAS, Gainesville, Florida, USA, pp. 41-59.* http://www.imok.ufl.edu/programs/veg-hort/tomato-institute/, *with permission.*

Criniviruses contain a linear single-stranded, positive-sense RNA genome and are transmitted by whiteflies in a semi-persistent manner. Two of these, tomato chlorosis virus (ToCV) and tomato infectious chlorosis virus (TICV) are wide spread and can cause considerable damage to tomato (Wintermantle, 2010). ToCV is most efficiently transmitted by *B. tabaci* and *Trileurodes abutiloneus* Haldeman, and less so by *T. vaporariorum*, whereas TICV is transmitted only by *T. vaporariorum*. Symptoms of TICV and ToCV in tomato are very similar and include interveinal yellowing, necrotic flecking, rolling, and thickening of older leaves and yield loss due primarily to reduction of photosynthetic area resulting in reduced size and number of fruit Arnó et al. (2009).

Sources of resistance to ToCV and TICV have been identified in wild tomato species but are not yet available in commercial cultivars. Therefore, vector control is critical to slow spread and severity of both diseases, although effectiveness is limited by rapid transmission and short persistence of the virus (Tzanetakis et al., 2013). Criniviruses have broad host ranges compared to Begomoviruses, with ToCV known to infect at least 24 species from 7 taxonomic families (Wintermantel and Wisler, 2006). Therefore, reducing primary spread from weed or other infected crops is an important management practice.

Tomato torrado virus (ToTV; Secoviridae) is the type member of an emerging virus group (van der Vlugt et al., 2015). Symptoms include necrosis of leaves and fruit and reduced plant growth. ToTV is transmitted semi-persistently by *B. tabaci*, *T. vaporariorum*, and *T. abutiloneus*. A growing number of torrado viruses of tomato have been described since the discovery of ToTV in 2004, including tomato marchitez virus, tomato chocolate spot virus, and tomato necrotic dwarf virus. There is little information available on crop loss or management of these viruses.

4.8 Integrated Pest Management

From the discussion above, it is clear that growers have a number of options for managing whiteflies and whitefly-vectored viruses. There are additional strategies that have been evaluated which do not fit into the categories previously mentioned. One such idea is to use Acibenzolar-S-Methyl (ASM), the inducer of systemic acquired resistance (SAR) to improve performance of insecticides. Fanigliulo et al. (2015) found that combining ASM and the insecticide cyantraniliprole provided the best defense against infection and disease development of TYLCV. Reduction in TYLCV disease severity was also demonstrated with the application of the SAR inducer, eugenol, to tomato (Wang and Fan, 2014). Another study applied exogenous SA to tomato plants which repelled whiteflies (Shi et al., 2016). Studies evaluating the impacts of plant chemicals have also shown some interesting results. Li et al. (2014) found that (E)-2-hexenal, 3-hexen-1-ol attracted *B. tabaci* adults, while limonene deterred whiteflies from settling. The authors suggested these compounds might be useful in a "push–pull" strategy for whiteflies. Another study with limonene, citral, and olive oil mixture lowered *B. tabaci* adult settling, and was effective in a push–pull strategy using yellow sticky cards as the "pull" part of the strategy (Du et al., 2016).

Similar to aphids, studies to control whiteflies with vermicomposts have been conducted. These are materials produced by microorganisms during the breakdown of waste products of earthworms (Arancon et al., 2007). Applying vermicompost as solid fertilizer and in aqueous extracts resulted in substantial reductions in adult *T. vaporariorum* (Foroushani et al., 2016). They found that mortality of the second nymphal instar occurred in vermicompost-treated plants.

5. ECONOMICS OF MANAGEMENT STRATEGIES

Whiteflies damage tomatoes by (1) extracting phloem sap, thereby weakening the plant and directly reducing yield; (2) excreting honeydew over the fruit and foliage that serves as a source for sooty mold fungus which reduces photosynthesis and must be washed off of fruit before marketing; (3) injecting saliva that in the case of *B. tabaci* (MEAM1) causes TIR; and (4) vectoring plant pathogens. While all of these types of damages can be present in the same tomato field, more often it is one or two of the damages that must be addressed by the grower to prevent economic loss. The reduction of whitefly numbers is an appropriate goal for all of these damage types, yet there are times when additional objectives may be more important. For example, preventing viruliferous whiteflies from coming into a field, or controlling them outside the target crop may be the best strategy to prevent the field from becoming infected by whitefly-borne viruses. The myriad of damage possibilities by whiteflies complicate the choices faced by growers and make it difficult to predict the economic outcomes of employing certain management strategies. However, in one study, Lindquist et al. (1972a) found that spraying tomato plants in the greenhouse every 5 days with endosulfan increased gross income by ca. $5000/acre. With our current insecticide formulations, we do not typically treat tomatoes every 5 days, but the return on investment demonstrated by Lindquist et al. (1972a) still makes insecticides the number one go-to strategy for whitefly management. Particular emphasis has been placed on the nicotinic receptor antagonists (see Section 4.2.2), which are applied to the soil and provide control by being transported in the phloem where whiteflies feed. These materials provide cost-effective control of whiteflies. In addition, due to the nature of their application to the soil, as opposed to spraying the foliage, they are compatible with natural enemies, thereby enabling a more integrated approach. This is particularly important, since a number of commercial companies produce natural enemies. This is resulting in more biological control applications in tomatoes, particularly in greenhouse situations. In addition, increased production of organically grown tomatoes favors a more IPM approach with less dependence on synthetic insecticides. Concomitant with this movement toward fewer insecticides are the advances being made in plant resistance against whiteflies and the viruses they transmit. The development of resistant varieties holds great economic potential for dealing with whiteflies on tomatoes in the future.

6. FUTURE PROSPECTS AND CONCLUSION

Trends in fresh-market tomato production in recent years have moved toward more production in greenhouses. The driving forces for this trend are improved control over environmental conditions which have resulted in higher quality and the ability to have more precision on the use of IPM strategies, particularly biological control. At the same time, greenhouse producers and especially growers of open-field tomatoes still rely on pesticides because of their relative ease of use and immediate kill following application. With increasing global warming, whitefly problems are likely to be exacerbated; Guo et al. (2012) noted that *B. tabaci* can survive long-term high temperature stress. The neonicotinoids continue to play a key role in whitefly control, and new synthetic and natural products are being developed. As mentioned above, there are a number of other creative strategies that are being evaluated for whitely control, and plant resistance is a fertile area of research. The more we integrate different types of controls, particularly within the context of other tomato pests that have been discussed in this book, the better chance we have of developing economically and environmentally sustainable pest management programs for whiteflies in tomato.

REFERENCES

Abd Rabou, S., Simmons, A.M., 2012a. Effect of three irrigation methods on incidences of *Bemisia tabaci* (Hemiptera: Aleyrodidae) and some whitefly-transmitted viruses in four vegetable crops. Trends in Entomology 8, 21–26.

Abd Rabou, S., Simmons, A.M., 2012b. Some cultural strategies to help manage *Bemisia tabaci* (Hemiptera: Aleyrodidae) and whitefly-transmitted viruses in vegetable crops. African Entomology 20, 371–379.

Abhary, M., Patil, B.L., Fauquet, C.M., 2007. Molecular biodiversity, taxonomy, and nomenclature of Tomato yellow leaf curl-like viruses. In: Czosnek, H. (Ed.), Tomato Yellow Leaf Curl Virus Disease: Management, Molecular Biology, Breeding for Resistance. Springer, Dordrecht, The Netherlands, pp. 85–118.

Adamson, R.M., Tonks, N.V., Maas, E.F., 1972. Yields of greenhouse tomatoes treated with Naled for control of the greenhouse whitefly. Journal of Economic Entomology 65, 1205.

Agnello, A.M., 2002. Petroleum-derived spray oils: chemistry, history, refining and formulation. In: Beattie, G.A.C., Watson, D.M., Stevens, M.L., Rae, D.J., Spooner-Harts, R.N. (Eds.), Spray Oils beyond 2000: Sustainable Pest and Disease Management. University of Western Sydney, Penrith, New South Wales, Australia, pp. 2–18.

Ahmad, M., 2007. Potentiation/antagonism of pyrethroids with organophosphate insecticides in *Bemisia tabaci* (Homoptera: Aleyrodidae). Journal of Economic Entomology 100, 886–893.

Ahman, I., Ekbom, B.S., 1981. Sexual behavior of the greenhouse whitefly (*Trialeurods vaporariorum*) – orientation and courtship. Entomologia Experimentalis et Applicata 29, 330–338.

Ahsan, M.I., Hossain, M.S., Parvin, S., Karim, Z., 2005. Effect of varieties and planting dates on the incidence of aphid and white fly attack on tomato. International Journal of Sustainable Agricultural Technology 1, 26–30.

Al-Musa, A., 1982. Incidence, economic importance and control of tomato yellow leaf curl in Jordan. Plant Disease 66, 561–563.

Al-Shihi, A.A., Al-Sadi, A.M., Al-Said, F.A., Ammara, U.E., Deadman, M.L., 2016. Optimising the duration of floating row cover period to minimise the incidence of tomato yellow leaf curl disease and maximise yield of tomato. Annals of Applied Biology 168, 328–336.

Al-Zyoud, F., 2007. Prey species preference of the predator *Serangium parcesetosum* Sicard (Col., Coccinellidae) and its interaction with another natural enemy. Pakistan Journal of Biological Sciences 10, 2159–2165.

Al-Zyoud, F., Sengonca, C., 2004. Prey consumption preferences of *Serangium parcesetosum* Sicard (Col., Coccinelidae) for different prey stages, species and parasitized prey. Journal of Pest Science 77, 197–204.

Al-Zyoud, F., Sengonca, C., Al-Abbadi, S., 2007. Evaluation of *Serangium parcesetosum* (Col.: Coccinellidae) for biological control of *Bemisia tabaci* under greenhouse conditions. Journal of Pest Science 80, 85–92.

Al-Zyoud, F., Tort, N., Sengonca, C., 2005. Influence of host plant species of *Bemisia tabaci* (Genn.) (Hom., Aleyrodidae) on some of the biological and ecological characteristics of the entomophagous *Serangium parcesetosum* Sicard (Col., Coccinellidae). Journal of Pest Science 78, 25–30.

Alomar, O., Riudavets, J., Castane, C., 2006. *Macrolophus caliginosus* in the biological control of *Bemisia tabaci* on greenhouse melons. Biological Control 36, 154–162.

Antignus, Y., 2000. Manipulation of wavelength – dependent behavior of insects: an IPM tool to impede insects and restrict epidemics of insect-borne viruses. Virus Research 71, 213–220.

Antignus, Y., Lapidot, M., Hadar, D., Messika, Y., Cohen, S., 1998. Ultraviolet-absorbing screens serve as optical barriers to protect crops from virus and insect pests. Journal of Economic Entomology 91, 1401–1405.

Antignus, Y., Mor, N., Joseph, R.B., Lapidot, M., Cohen, S., 1996. Ultraviolet-absorbing plastic sheets protect crops from insect pests and from virus diseases vectored by insects. Environmental Entomology 25, 919–924.

Antignus, Y., Nestel, D., Cohen, S., Lapidot, M., 2001. Ultraviolet-deficient greenhouse environment affects whitefly attraction and flight-behavior. Environmental Entomology 30, 394–399.

Antonious, G., Snyder, J., 2008. Tomato leaf crude extracts for insects and spider mite control. In: Preedy, V.R., Watson, R.R. (Eds.), Tomatoes and Tomato Products: Nutritional, Medicinal and Therapeutic Properties. Science Publishers, USA, pp. 269–297.

Arias, R., Hilje, L., 1993. Uso del frijol como cultivo trampa y de un aceite agrícola para disminuir la incidencia de virosis transmitida por *Bemisia tabaci* (Gennadius) en el tomate. Manejo Integrado de Plagas 27, 27–35.

Arnó, J., Albajes, R., Gabarra, R., 2006. Within-plant distribution and sampling of single and mixed infestations of *Bemisia tabaci* and *Trialeurodes vaporariorum* (Homoptera: Aleyrodidae) in winter tomato crops. Journal of Economic Entomology 99, 331–340.

Arnó, J., Gabarra, R., Liu, T.X., Simmons, A.M., Gerling, D., 2010. Natural enemies of *Bemisia tabaci*: predators and parasitoids. In: Stansly, P.A., Naranjo, S.E. (Eds.), *Bemisia*: Bionomics and Management of a Global Pest. Springer, Dordrecht, Heidelberg, London, New York, pp. 385–421.

Arnó, J., Roig, J., Riudavets, J., 2008. Evaluation of *Orius majusculus* and *O. laevigatus* as predators of *Bemisa tabaci* and estimation of their prey preference. Biological Control 44, 1–6.

Arnó, J., Gabarra, I., Ampert, R., Estopa, M., Gorman, K., Peterschmitt, M., Bonato, O., Vosman, B., Hommes, M., Albajes Garcia, R., 2009. Implementation of IPM Programs on European Greenhouse Tomato Production Areas: Tools and Constraints. University of Lleida, Lleida, Spain, p. 46.

Arancon, N.Q., Edwards, C.A., Oliver, T.J., Byrne, R.J., 2007. Suppression of two-spotted spider mite (*Tetranychus urticae*), mealy bugs (*Pseudococcus* sp.) and aphid (*Myzus persicae*) populations and damage by vermicomposts. Crop Protection 26, 26–39.

Asmita, S., Ukey, S.P., 2016. Seed treatment with botanicals against tomato jassids and whiteflies. The Indian Journal of Entomology 78, 229–232.

Baldin, E.L.L., Aguiar, G.P., Fanela, T.L.M., Soares, M.C.E., Groppo, M., Crotti, A.E.M., 2015. Bioactivity of *Pelargonium graveolens* essential oil and related monoterpenoids against sweet potato whitefly, *Bemisia tabaci* biotype B. Journal of Pest Science 88, 191–199.

Baldin, E.L.L., Crotti, A.E.M., Wakabayashi, K.A.L., Silva, J.P.G.F., Aguiar, G.P., Souza, E.S., Veneziani, R.C.S., Groppo, M., 2013. Plant-derived essential oils affecting settlement and oviposition of *Bemisia tabaci* (Genn.) biotype B on tomato. Journal of Pest Science 86, 301–308.

Bao Fundora, L., Ramirez Romero, R., Sanchez Hernandez, C.V., Sanchez-Martinez, J., Desneux, N., 2016. Intraguild predation of *Geocoris punctipes* on *Eretmocerus eremicus* and its influence on the control of the whitefly *Trialeurodes vaporariorum*. Pest Management Science 72, 1110–1116.

Barati, R., Golmohammadi, G., Ghajarie, H., Zarabi, M., Mansouri, R., 2014. Efficiency of some herbal pesticides on reproductive parameters of silverleaf whitefly, *Bemisia tabaci* (Gennadius) (Hemiptera: Aleyrodidae). Archives of Phytopathology and Plant Protection 47, 212–221.

Barry, J.D., Portillo, H.E., Annan, I.B., Cameron, R.A., Clagg, D.G., Dietrich, R.F., Watson, L.J., Leighty, R.M., Ryan, D.L., McMillan, J.A., Swain, R.S., Kaczmarczyk, R.A., 2015. Movement of cyantraniliprole in plants after foliar applications and its impact on the control of sucking and chewing insects. Pest Management Science 71, 395–403.

Bemis, F.E., 1904. The aleyrodids, or mealy-winged flies, of California with references to other American species. Proceedings of the United States Natural Museum 27, 471–537.

Ben-Yakir, D., Antignus, Y., Offir, Y., Shahak, Y., 2012. Colored shading nets impede insect invasion and decrease the incidences of insect-transmitted viral diseases in vegetable crops. Entomologia Experimentalis et Applicata 144, 249–257.

Berlinger, M.J., Dahan, R., Cohen, S., 1983. Use of plastic covering to prevent the spread of tomato yellow leaf curl virus in greenhouses. Hasadeh 63, 1862–1865.

Bernays, E.A., 1999. When host choice is a problem for a generalist herbivore: experiments with the whitefly *Bemisia tabaci*. Ecological Entomology 24, 260–267.

Bethke, J.A., Paine, T.D., 1991. Screen hole size and barriers for exclusion of insect pests of glasshouse crops. Journal of Entomological Science 26, 169–177.

Bethke, J.A., Paine, T.S., Nuessly, G.S., 1991. Comparative biology, morpho-metrics, and development of two populations of *Bemisia tabaci* (Homoptera: Aleyrodidae) on cotton and poinsettia. Annals of the Entomological Society of America 84, 407–411.

Blackmer, J.L., Byrne, D.N., 1993a. Flight behaviour of *Bemisia tabaci* in a vertical flight chamber: effect of time of day, sex, age and host quality. Physiological Entomology 18, 223–232.

Blackmer, J.L., Byrne, D.N., 1993b. Environmental and physiological factors influencing phototactic flight of *Bemisia tabaci*. Physiological Entomology 18, 336–342.

Bleeker, P.M., Diergaarde, P.J., Ament, K., Guerra, J., Weidner, M., Schütz, S., de Both, M.T.J., Harang, M.A., Schuurinck, R.C., 2009. The role of specific tomato volatiles in tomato-whitefly interaction. Plant Physiology 151, 925–935.

Bleeker, P.M., Mirabella, R., Diergaarde, P.J., VanDoorn, A., Tissier, A., Kant, M.R., Prins, M., de Vos, M., Haring, M.A., Schuurink, R.C., 2012. Improved herbivore resistance in cultivated tomato with the sesquiterpene biosynthetic pathway from a wild relative. Proceedings of the National Academy of Sciences 109, 20124–20129.

Böckmann, E., Hommes, M., Meyhöfer, R., 2015. Yellow traps reloaded: what is the benefit for decision making in practice? Journal of Pest Science 88, 439–449.

Bonato, O., Couton, L., Fargues, J., 2006. Feeding preference of *Macrolophus caliginosus* (Heteroptera: Miridae) on *Bemisia tabaci* and *Trialeurodes vaporariorum* (Homoptera: Aleyrodidae). Journal of Economic Entomology 99, 1143–1151.

Borisade, O.A., 2015. Rearing tomato whitefly and field evaluation of modified and unmodified conidia of *Beauveria bassiana*, *Isaria farinosa*, *Metarhizium anisopliae* and low rates of chlorpyrifos under tropical conditions. African Crop Science Journal 23, 177–195.

Boykin, L.M., 2014. *Bemisia tabaci* nomenclature: lessons learned. Pest Management Science 70, 1454–1459.

Breene, R.G., Meagher Jr., R.L., Nordlund, D.A., Wang, Y.T., 1992. Biological control of *Bemisia tabaci* (Homoptera: Aleyrodidae) in a greenhouse using *Chrysoperla rufilabris* (Neuropteran: Chrysopidae). Biological Control 2, 9–14.

Buckner, J.S., Jones, W.A., 2005. Transfer of methyl-branched hydrocarbons from the parasitoid, *Eretmocerus mundus*, to silverleaf whitefly nymphs during oviposition. Comparative Biochemistry and Physiology Part A: Molecular and Integrative Physiology 140, 59–65.

Burnett, T., 1949. The effect of temperature on an insect host-parasite population. Ecology 30, 113–134.

Butler Jr., G.D., Henneberry, T.J., Clayton, T.E., 1983. *Bemisia tabaci* (Homoptera: Aleyrodidae): development, oviposition and longevity in relation to temperature. Annals of the Entomological Society of America 76, 310–313.

Butler Jr., G.D., Henneberry, T.J., Stansly, P.A., Schuster, D.J., 1993. Insecticidal effect of selected soaps, oils, and detergents on the sweetpotato whitefly. Florida Entomologist 76, 162–167.

Byrne, D.N., Bellows Jr., T.S., 1991. Whiteflies biology. Annual Review of Entomology 36, 431–457.

CABI, 2017a. Center for Agriculture and Biosciences International, Invasive Species Compendium. *Bemisia tabaci* (tobacco whitefly) http://www.cabi.org/isc/datasheet/8927.

CABI, 2017b. Center for Agriculture and Biosciences International, Invasive Species Compendium. *Trialeurodes vaporariorum* (greenhouse whitefly) http://www.cabi.org/isc/datasheet/54660.

Calvitti, M., Buttarazzi, M., 1995. Determination of biological and demographic parameters of *Trialeurodes vaporariorum* Westwood (Homoptera Aleyrodidae) on two host plant species: zucchini (*Cucurbita pepo*) and tomato (*Lycopersicon esculentum*). Redia 78, 29–37.

Calvo, F.J., Bolckmans, K., Belda, J.E., 2009a. Development of a biological control-based integrated pest management method for *Bemisia tabaci* for protected sweet pepper crops. Entomologia Experimentalis et Applicata 133, 9–18.

Calvo, F.J., Bolckmans, K., Belda, J.E., 2012a. Release rate for a pre-plant application of *Nesidiocoris tenuis* for *Bemisia tabaci* control in tomato. BioControl 57, 809–817.

Calvo, F.J., Bolckmans, K., Stansly, P.A., Urbaneja, A., 2009b. Predation by *Nesidiocoris tenuis* on *Bemisia tabaci* and injury to tomato. BioControl 54, 237–246.

Calvo, F.J., Lorente, M.J., Stansly, P.A., Belda, J.E., 2012b. Preplant release of *Nesidiocoris tenuis* and supplementary tactics for control of *Tuta absoluta* and *Bemisia tabaci* in greenhouse tomato. Entomologia Experimentalis et Applicata 143, 111–119.

Calvo, F.J., Torres-Ruiz, A., Velazquez-Gonzalez, J.C., Rodriguez-Leyva, E., Lomeli-Flores, J.R., 2016. Evaluation of *Dicyphus hesperus* for biological control of sweet potato whitefly and potato psyllid on greenhouse tomato. BioControl 61, 415–424.

Cao, F., Liu, W., Fan, Z., Wan, F., Cheng, L., 2008. Behavioral responses of *Bemisia tabaci* biotype B to three host plants and their volatiles. Acta Entomologica Sinica 51, 830–838.

Castane, C., Zapata, R., 2005. Rearing the predatory bug *Macrolophus caliginosus* on a meat-based diet. Biological Control 34, 66–72.

Channarayappa, C., Shrivashankar, G., Muniyappa, V., Frist, R.H., 1992. Resistance of *Lycopersicon* species to *Bemisia tabaci*, a tomato leaf curl virus vector. Canadian Journal of Botany 70, 2184–2192.

Chavan, S.M., Kumar, S., Arve, S.S., 2013. Population dynamics and development of suitable pest management module against major insect pests of tomato (*Solanum lycopersicum*). Journal of Applied Horticulture 15, 150–155.

Chavan, R.D., Yeotikar, S.G., Gaikwad, B.B., Dongarjal, R.P., 2015. Management of major pests of tomato with biopesticides. Journal of Entomological Research 39, 213–217.

Chen, H., Lin, C., Tsai, W., Kenyon, L., Chan, M., Yen, J., Chang, S., Pena, R., Schafleitner, R., 2016. Resistance to viral yellow leaf curl in tomato through RNAi targeting two *Begomovirus* species strains. Journal of Plant Biochemistry and Biotechnology 25, 199–207.

Cheng, C., Junqi, J., Xiaofei, X., Ying, W., Guiting, L., 2014. Predatory mites control *Bemisia tabaci* on tomato plants in greenhouse. Anhui Nongye Daxue Xuebao 41, 685–689.

Chiel, E., Gerling, D., Steinberg, S., Klapwijk, J., Bolckmans, K., Zchori-Fein, E., 2012. Contagious sterility in the parasitoid wasp *Eretmocerus mundus* (Hymenoptera: Aphelinidae). Biocontrol Science and Technology 22, 61–66.

Choi, Y., Hwang, I., Lee, G., Kim, G., 2016. Control of *Bemisia tabaci* Genn. (Hemiptera: Aleyrodidae) adults on tomato plants using trap plants with systemic insecticide. Korean Journal of Applied Entomology 55, 109–117.

Choi, Y., Seo, J., Whang, I., Kim, G., Choi, B., Jeong, T., 2015. Effects of eggplant as a trap plant attracting *Bemisia tabaci* Genn. (Hemiptera: Aleyrodidae) adults available on tomato greenhouses. Korean Journal of Applied Entomology 54, 311–316.

Chun-Li, C., Jun-Rui, Z., Xian-Ju, K., Feng, G., 2015. Effect of previous infestation of tomatoes by *Bemisia tabaci* or *Trialeurodes vaporariorum* on the fitness of these whitefly species. Chinese Journal of Applied Entomology 52, 104–112.

Clayberg, C.C., Kring, J.B., 1974. Breeding tomatoes resistant to potato aphid and whitefly. HortScience 9, 297.

Cock, M.J.W., 1986. *Bemisia Tabaci* – A Literature Survey on the Cotton Whitefly with an Annotated Bibliography. CAB International Institute of Biological Control, UK.

Cockerell, T.D.A., 1902. The classification of the Aleyrodidae. Proceedings of the Acadamy of Natural Sciences of Philadelphia 54, 279–283.

Cohen, S., Kern, J., Harpaz, I., Ben-Joseph, R., 1988. Epidemiological studies of the tomato yellow leaf curl in the Jordan Valley. Phytoparasitica 16, 259–270.

Condron, C.H., Neiswander, R.B., Wessel, R.D., 1962. Toxicity of Dibrom vapors to greenhouse insects. Journal of Economic Entomology 55, 221–224.

Coombe, P.E., 1982. Visual behavior of the greenhouse whitefly, *Trialeurodes vaporariorum*. Physiological Entomology 7, 243–251.

Costa, H.S., Robb, K.L., 1999. Effects of ultraviolet-absorbing greenhouse plastic films on flight behavior of *Bemisia argentifolii* (Homoptera: Aleyrodidae) and *Frankliniella occidentalis* (Thysanoptera: Thripidae). Journal of Economic Entomology 92, 557–562.

Coudriet, D.L., Prabahker, N., Kishaba, A.N., Meyerdirk, D.E., 1985. Variation in development rate on different hosts and overwintering of the sweetpotato whitefly *Bemisia tabaci* (Homoptera: Aleyrodidae). Environmental Entomology 14, 516–519.

Csizinszky, A.A., Schuster, D.J., Kring, J.B., 1995. Color mulches influence yield and insect pest populations in tomatoes. Journal of the American Society for Horticultural Science 120, 778–784.

Csizinszky, A.A., Schuster, D.J., Polston, J.E., 1999. Effect of ultraviolet-reflective mulches on tomato yields and on the silverleaf whitefly. HortScience 34, 911–914.

Curry, J.P., Pimentel, D., 1971. Evaluation of tomato varieties for resistance to greenhouse whitefly. Journal of Economic Entomology 64, 1333–1334.

Cuthbertson, A.G.S., Vänninen, I., 2015. The importance of maintaining protected zone status against *Bemisia tabaci*. Insects 6, 432–441.

Cuthbertson, A.G.S., Walters, K.F.A., Deppe, C., 2005a. Compatibility of the entomopathogenic fungus *Lecanicillium muscarium* and insecticides for eradication of sweetpotato whitefly, *Bemisia tabaci*. Mycopathologia 160, 35–41.

Cuthbertson, A.G.S., Walters, K.F.A., Northing, P., 2005b. The susceptibility of immature stages of *Bemisia tabaci* to the entomopathogenic fungus *Lecanicillium muscarium* on tomato and verbena foliage. Mycopathologia 159, 23–29.

Cuthbertson, A.G.S., Walters, K.F.A., Northing, P., Luo, W., 2007. Efficacy of the entomopathogenic nematode, *Steinernema feltiae*, against sweetpotato whitefly *Bemisia tabaci* (Homoptera: Aleyrodidae) under laboratory and glasshouse conditions. Bulletin of Entomological Research 97, 9–14.

Czosnek, H., Sharma, P., Gaur, R.K., 2010. Management of tomato yellow leaf curl disease: a case study for emerging geminviral diseases. In: Sharma, P., Guar, R.K., Ikegami, M. (Eds.), Emerging Geminiviral Diseases and their Management. Nova Science Publishers, New York, USA, pp. 37–57.

De Ponti, O.M.B., Pet, G., Hogenboom, N.G., 1975. Resistance to the glasshouse whitefly (*Trialeurodes vaporariorum* Westw.) in tomato (*Lycopersicon esculentum* Mill.) and related species. Euphytica 24, 645–649.

De Resende, J.T., Maluf, W.R., das Gracas Cardoso, M., Gonçalves, L.D., Faria, M.V., do Nascimento, I.R., 2009. Resistance of tomato genotypes to the silverleaf whitefly mediated by acylsugars. Horticultura Brasileira 27, 345–348.

Deligeorgidis, P.N., Ipsilandis, C.G., Vaiopoulou, M., Kaltsoudas, G., Sidiropoulos, G., 2005. Predatory effect of *Coccinella septempunctata* on *Thrips tabaci* and *Trialeurodes vaporariorum*. Journal of Applied Entomology 129, 246–249.

Denholm, I., Cahill, M., Byrne, F.J., Devonshire, A.L., 1996. Progress with documenting and combating insecticide resistance in *Bemisia*. In: Gerling, D., Mayer, R.T. (Eds.), *Bemisia* 1995: Taxonomy, Biology, Damage, Control and Management. Intercept, Ltd., Andover, UK, pp. 577–603.

Diabate, D., Gnago, J.A., Koffi, K., Tano, Y., 2014. The effect of pesticides and aqueous extracts of *Azadirachta indica* (A. Juss) and *Jatropha carcus* L. on *Bemisia tabaci* (Gennadius) (Homoptera: Aleyrididae) and *Helicoverpa armigera* (Hübner) (Lepidoptera: Noctuidae) found on tomato plants in Côte d'Ivoire. Journal of Applied Biosciences 80, 7132–7143.

Dias, D.M., Resende, J.T.V., Marodin, J.C., Matos, R., Lustosa, I.F., 2016. Acyl sugars and whitefly (*Bemisia tabaci*) resistance in segregating populations of tomato genotypes. Genetics and Molecular Research 15, 1–11.

Dinsdale, A., Cook, L., Riginos, C., Buckley, Y.M., De Barro, P., 2010. Refined global analysis of *Bemisia tabaci* (Hemiptera: Sternorrhyncha: Aleyrodoidea: Aleyrodidae) mitochondrial cytochrome oxidase 1 to identify species level genetic boundaries. Annals of the Entomological Society of America 103, 196–208.

Dodds, P.N., Rathjen, J.P., 2010. Plant immunity: towards an integrated view of plant-pathogen interactions. Nature Reviews Genetics 11, 539–548.

Drost, Y.C., Elmula, A.F., Posthuma-Doodeman, C.J.A.M., van Lenteren, J.C., 1996. Development of criteria for evaluation of natural enemies in biological control: bionomics of different parasitoids of *Bemisia argentifolii*. OILB SROP Bulletin 19, 31–34.

Du, W., Han, X., Wang, Y., Qin, Y., 2016. A primary screening and applying of plant volatiles as repellents to control whitefly *Bemisia tabaci* (Gennadius) on tomato. Scientific Reports 6, 22140.

Eggenkamp-Rotteveel Mansfeld, M.H., van Lenteren, J.C., Ellenbroek, J.M., Woets, J., 1982. The parasite-host relationship between *Encarsia formosa* (Hym., Aphelinidae) and *Trialeurodes vaporariorum* (Hom., Aleyrodidae). XII. Population dynamics of parasite and host in a large, commercial glasshouse and test of the parasite-introduction method used in the Netherlands. Zeitschrift für Angewandte Entomologie 93 (113–130), 258–279.

Ekbom, B.S., 1981. Efficiency of the predator *Anthocoris nemorum* (Het.: Anthorcoridae) against the greenhouse whitefly, *Trialeurodes vaporariorum* (Hom.: Aleyrodidae). Zeitschrift für angewandte Entomologie 92, 26–34.

Ekbom, B.S., Rumei, X., 1996. Sampling and spatial patterns of whiteflies. In: Gerling, D. (Ed.), Whiteflies: their Bionomics, Pest Status, and Management. Intercept, Andover, Hants, UK, pp. 107–121.

El-Shafie, H.A.F., Abdelraheem, B.A., 2012. Field evaluation of three biopesticides for integrated management of major pests of tomato, *Solanum lycopersicum* L. in Sudan. Agriculture and Biology Journal of North America 3, 340–344.

El-Gendi, S.S., Adam, K.M., Bachatly, M.A., 1997. Effect of the planting date of tomato on the population density of *Bemisia tabaci* (Genn.) and *Heliothis armigera* (HB), viral infection and yield. Arab Universities Journal of Agricultural Science 5, 135–144.

El-Serwiy, S.A., Ali, A.A., Razoki, I.A., 1987. Effect of intercropping of some host plants with tomato on population density of tobacco whitefly, *Bemisia tabaci* (Genn.), and the incidence of tomato yellow leaf curl virus (TYLCV) in plastic houses. Journal of Agriculture and Water Resources Research, Plant Production 6, 79–81.

Elbaz, M., Hanson, P., Fgaier, S., Laarif, A., 2016. Evaluation of tomato entries with different combinations of resistance genes to tomato yellow leaf curl disease in Tunisia. Plant Breeding 135, 525–530.

Emilie, D., Mallent, M., Menut, C., Chandre, F., Martin, T., 2015. Behavioral response of *Bemisia tabaci* (Hemiptera: Aleyrodidae) to 20 plant extracts. Journal of Economic Entomology 108, 1890–1901.

ERS, 2017. U.S. Tomato Statistics (92010). Economic Research Service (ERS), U.S. Department of Agriculture (USDA). Updated July 2010 http://usda.mannlib.cornell.edu/MannUsda/viewDocumentInfo.do?documentID=1210.

Esashika, D.A.S., Michereff-Filho, M., Bastos, C.S., Inoue-Nagata, A.K., Dias, A.M., Ribeiro, M.G.P.M., 2016. Suscetibilidade de adultos de *Bemisia tabaci* biótipo B a inseticidas. Horticultura Brasileira 34, 189–195.

Espinel, C., Denis-Lozano, M., Villamizar, L.R., Grijalbab, E., Marina-Cotes, A.P., 2008. IPM strategy for the control of *Bemisia tabaci* (Hemiptera: Aleyrodidae) on melon and tomato. Revista Colombiana de Entomología 34, 163–168.

Evans, G.A., 2007. Parasitoids (Hymenoptera) Associated with Whiteflies (Aleyrodidae) of the World. U.S. Department of Agriculture, Beltsville, MD. Version 070202 http://www.sel.barc.usda.gov:8080/1WF/parasitoidcatalog.

Fancelli, M., Vendramim, J.D., Lourenção, A.L., Dias, C.T.S., 2003. Atratividade e preferência para oviposição de *Bemisia tabaco* (Gennadius) (Hemiptera: Aleyrodidae) biótipo B em genotipos de tomateiro. Neotropical Entomology 32, 319–328.

Fanela, T.L.M., Baldin, E.L.L., Pannuti, L.E.R., Cruz, P.L., Crotti, A.E.M., Takeara, R., Kato, M.J., 2016. Lethal and inhibitory activities of plant-derived essential oils against *Bemisia tabaci* Gennadius (Hemiptera: Aleyrodidae) biotype B in tomato. Neotropical Entomology 45, 201–210.

Fanigliulo, A., Viggiano, A., Crescenzi, A., Zingariello, E., Liguori, R., 2015. Control of tomato yellow leaf curl disease in tomato. Acta Horticulturae 1069, 191–196.

Fargues, J., Smits, N., Rougier, M., Boulard, T., Ridray, G., Lagier, J., Jeannequin, B., Fatnassi, H., Mermier, M., 2005. Effect of microclimate heterogeneity and ventilation system on entomopathogenic hyphomycete infection of *Trialeurodes vaporariorum* (Homoptera: Aleyrodidae) in Mediterranean greenhouse tomato. Biological Control 32, 461–472.

Faria, M., Wraight, S.P., 2001. Biological control of *Bemisia tabaci* with fungi. Crop Protection 20, 767–778.

Figuls, M., Castane, C., Gabarra, R., 1999. Residual toxicity of some insecticides on the predatory bugs *Dicyphus tamaninii* and *Macrolophus caliginosus*. BioControl 44, 89–98.

Firdaus, S., van Heusden, A.W., Hidayati, N., Supena, E.D.J., Visser, R.G.F., Vosman, B., 2012. Resistance to *Bemisia tabaci* in tomato wild relatives. Euphytica 187, 31–45.

Foroushani, A.P., Poorjavad, N., Haghigh, M., Khajehali, J., 2016. Effect of solid and aqueous extract of vermicompost on growth characteristics of tomato and greenhouse whitefly (*Trialeurodes vaporariorum*). Journal of Science and Technology of Greenhouse Culture 7, 35–46.

Fransen, J.J., Winkelman, K., van Lenteren, J.C., 1987. The differential mortality at various life stages of the greenhouse whitefly, *Trialeurodes vaporariorum* (Homoptera: Aleyrodidae), by infection with the fungus *Aschersonia aleyrodis* (Deuteromycotina: Coelomycetes). Journal of Invertebrate Pathology 50, 158–165.

Freitas, J.A., Maluf, W.R., Graças Cardoso, M., Gomes, L.A.A., Bearzotti, E., 2002. Inheritance of foliar zingiberene contents and their relationship to trichome densities and whitefly resistance in tomatoes. Euphytica 127, 275–287.

Friedmann, M., Lapidot, M., Cohen, S., Pilowsky, M., 1998. A novel source of resistance to tomato yellow leaf curl virus exhibiting a symptomless reaction to viral infection. Journal of the American Society of Horticultural Sciences 123, 1004–1006.

Gabarra, R., Zapata, R., Castañé, C., Riudavets, J., Arnó, J., 2006. Releases of *Eretmocerus mundus* and *Macrolophus caliginosus* for controlling *Bemisia tabaci* on spring and autumn greenhouse tomato crops. OILB/SROP Bulletin 29, 71–76.

Gelman, D.B., Gerling, D., Blackburn, M.B., Hu, J.S., 2005. Host-parasite interactions between whiteflies and their parasitoids. Archives of Insect Biochemistry and Physiology 60, 209–222.

Gennadius, P., 1889. Disease of tobacco plantations in the Trikonia. The aleyrodid of tobacco. Ellenike Georgia 5, 1–3.

Gerk, A.O., Vilela, E.F., Eiras, A.E., 1995. Biometry and life cycle of the whitefly, *Trialeurodes vaporariorum* (West.) and orientation aspects of its parasitoid *Encarsia formosa* Gahan. Anais da Sociedade Entomológica do Brasil 24, 89–97.

Gerling, D., 1966. Studies with whitefly parasites of Southern California. I. *Encarsia pergandiella* howard (Hymenoptera: Aphelinidae). The Canadian Entomologist 98, 707–724.

Gerling, D., 1990a. Natural enemies of whiteflies: predators and parasitoids. In: Gerling, D. (Ed.), Whiteflies: Their Bionomics, Pest Status and Management. Intercept, Andover, UK, pp. 147–185.

Gerling, D., 1990b. Whiteflies: Their Bionomics, Pest Status, and Management. Intercept, Andover, Hants, UK.

Gerling, D., Alomar, O., Arnó, J., 2001. Biological control of *Bemisia tabaci* using predators and parasitoids. Crop Protection 20, 779–799.

Gerling, D., Blackburn, M.B., 2013. Immature development of *Eretmocerus mundus* (Hymenoptera: parasitoids. Aphelinidae). Arthropod Structure & Development 42, 309–314.

Gerling, D., Mayer, R.T., 1996. *Bemisia*: 1995. Taxonomy, Biology, Damage, Control, and Management. Intercept, Andover, Hants, UK.

Gerling, D., Quicke, D.L.J., Orion, T., 1998. Oviposition mechanisms in the whitefly parasitoids *Encarsia transvena* and *Eretmocerus mundus*. Biocontrol 43, 289–297.

Gerling, D., Tremblay, E., Orion, T., 1991. Initial stages of the vital capsule formation in the *Eretmocerus-Bemisia tabaci* association. Redia 74, 411–415.

Ghahari, H., Abd-Rabou, S., Zahradnik, J., Ostovan, H., 2013. Annotated catalogue of whiteflies (Hemiptera: Sternorrhyncha: Aleyrodidae) from Arasbaran, Northwestern Iran. International Journal of Nematology and Entomology 1, 42–52.

Gilbertson, R.L., Batuman, O., Webster, C.G., Adkins, S., 2015. Role of the insect supervectors *Bemisia tabaci* and *Frankliniella occidentalis* in the emergence and global spread of plant viruses. Annual Review of Virology 2, 67–93.

Gilbertson, R.L., Rojas, M.R., Kon, T., Jaquez, J., 2007. Introduction of Tomato yellow leaf curl virus into the Dominican Republic: the development of a successful integrated pest management strategy. In: Czosnek, H. (Ed.), Tomato Yellow Leaf Curl Disease. Springer, Dordrecht, The Netherlands, pp. 279–303.

Gilbertson, R.L., Rojas, M., Natwick, E., 2011. Development of integrated pest management (IPM) strategies for whitefly (*Bemisia tabaci*)-transmissible geminiviruses. In: Thomson, W.M.O. (Ed.), The Whitefly, *Bemisia tabaci* (Homoptera: Aleyrodidae) Interaction with Geminivirus-infected Host Plants. Springer, Berlin, The Netherlands, pp. 323–356.

Gill, R.J., 1990. The morphology of whiteflies. In: Gerling, D. (Ed.), Whiteflies: Their Bionomics, Pest Status and Management. Intercept, Andover, UK, pp. 13–46.

Gillespie, D., Brodeur, J., Cloutier, C., Goettel, M., Jaramillo, P., Labbe, R., Roitberg, B., Thompson, C., VanLaerhoven, S., 2006. Combining pathogens and predators of insects in biological control. OILB/SROP Bulletin 29, 3–8.

Glas, J.J., Schimmel, B.C., Alba, J.M., Escobar-Bravo, R., Schuurink, R.C., Kant, M.R., 2012. Plant glandular trichomes as targets for breeding or engineering of resistance to herbivores. International Journal of Molecular Sciences 13, 17077–17103.

Goettel, M.S., Hajek, A.E., Siegel, J.P., Evans, H.C., 2001. Safety of fungal biocontrol agents. In: Butt, T.M., Jackson, C.W., Magan, N. (Eds.), Fungi as Biocontrol Agents: Progress, Problems and Potential. CABI, Wallingford, UK, pp. 347–376.

Gogo, E.O., Saidi, M., Ochieng, J.M., Martin, T., Baird, V., Ngouajio, M., 2014. Microclimate modification and insect pest exclusion using agronet improve pod yield and quality of French bean. HortScience 49, 1298–1304.

Greer, L., Dole, J.M., 2003. Aluminum foil, aluminium-painted, plastic, and degradable mulches increase yields and decrease insect-vectored viral diseases of vegetables. HortTechnology 13, 276–284.

Guo, J., Cong, L., Wan, F., 2013. Multiple generation effects of high temperature on the development and fecundity of *Bemisia tabaci* (Gennadius) (Hemiptera: Aleyrodidae) biotype B. Insect Science 20, 541–549.

Guo, J., Cong, L., Zhou, Z., Wan, F., 2012. Multi-generation life tables of *Bemisia tabaci* (Gennadius) biotype B (Hemiptera: Aleyrodidae) under high-temperature stress. Environmental Entomology 41, 1672–1679.

Guo, J., Ye, G., Dong, S., Liu, S., Leulier, F., 2010. An invasive whitefly feeding on a virus-infected plant increased its egg production and realized fecundity. PLoS One 5, e11713. http://dx.doi.org/10.1371/journal.pone.0011713.

Gusmao, M.R., Picanco, M.C., Guedes, R.N.C., Galvan, T.L., Pereira, E.J.G., 2006. Economic injury level and sequential sampling plan for *Bemisia tabaci* in outdoor tomato. Journal of Applied Entomology 130, 160–166.

Gusmao, M.R., Picanco, M.C., Zanuncio, J.C., Silva, D.J.H., Barrigossi, J.A.F., 2005. Standardised sampling plan for *Bemisia tabaci* (Homoptera: Aleyrodidae) in outdoor tomatoes. Scientia Horticulturae 103, 403–412.

Hai Lin, Y., Xiang Yong, L., Li Meng, Z., Yan Qiong, Y., Xue Qing, Z., 2014. Optimization for *Bemisia tabaci* egg development conditions using orthogonal design. South China Journal of Agricultural Sciences 45, 1970–1975.

Hargreaves, E., 1915. The life history and habits of the greenhouse whitefly. Annals of Applied Biology 1, 303–334.

Haydar, M.F., Aly, F.A., 1990. A Simple Approach for the Management of Whitefly-Borne Virus Diseases on Tomatoes, vol. 41. Cairo University Faculty of Agriculture Bulletin, pp. 649–664.

Heinz, K.M., Zalom, F.G., 1995. Variation in trichome-based resistance to *Bemisia argentifolii* (Homoptera: Aleyrodidae) oviposition on tomato. Journal of Economic Entomology 88, 1494–1502.

Henneberry, T.J., Jech, L., Hendrix, D.L., Steele, T., 1999. *Bemisia argentifolii* (Homoptera: Aleyrodidae). Factors affecting adult and nymph honeydew production. Southwestern Entomologist 24, 207–231.

Hilje, L., Costa, H.S., Stansly, P.A., 2001. Cultural practices for managing *Bemisa tabaci* and associated viral diseases. In: Naranjo, S., Ellsworth, P. (Eds.), Special Issue: Challenges and Opportunities for Pest Management of *Bemisia Tabaci* in the New CenturyCrop Protection, vol. 20, pp. 801–812.

Hilje, L., Stansly, P.A., 2008. Living ground covers for management of *Bemisia tabaci* (Gennadius) (Homoptera: Aleyrodidae) and tomato yellow mottle virus (ToYMoV) in Costa Rica. Crop Protection 27, 10–16.

Hirose, Y., Mitsunaga, T., Yano, E., Goto, C., 2009. Effects of sugars on the longevity of females *Eretmocerus eremicus*, and *Encarsia formosa*, parasitoids of *Bemisia tabaci* and *Trialeurodes vaporariorum*, as related to their honeydew feeding and host feeding. Applied Entomology and Zoology 44, 175–181.

Hoddle, M.S., 2004. Biological control of whiteflies on ornamental crops. In: Heinz, K., Van Driesche, R.G., Parrella, M.P. (Eds.), Biocontrol in Protected Culture. Ball Publishing, Batavia, IL, USA, pp. 149–170.

Hoddle, M.S., Van Driesche, R.G., Sanderson, J.P., 1997a. Biological control of *Bemisia argentifolii* (Homoptera: Aleyrodidae) on poinsettia with inundative releases of *Encarsia formosa* (Hymenoptera: Aphelinidae): are higher release rates necessarily better? Biological Control 10, 166–179.

Hoddle, M.S., Van-Driesche, R.G., Sanderson, J.P., 1998. Biology and use of the whitefly parasitoid *Encarsia formosa*. Annual Review of Entomology 43, 645–669.

Hoddle, M.S., Van Driesche, R.G., Sanderson, J.P., Minkenberg, O.P.J.M., 1997b. Biological control of *Bemisia argentifolii* (Homoptera: Aleyrodidae) on poinsettia with inundative releases of *Encarsia formosa* Beltsville strain (Hymenoptera: Aphelinidae): do release rates affect parasitism? Bulletin of Entomological Research 88, 47–58.

Hokkanen, H.M.T., 1991. Trap cropping in pest management. Annual Review of Entomology 36, 119–138.

Horowitz, A.R., Denholm, I., Morin, S., 2007. Resistance to insecticides in the TYLCV vector, *Bemisia tabaci*. In: Czosnek, H. (Ed.), Tomato Yellow Leaf Curl Virus Disease: Management, Molecular Biology, Breeding for Resistance. Springer, Dordrecht, The Netherlands, pp. 305–325.

Horowitz, A.R., Gorman, K., Ross, G., Denholm, I., 2003. Inheritance of pyriproxyfen resistance in the whitefly, *Bemisia tabaci* (Q biotype). Archives of Insect Biochemistry and Physiology 54, 177–186.

Hosseini, R.S., Madadi, H., Hosseini, M., Delshad, M., Dashti, F., 2015. Nitrogen in hydroponic growing medium of tomato affects the demographic parameters of *Trialeurodes vaporariorum* (Westwood) (Hemiptera: Aleyrodidae). Neotropical Entomology 44, 643–650.

Huang, Z., Ren, S.X., Yao, S.L., 2006. Life history of *Axinoscymnus cardilobus* (Col., Coccinellidae), a predator of *Bemisia tabaci* (Hom., Aleyrodidae). Journal of Applied Entomology 130, 437–441.

Hunter, W.B., Hiebert, E., Webb, S.E., Polston, J.E., Tsai, J.H., 1996. Precibarial and cibarial chemosensilla in the whitefly *Bemisia tabaci*. International Journal of Insect Morphology and Embryology 25, 295–304.

Hussey, N.W., Gurney, B., 1958. Report of the Glasshouse Crops Research Institute. 1957. Littlehampton, UK.

Hussey, N.W., Read, W.H., Hesling, J.J., 1969. The Pests of Protected Cultivation. The Biology and Control of Glasshouse and Mushroom Pests. Edward Arnold, London, p. 404.

Inglis, G.D., Goettel, M.S., Butt, T.M., Strasser, H., 2001. Use of hyphomycetous fungi for managing insect pests. In: Butt, T.M., Jackson, C.W., Magan, N. (Eds.), Fungi as Biocontrol Agents. Progress, Problems and Potential. CABI, Wallingford, UK, pp. 1–8.

Ioannou, N., 1987. Cultural management of tomato yellow leaf curl disease in Cyprus. Plant Pathology 36, 367–373.

Ishaaya, I., Mendelson, Z., Ascher, K.S., Casida, J.E., 1987. Cypermethrin synergism by pyrethroid esterase inhibitors in adults of the whitefly *Bemisia tabaci*. Pesticide Biochemistry and Physiology 28, 155–162.

Ishaaya, I., Horowitz, A.R., 1992. Novel phenoxy juvenile hormone analog (pyriproxyfen) suppresses embryogenesis and adult emergence of sweetpotato whitefly (Homoptera: Aleyrodidae). Journal of Economic Entomology 85, 2113–2117.

James, R.R., Buckner, J.S., Freeman, T.P., 2003. Cuticular lipids and silverleaf whitefly stage affect conidial germination of *Beauveria bassiana* and *Paecilomyces fumosoroseus*. Journal of Invertebrate Pathology 84, 67–74.

James, D.G., Price, T.S., 2002. Fecundity in twospotted spider mite (Acari: Tetranychidae) is increased by direct and systemic exposure to imidacloprid. Journal of Economic Entomology 95, 729–732.

Jamuna, B., Bheemanna, M., Hosamani, A.C., Govindappa, M.R., Nadagouda, S., 2016. Biology of whitefly *Bemisia tabaci* (Gennadius) on tomato. Journal of Experimental Zoology India 19, 475–477.

Jauset, A.M., Sarasua, M.J., Avilla, J., Albajes, R., 2000. Effect of nitrogen fertilization level applied to tomato on the greenhouse whitefly. Crop Protection 19, 255–261.

Jervis, M.A., Heimpel, G.E., 2005. Phytophagy. In: Jervis, M.A. (Ed.), Insects as Natural Enemies. Springer, Dordrecht, The Netherlands, pp. 525–550.

Ji, Y., Scott, J.W., Hanson, P., Graham, E., Maxwell, D.P., 2007. Sources of resistance, inheritance, and location of genetic loci conferring resistance to members of the tomato-infecting begomoviruses. In: Czosnek, H. (Ed.), Tomato Yellow Leaf Curl Virus Disease: Management, Molecular Biology, Breeding for Resistance. Springer, Dordrecht, The Netherlands, pp. 343–362.

Johnson, M.W., Caprio, L.C., Coughlin, J.A., Tabashnik, B.E., Rosenheim, J.A., Welter, S.C., 1992. Effect of *Trialeurodes vaporarorum* (Homoptera: Aleyrodidae) on yield of fresh market tomatoes. Journal of Economic Entomology 85, 2370–2376.

Jones, D.R., 2003. Plant viruses transmitted by whiteflies. European Journal of Plant Pathology 109, 195–219.

Junaid, Z., Freed, S., Khan, B.A., Farooq, M., 2016. Effectiveness of *Beauveria bassiana* against cotton whitefly, *Bemisia tabaci* (Gennadius) (Aleyrodidae: Homoptera) on different host plants. Pakistan Journal of Zoology 48, 91–99.

Juraske, R., Castells, F., Vijay, A., Muñoz, P., Antón, A., 2009. Uptake and persistence of pesticides in plants: measurements and model estimates for imidacloprid after foliar and soil application. Journal of Hazardous Materials 165, 683–689.

Kakimoto, K., Inoue, H., Yamaguchi, T., Ueda, S., Honda, K., Yano, E., 2007. Host plant effect on development and reproduction of *Bemisia argentifolii* Bellows and Perring (*B. tabaci* [Gennadius] B-biotype) (Homoptera: Aleyrodidae). Applied Entomology and Zoology 42, 63–70.

Kakutani, K., Matsuda, Y., Nonomura, T., Toyoda, H., Kimbara, J., Osamura, K., Kusakari, S., 2012. Practical application of an electric field screen to an exclusion of flying insect pests and airborne fungal conidia from greenhouses with a good air penetration. Journal of Agricultural Science 4, 51–60.

Kaloshian, I., Walling, L.L., 2005. Hemipterans as plant pathogens. Annual Review of Phytopathology 43, 491–521.

Kaloshian, I., Walling, L.L., 2015. Hemipteran and dipteran pests: effectors and plant host immune regulators. Journal of Integrative Plant Biology 58, 350–361.

Kapongo, J.P., Shipp, L., Kevan, P., Broadbent, B., 2008a. Optimal concentration of *Beauveria bassiana* vectored by bumble bees in relation to pest and bee mortality in greenhouse tomato and sweet pepper. BioControl 53, 797–812.

Kapongo, J.P., Shipp, L., Kevan, P., Sutton, J.C., 2008b. Co-vectoring of *Beauveria bassiana* and *Clonostachys rosea* by bumble bees (*Bombus impatiens*) for control of insect pests and suppression of grey mould in greenhouse tomato and sweet pepper. Biological Control 46, 508–514.

Karanja, J., Poehling, H.M., Pallmann, P., 2015. Efficacy and dose response of soil-applied neem formulations in substrates with different amounts of organic matter, in the control of whiteflies, *Aleyrodes proletella* and *Trialeurodes vaporariorum* (Hemiptera: Aleyrodidae). Journal of Economic Entomology 108, 1182–1190.

Karatolos, N., Williamson, M.S., Denholm, I., Gorman, K., Nauen, R., 2012. Resistance to spiromesifen in *Trialeurodes vaporariorum* is associated with a single amino acid replacement in its target enzyme acetyl-coenzyme A carboxylase. Insect Molecular Biology 21, 327–334.

Kim, J.S., Je, Y.H., Choi, J.Y., 2010. Complementary effect of Phloxine B on the insecticidal efficacy of *Isaria fumosorosea* SFP-198 wettable powder against greenhouse whitefly, *Trialeurodes vaporariorum* West. Pest Management Science 66, 1337–1343.

Kim, J.K., Park, J.J., Pak, C.H., Park, H., Cho, K., 1999. Implementation of yellow sticky trap for management of greenhouse whitefly in cherry tomato greenhouse. Journal of the Korean Society for Horticultural Science 40, 549–553.

Kim, J.K., Park, J.J., Park, H., Cho, K., 2001. Unbiased estimation of greenhouse whitefly, *Trialeurodes vaporariorum*, mean density using yellow sticky trap in cherry tomato greenhouses. Entomologia Experimentalis et Applicata 100, 235–243.

Kisha, J.S.A., 1981. Observations on the trapping of the whitefly *Bemisia tabaci* by glandular hairs on tomato leaves. Annals of Applied Biology 97, 123–127.

Koike, M., Higashio, T., Komori, A., Akiyama, K., Kishimoto, N., Masuda, E., Sasaki, M., Yoshida, S., Tani, M., Kuramoti, K., Sugimoto, M., 2004. *Verticillium lecanii* (*Lecanicillium* spp.) as epiphyte and its application to biological control of arthropod pests and diseases. OILB/SROP Bulletin 27, 41–44.

Krueger, H.R., Lindquist, R.K., Mason, J.F., Spadafora, R.R., 1973. Application of methomyl to greenhouse tomatoes: greenhouse whitefly control and residues in foliage and fruits. Journal of Economic Entomology 66, 1223–1224.

Kumar, P., Poehling, H.M., 2006a. Persistence of soil and foliar azadirachtin treatments to control sweetpotato whitefly *Bemisia tabaci* Gennadius (Homoptera: Aleyrodidae) on tomatoes under controlled (laboratory) and field (netted greenhouse) conditions in the humid tropics. Journal of Pest Science 79, 189–199.

Kumar, P., Poehling, H.M., 2006b. UV-blocking plastic films and nets influence vectors and virus transmission on greenhouse tomatoes in the humid tropics. Environmental Entomology 35, 1069–1082.

Lambert, L., Chouffot, T., Tureotte, G., Lemieux, M., Moreau, J., 2005. Biological control of greenhouse whitefly (*Trialeurodes vaporariorum*) on interplanted tomato crops with and without supplemental lighting using *Dicyphus hesperus* (Quebec, Canada). OILB/SROP Bulletin 28, 175–178.

Lapidot, M., Legg, J.P., Wintermantel, W.M., Polston, J.E., 2014. Management of whitefly-transmitted viruses in open-field production systems. In: Lobenstein, G., Katis, N. (Eds.), Advances in Virus Research: Control of Plant Virus Diseases: Seed-Propagated Crops. Elsevier, Inc., Oxford, UK, pp. 147–206.

Lapidot, M., Polston, J.E., 2010. Biology and epidemiology of *Bemisa*-vectored viruses. In: Stansly, P.A., Naranjo, S.E. (Eds.), *Bemisia*: Bionomics and Management of a Global Pest. Springer, Dordrecht, Heidelberg, London, New York, pp. 227–231.

Las, A., 1979. Male courtship persistence in the greenhouse whitefly, *Trialeurodes vaporariorum* Westwood (Homoptera: Aleyrodidea). Behavior 72, 107–125.

Leckie, B.M., D'Ambrosio, D.A., Chappell, T.M., Halitschke, R., De Jong, D.M., Kessler, A., Kennedy, G.G., Mutschler, M.A., 2016. Differential and synergistic functionality of acylsugars in suppressing oviposition by insect herbivores. PLoS One 11, e0153345.

Leckie, B.M., De Jong, D.M., Mutschler, M.A., 2012. Quantitative trait loci increasing acylsugars in tomato breeding lines and their impacts on silverleaf whiteflies. Molecular Breeding 30, 1621–1634.

Lee, D.H., Nyrop, J.P., Sanderson, J.P., 2010. Effect of host experience of the greenhouse whitefly, *Trialeurodes vaporariorum*, on trap cropping effectiveness. Entomologia Experimentalis et Applicata 137, 193–203.

Lee, M., Yoon, C.S., Yun, T.Y., Kim, H.S., Yoo, J.K., 2002. Selection of a highly virulent *Verticillium lecanii* strain against *Trialeurodes vaporariorum* at various temperatures. Journal of Microbiology and Biotechnology 12, 145–148.

Legaspi, J.C., Legaspi Jr., B.C., Simmons, A.M., Soumare, M., 2008. Life table analysis for immature and female adults of the predatory beetle, *Delphastus catalinae*, feeding on whiteflies under three constant temperatures. Journal of Insect Science 8, 7.

Legaspi, J.S., Nordlund, D.A., Legaspi Jr., B.C., 1996. Tri-trophic interactions and predation rates in *Chrysoperla* spp. attacking the silverleaf whitefly. Southwestern Entomologist 21, 33–42.

Legaspi, J.C., Simmons, A.M., Legaspi Jr., B.C., 2006. Prey preference by *Delphastus catalinae* (Coleoptera: Coccinellidae) on *Bemisia argentifolii* (Homoptera: Aleyrodidae): effects of plant species and prey stages. Florida Entomologist 89, 218–222.

Lei, H., Tjallingi, W.F., van Lenteren, J.C., 1998. Probing and feeding characteristics of the greenhouse whitefly in association with host-plant acceptance and whitefly strains. Entomologia Experimentalis et Applicata 88, 73–80.

Leibman, D., Prakash, S., Wolf, D., Zelcer, A., Anfoka, G., Haviv, S., Brumin, M., Gaba, V., Arazi, T., Lapidot, M., Gal-On, A., 2015. Immunity to tomato yellow leaf curl virus in transgenic tomato is associated with accumulation of transgene small RNA. Archives of Virology 160, 2727–2739.

Li, T.Y., Maschwitz, U., 1985. Sexual behavior in the whitefly *Trialeurodes vaporariorum* Westw. Acta Entomologica Sinica 28, 233–236.

Li, Y., Zhong, S., Qin, Y., Zhang, S., Gao, Z., Dang, Z., Pan, W., 2014. Identification of plant chemicals attracting and repelling whiteflies. Arthropod-Plant Interactions 8, 183–190.

Lima, I.P., Resende, J.T.V., Oliveira, J.R.F., Faria, M.V., Dias, D.M., Resende, N.C.V., 2016. Selection of tomato genotypes for processing with high zingiberene content, resistant to pests. Horticultura Brasileira 34, 387–391.

Lin, C., DerChung, W., JihZu, Y., BingHuei, C., ChinLing, W., WenHsiung, K., 2009. Control of silverleaf whitefly, cotton aphid and Kanzawa spider mite with oil and extracts from seeds of sugar apple. Neotropical Entomology 38, 531–536.

Lindquist, R.K., Bauerle, W.L., Spadafora, R.R., 1972a. Effect of the greenhouse whitefly on yields of greenhouse tomatoes. Journal of Economic Entomology 65, 1406–1408.

Lindquist, R.K., Krueger, H.R., Spadafora, R.R., Mason, J.F., 1972b. Application of aldicarb to greenhouse tomatoes: plant growth, fruit yields, greenhouse whitefly control and residues in fruits. Journal of Economic Entomology 65, 862–864.

Liu, S.S., De Barro, P.J., Xu, J., Luan, J.B., Zang, L.S., Ruan, Y.M., Wan, F.H., 2007. Asymmetric mating interactions drive widespread invasion and displacement in a whitefly. Science 318, 1769–1772.

Liu, T.X., 2005. Life history and life table analysis of the whitefly predator *Delphastus catalinae* (Coleoptera: Coccinellidae) on collards. Insect Science 12, 129–135.

Liu, T.X., Stansly, P.A., 1999. Searching and feeding behavior of *Nephaspis oculatus* and *Delphastus catalinae* (Coleoptera: Coccinellidae), predators of *Bemisia argentifolii* (Homoptera: Aleyrodidae). Environmental Entomology 28, 901–906.

Liu, T.X., Stansly, P.A., 2000. Insecticidal activity of surfactants and oils against silverleaf whitefly (*Bemisia argentifolii*) nymphs (Homoptera: Aleyrodidae) on collards and tomato. Pest Management Science 56, 861–866.

Liu, T.X., Stansly, P.A., 2004. Lethal and sublethal effects of two insect growth regulators on adult *Delphastus catalinae* (Coleoptera: Coccinellidae), a predator of whiteflies (Homoptera: Aleyrodidae). Biological Control 30, 298–305.

Liu, T.X., Stansly, P.A., Gerling, D., 2015. Whitefly parasitoids: distribution, life history, bionomics, and utilization. Annual Review of Entomology 60, 273–292.

Liu, T.X., Stansly, P.A., Hoelmer, K.A., Osborne, L.S., 1997. Life history of *Nephaspis oculatus* (Coleoptera: Coccinellidae), a predator of *Bemisia argentifolii* (Homoptera: Aleyrodidae). Annals of the Entomological Society of America 90, 776–782.

Lloyd, L., 1922. The control of the greenhouse white fly (*Asterochiton vaporariorum*) with notes on its biology. Annals of Applied Biology 9, 1–32.

López-Avila, A., 1986. Taxonomy and biology. In: Cock, M.J. (Ed.), *Bemisia Tabaci*: A Literature Survey on the Cotton Whitefly with an Annotated Bibliography. International Institute of Biological Control, Chamaleon Press, London, UK, pp. 3–11.

Lorenzo, M.E., Grille, G., Basso, C., Bonato, O., 2016. Host preference and biotic potential of *Trialeurodes vaporariorum* and *Bemisia tabaci* (Hemiptera: Aleyrodidae) in tomato and pepper. Arthropod-plant Interactions 10, 293–301.

Lucatti, A.F., Alvarez, A.E., Machado, C.R., Gilardon, E., Gilardón, E., 2010. Resistance of tomato genotypes to the greenhouse whitefly *Trialeurodes vaporariorum* (West.) (Hemiptera: Aleyrodidae). Neotropical Entomology 39, 792–798.

Lucatti, A.F., Meijer-Dekens, F.R., Mumm, R., Visser, R.G., Vosman, B., van Heusden, S., 2014. Normal adult survival but reduced *Bemisia tabaci* oviposition rate on tomato lines carrying an introgression from *S. habrochaites*. BMC Genetics 15, 142.

Lykouressis, D.P., Perdikis, D.C., Konstantinou, A.D., 2009. Predation rates of *Macrolophus pygmaeus* (Hemiptera: Miridae) on different densities of eggs and nymphal instars of the greenhouse whitefly *Trialeurodes vaporariorum* (Homoptera: Aleyrodidae). Entomologia Generalis 32, 105–112.

Maliepaard, C., Bas, N.J., Van Heusden, S., Kos, J., Pet, G., Verkerk, R., Vrielink, R., Zabel, P., Lindhout, P., 1995. Mapping of QTLs for glandular trichome densities and *Trialeurodes vaporariorum* (greenhouse whitefly) resistance in an F_2 from *Lycopersicon esculentum* X *Lycopersicon hirsutum* f. *glabratum*. Heredity 75, 425–433.

Mansour, A., Al-Musa, A., 1992. Tomato yellow leaf curl: host-range and virus-vector relationships. Plant Pathology 41, 122–125.

Martin, N.A., Ball, R.D., Noldus, L., van Lenteren, J.C., 1991. Distribution of greenhouse whitefly *Trialeurodes vaporariorum* (Homoptera, Aleyrodidae) and *Encarsia formosa* (Hymenoptera, Aphelinidae) in a greenhouse tomato crop – implications for sampling. New Zealand Journal of Crop and Horticultural Science 19, 283–290.

Martin, J.H., Mound, L.A., 2007. An Annotated Check List of the World's Whiteflies (Insecta: Hemiptera: Aleyrodidae). Zootaxa 1492. Magnolia Press, Aukland, New Zealand.

Martinez de Ilarduya, O., Nombela, G., Hwang, C., Williamson, V.M., Muniz, M., Kaloshian, I., 2004. *Rme1* is necessary for *Mi*-1-mediated resistance and acts early in the resistance pathway. Molecular Plant-microbe Interactions 17, 55–61.

Maynard, D.N., Cantliffe, D.J., 1989. Squash Silverleaf and Tomato Irregular Ripening: New Vegetable Disorders in Florida. Florida Cooperative Extension Service, IFAS VC-37. , p. 4.

McCollum, T.G., Stoffella, P.J., Powell, C.A., Cantliffe, D.J., Hanif-Khan, S., 2004. Effects of silverleaf whitefly feeding on tomato fruit ripening. Postharvest Biology and Technology 31, 183–190.

McDaniel, T., Tosh, C.R., Gatehouse, A.M.R., George, D., Robson, M., Brogan, B., 2016. Novel resistance mechanisms of a wild tomato against the glasshouse whitefly. Agronomy for Sustainable Development 36, 14.

McKenzie, C.L., Kumar, V., Palmer, C.L., Oetting, R.D., Osborne, L.S., 2014. Chemical class rotations for control of *Bemisia tabaci* (Hemiptera: Aleyrodidae) on poinsettia and their effect on cryptic species population composition. Pest Management Science 70, 1573–1587.

Metcalf, R.L., Metcalf, R.A., 1993. Destructive and Useful Insects: Their Habits and Control, fifth ed. McGraw-Hill, Inc., New York, USA.

Messelink, G., van Steenpaal, S.E.F., Ramakers, P., 2006. Evaluation of phytoseiid predators for control of western flower thrips on greenhouse cucumber. BioControl 51, 753–768.

Miles, H.W., 1927. On the control of glasshouse insects with calcium cyanide. Annals of Applied Biology 14, 240–246.

Milligan, S.B., Bodeau, J., Yaghoobi, J., Kaloshian, I., Zabel, P., Williamson, V.M., 1998. The root knot nematode resistance gene Mi from tomato is a member of the leucine zipper, nucleotide binding, leucine-rich repeat family of plant genes. The Plant Cell 10, 1307–1319.

Mohamed, E.F., 2010. Interaction between some viruses which attack tomato (*Lycopersicon esculentum* Mill.) plants and their effect on growth and yield of tomato plants. Journal of American Science 6, 311–320.

Mohd Rasdi, Z., Fauziah, I., Wan Mohamad, W.A.K., Abdul, S., Rahman, S.R., Che Salmah, M.R., Kamaruzaman, J., 2009. Biology of *Macrolophus caliginosus* (Heteroptera: Miridae) predator of *Trialeurodes vaporariorum* (Homoptera: Aleyrodidae). International Journal of Biology 1, 63–70.

Moriones, E., Navas-Castillo, J., 2000. Tomato yellow leaf curl virus, an emerging virus complex causing epidemics worldwide. Virus Research 71, 123–134.

Moreno-Delafuente, A., Garzo, E., Moreno, A., Fereres, A., 2013. A plant virus manipulates the behavior of its whitefly vector to enhance its transmission efficiency and spread. PLoS One 8, e61543.

Moreno-Ripoll, R., Agusti, N., Berruezo, R., Gabarra, R., 2012a. Conspecific and heterospecific interactions between two omnivorous predators on tomato. Biological Control 62, 189–196.

Moreno-Ripoll, R., Gabarra, R., King, R.A., Agusti, N., Symondson, W.O., Agustí, N., 2012b. Trophic relationships between predators, whiteflies and their parasitoids in tomato greenhouses: a molecular approach. Bulletin of Entomological Research 102, 415–423.

Moreno-Ripoll, R., Gabarra, R., Symondson, W.O., King, R.A., Agusti, N., Agustí, N., 2014. Do the interactions among natural enemies compromise the biological control of the whitefly *Bemisia tabaci*? Journal of Pest Science 87, 133–141.

Morrill, A.W., 1905. The greenhouse white-fly (*Aleyrodes vaporariorum* Westw.). United States Department of Agriculture Bureau of Entomology Circular 57, 1–9.

Mound, L.A., 1962. Studies on olfaction and color sensitivity of *Bemisia tabaci*. Entomologia Experimentalis et Applicata 5, 99–104.

Mound, L.A., Halsey, S.H., 1978. Whitefly of the World. Wiley, New York, USA, p. 340.

Muigai, S.G., Bassett, M.J., Schuster, D.J., Scott, J.W., 2003. Greenhouse and field screening of wild *Lycopersicon* germplasm for resistance to the whitefly *Bemisia argentifolii*. Phytoparasitica 31, 27–38.

Muigai, S.G., Schuster, D.J., Bassett, M.J., Scott, J.W., McAuslane, H.J., 2002. Mechanisms of resistance in *Lycopersicon* germplasm to the whitefly *Bemisia argentifolii*. Phytoparasitica 30, 347–360.

Mutisya, S., Saidi, M., Opiyo, A., Ngouajio, M., Martin, T., 2016. Synergistic effects of agronet covers and companion cropping on reducing whitefly infestation and improving yield of open field-grown tomatoes. Agronomy 6, 42.

Nakano, R., Tsuchida, Y., Doi, M., Ishikawa, R., Tatara, A., 2016. Control of *Bemisia tabaci* (Gennadius) on tomato in greenhouses by a combination of *Nesidiocoris tenuis* (Reuter) and banker plants. Annual Report of the Kansai Plant Protection Society 58, 65–72.

Naranjo, S.E., 2001. Conservation and evaluation of natural enemies in IPM systems for *Bemisia tabaci*. Crop Protection 20, 835–852.

Naranjo, S.E., Castle, S.J., De Barro, P.J., Liu, S.S., 2010. Population dynamics, demography, dispersal and spread of *Bemisis tabaci*. In: Stansly, P.A., Naranjo, S.E. (Eds.), *Bemisia*: Bionomics and Management of Global Pest. Springer, New York, pp. 185–226.

Naranjo, S.E., Hagler, J.R., 1998. Characterizing and estimating the impact of heteropteran predation. In: Coll, M., Ruberson, J. (Eds.), Predatory Heteroptera: Their Ecology and Use in Biological Control, Thomas Say Symposium Proceedings. Entomological Society of America, Lanham, Maryland, USA, pp. 170–197.

Nauen, R., Ghanim, M., Ishaaya, I., 2014. Whitefly special issue organized in two parts. Pest Management Science 70, 1438–1439.

Nauen, R., Reckmann, U., Armborst, S., Stupp, H.P., Elbert, A., 1999. Whitefly-active metabolites of midacloprid: biological efficacy and translocation in cotton plants. Pesticide Science 55, 265–271.

Nauen, R., Reckmann, U., Thomzik, J., Thielert, W., 2008. Biological profile of spirotetramat (Movento®) – a new two-way systemic (ambimobile) insecticide against sucking pest species. Bayer CropScience Journal 61, 245–278.

Nava-Camberos, U., Riley, D.G., Harris, M.K., 2001. Temperature and host plant effects on development, survival, and fecundity of *Bemisia argentifolii* (Homoptera: Aleyrodidae). Environmental Entomology 30, 55–63.

Nguyen-Dang, L., Vankosky, M., VanLaerhoven, S., 2016. The effects of alternative host plant species and plant quality on *Dicyphus hesperus* populations. Biological Control 100, 94–100.

Noemi Lopez, S., Arce Rojas, F., Villalba Velasquez, V., Cagnotti, C., 2012. Biology of *Tupiocoris cucurbitaceus* (Hemiptera: Miridae), a predator of the greenhouse whitefly *Trialeurodes vaporariorum* (Hemiptera: Aleyrodidae) in tomato crops in Argentina. Biocontrol Science and Technology 22, 1107–1117.

Nombela, G., Beitia, F., Muñiz, M., 2000. Variation in tomato host response to *Bemisia tabaci* (Hemiptera: Aleyrodidae) in relation to acyl sugar content and presence of the nematode and potato aphid resistance gene Mi. Bulletin of Entomological Research 90, 161–167.

Nombela, G., Beitia, F., Muniz, M., 2001. A differential interaction study of *Bemisia tabaci* Q-biotype on commercial tomato varieties with or without the Mi resistance gene, and comparative host responses with the B-biotype. Entomologia Experimentalis et Applicata 98, 339–344.

Nombela, G., Muniz, M., Kaloshian, I., 2003a. A mutation in the Rme1 tomato locus reduces Mi-1.2-mediated resistance to whitefly *Bemisia tabaci*. OILB/SROP Bulletin 26, 57–59.

Nombela, G., Williamson, V.M., Muñiz, M., 2003b. The root-knot nematode resistance gene Mi-1.2 of tomato is responsible for resistance against the whitefly *Bemisia tabaci*. Molecular Plant-Microbe Interactions 16, 645–649.

Nomikou, M., Janssen, A., Sabelis, M.W., 2003. Phytoseiid predators of whiteflies feed and reproduce on non-prey food sources. Experimental and Applied Acarology 31, 15–26.

Nomikou, M., Janssen, A., Schraag, R., Sabelis, M.W., 2001. Phytoseiid predators as potential biological control agents for *Bemisia tabaci*. Experimental and Applied Acarology 25, 271–291.

Nomikou, M., Janssen, A., Schraag, R., Sabelis, M.W., 2002. Phytoseiid predators suppress populations of *Bemisia tabaci* on cucumber plants with alternative food. Experimental and Applied Acarology 27, 57–68.

Nomikou, M., Janssen, A., Schraag, R., Sabelis, M.W., 2004. Vulnerability of *Bemisia tabaci* immatures to phytoseiid predators: consequences for oviposition and influence of alternative food. Experimental and Applied Acarology 110, 95–102.

Nonomura, T., Matsuda, Y., Kakutani, K., Kimbara, J., Osamura, K., Kusakari, S., Toyoda, H., 2012. An electric field strongly deters whiteflies from entering window-open greenhouses in an electrostatic insect exclusion strategy. European Journal of Plant Pathology 134, 661–670.

Nonomura, T., Matsuda, Y., Kakutani, K., Takikawa, Y., Kimbara, J., Osamura, K., Kusakari, S., Toyoda, H., 2014. Prevention of whitefly entry from a greenhouse entrance by furnishing an airflow-oriented pre-entrance room guarded with electric field screens. Journal of Agricultural Science 6, 172–184.

Noyes, J.S., 2012. Universal Chalcidoidea Database. Natural History Museum, London. http://www.nhm.ac.uk/chalcidoids.

Nzanza, B., Mashela, P.W., 2012. Control of whiteflies and aphids in tomato (*Solanum lycopersicum* L.) by fermented plant extracts of neem leaf and wild garlic. African Journal of Biotechnology 11, 16077–16082.

Okadome, K., Amano, H., 2014. Wavelengths of reflected light that attract greenhouse whitefly *Trialeurodes vaporariorum* (Hemiptera: Aleyrodidae) to board sticky traps using sharp-cut filters. Japanese Journal of Applied Entomology and Zoology 58, 197–201.

Okamoto, M., Kita, T., Okuda, H., Tanaka, T., Nakashima, T., 1994. Effects of aging on acute toxicity of nicotine in rats. Pharmacology and Toxicology 75, 1–6.

Oriani, M.A.D.G., Vendramim, J.D., 2010. Influence of trichomes on attractiveness and ovipositional preference of *Bemisia tabaci* (Genn.) B biotype (Hemiptera: Aleyrodidae) on tomato genotypes. Neotropical Entology 39, 1002–1007.

Orozco Munoz, A., Villalba Velasquez, V., Noemi Lopez, S., 2012. Development of *Tupiocoris cucurbitaceus* (Hemiptera: Miridae) on *Bemisia tabaci* (Hemiptera: Aleyrodidae) in several vegetables. Fitosanidad 16, 147–153.

Ozores-Hampton, M., Stansly, P.A., McAvoy, E., 2013. Evaluation of round and roma-type tomato varieties and advanced breeding lines resistant to Tomato yellow leaf curl virus in Florida. HortTechnology 23, 689–698.

Palacios, S.M., Carpinella, M.C., Mangeaud, A., Valladares, G., Defago, M.T., 2008. Effect of *Melia azedarach* fruit extract on *Trialeurodes vaporariorum* in organic crops under greenhouse conditions. Biopesticides International 4, 121–127.

Pan, H., Preisser, E.L., Chu, D., Wang, S., Wu, Q., Carriére, Y., Zhou, X., Zhang, Y., 2015. Insecticides promote viral outbreaks by altering herbivore competition. Ecological Applications 25, 1585–1595.

Pang, S., Wang, L., Hou, Y., Shi, Z., 2011. Interspecific interference competition between *Encarsia formosa* and *Encarsia sophia* (Hymenoptera: Aphelinidae) in parasitizing *Bemisia tabaci* (Hemiptera: Aleyrodidae) on five tomato varieties. Insect Science 18, 92–100.

Papayiannis, L.C., Katis, N.I., Idris, A.M., Brown, J.K., 2011. Identification of weed hosts of tomato yellow leaf curl virus in Cyprus. Plant Disease 95, 120–125.

Pappas, M.L., Migkou, F., Broufas, G.D., 2013. Incidence of resistance to neonicotinoid insecticides in greenhouse populations of the whitefly, *Trialeurodes vaporariorum* (Hemiptera: Aleyrodidae) from Greece. Applied Entomology and Zoology 38, 373–378.

Park, M.K., Kim, J.G., Song, Y.H., Lee, J., Shin, K., Cho, K., 2009. Effect of nitrogen levels of two cherry tomato cultivars on development, preference and honeydew production of *Trialeurodes vaporariorum* (Hemiptera: Aleyrodidae). Journal of Asia-Pacific Entomology 12, 227–232.

Park, J., Lee, J., Shin, K., Lee, S.E., Cho, K., 2011. Geostatistical analysis of the attractive distance of two different sizes of yellow sticky traps for greenhouse whitefly, *Trialeurodes vaporariorum* (Westwood) (Homoptera: Aleyrodidae), in cherry tomato greenhouses. Australian Journal of Entomology 50, 144–151.

Parker, T., 1928. The use of tetrachloroethane for commercial glasshouse fumigation. Annals of Applied Biology 15, 251–257.

Parolin, P., Bresch, C., Ottenwalder, L., Ion Scotta, M., Brun, R., 2013. False yellowhead (*Dittrichia viscosa*) causes over infestation with the whitefly pest (*Trialeurodes vaporariorum*) in tomato crops. International Journal of Agricultural Policy and Research 1, 311–318.

Parolin, P., Bresch, C., Poncet, C., Suay-Cortez, R., Van Oudenhove, L., 2015. Testing basil as banker plant in IPM greenhouse tomato crops. International Journal of Pest Management 61, 235–242.

Parr, W.J., Gould, H.J., Jessop, H.H., Ludlam, F.A.B., 1976. Progress towards a biological control programme for glasshouse whitefly (*Trialeurodes vaporariorum*) on tomatoes. Annals of Applied Biology 83, 349–363.

Paulson, G.S., Beardsley, J.W., 1985. Whitefly (Hemiptera, Aleyrodidae) egg pedicel insertion into host plant stomata. Annals of the Entomological Society of America 78, 506–508.

Pedigo, L.P., Rice, M.E., 2006. Entomology and Pest Management, fifth ed. Pearson Education Inc., Upper Saddle River, New Jersey, USA.

Pei-Xiang, Z., Mei-Chang, L., Yu-Chuan, Q., Jian-Jun, X., Yun-Hong, L., 2011. Control effects of whitefly by intercropping celery in greenhouse. Chinese Journal of Applied Entomology 48, 375–378.

Peralta, L., Hilje, L., 1993. Un intento de control de *Bemisia tabaci* con insecticidas sistemicos incorporados a la vainica como cultivo trampa, más aplicaciones de aceite en el tomate. Manejo Integrado de Plagas (Costa Rica) 30, 21–23.

Pereira Carvalho, R.C., Diaz Pendon, J.A., Fonseca, M.E.N., Boiteux, L.S., Fernandez-Munoz, R., Moriones, E., Resende, R.O., 2015. Recessive resistance derived from tomato cv. Tyking-limits drastically the spread of Tomato yellow leaf curl virus. Viruses 7, 2518–2525.

Perring, T.M., 1996. Biological differences of two species of *Bemisia* that contribute to adaptive advantage. In: Gerling, D., Mayer, R.T. (Eds.), *Bemisia* 1995: Taxonomy, Biology, Damage Control and Management. Intercept Ltd., Andover, Hants, UK, pp. 3–16.

Perring, T.M., 2001. The *Bemisia tabaci* species complex. Crop Protection 20, 725–737.

Perring, T.M., Cooper, A., Kazmer, D.J., 1992. Identication of the poinsettia strain of *Bemisia tabaci* (Homoptera, Aleyrodidae) on broccoli by electrophoresis. Journal of Economic Entomology 85, 1278–1284.

Perring, T.M., Cooper, A.D., Rodriguez, R.J., Farrar, C.A., Bellows, T.S., 1993. Identification of a whitefly species by genomic and behavioral studies. Science 259, 74–77.

Perring, T.M., Symmes, E.J., 2006. Courtship behavior of *Bemisia argentifolii* (Hemiptera: Aleyrodidae) and whitefly mate recognition. Annals of the Entomological Society of America 99, 598–606.

Picó, B., Díez, M.J., Nuez, F., 1996. Viral diseases causing the greatest economic losses to the tomato crop. II. The Tomato yellow leaf curl virus—a review. Scientia Horticulturae 67, 151–196.

Pinto-Zevallos, D.M., Vanninen, I., 2013. Yellow sticky traps for decision-making in whitefly management: what has been achieved? Crop Protection 47, 74–84.

Polston, J.E., Anderson, P.K., 1997. The emergence of whitefly transmitted geminiviruses in tomato in the Western Hemisphere. Plant Disease 81, 1358–1369.

Polston, J.E., De Barro, P., Boykin, L.M., 2014. Transmission specificities of plant viruses with the newly identified species of the *Bemisia tabaci* species complex. Pest Management Science 70, 1547–1552.

Polston, J.E., Schuster, D.J., Taylor, J.E., 2009. Identification of weed reservoirs of Tomato yellow leaf curl virus in Florida. In: Simone, E., Snodgrass, C., Ozores-Hampton, M. (Eds.), Florida Tomato Institute Proceedings. Naples, Florida, USA. University of Florida Institute of Food and Agricultural Sciences, Gainesville, Florida, USA, pp. 32–33.

Powell, C.A., Stoffella, P.J., 1998. Control of tomato irregular ripening with imidacloprid. HortScience 33, 283–284.

Prabhaker, N., Castle, S.J., Naranjo, S.E., Toscano, N.C., Morse, J.G., 2011. Compatibility of two systemic neonicotinoids, imidacloprid and thiamethoxam, with various natural enemies of agricultural pests. Journal of Economic Entomology 104, 773–781.

Prasanna, H.C., Sinha, D.P., Rai, G.K., Krishna, R., Kashyap, S.P., Singh, N.K., Singh, M., Malathi, V.G., 2015. Pyramiding Ty-2 and Ty-3 genes for resistance to monopartite and bipartite Tomato leaf curl viruses of India. Plant Pathology 64, 256–264.

Pratissoli, D., de Carvalho, J.R., Pastori, P.L., Oliveira de Freitas Bueno, R.C., Zago, H.B., 2015. Incidence of leaf miner and insect vectors for pest management systems in the tomato. Revista Ciencia Agronomica 46, 607–614.

Prota, N., Bouwmeester, H.J., Jongsma, M.A., 2014. Comparative antifeedant activities of polygodial and pyrethrins against whiteflies (*Bemisia tabaci*) and aphids (*Myzus persicae*). Pest Management Science 70, 682–688.

Qiu, Y.T., van Lenteren, J.C., Drost, Y.C., Posthuma-Doodeman, C.J.A.M., 2004. Life-history parameters of *Encarsia formosa, Eretmocerus eremicus* and *E. mundus*, aphelinid parasitoids of *Bemisia argentifolii* (Hemiptera: Aleyrodidae). European Journal of Entomology 101, 83–94.

Quaintance, A.L., 1900. Contributions toward a monograph of the American Aleurodidae. Bulletin of the United States Department of Agriculture Entomology Technician 8, 9–43.

Quaintance, A.L., Baker, A.C., 1913. Classification of the Aleyrodidae. Part I. United States Department of Agriculture Technical Series 27, 1–93.

Qureshi, M.S., Midmore, D.J., Syeda, S.S., Playford, C.L., 2007. Floating row covers and pyriproxyfen help control silverleaf whitefly *Bemisia tabaci* (Gennadius) Biotype B (Homoptera: Aleyrodidae) in zucchini. Australian Journal of Entomology 46, 313–319.

Rajasri, M., Lakshmi, K.V., Reddy, K.L., 2009. Management of whitefly transmitted tomato leaf curl virus using guard crops in tomato. Indian Journal of Plant Protection 37, 101–103.

Ren, S.X., Stansly, P.A., Liu, T.X., 2002. Life history of the whitefly predator *Nephaspis oculatus* (Coleoptera: Coccinellidae) at six constant temperatures. Biological Control 23, 262–268.

Richardson, H., Bulger, J.W., Busbey, R.L., Nelson, R.H., Weigel, C.A., 1943. Studies on nicotine fumigation in greenhouses. U.S. Department of Agriculture Circular 684, 1–15.

Riddick, E.W., Simmons, A.M., 2014. Do plant trichomes cause more harm than good to predatory insects? Pest Management Science 70, 1655–1665.

Rivas, F., Nunez, P., Jackson, T., Altier, N., Nuñez, P., 2014. Effect of temperature and water activity on mycelia radial growth, conidial production and germination of *Lecanicillium* spp. isolates and their virulence against *Trialeurodes vaporariorum* on tomato plants. BioControl 59, 99–109.

Roberts, P.A., Thomason, I.J., 1986. Variability in reproduction of isolates of *Meloidogyne incognita* and *M. javanica* on resistant tomato genotypes. Plant Disease 70, 547–551.

Rodriguez, A.E., Tingey, W.M., Mutschler, M.A., 1993. Acylsugars of *Lycopersicon pennellii* deter settling and feeding of the green peach aphid (Homoptera: Aphididae). Journal of Economic Entomology 86, 34–39.

Rodriguez-Lopez, M.J., Garzo, E., Bonani, J.P., Fernandez-Munoz, R., Moriones, E., Fereres, A., 2012. Acylsucrose producing tomato plants forces *Bemisia tabaci* to shift oviposition site. PLoS One 7, e33064.

Russell, L.M., 1948. The North American Species of Whiteflies of the Genus *Trialeurodes*. United States Department of Agriculture Miscellaneous Publication No. 635, p. 85.

Ryckewaert, P., Alauzet, C., 2002. The natural enemies of *Bemisia argentifolii* in Martinique. BioControl 47, 115–126.

Sachin, U.S., Kumari, J.H.S., Kumari, S., 2016. Bio-efficacy of selected insecticides against adult whitefly, *Trialeurodes vaporariorum* (Westwood) under field condition of Chikkamagalura Taluk. Journal of Experimental Zoology India 19, 1069–1075.

Saikia, A.K., Muniyappa, V., 1989. Epidemiology and control of Tomato leaf curl virus in Southern India. Tropical Agriculture (Trinidad) 66, 350–354.

Saito, T., 2005. Preliminary experiments to control the silverleaf whitefly with electrostatic spraying of a mycoinsecticide. Applied Entomology and Zoology 40, 289–292.

Salas, J., Mendoza, O., 1995. Biology of the sweetpotato whitefly (Homoptera: Aleyrodidae) on tomato. Florida Entomologist 78, 154–160.

Salati, R., Nahkla, M.K., Rojas, M.R., Guzman, P., Jaquez, J., Maxwell, D.P., Gilbertson, R.L., 2002. Tomato yellow leaf curl virus in the Dominican Republic: characterization of an infectious clone, virus monitoring in whiteflies, and identification of reservoir hosts. Phytopathology 92, 487–496.

Sanchez, J.A., Gillespie, D.R., McGregor, R.R., 2003. The effects of mullein plants (*Verbascum thapsus*) on the population dynamics of *Dicyphus hesperus* in tomato greenhouses. Biological Control 28, 313–319.

Sanchez-Campos, S., Navas-Castillo, J., Monci, F., Diaz, J.A., Moriones, E., 2000. *Mercurialis ambigua* and *Solanum luteum*: two newly discovered natural hosts if tomato yellow leaf curl geminiviruses. European Journal of Plant Pathology 106, 391–394.

Schmidt, R.A., 2014. Leaf structures affect predatory mites (Acari: Phytoseiidae) and biological control: a review. Experimental and Applied Acarology 62, 1–17.

Schuster, D.J., 2001. Relationship of silverleaf whitefly population density to severity of irregular ripening of tomato. HortScience 36, 1089–1090.

Schuster, D.J., 2002. Action threshold for applying insect growth regulators to tomato for management of irregular ripening caused by *Bemisia argentifolii* (Homoptera: Aleyrodidae). Journal of Economic Entomology 95, 372–376.

Schuster, D.J., 2003. Preference of *Bemisia argentifolii* (Homoptera: Aleyrodidae) for selected vegetable hosts. Journal of Agricultural and Urban Entomology 20, 59–67.

Schuster, D.J., 2004. Squash as a trap crop to protect tomato from whitefly-vectored tomato yellow leaf curl. International Journal of Pest Management 50, 281–284.

Schuster, D.J., Mueller, T.F., Kring, J.B., Price, J.F., 1990. Relationship of the sweetpotato whitefly to a new tomato fruit disorder in Florida. HortScience 25, 1618–1620.

Schuster, D.J., Stansly, P.A., Polston, J.E., 1996. Expressions of plant damage from *Bemisia*. In: Gerling, G., Mayer, R.T. (Eds.), *Bemisia* 1995: Taxonomy, Biology, Damage Control and Management. Intercept, Ltd., Andover, UK, pp. 153–165.

Schuster, D.J., Stansly, P.A., Polston, J.E., Gilreath, P.R., McAvoy, E., 2007. Management of Whiteflies, Whitefly-Transmitted Plant Virus, and Insecticide Resistance for Vegetable Production in Southern Florida. ENY-735, IFAS Extension. University of Florida, Gainesville, Florida, USA.

Schuster, D.J., Thompson, S., Ortega, L.D., Polston, J.E., 2009. Laboratory evaluation of products to reduce settling of sweetpotato whitefly adults. Journal of Economic Entomology 102, 1482–1489.

Scopes, N.E.E., Biggerstaff, S.M., 1971. The production, handling and distribution of the whitefly *Trialeurodes vaporariorum* and its parasite *Encarsia formosa* for use in biological control programmes in glasshouses. Plant Pathology 20, 111–116.

Scott, J.W., Stevens, M.R., Barten, J.H.M., Thome, C.R., Polston, J.E., Schuster, D.J., Serra, C.A., 1996. Introgression of resistance to whitefly-transmitted geminiviruses from *Lycopersicon chilense* to tomato. In: Gerling, D., Mayer, R.T. (Eds.), *Bemisia* 1995: Taxonomy, Biology, Damage, Control and Management. Intercept, Ltd., Andover, UK, pp. 357–367.

Selby, T.P., Lahm, G.P., Stevenson, T.M., Hughes, K.A., Cordova, D., Annan, I.B., Barry, J.D., Benner, E.A., Currie, M.J., Pahutski, T.F., 2013. Discovery of cyantraniliprole, a potent and selective anthranilic diamide ryanodine receptor activator with cross-spectrum insecticidal activity. Bioorganic and Medicinal Chemistry Letters 23, 6341–6345.

Sengonca, C., Al-Zyoud, F., Blaeser, P., 2004. Life table of the entomophagous ladybird *Serangium parcesetosum* Sicard (Coleoptera: Coccinellidae) by feeding on *Bemisia tabaci* (Genn.) (Homoptera: Aleyrodidae) as prey at two different temperatures and plant species. Zeitschrift für Pflanzenkrankheiten und Pflanzenschutz 111, 598–609.

Sengonca, C., Al-Zyoud, F., Blaeser, P., 2005. Prey consumption by larval and adult stages of the entomophagous ladybird *Serangium parcesetosum* Sicard (Col., Coccinellidae) of the cotton whitefly, *Bemisia tabaci* (Genn.) (Hom., Aleyrodidae), at two different temperatures. Journal of Pest Science 78, 179–186.

Sharaf, N., Batta, Y., 1985. Effect of some factors on the relationship between the whitefly *Bemisia tabaci* Genn. (Homoptera, Aleyrodidae) and the parasitoid *Eretmocerus mundus* Mercet (hymenoptera, Aphelinidae). Zeitschrift fur Angewandte Entomologie 99, 267–276.

Shefali, M., Jagadeesh, K.S., Krishnaraj, P.U., Prem, S., 2014. Biocontrol of tomato leaf curl virus (ToLCV) in tomato with chitosan supplemented formulations of *Pseudomonas* sp. under field conditions. Australian Journal of Crop Science 8, 347–355.

Shi, X., Chen, G., Tian, L., Peng, Z., Xie, W., Wu, Q., Wang, S., Zhou, X., Zhang, Y., 2016. The salicylic acid-mediated release of plant volatiles affects the host choice of *Bemisia tabaci*. International Journal of Molecular Sciences 17, 1048.

Shojai, M., Ostovan, H., Zamanizadeh, H., Labbafi, Y., Nasrollahi, A., Ghasemzadeh, M., Rajabi, M.Z., 2003. The management of cucumber and tomato intercrops and implementation of non-chemical control of pests and diseases for organic crop production in the greenhouse. Journal of Agricultural Sciences – Islamic Azad University 9, 1–39.

Simmons, A.M., Legaspi, J.C., 2004. Survival and predation of *Delphastus catalinae* (Coleoptera: Coccinellidae), a predator of whiteflies (Homoptera: Aleyrodidae), after exposure to a range of constant temperatures. Environmental Entomology 33, 839–843.

Simmons, A.M., Legaspi, J.C., 2007. Ability of *Delphastus catalinae* (Coleoptera: Coccinellidae), a predator of whiteflies (Homoptera: Aleyrodiade), to survive mild winters. Journal of Entomological Science 42, 163–173.

Simmons, A.M., Legaspi, J.C., Legaspi, B.C., 2008. Response of *Delphastus catalinae* (Coleoptera: Coccinellidae), a predator of whiteflies (Homoptera: Aleyrodidae), to relative humidity: oviposition, hatch, and immature survival. Annals of the Entomological Society of America 101, 378–383.

Singh, P., Beattie, G.A.C., Clift, A.D., Watson, D.M., Furness, G.O., Tesoriero, L., Rajakulendran, V., Parkes, R.A., Scanes, M., 2000. Petroleum spray oils and tomato integrated pest and disease management in southern Australia. General and Applied Entomology 29, 69–93.

Sivinski, J., Wahl, D., Holler, T., Al Dobai, S., Sivinski, R., 2011. Conserving natural enemies with flowering plants: estimating floral attractiveness to parasitic Hymenoptera and attraction's relationship to flower and plant morphology. Biological Control 58, 208–214.

Smith, F.F., 1970. Identifying and controlling the greenhouse whitefly. American Vegetable Grower 18, 41–42.

Smith, F., Fulton, R.A., Brierly, P., 1947. Use of DDT and H.E.T.P. as aerosols in greenhouses. Agricultural Chemicals 2, 28–31.

Smith, H.A., Giurcanu, M.C., 2013. Residual effects of new insecticides on egg and nymph densities of *Bemisia tabaci* (Hemiptera: Aleyrodidae). The Florida Entomologist 96, 504–511.

Smith, H.A., Giurcanu, M.C., 2014. New insecticides for management of tomato yellow leaf curl, a virus vectored by the silverleaf whitefly, *Bemisia tabaci*. Journal of Insect Science 14, 183.

Smith, H.A., Koenig, R.L., McAuslane, H.J., McSorley, R., 2000. Effect of silver reflective mulch and a summer squash trap crop on densities of immature *Bemisia argentifolii* (Homoptera: Aleyrodidae) on organic bean. Journal of Economic Entomology 93, 726–731.

Smith, S.F., Krischik, V.A., 1999. Effects of systemic imidacloprid on *Coleomegilla maculata* (Coleoptera: Coccinellidae). Environmental Entomology 28, 1189–1195.

Smith, H.A., McSorley, R., Sierra Izaguirre, J.A., 2001. The effect of intercropping common bean with poor hosts and non-hosts on numbers of immature whiteflies (Homoptera: Aleyrodidae) in the Salamá valley, Guatemala. Environmental Entomology 30, 89–100.

Smith, H.A., Nagle, C.A., Evans, G.A., 2014. Densities of eggs and nymphs and percent parasitism of *Bemisia tabaci* on common weeds in west central Florida. Insects 5, 860–876.

Smith, H.A., Seijo, T.E., Vallad, G.E., Peres, N.A., Druffel, K.L., 2015. Evaluating weeds as hosts of tomato yellow leaf curl virus. Environmental Entomology 44 (4), 1101–1107.

Soliman, M.M.M., Tarasco, E., 2008. Toxic effects of four biopesticides (Mycotal, Vertalec, Vertemic and neem Azal-t/S) on *Bemisia tabaci* (Gennadius) and *Aphis gossypii* (Glover) on cucumber and tomato plants in greenhouses in Egypt. Entomologica 41, 195–217.

Song, J.H., Lee, K.J.H., Yang, Y.T., Lee, S.C., 2014. Sampling plan for *Bemisia tabaci* adults by using yellow-color sticky traps in tomato greenhouses. Korean Journal of Applied Entomology 53, 375–380.

Spasova, P., Khristova, E., Elenkov, E.S., 1980. Pathogenicity of various species of fungi of the genus *Aschersonia* to larvae of the greenhouse whitefly (*Trialeurodes vaporariorum* Westw.) on tomatoes and cucumbers. Gradinarska i Lozarska Nauka 17, 70–75.

Speyer, E.R., Owen, O., 1926. The fumigation of tomato houses with hydrocyanic acid gas. Annals of Applied Biology 13, 144–147.

Stacey, D.L., 1977. 'Banker' plant production of *Encarsia formosa* Gahan and its use in the control of glasshouse whitefly on tomatoes. Plant Pathology 26, 63–66.

Stansly, P.A., 1996. Seasonal abundance of silverleaf whitefly in southwest Florida vegetable fields. Proceedings of the Florida State Horticultural Society 108, 234–242.

Stansly, P.A., Calvo, J., Urbaneja, A., 2005a. Augmentative biological control of *Bemisia tabaci* biotype "Q" in greenhouse pepper using *Eretmocerus* spp. (Hym. Aphelinidae). Crop Protection 24, 829–835.

Stansly, P.A., Calvo, J., Urbaneja, A., 2005b. Release rates for control of *Bemisia tabaci* (Homoptera: Aleyrodidae) biotype "Q" with *Eretmocerus mundus* (Hymenoptera: Aphelinidae) in greenhouse tomato and pepper. Biological Control 35, 124–133.

Stansly, P.A., Kostyk, B.C., 2013. Soil applications of imidacloprid and cyantraniliprol for control of sweetpotato whitefly on staked tomatoes, 2012. Arthropod Management Tests 38, 1–3.

Stansly, P.A., Kostyk, B.C., 2016a. Control of sweetpotato whitefly with insecticides applied to soil and foliage of TYLCV tolerant and susceptible staked tomatoes. Arthropod Management Tests 39, 1–5.

Stansly, P.A., Kostyk, B.C., 2016b. Control of sweetpotato whitefly with foliar insecticides on staked tomatoes, fall 2013. Arthropod Management Tests 40, 1–3.

Stansly, P.A., Liu, T.X., Schuster, D.J., 2002. Effects of horticultural mineral oils on a polyphagous whitefly, its plant hosts and its natural enemies. In: Beattie, G.A.C., Watson, D.M., Stevens, M.L., Rae, D.J., Spooner-Harts, R.N. (Eds.), Spray Oils beyond 2000: Sustainable Pest and Disease Management. University of Western Sydney, Penrith New South Wales, Australia, pp. 120–133.

Stansly, P.A., Liu, T.X., Schuster, D.J., Dean, D.E., 1996. Role of biorational insecticides in management *of Bemisia*. In: Gerling, D., Mayer Jr., R.T. (Eds.), *Bemisia* 1995: Taxonomy, Biology, Damage Control and Management. Intercept, Ltd., Andover, UK, pp. 605–615.

Stansly, P.A., Natwick, E.T., 2010. Integrated systems for managing *Bemisia tabaci* in protected and open field agriculture. In: Stansly, P.A., Naranjo, S.E. (Eds.), *Bemisia*: Bionomics and Management of a Global Pest. Springer, Dordrecht, Heidelberg, London, New York, pp. 467–497.

Stansly, P.A., Naranjo, S.E., 2010. *Bemisia*: Bionomics and Management of a Global Pest. Springer, New York, USA.

Stansly, P.A., Sanchez, P.A., Rodriguez, J.M., Canizares, F., Nieto, A., Leyva, M.L., Fajardo, M., Suarez, V., Urbaneja, A., 2004. Prospects for biological control of *Bemisia tabaci* (Homoptera, Aleyrodidae) in greenhouse tomatoes of southern Spain. Crop Protection 23, 701–712.

Stansly, P.A., Schuster, D.J., 1990. Update on sweetpotato whitefly. In: Stall, W.M. (Ed.), Proceedings of Florida Tomato Institute, Vegetable Crops Special Series SS-VEC-001. IFAS, Gainesville, Florida, USA, pp. 41–59. http://www.imok.ufl.edu/programs/veg-hort/tomato-institute/.

Su, Q., Oliver, K.M., Xie, W., Wu, Q., Wang, S., Zhang, Y., 2015. The whitefly-associated facultative symbiont *Hamiltonella defensa* suppresses induced plant defences in tomato. Functional Ecology 29, 1007–1018.

Sun, Y.X., Liu, T.X., 2016. Effectiveness of imidacloprid in combination with a root nitrogen fertilizer applied to tomato seedlings against *Bemisia tabaci* (Hemiptera: Aleyrodidae). Crop Protection 80, 56–64.

Syed, A.N., Ashfaq, M., Khan, S., 2005. Comparison of development and predation of *Chrysoperla carnea* (Neuroptera: Chrysopidae) on different densities of two hosts (*Bemisia tabaci*, and *Amrasca devastans*). Pakistan Entomologist 27, 41–44.

Takikawa, Y., Matsuda, Y., Kakutani, K., Nonomura, T., Kusakari, S., Okada, K., Kimbara, J., Osamura, K., Toyoda, H., 2015. Electrostatic insect sweeper for eliminating whiteflies colonizing host plants: a complementary pest control device in an electric field screen-guarded greenhouse. Insects 6, 442–454.

Takikawa, Y., Matsuda, Y., Nonomura, T., Kakutani, K., Okada, K., Morikawa, S., Shibao, M., Kusakari, S., Toyoda, H., 2016. An electrostatic nursery shelter for raising pest and pathogen free tomato seedlings in an open-window greenhouse environment. Journal of Agricultural Science (Toronto) 8, 13–25.

Tanaka, N., Matsuda, Y., Kato, E., Kokabe, K., Furukawa, T., Nonomura, T., Ken-ichiro, H., Kusakari, S., Imura, T., Kimbara, J., Toyoda, H., 2008. An electric dipolar screen with oppositely polarized insulators for excluding whiteflies from greenhouses. Crop Protection 27, 215–221.

Tay, W.T., Evans, G.A., Boykin, L.M., DeBarro, P.J., 2012. Will the real *Bemisia tabaci* please stand up? PLoS One 7, e50550.

Thomson, W.M.O., 2011. The Whitefly, *Bemisia Tabaci* (Homoptera: Aleyrodidae) Interaction with Geminivirus-infected Host Plants: *Bemisia Tabaci*, Host Plants and Geminiviruses. Springer, Berlin, The Netherlands.

Tomizawa, M., Casida, J.E., 1999. Minor structural changes in nicotinoid insecticides confer differential subtype selectivity for mammalian nicotinic acetylcholine receptors. British Journal of Pharmacology 127, 115–122.

Toscano, L.C., Boiça Jr., A.L., Maruyama, W.I., 2002. Nonpreference of whitefly for oviposition in tomato genotypes. Scientia Agricola 59, 677–681.

Trottin-Caudal, Y., Chabrière, C., Fournier, C., Leyre, J.M., 2006. Current situation of *Bemisia tabaci* in protected vegetable crops in the South of France. OILB/SROP Bulletin 29, 53–58.

Tsueda, H., Taguchi, Y., Katsuyama, N., 2007. Lethal high temperatures for the sweetpotato whitefly (*Bemisia tabaci* (Gennadius) B Biotype) and control effects under greenhouse conditions using solar radiation. Japanese Journal of Applied Entomology and Zoology 51, 197–204.

Tsueda, H., Tsuduki, T., Tsuchida, K., 2014. Factors that affect the selection of tomato leaflets by two whiteflies, *Trialeurodes vaporariorum* and *Bemisia tabaci* (Homoptera: Aleyrodidae). Applied Entomology and Zoology 49, 561–570.

Tzanetakis, I.E., Martin, R.R., Wintermantel, W.M., 2013. Epidemiology of criniviruses: an emerging problem in world agriculture. Frontiers in Microbiology 4, 193–207.

UC-IPM, 2016. UC Pest Management Guidelines – Tomato Whiteflies. University of California Agriculture and Natural Resources, University of California. http://ipm.ucanr.edu/PMG/r783301211.html.

Ucko, O., Cohen, S., Ben-Joseph, R., 1998. Prevention of virus epidemics by a crop free period in the Arava region of Israel. Phytoparasitica 26, 313–321.

Ujváry, I., 1999. Nicotine and other insecticidal alkaloids. In: Yamamoto, I., Casida, J. (Eds.), Nicotinoid Insecticides and the Nicotinic Acetylcholine Receptor. Springer, Tokyo, Japan, pp. 29–69.

Urbaneja, A., Tapia, G., Stansly, P., 2005. Influence of host plant and prey availability on developmental time and survivorship of *Nesidiocoris tenuis* (Het.: Miridae). Biocontrol Science and Technology 15, 513–518.

van der Vlugt, R.A.A., Verbeek, M., Dullemans, A.M., Wintermantel, W.M., Cuellar, W.J., Fox, A., Thompson, J.R., 2015. Torradoviruses. Annual Review of Phytopathology 53, 485–512.

van Kleeff, P.J.M., Galland, M., Schuurink, R.C., Bleeker, P.M., 2016. Small RNAs from *Bemisia tabaci* are transferred to *Solanum lycopersicum* phloem during feeding. Frontiers in Plant Science 7, 1759.

van Lenteren, L.C., Noldus, L.P.I.I., 1990. Whitefly-plant relationships: behavioural and ecological aspects. In: Gerling, D. (Ed.), Whiteflies: Their Bionomics. Pest Status and Management. Intercept, Andover, UK, pp. 47–89.

van Lenteren, J.C., van Roermund, H.J.W., Sutterlin, S., 1996. Biological control of greenhouse whitefly (*Trialeurodes vaporariorum*) with the parasitoid *Encarsia formosa*: how does it work? Biological Control 6, 1–10.

van Lenteren, J.C., van Vianen, A., Gast, H.F., Kortenhoff, A., 1987. The parasite-host relationship between *Encarsia formosa* Gahan (hymenoptera: Aphelinidae) and *Trialeurodes vaporariorum* (Westwood) (Homoptera: Aleyrodidae). Journal of Applied Entomology 103, 69–84.

Vavrina, C.S., Stansly, P.A., Liu, T.X., 1995. Household detergent on tomato: phytotoxicity and toxicity to silverleaf whitefly. HortScience 30, 1406–1409.

Verma, A.K., Mitra, P., Saha, A.K., Ghatak, S.S., Bajpai, A.K., 2013. Effect of trapcrops on the population of the whitefly *Bemisia tabaci* (Genn.) and the diseases transmitted by it. Bulletin of Indian Academy of Sericulture, Unit of Association for Development of Sericulture 17, 37–44.

Vestergaard, S., Cherry, A., Keller, S., Goettel, M., 2003. Safety of hyphomycete fungi as microbial control agents. In: Hokkanen, H.M.T., Hajek, A.E. (Eds.), Environmental Impacts of Microbial Insecticides: Needs and Methods for Risk Assessment. Springer, Dordrecht, The Netherlands, pp. 35–62.

Vet, L.E.M., van Lentern, J.C., Woets, J., 1980. The parasite-host relationship between *Encarsia formosa* and *Trialeurodes vaporariorum*. IX. A review of the biological control of the greenhouse whitefly with suggestions for future research. Zeitschrift für Angewandte Entomologie 90, 26–51.

Vidal, C., Fargues, J., Rougier, M., Smits, N., 2003. Effect of air humidity on the infection potential of hyphomycetous fungi as mycoinsecticides for *Trialeurodes vaporariorum*. Biocontrol Science and Technology 13, 183–198.

Wagner, T.L., 1995. Temperature-dependent development, mortality, and adult size of sweet-potato whitefly biotype-B (Homoptera, Aleyrodidae) on cotton. Environmental Entomology 24, 1179–1188.

Walker, G.P., Perring, T.M., Freeman, T.P., 2010. Life history, functional anatomy, feeding, and mating behavior. In: Stansly, P.A., Naranjo, S.E. (Eds.), *Bemisia*: Bionomics and Management of a Global Pest. Springer, New York, USA, pp. 109–160.

Wang, C., Fan, Y., 2014. Eugenol enhances the resistance of tomato against tomato yellow leaf curl virus. Journal of the Science of Food and Agriculture 94, 677–682.

Wang, L.D., Huang, J., You, M.S., Guan, X., Liu, B., 2007. Toxicity and feeding deterrence of crude toxin extracts of *Lecanicillium (Verticillium) lecanii* (Hyphomycetes) against sweet potato whitefly, *Bemisia tabaci* (Homoptera: Aleyrodidae). Pest Management Science 63, 381–387.

Wang, Z., KaiYi, W., ShuiFa, Z., ZhongQiang, L., CuiXia, M., 2014. Whiteflies counting with K-means clustering and ellipse fitting. Transactions of the Chinese Society of Agricultural Engineering 30, 105–112.

Wang, K.H., Tsai, J.H., 1996. Temperature effect on development and reproduction of silverleaf whitefly (Homoptera: Aleyrodidae). Annals of the Entomological Society of America 89, 375–384.

Williamson, M.V., Ho, J.Y., Wu, F.F., Miller, N., Kaloshian, I., 1994. A PCR-based marker tightly linked to the nematode resistance gene, *Mi*, in tomato. Theoretical and Applied Genetics 87, 757–763.

Wilson, G.F., 1931. Biological control of the greenhouse whitefly. Gardner's Chronical 1931, 15–17.

Wintermantle, 2010. Transmission efficiency and epidemiology of criniviruses. In: Stansly, P.A., Naranjo, S.E. (Eds.), *Bemisia*: Bionomics and Management of a Global Pest. Springer, Dordrecht, Heidelberg, London, New York, pp. 319–331.

Wintermantel, W.M., Wisler, G.C., 2006. Vector specificity, host range, and genetic diversity of Tomato chlorosis virus. Plant Disease 90, 814–819.

Woets, J., 1978. Development of an introduction scheme for *Encarsia forrnosu* Gahan (hymenoptera: Aphelinidae) in greenhouse tomatoes to control the greenhouse whitefly, *Trialeurodes vaporariorum* (Westwood). (Homoptera: Aleyrodidae). In: Proceedings of the International Symposium on crop protectionMededelingen Faculteit Landbouwwetenschappen Rijksuniversiteit Gent, vol. 43, pp. 379–385.

Xiao, Y., Chen, J., Cantliffe, D., Mckenzie, C., Houben, K., Osborne, L.S., 2011. Establishment of papaya banker plant system for parasitoid, *Encarsia sophia* (Hymenoptera: Aphilidae) against *Bemisia tabaci* (Hemiptera: Aleyrodidae) in greenhouse tomato production. Biological Control 58, 239–247.

Xie, M., Xie, W., Wu, G., 2011. Effects of temperature on the growth and reproduction characteristics of *Bemisia tabaci* B-biotype and *Trialeurodes vaporariorum*. Journal of Applied Entomology 135, 252–257.

Yamamoto, I., 1999. Nicotine to nicotinoids: 1962 to 1997. In: Yamamoto, I., Casida, J. (Eds.), Nicotinoid Insecticides and the Nicotinic Acetylcholine Receptor. Springer, Tokyo, Japan, pp. 3–27.

Yang, T.C., Chi, H., 2006. Life tables and development of *Bemisia argentifolii* (Homoptera: Aleyrodidae) at different temperatures. Journal of Economic Entomology 99, 691–698.

Youssef, H.I., Hady, S.A., Ismail, H., 2001. Intercropping pattern against *Bemisia tabaci* (Gennadius) infestation (Homoptera: Aleyrodidae). Annals of Agricultural Science, Moshtohor 39, 651–654.

Zalom, F.G., Lanini, W.T., Miyao, G., Davis, R.M., 2001. A continuum of integrated pest management practices in processing tomatoes. In: Proceedings of the Seventh International Symposium on the Processing Tomato, Acta Horticulturae, vol. 542, pp. 55–62.

Zeshan, M.A., Khan, M.A., Ali, S., Arshad, M., 2015. Correlation of conducive environmental conditions for the development of whitefly, *Bemisia tabaci* population in different tomato genotypes. Pakistan Journal of Zoology 47, 1511–1515.

Zhang, G., Li, D., Liu, T., Wan, F., Wang, J., 2011. Interspecific interactions between *Bemisia tabaci* biotype B and *Trialeurodes vaporariorum* (Hemiptera: Aleyrodidae). Environmental Entomology 40, 140–150.

Zhou, F., AiMin, Y., XueHao, C., Feng, Y., JiDe, F., 2014. Study on the ecological control technology against *Bemisia tabaci* (Gennadius) on facility vegetables. Journal of Yangzhou University, Agriculture and Life Sciences 35, 75–79.

Zimmermann, G., 2008. The entomopathogenic fungi *Isaria farinosa* (formerly *Paecilomyces farinosus*) and the *Isaria fumosorosea* species complex (formerly *Paecilomyces fumosoroseus*): biology, ecology and use in biological control. Biocontrol Science and Technology 18, 865–901.

Chapter 5

Mites: Biology, Ecology, and Management

Gerald E. Brust[1], Tetsuo Gotoh[2]

[1]CMREC-UMF, University of Maryland, Upper Marlboro, MD, United States; [2]Ibaraki University, Ami, Japan

1. INTRODUCTION

Mites are a diverse group of arthropods compared to insects, and are among the most damaging pests of tomatoes worldwide (Jeppson et al., 1975). Spider mites (family Tetranychidae) are especially troublesome, as an outbreak of these pests can lead to serious losses and even total failure of the tomato crop. Within the spider mite family the most widespread and abundant tomato pest is the two-spotted spider mite (TSSM), *Tetranychus urticae* Koch (Acari: Tetranychidae). One reason this mite is such a ubiquitous pest is that the "species" is probably a complex of biologically variable species (Bolland et al., 1998). Much controversy exists regarding the taxonomic placement of TSSM, and there are approximately 65 synonyms included under this one species. One such problematic "species" of *Tetranychus* is *Tetranychus cinnabarinus* (Boisduval), the carmine spider mite, which is considered a red biotype of *T. urticae* in some studies (Dupont, 1978; Auger et al., 2013) and a separate species in others (Brandenburg and Kennedy, 1981; Kuang and Cheng, 1990). Thus a comprehensive presentation of all the various synonyms of TSSM is not possible in this chapter and the most common descriptions of *T. urticae* will be used. Another species closely related to *T. urticae* is *Tetranychus evansi* Baker & Pritchard, the red tomato spider mite. This mite species has spread rapidly throughout the tropical and subtropical areas of the world during the last several decades, and has become a major pest on tomato (Gotoh et al., 2003; Navajas et al., 2013). In the family Eriophyidae (superfamily Eriophyoidea) there are two major pest species of tomato, the tomato russet mite (TRM) *Aculops lycopersici* (Massee) and the tomato erineum mite *Aceria lycopersici* (Wolffenstein). *Ace. lycopersici* is closely related to the TRM and will be mentioned only when it differs substantially from *Acu. lycopersici*. The broad mite, *Polyphagotarsonemus latus* (Banks) (family Tarsonemidae) is known as citrus silver mite or yellow tea mite. This species is found throughout the tropics and subtropics, and is mostly a problem in tomato greenhouse operations.

2. TWO-SPOTTED SPIDER MITE, *TETRANYCHUS URTICAE* KOCH

2.1 Identification

Female *T. urticae* are 0.4 mm long and broadly oval with 12 pairs of dorsal setae, whereas the male is slightly wedge-shaped with a narrow caudal end (Boudreaux and Dosse, 1963; Bolland et al., 1998). Female mites are approximately 50% larger than males. The color of the adult mite varies; usually they are greenish-yellow or almost translucent green, although they can also be brown or orange-red (Boudreaux and Dosse, 1963; English-Loeb, 1990). Overwintering females are consistently orange to orange-red (Fig. 5.1A). Two large dark spots (consisting of food or bodily waste) on either sides of the idiosoma can usually be seen through the body wall (from which the species gets its common name). Mites that have recently molted may not have these spots (Shih et al., 1976; English-Loeb, 1990).

T. urticae eggs are about 0.15 mm in diameter, round, and translucent when first laid (Fig. 5.1B) but become white as hatching approaches. Most often the eggs are laid on the underside of leaves. Immature mites have the same general color and shape as adults, although the larvae have only 6 legs.

2.2 Biology

The life cycle for *T. urticae* consists of egg, larva, protonymph, deutonymph, and adult stages. Because spider mites are arrhenotokous (unmated females produce haploid eggs that develop into males while mated females produce diploid eggs that develop into females), this can lead to circumstances where there are very few males in the population. When this circumstance

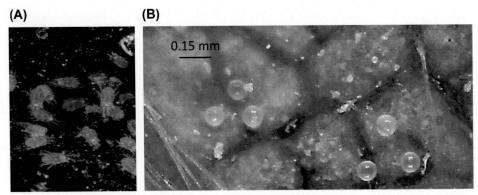

FIGURE 5.1 (A) Two-spotted spider mite overwintering female adult. (B) Two-spotted spider mite eggs.

occurs the female may mate with her sons leading to inbreeding. Inbreeding can further occur by mating of brothers and sisters. Despite this potential for inbreeding, spider mites have considerable genetic diversity which allows them to adapt to new conditions and environments (Helle and Sabelis, 1985; Sabelis, 1991; Krips et al., 1998; Tommaso et al., 2007). Females oviposit fertilized and unfertilized eggs with the sex ratio of mated females being 3:1 female to male (Shih et al., 1976; Gotoh, 1997; Rotem and Agrawal, 2003). Mated females can lay 20–30 eggs/day for a total of more than 200 over their 7–18 day life span depending on abiotic factors. Eggs, usually attached to fine silk webbing, hatch in approximately 3 days. After hatching, the first immature stage (larva) emerges with three pairs of legs, while all the nymphal and adult stages have four pairs. After each larval and nymphal stage there is a brief quiescent period, the chrysalis, during which the mite appears to be attached to the plant surface. Male mites generally mature more quickly than females and the adult males can be often seen guarding a deutonymphs chrysalis female so that they can mate with her as soon as she molts. At times a number of males will guard the same female deutonymph chrysalis, but usually only the most aggressive male succeeds in mating. This guarding behavior is often driven by the female who releases a sex pheromone at times resulting in one female mating with several males (Helle and Sabelis, 1985; Gotoh, 1997; Krips et al., 1998). The female mite has a pre-ovipositional period of approximately 24 h during which she often disperses. Development is most rapid during hot, dry weather (Sabelis, 1991; Rotem and Agrawal, 2003; Tommaso et al., 2007). A single generation requires 5–20 days to reach adulthood before producing the next generation. There are many overlapping generations per year and only mated females overwinter in temperate regions.

T. urticae undergoes diapause in temperate regions when climatic conditions become too cold or hot. Diapause is indicated by a change in color to orange or reddish-orange. Diapause is induced by several factors that include food quality, temperature, and day length (Shih et al., 1976; Veerman, 1985). Once in a hibernal (winter) diapause, females need a period of low temperature and short day lengths before diapause can be broken. How long and how deeply a population of TSSM stays in diapause is determined by their specific climatic conditions (Rotem and Agrawal, 2003; Kawakami et al., 2009). This can produce site situations that will cause some mite populations to emerge from diapause earlier or later than other local populations resulting in a slow but steady release of spider mite females throughout the early part of the season. Female *T. urticae* may also stop reproduction during the cooler winter months in temperate production areas. Reproduction in greenhouses, tropical, or subtropical environments typically occurs year round. *T. urticae* often uses aerial dispersal (see below) from tomato fields in the autumn to move to overwintering hosts such as chickweed, *Stellaria media* (L.) Vill. Females can also hibernate in ground litter or under the bark of trees or shrubs (Shih et al., 1976; Sabelis, 1991; Meck et al., 2009).

T. urticae is able to produce silk from glands located on the apex of their pedipalps. How well an individual mite survives can be greatly influenced by the mite's ability to produce silk, as silk is used for web building, protection, and communication. *T. urticae* also uses silk as a social cue in selecting their microhabitat with group behavior being shaped by the individuals' response to social cues, such as the amount of silk already present on a leaf. When mite population begins to increase on the undersides of leaves, the combined production of silk by the adults can serve as protection from some natural enemies and increase humidity levels preventing desiccation (Gerson, 1985). This silken "roof" also prevents pesticides from the leaf surface where mites are located. However, the presence of TSSM silk can be a signal for some specialist predators such as phytoseiid mites (family Phytoseiidae) and ladybird beetles to search more intensely when it is encountered (van de Vrie et al., 1972). *T. urticae* also use their silk for escape. Mites use the silk to spin down when they encounter a pesticide residue, especially pyrethroids, allowing them to avoid contact with the control treatment (Margolies and Kennedy, 1988). Many Tetranychidae use silk to disperse via a process called ballooning; TSSM disperses in two different ways using their silken threads. One way is when only mated females that have not laid eggs form large silken assembles at the top of a plant. At the apex of the plant, they are picked up by the wind for dispersal (Gerson, 1985). The second is when overcrowding occurs on a poor host plant. The mites collect on branch tips in masses of male and female mites, spin down on silk, and are picked up by the wind or animals (Clotuche et al., 2011). *T. urticae* can also use the silk as anchors and possibly as interference to other mites.

2.3 Distribution, Host Range, and Seasonal Occurrence

T. urticae is a cosmopolitan spider mite. Although considered a temperate zone species, it is also found throughout sub-tropical regions (Sabelis, 1985a,b). *T. urticae* is reported on over 300 host plants worldwide, including over 100 cultivated species such as cotton, corn, soybean, tomato, pepper, and numerous fruit and ornamental species (Bolland et al., 1998). A number of native plant species, including cutleaf evening primrose, violets, chickweed, clovers, pokeweed, wild mustard, and blackberry are common hosts from which infestations spread to nearby crops (Migeon et al., 2010). *T. urticae* can be found on crop plants from early spring until the first killing frost in temperate areas of the world. Their populations begin to increase gradually in the early part of the growing season and rapidly at higher temperatures. Temperatures at or greater than 26.5°C with low rain fall and low to moderate humidity are prime environmental conditions for TSSM populations to rapidly increase to damaging levels (Gerson, 1985).

2.4 Damage, Losses, and Economic Thresholds

Two-spotted spider mites have stylet-like chelicerae used for piercing host plants. This releases cellular content of the epidermal cells which the mite sucks up using its rostrum. The mite's feeding causes the mesophyll cells in the area to collapse creating very small white chlorotic spots on the leaves where they have removed the chlorophyll (Fig. 5.2) (Sabelis, 1985a,b).

FIGURE 5.2   White speckles on tomato leaf from two-spotted spider mite feeding.

As feeding damage progresses a stippled appearance of the foliage is evident. After several days of heavy mite feeding, necrotic spots begin to develop on leaf tissue and leaves will turn yellow or gray and collapse. Both *T. urticae* and *T. cinnabarinus* have been found to cause an unusual hyper-necrotic response in tomato that involves premature chlorosis of infested leaflets that consequently wilt and die (Foster and Barker, 1978; Szwejda, 1993). Although not common, TSSM feeding damage on tomato flowers causes a browning and withering of the petals. Mites will feed directly on the tomato fruit, usually at the stem-end around the cap area (Meck et al., 2009). This feeding damage is rough to touch and has small depressed areas where the mites have removed chlorophyll and the cells have collapsed. Another fruit problem caused by TSSM is gold flecking, which appears as yellow or gold spots scattered over the surface of the fruit as it ripens. When this flecking is severe it can reduce the market value of the fruits. The flecks are only in the epidermal layer of the fruit and do not penetrate beyond this (Brust, 2014). These flecks have been determined to be calcium oxalate crystals (Den Outer and Van Veenendaal, 1988). Gold fleck is thought to be a response to certain stresses the plant encounters during the season, such as high temperatures and humidity or TSSM or thrips feeding (De kreij et al., 1992; Ghidiu et al., 2006; Brust, 2014), or too high a level of calcium in the fruit as it is ripening (Den Outer and Van Veenendaal, 1988). Of all the possible causes, TSSM seems to be the most important in causing this fruit ripening problem in temperate regions (Brust, 2014).

T. urticae is also implicated in the transmission of several viruses that include potato virus Y, tobacco mosaic virus, and tobacco ringspot virus. The mite does not actually inject the virus into the plant, instead excretes the virus onto the leaf surface and allows entry of the virus into the plant through feeding damage (Oldfield, 1970; Jeppson et al., 1975).

T. urticae, and most probably its species complex, is responsible for 10–50% yield losses in an average tomato production season. This range is so large because mite infestations can be severe in some areas of a field and almost nonexistent in others. Environmental conditions and management programs (excessive early season insecticide applications) influence the severity of TSSM outbreaks and potential yield loss (Wilkerson et al., 2005). Yield loss is not only due to a reduction in tonnage of fruit, but also quality and size and therefore marketable yield (Oldfield, 1970; Metcalf and Metcalf, 1993; Meck, 2010). Crop losses can occur when about 30% of the tomato leaf surface is damaged by spider mite feeding. In a study by Meck (2010) on tomatoes in North Carolina (United States), it was found that economic thresholds were very low at 1–2 mites/tomato leaflet. This threshold is extremely low and probably not practical for most tomato operations. Sampling for mites in a tomato field has shown that mite populations were highly aggregated and the number of samples required for just 60% precision was too large to be practical (Lange and Bronson, 1981; Park and Lee, 2007; Meck, 2010). Jayasinghe and Mallik (2010) in Thihagoda, Sri Lanka found that the middle developmental stage of tomato was the most critical period for mite damage and accounted for more than 50% of the total yield loss compared with early or late infestations. They developed an economic injury level (EIL) based on initial number of mites released on the plant and the number of days mites fed on the plant. While this EIL is a good place to start in understanding the relationship between mite numbers, feeding duration, and yield reduction, it is not practical at this time because it is impossible to know when and how many mites were initially there on a tomato plant and how long they had been feeding. Therefore, this EIL does not lend itself to commercial use.

3. TOMATO RED SPIDER MITE, *TETRANYCHUS EVANSI* BAKER & PRITCHARD

3.1 Identification

T. evansi belongs to "Group 5" of the genus *Tetranychus* (Flechtmann and Knihinicki, 2002). To an untrained eye *T. evansi* looks very similar to *T. urticae* with a few exceptions. The body is from pale to dark orange in color and legs are also orange. Empodia (a lobe or spine between the two claws at the end of the tarsus) I and II of females have three pairs of proximoventral hairs and a tiny mediodorsal spur. All four proximal tactile setae on female tarsus I are nearly in line with the proximal set of duplex setae. The aedeagus (penis) is upturned distally and the dorsal margin is slightly convex (Baker and Pritchard, 1960).

3.2 Biology

The ability to predict population outbreaks of a new pest such as *T. evansi* greatly depends on its capacity to develop at different temperatures. Therefore, it is important to know the temperature requirements of a target species to understand its potential distribution and population dynamics (Ullah and Gotoh, 2013). The r_m-value (intrinsic rate of natural increase) of a species is used to describe the growth and adaptation of a population to certain environmental conditions (Birch, 1948). Life history components, such as developmental rates, oviposition, survival rates, and the offspring proportion of females determine the r_m-value (Sabelis, 1985b). The r_m-value can be used to forecast how much of a potential problem a mite species may pose in an agricultural system (Margolies and Wrensch, 1996; Gotoh et al., 2004).

Sabelis (1991) conducted a broad review of the life history parameters of tetranychid mites and found the r_m-values for these mites range from 0.219 to 0.336/day at approximately 25°C. The r_m-values of *T. evansi* fall within this range. Gotoh et al. (2004) conducted studies on temperature-dependent developmental and reproductive traits of seven *T. evansi* strains collected from Brazil, France, Japan, Kenya, Canary Islands, Spain, and Taiwan. Temperature had a significant effect on the egg to female adult duration among the seven strains studied. The developmental times of the *T. evansi* strains ranged from 41.0 to 45.1 days at 15°C to 5.5–6.5 days at 40°C, and 9.7–10.5 days at 25°C. The strain from Brazil significantly differed from the France, Japan, Kenya, Canary Islands, and Taiwan strains in developmental duration. However, the overall life history parameters were very close among the seven *T. evansi* strains.

Sarmento et al. (2011a,b) showed that *T. evansi* suppresses the proteinase inhibitor (PI) activity in tomato plants by about 33% in comparison with the PI in non-infested tomato plants. This suppression resulted in higher oviposition rates (about twofold) on leaves that were attacked previously by conspecifics than on leaves that were not attacked, or on leaves that were previously attacked by heterospecifics (*T. urticae*). In *T. urticae*, oviposition rates were highest on leaves obtained from a "clean" plant followed by non-attacked leaves from an attacked plant. PIs are involved in induced plant defense and hamper the action of digestive proteinases present in the herbivore gut (Sarmento et al., 2011a). Oviposition rates of *T. evansi* were drastically reduced by 30% on leaves attacked previously by *T. urticae* compared with ones attacked by *T. evansi* (Sarmento et al., 2011b).

3.3 Distribution, Host Range, and Seasonal Occurrence

T. evansi was first described as a new species based on specimens from tomato (*Lycopersicon esculentum* Miller) in the Mauritius Islands (Baker and Pritchard, 1960). However, it is considered to have originated from Brazil, from where it was first described as *Tetranychus marianae* McGregor (McGregor, 1950; Silva, 1954; Gutierrez and Etienne, 1986; de Moraes et al., 1987; Furtado et al., 2007; Navajas et al., 2013). Until the 1960s, it was known only from Mauritius Islands, Brazil, and Florida (United States). By the mid-1980s only a few other countries, such as Reunion Island, Rodrigues Island, Seychelles, Cuba, Puerto Rico, and Zimbabwe, reported it (Gutierrez, 1974; de Moraes et al., 1987; Furtado et al., 2007). Since the 1990s it has spread throughout the world (Fig. 5.3).

This species distribution is throughout the tropical to subtropical regions all over the world and is active year round (Migeon et al., 2009; Navajas et al., 2013). In Japan, *T. evansi* was first described as *Tetranychus takafujii* Ehara and Ohashi in 2001 at Kyoto and Osaka prefectures, where it specialized on solanaceous plants (nightshades, *Solanum nigrum* L., *Solanum carolinense* L., *Solanum melongena* L.). In the next year, it was found in Tokyo and Hyogo (Fig. 5.4). *T. evansi* was observed at 10 prefectures starting in 2008 from *S. nigrum*, *S. carolinense*, *Solanum photeinocarpum* Nakamura et Odashima, *S. melongena* to tomato.

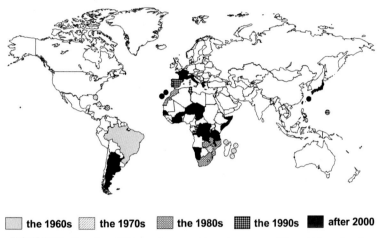

▨ the 1960s   ▨ the 1970s   ▨ the 1980s   ▦ the 1990s   ■ after 2000

FIGURE 5.3   Invasion of *Tetranychus evansi* to various areas of the world per decade.

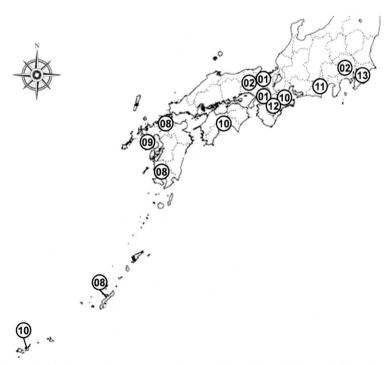

FIGURE 5.4   Observed year of *Tetranychus evansi* at prefectures of Japan. Numerals in *circles* indicate the year of the 2000s.

To date, more than 110 host species of *T. evansi* have been reported in both outdoor crops and greenhouses. Several *T. evansi* strains can attack different plant families such as Malyaceae (okra), Fabaceae, and Rosacea (Bolland et al., 1998; Navajas et al., 2013).

3.4 Damage, Losses, and Economic Thresholds

T. evansi infests abaxial and adaxial surfaces and causes graying or browning of the leaves, which eventually wither and drop from the plant. Symptoms and losses are similar to *T. urticae*. This species is considered an important pest of all solanaceous plants, such as tomato, potato, and eggplant in Africa, and can cause up to 90% yield loss in southern Africa (Sarr et al., 2002; Saunyama and Knapp, 2003).

4. TOMATO RUSSET MITE, *ACULOPS LYCOPERSICI* (MASSEE)

4.1 Identification

The TRM is minute and difficult to see even with a 15× hand lens. It has a spindle-shaped body that is 150–200 μm long with two pairs of legs at the broader head end and long hairs on the narrower posterior end (Fig. 5.5).

The prodorsal shield has intricate microscopic sculpturing and lobes (Lamb, 1953; Kay, 1986). A morphologically detailed description of the adult can be found in Keifer (1940), Bailey and Keifer (1943), and Perring and Farrar (1986). Adults are transparent and tan-yellow or pink. Eggs are translucent white when laid, but turn yellow as they mature (Bailey and Keifer, 1943; Kay, 1986). In approximately 2 days, eggs hatch and larvae are off-white with two pairs of legs. Nymphs are yellowish and slightly larger than the larva. Larvae and nymphs remain close to where they originally hatched and tend to congregate near leaf edges (Keifer, 1940; Davis, 1964; Kay, 1986). The two pairs of legs have distinctive claw-like structures located at the distal end of the tarsi called feather-claws (Keifer, 1940, 1952).

4.2 Biology

The TRM has a very fast life cycle of 6–7 days (egg to adult), and females have a 22 days longevity while males are around 16 days (Anderson, 1954; Carmona, 1964; Anonymous, 1972). Abou-Awad (1979) found that after each larval and nymphal stage, there is a quiescent stage similar to *T. urticae*. Populations of TRM increase rapidly at approximately 26°C and 30% humidity, although slightly higher temperatures and humidities have resulted in marginally faster developmental times (Bourne, 1952; Lamb, 1953; Davis, 1964; Abou-Awad, 1979). High humidity levels appear to slow the development of the mite only at high temperatures, but favor its development at lower temperatures (Rice and Strong, 1962; Perring and Farrar, 1986; Fischer and Mourrut-Salesse, 2005). Development was slowed when temperatures fluctuated around a mean compared with conditions at a constant temperature (Bourne, 1952; Lamb, 1953). Photoperiod was found to have no effect on TRM development (Davis, 1964). Temperatures up to 30°C are favorable for mite survival, but above 30°C survival is reduced by 22%. Females lay between 15 and 55 eggs over their life time (Monkman, 1992). Non-fertilized females produce only male eggs and fertilized females produce both male and female eggs (Flechtmann, 1974, 1977). Eggs hatch in

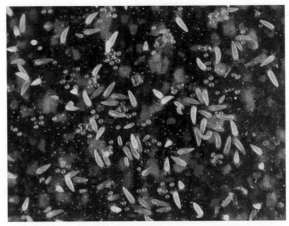

FIGURE 5.5 Tomato russet mite adults, nymphs, and eggs.

45–50 h at 21°C. The larval stage lives for 1 day while the nymphal stage for 2 days (Carmona, 1964; Osman, 1975; Abou-Awad, 1979; Fischer and Mourrut-Salesse, 2005). Females start laying eggs as soon as they mature. Adults have been found to overwinter on nightshade weeds and petunias but are not reproductively active at this time (Bailey and Keifer, 1943; Kay, 1986). Female mites disperse from dying host plants or overwintering sites by moving to the top of a plant and standing on their tapered posterior end allowing the wind to blow them away from the plant (Sloan, 1938; Keifer, 1952, 1966). At times several TRM females can form "dispersal chains" and they move as a group from the plant. Nightshade plants (*Solanum* spp.), *Datura* spp. and other solanaceous plants as well as *Convolvulus arvensis* L. can all serve as alternative hosts and sources of infestation (Zalom et al., 1986; Monkman, 1992).

4.3 Distribution, Host Range, and Seasonal Occurrence

The TRM is worldwide in distribution and occurs in all regions where tomatoes and other solanaceous crops are grown. There are few areas of the world where this pest is not present, with the exceptions being the furthest northern and southern latitudes, although it was reported in Antarctica (Lindquist et al., 1996; Máca, 2012).

Host range studies of TRM found it to be capable of reproducing on plants belonging to several solanaceous genera, as well as on field bindweed, morning glory, jimsonweed, sunberry, wild goose berry, and Chinese thorn apple (Perring and Farrar, 1986; Lindquist et al., 1996). By far the preferred host of TRM is tomato. There are some hosts (tobacco, some species of *Convolvulus*) listed in the literature that have been cited as hosts for TRM, but further studies demonstrated that those plants could not support reproduction. More detailed host lists can be found in Perring and Farrar (1986), Lindquist et al. (1996) and CABI (2013).

Tomato russet mites are a major pest in the tropical parts of the world year round. In subtropical areas they are a major pest during the summer and fall if weather conditions are warm (Lamb, 1953). In temperate areas TRM is a pest during the summer months, and will cause severe crop damage under the suitable environmental conditions of heat and humidity.

4.4 Damage, Losses, and Economic Thresholds

TRMs have piercing-sucking mouthparts and damage is usually found first on the lower leaves of a plant. TRM usually feeds on the undersides of leaves, but can also be found feeding on the stems of leaves and flowers. TRM removes cell contents from leaves, stems, and fruit cells. This feeding can produce a greasy appearance, which later becomes bronzed. Leaves may turn yellow, curl upwards, and dry out. Damage on tomatoes often starts on the lower leaves of a plant and moves upward. This feeding damage can be confused with nutritional deficiencies, plant disease, or water stress. Royalty and Perring (1988) and Petanović and Kielkiewicz (2010) showed that TRM feeding on tomato leaves resulted in the upper and lower epidermal cells being destroyed, although parenchyma cells were not damaged. Guard cells were also damaged, thereby inhibiting gas exchange and photosynthesis. Research also suggested that as TRM density increased, the feeding activity of each individual mite was intensified. *Acu. lycopersici* can cause serious reductions in yield in tomato crops, but usually only when young plants are exposed to attack.

Ace. lycopersici's damage differs from TRM. Their feeding gives the feeding area a "fuzzy, hair like, or white mold" appearance (Massee, 1937; Jeppson et al., 1975). The mite's feeding also causes misshapen leaves, curled leaf edges, and necrotic spots.

In some cases there have been losses up to 65% when young plants have become heavily infested shortly after transplanting (Eschiapati et al., 1975; Oliveira et al., 1982). Early infestations (25–30 days after transplanting (AT)) have led to 60% yield reductions versus later infestations (55–70 days AT) leading to only a 5–10% reduction in yield (Eschiapati et al., 1975; Estebanes-Gonzalez and Rodriguez-Navarro, 1991; Brust, 2013). Any stress on a tomato plant (drought, nutritional, disease, etc.) can lead to a TRM population increase on those stressed plants compared with plants that are not stressed (Estebanes-Gonzalez and Rodriguez-Navarro, 1991). When damage to lower plant parts increases, the mites move up to younger foliage. As plants begin to die, mites can move to the highest parts of the plant where they will be picked up by the wind. Although not common, *Acu. lycopersici* can be infected by the fungal pathogen, *Hirsutella thompsonii* F.E. Fisher (Cabrera, 1984).

Because TRM is so difficult to monitor using conventional methods and is so devastating in its damage to tomato plants, evaluating their population densities and the related yield losses has proved to be difficult. Thus thresholds for this mite in tomato have not been reliably determined. Growers should examine young tomato plants closely looking for the earliest symptoms of feeding from the mite. Look for bronzing on lower leaves and check these leaves and those immediately above them for mites. Once the initially damaged plants are found the area should be marked and checked again in a few days to see if mites and damage are increasing.

5. BROAD MITE, *POLYPHAGOTARSONEMUS LATUS* (BANKS)

5.1 Identification

Female mites are about 0.2 mm long and oval in shape. Their bodies are swollen in profile and transparent, light yellow to pale brown or green, and waxy with a faint, median stripe that forks near the back end of the body (Fig. 5.6) (Gerson, 1992; de Coss-Romero and Peña, 1998; Montasser et al., 2011).

Males are small (0.11 mm) and have relatively long legs. They are similar in color and lack the median stripe found in the females. The fourth pair of legs on the female is reduced to a slender long hair extending from the tip. The fourth pair of legs of males ends in strong claws that are used to pick up the female nymph and place her at right angles to the male's body for transport and later mating (Peña et al., 1996; Ferreira et al., 2006).

Broad mite eggs are elliptical, gleaming, and colorless, about 0.08 mm long, and are covered with 30–40 scattered white knobs on the upper surface of each egg (Fig. 5.7) (Montasser et al., 2011). The distinctive egg is a key identifying characteristic to use when confirming that plant damage is being caused by broad mites. When eggs hatch, larvae emerge that are roughly 0.05–0.1 mm long and due to minute ridges found on their skin appear white, but later become transparent. Larvae have three pairs of legs. After 2–3 days, the larva becomes a quiescent "pupa" that appears as a clear immobile engorged body that is pointed at both ends (Gerson, 1992; de Coss-Romero and Peña, 1998). The "pupa" is about 0.08 mm long. Once they molt from this stage they become adults, which have four pairs of legs.

5.2 Biology

Broad mites have a somewhat modified four stage life cycle: egg, larva, quiescent pupa (some researchers call this a nymphal or even a larval stage), and adult. Non-mated females lay male eggs; mated females lay female or male eggs at a 4:1 ratio

FIGURE 5.6 Female adult broad mite on tomato leaf.

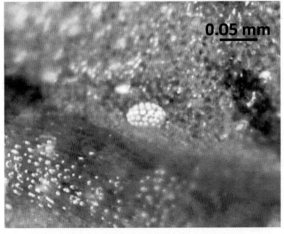

FIGURE 5.7 One broad mite egg on tomato leaf.

(Ewing, 1939; Jones and Brown, 1983; Li et al., 1985; Ferreira et al., 2006). Female broad mites lay 1–6 eggs/day, 35–70 eggs over a 7–12-day period (Brown and Jones, 1983; Gerson, 1992; de Coss-Romero and Peña, 1998). Reproduction usually does not occur below 14°C or above 34°C. Males live 5–9 days while females live 8–13 days (Ewing, 1939; Brown and Jones, 1983; Li et al., 1985; Montasser et al., 2011). The eggs hatch in 2 or 3 days and emerge as larvae (Li et al., 1985; Gerson, 1992). Larvae are slow moving and do not disperse far. In 2 or 3 days, the larvae develop into a quiescent pupal stage. Quiescent female pupae become attractive to the males who pick them up and carry them to new foliage using their specialized appendages. The males then wait for the adult female to emerge at which time they immediately mate with her (Jones and Brown, 1983; Li et al., 1985; Gerson, 1992; Peña and Campbell, 2005). Although females are very active, males account for much of the dispersal of the broad mite population as they carry the quiescent female pupa to new areas on the plant (Gerson, 1992). It has been found that adult broad mites under certain circumstances disperse by phoresy (method of dispersal in which the mites cling to the surface of another arthropod animal to be carried to a new site) using green peach aphids and several whitefly species (Fan and Petitt, 1998; Ferreira et al., 2006).

5.3 Distribution, Host Range, and Seasonal Occurrence

Broad mites are worldwide in distribution in both field and protected areas. *P. latus* has a wide host range in tropical and sub-tropical areas including 60 families of plants (Li et al., 1985; Grinberg et al., 2005; Ferreira et al., 2006; Alagarmalai et al., 2009). In temperate climates it is a greenhouse pest throughout the year and a crop pest during the summer. Broad mites infest a great many ornamental plants such as African violet, azalea, begonia, Cannabis, chrysanthemums, cyclamen, dahlia, gerbera, gloxinia, jasmine, impatiens, lantana, marigold, snapdragon, verbena, and zinnia as well as crops such as apple, avocado, cantaloupe, castor, chili, citrus, coffee, cotton, eggplant, guava, papaya, pear, potato, sesame, string or pole beans, mango, tea, tomato, and watermelon (Rice and Strong, 1962; Peña and Campbell, 2005; Grinberg et al., 2005; Ferreira et al., 2006).

Broad mite is a major pest in the tropical parts of the world year round. In subtropical areas it is a major pest during the summer and fall if weather conditions are warm and wet (Li et al., 1985; Ferreira et al., 2006). In temperate areas it is a pest during the summer months, but under the right environmental conditions of heat and humidity it can cause severe damage—similar to TRM.

5.4 Damage, Losses, and Economic Thresholds

The dramatic effects of broad mite infestations and feeding become evident by the appearance of tissue damage and changes in plant morphology (Alagarmalai et al., 2009). The earlier the plants are infested by broad mites the greater the damage and reduction in yields (de Coss-Romero and Peña, 1998). Infested plants show growth inhibition and a decrease in leaf numbers and leaf area (Schoonhoven et al., 1978; Hill, 1983; Alagarmalai et al., 2009). Feeding by broad mites may cause leaves to bronze and thicken, become brittle, corky, or cupped downward, and narrower than normal because the mesophyll cells in the area collapse (Fig. 5.8).

FIGURE 5.8 Leaf deformation on pepper from broad mite feeding.

Young stem growth may be distorted and stunted, and heavy feeding causes tomato flowers to die and drop off. Severely damaged plants may die. Broad mites cause damage even at low densities because they secrete a plant growth regulator–like toxin when they feed (Cross and Bassett, 1982; Nemesthoty et al., 1982; Peña, 1999). Mite damage can be confused with other plant problems such as viral diseases, micronutrient deficiencies, or herbicide injury. Even after control, damage may remain for weeks afterward (Schoonhoven et al., 1978; Nemesthoty et al., 1982). These factors make it challenging to make proper evaluations of control tactics, often leading to reports of control failures.

The broad mite's minute size and ability to damage plants at very low densities generally results in plant injury serving as the first indication of an infestation. When damage is noted, terminals of symptomatic plants should be examined under magnification to verify the presence of broad mites (look for their eggs). Damage will usually start in small clumps in a field and can spread rapidly. As few as five broad mites on young plants may cause substantial plant damage leading to reduced fruit production. Tomatoes, however, are less susceptible to broad mite feeding compared to sweet pepper or cucumber, and can tolerate 5–10 broad mites on a small plant (Schoonhoven et al., 1978; Nemesthoty et al., 1982; Peña, 1999).

6. MANAGEMENT

There are no reports on pest management practices for *T. evansi*, because there are so few studies on its population dynamics and its predators on solanaceous crops. Therefore, this section will be a general discussion of mite management unless there are specific species to highlight.

6.1 Monitoring

Fields should be regularly inspected for mites to prevent population build up. However, the aggregated distribution of mites in the early infestation makes scouting for them challenging and a poor tool for predictive purposes (Lange and Bronson, 1981; Park and Lee, 2007; Meck, 2010). Thus, scouting for mite feeeding damage is a better predictive tool. It is very difficult to control mite populations once they are established. Because damage symptoms can resemble nutrient deficiencies or plant disease, the presence of mites must be confirmed. There are methods to monitor mites used by researchers, but these are time consuming and labor intensive, and are not amenable to commercial grower situations.

6.2 Cultural Control

Basic crop sanitation helps in the control of all the mite species by keeping the area around the crop free of mite-weed hosts, which will reduce the sources of infestation (Kay, 1986). Some weeds are better at sustaining mite pests than others and these weeds, such as plantains, black nightshade, or solanaceous weeds should be targeted. Once harvest is complete, crop residues should be destroyed thus removing a breeding ground for the mites. Pruning back the affected plants and removing infested leaves will reduce pest numbers. For TSSM, dust management is particularly important. Dust on foliage makes it easier for mites to become a problem, although the mechanism of this effect has not been conclusively determined. There is evidence to show that dust interferes with mite predators and that the dust makes the foliage more hospitable for the mites. In many areas of the world there are periods of dry weather during the production of the tomato crop and the production of dust will be difficult to stop. Growing some type of cover crop around the field or between the field and an area of dust production can greatly reduce the amount of dust that is deposited on the tomato plant.

Tomato plants should be kept as healthy as possible by feeding, mulching, and watering at the proper levels (Varela et al., 2003). Water-stressed plants increase spider mite populations compared with the same plants without water stress (English-Loeb, 1990). Excessive nitrogen applications can cause succulent tomato plant growth, which can stimulate TSSM reproduction (van de Vrie et al., 1972).

6.3 Host-Plant Resistance

6.3.1 *Tetranychus evansi*

Some tomato varieties contain zingiberene, a sesquiterpene present in glandular trichomes of *Lycopersicon hirsutum* Dunal (ex. varieties PI-127826 and PI134414) (Maluf et al., 2001, 2007). Zingiberene content is markedly influenced by the density of glandular trichomes in tomato leaflets, especially type IV trichome density (see Luckwill (1943) for trichome classification). Higher densities of glandular trichomes (especially type IV and VI) decreased the distances walked by mites on the leaf surface (Maluf et al., 2001, 2007). Therefore, selection of tomato varieties with greater densities of glandular trichomes can be an effective tool to obtain tomato genotypes with greater levels of resistance (including repellency) to *T. evansi* (Maluf et al., 2007) and *T. urticae* (Alba et al., 2009).

Fecundity and longevity of *T. evansi* on resistant varieties were negatively correlated with the density of trichome types I and IV, while these values were high for the susceptible variety "Money Maker" (Murungi et al., 2009). Onyambus et al. (2011) reported that *T. evansi* did not lay eggs and showed very short adult longevity on two (PI134417 and LA2285) out of eight tomato varieties that had trichome types IV and VI. Thus, it is possible to develop resistant tomato varieties against *T. evansi* and possibly other mite species (Onyambus et al., 2011).

Another possible resistance factor of tomato genotypes to *T. evansi* and other tetranychids is acyl sugars present in tomato leaflets (Resende et al., 2008; Alba et al., 2009; Maluf et al., 2010). The greater the concentration of acyl sugars the slower the movement of *T. evansi* on leaflets (2.6 vs. 12.4 mm after 60 min) (Kant et al., 2004; Resende et al., 2008).

6.3.2 Aculops lycopersici

In greenhouse and field studies conducted in Kenya, differences were observed in the levels of susceptibility to *Acu. lycopersici* among 12 varieties of tomato (Kamau et al., 1992). Plant damage from foliar feeding was tolerated better on the varieties Early Stone Improved and Beauty compared to the other 10 varieties. The variety that was most damaged by TRM feeding was Red Cloud, but the percentage yield loss was greatest on Hybrid Beefmaster and Oxheart, and lowest on Roma and Money Maker.

6.4 Biological Control

There has been much work done on the biocontrol of *T. urticae*, but not as much on the other mite pests. With any of the mites, conservation of natural enemies is the key management program a grower can undertake. This entails altering some agricultural practices to increase the numbers and activity of these predators, i.e., predatory beetles in the family Staphylinidae and Coccinellidae, lacewings, predatory thrips, *Orius* spp., mirid bugs, and predatory flies in the family Cecidomyiidae. This might include changing the type and rates of chemical pesticide applications, altering irrigation practices, growing cover crops and companion plants that help natural enemies, reducing dust production near the crop, etc.

Much work has been done on augmentative control, which involves the mass rearing of and release of TSSM predators. The most commonly available predatory mites that have proved most efficacious for TSSM control are in the family Phytoseiidae. There are many different species commercially available. Some phytoseiid mites are specialists such as *Phytoseiulus persimilis* Athias-Henriot which feed exclusively on spider mites such as *T. urticae* and *T. cinnabarinus*. *P. persimilis* can rapidly increase in numbers in response to surging spider mite populations and is good to use if a large mite outbreak is occurring (Nihoul, 1994; Drukker et al., 1997). However, *P. persimilis* is susceptible to population crashes as its food source (spider mites) becomes scarce, and at times has problems moving on tomato plants because of the trichomes on the leaves. *P. persimilis* performs best when released on plants already infested with TSSM (Nihoul, 1994; Drukker et al., 1997). Combining *P. persimilis* with the predatory gall midge *Feltiella acarisuga* (Vallot) is particularly useful on tomato as the trichomes do not affect *F. acarisuga* and neither predator appears to interfere to any great extent with the other (Drukker et al., 1997).

Although *P. persimilis* is available in Europe, Africa, and Asia, other mites may be better suited for particular situations (Stoll, 1988; Seif et al., 2001). The mite predators *Neoseiulus californicus* (McGregor) and *Neoseiulus cucumeris* (Oudemans) are more adaptable generalist feeders (TSSM and broad mites), utilizing alternate food sources including pollen, which allows them to be used in the early part of the season to suppress early mite (TSSM and broad mite) outbreaks (McMurtry and Croft, 1997; Weintraub et al., 2003). *Galendromus occidentalis* (Nesbitt) tolerates high temperatures and low humidities (McMurtry and Croft, 1997). *Amblyseius andersoni* (Chant) is popular in Europe where it is available in slow release sachets embedded in long ribbons easily applied to crops. How many predatory mites to release depends on many things, such as the stage of the crop, cultivar, environment, area to be covered, release rates recommended, density of the pest. The release rate can be very site-specific and therefore growers need to work closely with the commercial biocontrol producer and university personnel to determine the best timing and density of a release for the best control potential (McMurtry and Croft, 1997).

Biological control agents for *T. evansi* were reviewed in detail by Navajas et al. (2013). Gutierrez and Etienne (1986) searched for natural enemies of *T. evansi* in South America. A few predators were found to be promising, such as the phytoseiid mite, *Phytoseiulus longipes* Evans (Badii, 1981; Furtado et al., 2006, 2007; Ferreira et al., 2006). Furtado et al. (2006) concluded that out of the 28 phytoseiid species surveyed on solanaceous plants, *P. longipes* is the most capable predator to control *T. evansi*, because it could easily move about on tomato leaves infested by *T. evansi* without being hampered by trichomes or the profuse webbing produced by the mites. The total fecundity of this predatory strain while feeding on *T. evansi* was similar to when it fed on *T. urticae* (Furtado et al., 2007). Another strain of *P. longipes* collected in Morocco resulted in a very low oviposition rate (0.1 eggs/day) when fed on *T. evansi*, compared with 2.7 eggs/day when fed on *T. urticae* (de Moraes and McMurtry, 1987). Similar data were obtained in a Chilean strain, in which 89% of immatures did not complete their development when fed on *T. evansi*, although more than 90% of immatures reached adulthood when fed on *T. urticae* (Ferreira et al., 2006). Thus, there are conspicuous differences in feeding habits among *P. longipes* strains, although all strains belong to the same species (Furtado et al., 2007). The spider mite predators *P. persimilis* and *N. californicus* did not survive well when fed only *T. evansi* and probably should not be used for its management (de Moraes and McMurtry, 1987; Migeon et al., 2009).

When using augmentative release as a biocontrol strategy an important consideration is the quality of commercially available natural enemies. At times the quality of the shipment may be reduced due to poor production or shipping/handling preparations. Growers should evaluate the quantity and quality of each shipment of natural enemies as soon as the shipment arrives. Growers should calibrate their application rate by counting the number of active natural enemies and contact the supplier immediately if natural enemy quality is unsatisfactory.

While there are fewer studies on biocontrol of TRM and broad mites than the spider mites, there are predators that show potential. However, *P. persimilis*, which is a very important specialist predator of the TSSM, does not prey on TRM. *Amblyseius fallacis* (Garman) appears to survive and reproduce on and control both TSSM and TRM (Brodeur et al., 1997). There also may be potential for the use of *Homeopronematus anconai* (Baker) as a biological control agent as studies indicate that this predator can be effective in controlling TRM in tomato (Hessein and Perring, 1986, 1988; Perring and Farrar, 1986). The predatory mite *Amblyseius swirskii* Athias-Henriot was found to be a good predator of TRM and broad mite under laboratory and field conditions (van Maanen et al., 2010). Under field conditions *A. swirskii* was able to survive on maize pollen when broad mite densities became low (Momen and Abdel-Khalek, 2008; van Maanen et al., 2010; Onzo et al., 2012). *N. cucumeris* was found to control broad mite populations, but needed to be released on every plant or on every other plant; this release method resulted in controls equal to chemical applications (Schoonhoven et al., 1978; Waterhouse and Norris, 1987; Weintraub et al., 2003). Lower release rates resulted in poor control. Other studies have shown that *Amblyseius largoensis* (Muma) has potential as a biological control of broad mites (Rodríguez et al., 2011). While there have been many studies examining other mite predators for control of mite pests on tomato under laboratory or greenhouse conditions, when moved to the field, few, if any, have been successful enough to warrant further study.

Other "biocontrols" include biopesticides such as entomopathogenic fungi. These fungi include *Neozygites floridana* (Weiser & Muma) (Humber et al., 1981; Maniania et al., 2008; Wekesa et al., 2008, 2010, 2011; Duarte et al., 2009; Ribeiro et al., 2009), *Beauveria bassiana* (Balsamo-Crivelli) Vuillemin, and *Metarhizium anisopliae* (Metchnikoff) Sorokin (Wekesa et al., 2005, 2006; Bugeme et al., 2008). de Faria and Wraight (2007) provided a list of commercially available species of the mycoacaricides for controlling tetranychid mites, in which 17 formulations from five fungal species were listed: *B. bassiana*, *H. thompsonii*, *Isaria fumosorosea* Wize, *Lecanicillium muscarium* (=*Verticillium lecanii*) R. Zare & W. Gams, and *M. anisopliae* (see also Maniania et al., 2008).

N. floridana has an optimal sporulation at 25°C and optimal germination between 25°C and 29°C, with the LC_{50} ranging from 3.16 to 3.47 days among three isolates at 29°C (Wekesa et al., 2010). Similarly, three of five isolates were pathogenic to *T. evansi*, and infestation rates by the fungal isolates on the mites were greater than 50% (Ribeiro et al., 2009). When both *N. floridana* and the phytoseiid mite *P. longipes* were simultaneously released on *T. evansi* infested leaflets, egg predation by *P. longipes* was greatly reduced because the predators spent increased time in self grooming, removing capilliconidia (from *N. floridana*) that were attached to their bodies (Wekesa et al., 2007). Other than the extra grooming there were no other effects of *N. floridana* on *P. longipes*.

For *B. bassiana* and *M. anisopliae*, optimal fungal germination was observed at 25°C and 30°C (Bugeme et al., 2008). Adults and deutonymphs of *T. evansi* were more susceptible to fungal infection than larval and protonymphal stages. While *T. evansi* eggs were found to be vulnerable to fungal infection, their mortality was dose-dependent (Wekesa et al., 2006). Under laboratory conditions *M. anisopliae* was pathogenic to adult females of *T. evansi*, causing mortality between 22% and 78% (Bugeme et al., 2008). All fungal isolates of *B. bassiana* showed similar results (Bugeme et al., 2008).

Chromobacterium subtsugae Martin is a naturally occurring, recently discovered bacterium, which produces a number of compounds that contribute to the formation of several complex modes of action, creating a biopesticide that is highly active against mites (Palmer and Vea, 2012). Control of mites using *C. subtsugae* is achieved by complex unique combinations of repellency, oral toxicity, reduced egg hatching, and reduced fecundity. This bacterium has also been shown to have only minor effects on most beneficials (Palmer and Vea, 2012).

6.5 Chemical Control

By far the most common and effective method for controlling mite pests on tomatoes is by using synthetic chemicals. However, the non-judicious use of chemical compounds against other pests such as caterpillars or thrips can cause flare-up of mites, especially

TSSM. Only the common name of the chemical miticide will be provided and the chemicals presented here are not meant to be an exhaustive list of control options, but a general guide to the chemicals that are currently available and efficacious.

There are eight acaricides that consistently have shown efficacy on TSSM, TRM, and broad mites for many years: abamectin, acequinocyl, bifenazate, chlorfenapyr, etoxazole, fenpyroximate, milbemectin, and spiromesifen (Jensen and Mingochi, 1988; Liu, 2005; Walgenbach and Schoof, 2005, 2009, 2010a,b, 2011; Schuster, 2006; Schuster et al., 2009; Layton et al., 2011; Stansly and Kostyk, 2012, 2013). Abamectin serves as a good rescue miticide and offers residual control of TSSM (Walgenbach and Schoof, 2005, 2009). Bifenazate controls TSSM with less toxicity to predaceous mites and beneficial arthropods. Spiromesifen and fenpyroximate are slow acting contact controls that can take 2–5 days for results to be observed (CABI, 2005; Schuster, 2006; Schuster et al., 2009; Layton et al., 2011). Etoxazol is a translaminar growth-regulating (molt inhibitor) miticide that is specific to plant feeding mites. It kills the eggs and nymphs of TSSM and sterilizes the female adult, but does not adversely affect mite predators (Haji et al., 1988; CABI, 2005).

Experiments in Iran with six acaricides showed that bromopropylate, sulfur, and karathane were the most effective compounds for TRM (Baradaran-Anaraki and Daneshvar, 1992) while lambda-cyhalothrin or fluvalinate was found to give 98% control of TRM (Cheremushkina et al., 1991). Flubenzimine and abamectin were found to be the most effective treatments for TRM control followed by dicofol (dicofol is no longer used in many countries because of its toxicity to the environment) and dicofol plus tetradifon (Haji et al., 1988). Fourteen acaricides were tested against *Acu. lycopersici* on tomatoes in Australia (Kay and Shepherd, 1988). The most effective chemicals in controlling an established infestation were flubenzimine, cyhexatin, and azocyclotin. For best control of TRM, applications were needed every 7–10 days. Also, flubenzimine and cyhexatin were the most effective preventative treatments for TRM (Kay and Shepherd, 1988). Royalty and Perring (1988) found that judicious applications of avermectin B could provide good control of TRM while conserving its predator, *H. anconai*. Pesticides and spray oils were tested against broad mites with the most efficacious being: abamectin, endosulfan, fenpyroximate, pyridaben, tebufenpyrad, dicofol, petroleum-based spray oils, and canola oil (Herron et al., 1996; Peña, 1988).

There are three reports on chemical control of *T. evansi*; Blair (1989), Gotoh et al. (2011) and Nyoni et al. (2011). In one study pesticides were foliar-applied and five were soil-applied to determine control efficacy for *T. evansi* in Zimbabwe with 48% of the chemicals tested being effective (Blair, 1989). A few chemicals such as binapacryl, cyhexatin, and dicofol are no longer used in some countries because of their toxicity to humans and environment. Out of 23 chemicals tested, 14 and 19 were effective to adult females and eggs of the Zimbabwean *T. evansi* strain, respectively (Blair, 1989).

Nine strains originating from Brazil, France, Kenya, Spain, Canary Island, Taiwan, and Japan (Kagoshima, Osaka, and Tokyo) were examined by Gotoh et al. (2011). All 11 chemicals tested were found to be effective for the control of *T. evansi* adults. Four strains used by Nyoni et al. (2011) were collected from Malawi and France. The LC_{50} values of four chemicals to adult females were variable. Only abamectin was considered to be effective to all four *T. evansi* strains tested (Nyoni et al., 2011). For bifenthrin, the LC_{50} values exceeded the recommended concentration in all four strains tested, and the LC_{50} values of two Malawian strains (1858–3560 mg/L) also were 20- to 39-fold greater than those of the two French strains (92.0–134.6 mg/L). For chlorpyriphos and fenpyroximate, LC_{50} values were similar among four strains (Nyoni et al., 2011).

Based on these studies, we recommend that the following chemicals should not be used to manage *T. evansi*: acephate, amitraz, bifenthrin, chlorpyriphos, cyflumetofen, dimethoate, flucythrinate, flufenoxuron, hexythiazox, and tetradifon because a marked resistance has already been documented in several studies. The present knowledge on chemical control suggests that 23 of the 35 acaricides could reduce *T. evansi* populations.

Calendar applications (spraying chemicals at regular predefined intervals rather than in response to pest populations) of chemicals increase the risk of resistance development in spider mites. This also applies to TRM and broad mites as well, but spider mites have a remarkable potential for rapid development of resistance (Cross and Bassett, 1982; Croft and Van de Baan, 1988; Peña, 1999; van Leeuwen et al., 2009; Ullah and Gotoh, 2013). Repeated use of acaricides with the same or similar modes of action should be avoided as this may lead to increased resistance. Development of resistance can be delayed by rotating acaricides; i.e., successively using acaricides with different modes of action, or by using mixtures of several compatible acaricides (Georghio, 1980; Ives et al., 2011; Ullah and Gotoh, 2013).

Other chemical controls include various "horticultural oils" that either are refined petroleum products (mineral oils) or plant-derived oils that can effectively reduce mite populations on tomatoes as long as coverage of the plant is thorough. These horticultural oils differ in their mode of action. One of the most direct pathways is by obstructing the air passages that mites use to breathe, causing them to suffocate. In other cases, the oils may interact with the fatty acids of the pest and degrade their lipid/waxy layer that normally keeps them from drying out, or the oils may interfere with the insect's normal metabolic processes (Hamilton, 1993; Taverner, 2002). The petroleum oils have their impurities removed that are associated with plant injury. These refined, purified oils are mixed with an emulsifier, which then allows the oil and water to mix (Cranshaw and Baxendale, 2013). Horticultural oils when applied thoroughly to plants give results that are similar to the best synthetic chemical compounds. However, it is difficult to apply the oils to the underside of the foliage or in tightly folded leaves where most of the mites are located. This inability to effectively apply the oils to tomato plants can result in spotty or reduced control of the mites. Though usually not as reliable as petroleum-based oils, some of the common plant-derived oils that have shown efficacy on mites include: caraway seed, citronella java, lemon eucalyptus, pennyroyal, peppermint, rosemary, sesame, thyme and cinnamon (Won-Il et al., 2004), Neem (Makundi and Kashenge, 2002), and Neem extract (http://www.infonet-biovision.org/PlantHealth/Plant-extract-Neem). Sulfur dust or wettable sulfur can be used although research shows that the efficacy of the sulfur products is especially variable (CABI, 2005).

7. ECONOMICS OF MANAGEMENT STRATEGIES

For any of the mite pests discussed in this chapter, cultural control is necessary to reduce mite infestations. This includes making sure that mites are not introduced to the field with transplants. Once mite numbers are causing damage, the most economical control is the judicial use of a chemical miticide, either synthetic or organic. The key is to find the mite infestations early enough so that horticultural oils or natural enemies can be utilized to their best extent. Discovering infestations later when mite populations are widespread and yield reducing will necessitate the use of several synthetic chemical applications. While endemic predators of mites should be encouraged in the tomato system, inundative or augmentative releases of predatory mites in a field situation may not be as economical as chemical applications.

8. FUTURE PROSPECTS

Proper, timely, and cost-effective monitoring of mites in a tomato field is one of the most important priorities for developing good IPM programs. As it stands now, there is no reliable cost-effective method to detect low, scattered mite densities, which would allow growers to use more options for control. Biological control using augmentative methods may be cost prohibitive for most small-land holders, therefore new mass-rearing methods are needed to reduce the costs. The application of biocontrol agents is still relatively primitive and more work is needed to examine how it could be done more efficiently and cost-effectively. Some tomato cultivars have shown moderate to high levels of resistance to mites and the factors that induce resistance are inherited, thus breeding for some levels of resistance to mites should be a priority for future integrated management programs.

In Japan *T. evansi* never causes much damage to nightshade plants. Analysis of the cause of this phenomenon may lead to a new control method. On the other hand, Furtado et al. (2007) pointed out that natural enemies can control *T. evansi* on nightshade, although they did not show any data on population dynamics. For confirming why *T. evansi* does not increase its population density and damage nightshade plants, we need to examine and compare patterns of population dynamics on solanaceous crops and weeds. Additionally, there is little information on the distribution of *T. evansi* in China and the United States. This information would be helpful in estimating the invasion route from South America to Pacific Rim counties.

Fungal and bacterial endophytes are microscopic organisms that are present in all plant species. These organisms have been found at times to reduce pest feeding on plant hosts through the production of inhibitory chemicals. In some grass species these fungal endophytes have been utilized as biocontrols by successfully reducing insect pest populations and damage while increasing host growth. The utilization of endophytes in broadleaf or vegetable plants is just at the beginning stages. Research is needed to identify which endophytes are present in vegetables and how they can be best utilized for control of mite pests within an integrated pest management system.

REFERENCES

Abou-Awad, B.A., 1979. The tomato russet mite, *Aculops lycopersici* (Massee) (Acari: Eriophyidae) in Egypt. Anzeiger für Schädlingskunde, Pflanzenschutz, Umweltschutz 52, 153–156.

Alagarmalai, J., Grinberg, M., Perl-Treves, R., Soroker, V., 2009. Host selection by the herbivorous mite *Polyphagotarsonemus latus* (Acari: Tarsonemidae). Journal of Insect Behavior 22, 375–387.

Alba, J.M., Montserrat, M., Fernández-Muñoz, R., 2009. Resistance to the two-spotted spider mite (*Tetranychus urticae*) by acylsucroses of wild tomato (*Solanum pimpinellifolium*) trichomes studied in a recombinant inbred line population. Experimental and Applied Acarology 47, 35–47.

Anderson, L.D., 1954. The tomato russet mite in the United States. Journal of Economic Entomology 47, 1001–1005.

Anonymous, 1972. Tomato mite. Tasmanian Journal of Agriculture 43, 244–245.

Auger, P., Migeon, A., Ueckermann, E.A., Tiedt, L., Navarro, M., 2013. Evidence for synonymy between *Tetranychus urticae* and *Tetranychus cinnabarinus* (Acari, Prostigmata, Tetranychidae): review and new data. Acarologia 53, 383–415.

Badii, M.H., 1981. Experiments on the Dynamics of Predation of *Phytoseiulus longipes* on the Prey *Tetranychus pacificus* (Acarina: Phytoseiidae, Tetranychidae) (Ph.D. thesis). University of California, Riverside, USA, p. 153.

Bailey, S.F., Keifer, H.H., 1943. The tomato russet mite, *Phyllocoptes destructor* Keifer: its present status. Journal of Economic Entomology 36, 706–712.

Baker, E.W., Pritchard, A.E., 1960. The tetranychoid mites of Africa. Hilgardia 29, 455–574.

Baradaran-Anaraki, P., Daneshvar, H., 1992. Studies on the biology and chemical control of tomato russet mite, *Aculops lycopersici* (Acari: Eriophyidae), in Varanin. Applied Entomology and Phytopathology 59, 25–27.

Birch, L.C., 1948. The intrinsic rate of natural increase of an insect population. Journal of Animal Ecology 17, 15–26.

Blair, B.W., 1989. Laboratory screening of acaricides against *Tetranychus evansi* Baker and Pritchard. Crop Protection 8, 212–216.

Bolland, H.R., Gutierrez, J., Flechtmann, C.H.W., 1998. World Catalogue of the Spider Mite Family (Acari: Tetranychidae). Koninklijke Brill NV, Leiden, The Netherlands, p. 392.

Boudreaux, H.B., Dosse, G., 1963. Concerning the names of some common spider mites. Advances in Acarology 1, 350–364.

Bourne, A.I., 1952. Tomato russet mite (*Vasates destructor*) – Massachusetts. Cooperative Economic Insect Report 2, 179.

Brandenburg, R., Kennedy, G.G., 1981. Differences in dorsal integumentary lobe densities between *Tetranychus urticae* Koch and *Tetranychus cinnabarinus* (Boisduval) (Acarina: Tetranychidae) from Northeastern North Carolina. International Journal of Acarology 7, 231–234.

Brodeur, J., Bouchard, A., Turcotte, G., 1997. Potential of four species of predatory mites as biological control agents of the tomato russet mite, *Aculops lycopersici* (Massee) (Eriophyidae). The Canadian Entomologist 129, 1–6.

Brown, R.D., Jones, V.P., 1983. The broad mite on lemons in Southern California. California Agriculture 37, 21–22.

Brust, G.E., 2013. Tomato Russet Mite Damage to Tomato Plants. University of Maryland Extension, College Park, Maryland, USA. http://extension.umd.edu-learn/insect-pests-tomato.

Brust, G.E., 2014. Gold Fleck on Tomatoes Caused by Many Things. University of Maryland Extention, College Park, Maryland, USA. http://extension.umd.edu/learn/gold-flecking-tomato-caused-many-things.

Bugeme, D.M., Maniania, N.K., Knapp, M., Boga, H.I., 2008. Effect of temperature on virulence of *Beauveria bassiana* and *Metarhizium anisopliae* isolates to *Tetranychus evansi*. Experimental and Applied Acarology 46, 275–285.

CABI, 2005. Crop Protection Compendium, 2005 ed. CAB International Publishing, Wallingford, UK. www.cabi.org.

CABI, 2013. Invasive Species Compendium: *Aculops lycopersici*. http://www.cabi.org/isc/datasheet/56111.

Cabrera, R.I., 1984. The mite *Vasates destructor*, a new host of the fungus *Hirsutella thompsonii*. Ciencia y Tecnica en la Agricultura Protección de Plantas 7, 69–79.

Carmona, M.M., 1964. Contribuicao para o conhecimento dos acaros das plantas cultivadas im Portugal. Agronomia Lusitana 22, 221–230.

Cheremushkina, N.P., NKh, A.A., Makarenkova, A., Golyshin, N.M., 1991. The rust mite of tomato. Zashchita Rastenii 11, 44–45.

Clotuche, G., Mailleux, A., Fernández, A.A., Deneubourg, J., Detrain, C., Hance, T., 2011. The formation of collective silk balls in the spider mite *Tetranychus urticae* Koch. PLoS One 6 (4), e18854.

Cranshaw, W.S., Baxendale, B., 2013. Insect Control: Horticultural Oils. Fact Sheet No. 5.569 Colorado State University Extension, Fort Collins, Colorado, USA. http://www.ext.colostate.edu/pubs/insect/05569.pdf.

Croft, B.A., Van de Baan, H.E., 1988. Ecological and genetic factors influencing evolution of pesticide resistance in tetranychid and phytoseiid mites. Experimental and Applied Acarology 4, 277–300.

Cross, J.V., Bassett, P., 1982. Damage to tomato and aubergine by broad mite, *Polyphagotarsonemus latus* (Banks). Plant Pathology 31, 391–393.

Davis, R., 1964. Autecological studies of *Rhynaeus breitlowi* Davis. Florida Entomologist 47, 113.

de Coss-Romero, M., Peña, J.E., 1998. Relationship of broad mite (Acari: Tarsonemidae) to host phenology and injury levels in *Capsicum annuum*. Florida Entomologist 81, 515–526.

de Faria, M.R., Wraight, S.P., 2007. Mycoinsecticides and mycoacaricides: a comprehensive list with worldwide coverage and international classification of formulation types. Biological Control 43, 237–256.

De kreij, C., Janse, J., Van Goor, B.J., Van Doesburg, J.D.J., 1992. The incidence of calcium oxalate crystal in fruit walls of tomato (*Lycopersicon esculenum* Mill.) as affected by humidity phosphate and calcium supply. Journal of Horticultural Science 67, 45–50.

de Moraes, G.J., McMurtry, J.A., 1987. Effect of temperature and sperm supply on the reproductive potential of *Tetranychus evansi* (Acari: Tetranychidae). Experimental and Applied Acarology 3, 95–107.

de Moraes, G.J., McMurtry, J.A., Baker, E.W., 1987. Redescription and distribution of the spider mites *Tetranychus evansi* and *T. marianae*. Acarologia 28, 333–343.

Den Outer, R.W., Van Veenendaal, W.L.H., 1988. Gold specks and crystals in tomato fruits (*Lycopersicon esculenum* Mill.). Journal of Horticultural Science 63, 645–649.

Drukker, B., Janssen, A., Ravensberg, W., Sabelis, M.W., 1997. Improved capacity of the mite predator *Phytoseiulus persimilis* (Acari: Phytosiidae) on tomato. Experimental and Applied Acarology 21, 507–518.

Duarte, V.S., Silva, R.A., Wekesa, V.W., Rizzato, F.B., Dias, C.T.S., Delalibera Jr., I., 2009. Impact of natural epizootics of the fungal pathogen *Neozygites floridana* (Zygomycetes: Entomophthorales) on population dynamics of *Tetranychus evansi* (Acari: Tetranychidae) in tomato and nightshade. Biological Control 51, 81–90.

Dupont, L.M., 1978. On gene flow between *Tetranychus urticae* Koch, 1836 and *Tetranychus cinnabarinus* (Boisduval) Boudreaux, 1956 (Acari: Tetranychidae): sononomy between the two species. Entomologia Experimentalis et Applicata 25, 297–303.

English-Loeb, G.M., 1990. Plant drought stress and outbreaks of spider mites: a field test. Ecology 71, 1401–1411.

Eschiapati, D., Oliveira, C.A.L. de, Velho, D., Sponchiado, O.J., 1975. Efieto da epoca de infestacao do microacaro, *Aculops* sp., na cultura do tomateiro. Ciencia e Cultura 27, 1336–1337.

Estebanes-Gonzalez, M.L., Rodriguez-Navarro, S., 1991. Observations on some mites of the families Tetranychidae, Eriophyidae, Acaridae and Tarsonemidae (Acari), in horticultural crops from Mexico. Folia Entomológica Mexicana 83, 199–212.

Ewing, H.E., 1939. A Revision of the Mites of the Subfamily Tarsoneminae of North America, the West Indies, and the Hawaiian Islands. United States Department of Agriculture, Washington, DC. Technical Bulletin 653, 64.

Fan, Y.Q., Petitt, F.L., 1998. Dispersal of the broad mites, *Polyphagotarsonemus latus* (Acari: Tarsonemidae) on *Besmisia argentifolii* (Homoptera: Aleyrodidae). Experimental and Applied Acarology 22, 411–415.

Ferreira, R.C., de Oliveira, J.V., Haji, F.P., Gondim, M.G., 2006. Biologia, exigências térmicas e tabela de vida de fertilidade do ácaro-branco *Polyphagotarsonemus latus* (Banks) (Acari: Tarsonemidae) em videira (*Vitis vinifera* L.) cv. Itália. Neotropical Entomology 35, 126–132.

Fischer, S., Mourrut-Salesse, J., 2005. Tomato russet mite in Switzerland (*Aculops lycopersici*: Acari, Eriophyidae). Revue Suisse de Viticulture Arboriculture Horticulture 37, 227–232.

Flechtmann, C.H.W., 1974. Phytophagous mites of economic importance in Brazil. In: Piffl, E. (Ed.), Proceedings of the 4th International Congress of Acarology. Akademiai Kiado, Budapest, Hungary, pp. 185–187.

Flechtmann, C.H.W., 1977. Acaros de Importancia Agricola, pp. 112–114 Sao Paulo, Livraria, Brazil.

Flechtmann, C.H.W., Knihinicki, D.K., 2002. New species and new record of *Tetranychus* Dufour from Australia, with a key to the major groups in this genus based on females (Acari: Prostigmata: Tetranychidae). Australian Journal of Entomology 41, 118–127.

Foster, G.N., Barker, J., 1978. A new biotype of red spider mite (*Tetranychus urticae* (Koch)) causing atypical damage to tomatoes. Plant Pathology 27, 47–48.

Furtado, I.P., Moraes, G.J., Kreiter, S., Knapp, M., 2006. Search for effective natural enemies of *Tetranychus evansi* in south and Southeast Brazil. Experimental and Applied Acarology 40, 157–174.

Furtado, I.P., Moraes, G.J., Kreiter, S., Tixier, M.S., Knapp, M., 2007. Potential of a Brazilian population of the predatory mite *Phytoseiulus longipes* of a biological control agent of *Tetranychus evansi* (Acari: Phytoseiidae, Tetranychidae). Biological Control 42, 139–147.

Georghio, G.P., 1980. Insecticide resistance and prospects for its management. Residue Reviews 76, 131–145.

Gerson, U., 1985. Webbing. In: Helle, W., Sabelis, M.W. (Eds.), Spider Mites: Their Biology, Natural Enemies and Control. Elsevier, Amsterdam, The Netherlands, pp. 223–231.

Gerson, U., 1992. Biology and control of the broad mite, *Polyphagotarsonemus latus* (Banks) (Acari: Tarsonemidae). Experimental and Applied Acarology 13, 163–178.

Ghidiu, G.M., Hitchner, E.M., Funderburk, J.E., 2006. Gold fleck damage to tomato fruit caused by feeding of *Frankliniella occidentalis* (Thysanoptera: Thripidae). Florida Entomologist 89, 279–281.

Gotoh, T., 1997. Annual life cycles of populations of the two-spotted spider mite, *Tetranychus urticae* Koch (Acari: Tetranychidae) in four Japanese pear orchards. Applied Entomology and Zoology 32, 207–216.

Gotoh, T., Fujiwara, S., Kitashima, Y., 2011. Susceptibility to acaricides in nine strains of the tomato red spider mite *Tetranychus evansi* (Acari: Tetranychidae). International Journal of Acarology 37, 93–102.

Gotoh, T., Ishikawa, Y., Kitashima, Y., 2003. Life-history traits of the six *Panonychus* species from Japan (Acari: Tetranychidae). Experimental and Applied Acarology 29, 241–252.

Gotoh, T., Suwa, A., Kitashima, Y., Rezk, H.A., 2004. Developmental and reproductive performance of *Tetranychus pueraricola* Ehara and Gotoh (Acari: Tetranychidae) at four constant temperatures. Applied Entomology and Zoology 39, 675–682.

Grinberg, M., Treves, R.P., Palevsky, E., Shomer, I., Soroker, V., 2005. Interaction between cucumber plants and the broad mite, *Polyphagotarsonemus latus*: from damage to defense gene expression. Entomologia Experimentalis et Applicata 115, 134–144.

Gutierrez, J., 1974. Les especes du genre *Tetranychus* Dufour (Acariens: Tetranychidea) ayant une incidence economique a Madagascar et dans les iles voisines. Acarologia 16, 258–270.

Gutierrez, J., Etienne, J., 1986. Les Tetranychidae de l'ile de la Reunion et quelques-uns de leurs predateurs. L'Agronomie Tropicale 41, 84–91.

Haji, F.N.P., Morps, G.J.de., Lacerda, C.A.de., Neto, R.S., 1988. Chemical control of the tomato russet mite *Aculops lycopersici* (Massee, 1937). Anais da Sociedade Entomologica do Brasil 17, 437–442.

Hamilton, R.J., 1993. Structure and general properties of mineral and vegetable oils used as spray adjuvants. Pesticide Science 37, 141–146.

Helle, W., Sabelis, M.W. (Eds.), 1985. Spider Mites: Their Biology, Natural Enemies and Control. Elsevier, Amsterdam, The Netherlands, p. 458.

Herron, G., Jiang, L., Spooner-Hart, R., 1996. A laboratory-based method to measure relative pesticide and spray oil efficacy against broad mite, *Polyphagotarsonemus latus* (Banks) (Acari: Tarsonemidae). Experimental and Applied Acarology 20, 492–502.

Hessein, N.A., Perring, T.M., 1986. Feeding habits of the Tydeidae with evidence of *Homeopronematus anconai* (Acari: Tydeidae) predation on *Aculops lycopersici* (Acari: Eriophyidae). International Journal of Acarology 12, 215–221.

Hessein, N.A., Perring, T.M., 1988. The importance of alternate foods for the mite *Homeopronematus anconai* (Acari: Tydeidae). Annals of the Entomological Society of America 81, 488–492.

Hill, D.S., 1983. *Polyphagotarsonemus latus* (Banks). In: Agricultural Insect Pests of the Tropics and their Control. Cambridge University Press, UK, p. 504.

Humber, R.A., Moraes, G.J., dos Santos, J.M., 1981. Natural infection of *Tetranychus evansi* [Acarina: Tetranychidae] by a *Triplosporium* sp [Sygomycetes: Entomophthorales] in Northeastern Brazil. Entomophaga 26, 421–425.

Ives, A.R., Glaum, P.R., Ziebarth, N.L., Andow, D.A., 2011. The evolution of resistance to two-toxin pyramid transgenic crops. Ecological Applications 21, 503–515.

Jayasinghe, G.G., Mallik, B., 2010. Growth stage based economic injury levels for two spotted spider mite, *Tetranychus urticae* Koch (Acari, Tetranychidae) on tomato *Lycopersicon esculentum* Mill. Tropical Agricultural Research 22, 54–65.

Jensen, A., Mingochi, D.S., 1988. Chemical control of red spider mite (*Tetranychus urticae* Koch) on tomatoes in Zambia. Acta Horticulturae 218, 275–280.

Jeppson, L.R., Keifer, H.H., Baker, E.W., 1975. Mites Injurious to Economic Plants. Berkeley: University California Press, USA.

Jones, V.P., Brown, R.D., 1983. Reproductive responses of the broad mite, *Polyphagotarsonemus latus* (Acari: Tarsonemidae), to constant temperature-humidity regimes. Annals of the Entomological Society of America 76, 466–469.

Kamau, A.W., Mueke, J.M., Khaemba, B.M., 1992. Resistance of tomato varieties to the tomato russet mite, *Aculops lycopersici* (Massee) (Acarina: Eriophyidae). International Journal of Tropical Insect Science 13, 351–356.

Kant, M.R., Ament, K., Sabelis, M.W., Haring, M.A., Schuurink, R.C., 2004. Differential timing of spider mite-induced direct and indirect defenses in tomato plants. Plant Physiology 135, 483–495.

Kawakami, Y., Numata, H., Ito, K., Gotoh, S.G., 2009. Dominant and recessive inheritance patterns of diapause in the two-spotted spider mite *Tetranychus urticae*. Journal of Heredity 101, 20–25.

Kay, I.R., 1986. Tomato russet mite: A serious pest of tomatoes. Queensland Agricultural Journal 112, 231–232.

Kay, I.R., Shepherd, R.K., 1988. Chemical control of the tomato russet mite on tomatoes in the dry tropics of Queensland. Queensland Journal of Agricultural and Animal Sciences 45, 1–8.

Keifer, H.H., 1940. Eriophyid studies X. Bulletin of the California Department of Agriculture 29, 21–46.

Keifer, H.H., 1952. The eriophyid mites of California. Bulletin of the California Insect Survey, 2. University of California Press, Berkeley and Los Angeles, USA.

Keifer, H.H., 1966. *Aculops*, New Genus, vol. 21. California Department of Agriculture Eriophyid studies, p. 9.

Krips, O.E., Witu, A., Willems, P.E.L., Dicke, M., 1998. Intrinsic rate of population increase of the spider mite *Tetranychus urticae* on the ornamental crop gerbera: intraspecific variation in host plant and herbivore. Entomologia Experimentalis et Applicata 89, 159–168.

Kuang, H.Y., Cheng, L.S., 1990. Studies on the differentiation of two sibling species, *Tetranychus cinnabarinus* (Boisduval) and *T. urticae* Koch. Acta Entomologica Sinica 33, 109–116.

Lamb, K.P., 1953. A revision of the gall-mites (Acarina, Eriophyidae) occurring on tomato (*Lycopersicum esculentum* Mill.), with a key to the Eriophyidae recorded from solanaceous plants. Bulletin of Entomological Research 44, 343–350.

Lange, W.H., Bronson, L., 1981. Insect pests of tomatoes. Annual Review of Entomology 26, 345–371.

Layton, B.M., Zhao, Y., Gu, M., 2011. Efficacy of selected insecticides for spider mite control on greenhouse grown tomatoes 2010. Arthropod Management Tests 36, e86.

Li, L.S., Li, Y.R., Bu, G.S., 1985. The effect of temperature and humidity on the growth and development of the broad mite, *Polyphagotarsonemus latus*. Acta Entomologica Sinica 28, 181–187.

Lindquist, E.E., Sabelis, M.W., Brun, J., 1996. Eriophyoid Mites – Their Biology, Natural Enemies and Control. Elsevier, Amsterdam, The Netherlands, pp. 137–156.

Liu, L.U., 2005. Efficacy of selected insecticides against pepper pests on tomato, fall 2003. Arthropod Management Tests 30, e47.

Luckwill, L.C., 1943. The Genus *Lycopersicon*: An Historical, Biological and Taxonomic Survey of the Wild and Cultivated Tomatoes. Aberdeen University Studies, No. 120. Aberdeen University Press, Aberdeen, Scotland.

Máca, J., 2012. Nové nálezy hálek (zoocecidií) v jizních Cechách (New records of galls (zoocecidia) in (Czech Republic)). Sborník Jihoceského Muzea v Ceských Budejovicích, Prírodní Vedy 52, 197–208.

Makundi, R.H., Kashenge, S., 2002. Comparative efficacy of neem, *Azadirachta indica*, extract formulations and the synthetic acaricide, Amitraz (Mitac), against the two spotted spider mites, *Tetranychus urticae* (Acari: Tetranychidae), on tomatoes, *Lycopersicum esculentum*. Zeitschrift fur Pflanzenkrankheiten und Pflanzenschutz 109, 57–63.

Maluf, W.R., Campos, G.A., Cardoso, M.G., 2001. Relationships between trichome types and spider mite (*Tetranychus evansi*) repellence in tomatoes with respect to foliar zingiberene contents. Euphytica 121, 73–80.

Maluf, W.R., Inoue, I.F., Ferreira, R.P.D., Gomes, L.A.A., Castro, E.M., Cardoso, M.G., 2007. Higher glandular trichome density in tomato leaflets and repellence to spider mites. Pesquisa Agropecuaria Brasileira 42, 1227–1235.

Maluf, W.R., Maciel, G.M., Gomes, L.A.A., Cardoso, M.G., Goncalves, L.D., Silva, E.C., Knapp, M., 2010. Broad-spectrum arthropod resistance in hybrids between high- and low-acylsugar tomato lines. Crop Science 50, 439–450.

Maniania, N.K., Bugeme, D.M., Wekesa, V.W., Delalibera Jr., I., Knapp, M., 2008. Role of entomopathogenic fungi in the control of *Tetranychus evansi* and *Tetranychus urticae* (Acari: Tetranychidae), pests of horticultural crops. Experimental and Applied Acarology 46, 259–274.

Margolies, D.C., Kennedy, G.G., 1988. Fenvalerate-induced aerial dispersal by the twospotted spider mite. Entomologia Experimentalis et Applicata 46, 233–240.

Margolies, D.C., Wrensch, D.L., 1996. Temperature-induced changes in spider mite fitness: offsetting effects of development time, fecundity, and sex ratio. Entomologia Experimentalis et Applicata 78, 111–118.

Massee, M.A., 1937. An Eriophyid mite injurious to tomato. Bulletin of Entomological Research 28, 403–406.

McGregor, E.A., 1950. Mites of the family Tetranychidae. American Midland Naturalist 44, 257–420.

McMurtry, J.A., Croft, B.A., 1997. Life-styles of phytoseiid mites and their roles in biological control. Annual Review of Entomology 42, 291–321.

Meck, E.D., 2010. Management of the Twospotted Spider Mite *Tetranychus urticae* (Acari: Tetranychidae) in North Carolina Tomato Systems (Ph.D. thesis). North Carolina State University, Raleigh, North Carolina, USA.

Meck, E.D., Walgenbach, J.F., Kennedy, G.G., 2009. Effect of vegetation management on autumn dispersal of *Tetranychus urticae* (Acari: Tetranychidae) from tomato. Journal of Applied Entomology 133, 742–748.

Metcalf, R.L., Metcalf, R.A., 1993. Destructive and useful insects. In: Insects Injurious to Vegetable Gardens and Truck Crops. fifth ed. McGraw-Hill, Inc., New York, USA.

Migeon, A., Ferragut, F., Escudero-Colomar, L., Fiaboe, K., Knapp, M., Moraes, G.J., Ueckermann, E., Navajas, M., 2009. Modelling the potential distribution of the invasive tomato red spider mite, *Tetranychus evansi* (Acri: Tetranychidae). Experimental and Applied Acarology 48, 199–212.

Migeon, A., Nouguier, E., Dorkeld, F., 2010. Spider mites web: a comprehensive database for the Tetranychidae. Trends in Acarology 557–560.

Momen, F.M., Abdel-Khalek, A., 2008. Effect of the tomato rust mite *Aculops lycopersici* (Acari: Eriophyidae) on the development and reproduction of three predatory phytoseiid mites. International Journal of Tropical Insect Science 28, 53–57.

Monkman, K.D., 1992. Tomato russet mite-Acari: Eriophyoidea. Monthly Bulletin Department of Agriculture, Fisheries and Parks Bermuda 63, 19–21.

Montasser, A.A., Taha, A.A., Hanafy, A.R.I., Hassan, G.M., 2011. Biology and control of the broad mite *Polyphagotarsonemus latus* (Banks, 1904) (Acari: Tarsonemidae). International Journal of Environmental Science and Technology 1, 26–34.

Murungi, L.K., Knapp, M., Masinde, P.W., Onyambu, G., Gitonga, L., Agong, S.G., 2009. Host-plant acceptance, fecundity and longevity of *Tetranychus evansi* (Acari: Tetranychidae) on selected tomato accessions. African Journal of Horticultural Science 2, 79–91.

Navajas, M., de Moraes, G.J., Auger, P., Migeon, A., 2013. Review of the invasion of *Tetranychus evansi*: biology, colonization pathways, potential expansion and prospects for biological control. Experimental and Applied Acarology 59, 43–65.

Nemesthoty, K., Volcsansky, E., Simon, N., 1982. Influence of damage of the mites *Tarsonemus pallidus* and *Polyphagotarsonemus latus* Banks (Acari: Tarsonemidae) on the morphological properties of fashedera and hedera leaves. Novenyvedelem 10, 437–442.

Nihoul, P., 1994. Phenology of glandular trichomes related to entrapment of *Phytoseiulus persimilis* A.H. in the glasshouse tomato. Journal of Horticultural Science 69, 123–129.

Nyoni, B.N., Gorman, K., Mzilahowa, T., Williamson, M.S., Navajas, M., Field, L.M., Bass, C., 2011. Pyrethroid resistance in the tomato red spider mite, *Tetranychus evansi*, is associated with mutation of the para-type sodium channel. Pest Management Science 67, 891–897.

Oldfield, G.N., 1970. Mite transmission of plant viruses. Annual Review of Entomology 15, 343–380.

Oliveira, C.A.L., Eschiapapti, D., Velho, D., Sponchiado, O.J., 1982. Quantitative losses caused by the tomato russet mite, *Aculops lycopersici* (Massee) in field tomato crop. Ecossistema 7, 14–18.

Onyambus, G.K., Maranga, R.O., Gitonga, L.M., Knapp, M., 2011. Host plant resistance among tomato accessions to the spider mite *Tetranychus evansi* in Kenya. Experimental and Applied Acarology 54, 385–393.

Onzo, A., Houedokoho, A.F., Hanna, R., 2012. Potential of the predatory mite, *Amblyseius swirskii* to suppress the broad mite, *Polyphagotarsonemus latus* on the gboma eggplant, *Solanum macrocarpon*. Journal of Insect Science 12, 1–11.

Osman, A.A., 1975. Efficiency of some fungicides in the control of the eriophyid mite, *Vasates lycopersici* (Massee) in Egypt (Acarina: Eriophyidae). Bulletin of the Entomological Society of Egypt Economic Series 9, 115–118.

Palmer, C., Vea, E., 2012. IR-4 Ornamental Horticulture Program Mite Efficacy: A Literature Review. http://ir4.rutgers.edu/Ornamental/SummaryReports/MiteEfficacyDataSummary2012.pdf.

Park, Y.L., Lee, J.H., 2007. Seasonal dynamics of economic injury levels for *Tetraychus urticae* Koch (Acari: Tetranychidae) on *Cucumis sativus* L. Journal of Applied Entomology 131, 588–592.

Peña, J.E., 1999. Broad mite: damage and control. In: Needham, G.R., Mitchell, R., Horn, D.J., Welbourn, W.C. (Eds.), Acarology IX. Ohio Biological Survey, pp. 265–270.

Peña, J.E., Campbell, C.W., 2005. Broad Mite. Department of Entomology and Nematology, Florida Cooperative Extension Service, Publication No. ENY618. Institute of Food and Agricultural Sciences, University of Florida, USA.

Peña, J.E., 1988. Chemical control of broad mites (Acarina: Tarsonemidae) in limes (*Citrus latifolia*). Proceedings of Florida State Horticultural Society 101, 247–249.

Peña, J.E., Osborne, L.S., Duncan, R.E., 1996. Potential of fungi as biocontrol agents of *Polyphagotarsonemus latus*. Entomophaga 41, 27–36.

Perring, T.M., Farrar, C.A., 1986. Historical perspective and current world status of the tomato russet mite (Acari: Eriophyidae). Miscellaneous Publications of the Entomological Society of America 63, 1–9.

Petanović, R., Kielkiewicz, M., 2010. Plant–eriophyoid mite interactions: cellular biochemistry and metabolic responses induced in mite-injured plants. In: Ueckermann, E.U. (Ed.), Eriophyoid Mites: Progress and Prognoses. Springer Science+Business Media B.V., The Netherlands, pp. 61–80.

Resende, J.T.V., Maluf, W.R., Cardoso, M.G., Faria, M.V., Gonalves, L.D., Nascimento, I.R., 2008. Resistance of tomato genotypes with high level of acylsugars to *Tetranychus evansi* Baker and Pritchard. Scientia Agricola, Piracicaba 65, 31–35.

Ribeiro, A.E., Gondim Jr., M.G., Calderan, E., Delalibera Jr., I., 2009. Host range of *Neozygites floridana* isolates (Zygomycetes: Entomophthorales) to spider mites. Journal of Invertebrate Pathology 102, 196–202.

Rice, R.E., Strong, F.E., 1962. Bionomics of the tomato russet mite, *Vasates lycopersici* (Massee). Annals of the Entomological Society of America 55, 431–435.

Rodrıguez, H., Ramos, M., Montoya, A., Rodrıguez, Y., Chico, R., Miranda, I., Depestre, T., 2011. Development of *Amblyseius largoensis* as biological control agent of the broad mites (*Polyphagotarsonemus latus*). Biotecnología Aplicada 28, 171–175.

Rotem, K.A., Agrawal, A.A., 2003. Density dependent population growth of the two-spotted spider mite, *Tetranychus urticae*, on the host plant *Leonurus cardiac*. Nordic Society Oikos 103, 559–565.

Royalty, R.N., Perring, T.M., 1988. Morphological analysis of damage to tomato leaflets by tomato russet mite (Acari: Eriophyidae, Tydeidae). Journal of Economic Entomology 81, 816–820.

Sabelis, M.W., 1985a. Capacity for population increase. In: Helle, W., Sabelis, M.W. (Eds.), World Crop Pests: Spider Mites, Their Biology, Natural Enemies and Control, vol. 1B. Elsevier, Amsterdam, The Netherlands, pp. 35–41.

Sabelis, M.W., 1985b. Reproductive strategy. In: Helle, W., Sabelis, M.W. (Eds.), Spider Mites: Their Biology, Natural Enemies and Control, vol. 1B. Elsevier, Amsterdam, The Netherlands, pp. 265–278.

Sabelis, M.W., 1991. Life-history evolution of spider mites. In: Schuster, R., Murphy, P.W. (Eds.), The Acari: Reproduction, Development and Life-History Strategies. Chapman and Hall, London, UK, pp. 23–49.

Sarmento, R.A., Lemos, F., Bleeker, P.M., Schuurink, R.C., Pallini, A., Oliveira, M.G., Lima, E.R., Kant, M., Sabelis, M.W., Janssen, A., 2011a. A herbivore that manipulates plant defense. Ecology Letters 14, 229–236.

Sarmento, R.A., Lemos, F., Dias, C.R., Kikuchi, W.T., Rodrigues, J.C.P., Pallini, A., Sabelis, M.W., Janssen, A., 2011b. A herbivorous mite down-regulates plant defense and produces web to exclude competitors. PLoS One 6, e23757.

Sarr, I., Knapp, M., Ogol, C.K.P., Baumgärtner, J., 2002. Impact of predators on *Tetranychus evansi* Baker and Pritchard populations and damage on tomatoes (*Lycopersicon esculentum* Mill.) in Kenya. In: XI International Congress of Acarology (Abstract Book), September 18–23, Merida, Mexico, p. 271.

Saunyama, I.G.M., Knapp, M., 2003. Effect of pruning and trellising of tomatoes on red spider mite incidence and crop yield in Zimbabwe. African Crop Science Journal 11, 269–277.

Schoonhoven, A.V., Piedrahita, J., Valderrama, R., Galvez, G., 1978. Biologia, daño y control del acaro tropical *Polyphagotarsonemus latus* (Banks) (Acarina: Tarsonemidae) en frijol. Turrialba 28, 77–80.

Schuster, D.A., 2006. Control of the broad mite on bell pepper, Spring 2005. Arthropod Management Tests 31 (1), e45.

Schuster, D.A., Shurtleff, S., Kalb, S., 2009. Broad mite control on tomato, fall 2007. Arthropod Management Tests 34, e46.

Seif, A.A., Varela, A.M., Loehr, B., Michalik, S., 2001. A Guide to IPM in French Beans Production with Emphasis on Kenya. ICIPE Science Press, Nairobi, Kenya, p. 88.

Shih, C.T., Poe, S.L., Cromroy, H.L., 1976. Biology, life table and intrinsic rate of increase of *Tetranychus urticae*. Annals of the Entomological Society of America 69, 362–364.

Silva, P., 1954. Um novo acaro nocivo ao tomaeiro na Bahia (*Tetranychus marianae* McGregor, 1950-Acaruba). Boletim do Instituto Biologico da Bahia 1, 18–37.

Sloan, W.J.S., 1938. Mite injury of tomatoes. Queensland Agricultural Journal 50, 370–371.

Stansly, P.A., Kostyk, B., 2012. Control of broad mite on 'Jalepeno' pepper with oberon and movento, 2010. Arthropod Management Tests 37 (1), e42.

Stansly, P.A., Kostyk, B., 2013. Control of broad mite on 'Jalepeno' pepper with tolfenpyrad and industry standard, 2011. Arthropod Management Tests 38, e44.

Stoll, G., 1988. Natural Crop Protection in the Tropics. Agrecole. c/o Ökozentrum, CH-4438, Langenbruck, Switzerland.

Szwejda, J., 1993. Injury symptoms and control of two spider mite species: *Tetranychus urticae* and *T. cinnabarinus* occurring on cucumbers and tomatoes. Materialy Sesji Instytutu Ochrony Roslin 33, 128–135.

Taverner, P., 2002. In: Beattie, G.A.C. (Ed.), Spray Oils beyond 2000: Sustainable Pest and Disease Management. University of Western Sydney, Sydney, Australia, pp. 125–136.

Tommaso, A., Sílvia, A., Josep, J., 2007. Estimating the intrinsic rate of increase of *Tetranychus urticae*: which is the minimum number of immature individuals to consider. Experimental and Applied Acarology 41, 55–59.

Ullah, M.S., Gotoh, T., 2013. Laboratory-based toxicity of some acaricides to *Tetranychus macfarlanei* and *Tetranychus truncatus* (Acari: Tetranychidae). International Journal of Acarology 39, 244–251.

van de Vrie, M., McMurtry, J.A., Huffaker, C.B., 1972. Ecology of tetranychid mites and their natural enemies: a review. III. Biology, ecology and pest status and host plant relations of tetranychids. Hilgardia 41, 343–432.

van Leeuwen, T., Vontas, J., Tsagkarakou, A., Tirry, L., 2009. Mechanisms of acaricide resistance in the two-spotted spider mite *Tetranychus urticae*. In: Ishaaya, I., Horowitz, A.R. (Eds.), Biorational Control of Arthropod Pests. Springer, Dordrecht, Netherlands, pp. 347–393.

van Maanen, R., Vila, E., Sabelis, M.W., Janssen, A., 2010. Biological control of broad mites (*Polyphagotarsonemus latus*) with the generalist predator *Amblyseius swirskii*. Experimental and Applied Acarology 52, 29–34.

Varela, A.M., Seif, A.A., Loehr, B., 2003. A Guide to IPM in Tomato Production in Eastern and Southern Africa. ICIPE Science Press, Nairobi, Kenya.

Veerman, A., 1985. Diapause. In: Helle, W., Sabelis, M.W. (Eds.), Spider Mites: Their Biology, Natural Enemies and Control, vol. 1. Elsevier, Amsterdam, Netherlands, pp. 279–316.

Walgenbach, J.F., Schoof, S.C., 2005. Insect control on staked tomatoes, 2004. Arthropod Management Tests 30, e91.

Walgenbach, J.F., Schoof, S.C., 2009. Twospotted spider mite control on tomatoes, 2008. Arthropod Management Tests 34, e89.

Walgenbach, J.F., Schoof, S.C., 2010a. Twospotted spider mite control on tomatoes, 2009. Arthropod Management Tests 35, e45.

Walgenbach, J.F., Schoof, S.C., 2010b. Tomato chemigation study, 2009. Arthropod Management Tests 35, e44.

Walgenbach, J.F., Schoof, S.C., 2011. Tomato chemigation study. Arthropod Management Tests 36, e90.

Waterhouse, D.F., Norris, K.R., 1987. *Polyphagotarsonemus latus* (Banks). In: Biological Control Pacific Prospects. Inkata Press, Melbourne, Australia, p. 454.

Weintraub, P.G., Kleitman, S., Mori, R., Shapiro, N., Palevsky, E., 2003. Control of the broad mite (*Polyphagotarsonemus latus* (Banks)) on organic greenhouse sweet peppers (*Capsicum annuum* L.) with the predatory mite, *Neoseiulus cucumeris* (Oudemans). Biological Control 27, 300–309.

Wekesa, V.W., de Moraes, G.J., Knapp, M., Delalibera Jr., I., 2007. Interactions of two natural enemies of *Tetranychus evansi*, the fungal pathogen *Neozygites floridana* (Zygomycetes: Entomophthorales) and the predatory mite, *Phytoseiulus longipes* (Acari: Phytoseiidae). Biological Control 41, 408–414.

Wekesa, V.W., Knapp, M., Delalibera Jr., I., 2008. Side-effects of pesticides on the life cycle of the mite pathogenic fungus *Neozygites floridana*. Experimental and Applied Acarology 46, 287–297.

Wekesa, V.W., Knapp, M., Maniania, N.K., Boga, H.I., 2006. Effects of *Beauveria bassiana* and *Metarhizium anisopliae* on mortality, fecundity and egg fertility of *Tetranychus evansi*. Journal of Applied Entomology 130, 155–159.

Wekesa, V.W., Maniania, N.K., Knapp, M., Boga, H.I., 2005. Pathogenicity of *Beauveria bassiana* and *Metarhizium anisopliae* to the tobacco spider mite *Tetranychus evansi*. Experimental and Applied Acarology 36, 41–50.

Wekesa, V.W., Moraes, G.J., Ortega, E.M.M., Delalibera Jr., I., 2010. Effect of temperature on sporulation of *Neozygites floridana* isolates from different climates and their virulence against the tomato red spider mite, *Tetranychus evansi*. Journal of Invertebrate Pathology 103, 36–42.

Wekesa, V.W., Vital, S., Silva, R.A., Ortega, E.M.M., Klingen, I., Delalibera Jr., I., 2011. The effect of host plants on *Tetranychus evansi*, *Tetranychus urticae* (Acari: Tetranychidae) and on their fungal pathogen *Neozygites floridana* (Entomophthorales: Neozygitaceae). Journal of Invertebrate Pathology 107, 139–145.

Wilkerson, J.L., Webb, S.E., Capinera, J.L., 2005. Vegetable Pests II: Acari-Hemiptera-Orthoptera-Thysanoptera. University of Florida/IFAS Extension Service CD-ROM, Florida, USA. SW 181.

Won-Il, C., Lee, S.G., Park, H.M., Ahn, Y.J., 2004. Toxicity of plant essential oils to *Tetranychus urticae* (Acari: Tetranychidae) and *Phytoseiulus persimilis* (Acari: Phytoseiidae). Journal of Economic Entomology 97, 553–558.

Zalom, F.G., Kitzmiller, J., Wilson, L.T., Gutierrez, P., 1986. Observation of tomato russet mite (Acari: Eriophyidae) damage symptoms in relation to tomato plant development. Journal of Economic Entomology 79, 940–942.

Chapter 6

Lepidopterous Pests: Biology, Ecology, and Management

Alvin M. Simmons[1], Waqas Wakil[2], Mirza A. Qayyum[2,3], Srinivasan Ramasamy[4], Thomas P. Kuhar[5], Christopher R. Philips[6]

[1]USDA, ARS, Charleston, SC, United States; [2]University of Agriculture, Faisalabad, Pakistan; [3]Muhamamd Nawaz Sharif University of Agriculture, Multan, Pakistan; [4]AVRDC – The World Vegetable Center, Tainan, Taiwan; [5]Virginia Tech, Blacksburg, VA, United States; [6]University of Minnesota, Grand Rapids, MN, United States

1. INTRODUCTION

Tomato (*Solanum lycopersicum* L.) is a host for a diverse array of arthropod pests including several species of Lepidoptera. Some of these pests are more problematic and widespread than others. This chapter concerns nine of the particularly problematic lepidopterans (*Helicoverpa armigera* (Hübner), *Helicoverpa zea* (Boddie), *Keiferia lycopersicella* (Walsingham), *Manduca quinquemaculata* (Haworth), *Manduca sexta* (L.), *Phthorimaea operculella* (Zeller), *Spodoptera exigua* (Hübner), *Spodoptera litura* (F.), and *Tuta absoluta* (Meyrick)) attacking tomato. These species are diverse in their feeding nature, attacking foliage, stems, immature, and ripe fruits. Moreover, some are polyphagous while others are known to feed only on a few species. Nevertheless, all of these insects have host plants other than tomato. Because some of these pests are more adaptive, they tend to be more damaging in some regions. Several tools for an integrated pest management (IPM) approach are available to relieve the problems on tomato from these lepidopterans, and some of these tools are well-suited for use against multiple lepidopterous pests.

2. TOMATO FRUITWORM, *HELICOVERPA ARMIGERA* (HÜBNER)

H. armigera (tomato fruitworm, old world bollworm, African cotton bollworm, tomato worm, or corn earworm) is a polyphagous and highly mobile insect pest which continues to spread to new environments around the world.

2.1 Description and Biology

Adult *H. armigera* males are usually pale yellow with an olive green color, while adult females are reddish-brown. The female moths typically emerge first during the season, and live longer than males. Pearson (1958) reported that longevity varied from 1 to 23 days for males and 5–28 days for females, while Bhatt and Patel (2001) reported a lifespan of about 51 days for males and 54 days for females. Females lay eggs singly and scattered, usually on or near leaflets, floral buds, or young fruit. They prefer to oviposit on hairy surfaces of plants with peak egg laying before or during host flowering (King, 1994). A female can lay 730 to 1702 eggs, with a maximum of 4394 eggs within 10–23 days (King, 1994; Fowler and Lakin, 2001). The eggs hatch in 3 days at 25°C and may take 10–11 days at low temperatures (CABI, 2016a). Upon hatching, neonate larvae are creamy white with dark brown or black heads with prominent spines on the body. Older larvae vary in color from pale green to brown to black with lateral stripes on the body (Fig. 6.1A). The larval period is about 15–25 days; there are six instars. Later larval instars are found singly on fruit or on other plant parts, as they are known to be cannibalistic (Twine, 1971; Kakimoto et al., 2003). Pupation occurs in the soil, and lasts about 10–14 days. Pupae are dark brown.

Sustainable Management of Arthropod Pests of Tomato. http://dx.doi.org/10.1016/B978-0-12-802441-6.00006-1

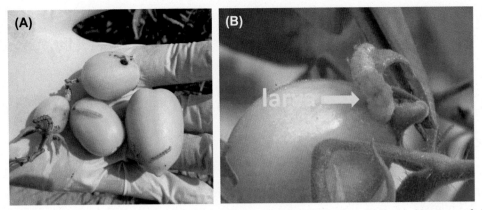

FIGURE 6.1   (A) Larvae *Helicoverpa armigera* on tomato fruits and (B) larva feeding with head inside a tomato fruit.

2.2 Distribution, Host Range, and Seasonal Occurrence

H. armigera damages and feeds on at least 180 cultivated and wild plant species in 45 families (Venette et al., 2003a). Its distribution is expanding and includes at least 145 countries and territories (51 in Africa, 42 in Asia, 29 in Europe, 20 in Oceania, and 3 in South America) (Sullivan and Molet, 2014). In late 2014, adults were collected in traps in several locations in Puerto Rico (USDA-APHIS, 2014). Subsequently, a male specimen was captured in a pheromone trap in the state of Florida in the United States in June 2015 (Hayden and Brambila, 2015). Beyond three males being detected in traps in June and July in Forida in 2015, no additional *H. armigera* specimens have been detected after nearly a year of continuous extensive trapping, which suggests that this was an isolated regulatory incident on the United States mainland (USDA-APHIS, 2016). Yet, as illustrated by its spread around the world, there is concern about populations of this pest becoming established in North America. *H. armigera* can have continuous populations throughout the year in tropical areas (Fitt, 1989). Yet, its distribution range is extended by its migration and overwintering abilities (Fitt, 1989; Gregg et al., 1993). Depending on certain climatic conditions, it may spend parts of its seasonal cycle in diapause. For example, diapause occurs when larvae are exposed to day lengths of 11.5–12.5 h and low temperatures (19–23°C) or extended periods of hot and dry conditions (≥35°C) (King, 1994; Zhou et al., 2000; Shimizu and Fujisaki, 2002). Pupae can also enter a summer diapause during prolonged hot (above 37°C) and dry conditions (Nibouche, 1998). In addition, facultative pupal diapause can be induced in cooler areas (15–23°C) with shortened day lengths (11–14 h/day) (CABI, 2016a).

2.3 Damage, Economic Thresholds, and Losses

H. armigera is a major economic pest of many agricultural and horticultural crops (Torres-Villa et al., 1996). More damage has been observed in tropical regions compared with temperate regions, which may be a result of more generations in the tropics (USDA-APHIS, 2014). Neonate larvae feed on the surfaces of leaves or floral buds, but older larvae prefer to feed on fruits. The larvae make holes in fruits and then thrust their heads inside to feed (Fig. 6.1B). The holes are circular and often surrounded by fecal pellets. The larva can feed on the inner contents and may hollow out the fruit. Severely damaged fruits may rot and drop to the ground. This pest was estimated to cause 5–55% loss in most tomato-producing areas in the world (Kashyap and Batra, 1987). Global losses from this pest can be in excess of $2 billion annually (Hayden and Brambila, 2015).

2.4 Management

2.4.1 Crop Rotation

Although long distance host location by *H. armigera* is not well understood, field and laboratory studies have demonstrated the ability of the adults to discriminate between hosts and non-hosts within a short range (Jallow et al., 2004). The size of local populations of *H. armigera* is important because of the immediate threat to local crops, and crop rotation can help to mitigate attack from this pest. Some host plants such as corn (maize) (*Zea mays* L.) and okra (*Abelmoschus esculentus* (L.) Moench) have contrasting assets because they are good for attraction and attack by *H. armigera*, but the hosts can be relatively inferior for survival and fecundity of the offspring compared with some other hosts (Jallow et al., 2001). In addition to tomato, crops such as chickpea (*Cicer arietinum* L.) or cotton (*Gossypium* spp.) are particularly good hosts for building up populations of *H. armigera* in response to favorable life-table parameters (Feng et al., 2010; Razmjou et al., 2014). If growers plant tomato after a previous tomato crop or follow with other host plants, e.g., chickpea, corn, or cotton, damage will be higher from the emerging *H. armigera* than the previous crop. The tomato crop can be rotated with a non-host cereal crop, cucurbit, or cruciferous vegetable to reduce infestation and damage to tomato by *H. armigera*.

2.4.2 Trap Crops

Planting African marigold (*Tagetes erecta* L.) longitudinally as a trap crop on both sides of 10 or 15 rows of tomato has been shown to reduce the incidence of *H. armigera* (Srinivasan et al., 1994). Tropical soda apple (*Solanum viarum* Dunal) is an excellent ovipositional attractant that does not support the development of *H. armigera* larvae, thus it serves as a "dead-end" trap crop (Srinivasan et al., 2013).

2.4.3 Host-Plant Resistance

Planting *H. armigera*-resistant tomato cultivars would reduce pest damage, however, commercial tomato cultivars with appreciable levels of resistance are not available. Germplasm screening at AVRDC—The World Vegetable Center in Taiwan—revealed the presence of high levels of *H. armigera* resistance only in wild *Solanum* species, particularly *Solanum habrochaites* S. Knapp & D.M. Spooner and *Solanum pennellii* Correll. Efforts on introgressing resistance from wild species into cultivated tomato resulted in resistant accessions, but with small fruits (Talekar et al., 2006). Leaves and fruits of transgenic tomato plants, that were transformed using a synthetic Cry1Ac gene coding for an insecticidal crystal protein of *Bacillus thuringiensis* (*Bt*) Berliner and highly specific to *H. armigera*, were demonstrated to be highly resistant to the larvae of *H. armigera* (Mandaokar et al., 2000). Among other genetic studies, the GroEL homolog XnGroEL protein of *Xenorhabdus nematophila* in tomato (Kumari et al., 2015) and host-induced RNA interference (HI-RNAi) by chitinase genes were demonstrated to enhance resistance of tomato to *H. armigera* (Mamta et al., 2015).

2.4.4 Pheromone-Mediated Management Tools

Traps using *H. armigera* sex pheromone can be used to monitor, mass trap, or disrupt male moths during the mating period. Traps baited with *H. armigera* pheromone lure adult males, and the data can be used to predict population buildup in the field. However, mass trapping may have limited direct effect when there are multiple host species in or near tomato production systems in the tropics. High concentrations of pheromone can be placed in the field to permeate the air and overwhelm the males making it very difficult for the males to locate a receptive female (AVRDC, 1988). Failure to mate results in no egg production or in only non-viable eggs. Thus, the population is reduced.

2.4.5 Biological Control

A large suite of natural enemies (at least 290 parasitoids, predators, and pathogens) attack *H. armigera* (Ahmad, 2003). The parasitoids (at least 163 species) attack eggs, larvae, and pupae of *H. armigera*; they are represented by two families of Diptera (37 species of Sarcophagidae and Tachinidae) while the other species are in eight families (primarily Braconidae, Ichneumonidae, and Trichogrmmatidae) in the order Hymenoptera (Romeis and Shanower, 1996; Ahmad, 2003). The predators (at least 127 species) are generalist in 7 orders of arthropods (Araneae, Coleoptera, Dermaptera, Hemiptera, Hymenoptera, Mantodea, and Neuroptera) with most of the species occurring in Hemiptera and Coleoptera (Romeis and Shanower, 1996; Ahmad, 2003). The predators primarily attack eggs and larvae; three species of Carabidae and Labiduridae also attack the pupal stage, and two species of spiders have been reported to prey on the adult stage of *H. armigera* (Romeis and Shanower, 1996; Ahmad, 2003). The pathogens (at least 10 viral, fungal, bacterial, protozoa, and nematode species) attack the larval stage of *H. armigera* (Ahmad, 2003). The incidence of attack and distribution of the natural enemies varies by region and can be affected by host plant and climate (van den Berg et al., 1988; Romeis and Shanower, 1996). Egg parasitoids (e.g., *Trichogramma pretiosum* Riley) and larval parasitoids (e.g., *Campoletis chlorideae* Uchida) can be conserved and/or released in tomato fields at regular intervals to curb the buildup of *H. armigera* (Romeis and Shanower, 1996).

2.4.6 Biopesticides

H. armigera has developed resistance to multiple classes of insecticides (Qayyum et al., 2015a). However, commercially available biopesticides based on *B. thuringiensis* (Bibi et al., 2013), *H. armigera* nucleopolyhedrovirus

(HaNPV) (Jayewar and Sonkamble, 2015), and neem (*Azadirachta indica* A. Juss.) (Yadav et al., 2015) can be used against *H. armigera*. These microbial control agents in different combinations exhibited enhanced efficacy against *H. armigera* (Wakil et al., 2012, 2013; Qayyum et al., 2015b). Endophytic fungi also has potential to control *H. armigera* (Qayyum et al., 2015c).

3. TOMATO FRUITWORM, *HELICOVERPA ZEA* (BODDIE)

H. zea is an important polyphagous insect pest of tomato and other crops, especially in the United States. Depending on the host, its common names include tomato fruitworm, corn earworm, sorghum headworm, vetchworm, and cotton bollworm.

3.1 Description and Biology

H. zea adults are medium-sized moths with a wingspan of 32–45 mm. The color of the adults varies, but the forewings generally appear yellowish-brown with a dark spot near the center (Fig. 6.2A). On the hind wings, a dark band flows toward the wing margins, but the wing margin and remaining part appears creamy white (Gilligan and Passoa, 2014). Eggs are pale green, turning yellow near hatching, and appear as a flattened sphere with several ridges originating from the top center. Larvae of *H. zea* and *H. armigera* appear identical. The larvae (Fig. 6.2B) range in colors, including brown, green, pinkish, yellowish, or blackish from early to late instars (Flood et al., 2005). They vary in size from 1.5 mm in neonates to 25 mm toward maturity. The head capsule appears orange or light brown with white net–like markings. Larvae usually have a broad, dark band running laterally down their back located above the spiracles and a lighter band below. Pupation occurs in the soil, and the brownish pupae are 18–19 mm long (Quaintance and Brues, 1905). A female lays 35 eggs a day, with 500–3000 eggs in its lifetime (Capinera, 2001). Neonates emerge from eggs after 3–4 days and start feeding on tender foliage, but increasingly prefer fruiting structures as they mature. There are generally 5–6 larval instars, with the larval period lasting 2–3 weeks under field conditions (Capinera, 2001). In general, a single larva invades a fruiting structure. Pupae can overwinter in the fall. Depending on environment, the life cycle of *H. zea* is completed in about 30 days with 1–7 generations in a year (Capinera, 2001).

3.2 Distribution, Host Range, and Seasonal Occurrence

H. zea is widely distributed in North, Central, and South America, the Caribbean Basin, as well as Hawaii and some other Pacific Islands (Hardwick, 1965). High levels of crop damage are observed late in the season in the United States (Flint, 1998). Weather regimes and wind patterns are the most influential components for migration of the adults (Sandstrom et al., 2007). The moths can travel several hundred kilometers in a single night (Westbrook, 2008). They are active year round in tropical and subtropical climates and are multivoltine in temperate climates. There is one generation per year in the northern areas of North America (e.g., southern Canada, Minnesota, and Wisconsin) (Capinera, 2007), two–three generations in the northeastern and northwestern United States (e.g., Maryland and Oregon) (Coop et al., 1993), three–four generations in the South Central United States (e.g., Arkansas and Kentucky) (Isely, 1935), and between four and seven generations in the southern United States (e.g., Louisiana, Florida, and Texas) (Capinera, 2007). Larvae from the mobile adults damage several crops during the multiple generations each year (Westbrook and Lopez,

FIGURE 6.2 (A) Adult *Helicoverpa zea* and (B) larva on tomato fruit and hole damage from feeding.

2010), but it prefers corn (Neunzig, 1969). Over 150 plant species are hosts for *H. zea* (Gilligan and Passoa, 2014). This includes species within the families Fabaceae, Malvaceae, Poaceae, and Solanaceae (Camelo et al., 2011) and over 30 species of cultivated plants (Blanco et al., 2007).

3.3 Damage, Economic Thresholds, and Losses

Heavy and frequent losses caused by *H. zea* occur in fields and gardens, and sometimes damage also occur in trees, ornamentals, and greenhouse crops. Larvae feed on many plant parts, preferring flowers and fruits; mature larvae (Fig. 6.2B) can feed on multiple fruits (Capinera, 2001). Late plantings may result in 100% fruit damage. Locating 7 damaged fruits out of 100 randomly sampled fruits indicates a need for immediate insecticide treatment (Flint, 1998). An action threshold regimen with minimum fruit damage, reduced number of insecticide sprays, and maximum yield returns was determined when the first insecticidal application was made when 10% of the sampled tomato plants were infested with *H. zea* eggs, and the second spray was made when more than 3 damaged fruits/100 market-sized unripe fruits were observed (Kuhar et al., 2006). The high fecundity rate, polyphagous feeding behavior, high mobility, seasonal migration, and facultative pupal diapause of *H. zea* contribute to its high pest status.

3.4 Management

Since the mid-19th century, there has been a quest to protect field crops and gardens from *H. zea* in the United States Control strategies to manage *H. zea* fall into two broad categories: overall pest reduction and crop-targeted protection from pests. IPM has been the preferred tool to control this pest (Bottrell, 1979).

3.4.1 Trapping

H. zea is often monitored for population indices and expected infestation using a four-component sex pheromone (Klun et al., 1980). Light traps can also be used to provide information on infestation (Walgenbach et al., 1989), but this tool is labor-intensive and often rejected by growers. Instead, pheromone traps are the more accepted trap used by growers due to species specificity and its feasibility of use. Pheromone traps also revealed that the wild tomato species (*S. habrochaites* = *Lycopersicon hirsutum* Dunal) are more preferred by *H. zea* for oviposition than commercial tomato (Campbell, 1990). Advanced technologies have led to careful observations of the movement of adult *H. zea* using aerial collections, digital imaging, and radar (Westbrook, 2008) which may aid in population monitoring of *H. zea* in a given region.

3.4.2 Cultural Control

Intercropping corn into tomato fields effectively diverted the pest infestation from tomato to corn borders resulting in a reduced population level of *H. zea* in tomato fields; Sugar, Jean, and Java varieties of corn served as good trap crops (Rhino et al., 2014). Crop rotation, disking, and culling plants immediately after harvest, and deep plowing can help suppress populations of *H. zea* (Flint, 1998). To avoid the synchrony between oviposition and susceptible plant stage, timely or early planting help control *H. zea* (Hardwick, 1965; Flood et al., 2005).

3.4.3 Physical Control

Abiotic factors (e.g., rain and wind) can impact the survival of *H. zea* (Nuessly and Sterling, 1994). At 10% relative humidity (RH), the survival of *H. zea* eggs was reduced by up to 50% when the exposure temperature was raised from 30 to 40°C for 4 h; 100% mortality occurred within 8 h at 40°C and 10% RH (Fye and Surber, 1971). Thus, hot ambient temperatures can reduce egg survival. Abiotic factors that affect the populations can be used to help model and predict populations of *H. zea*.

3.4.4 Host-Plant Resistance

Host-plant resistance to *H. zea* was first reported in sweet corn (*Z. mays*) (Collins and Kempton, 1917). Yet, tomato plants can respond to *H. zea* oviposition as a signal to restrict subsequent herbivory, and larval emergence induces an increased level of plant defense against herbivores (Kim et al., 2012). Oviposition on tomato plants resulted in a higher expression of pin2, a gene encoding defense proteins against herbivores, and feeding by *H. zea* is the source of cross-resistance (Kim et al., 2012). Feeding by *H. zea* can enhance tomato plant resistance to the potato aphid (*Macrosiphum euphorbiae* Thomas) and the two-spotted spider mite (*Tetranychus urticae* Koch) (Stout et al., 1998). The extract of *Clitoria ternatea* L. has been introduced as "Sero-X," a formulation that could be used effectively against *Helicoverpa* spp. and sucking insect pests of vegetables (Mensah et al., 2014). The treatment of methyl JA (methyl jasmonic acid) to a susceptible tomato mutant reinstated its normal wound response to herbivory, enhanced resistance to *H. zea*, and increased leaf trichome density (Tian et al., 2014).

3.4.5 Biological Control

Insects in the genus *Trichogramma* are the most efficient parasitoids of *H. zea* (Vargas and Nishida, 1980). For example, *T. pretiosum* were shown to destroy 40% of *H. zea* eggs in tomato fields in southern California, and 80% of the eggs in the Sacramento Valley, California, United States (Flint, 1998). Important *H. zea* larval parasitoids include *Hyposoter exiguae* (Viereck) (Flint, 1998), *Archytas marmoratus* (Townsend) (Gross, 1990), *Campoletis* spp. (Capinera, 2001), *Cotesia* spp. (Tipping et al., 2005), *Eucelatoria bryani* Sabrosky and *Eucelatoria robentis* (Coquillett) (Reitz, 1996), and *Microplitis cropeipes* (Cresson) (Herbert et al., 1993).

Several generalist predators feed on eggs and larvae of *Helicoverpa* spp. In Kentucky (United States), Nabidae was the major predator group reported to feed on *Helicoverpa* in soybean (about 50% predation of eggs and larvae). *Coleomegilla maculata* DeGeer resulted in about 45% predation of eggs and larvae in corn and other predators such as *Geocoris punctipes* (Say) in soybean, and *Orius insidiosus* (Say) and *Lygus lineolaris* Palisot de Beauvois in sweet corn caused ≥10% egg predation (Pfannenstiel and Yeargan, 2002).

Entomopathogenic nematodes in the genus *Steinernema* can infect pupal and pre-pupal *H. zea* (Raulston et al., 1992). *Steinernema riobravis* Cabanillas, Poinar and Raulston, caused pupal mortality against *H. zea* in the southern United States (Feaster and Steinkraus, 1996). Nuclear polyhedrosis viruses have also been evaluated as biocontrol agents against *H. zea* on several crops (Washburn et al., 2001). Higher insecticidal activity of NPV of *Anagrapha falcifera* Kirby (AfNPV) against *H. zea* was observed on tomato and corn than on collard (*Brassica oleracea* L. var. acephala de Condolle), snap bean (*Phaseolus vulgaris* L.), and cotton (Farrar and Ridgway, 2000). *B. thuringiensis* can control 40–60% of larval population of *H. zea*, especially when applied for young larvae (Flint, 1998). *B. thuringiensis* dust (Dipel) and liquid formulations (Thuricide) are generally effective and provide long-term *H. zea* control on homegrown tomato (Williamson, 2014). A concealed feeding habit protects *H. zea* against foliar insecticides; however, transgenic crops lack this issue as toxins are expressed within all plant tissues (Sims et al., 1996).

3.4.6 Autocidal Control

The inherited sterility (IS) technique can be used to effectively control the population of *H. zea*. The long-term release of irradiated *H. zea* (100 Gy) reduced or delayed the occurrence of wild males, and the resulting F_1 sterility was detrimental to the overall population growth of *H. zea* (Carpenter and Gross, 1993). The compatibility of F_1 sterility with other management methods against *H. zea* has been confirmed in several studies. For instance, Hamm and Carpenter (1997) successfully used nuclear polyhedrosis viruses with F_1 sterility, and host-plant resistance was used with F_1 sterility by Carpenter and Wiseman (1992) to control *H. zea*. F_1 progeny from irradiated males and wild females was also a suitable host for an *H. zea* larval parasitoid, *A. marmoratus* (Mannion et al., 1995).

3.4.7 Chemical Control

Insecticide management for *H. zea* is focused on the early instars. Long-term use of pyrethroids has resulted in populations of *H. zea* resistant to several insecticides (Flood and Rabaey, 2007). Insecticides with low-to-moderate control levels against *H. zea* on tomato include carbaryl, endosulfan, indoxacarb, methomyl, and methoxyfenozide, but spinosad provided the best control among these (Weinzierl, 2007). Indoxacarb (at 2.4–6.0 g ai/100 L), deltamethrin (at 21.5 g ai/100 L), and triflumuron (at 15 g ai/100 L) all provided up to 80% control of *H. zea* on tomato crops (Martinelli et al., 2003). When the systemic chlorantraniliprole was used as a drip chemigation treatment, the percentage of tomato damaged by *H. zea* was reduced when compared to a foliar application (Kuhar et al., 2010). The safety of an anthranilic diamide formulation (cyazypyr 10% OD) to field populations of natural enemies of tomato insect pests was reported recently (Mandal, 2012). The effects of six insecticides belonging to different groups (lambda-cyhalothrin, spinosad, thiodicarb, chlorfenapyr, indoxacarb, and emamectin benzoate) on *H. zea* were determined on cotton. Spinosad and thiodicarb effectively controlled *H. zea* and usage rates of lambda-cyhalothrin, spinosad, and thiodicarb to control *H. zea* infestation were reduced (Brickle et al., 2001). A study of two silicon minerals, wollastonite and olivine (acidulated and non-acidulated), revealed an effect of olivine on growth of *H. zea* compared to wollastonite which suggests an opportunity for future research to identify the safety of this mineral as a dust insecticide against *Helicoverpa* spp. (Stanley et al., 2014).

4. TOMATO PINWORM, *KEIFERIA LYCOPERSICELLA* (WALSINGHAM)

K. lycopersicella, tomato pinworm, is a major pest of tomato in tropical and subtropical regions of the United States.

4.1 Description and Biology

Elmore and Howland (1943) and Capinera (2001) provided excellent reviews of the biology of *K. lycopersicella*. Development from oviposition to adult requires 456 DD at a threshold temperature of 9.5°C (Weinberg and Lange, 1980). Eggs are laid singly or in groups of two or three on host-plant foliage. Eggs typically hatch in 4–7 days (Capinera, 2001). Newly deposited eggs are pale yellow, turning orange before hatching. There are four larval instars. Capps (1946) provided a key to identify larvae of *K. lycopersicella* and other species with which it might be confused. The first instar spins silk over itself and tunnels into the leaf. First and second instars are less than 1 mm long, yellowish-gray with a dark head capsule, and are typically found inside leaf mines, where they deposit their feces in a pile at the entrance (Capinera, 2001). As the larva matures, it develops darker pigmentation (first orange, then purple) on the dorsum of the abdomen (Capps, 1946). This irregular pigmented region contains two light-colored circular spots and two elongated light-colored sections that can be used as characters to identify this species (Capps, 1946). The third and fourth instars feed from within tied or folded leaves or bore into stems or fruits. Mature larvae pupate near the soil surface in loosely woven cells covered with soil particles. Pupae are initially green and then turn brown as they age. Adults emerge in 2–4 weeks as small gray moths with a wingspan of 10 mm (Capinera, 2001). Mating begins soon after adult eclosion and usually occurs at night (McLaughlin et al., 1979).

4.2 Distribution, Host Range, and Seasonal Occurrence

Apparently native to Central America, *K. lycopersicella* was first found in California, United States in 1923 (Capinera, 2001). It occurs in warm regions of Central, South, and North America as well as the Caribbean Islands and Hawaii. Populations also occur in greenhouses throughout North America. This pest can feed and develop on many species within the family Solanaceae including potato (*Solanum tuberosum* L.), eggplant (*Solanum melongena* L.), and several nightshades and nettles (*Solanum* spp.), but it strongly prefers tomato (Batiste and Olson, 1973; Schuster, 1989). It can complete up to eight generations per year. In warm regions, *K. lycopersicella* can complete a generation in 30 days (Lin and Trumble, 1985). It does not undergo diapause.

4.3 Damage, Economic Thresholds, and Losses

Small larvae of *K. lycopersicella* feed on leaves, causing blotch-type mines, while large larvae tie leaves and bore into fruit (Fig. 6.3). Infestation levels and damage can reach 100% in southern California, South Florida, or in greenhouses if immediate control measures are not taken (Capinera, 2001; Natwick et al., 2014). It causes indirect damage to tomato plants by mining and folding leaves and causes direct damage by boring into fruit leaving a small "pin-size" hole. Secondary damage may also result when plant tissues become infected by pathogens, usually resulting in plant death or fruit rot. Fruit damage

FIGURE 6.3 Larva of *Keiferia lycopersicella* on tomato fruit and feeding damage. *Adapted from Alton N. Sparks Jr., University of Georgia, Bugwood.org.*

is usually greatest in the lower canopy of the plant (Wellik et al., 1979). A density of about one larva per plant can cause economic damage to tomato (Capinera, 2001).

4.4 Management

Although damage by this pest can be quite severe, an IPM program based on intensive sampling, parasitoid releases, use of mating disruption, and applications of selective insecticides can significantly reduce *K. lycopersicella* densities and concomitant feeding damage (Trumble and Alvarado-Rodriguez, 1993).

4.4.1 Monitoring and Trapping

The sex pheromone of *K. lycopersicella* was identified as a 96:4 mixture of [E]-4:[Z]-4-tridecenyl acetate (Charlton et al., 1991). Pheromone-baited delta traps or other types of traps can be used to monitor *K. lycopersicella* adults to time insecticide applications or when to start visual sampling (van Steenwyk et al., 1983). Although van Steenwyk et al. (1983) suggested an action threshold of 10 months per trap per night, pheromone trap catch density is probably too variable to base control decisions, but it can be used as an indicator of mating activity and the need for field scouting. California tomato production guidelines recommend that traps be installed at planting at the density of one per 4 ha and be checked twice per week until harvest (Natwick et al., 2014). Plant inspections for larvae should begin as soon as moths are caught in traps. Because leaf mines and folded leaf shelters are conspicuous, visual inspections can be used to sample for *K. lycopersicella* larvae. Wellik et al. (1979) suggested that the lower canopy stratum be sampled. Natwick et al. (2014) recommended inspecting foliage in California tomatoes as soon as seedlings are well established. Treatment is advised when an average of 1–2 larvae per 2-m row section is found.

4.4.2 Sanitation, Regulatory, and Cultural Control

Sanitation is an important first step in reducing pest populations. Infestations, particularly in colder-climate regions, often result from the movement or shipment of pinworms on transplants, or in picking containers, crates, or infested fruit regions or greenhouses with established populations of *K. lycopersicella*. Outbreaks may also result from populations perpetuated on plants left in fields after harvest or left in seed flats or compost heaps. Pena and Waddill (1985) reported very high infestation levels of pinworms in volunteer (self-sown) tomatoes growing in fields in South Florida that were harvested previously, then abandoned or mowed. Thus, infested plants and culled tomatoes should be destroyed after harvest to eliminate or reduce breeding populations (Capinera, 2001). Tomatoes planted near infested greenhouses are also susceptible to pest outbreaks. In the United States, a combination of crop rotation with a host-free period has become an essential IPM component for reducing *K. lycopersicella* populations in tomato crops; these strategies are most effective when practiced over large areas (UC IPM, 2016).

4.4.3 Mating Disruption

Pheromones can be used to confuse males for mating disruption (van Steenwyk and Oatman, 1983). A mating disruption study in tomato fields in California revealed that pinworm damage was reduced from 12% to 65% in insecticide control fields to less than 5% in pheromone-inundated fields (Jimenez et al., 1988). The pheromone was also applied to a hard matrix formed into a hanging spiral for air dispersal (Wyenandt et al., 2016). Fewer of these lures are needed per area for effective mating disruption.

4.4.4 Biological Control

Over 20 species of parasitoids were found attacking *K. lycopersicella* on tomato (Oatman and Platner, 1989). These parasitoids include members of: Braconidae (*Agaths* sp., *Pseudapanteles dignus* (Muesebeck), *Bracon gelechiae* Ashmead, *Bracon* spp., *Chelonus blackburni* Cameron, *Chelonus phthorimaeae* Gahan, *Orgilus* spp., *Parahormius pallidipes* (Ashmead)), Ichneumonidae (*Campoplex phthorimaeae* (Cushman), *Campoplex* sp., *Pristomerus hawaiiensis* Perkins, *Pristomerus spinator* (F.), *Trathala flavoorbitalis* (Cameron)), Eulophidae (*Elasmus nigripes* Howard, *Sympiesis stigmatipennis* Girault), Pteromalidae (*Zatropis* sp.), Bethylidae (*Goniozus* sp.), and Trichogrammatidae (*T. pretiosum*). *P. dignus* is an important parasitoid and, along with other braconids, ichneumonids, and eulophids, up to 70% parasitism may result in California and Florida (Oatman et al., 1979; Pena and Waddill, 1983). *P. dingus* and *Apanteles scutellaris* Muesebeck were combined to reduce the population of *K. lycopersicella* by 50–60% (Poe et al., 1975). Another important parasitoid is *T. pretiosum* (egg parasitoid); it attacks the egg stage of *K. lycopersicella* (Pena and Waddill, 1983; Medeiros et al., 2009a).

4.4.5 Chemical Control

Insecticides used to control *K. lycopersicella* can be intensive. Up to 16 applications of insecticides were made against this pest on cherry tomatoes in southern California in the 1980s (Jimenez et al., 1988). Populations of first and second instars are controlled effectively with insecticides, but later instars are often protected by leaf folds or fruits, and are more difficult to control with insecticides (Poe, 1973). There are many narrow-spectrum insecticides that provide effective control of *K. lycopersicella* including diamides, spinosyns, and avermectins. These insecticides provide effective control while minimizing the impact on non-target organisms. *B. thuringiensis* has been demonstrated to be comparable to numerous synthetic insecticides for efficacy against *K. lycopersicella* (Seal and McCord, 1996). There has been a decrease in its susceptibility to several broad-spectrum insecticides including pyrethroids and carbamates in Florida (Schuster, 1989). If broad-spectrum insecticides are used, it becomes necessary to treat throughout the season until final harvest, which may also result in secondary pest outbreaks such as *Liriomyza* leafminers.

5. TOMATO HORNWORM, *MANDUCA SEXTA* (L.), AND *MANDUCA QUINQUEMACULATA* (HAWORTH)

The tobacco hornworm (*M. sexta*) and tomato hornworm (*M. quinquemaculata*) are large size lepidopterans which have similar appearance (adults in Fig. 6.4A and B), biology, ecology, and damage. They both attack tomato and other plants. Two common names for adult *M. quinquemaculata* are five-spotted hawk moth and Carolina sphinx. Another common name for larva *M. sexta* is goliath worm.

5.1 Description and Biology

M. sexta is often used as a teaching and research model for many aspects of biology because of its large size and relative ease of rearing. Eggs are deposited on foliage of host plants, and females can deposit up to 1400 eggs (Yamamoto, 1968). The eggs hatch in 6–7 days while the larval stage lasts 20–30 days depending on temperature (Capinera, 2001). Larvae of *M. quinquemaculata* have eight "V-shaped" whitish patterns along their sides (Fig. 6.4C) while larvae of *M. sexta* have seven whitish lines along their sides (Fig. 6.4D). Larvae for both species have a prominent pointed structure or "horn" arising

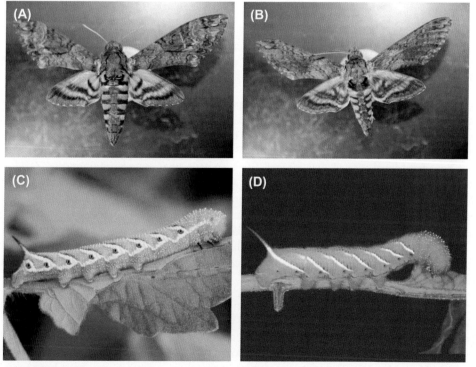

FIGURE 6.4 (A) Adult *Manduca quinquemaculata*; (B) adult *Manduca sexta*; (C) larva of *Manduca quinquemaculata* feeding on tomato foliage; and (D) larva of *Manduca sexta* feeding on tomato foliage. *Photos (C,D) by B. Merle Shepard, with permission.*

dorsally from the terminal abdominal segment. There are usually five larval instars, but sometimes there are six, and may reach up to 81 mm long (Capinera, 2001). It is well known that *M. sexta* is tolerant to nicotine. Moreover, *M. sexta* can tolerate many other alkaloids (Wink and Theile, 2002). Although some other lepidopterans can tolerate nicotine and other alkaloids (Wink and Theile, 2002), the assessment of *M. quinquemaculata* to these compounds has not been reported.

5.2 Distribution, Host Range, and Seasonal Occurrence

A recent phylogenetic analysis suggests that *M. sexta* and *M. quinquemaculata* are closely related, with an ancestral center of diversification in Central America (Kawahara et al., 2013). The distribution, host range, and seasonal occurrence are similar between *M. sexta* and *M. quinquemaculata*. The distribution for *M. sexta* extends from southern Canada through the United States, Caribbean, Central America, and into South America, but tends to have lower incidences in the northeastern United States and Canada (Capinera, 2001). However, populations of *M. quinquemaculata* are found throughout the United States, in southern Canada, and northern Mexico (BAMONA, http://www.butterfliesandmoths.org/species/Manduca-quinquemaculata). Both species are reported to feed only on solanaceous species. Tobacco (*Nicotiana* spp.) and tomato are common hosts. They also feed on wild and cultivated species including eggplant and potato (Lawson, 1959; Capinera, 2001). These two species of *Manduca* overwinter as pupae and undergo one–four generations per year depending on temperature in a given region (Capinera, 2001).

5.3 Damage, Economic Thresholds, and Losses

Both species feed on leaves and can defoliate plants in a relatively short period of time. Most defoliation happens during feeding by the last two larval instars. Home gardens and small-hectare production receive the greatest amount of damage. Treatments for other pests in tomato often eliminate the threat from *M. quinquemaculata* and *M. sexta*. *M. sexta* is a common pest of tomato throughout the United States (Flood et al., 2005). It has long been recognized as an important crop pest in the United States. Over 60 years ago, it was reported as being the source of the primary loss (estimated as $25,530,000) to tobacco which was the greatest economic crop in the southeastern United States at that time (Madden and Chamberlin, 1945). There are limited reports on economic threshold information on tomato for both *M. sexta* and *M. quinquemaculata*. Because of the voracious appetite of the larvae, the Ministry of Agriculture, Food and Rural Affairs (2009) Ontario, Canada supports a treatment threshold of one larva per 30 plants for infestation by either *M. sexta* or *M. quinquemaculata*.

5.4 Management

Various management strategies can help to control populations of *M. sexta* and *M. quinquemaculata*.

5.4.1 Trapping

Although adult *M. sexta* and *M. quinquemaculata* can be captured using light traps, this is not a practical method to control these pests (Capinera, 2001). A good understanding of the response of these species to odors may help in the development of improved trapping strategies. A pheromone blend consisting of (Z)-9-hexadecenal, (Z)-11-hexadecenal, (E)-11-hexadecenal, hexadecanal, (E,Z)-10,12-hexadecadienal, (E,E)-10,12-hexadecadienal, (E,E,Z)-10,12,14-hexadecatrienal, (E,E,E)-10,12,14-hexadecatrienal, (Z)-11-octadecenal, (Z)-13-octadecenal, octadecanal, and (Z,Z)-11,13-octadecadienal was found to be attractive to male *M. sexta* (Tumlinson et al., 1989). Gas chromatography coupled with electroantennographic detection revealed nine active compounds ((Z)-3-hexenyl acetate, nonanal, decanal, phenylacetaldehyde, methyl salicylate, benzyl alcohol, geranyl acetone, (E)-nerolidol, and one unidentified compound) from headspace volatiles of tomato leaves and three other host plants (*Capiscum annuum* L., *Datura wrightii* Regel, and *Proboscidea parviflora* (Woot.) Woot. & Standl.) (Fraser et al., 2003). Researchers have examined the receptors of *M. sexta* and have identified an alternate polyadenylation signal that appears to be associated with transcript processing (Patch et al., 2009; Große-Wilde et al., 2010; Garczynski et al., 2011).

5.4.2 Cultural Control

The type of fertilizer used on tomatoes may have an effect on insect populations. Damage to tomato from infestation by *M. quinquemaculata* was demonstrated by Yardim et al. (2006) to be suppressed by food waste vermicomposts as substitutes to a soilless medium, MetroMix 360. They speculated that the decrease in damage may have been from a more balanced release of fertilizer and an elevated level of phenolic compounds in the plants. Larvae, notably the later instars, can be easily observed, and manually removed from tomato plants in gardens and small fields. Field cultivation can result in over 90% mortality of pupae that have burrowed in the soil (Capinera, 2001).

5.4.3 Host-Plant Resistance

Transgenic tomato plants expressing a protein from *Bacillus thuringinsi*s var *kurstaki* (HD-1) sustained little damage from *M. sexta* while control plants were heavily damaged from feeding (Delannay et al., 1989). These transgenic plants were also protected from damage by *K. lycopersicella* and *M. quinquemaculata*. In related field trials, damage to non-transformed tomato plants resulted in 100% defoliation from *M. sexta* while transgenic tomato plants received only minor damage and were comparable to plants treated with insecticide (Fischhoff et al., 1987). Plant inbreeding may result in hosts being more attractive to hornworms. In a study with *Solanum carolinense* L., Kariyat et al. (2014) demonstrated the attraction and acceptance of this host on *M. sexta* feeding. Jasmonate-dependent depletion of soluble sugars of *Nicotiana attenuata* Torr. ex S. Watson resulted in the plant being more susceptible to attack by *M. sexta* larvae (Machado et al., 2015). Throughout larval development, jasmonates reduced the constitutive and herbivore-induced concentration of glucose and fructose in leaves of the plant, and the growth of the larvae increased (Machado et al., 2015). In cultivated tomato, the plant was demonstrated to develop increased tolerance after prior feeding by *M. sexta* (Korpita et al., 2014). The previously damaged plant grew thicker stems, had reduced chlorophyll, and produced more leaves than a mechanically damaged treatment.

5.4.4 Biological Control

Natural populations of parasitoids are important in suppressing *M. sexta* and *M. quinquemaculata*. Lawson (1959) provided a list, although incomplete, of numerous natural enemies of *M. sexta* and *M. quinquemaculata* in tobacco focused on eggs and larvae. Lawson (1959) noted that the most important species in North Carolina, United States, were a hemipteran egg predator (*Jalysus spinosus* (Say)), a vespid larvae predator (*Polistes fuscatus* (F.)), and a braconid parasitoid (*Apanteles congregates* (Say)). Because of the adverse effects of nicotine and other compounds on parasitoids, the success of the braconid parasitoid *Cotesia congregata* (Say) may be less in regions with high incidence of tobacco versus other solanaceous host plants (Kester and Barbosa, 1991). Using *M. sexta* as a model for isolating insecticidal bacteria, 14 isolates were identified as belonging to the *Bacillus cereus* Frankland and Frankland group, two were identified as psychrotrophic *Bacillus weihenstephanensis* Lechner, three were *Lysinibacillus fusiformis* (Ahmed), and two others were identified as *Enterococcus faecalis* (Andrewes and Horder) (Martin et al., 2008). The predator *Podisus maculiventris* (Say) was demonstrated to have a reduction in prey consumption of *M. sexta* and reduced non-consumptive impacts on resistant tomato compared to susceptible tomato (Kaplan and Thaler, 2010). In a field study with *M. sexta* and *M. quinquemaculata*, mortality from parasitism by *C. congregata* and mortality for a combination of parasitism and predation varied based on leaf location by *M. sexta*, but not for *M. quinquemaculata* (Kester et al., 2002).

5.4.5 Chemical Control

Both *M. sexta* and *M. quinquemaculata* are susceptile to a wide range of conventional insecticides. However, they can also be controlled by more environmentally friendly insecticides such as spinosids (Herzog et al., 2002) and *B. thuringiensis* (Martin et al., 2008). Differences in binding of different classes of insecticides by hemolymph proteins of *M. sexta* have been demonstrated (Helling et al., 1986). To aid in the control without affecting non-target species, *M. sexta* was selectively killed when it was fed species-specific dsRNA targeting vATPase transcripts (Whyard et al., 2009).

6. POTATO TUBERWORM, *PHTHORIMAEA OPERCULELLA* (ZELLER)

P. operculella, commonly known as potato tuberworm, is a key pest to solanaceous crops with cosmopolitan and oligophagous pest status.

6.1 Description and Biology

P. operculella (Fig. 6.5B) adult moths are 10mm long with a wingspread of 12mm. The forewings of males have two–three dark spots while females have x-shaped spots (Rondon et al., 2007). The edges of both pairs of wing are fringed. Some 16–20h after emergence, adults seek mates and copulate for 85–200min (Makee and Saour, 2001). Moths are not active during the day; therefore, oviposition occurs at night (Traynier, 1975). They generally oviposit on foliage, but in the absence of foliage, they may oviposit on soil or plant debris (Traynier, 1975; Rondon et al., 2007). The eggs are spherical and white or yellowish and are either deposited singly or in batches of 2–20 (Rondon et al., 2007). A female can lay up to 290 eggs which hatch

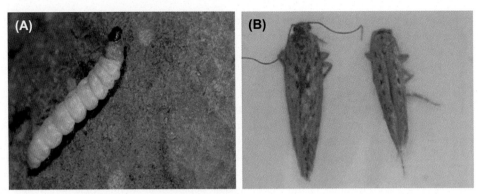

FIGURE 6.5 (A) Larva and (B) adult *Phthorimaea operculella. Rondon, S.I., 2010. The potato tuberworm: a literature review of its biology, ecology, and control. American Journal of Potato Research 87, 149–166.*

within 5–34 days, with an upper critical temperature of 36°C (Trivedi and Rajagopal, 1992). Newly hatched larvae are gray or yellow-to-white (0.1–0.2 cm in length); toward maturity (Fig. 6.5A), the color changes to pink and greenish with an average size of 1.5–2 cm (Maharjan and Jung, 2012). Sexual dimorphism starts to appear in the third larval instar, and two elongated yellowish testes are present on the fifth and sixth abdominal segment in the fourth larval instar, which differentiate males from females (Chauha and Verma, 1991). It takes the larvae 13–33 days to pupate depending on temperature (Trivedi and Rajagopal, 1992). The pre-pupae spin silken cocoons and pupate on the soil surface or in leaf debris. Pupae are generally smooth, brown, and 0.84 cm long (Rondon, 2010). Pupation lasts 5–7 days in optimum conditions (Krsteska and Stojanoski, 2012).

6.2 Distribution, Host Range, and Seasonal Occurrence

P. operculella is widely distributed and may have originated in western South America (Balachowsky and Real, 1966). Reports of potato-tuber damage in Australia and New Zealand introduced *P. operculella* as an important pest to the world in the mid-1800s (Berthon, 1855). It has dispersed to more than 90 countries, particularly tropical and subtropical countries in Asia, Africa, Australia, and Central, South, and North America (Flint, 1986; Rothschild, 1986; Cisneros and Gregory, 1994). It has distinct geographical populations across the United States (Medina et al., 2010). The adults can fly non-stop for more than 5 h which can cover up to 10 km in a single flight (Foley, 1985). Each stage can overwinter above 0°C and recolonize host crops in the spring (Coll et al., 2000; Krsteska and Stojanoski, 2012). Larvae are oligophagous on plants of solanaceous family and few additional families. It prefers to feed on potato, tomato, eggplant, pepper (*C. annuum*), tobacco, and wild solanaceous plants (Kroschel and Lacey, 2009). Other hosts include ground cherry (*Physalis pruniosa* L.) and jimson weed (*Datura stramonium* L.) (Das and Raman, 1994).

6.3 Damage, Economic Thresholds, and Losses

Larvae of *P. operculella* cause principle damage by boring into tomato fruits, making them unmarketable. The presence of larvae during harvest can result in the transfer of infested fruits to storage facilities where they can increase losses by infesting other fruits (Kroschel, 1995). Although direct damage is not always severe, under high infestation levels, larvae tunnel into the fruit and spin webbing that may be filled with frass (Ferro and Boiteau, 1993). In temperate regions, populations build after mild winter hibernations and increase with occasional pest outbreaks, resulting in significant economic losses (DeBano et al., 2010). There is no reported economic threshold for *P. operculella* (Rivera, 2011).

6.4 Management

Pest management practices are focused at minimizing the number of females so that there can be a source reduction for the next generations.

6.4.1 Trapping (Monitoring/Mass Trapping)

To estimate populations of adult *P. operculella*, pheromone traps are well suited (Subchev et al., 2013). A gland occurring as a dorsal invagination, anterior to the last abdominal segment is the sex pheromone source. The pheromone, a mixture of *cis*-7, *cis*-11-tridecadien-1-ol acetate, is responsible for sex communication between sexes (Fouda et al., 1975). Another compound, *trans*-4, *cis*-7-tridecadien-1-ol acetate, is also active (Roelofs et al., 1975). Numerous blends of chemicals are available for mass trapping, monitoring, and mating disruption of *P. operculella*. Mass trapping efficiently works in fields and in storage facilities against this pest (Raman, 1988). Prolonged placement of pheromone sachets in sticky traps provides additional benefits when compared to water-pan traps (Rivera, 2011). Moths captured in traps do not provide quantitative information on density, but aid in estimating the population trend in an area (Rivera, 2011). Pheromone traps can be geo-referenced to obtain geo-statistical maps to assist pest managers in monitoring an area and predicting if and when treatment is needed. Recent research has proved georeferencing of pheromone traps as a valuable technology to describe the phenology and distribution of *P. operculella* (Masetti et al., 2015).

6.4.2 Cultural Control

Cultural control is vital in managing *P. operculella* infestations in the field (Das and Raman, 1994). Intercropping tomato with any nonpreferred crop can reduce the incidence of *P. operculella*. For example, 80–91% reduction in tomato damage was reported in Egypt by intercropping with onion and garlic; this resulted in an increase of 114–207% and 104–284% fruit yield, respectively (Afifi et al., 1990). The survival of *P. operculella* greatly depends on the presence of alternate hosts nearby, which helps to ensure its reproduction and perpetuation. Therefore, non-host crops surrounding the primary crop can disrupt this insect's seasonal biology (Broodryk, 1971; Das and Raman, 1994).

6.4.3 Host-Plant Resistance

A lower performance of *P. operculella* on some non-preferred and less susceptible cultivars can be used in IPM programs (Golizadeh et al., 2014). Several traits have been located in tomato plants for resistance against *P. operculella*. Trichome-based resistance of *Solanum* species has been found where type V and VI trichomes were associated with larval mortality (Duffey, 1986). Larval mortality in *P. operculella* is linearly related to high densities of type VI and low densities of type V trichomes (Simmons et al., 2006). Egg laying capacity of *P. operculella* is also affected adversely by glandular pubescence (Gurr and McGrath, 2001). Plant volatile compounds such as hexanal, octanal, and 1-octen-3-ol have been shown to deter oviposition, leading to reduced egg laying by *P. operculella* (Anfora et al., 2014).

6.4.4 Biological Control

A reduction in *P. operculella* damage was seen in Argentina and Brazil due to natural parasitism (Lloyd, 1972). Many of these parasitoids including *Apanteles subandinus* Blanchard, *B. gelechiae*, *Copidosoma koehleri* Blanchard, and *Orgilus Lepidus* Muesebeck were exported to other countries and have become successful biocontrol agents (Sankaran and Girling, 1980). An ichneumonid, *Diadegma turcator* Aubert (=*Diadegma pulchripes* (Kokujev)), in Sardinia, accounted for 65% parasitism (Ortu and Floris, 1989). A combination of ichneumonids and braconids was even more effective (Pucci et al., 2003). Parasitic wasps, *D. pulchripes*, *Temelucha decorata* (Gravenhorst), *B. gelechiae*, and two unidentified braconids emerged from field collected *P. operculella* larvae (Coll et al., 2000). The most abundant predators observed for *P. operculella* in a field in Israel were *Coccinella septempunctata* (L.), *Chrysoperla carnea* (Stephens), *Orius albidipennis* Reuter, and four species of ants; overall, parasitism was 40% and predation was 79% (Coll et al., 2000).

A foliar application of *B. thuringiensis kurstaki* (*Btk*) against *P. operculella* resulted in a control level similar to a soil application of parathion and carbaryl (Awate and Naik, 1979). A nearly 70% reduction in larval density and fruit damage was observed after a foliar application of *Btk* in tomato (Broza and Sneh, 1994). Two common insect pathogenic species, *Beauveria bassiana* (Balsamo-Crivelli) Vuillemin and *Metarhizium anisopliae* (Metchnikoff) Sorokin have also been reported to kill early instars of *P. operculella* (Sewify et al., 2000; Sun et al., 2004). Entomopathogenic nematodes (*Steinernema* spp. and *Heterorhabditis* spp.) are candidates for the biocontrol of *P. operculella* larvae (Wraight et al., 2007). Viral formulations can be effective based on several reports of *Baculovirus phthorimaea* (Fano and Winter, 1997) and granulosis virus (PoGV) (Kroschel et al., 1996). PoGV has a worldwide distribution (Zeddam et al., 1999), and its field studies revealed 35–40% reduction in larval population of *P. operculella* (Kroschel, 1995; Laarif et al., 2003).

6.4.5 Autocidal Control

The use of sterile insect techniques highlighted several possibilities to suppress *P. operculella* using classical and modern approaches. In developing an effective program using sterility as a control method against *P. operculella*, information including sex ratio, age, and weight of the adults are useful. Weight of both sexes had an impact on the number of times they mate and mating success (Makee and Saour, 2001). When the larvae of different ages were exposed to doses ≥100 Gy, only 13–35% pupated, but no adult emergence was observed (Saour and Makee, 2004).

6.4.6 Chemical Control

The chemical control of *P. operculella* is a common approach, but a sustainable cropping system would be better served if treatment with chemicals was a secondary option because of the latent feeding of the larvae and the development of strong resistance toward many of the traditional chemistries including organophosphates, carbamates, and pyrethroids (Doğramaci and Tingey, 2008; El-Kedy, 2011). Pyrethroids, carbamates, and organophosphates have been used against this pest and indoxacarb, novaluron, and spinosad are three of the newer compounds used (Lawrence, 2009). Use of imidacloprid and acetamiprid has been found unsatisfactory for the control of *P. operculella* in tomato and tobacco (Lawrence, 2009; Vaneva-Gancheva and Dimitrov, 2013). However, Ali (2012) evaluated a broad range of new chemistry insecticides in Sudan in tomatoes and suggested abamectin and lufenuron as the most effective formulations in controlling *P. operculella*. Difenoconazole, imidacloprid, pirimicarb, and thiodicarb were found not to interfere with *O. lepidus*, a braconid parasitoid of *P. operculella* (Symington, 2003). Recent insecticide recommendations for the management of *P. operculella* include flubendiamide and chlorantraniliprole in many cropping systems (Rivera, 2011).

6.4.7 Other Control Tactics

The use of phenological-based studies linked with geographic information systems (GIS) helps to devise good predictive models determining the pests' life history, i.e., development, mortality, and reproduction in different regions of the world (Sporleder et al., 2008). Synergistic interaction against *P. operculella* has been reported by integrating lethal extracts of *Atropa belladonna* L. and *Hyoscyamus niger* L. leaves with entomopathogenic fungi (*B. bassiana*) and bacteria (*B. thuringiensis*) (Sabbour and Ismail, 2002).

7. BEET ARMYWORM, *SPODOPTERA EXIGUA* (HÜBNER)

S. exigua (beet armyworm, small mottled willow moth, or asparagus fern caterpillar) is a polyphagous insect pest of wide distribution.

7.1 Description and Biology

S. exigua adults are medium-sized moths (25–30 mm). Forewings are mottled gray and brown with a light-colored, bean-shaped spot and a pattern of irregular bands. Hind wings are much more gray or whitish with a dark outline. Adults live for approximately 10 days (Srinivasan et al., 2010). After emergence, adults mate and oviposition starts 2–3 days later, lasting for 3–7 days. Females prefer upper tender foliage for oviposition, particularly the lower surface of leaves, tip of the branch, and flowers near blossoming. A female lays 500–600 eggs in clusters of 100–150 eggs, covering the eggs in brown hairs; the eggs hatch in 3–5 days (Srinivasan et al., 2010). First and second instar larvae are pale green to yellowish and later instars are darker (Fig. 6.6). Upon hatching, larvae are 1 mm, but reach 22.3 mm in length as fifth instars (Wilson, 1934). The larvae develop in 2–3 weeks (depending on temperature), and there are generally five instars, but more can develop (Capinera, 2001; Srinivasan et al., 2010). Pupae are light brown, and pupation occurs in the soil after the larvae makes a chamber from sand and soil particles. Pupae are 15–22 mm, and the pupal duration varies from

FIGURE 6.6 Larvae of *Spodoptera exugia* feeding on tomato leaf.

7 to 11 days (Srinivasan et al., 2010). The pupal chamber helps protect *S. exigua* from chill injury (Zheng et al., 2011a). The life cycle of *S. exigua* may take from 24 (Wilson, 1934) to 126 days (Campbell and Duran, 1929) depending on temperature. In temperature-developmental rate studies, no development was observed at 12 and 36°C (Karimi-Malati et al., 2014). Developmental time (16.9 day) was slowest at 30°C, and mean generation time was shortest (15.1 day) at 33°C (Karimi-Malati et al., 2014).

7.2 Distribution, Host Range, and Seasonal Occurrence

S. exigua originated from Southeast Asia. The migratory capacity of *S. exigua* facilitated its geographical distribution (Kimura, 1991; Westbrook, 2008). It is widely distributed in temperate (Zheng et al., 2011b), tropical, and subtropical areas of the world (Liburd et al., 2000) and has occasional outbreaks worldwide. It has been known in North America since 1876 (Capinera, 2001). It does not diapause, but can overwinter in Arizona, Florida, and Texas in the United States with reduced winter populations (Liburd et al., 2000). It is found in the southern half of the United States. during the summer season (Capinera, 2001). This pest previously was reported in 67 countries; about 70% of these were in Asia and Africa (CAB, 1972). It is now found in at least 101 countries, with expansions expected in northern Europe and South America. Thus the geographical range should be revised to between 64°N and 45°S latitude (Zheng et al., 2011b). Host-plant species from more than 18 families (Pearson, 1982) have been reported, including tomato (Srinivasan et al., 2010) and several other agriculturally important plants.

7.3 Damage, Economic Thresholds, and Losses

S. exigua damages field, vegetable, and floral crops (Suenaga and Tanaka, 1997). In the southeastern United States, it has evolved as a key pest in tomato (Liburd et al., 2000). Heavy losses have been reported from several parts of the world (Zalom et al., 1986; Yee and Toscano, 1998). Damage to foliage is the major concern as it leads to reduced productivity. Larvae feed on fruits and foliage and they can be serious defoliators in flowering crops. Newly hatched larvae feed gregariously but become solitary and highly active with maturity (Capinera, 2001). Feeding by young larvae may leave behind skeletonized leaves. After fruit set, larvae start feeding on fruits, reducing the value of the fresh market crop. In addition, larval contamination affects the value of processing tomatoes (Zalom and Jones, 1994). One *S. exigua* larva per 40 plants caused 5% tomato yield loss in field trials (Riley, 2006). Taylor and Riley (2008) estimated the economic injury level (EIL) as 1 larva/20 plants during the early season, 1 larva/9 plants during the late season, and 1 larva/17 plants for the season-long EIL. An impact of 1 larva/6 plants (averaged across the growing season), resulted in a 5% reduction in yield (Taylor and Riley, 2008).

7.4 Management

Short-term pest management practices such as insecticides tend to provide immediate pest relief, while long-term management practices such as plant manipulation may not result in immediate control. Yet, safe and reliable short-term and long-term practices can be complementary in the long-term management of armyworm populations.

7.4.1 Trapping (Monitoring/Mass Trapping)

Pheromone trapping is the most reliable way to capture and monitor adult *S. exigua*. The sex pheromones of female *S. exigua* have been identified (Z9, E12-14:Ac and Z9-14:OH) (Dong and Du, 2003), and they are used to monitor populations of this pest. A 6-year study (1994–2000) using pheromone traps against *S. exigua* suggested that because its widescale immigration starts in mid-July each year, this date may be used as a standard to determine when immediate treatment is needed to prevent economic crop damage in the Mississippi Delta of the United States (Adamczyk et al., 2003). Trap data also defined the activity period of male *S. exigua*; 15% were captured during the early hours of night, with peak capture during the last 4 h of the night and minimum catches per trap at sunrise (Cheng et al., 2015). Pheromone traps can disrupt mating and thus be used to control the pest (Kerns, 2000). For example, a synthetic sex pheromone was very effective in disrupting female–male communication and controlling *S. exigua* in onion fields in Kyoto, Japan (Yoshiyasu et al., 1995). The use of puffers is another strategy to be used with pheromone to get some level of control through mating disruption (Shorey and Gerber, 1996). A limitation associated with pheromone trapping of *S. exigua* is that adult moth catches per traps do not depict actual larval damage in the field. Monitoring of the pest can be done by inspecting egg masses, larval count, and hits (actively feeding group of larvae on a plant).

7.4.2 Mechanical Control

In garden plots and small farms, large numbers of larvae causing obvious feeding injury can be handpicked from the plants (Srinivasan et al., 2010). However, this is unrealistic on large-scale farms because such labor-intensive practices are often uneconomic for commercial growers. In the same fashion, egg masses can be handpicked from plants (Srinivasan et al., 2010).

7.4.3 Host-Plant Resistance

Low survival by *S. exigua* has been identified in several accessions of tomato (Eigenbrode and Trumble, 1993). Methyl ketones and the sesquiterpene hydrocarbon zingiberene can confer resistance to *S. exigua* (Eigenbrode et al., 1996). The enzymes, polyphenol oxidases (PPOs), in tomato, play an important role in resistance to insects. Larvae of *S. exigua* on leaves of tomato genotypes with suppressed PPO fed more and had higher weight than those on tomatoes which overexpressed PPO (Bhonwong et al., 2009). Among various plant species used to study the life history of *S. exigua*, long bean (*Vigna unguiculata* subsp. *sesquipedalis* L.) was the most suitable host plant, allowing the fastest development with the fewest number of larval instars, and the highest survival rate (Azidah and Sofian-Azirun, 2006). Based on the findings that jasmonic acid (JA) was necessary for the formation of glandular trichomes and related metabolites as defense traits in tomato (Bosch et al., 2014a), the role of JA in plant defense against *S. exigua* was studied. JA-deficient tomato was the preferred host plant by *S. exigua* over wild-type tomato plants for oviposition and feeding (Bosch et al., 2014b). Elicitors exclusively present in the oral secretion of *S. exigua* were mainly responsible for these pronounced results (Bosch et al., 2014b).

7.4.4 Biological Control

Several natural enemies are important in the management of *S. exigua* including parasitoids (*Meteorus autographae* Muesebeck (Ruberson et al., 1993), *Cotesia marginiventris* (Cresson) (Henneberry et al., 1991), *Lespesia archippivora* (Riley) (Eveleens et al., 1973) and *Chelonus insularis* Cresson (Pearson, 1982)), and egg and larval predators (e.g., *C. carnea* (Hogg and Gutierrez, 1980) and *Orius tristicolor* (White) (Eveleens et al., 1973)). Most of the parasitoids are in the orders Diptera (Tachinidae) and Hymenoptera (Braconidae and Ichneumonidae), while the major predators are from the orders Hemiptera, Hymenoptera, Coleoptera, Neuroptera, and Dermaptera (Ruberson et al., 1994).

The use of pathogens is another promising strategy for the control of *S. exigua*. Field trials on vegetable crops in Thailand with a formulation of nuclear polyhedrosis virus (Spod-X) of *S. exigua*, showed Spod-X as a cost effective and long-term management tool for *S. exigua* (Kolodny-Hirsch et al., 1997). For the performance of *S. exigua* nuclear polyhedrosis virus (SeNPV), a high RH level (90–100%) reduced the LD_{50} value of SeNPV against the larvae (Li et al., 2014). As an integrated approach, the interaction between the larval braconid parasitoids *Meteorus pulchricornis* (Wesmael) (Guo et al., 2013), *Microplitis bicoloratus* Xu and He (Cai et al., 2012), and *Microplitis pallidipes* Szepligeti (Jiang et al., 2011), and SeNPV enhanced the efficacy of the natural enemies. The addition of optical brighteners to nuclear polyhedrosis virus of *S. exigua* reduced the LC_{50} and LT_{50} values of the pathogen, and retarded larval development of *S. exigua* (Shapiro and Argauer, 2001).

B. bassiana (Wraight et al., 2010; Al-Kherb, 2014), *M. anisopliae* (Al-Kherb, 2014), and *Nomuraea rileyi* (Farlow) Samson (Lee et al., 2012) are the major reported fungal strains with good insecticidal potential against *S. exigua*. The role of secondary metabolites of the endophytic fungus, *Alternaria alternata* (Fr.) Keissl., as an immunosuppressive agent against *Spodoptera* sp. was reported by Kaur et al. (2015). Bacterial formulations, particularly those containing *B. thuringiensis*, also offer an enhanced effect to control *S. exigua* in vegetables and cash crops (Siebert et al., 2012). A bioassay using five *Bt* Vip3 proteins determined that *S. exigua* was most susceptible to Vip3Aa and Vip3Ae (Ruiz de Escudero et al., 2014).

7.4.5 Autocidal Control

Based on the concept of Knipling (1992) and Carpenter and Gross (1993) that F_1 sterility would be more effective if integrated with other pest control tactics, Carpenter et al. (1996) combined two approaches, i.e., F_1 sterility and control with a parasitoid. They found that the application of the parasitoid *C. marginiventris* and F_1 sterility was a compatible strategy that can be used effectively for an *S. exigua* management program.

7.4.6 Chemical Control

Insecticide treatments of *S. exigua* are generally started at the first detection of eggs, and are most effective when applied before the third instar moves into the fruit (Sparks, 2014). Resistance has developed in *S. exigua* against carbamates

(Mascarenhas et al., 1998), organophosphates (Ahmad and Arif, 2010), pyrethroids (Shimada et al., 2005), and spinosad (Wang et al., 2002). The level of resistance varies among geographical populations of *S. exigua* (Osorio et al., 2008; Che et al., 2013). Based on the results of studies showing rapid feeding cessation and decline in the feeding damage by *S. exigua*, Lai and Su (2011) further investigated the insecticidal role of chlorantraniliprole, an anthranilic diamide with a mode of action belonging to class 28 (IRAC, 2016). Their findings suggested that this chemical provided an exceptional effect against the neonate larvae (Lai and Su, 2011). Yet, there may be a tendency of *S. exigua* to develop resistance against chlorantraniliprole with successive exposures up to 23 generations (Lai and Su, 2011). New chemistries such as emamectin benzoate (Bengochea et al., 2014), spinotram and indoxacarb (Huang et al., 2011), and in some situations, spinosad (Moadeli et al., 2014) may be a preferred option against different stages of *S. exigua*. Plant-based products can also be used against *S. exigua* (Juárez et al., 2014; Ntalli et al., 2014). Huang et al. (2014) recently reported on the effect of the essential oil of *Pogostemon cablin* (Blanco) Benth and demonstrated that it offers good insecticidal potential, particularly antifeedant, larvicidal, growth inhibitory, and pupicidal effects against *S. exigua*. The extracts of eight Chinese medicinal plant species exhibited good antifeedant activity against third instar *S. exigua* larvae in laboratory trials (Feng et al., 2012). Insect growth regulators, including several commercial formulations such as tebufenozide (Liu et al., 2008) and methoxyfenozide (Aguire et al., 2013), have been extensively studied and demonstrated efficacious against *S. exigua*.

7.4.7 Integrated Pest Management

As an integrated approach with the combined use of microbial agents (*B. bassiana*, *M. anisopliae*, and *B. thuringiensis*, formulated as Dipel 2×) (Mesbah et al., 2004) and microbial formulations (two fungal Biosect and Biover, one *Bt* formulation Dipel 2×) with a chemical insecticide (Lannate) (Zaki and Abdel-Raheem, 2010) were shown to control *S. exigua*. Liburd et al. (2000) proposed an IPM program for tomato where they used seven *Bt* commercial formulations, SeNVP, entomophagous nematodes (*Steinernema carpocapsae* (Weiser)), three chemical insecticides (methomyl, fenpropathrin, chlorpyrifos), and neem plant extract against three *Spodoptera* species including *S. exigua*. The pest populations were below EIL in treated plots, thus the treatments resulted in reduced fruit injuries and higher marketable yields of the tomatoes (Liburd et al., 2000).

8. COMMON ARMYWORM, *SPODOPTERA LITURA* (F.)

The common armyworm, *S. litura* is a polyphagous and highly mobile insect.

8.1 Biology

Adults of *S. litura* are very similar in size and color to *Spodoptera ornithogalli* Guenée (yellow-striped armyworm, found in North and Central America) and *Spodoptera littoralis* Boisduval (African cotton worm, found in the Mediterranean, Middle East, and Africa) (Mochida, 1973; Venette et al., 2003b). However, *S. litura* is the predominant species on tomato in tropical South Asia and Southeast Asia, and the geographical distribution among *S. litura* and *S. ornithogalli* or *S. littoralis* does not overlap in this region. The adults are usually brown; the forewings have numerous crisscross streaks on a cream or brown background (Fig. 6.7B). The hind wings are white with a brown patch along the border. The eggs are laid in groups of 200–300, and covered with brown hairs from the body of the female. Eggs hatch in 3–5 days. The young larvae feed in groups, but older larvae feed individually. The mature larvae (Fig. 6.7A) are green, pale greenish brown, or black, with

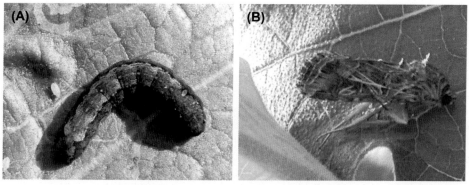

FIGURE 6.7   (A) Larva and (B) adult of *Spodoptera litura*.

prominent black spiracles. The body may have transverse and longitudinal gray and yellow bands. Morphometric analyses of *S. litura* reared on tomato in the laboratory supports that has six larval instars on this host (Vashisth and Chandel, 2013). The larvae are nocturnal and feed actively during the night, but hide during the day in cracks and crevices in the soil and in plant debris. The larval period lasts 15–30 days and pupation takes place in the soil. Pupae are shiny reddish-brown. The duration of the pupal period varies from 1–3 weeks. Populations based on trap captures and larvae per leaf counts positively correlate with the weather (Monobrullah et al., 2007).

8.2 Damage, Economic Thresholds, and Losses

S. litura causes serious damage to many economically important crops, and feeds on over 100 plant species (Garad et al., 1984; Maree et al., 1999; Ahmad et al., 2013). The preference of this pest is less on tomato than some other hosts such as cotton. Larvae of *S. litura* feed on leaf surfaces or on whole leaves, leaving only the main veins. They rarely feed on immature tomato fruits, but this type of feeding causes irregular holes in the fruit. Sometimes, the larvae cut tomato seedlings at the soil level. In a year-round field study on tomato, 80% plant infestation was observed in India (Monobrullah et al., 2007). However, there is a dearth of published data on damage threshold for this lepidopterous pest on tomato.

8.3 Management

Most of the pest management practices for *S. litura* are similar to managing *H. armigera*. Castor (*Ricinus communis* L.) and taro (*Colocasia esculenta* (L.) Schott) can be grown as trap crops along the field border to attract the egg-laying female adult moths. Since eggs are laid in masses, the egg masses and young larvae can be handpicked and destroyed in small-scale production. In addition, high rates of *S. litura* parasitism by *Microplitis prodeniae* Rao and Kurian and *C. chlorideae* were observed when taro was used as a trap crop (Zhou et al., 2010). Over 100 parasitoids and predators have been recorded as natural enemies of *S. litura* (Rao et al., 1993). A study based on trapping with sex pheromones across five Southeast Asian countries suggests that natural enemies are important in lowering population levels of *S. litura* in locations where the adults are found year round, compared to locations where they are absent during the winter period (Tojo et al., 2008). Sex pheromones of *S. litura* are commercially available in many countries and can be used for monitoring and mass trapping. *S. litura* is more susceptible to entomopathogenic fungi in the earlier instars than the later instars (Asi et al., 2013). *S. litura* nuclear-polyhedrovirus (SlNPV) is commercially available in some countries, and can be used to replace chemical pesticides. The expression of *Bt*-Cry toxins through transgenic technology is an ongoing approach for protecting tomato plants against lepidopterous pests. A transgenic tomato line, Ab25 E, expressing Cry1Ab gene of *B. thuringinsis* was recently selected, and feeding on leaves of this plant by *S. litura* resulted in 100% larval mortality (Koul et al., 2014). Irradiated male *S. litura* with substerilizing gamma doses (either 100 or 130 Gy) resulted in little or no degradation of evaluated flight, behavior, or mating parameters (Seth et al., 2016). Thus, IS may offer a role in an IPM program for *S. litura*.

9. TOMATO LEAFMINER, *TUTA ABSOLUTA* (MEYRICK)

The tomato leafminer, *T. absoluta*, has become an important invasive pest of tomatoes. It is one of the most important worldwide pests of tomato.

9.1 Identification and Biology

Adult *T. absoluta* moths (Fig. 6.8) are small, about 1 cm long, mottled gray with long filiform antennae. The adults live about 30 days (Vargas, 1970; Estay, 2000). The adults are active at night and rest among vegetation during the day (Fernández and Montagne, 1990). Female moths can oviposit for more than 20 days and deposit as many as 260 eggs in their lifetime (Fernández and Montagne, 1990; CABI, 2016b). Eggs are small and oval, starting out as creamy white and turning yellow to yellow-orange (Estay, 2000). As the eggs darken, an outline of the larval head capsule can be seen; this is called the blackhead stage of egg development (Vargas, 1970). Larvae are dorsoventrally flattened, and their color changes from creamy white to dark green as they progress through four instars (Estay, 2000). Late instars may appear pink before pupation. Upon eclosion, larvae penetrate and feed on apical buds, flowers, new fruit, leaves, or stems of solanaceous plants (Pastrana, 2004). Mature larvae build a silken cocoon where pupation occurs. Newly formed pupae are dark with a greenish tint and then darken to brown over time (Estay, 2000).

Reproductive rates of *T. absoluta* vary depending on environmental conditions with a potential maximum of 12 generations a year (EPPO, 2005). The lower developmental temperature is approximately 8°C. At a constant 25°C, *T. absoluta*

FIGURE 6.8   Adult of *Tuta absoluta. Photo by Aziz Ajlan, with permission.*

completes a generation in about 28 days (Vargas, 1970). However, development is highly temperature dependent with the complete life cycle taking 76 and 24 days at 14 and 27°C, respectively (Barrientos et al., 1998). It is unknown whether this species undergoes diapause as it has been shown to overwinter in the egg, pupal, and adult stages (EPPO, 2005).

9.2 Distribution, Host Range, and Seasonal Occurrence

T. absoluta originated in South America and it has become a major threat to tomato production worldwide (Desneux et al., 2010). *T. absoluta* is a primary pest of tomato in much of South America and was also detected in eastern Spain in 2006 (Urbaneja et al., 2008). Since then it has spread rapidly throughout Europe and the Mediterranean Basin, including areas in North Africa and the Middle East (Desneux et al., 2010; Kýlýc, 2010; Seplyarsky et al., 2010); and the likelihood of its spread to new regions is expected. In October 2014, *T. absoluta* was reported for the first time in India (ICAR, 2010). As of mid-2016, *T. absoluta* was not known to occur in the United States. It attacks a wide range of hosts; in addition to tomato, cultivated host plants include potato, eggplant, pepper, and bean (*P. vulgaris*), representing a shift from Solanaceae to Fabaceae (Bloem and Spaltenstein, 2011). In addition to cultivated hosts, *T. absoluta* utilizes other wild solanaceous plants including *Solanum nigrum* L., *Solanum elaeagnifolium* Cav., *D. stramonium, Datura quercifolia* Kunth (=*Datura ferox L.*), and *Nicotiana glauca* Graham (EPPO, 2005).

9.3 Damage, Economic Thresholds, and Losses

Where it has become established, *T. absoluta* is one of the most damaging insect pests of tomato (Miranda et al., 1998; Estay, 2000; EPPO, 2005; Desneux et al., 2010). Without adequate controls, infestations of *T. absoluta* can result in 90–100% loss of field-produced tomatoes (Estay, 2000; Vargas, 1970). *T. absoluta* is a major limiting factor for tomato production in South America (Ferrara et al., 2001). Plant injury consists of mine-formation within the mesophyll by feeding larvae, thus affecting the plant's photosynthetic capacity, and resulting in lower fruit yield (Desneux et al., 2010). Heavy infestations can lead to complete defoliation leaving only a skeletonized leaf and frass. The most significant losses occur from fruit attack (Vargas, 1970), which can further be colonized by pathogens causing fruit rot. *T. absoluta* attacks leaves, flowers, stems and fruits, and in the absence of control measures, growers can lose their entire crop (Apablaza, 1992). Infestations are fairly easy to observe due to the high number of mines, galleries, and dark frass. However, economic thresholds based on fruit damage or infestation levels have not been established. While all plants are vulnerable to attack, damage is more severe in young plants. Thus, control measures are most critical for younger plants.

9.4 Management

The threat of *T. absoluta* has led to extensive insecticide use by tomato growers. This highlights the need for a comprehensive IPM program to keep *T. absoluta* densities below economically damaging levels, and to slow its spread (Arno and Gabarra, 2011). The best strategy for long-term control of *T. absoluta* is to prevent outbreaks by using numerous practices that complement each other and lead to a more sustainable pest-management program. These methods include biological, biotechnological, cultural, and chemical controls. The use of pesticides is most suited only when preventive measures are inadequate and careful field monitoring indicates that economic loss is likely.

9.4.1 Monitoring and Trapping

Monitoring for the presence and density is a critical step in improving the management of *T. absoluta*. Lures that contain two major components of *T. absoluta* sex pheromone are commercially available, and can be used in traps (Attygalle et al., 1996). Adult captures in pheromone traps highly correlate with densities of leaf mines and larvae in the field, leading to the establishment of treatment thresholds based on trap catch (Ferrara et al., 2001; Salas, 2004). However, these thresholds vary in different locations. For example, in some parts of South America, thresholds have been set at 45 ± 19.5 adults/trap/day, whereas, in Chile, thresholds are set at more than twice that number—100 adults/trap/day (Benvenga et al., 2007). Monitoring for *T. absoluta* in the field requires intensive training and specialized manual labor. Scouting for this pest and interpreting infestation levels based on visual inspection is extremely difficult and time-consuming. While some thresholds are based on levels of defoliation, others are based on egg and larval counts (Benvenga et al., 2007; Desneux et al., 2010).

9.4.2 Cultural Control

Sanitation is a key first step in preventing initial infestations and spread of *T. absoluta*. Plants should be routinely examined for any evidence of infestation and all infested materials should be disposed of carefully (Bloem and Spaltenstein, 2011). The incorporation of crop residues after harvest effectively interrupts the life cycle of *T. absoluta* by killing the immature stages, and all equipment used in fields with known infestations should be thoroughly cleaned by high-pressure washing or steam. In addition, solanaceous weeds growing within 50 m of infested fields or processing areas should be removed (Bloem and Spaltenstein, 2011). Other cultural methods to decrease *T. absoluta* densities include non-solanaceous crop rotations, and proper fertilization and irrigation. In field tomatoes, a 6-week host-free period is recommended and can be adjusted based on climate (Bloem and Spaltenstein, 2011; Urbaneja et al., 2013). When possible, growers should avoid early and late-season tomato crops if *T. absoluta* is known to be present. Also, in field situations, conventional and center-pivot irrigation are favored over soil irrigation because these methods disturb eggs, larvae, and pupae, and can increase mortality in field populations (Bloem and Spaltenstein, 2011).

9.4.3 Biological Control

Biological control is an important component in the integrated management of *T. absoluta* (Haji et al., 2002). Almost 50 natural enemies have been reported in its native range in South America (Desneux et al., 2010). The eggs are parasitized primarily by *Trichogramma* spp. (family Trichogrammatidae) wasps, but they are also parasitized by wasps in the families Encyrtidae and Eupelmidae. Augmentative releases of *Trichogramma* spp., especially *T. pretiosum* (Medeiros et al., 2009a), have been successful in controlling *T. absoluta* in Colombia (Salas, 2001) and Brazil (Parra and Zucchi, 2004), but unsuccessful in Chile (Taco et al., 1998). The importance of predators attacking *T. absoluta* in South America has been acknowledged, but a complete understanding of their ecology and overall impact on *T. absoluta* mortality is needed (Desneux et al., 2010). Miranda et al. (1998) reported that the highest predation of *T. absoluta* occurred during the larval stage, with anthocorid bugs (Anthocoridae), coccinelid beetles (Coccinellidae), and to a lesser extent, predacious thrips (Phlaeothripidae) accounting for most mortality. Conversely, higher levels of egg mortality were attributed to parasitoids, but predators still played a large role. Bacci et al. (2008) found that the most important mortality factor for the larvae of *T. absoluta* was predation, particularly by *Protonectarina sylveirae* (Saussure) (Vespidae) as well as anthocorid and mirid bugs. Also, several researchers reported a decrease in populations of *T. absoluta* related to an increase in predator populations when conservative biological controls, such as companion plants, were applied (Medeiros et al., 2009b).

9.4.4 Chemical Control

Insecticides are the primary tool used to control *T. absoluta* (Bielza, 2010), which has led to many insecticide resistant populations. For example, resistance to cartap, abamectin, and permethrin in Brazil (Siqueira et al., 2000, 2001), and to pyrethroids in Chile (Salazar and Araya, 1997) and Argentina (Lietti et al., 2005) have been reported. Yet, insecticide use for *T. absoluta* management remains high. There are clear differences in efficacy among commercial insecticides on populations of *T. absoluta* (Braham and Hajji, 2012; Hanafy and El-Sayed, 2013). Among three bioinsecticides (spinetoram, spinosad, and emamectin) evaluated in tomato fields in Egypt, spinetoram was the most effective and Emamectin was the least (by about twofold) effective (Hanafy and El-Sayed, 2013). In that same study, spinetoram resulted in greater efficacy than the standard insecticides Pyridalyl and Indoxcarb. In numerous field trials in Tunisia among a suite of biorational insecticides (including spinosad, kaolin clay, and plant extracts such as neem) and synthetic insecticides, efficacy was high (around 90% mortality) for spinosad and emamectin benzoate, but was low for triflumuron, diafenthiuron, and the plant extract Tutafort (Hanafy and El-Sayed, 2013). Treatment of tomato for *T. absoluta* control with either azadirachtin or chlorantraniliprole resulted in no adverse impact on adult survival or egg hatch of predatory bugs (*Amphiareus constrictus* (Stål) and *Blaptostethus pallescens* Poppius). However, nymphal survival of the predators was reduced (Gontijo et al., 2015). As novel insecticides become available, it is important that they are used judiciously, combining them with all other available tools for *T. absoluta* management and rotating modes of action to hinder the development of insecticide resistance.

10. ECONOMICS OF MANAGEMENT STRATEGIES

Globally, losses from insect pests, weeds, and diseases average 35% of potential crop yields for six major crops (Oerke, 2006). Losses to preharvest pests are similar in vegetable crops (Dhaliwal et al., 2010). With commercial vegetables being valued at over $20 billion in the United States (USDA, 2014), losses and pest control costs can exceed $10 billion in some years. Worldwide, approximately 30% of all insecticides used are applied to vegetables and fruits (Shelton et al., 2008). The world's leading producers of tomatoes are China and the United States. Fresh and processed tomatoes account for more than $2 billion in annual farm cash receipts. Because the value of fresh market tomatoes is dependent on crop quality, cosmetic damage must be kept to a minimum. As a result, over 95% of the United States tomato crop receives insecticide applications (Gianessi, 2009). These applications add tremendous costs to producers, but are often viewed as the only effective means to manage certain pests, including most lepidopterans.

While costs greatly vary by locations, many studies have found that even with the increased overall production cost, the use of synthetic pesticides can lead to increases in net profits. In Turkey, the average cost of pesticide applications in processed and fresh-market tomatoes ranges 6–6.5% ($73–$82/ha) of the total production cost (Engindeniz and Cosar, 2013). Nevertheless, these expenditures led to net profits of $235 and $507/ha, respectively (Engindeniz and Cosar, 2013). In the United States, pesticide costs represent 1–5% of the total costs of tomato production (Miyao et al., 2008; VanSickle et al., 2009). For example, in the state of Virginia, insecticide costs ($93/ha) represented about 5% of the total cost of tomato production, but prevented losses lead to increased profits on $1214–$2785/ha (Nault and Speese, 2002). Additional studies in the southeastern United States found profits similar to those in Turkey with insecticide applications increasing net profit by $451–$493/ha (Walgenbach and Estes, 1992; Zehnder et al., 1994).

While there is no doubt pesticides can help to manage pests and maintain profits, it is also clear that a balanced integrated approach can be more sustainable and lead to higher revenues. Evaluation of an integrated management program in Mexico found that net profits increased as much as US $95/ha when compared to a conventional treatment approach, even though in some cases, lepidopteran pest abundance was higher (Trumble and Alvarado-Rodriguez, 1993). In Australia, these increases were more modest, but growers using integrated management approaches saw on average, a $15.88/ha increase in profits than growers using conventional methods (Grinter, 1996). Not only can an integrated approach reduce costs and increase profits, it can also slow the development of resistance; all of the pests described in this chapter display some level of resistance to multiple active ingredients (APRD, 2016).

11. SUMMARY

The lepidopteran pest complex presents many challenges for tomato growers worldwide. The effort to develop more sustainable management programs for these pests has been impeded, primarily because this crop has a low tolerance for insect damage and effective insecticides are available. However, with increased incidence of insecticide resistance and concerns for the interest of human safety, environmental protection, and sustainability, there is a need for a truly integrated approach. It is clear that integrated approaches can maintain high yields, and increase profit, thereby leading to an overall increase in the sustainable management of lepidopteran pests in tomato. Although synthetic insecticide use remains the predominant control strategy in commercial tomato production, there are ample opportunities to increase the use of more sustainable approaches. However, because of indiscriminate insecticide use, there are concerns and limitations to the broader use of biological control agents in tomato production systems.

As new information becomes available and new programs are developed, growers should include multiple tactics that work together and can be implemented over wide areas. Growers need a truly integrated approach that incorporates host-plant resistance, biological controls, cultural controls, behavior control, sanitation, and judicious use of insecticides such as selecting insecticides that minimally impact natural enemies. When used alone, these strategies will not work sufficiently. When used together, however, they can complement each other and lead to improved pest suppression, reduced insecticide resistance, and a more sustainable pest-management program. Further studies are needed to identify the most effective components of an environmentally sound management program. Refinement of IPM techniques and educating growers, extension entomologists, and other IPM practitioners are critical in advancing the sustainable management programs of lepidopteran pests.

REFERENCES

Adamczyk, J.J., Williams, M.R., Reed, J.T., Hubbard, D.W., Hardee, D.D., 2003. Spatial and temporal occurrence of beet armyworm (Lepidoptera: Noctuidae) moths in Mississippi. Florida Entomologist 86, 229–232.

Afifi, F.M.L., Haydar, M.F., Omar, H.I.H., 1990. Effect of Different Intercropping Systems on Tomato Infestation with Major Insect Pests. *Bemisia tabaci* (Genn.) (Hemiptera: Aleyrodidae), *Myzus persicae* Sulzer (Homoptera: Aphididae) and *Phthorimaea operculella* Zeller (Lepidoptera: Gelechiidae), vol. 41. Bulletin of Faculty of Agriculture, University of Cairo, pp. 885–900.

Aguire, O.U., Martinez, A.M., Campos-Garcia, J., Hernandez, L.A., Figeuroa, J.I., Vinuela, E., Chavarrieta, J.M., Smagghe, G., Pineda, S., 2013. Foliar persistence and residual activity of methoxyfenozide against beet armyworm (Lepidoptera: Noctuidae). Insect Science 6, 734–742.

Ahmad, M., 2003. A checklist of natural enemies of *Helicoverpa armigera* (Hübner). Journal of Agricultural Research 41, 267–278.

Ahmad, M., Arif, M.I., 2010. Resistance of beet armyworm *Spodoptera exigua* (Lepidoptera: Noctuidae) to endosulfan, organophosphorus and pyrethroid insecticides in Pakistan. Crop Protection 29, 1428–1433.

Ahmad, M., Ghaffar, A., Rafiq, M., 2013. Host plants of leaf worm, *Spodoptera litura* (Fabricius), Lepidoptera: Noctuidae) in Pakistan. Asian Journal of Agricultural Biology 1, 23–28.

Ali, A.E., 2012. A note on evaluation of some insecticides for the control of potato tuber moth, *Phthorimaea operculella*, (Lepidoptera: Gelechiidae), on tomato. Sudan Journal of Agricultural Research 19, 121–124.

Al-Kherb, W.A., 2014. Virulence bio-assay efficiency of *Beauveria bassiana* and *Metarhizium anisopliae* for the biological control of *Spodoptera exigua* Hübner (Lepidoptera: Noctuidae) eggs and the 1st instar larvae. Australian Journal of Basic and Applied Sciences 8, 313–323.

Anfora, G., Vitagliano, S., Larsson, M.C., Witzgall, P., Tasin, M., Germinara, G.S., de Cristofaro, A., 2014. Disruption of *Phthorimaea operculella* (Lepidoptera: Gelechiidae) oviposition by the application of host plant volatiles. Pest Management Science 70, 628–635.

Apablaza, J., 1992. La polilla del tomate y su manejo. Tattersal 79, 12–13.

APRD, 2016. *Helicoverpa armigera*, Arthropod Pesticide Resistance Database. Michigan State University, Michigan, USA. http://www.pesticideresistance.org/display.php?page=species&arId=41.

Arno, J., Gabarra, R., 2011. Side effects of selected insecticides on the *Tuta absoluta* (Lepidoptera: Gelechiidae) predators *Macrolophus pygmaeus* and *Nesidiocoris tenuis* (Hemiptera: Miridae). Journal of Pest Science 84, 513–520.

Asi, M.R., Bashir, M.H., Afzal, M., Akram, M., 2013. Potential of entomopathogenic fungi for biocontrol of *Spodoptera litura* Fabricius (Lepidoptera: Noctuidae). Journal of Animal and Plant Sciences 23, 913–918.

Attygalle, A.B., Jham, G.N., Svatos, A., Frighetto, R.T.S., Ferrara, F.A., Vilela, E.F., Uchoa Fernandes, M.A., 1996. (3E,SZ,11Z)-3,8,11-tetradecatrienyl acetate, major sex pheromone component of the tomato pest *Scrobipalpuloides absoluta* (Lepidoptera: Gelechiidae). Bioorganic and Medicinal Chemistry 4, 305–314.

AVRDC – The World Vegetable Center (AVRDC), 1988. 1986 Progress Report. Asian Vegetable Research and Development Center, Shanhua, Taiwan.

Awate, B.G., Naik, L.M., 1979. Efficacies of insecticidal dusts applied to soil surface for controlling potato tuberworm (*Phthorimaea operculella* Zeller) in field. Journal of Maharashtra Agricultural Universities 4, 100.

Azidah, A.A., Sofian-Azirun, M., 2006. Life history of *Spodoptera exigua* (Lepidoptera: Noctuidae) on various host plants. Bulletin of Entomological Research 96, 613–618.

Bacci, L., Picanço, M.C., Sousa, F.F., Silva, E.M., Campos, M.R., Tomé, H.V.T., 2008. Inimigos naturais da traça do tomateiro. Horticultura Brasileira 26, 2808–2812.

Balachowsky, A.S., Real, P., 1966. La teigne de la pomme de terre. In: Balachowsky, A.S. (Ed.), Entomologie Appliquee a L'agriculture, Lepidopteres. vol. 2. Masson, Paris, France, pp. 371–381.

Barrientos, Z.R., Apablaza, H.J., Norero, S.A., Estay, P.P., 1998. Temperatura base y constant térmica de desarrollo de la polilla del tomate, *Tuta absoluta* (Lepidoptera: Gelechiidae). Ciencia e Investigación Agraria 25, 133–137.

Batiste, W.C., Olson, W.H., 1973. Laboratory evaluations of some solanaceous plants as possible hosts for tomato pinworm. Journal of Economic Entomology 66, 109–111.

Bengochea, P., Sánchez-Ramos, I., Saelices, R., Amor, F., del Estal, P., Viñuela, E., Adán, A., López, A., Budia, F., Medina, P., 2014. Is emamectin benzoate effective against the different tages of *Spodoptera exigua* (Hübner) (Lepidoptera, Noctuidae)? Irish Journal of Agricultural and Food Research 53, 37–49.

Benvenga, S.R., Fernandes, O.A., Gravena, S., 2007. Decision making for integrated pest management of the South American tomato pinworm based on sexual pheromone traps. Horticultura Brasileira 25, 164–169.

Berthon, C.H., 1855. On the potato grub of Tasmania. In: Papers and Proceedings of the Royal Society of Van Diemen's Land, vol. 3, Part 1. Walch and Sons and Huxtable and Deakin Booksellers, Hobart Town, UK, pp. 76–80.

Bhatt, N., Patel, R., 2001. Biology of chickpea pod borer, *Helicoverpa armigera*. Indian Journal of Entomology 63, 255–259.

Bhonwong, A., Stout, M.J., Attajarusit, J., Tantasawat, P., 2009. Defensive role of tomato polyphenol oxidases against cotton bollworm (*Helicoverpa armigera*) and beet armyworm (*Spodoptera exigua*). Journal of Chemical Ecology 35, 28–38.

Bibi, A., Ahmed, K., Ayub, N., Alam, S., 2013. Production of low cost *Bacillus thuringiensis* based-biopesticide for management of chickpea pod-borer *Helicoverpa armigera* (Huebn) in Pakistan. Natural Science 5, 1139–1144.

Bielza, P., 2010. La resistencia a insecticidas en *Tuta absoluta*. Phytoma España 217, 103–106.

Blanco, C.A., Terán-Vargas, A.P., López Jr., J.D., Kauffman, J.V., Wei, X., 2007. Densities of *Heliothis virescens* and *Helicoverpa zea* (Lepidoptera: Noctuidae) in three plant hosts. Florida Entomologist 90, 742–750.

Bloem, S., Spaltenstein, E., 2011. New Pest Response Guidelines: Tomato Leafminer (*Tuta absoluta*). United States Department of Agriculture, Animal and Plant Health Inspection Service-PPQ-EDP-Emergency Management, Riverdale, Maryland, USA.

Bosch, M., Berger, S., Schaller, A., Stintzi, A., 2014a. Jasmonate-dependent induction of polyphenol oxidase activity in tomato foliage is important for defense against *Spodoptera exigua* but not against *Manduca sexta*. BMC Plant Biology 14 (257), 1–15.

Bosch, M., Wright, L.P., Gershenzon, J., Wasternack, C., Hause, B., Schaller, A., Stintzi, A., 2014b. Jasmonic acid and its precursor 12-oxophytodienoic acid control different aspects of constitutive and induced herbivore defenses in tomato. Plant Physiology 166, 396–410.

Bottrell, D.G., 1979. Guidelines for Integrated Control of Maize Pests. FAO Plant Production and Protection Paper No. 18. Food and Agriculture Organization of the United Nations, Rome, Italy, p. 91.

Braham, M., Hajji, L., 2012. Management of *Tuta absoluta* (Lepidoptera, Gelechiidae) with insecticides on tomatoes. In: Perveen, F. (Ed.), Insecticides – Pest Engineering. InTech, Rijeka, Croatia, pp. 333–354. http://www.intechopen.com/books/insecticides-pest-engineering/management-of-tuta-absoluta-lepidoptera-gelechiidae-with-insecticides-on-tomatoes.

Brickle, D.S., Turnipseed, S.G., Sullivan, M.J., 2001. Efficacy of insecticides of different chemistries against *Helicoverpa zea* (Lepidoptera: Noctuidae) in transgenic *Bacillus thuringiensis* and conventional cotton. Journal of Economic Entomology 94, 86–92.

Broodryk, S.W., 1971. Ecological investigations on the potato tuber moth, *Phthorimaea operculella* (Zeller) (Lepidoptera: Gelechiidae). Phytophylactica 3, 73–84.

Broza, M., Sneh, B., 1994. *Bacillus thuringiensis* spp. *Kurstaki* as an effective control agent of lepidopteran pests in tomato fields in Israel. Journal of Economic Entomology 87, 923–928.

CAB (Commonwealth Agricultural Bureaux), 1972. *Spodoptera exigua* (Hübner). Distribution Maps of Plant Pests, Map. 302. CAB, Wallingford, UK.

CABI (Commonwealth Agricultural Bureaux International), 2016a. *Helicoverpa armigera* (Cotton Bollworm) – Datasheet. Invasive Species Compendium. Commonwealth Agricultural Bureau, International (CABI), Wallingford, UK. http://www.cabi.org/isc/datasheet/26757.

CABI (Commonwealth Agricultural Bureaux International), 2016b. *Tuta absoluta* – Datasheet. Invasive Species Compendium Commonwealth Agricultural Bureau, International (CABI), Wallingford, UK. http://www.cabi.org/isc/datasheet/49260.

Cai, Y., Fan, J., Sun, S., Wang, F., Yang, K., Li, G., Pang, Y., 2012. Interspecific interaction between *Spodoptera exigua* multiple nucleopolyhedrovirus and *Microplitis bicoloratus* (Hymenoptera: Braconidae: Microgastrina) in *Spodoptera exigua* (Lepidoptera: Noctuidae) larvae. Journal of Economic Entomology 105, 1503–1508.

Camelo, L., Adams, T.B., Landolt, P.J., Zack, R.S., Smithhisler, C., 2011. Seasonal patterns of capture of *Helicoverpa zea* (Boddie) and *Heliothis phloxiphaga* (Grote and Robinson) (Lepidoptera: Noctuidae) in pheromone traps in Washington state. Journal of the Entomological Society of British Columbia 108, 1–8.

Campbell, C.D., 1990. Aspects of Heliothis Management on Staked Tomatoes in Western North Carolina (M.S. thesis). North Carolina State University, Raleigh, North Carolina, USA, p. 77.

Campbell, R.E., Duran, V., 1929. Notes on the sugar-beet army worm in California. California Department of Agriculture Monthly Bulletin 18, 267–275.

Capinera, J.L., 2001. Handbook of Vegetable Pests. Academic Press, San Diego, California, USA, p. 729.

Capinera, J.L., 2007. Corn Earworm, *Helicoverpa* (=Heliothis) *zea* (Boddie) (Lepidoptera: Noctuidae). Department of Entomology and Nematology, Florida Cooperative Extension Service Publication No. EENY-145 (IN30200). University of Florida, Gainesville, Florida, USA.

Capps, H.W., 1946. Description of the larvae of *Keiferia penicula* Heim., with a key to the larvae of related species attacking eggplant, potato and tomato in the United States. Annals of the Entomological Society of America 39, 561–563.

Carpenter, J.E., Gross, H.R., 1993. Suppression of feral *Helicoverpa zea* (Lepidoptera: Noctuidae) populations following the infusion of inherited sterility from released substerile males. Environmental Entomology 22, 1084–1091.

Carpenter, J.E., Hidrayani, Sheehan, W., 1996. Compatibility of F1 sterility and a parasitoid, *Cotesia marginiventris* (Hymenoptera: Braconidae), for managing *Spodoptera exigua* (Lepidoptera: Noctuidae): acceptability and suitability of hosts. Florida Entomologist 79, 289–290.

Carpenter, J.E., Wiseman, B.R., 1992. Effects of inherited sterility and insect resistant dent-corn silks on *Helicoberpa zea* (Lepidoptera: Noctuidae) development. Journal of Entomological Science 27, 413–420.

Charlton, R.E., Wyman, J.A., McLaughlin, J.R., Du, J.W., Roelofs, W.L., 1991. Identification of sex-pheromone of tomato pinworm, *Keiferia lycopersicella* (Wals). Journal of Chemical Ecology 17, 175–183.

Chauha, U., Verma, L.R., 1991. Biology of potato tuber moth, *Phthorimaea operculella* (Zeller) with special reference to pupal eye pigmentation and adult sexual dimorphism. Entomon 16, 63–67.

Che, W., Shi, T., Wu, Y., Yang, Y., 2013. Insecticide resistance status of field populations of *Spodoptera exigua* (Lepidoptera: Noctuidae) from China. Journal of Economic Entomology 106, 1855–1862.

Cheng, W.J., Zheng, X.L., Wang, P., Zhou, L.L., Lei, C.L., Si, S.Y., Wang, X.P., 2015. The circadian rhythm of flight activity of *Spodoptera exigua* males in response to sex pheromone. Entomologia Experimentalis et Applicata 154, 154–160.

Cisneros, F., Gregory, P., 1994. Potato pest management. In: Proceedings of the Presidential Meeting on the "Impact of Genetic Variation on Sustainable Agriculture," Aspects of Applied Biology, vol. 39. Association of Applied Biologists, UK, pp. 113–124.

Coll, M., Gavish, S., Dori, I., 2000. Population biology of the potato tuber moth, *Phthorimaea operculella* (Lepidoptera: Gelechiidae), in two potato cropping systems in Israel. Bulletin of Entomological Research 90, 309–315.

Collins, G.N., Kempton, J.H., 1917. Breeding sweet corn for resistance to the corn earworm. Journal of Agricultural Research 11, 549–572.

Coop, L.B., Croft, B.A., Drapek, R.J., 1993. Model of corn earworm development and crop loss in sweet corn. Journal of Economic Entomology 86, 906–916.

Das, G.P., Raman, K.V., 1994. Alternate hosts of the potato tuber moth, *Phthorimaea operculella* (Zeller). Crop Protection 13, 83–86.

DeBano, S.J., Hamm, P.B., Jensen, A., Rondon, S.I., Landolt, P.J., 2010. Spatial and temporal dynamics of potato tuberworm (Lepidoptera: Gelechiidae) in the Columbia basin of the Pacific Northwest. Environmental Entomology 39, 1–14.

Delannay, X., LaVallee, B.J., Proksch, R.K., Fuchs, R.L., Sims, S.R., Greenplaate, J.T., Marrone, P.G., Dodson, R.B., Augustine, J.J., Layton, J.G., Fischhoff, D.A., 1989. Field performance of transgenic tomato plants expressing the *Bacillus thuringiensis* var. *kurstaki* insect control protein. Nature Biotechnology 7, 1265–1269.

Desneux, N., Wajnberg, E., Wyckhuys, K.A.G., Burgio, G., Arpaia, S., Narváez-Vasquez, C.A., González-Cabrera, J., Catalán Ruescas, D., Tabone, E., Frandon, J., Pizzol, J., Poncet, C., Cabello, T., Urbaneja, A., 2010. Biological invasion of European tomato crops by *Tuta absoluta*: ecology, geographic expansion and prospects for biological control. Journal of Pest Science 83, 197–215.

Dhaliwal, G.S., Jindal, V., Dhawan, A.K., 2010. Insect pest problems and crop losses: changing trends. Indian Journal of Ecology 37, 1–7.

Doğramaci, M., Tingey, W.M., 2008. Comparison of insecticide resistance in a North American field population and laboratory colony of potato tuberworm (Lepidoptera: Gelechiidae). Journal of Pest Science 81, 17–22.

Dong, S.L., Du, J.W., 2003. Chemical identification and field tests of sex pheromone of beet armyworm *Spodoptera exigua*. Acta Phytophylacica Sinica 29, 19–24.

Duffey, S.S., 1986. Plant glandular trichomes: their partial role in defense against insects. In: Juniper, B., Southwood, T.R.E. (Eds.), Insects and the Plant Surface. Edward Arnold, London, pp. 151–172.

Eigenbrode, S.D., Trumble, J.T., 1993. Antibiosis to beet armyworm (*Spodoptera exigua*) in *Lycopersicon* accessions. HortScience 28, 932–934.

Eigenbrode, S.D., Trumble, J.T., White, K.K., 1996. Trichome exudates and resistance to beet armyworm (Lepidoptera: Noctuidae) in *Lycopersicon hirsutum* f. typicum accessions. Environmental Entomology 25, 90–95.

El-Kedy, H., 2011. Insecticide resistance in potato tuber moth *Phthorimaea operculella* Zeller in Egypt. Journal of American Science 7, 263–266.

Elmore, J.C., Howland, A.F., 1943. Life History and Control of the Tomato Pinworm. United States Department of Agriculture, Technical Bulletin 841, p. 30.

Engindeniz, S., Cosar, G.O., 2013. An economic comparison of pesticide applications for processing and table tomatoes: a case study for Turkey. Journal of Plant Protection Research 53, 230–237.

EPPO, 2005. Data sheets on quarantine pests: *Tuta absoluta*. European and Mediterranean Plant Protection Organization Bulletin 35, 434–435.

Estay, P., 2000. Polilla del tomate Tuta absoluta (Meyrick). Informativo La Platina, 9. Instituto de Investigationes Agropecuarias, Centro Regional de Investigacion la Platina, Ministerio de Agricultura Santiago, Chile.

Eveleens, K.G., Bosch, R.V.D., Ehler, L.E., 1973. Secondary outbreaks of beet armyworm by experimental insecticide applications in cotton in California. Environmental Entomology 2, 497–503.

Fano, H., Winter, P., 1997. The Economics of Biological Control in Peruvian Potato Production. Social Science Department Working Paper Series, International Potato Center No. 1997-7. Centro Internacional de la Papa, Lima, Peru, p. 35.

Farrar Jr., R.R., Ridgway, R.L., 2000. Host plant effects on the activity of selected nuclear polyhedrosis viruses against the corn earworm and beet armyworm (Lepidoptera: Noctuidae). Environmental Entomology 29, 108–115.

Feaster, M.A., Steinkraus, D.C., 1996. Inundative biological control of *Helicoverpa zea* (Lepidoptera: Noctuidae) with the entomopathogenic nematode *Steinernema riobravis* (Rhabditida: Steinernematidae). Biological Control 7, 38–43.

Feng, H., Gould, F., Huang, Y., Jiang, Y., Wu, K., 2010. Modelling the population dynamics of cotton bollworm *Helicoverpa armigera* (Hübner) (Lepidoptera: Noctuidae) over a wide area in northern China. Ecological Modelling 221, 1819–1830.

Feng, X., Jiang, H., Zhang, Y., He, W., Zhang, L., 2012. Insecticidal activities of ethanol extracts from thirty Chinese medicinal plants against *Spodoptera exigua* (Lepidoptera: Noctuidae). Journal of Medicinal Plants Research 6, 1263–1267.

Fernández, S., Montagne, A., 1990. Biología del minador del tomate, *Scrobipalpula absoluta* (Meyrick) (Lepidoptera: Gelechiidae). Boletín de Entomología Venezolana 5, 89–99.

Ferrara, F.A.A., Vilela, E.F., Jham, G.N., Eiras, A.E., Picanco, M.C., Attygalle, A.B., Svatos, A., Frighetto, R.T.S., Meinwald, J., 2001. Evaluation of the synthetic major component of the sex pheromone of *Tuta absoluta* (Meyrick) (Lepidoptera: Gelechiidae). Journal of Chemical Ecology 27, 907–917.

Ferro, D.N., Boiteau, G., 1993. Management of inset pests. In: Rowe, R.C. (Ed.), Potato Health Management. American Phytopathological Society Press, Saint Paul, Minnesota, USA, pp. 103–115.

Fischhoff, D.A., Bowdish, K.S., Perlak, F.J., Marrone, P.G., McCormick, S.M., Niedermeyer, J.G., Dean, D.A., Kusano-Kretzmer, K., Mayer, E.J., Rochester, D.E., Rogers, S.G., Fraley, R.T., 1987. Insect tolerant transgenic tomato plants. Nature Biotechnology 5, 807–813.

Fitt, G.P., 1989. The ecology of *Heliothis* species in relation to agroecosystems. Annual Review of Entomology 34, 17–52.

Flint, M., 1986. Integrated Pest Management for Potatoes in the Western United States. Statewide Integrated Pest Management Program. University of California Agriculture and Natural Resources Publication No. 3316, Davis, California, USA, p. 146.

Flint, M., 1998. Integrated Pest Management for Tomatoes, Statewide Integrated Pest Management Program, fourth ed. University of California Agriculture and Natural Resources Publication No. 3274, Davis, California, USA, p. 118.

Flood, B., Foster, R., Hutchison, W.D., Pataky, S., 2005. Sweet corn. In: Foster, R., Flood, B. (Eds.), Vegetable Insect Management. Meister Media Worldwide Press, Willoughby, Ohio, USA, pp. 38–63.

Flood, B.R., Rabaey, T.L., 2007. Potential impact of pyrethroid resistance in *Helicoverpa zea* to the midwest processing industry: sweet corn and snap beans. Online Plant Health Progress. http://dx.doi.org/10.1094/PHP-2007-0719-06-RV.

Foley, D.H., 1985. Tethered flight of the potato moth, *Phthorimaea operculella*. Physiological Entomology 10, 45–51.

Fouda, H.G., Seiber, J.N., Bacon, O.G., 1975. A potent sex attractant for the potato tuberworm moth. Journal of Economic Entomology 68, 423–427.

Fowler, G., Lakin, K., 2001. Risk Assessment: The Old Bollworm, *Helicoverpa armigera* (Hubner), (Lepidoptera: Noctuidae). United States Department of Agriculture. Animal and Plant Health Inspection Service, Center for Plant Health Science and Technology (Internal Report), Raleigh, North Carolina, USA, p. 19.

Fraser, A.M., Mechaber, W.L., Hildebrand, J.G., 2003. Electroantennographic and behavioral responses of the sphinx moth *Manduca sexta* to host plant headspace VOLATILES. Journal of Chemical Ecology 29, 1813–1833.

Fye, R.E., Surber, D.E., 1971. Effects of several temperature and humidity regimens on eggs of six species of lepidopterous pests of cotton in Arizona. Journal of Economic Entomology 64, 1138–1142.

Garad, G.P., Shivpuje, P.R., Bilapte, G.G., 1984. Life fecundity tables of *Spodoptera litura* (F.) on different hosts. Proceedings of the National Academy of Sciences 93, 29–33.

Garczynski, S.F., Wanner, K.W., Unruh, T.R., 2011. Identification and initial characterization of the 3′ end of gene transcripts encoding putative members of the pheromone receptor subfamily in Lepidoptera. Insect Science 19, 64–74.

Gianessi, L., 2009. The Benefits of Insecticide Use: Tomatoes. Crop Life Foundation, Crop Protection Research Institute, Washington, DC, USA, p. 18.

Gilligan, T.M., Passoa, S.C., 2014. LepIntercept – An Identification Resource for Intercepted Lepidoptera Larvae. Identification Technology Program (ITP), updated February, 2014 Fort Collins Company, Colorado, USA. http://idtools.org/id/leps/lepintercept/.

Golizadeh, A., Esmaeili, N., Razmjou, J., Rafiee-Dastjerdi, H., 2014. Comparative life tables of the potato tuberworm, *Phthorimaea operculella*, on leaves and tubers of different potato cultivars. Journal of Insect Science 14, 42.

Gontijo, L.M., Celestino, D., Queiroz, O.S., Guedes, R.N.C., Picanço, M.C., 2015. Impacts of azadirachtin and chlorantraniliprole on the developmental stages of pirate bug predators (Hemiptera: Anthocoridae) of the tomato pinworm *Tuta absoluta* (Lepidoptera: Gelechiidae). Florida Entomologist 98, 59–64.

Gregg, P.C., Fitt, G.P., Coombs, M., Henderson, G.S., 1993. Migrating moths (Lepidoptera) collected in tower-mounted light traps in northern New South Wales, Australia: species composition and seasonal abundance. Bulletin of Entomological Research 83, 563–578.

Grinter, J.P., 1996. The economics of integrated pest management: processing tomatoes. In: 40th Annual Conference of the Australian Agricultural and Resource Economics Society, Sustainable Development Unit. University of Melbourne, Australia, pp. 11–16.

Gross, H.R., 1990. Field release and evaluation of *Archytas marmoratus* (Diptera: Tachinidae) against larvae of *Heliothis zea* (Lepidoptera: Noctuidae) in whorl stage corn. Environmental Entomology 19 (4), 1122–1129.

Große-Wilde, E., Stieber, R., Forstner, M., Krieger, J., Wicher, D., Hansson, B.S., 2010. Sex-specific odorant receptors of the tobacco hornworm *Manduca sexta*. Frontiers in Cellular Neuroscience 4, 22.

Guo, H.F., Fang, J.C., Zhong, W.F., Liu, B.S., 2013. Interactions between *Meteorus pulchricornis* and *Spodoptera exigua* multiple nucleopolyhedrovirus. Journal of Insect Science 13, 12.

Gurr, G.M., McGrath, D., 2001. Effect of plant variety, plant age and photoperiod on glandular pubescence and host-plant resistance to potato moth (*Phthorimaea operculella*) in *Lycopersicon* spp. Annals of Applied Biology 138, 221–230.

Haji, F.N.P., Prezotti, L., Carneiro, J.S., Alencar, J.A., 2002. *Trichogramma pretiosum* para o controle de pragas no tomateiro industrial. In: Parra, J.P.P., Botelho, P.S., Correa-Ferreira, B.S., Bento, J.M.S. (Eds.), Controle Biológico no Brasil: Parasitóides e Predadores. Manole, Sao Paulo, Brazil, pp. 477–494.

Hamm, J.J., Carpenter, J.E., 1997. Compatibility of polynuclear hydrosis viruses and inherited sterility for the control of corn earworm and fall armyworm (Lepidoptera: Noctuidae). Journal of Entomological Science 32, 48–53.

Hanafy, H.E.M., El-Sayed, W., 2013. Efficacy of bio-and chemical insecticides in the control of *Tuta absoluta* (Meyrick) and *Helicoverpa armigera* (Hubner) infesting tomato plants. Australian Journal of Basic and Applied Sciences 7, 943–948.

Hardwick, D.F., 1965. The corn earworm complex. Memoirs of the Entomological Society of Canada 40, 3–246.

Hayden, J.E., Brambila, J., 2015. *Helicoverpa armigera* (Lepidoptera: Noctuidae), the Old World Bollworm. Pest Alert No. FDACS-02039. Florida Department of Agriculture and Consumer Services, Florida, USA.

Helling, D.J., Browne, C., Guthrie, F.E., 1986. Binding of insecticides by hemolymph proteins of the tobacco hornworm, *Manduca sexta* (L.). Pesticide Biochemistry and Physiology 25, 125–132.

Henneberry, T.J., Vail, P.V., Pearson, A.C., Sevacherian, V., 1991. Biological control agents of noctuid larvae (Lepidoptera: Noctuidae) in the Imperial Valley of California. Southwestern Entomologist 16, 81–89.

Herbert, D.A., Zehnder, G.W., Speese, J., Powell, J.E., 1993. Parasitization and timing of diapause in Virginia *Microplitis croceipes* (Hymenoptera: Braconidae): implications for biocontrol of *Helicoverpa zea* (Lepidoptera: Noctuidae) in soybean. Environmental Entomology 22, 693–698.

Herzog, G.A., McPherson, R.M., Jones, D.C., Ottens, R.J., 2002. Baseline susceptibility of tobacco hornworms (Lepidoptera: Sphingidae) to acephate, methomyl and spinosad in Georgia. Journal of Entomological Science 37, 94–100.

Hogg, D.B., Gutierrez, A.P., 1980. A model of the flight phenology of the beet armyworm (Lepidoptera: Noctuidae) in Central California. Hilgardia 48, 1–36.

Huang, S.H., Xian, J.D., Kong, S.Z., Li, Y.C., Xie, J.H., Lin, J., Chen, J.N., Wang, H.F., Su, Z.R., 2014. Insecticidal activity of pogostone against *Spodoptera litura* and *Spodoptera exigua* (Lepidoptera: Noctuidae). Pest Management Science 70, 510–516.

Huang, X.P., Dripps, J.E., Quiñones, S., Min, Y.K., Tsai, T., 2011. Spinetoram, a new spinosyn insecticide for managing diamondback moth and other insect pests of crucifers. In: Srinivasan, R., Shelton, A.M., Collins, H.L. (Eds.), Proceedings of the Sixth International Workshop on Management of the Diamondback Moth and Other Crucifer Insect Pests, March 21–25, 2011. Kasetsart University, Kamphaeng Saen Campus, Nakhon Pathom, Thailand, pp. 222–227.

ICAR (Indian Council of Agricultural Research), 2010. Tuta Absoluta: A New Invasive Pest Alert. Indian Council of Agricultural Research, New Delhi, India. http://www.icar.org.in/en/node/8600.

IRAC, 2016. Insecticide Resistance Action Committee – Mode of Action Classification Scheme, Version 8.1. IRAC International MoA Working Group. http://www.irac-online.org/documents/moa-classification/?ext=pdf.

Isely, D., 1935. Relation of Hosts to Abundance of Cotton Bollworm. University of Arkansas Agricultural Experiment Station Bulletin No. 320, Fayetteville, Arkansas, USA, p. 30.

Jallow, M.F.A., Cunningham, J.P., Zalucki, M.P., 2004. Intra-specific variation for host plant use in *Helicoverpa armigera* (Hübner) (Lepidoptera: Noctuidae): implications for management. Crop Protection 23, 955–964.

Jallow, M.F.A., Matsumura, M., Suzuku, Y., 2001. Oviposition preference and reproductive performance of Japanese *Helicoverpa armigera* (Hübner) (Lepidoptera: Noctuidae). Applied Entomology and Zoology 36, 419–426.

Jayewar, N.E., Sonkamble, M.M., 2015. Evaluation of biopesticides against the capitulum borer *Helicoverpa armigera* on sunflower. Journal of Biopesticides 8, 93–97.

Jiang, J., Zeng, A., Ji, X., Wan, N., Chen, X., 2011. Combined effect of nucleopolyhedrovirus and *Microplitis pallidipes* for the control of the beet armyworm, *Spodoptera exigua*. Pest Management Science 67, 705–713.

Jimenez, M.J., Toscano, N.C., Flaherty, D.L., Ilic, P., Zalom, F.G., Kid, K., 1988. Controlling tomato pinworm by mating disruption. California Agriculture 26, 10–12.

Juárez, Z.N., Fortuna, A.M., Sánchez-Arreola, E., López-Olguín, J.F., Bach, H., Hernández, L.R., 2014. Antifeedant and phagostimulant activity of extracts and pure compounds from *Hymenoxys robusta* on *Spodoptera exigua* (Lepidoptera: Noctuidae) larvae. Natural Product Communications 9, 895–898.

Kakimoto, T., Fujisaki, K., Miyatake, T., 2003. Egg laying preference, larval dispersal, and cannibalism in *Helicoverpa armigera* (Lepidoptera: Noctuidae). Annals of the Entomological Society of America 96, 793–798.

Kaplan, I., Thaler, J.S., 2010. Plant resistance attenuates the consumptive and non-consumptive impacts of predators on prey. Synthesising Ecology 119, 1105–1113.

Karimi-Malati, A., Fathipour, Y., Talebi, A.A., Bazoubandi, M., 2014. Life table parameters and survivorship of *Spodoptera exigua* (Lepidoptera: Noctuidae) at constant temperatures. Environmental Entomology 43, 795–803.

Kariyat, R.R., Scanlon, S.R., Moraski, R.P., Stephenson, A.G., Mescher, M.C., De Moraes, C.M., 2014. Plant inbreeding and prior herbivory influence the attraction of caterpillars (*Manduca sexta*) to odors of the host plant *Solanum carolinense* (Solanaceae). American Journal of Botany 101, 376–380.

Kashyap, R.K., Batra, B.R., 1987. Influence of some crop management practices on the incidence of *Heliothis armigera* (Hubner) and yield of tomato (*Lycopersicon esculentum* Mill) in India. Tropical Pest Management 33, 166–169.

Kaur, H.P., Singh, B., Thakur, A., Kaur, A., Kaur, S., 2015. Studies on immunomodulatory effect of endophytic fungus *Alternaria alternata* on *Spodoptera litura*. Journal of Asia-Pacific Entomology 18, 67–75.

Kawahara, A.Y., Breinholt, J.W., Ponce, F.V., Haxaire, J., Xiao, L., Lamarre, G.P., Rubinoff, D., Kitching, I.J., 2013. Evolution of *Manduca sexta* hornworms and relatives: biogeographical analysis reveals an ancestral diversification in Central America. Molecular Phylogenetics and Evolution 68, 381–386.

Kerns, D.L., 2000. Mating disruption of beet armyworm (Lepidoptera: Noctuidae) in vegetables by a synthetic pheromone. Crop Protection 19, 327–334.

Kester, K.M., Barbosa, P., 1991. Behavioral and ecological constraints imposed by plants on insect parasitoids: implications for biological control. Biological Control 1, 94–106.

Kester, K.M., Peterson, S.C., Hanson, F., Jackson, D.M., Severson, R.F., 2002. The roles of nicotine and natural enemies in determining larval feeding site distributions of *Manduca sexta* L. and *Manduca quinquemaculata* (Haworth) on tobacco. Chemoecology 12, 1–10.

Kim, J., Tooker, J.F., Luthe, D.S., De Moraes, C.M., Felton, G.W., 2012. Insect eggs can enhance wound response in plants: a study system of tomato *Solanum lycopersicum* L. and *Helicoverpa zea* Boddie. PLoS One 7 (5), e37420.

Kimura, S., 1991. Immigration of the beet armyworm, *Spodoptera exigua* (Hübner), to northern coastal area of Akita Prefecture, 1990, in relation to atmospheric conditions. Annual Report of the Society of Plant Protection of North Japan 42, 148–151.

King, A.B.S., 1994. *Heliothis/Helicoverpa* (Lepidoptera: Noctuidae). In: Matthews, G.M., Tunstall, J.P. (Eds.), Insect Pests of Cotton. CABI, UK, pp. 39–106.

Klun, J.A., Plimmer, J.R., Bierl-Leonhardt, B.A., Sparks, A.N., Primiani, M., Chapman, O.L., Lee, G.H., Lepone, G., 1980. Sex pheromone chemistry of female corn earworm moth, *Heliothis zea*. Journal of Chemical Ecology 6, 165–175.

Knipling, E.F., 1992. Principles of Insect Parasitism Analyzed from New Perspectives: Practical Implications for Regulating Insect Populations by Biological Means. Agriculture Handbook No. 693. USDA-ARS, Washington, DC, USA, p. 349.

Kolodny-Hirsch, D.M., Sitchawat, T., Jansiri, T., Chenrchaivachirakul, A., Ketunuti, U., 1997. Field evaluation of the *Spodoptera exigua* (Lepidoptera: Noctuidae) nuclear polyhydrosis virus for control of beet armyworm on vegetable crops in Thailand. Biocontrol Science and Technology 7, 475–488.

Korpita, T., Gomez, S., Orians, C.M., 2014. Cues from a specialist herbivore increase tolerance to defoliation in tomato. Functional Ecology 28, 395–401.

Koul, B., Srivastava, S., Sanyal, I., Tripathi, B., Sharma, V., Amla, D.V., 2014. Transgenic tomato line expressing modified *Bacillus thuringiensis* cry1Ab gene showing complete resistance to two lepidopteran pests. SpringerPlus 3, 84.

Kroschel, J., 1995. Integrated Pest Management in Potato Production in the Republic of Yemen with Special Reference to the Integrated Biological Control of the Potato Tuber Moth (*Phthorimaea operculella* Zeller). Tropical Agriculture 8. Margraf Verlag, Weikersheim, Germany, p. 227.

Kroschel, J., Kaack, H.J., Fritsch, E., Huber, J., 1996. Biological control of the potato tuber moth (*Phthorimaea operculella* Zeller) in the Republic of Yemen using granulosis virus: propagation and effectiveness of the virus in field trials. Biocontrol Science and Technology 6, 217–226.

Kroschel, J., Lacey, L.A., 2009. Integrated pest management for the potato tuber moth, *Phthorimaea operculella* (Zeller) – a potato pest of global importance. In: Kroschel, J. (Ed.). Advances in Crop Research. Advances in Crop Research, vol. 10. Tropical Agriculture, Margraf Publishers, Weikersheim, Germany, p. 147.

Krsteska, V., Stojanoski, P., 2012. *Phthorimaea operculella* (Zeller, 1873), in the tobacco agro-ecosystem. In: Terzić, S. (Ed.), International Conference on BioScience: Biotechnology and Biodiversity – Step in the Future. The Fourth Joint UNS – PSU Conference, June 18–20, 2012, pp. 72–77 Novi Sad, Serbia.

Kuhar, T.P., Nault, B.A., Hitchner, E.M., Speese III, J., 2006. Evaluation of action threshold-based insecticide spray programs for tomato fruitworm management in fresh-market tomatoes in Virginia. Crop Protection 25, 604–612.

Kuhar, T.P., Walgenbach, J.F., Doughty, H.B., 2010. Control of *Helicoverpa zea* in tomatoes with chlorantraniliprole applied through drip chemigation. Online Plant Health Progress. http://dx.doi.org/10.1094/PHP-2009-0407-01-RS.

Kumari, P., Mahapatro, G.K., Banerjee, N., Sarin, N.B., 2015. Ectopic expression of GroEL from *Xenorhabdus nematophila* in tomato enhances resistance against *Helicoverpa armigera* and salt and thermal stress. Transgenic Research 24, 85–873.

Kýlýc, T., 2010. First record of *Tuta absoluta* in Turkey. Phytoparasitica 38, 243–244.

Laarif, A., Fattouch, S., Essid, W., Marzouki, N., Salah, H.B., Hammouda, M.H.B., 2003. Epidemiological survey of *Phthorimaea operculella* granulosis virus in Tunisia. European and Mediterranean Plant Protection Organization Bulletin 33, 335–338.

Lai, T., Su, J., 2011. Effects of chlorantraniliprole on development and reproduction of beet armyworm, *Spodoptera exigua* (Hübner). Journal of Pest Science 84, 381–386.

Lawrence, J.L., 2009. Damage Relationships and Control of the Tobacco Splitworm (Gelechiidae: *Phthorimaea operculella*) in Flue-Cured Tobacco (M.S. thesis). North Carolina State University, Raleigh, North Carolina, USA, p. 47.

Lawson, F.R., 1959. The natural enemies of the hornworms on tobacco (Lepidoptera: Sphingidae). Annals of the Entomological Society of America 52, 741–755.

Lee, W.W., Shin, T.Y., Ko, S.H., Choi, J.B., Bae, S.M., Woo, S.D., 2012. Entomopathogenic fungus *Nomuraea rileyi* for the microbial control of *Spodoptera exigua* (Lepidoptera: Noctuidae). The Korean Journal of Microbiology 48, 284–292.

Li, M., Liu, X.M., Wang, Y., Wang, X.P., Zhou, L.L., Si, S.Y., 2014. Effect of relative humidity on entomopathogens infection and antioxidant responses of the beet armyworm, *Spodoptera exigua* (Hübner). African Entomology 22, 651–659.

Liburd, O.E., Funderburk, J.E., Olson, S.M., 2000. Effect of biological and chemical insecticides on *Spodoptera* species (Lep., Noctuidae) and marketable yields of tomatoes. Journal of Applied Entomology 124, 19–25.

Lietti, M.M.M., Botto, E., Alzogaray, R.A., 2005. Insecticide resistance in Argentine populations of *Tuta absoluta* (Meyrick) (Lepidoptera: Gelechiidae). Neotropical Entomology 34, 113–119.

Lin, S.Y., Trumble, J.T., 1985. Influence of temperature and tomato maturation on development and survival of *Keiferia lycopersicella* (Lepidoptera: Gelechiidae). Environmental Entomology 14, 855–858.

Liu, W.W., Wei, M.U., Zhu, B.Y., Liu, F., 2008. Effects of tebufenozide on the biological characteristics of beet armyworm (*Spodoptera exigua* Hübner) and its resistance selection. Agricultural Sciences in China 7, 1222–1227.

Lloyd, D.C., 1972. Some South American Parasites of the Potato Tuber Moth *Phthorimaea operculella* (Zeller) and Remarks on Those in Other Continents. Technical Bulletin of the Commonwealth Institute of Biological Control No. 15. , pp. 35–49.

Machado, R.A.R., Arce, C.C.M., Feerrieri, A.P., Baldwin, I.T., Erb, M., 2015. Jasmonate-dependent depletion of soluble sugars compromises plant resistance to *Manduca sexta*. New Phytologist 207, 91–105.

Madden, A.H., Chamberlin, F.S., 1945. Biology of the Tobacco Hornworm in the Southern Cigar-tobacco District. USDA Technical Bulletin No. 896. , pp. 1–51.

Maharjan, R., Jung, C., 2012. Biological characteristics of potato tuber moth, *Phthorimaea operculella* (Zeller) (Lepidoptera: Gelechiidae), and its management relevant to Nepal and Korea. Korean Journal of Soil Zoology 16, 25–32.

Makee, H., Saour, G., 2001. Factors influencing mating success, mating frequency, and fecundity in *Phthorimaea operculella* (Lepidoptera: Gelechiidae). Environmental Entomology 30, 31–36.

Mandal, S.K., 2012. Bio-efficacy of Cyazypyr 10% OD, a new anthranilic diamide insecticide, against the insect pests of tomato and its impact on natural enemies and crop health. Acta Phytopathologica et Entomologica Hungarica 47, 233–249.

Mandaokar, A.D., Goyal, R.K., Shukla, A., Bisaria, S., Bhalla, R., Reddy, V.S., Chaurasia, A., Sharma, R.P., Altosaar, I., Kumar, P.A., 2000. Transgenic tomato plants resistant to fruit borer (*Helicoverpa armigera* Hubner). Crop Protection 19, 307–312.

Mannion, C.M., Carpenter, J.E., Gross, H.R., 1995. Integration of inherited sterility and a parasitoid, *Archytas marmoratus* (Diptera: Tachinidae), for managing *Helicoverpa zea* (Lepidoptera: Noctuidae): acceptability and suitability of hosts. Environmental Entomology 24, 1679–1684.

Maree, J.M., Kallar, S.A., Khuhro, R.D., 1999. Relative abundance of *Spodoptera litura* F. and *Agrotis ypsilon* Rott. on cabbage. Pakistan Journal of Zoology 31, 31–34.

Martin, P.A.E., Mongeon, E.A., Gundersen-Rindal, D.E., 2008. Microbial combinatorics: a simplified approach for isolating insecticidal bacteria. Biocontrol Science and Technology 18, 291–305.

Martinelli, S., Montagna, M.A., Picinato, N.C., Silva, F.M.A., Fernandes, O.A., 2003. Efficacy of indoxacarb in the control of vegetable pests. Horticultura – Brasileira 21, 501–505.

Mascarenhas, V.J., Graves, J.B., Leonard, B.R., Burris, E., 1998. Dosage-mortality responses of third instars of beet armyworm (Lepidoptera: Noctuidae) to selected insecticides. Journal of Agricultural Entomology 15, 125–140.

Masetti, A., Butturini, A., Lanzoni, A., De Luigi, V., Burgio, G., 2015. Area-wide monitoring of potato tuberworm (*Phthorimaea operculella*) by pheromone trapping in Northern Italy: phenology, spatial distribution and relationships between catches and tuber damage. Agricultural and Forest Entomology 17, 138–145.

McLaughlin, J.R., Antonio, A.Q., Poe, S.L., Minnick, D.R., 1979. Sex pheromone biology of the adult tomato pinworm, *Keiferia lycopersicella* (Walshingham). Florida Entomologist 62, 35–41.

Medeiros, M.A., Boas, G.L.V., Vilela, N.J., Carrijo, O.A., 2009a. A preliminary survey on the biological control of South American tomato pinworm with the parasitoid *Trichogramma pretiosum* in greenhouse models. Horticultura Brasileira 27, 80–85.

Medeiros, M.A., Sujii, E.R., Morais, H.C., 2009b. Effect of plant diversification on abundance of South American tomato pinworm and predators in two cropping systems. Horticultura Brasileira 27, 300–306.

Medina, R.F., Rondon, S.I., Reyna, S.M., Dickey, A.M., 2010. Population structure of *Phthorimaea operculella* (Lepidoptera: Gelechiidae) in the United States. Environmental Entomology 39, 1037–1042.

Mensah, R., Moore, C., Watts, N., Deseo, M.A., Glennie, P., Pitt, A., 2014. Discovery and development of a new semiochemical biopesticide for cotton pest management: assessment of extract effects on the cotton pest *Helicoverpa* spp. Entomologia Experimentalis et Applicata 152, 1–15.

Mesbah, I.I., Metwally, S.M., Abo-Attia, F.A., Bassyouni, A.M., Shalaby, G.A., 2004. Utilization of biological control agents for controlling some sugar beet insect pests at Kafr El-Sheikh region. Egyptian Journal of Biological Pest Control 14, 78–82.

Ministry of Agriculture, Food and Rural Affairs, 2009. Tomato or Tobacco Hornworm. Ontario CropIPM, Ministry of Agriculture, Food and Rural Affairs, Ontario, Canada. http://cropipm.creativedonkeys.com/english/tomatoes/insects/tomato-or-tobacco-hornworm.html#advanced.

Miranda, M.M.M., Picanço, M., Zanuncio, J.C., Guedes, R.N.C., 1998. Ecological life table of *Tuta absoluta* (Meyrick) (Lepidoptera: Gelechiidae). Biological Science and Technology 8, 597–606.

Miyao, G., Klonsky, K.M., Livingston, P., 2008. Sample Costs to Produce Processing Tomatoes. University of California Cooperative Extension Publication No. TM-SV-08-1, Department of Agricultural and Resource Economics. University of California, Davis, California, USA, p. 20.

Moadeli, T., Hejazi, M.J., Golmohammadi, G., 2014. Lethal effects of pyriproxyfen, spinosad, and indoxacarb and sublethal effects of pyriproxyfen on the 1st instars larvae of beet armyworm, *Spodoptera exigua* Hübner (Lepidoptera: Noctuidae) in the laboratory. Journal of Agricultural Science and Technology 16, 1217–1227.

Mochida, O., 1973. Two important insect pests, *Spodoptera litura* (F.) and *S. littoralis* (Boisd.) (Lepidoptera: Noctuidae), on various crops – morphological discrimination of the adult, pupal, and larval stages. Applied Entomology and Zoology 8, 205–214.

Monobrullah, M., Bharti, P., Shankar, U., Gupta, R.K., 2007. Trap catches and seasonal incidence of *Spodoptera litura* on cauliflower and tomato. Plant Protection Science 15, 73–76.

Mamta, Reddy, K.R., Rajam, M.V., 2015. Targeting chitinase gene of *Helicoverpa armigera* by host-induced RNA interference confers insect resistance in tobacco and tomato. Plant Molecular Biology 90, 281–292.

Natwick, E.T., Stoddard, C.S., Zalom, F.G., Trumble, J.T., Miyao, G., Stapleton, J.J., 2014. Insects and Mites in IPM Pest Management Guidelines: Tomato. Agriculture and Natural Resources, University of California Publication No. 3470. University of California, Davis, California, USA.

Nault, B.A., Speese III, J., 2002. Major insect pests and economics of fresh-market tomato in eastern Virginia. Crop Protection 21, 359–366.

Neunzig, H.H., 1969. The Biology of the Tobacco Budworm and the Corn Earworm in North Carolina with Particular Reference to Tobacco as a Host. North Carolina Agricultural Experiment Station Technical Bulletin No. 196. North Carolina State University, Raleigh, North Carolina, USA, p. 76.

Nibouche, S., 1998. High temperature induced diapause in the cotton bollworm *Helicoverpa armigera*. Entomologia Experimentalis et Applicata 87, 271–274.

Ntalli, N., Kopiczko, A., Radtke, K., Marciniak, P., Rosinski, G., Adamski, Z., 2014. Biological activity of *Melia azedarach* extracts against *Spodoptera exigua*. Biologia 69, 1606–1614.

Nuessly, G.S., Sterling, W.L., 1994. Mortality of *Helicoverpa zea* (Lepidoptera: Noctuidae) eggs in cotton as a function of oviposition sites, predator species and desiccation. Environmental Entomology 23, 1187–1202.

Oatman, E.R., Platner, G.R., 1989. Parasites of the potato tuberworm, tomato pinworm, and other closely-related gelechiids. Proceedings of the Hawaiian Entomological Society 29, 23–30.

Oatman, E.R., Wyman, J.A., Platner, G.R., 1979. Seasonal occurrence and parasitization of the tomato pinworm on fresh market tomatoes in southern California. Environmental Entomology 8, 661–664.

Oerke, E.-C., 2006. Crop losses to pests. Journal of Agricultural Science 144, 31–43.

Ortu, S., Floris, I., 1989. Preliminary study on the control of *Phthorimaea operculella* (Zeller) (Lepidoptera: Gelechiidae) on potato crops in Sardinia. Difesa Delle Piante 12, 81–88.

Osorio, A., Martínez, A.M., Schneider, M.I., Díaz, O., Corrales, J.L., Avilés, M.C., Smagghe, G., Pineda, S., 2008. Monitoring of beet armyworm resistance to spinosad and methoxyfenozide in Mexico. Pest Management Science 64, 1001–1007.

Parra, J.R.P., Zucchi, R.A., 2004. *Trichogramma* in Brazil: feasibility of use after twenty years of research. Neotropical Entomology 33, 271–281.

Pastrana, J.A., 2004. Los lepidopteros Argentinos – sus plantas hospederas y otros sustratos alimenticios. Sociedad Entomologica Argentina 334.

Patch, H.M., Velarde, R.A., Walden, K.K.O., Robertson, H.M., 2009. A candidate pheromone receptor and two odorant receptors of the hawkmoth *Manduca sexta*. Chemical Senses 34 (4), 305–316.

Pearson, A.C., 1982. Biology, Population Dynamics, and Pest Status of the Beet Armyworm (*Spodoptera exigua*) in the Imperial Valley of California (Ph.D. thesis). University of California, Riverside, California, USA, p. 564.

Pearson, E.O., 1958. Insect Pests of Cotton in Tropical Africa. Commonwealth Institute of Entomology, London, UK, p. 355.

Pena, J.E., Waddill, V., 1983. Larval and egg parasitism of *Keiferia lycopersicella* (Walsingham) (Lepidoptera: Gelechiidae) in southern Florida tomato fields. Environmental Entomology 12, 1322–1326.

Pena, J.E., Waddill, V.H., 1985. Influence of postharvest cultural practices on tomato pinworm population in southern Florida. Proceedings of the Florida State Horticultural Society 98, 251–254.

Pfannenstiel, R.S., Yeargan, K.V., 2002. Identification and diel activity patterns of predators attacking *Helicoverpa zea* (Lepidoptera: Noctuidae) eggs in soybean and sweet corn. Environmental Entomology 31, 232–241.

Poe, S.L., 1973. The Tomato Pinworm in Florida. State Department of Agriculture and Consumer Services Entomology, p. 5. AREC Research Report GC1973-2.

Poe, S.L., Crill, J.P., Everett, P.H., 1975. Tomato pinworm population management in semitropical agriculture. Proceedings of the Florida State Horticulture Society 88, 160–165.

Pucci, C., Spanedda, A.F., Minutoli, E., 2003. Field study of parasitism caused by endemic parasitoids and by the exotic parasitoid *Copidosoma koehleri* on *Phthorimaea operculella* in central Italy. Bulletin of Insectology 56, 221–224.

Qayyum, M.A., Wakil, W., Arif, M.J., Sahi, S.T., 2015b. *Bacillus thuringiensis* and nuclear polyhedrosis virus for the enhanced bio-control of *Helicoverpa armigera*. International Journal of Agriculture and Biology 17, 1043–1048.

Qayyum, M.A., Wakil, W., Arif, M.J., Sahi, S.T., Dunlap, C.A., 2015c. Infection of *Helicoverpa armigera* by endophytic *Beauveria bassiana* colonizing tomato plants. Biological Control 90, 200–207.

Qayyum, M.A., Wakil, W., Arif, M.J., Sahi, S.T., Saeed, N.A., Russell, D.A., 2015a. Multiple resistances against formulated organophosphates, pyrethroids, and newer-chemistry insecticides in populations of *Helicoverpa armigera* (Lepidoptera: Noctuidae) from Pakistan. Journal of Economic Entomology 108, 286–293.

Quaintance, A.L., Brues, C.T., 1905. The cotton bollworm. United States Department of Agriculture, Bureau of Entomology Bulletin 50, 1–112.

Raman, K.V., 1988. Control of potato tuber moth *Phthorimaea operculella* with sex pheromones in Peru. Agriculture, Ecosystems and Environment 21, 85–99.

Rao, G., Wightman, J., Ranga Rao, D., 1993. World review of the natural enemies and diseases of *Spodoptera litura* (F.) (Lepidoptera: Noctuidae). Insect Science and its Application 14, 273–284.

Raulston, J.R., Pair, S.D., Loera, J., Cabanillas, H.E., 1992. Prepupal and pupal parasitism of *Helicoverpa zea* and *Spodoptera frugiperda* (Lepidoptera: Noctuidae) by *Steinernema* sp. in cornfields in the lower Rio Grande valley. Journal of Economic Entomology 85, 1666–1670.

Razmjou, J., Naseri, B., Hemati, S.A.J., 2014. Comparative performance of the cotton bollworm, *Helicoverpa armigera* (Hübner) (Lepidoptera: Noctuidae) on various host plants. Journal of Pest Science 87, 29–37.

Reitz, S.R., 1996. Development of *Eucelatoria bryani* and *Eucelatoria rubentis* (Diptera: Tachinidae) in different instars of *Helicoverpa zea* (Lepidoptera: Noctuidae). Annals of Entomological Society of America 89, 81–87.

Rhino, B., Grechi, I., Marliac, G., Trebeau, M., Thibaut, C., Ratnadass, A., 2014. Corn as trap crop to control *Helicoverpa zea* in tomato fields: importance of phenological synchronization and choice of cultivar. International Journal of Pest Management 60, 73–81.

Riley, D.G., 2006. Beet armyworm threshold development in tomato. Agricultural Profitability and Sustainability. College of Agricultural and Environmental Sciences, University of Georgia, Athens, Georgia, USA. http://www.caes.uga.edu/Applications/ImpactStatements/index.cfm?referenceInterface=IMPACT_STATEMENT&subInterface=detail_main&PK_ID=462.

Rivera, M.J., 2011. The Potato Tuberworm, *Phthorimaea operculella* (Zeller), in the Tobacco, *Nicotiana tabacum* L., Agroecosystem: Seasonal Biology and Larval Behavior (M.S. thesis). North Carolina State University, Raleigh, North Carolina, USA, p. 93.

Roelofs, W.L., Kochansky, J.P., Carde, R.T., Kennedy, G.G., Henrick, C.A., Labovitz, J.N., Corbin, V.L., 1975. Sex pheromone of the potato tuberworm moth, *Phthorimaea operculella*. Life Sciences 17, 699–705.

Romeis, J., Shanower, T.G., 1996. Arthropod natural enemies of *Helicoverpa armigera* (Hübner) (Lepidoptera: Noctuidae) in India. Biocontrol Science and Technology 6, 481–508.

Rondon, S.I., 2010. The potato tuberworm: a literature review of its biology, ecology, and control. American Journal of Potato Research 87, 149–166.

Rondon, S.I., DeBano, S.J., Clough, G.H., Hamm, P.B., Jensen, A., Schreiber, A., Alvarez, J.M., Thornton, M., Barbour, J., Doğramaci, M., 2007. Biology and Management of the Potato Tuberworm in the Pacific Northwest. A Pacific Northwest Extension Publication No. PNW 594. Oregon State University, Washington State University and the University of Idaho, USA, p. 8.

Rothschild, G.H.L., 1986. The potato moth – an adaptable pest of short term cropping systems. In: Kitching, R.L. (Ed.), The Ecology of Exotic Animals and Plants. Wiley, Brisbane, Australia, pp. 144–162.

Ruberson, J.R., Herzog, G.A., Lambert, W.R., Lewis, W.J., 1994. Management of the beet armyworm (Lepidoptera: Noctuidae) in cotton: role of natural enemies. Florida Entomologist 77, 440–453.

Ruberson, J.R., Herzog, G.A., Lewis, W.J., 1993. Parasitism of the beet armyworm, *Spodoptera exigua*, in South Georgia cotton. In: Proceedings, Beltwide Cotton Conference, National Cotton Council, Memphis, Tennessee, USA, pp. 993–997.

Ruiz de Escudero, I., Banyuls, N., Bel, Y., Maeztu, M., Escriche, B., Muñoz, D., Caballero, P., Ferré, J., 2014. A screening of five *Bacillus thuringiensis* Vip3A proteins for their activity against lepidopteran pests. Journal of Invertebrate Pathology 117, 51–55.

Sabbour, M., Ismail, I.A., 2002. The combined effect of microbial control agents and plant extracts against potato tuber moth *Phthorimaea operculella* Zeller. Bulletin of the National Research Centre (Cairo) 27, 459–467.

Salas, J., 2001. Insectos Plagas del Tomate. Manejo Integrado. Maracay, Venezuela.

Salas, J., 2004. Capture of *Tuta absoluta* (Lepidoptera: Gelechiidae) in traps baited with its sex pheromone. Revista Colombiana de Entomologia 30, 75–78.

Salazar, E.R., Araya, J.E., 1997. Detección de resistencia a insecticidas en la polilla del tomate. Simiente 67, 8–22.

Sandstrom, M.A., Changnon, D., Flood, B.R., 2007. Improving our understanding of *Helicoverpa zea* migration in the Midwest: assessment of source populations. Online Plant Health Progress. http://dx.doi.org/10.1094/PHP-2007-0719-08-RV.

Sankaran, T., Girling, D., 1980. The current status of biological control of the potato tuber moth. Biocontrol News and Information 1, 207–211.

Saour, G., Makee, H., 2004. Susceptibility of potato tuber moth (Lepidoptera: Gelechiidae) to postharvest gamma irradiation. Journal of Economic Entomology 97, 711–714.

Schuster, D.R., 1989. Tomato Pinworm. Institute of Food and Agricultural Science Publication No. EENY-74 University of Florida, Gainesville, Florida, USA. http://entnemdept.ufl.edu/creatures/veg/tomato/tomato_pinworm.htm.

Seal, D.R., McCord, E., 1996. Management of the tomato pineworm, *Keiferia lycopersicella* (Walsingham) (Lepidoptera: Gelechiidae) in South Florida. Proceeding of the Florida State Horticultural Society 109, 196–200.

Seplyarsky, V., Weiss, M., Haberman, A., 2010. *Tuta absoluta* Povolny (Lepidoptera: Gelechiidae), a new invasive species in Israel. Phytoparasitica 38, 445–446.

Seth, R.K., Khan, Z., Rao, D.K., Zarin, M., 2016. Flight activity and mating behavior of irradiated *Spodoptera litura* (Lepidoptera: Noctuidae) males and their F1 progeny for use of inherited sterility in pest management approaches. Florida Entomologist 99, 119–130.

Sewify, G.H., Abol-Ela, S., Eldin, M.S., 2000. Effects of the entomopathogenic fungus *Metarhizium anisopliae* and granulosis virus (GV) combinations on the potato tuber moth *Phthorimaea operculella* (Zeller) (Lepidoptera: Gelechiidae). Bulletin of Faculty of Agriculture – University of Cairo 51, 95–106.

Shapiro, M., Argauer, R., 2001. Relative effectiveness of selected stilbene optical brighteners as enhancers of the beet armyworm (Lepidoptera: Noctuidae) nuclear polyhedrosis virus. Journal of Economic Entomology 94, 339–343.

Shelton, A.M., Fuchs, M., Shotkoski, F.A., 2008. Transgenic vegetables and fruits for control of insects and insect-vectored pathogens. In: Romeis, J., Shelton, A.M., Kennedy, G.G. (Eds.). Integration of Insect-resistant Genetically Modified Crops within IPM Programs, Progress in Biological Control. Integration of Insect-resistant Genetically Modified Crops within IPM Programs, Progress in Biological Control, vol. 5. Springer Science+Business Media, The Netherlands, pp. 249–272.

Shimada, K., Natsuhara, K.A., Oomori, Y., Miyata, T., 2005. Permethrin resistance mechanisms in the beet armyworm (Spodoptera exigua (Hubner). Journal of Pesticide Science 30, 214–219.

Shimizu, K., Fujisaki, K., 2002. Sexual differences in diapause induction of the cotton bollworm, *Helicoverpa armigera* (Hb.) (Lepidoptera: Noctuidae). Applied Entomology and Zoology 37, 527–533.

Shorey, H.H., Gerber, R.G., 1996. Disruption of pheromone communication through the use of puffers for control of beet armyworm (Lepidoptera: Noctuidae) in tomatoes. Environmental Entomology 25, 1401–1405.

Siebert, M.W., Nolting, S.P., Hendrix, W., Dhavala, S., Craig, C., Leonard, B.R., Stewart, S.D., All, J., Musser, F.R., Buntin, G.D., Samuel, L., 2012. Evaluation of corn hybrids expressing Cry1F, cry1A.105, Cry2Ab2, Cry34Ab1/Cry35Ab1, and Cry3Bb1 against southern United States insect pests. Journal of Economic Entomology 105, 1825–1834.

Simmons, A.T., Nicol, H.I., Gurr, G.M., 2006. Resistance of wild *Lycopersicon* species to the potato moth, *Phthorimaea operculella* (Zeller) (Lepidoptera: Gelechiidae). Australian Journal of Entomology 45, 81–86.

Sims, S.R., Greenplate, J.T., Stone, T.B., Caprio, M., Gould, F., 1996. Monitoring strategies for early detection of Lepidoptera resistance to *Bacillus thuringiensis* insecticidal proteins. In: Brown, T.M. (Ed.), Molecular Genetics and Ecology of Pesticide Resistance. American Chemical Society Symposium Series, Washington, DC, USA, pp. 229–242.

Siqueira, H.A.A., Guedes, R.N.C., Fragoso, D.B., Magalhaes, L.C., 2001. Abamectin resistance and synergism in Brazilian populations of *Tuta absoluta* (Meyrick) (Lepidoptera: Gelechiidae). International Journal of Pest Management 47, 247–251.

Siqueira, H.A.A., Guedes, R.N.C., Picanco, M.C., 2000. Cartap resistance and synergism in populations of *Tuta absoluta* (Lepidoptera: Gelechiidae). Journal of Applied Entomology 124, 233–238.

Sparks Jr., A.N., 2014. Insect Management: Commercial Tomato Production Handbook. Bulletin 1312, reviewed January, 2014. University of Georgia Cooperative Extension, College of Agricultural and Environmental Sciences, University of Georgia, Athens, Georgia, USA, pp. 31–35.

Sporleder, M., Simon, R., Juarez, H., Kroschel, J., 2008. Regional and seasonal forecasting of the potato tuber moth using a temperature-driven phenology model linked with geographic information systems. In: Kroschel, J., Lacey, L. (Eds.), Integrated Pest Management for the Potato Tuber Moth, *Phthorimaea operculella* Zeller – A Potato Pest of Global Proportion. Margraf Publishers, Weikersheim, Germany, pp. 15–30. Tropical Agriculture 20, Advances in Crop Research.

Srinivasan, K., Krishna Moorthy, P.N., Raviprasad, T.N., 1994. African marigold as a trap crop for the management of the fruit borer, *Helicoverpa armigera* on tomato. International Journal of Pest Management 40, 56–63.

Srinivasan, R., Su, F.C., Huang, C.C., 2013. Oviposition dynamics and larval development of *Helicoverpa armigera* on a highly preferred unsuitable host plant, *Solanum viarum*. Entomologia Experimentalis et Applicata 147, 217–224.

Srinivasan, R., Su, F.C., Mei-Ying, L., Hsu, Y., 2010. Inseect and mite pests on tomato: identification and management. In: Srinivasan, R. (Ed.), Safer Tomato Production Techniques, a Field Guide for Soil Fertility and Pest Management. AVRDC – The World Vegetable Center, Thailand, pp. 23–59.

Stanley, J.N., Baqir, H.A., McLaren, T.I., 2014. Effect on larval growth of adding finely ground silicon-bearing minerals (wollastonite or olivine) to artificial diets for *Helicoverpa* spp. (Lepidoptera: Noctuidae). Austral Entomology 53, 436–443.

Stout, M.J., Workman, K.V., Bostock, R.M., Duffey, S.S., 1998. Specificity of induced resistance in the tomato, *Lycopersicon esculentum*. Oecologia 113, 74–81.

Subchev, M., Toshova, T.B., Atanasova, D.I., Petrova, V.D., Miklós, T., 2013. Seasonal flight of the potato tuber moth, *Phthorimaea operculella* (Zeller) (Lepidoptera: Gelechiidae) in three regions in Bulgaria established by pheromone traps. Acta Phytopathologica et Entomologica Hungarica 48, 75–86.

Suenaga, H., Tanaka, A., 1997. Occurrence of beet armyworm, *Spodoptera exigua* (Hubner) (Lepidoptera: Noctuidae) on young growing stage of garden pea, *Pisum sativum* L. Japanese Journal of Applied Entomology and Zoology 41, 17–25.

Sullivan, M., Molet, T., 2014. CPHST Pest Datasheet for *Helicoverpa armegera*. United States Department of Agriculture-APHIS-PPQ-CPHST, p. 17. revised April 2014 http://www.aphis.usda.gov/plant_health/plant_pest_info/owb/downloads/owb-factsheet.pdf.

Sun, Y.X., Li, Z.Y., Gui, F.R., Yan, N.S., Chen, B., Xu, R.Q., He, S.Q., 2004. Susceptibility of *Phthorimaea operculella* (Zeller) larvae to eight isolates of *Beauveria*. Southwest China Journal of Agricultural Sciences 17, 627–629.

Symington, C.A., 2003. Lethal and sublethal effects of pesticides on the potato tuber moth, *Phthorimaea operculella* (Zeller) (Lepidoptera: Gelechiidae) and its parasitoid *Orgilus lepidus* Muesebeck (Hymenoptera: Braconidae). Crop Protection 22, 513–519.

Taco, E., Quispe, R., Bobadilla, D., Vargas, H., Jimenez, M., Morales, A., 1998. Resultados preliminares de un ensayo de control biológico de la polilla del tomate, *Tuta absoluta* (Meyrick), en el valle de Azapa. In: Sociedad. IX Congreso Latino-Americano de Horticultura, Santiago, Chile, pp. 54–56.

Talekar, N.S., Opena, R.T., Hanson, P., 2006. *Helicoverpa armigera* management: a review of AVRDC's research on host plant resistance in tomato. Crop Protection 25 (5), 461–467.

Taylor, J.E., Riley, D.G., 2008. Artificial infestations of beet armyworm, *Spodoptera exigua* (Lepidoptera: Noctuidae), used to estimate an economic injury level in tomato. Crop Protection 27, 268–274.

Tian, D., Peiffer, M., De Moraes, C.M., Felton, G.W., 2014. Roles of ethylene and jasmonic acid in systemic induced defense in tomato (*Solanum lycopersicum*) against *Helicoverpa zea*. Planta 9, 577–589.

Tipping, P.W., Holko, C.A., Bean, R.A., 2005. *Helicoverpa zea* (Lepidoptera: Noctuidae) dynamics and parasitism in Maryland soybeans. Florida Entomologist 88, 55–60.

Tojo, S., Mishima, H., Kamiwada, H., Ngakan, P.O., Chang, K.S., 2008. Variations in the occurrence patterns of male moths of the common cutworm, *Spodoptera litura* (Lepidoptera: Noctuidae) among Southeastern Asian countries, as detected by sex pheromone trapping. Applied Entomology and Zoology 43, 569–576.

Torres-Villa, L.M., Rodrigues, M., Lacasa, A., 1996. An unusual behaviour in *Helicoverpa armigera* Hubner (Lepidoptera: Noctuidae): pupation inside tomato fruits. Journal of Insect Behaviour 9, 981–984.

Traynier, R.M., 1975. Field and laboratory experiments on the site of oviposition by the potato moth *Phthorimaea operculella* (Zell.) (Lepidoptera, Gelechiidae). Bulletin of Entomological Research 65, 391–398.

Trivedi, T.P., Rajagopal, D., 1992. Distribution, biology, ecology and management of potato tuber moth, *Phthorimaea operculella* (Zeller) (Lepidoptera: Gelechiidae): a review. Tropical Pest Management 38, 279–285.

Trumble, J.T., Alvarado-Rodriguez, B., 1993. Development and economic-evaluation of an IPM program for fresh-market tomato production in Mexico. Agriculture Ecosystem and Environment 43, 267–284.

Tumlinson, J.H., Brennan, M.M., Doolittle, R.E., Mitchell, E.R., Brabham, A., Mazomenos, B.E., Baumhover, A.H., Jackson, D.M., 1989. Identification of a pheromone blend attractive to *Manduca sexta* (L.) males in a wind tunnel. Archives of Insect Biochemistry and Physiology 10, 255–271.

Twine, P.H., 1971. Cannibalistic behaviour of *Heliothis armigera* (Hübner) (Lepidoptera: Noctuidae). Queensland Journal of Agricultural and Animal Science 28, 153–157.

UC IPM, 2016. UC Management Guidelines for Tomato Pinworm on Tomato. University of California. http://ipm.ucanr.edu/PMG/r783300411.html.

Urbaneja, A., Desneux, N., Gabarra, R., Arnó, J., González-Cabrera, J., Mafra Neto, A., Stoltman, L., Pinto, A.de.S., Parra, J.R.P., 2013. Biology, ecology and management of the South American tomato pinworm, *Tuta absoluta*. In: Peña, J.E. (Ed.), Potential Invasive Pests of Agricultural Crops. CABI, Wallingford, UK, pp. 98–125.

Urbaneja, A., Montón, H., Vanaclocha, P., Mollá, O., Beitia, F., 2008. La polilla del tomate, *Tuta absoluta*, una nueva presa para los míridos *Nesidiocoris tenuis* y *Macrolophus pygmaeus*. Agricola Vergel 320, 361–367.

USDA, 2014. Crop Protection and Quarantine, Action Plan [2015–2020]. USDA Agricultural Research Service, National Program 304. USDA, ARS, Washington, DC, USA, p. 28.

USDA-APHIS, 2014. *Helicoverpa armegera*, Old World Bollworm. Technical Working Group Meeting, 18 November, 2014, Portland, Oregon, USA, p. 6.

USDA-APHIS, 2016. Old World Bollworm (Helicoverpa armigera) in Florida Deemed an Isolated Regulatory Incident. https://www.aphis.usda.gov/plant_health/plant_pest_info/owb/downloads/DA-2016-40.pdf.

van den Berg, H., Waage, J.K., Cock, M.J.W., 1988. Natural Enemies of *Helicoverpa armigera* in Africa: A Review. CAB International Institute of Biological Control, Chameleon Press Limited, London, UK, p. 81.

van Steenwyk, R.A., Oatman, E.R., 1983. Mating disruption of tomato pinworm (Lepidoptera: Gelechiidae) as measured by pheromone trap, foliage, and fruit infestation. Journal of Economic Entomology 76, 80–84.

van Steenwyk, R.A., Oatman, E.R., Wyman, J.A., 1983. Density treatment level for tomato pinworm (Lepidoptera: Gelechiidae) based pheromone trap catches. Journal of Economic Entomology 76, 440–445.

Vaneva-Gancheva, T., Dimitrov, Y., 2013. Chemical control of the potato tuber moth *Phthorimaea operculella* (Zeller) on tobacco. Bulgarian Journal of Agricultural Science 19, 1003–1008.

VanSickle, J.J., Smith, S., McAvoy, E., 2009. Production Budget for Tomatoes in the Manatee/Ruskin Area of Florida. Publication No. FE817, Food and Resource Economics Department, Florida Cooperative Extension Service. University of Florida, Gainesville, Florida, USA.

Vargas, H., 1970. Observaciones sobre la biología y enemigos naturales de la polilla del tomate, *Gnorismoschema absoluta* (Meyrick) (Lep. Gelechiidae). IDESIA 1, 75–110.

Vargas, R., Nishida, T., 1980. Life table of the corn earworm, *Heliothis zea* (Boddie), in sweet corn in Hawaii. Proceedings of the Hawaiian Entomological Society 23, 301–307.

Vashisth, S., Chandel, Y.S., 2013. Morphotrics of *Spodoptera litura* on tomato. Indian Journal of Plant Protection 41, 175–177.

Venette, R.C., Davis, E.E., Zaspel, J., Heisler, H., Larson, M., 2003a. Mini Risk Assessment: Old World Bollworm, *Helicoverpa Armigera* Hübner (Lepidoptera: Noctuidae). University of Minnesota, Saint Paul, Minnesota, USA. https://www.aphis.usda.gov/plant_health/plant_pest_info/owb/downloads/mini-risk-assessment-harmigerapra.pdf.

Venette, R.C., Davis, E.E., Zaspel, J., Heisler, H., Larson, M., 2003b. Mini Risk Assessment: Rice Cutworm, *Spodoptera litura* Fabricius (Lepidoptera: Noctuidae). University of Minnesota, Saint Paul, Minnesota, USA. http://www.aphis.usda.gov/plant_health/plant_pest_info/pest_detection/downloads/pra/sliturapra.pdf.

Wakil, W., Ghazanfar, M.U., Nasir, F., Qayyum, M.A., Tahir, M., 2012. Insecticidal efficacy of *Azadirachta indica*, nucleopolyhedrovirus and chlorantraniliprole single or combined against field populations of *Helicoverpa armigera* Hübner (Lepidoptera: Noctuidae). Chilean Journal of Agricultural Research 72, 52–62.

Wakil, W., Ghazanfar, M.U., Riasat, T., Kwon, Y.J., Qayyum, M.A., Yasin, M., 2013. Effect of interaction among *Metarhizium anisopliae*, *Bacillus thuringiensis* and chlorantraniliprole on the mortality and pupation of geographically distinct *Helicoverpa armigera* populations. Phytoparasitica 41, 221–234.

Walgenbach, J.F., Estes, E.A., 1992. Economics of insecticide use on staked tomatoes in western North Carolina. Journal of Economic Entomology 85, 888–894.

Walgenbach, J.F., Shoemaker, P.B., Sorensen, K.A., 1989. Timing pesticide applications for control of *Heliothis zea* (Boddie) (Lepidoptera: Noctuidae) *Alternaria solani* (Ell. and G. Martin) Sor., and *Phytophthora infestans* (Mont.) DeBary, on tomatoes in western North Carolina. Journal of Agricultural Entomology 6, 159–168.

Wang, W., Mo, J., Cheng, J., Zhuang, P., Tang, Z., 2002. Selection and characterization of spinosad resistance in *Spodoptera exigua* (Hübner) (Lepidoptera: Noctuidae). Pesticide Biochemistry and Physiology 84, 180–187.

Washburn, J.O., Wong, J.F., Volkman, L.E., 2001. Comparative pathogenesis of *Helicoverpa zea* S nucleopolyhedrovirus in noctuid larvae. Journal of General Virology 82, 1777–1784.

Weinberg, H.L., Lange, W.H., 1980. Developmental rate and lower temperature threshold of the tomato pinworm. Environmental Entomology 9, 245–246.

Weinzierl, R., 2007. Alternatives to pyrethroids for managing corn earworm in sweet corn, seed corn, tomatoes and peppers. Online Plant Health Progress. http://dx.doi.org/10.1094/PHP-2007-0719-05-RV.

Wellik, M.J., Slosser, J.E., Kirby, R.D., 1979. Evaluation of procedures for sampling *Heliothis zea* and *Keiferia lycopersicella* on tomatoes. Journal of Economic Entomology 72, 777–780.

Westbrook, J.K., 2008. Noctuid migration in Texas within the nocturnal aeroecological boundary layer. Integrative and Comparative Biology 48, 99–106.

Westbrook, J.K., Lopez Jr., J.D., 2010. Long-distance migration in *Helicoverpa zea*: what we know and need to know. Southwestern Entomologist 35, 355–360.

Whyard, S., Singh, A.D., Wong, S., 2009. Ingested double-stranded RNAs can act as species-specific insecticides. Insect Biochemistry and Molecular Biology 39, 824–832.

Williamson, J., 2014. Tomato Insect Pests, Home and Garden Information Center. Clemson University Cooperative Extension Service Publication No. HGIC 2218 Clemson University, South Carolina, USA. http://www.clemson.edu/extension/hgic/pests/pdf/hgic2218.pdf.

Wilson, J.W., 1934. The asparagus caterpillar: its life history and control. Florida Agricultural Experiment Station Bulletin 271, 1–26.

Wink, M., Theile, V., 2002. Alkaloid tolerance in *Manduca sexta* and phylogenetically related sphingids (Lepidoptera: Sphingidae). Chemoecology 12, 29–46.

Wraight, S.P., Ramos, M.E., Avery, P.B., Jaronski, S.T., Vandenberg, J.D., 2010. Comparative virulence of *Beauveria bassiana* isolates against lepidopteran pests of vegetable crops. Journal of Invertebrate Pathology 103, 186–199.

Wraight, S.P., Sporleder, M., Poprawski, T.J., Lacey, L.A., 2007. Application and evaluation of entomopathogens in potato. In: Lacey, L.A., Kaya, H.K. (Eds.), Field Manual of Techniques in Invertebrate Pathology, second ed. Springer, Dordrecht, The Netherlands, pp. 329–359.

Wyenandt, A., Hamilton, G., Kuhar, T., Sanchez, E., VanGessell, M., 2016. Mid-Atlantic Commercial Vegetable Production Recommendations, 2016–2017. Rutgers New Jersey Agricultural Experiment Station Cooperative Extension Publication No. E001. The State University of New Jersey, New Brunswick, New Jersey, USA, p. 412.

Yadav, S.S., Singh, B., Kumar, A., Vir, S., 2015. Evaluation of effective doses of neem based biopesticides against different stages of *Heliothis armigera* (Hubner). Journal of Global Biosciences 4, 2901–2910.

Yamamoto, R.T., 1968. Mass rearing of the tobacco hornworm. I. Egg production. Journal of Economic Entomology 61, 170–174.

Yardim, E.N., Arancon, Q.A., Edwards, C.A., Oliver, T.J., Byrne, R.J., 2006. Suppression of tomato hornworm (*Manduca quinquemaculata*) and cucumber beetles (*Acalymma vittatum*) and *Diabrotica undecimpunctata*) populations and damage by vermicomposts. Pedobiologia 50, 23–29.

Yee, W.L., Toscano, N.C., 1998. Laboratory evaluations of synthetic and natural insecticides on beet armyworm (Lepidoptera: Noctuidae) damage and survival on lettuce. Journal of Economic Entomology 91, 56–63.

Yoshiyasu, Y., Yamagishi, M., Katayama, J., Oyake, K., 1995. Control of the beet army worm *Spodoptera exigua* (Hübner), on Welsh onion by synthetic sex pheromone in Yodo district Kyoto. Scientific Reports of the Kyoto Prefectural University – Agriculture 47, 1–8.

Zaki, F.N., Abdel-Raheem, M.A., 2010. Use of entomopathogenic fungi and insecticide against some insect pests attacking peanuts and sugarbeet in Egypt. Archives of Phytopathology and Plant Protection 43, 1819–1828.

Zalom, F.G., Jones, A., 1994. Insect fragments in processed tomatoes. Journal of Economic Entomology 87, 181–186.

Zalom, F.G., Wilson, L.T., Hoffmann, M.P., 1986. Impact of feeding by tomato fruitworm, *Heliothis zea* (Boddie) (Lepidoptera: Noctuidae), and beet armyworm, *Spodoptera exigua* (Hubner) (Lepidoptera: Noctuidae), on processing tomato fruit quality. Journal of Economic Entomology 79, 822–826.

Zeddam, J.L., Pollet, A., Mangoendiharjo, S., Ramadhan, T.H., López-Ferber, M., 1999. Occurrence and virulence of a granulosis virus in *Phthorimaea operculella* (Lep., Gelechiidae) populations in Indonesia. Journal of Invertebrate Pathology 74, 48–54.

Zehnder, G.W., Sikora, E.J., Goodman, W.R., Hollingsworth, M.H., 1994. Insect scouting reduces insecticide use on tomatoes. Alabama Agricultural Experiment Station. Highlights of Agricultural Research 41 (3), 5.

Zheng, X.L., Cong, X.P., Wang, X.P., Lei, C.L., 2011a. Pupation behaviour, depth, and site of *Spodoptera exigua*. Bulletin of Insectology 64, 209–214.

Zheng, X.L., Cong, X.P., Wang, X.P., Lei, C.L., 2011b. A review of geographic distribution, overwintering and migration in *Spodoptera exigua* Hübner (Lepidoptera: Noctuidae). Journal of the Entomological Research Society 13, 39–48.

Zhou, X., Coll, M., Applebaum, S., 2000. Effect of temperature and photoperiod on juvenile hormone biosynthesis and sexual maturation in the cotton bollworm, *Helicoverpa armigera*: implications for life history traits. Insect Biochemistry and Molecular Biology 30, 863–868.

Zhou, Z.S., Chen, Z.P., Xu, Z.F., 2010. Potential of trap crops for integrated management of the tropical armyworm, *Spodoptera litura* in tobacco. Journal of Insect Science 10, 117.

Chapter 7

Psyllids: Biology, Ecology, and Management

Sean M. Prager[1], John T. Trumble[2]

[1]University of Saskatchewan, Saskatoon, SK, Canada; [2]University of California, Riverside, CA, United States

1. INTRODUCTION

The tomato/potato psyllid, *Bactericera cockerelli* (Šulc) has been recognized as a pest of solanaceous crop plants for nearly a century. In most locations, however, it was treated as a minor pest and only a concern due to occasional outbreaks. Moreover, the psyllid often was considered a threat to potatoes (*Solanum tuberosum* L.) and not tomatoes (*Solanum lycopersicum* L.). Recently, a range expansion accompanied by a change in the psyllid's genetic structure and the introduction of a new pathogen has changed this attitude.

The first report of *B. cockerelli* as a pest appeared in 1915. In the California (United States) State Horticulture Bulletin, Compere (1915) reported the psyllid (then *Paratrioza cockerelli* Šulc) as a pest damaging False Jerusalem Cherry (*Solanum capsicastrum* L.), an ornamental plant grown in northern California. This was followed by a report in which Essig (1917) identified the tomato psyllid as a pest of economic importance, but noted that severe outbreaks were rare. However, in the early 1920s psyllids were associated with a plant condition known as "psyllid yellows" (Binkley, 1929), and at this point the psyllid became a greater concern.

Psyllid yellows was first reported in Utah (United States) (Carter, 1939) before appearing in various Rocky Mountain States and California in the autumn of 1928 (Richards, 1928; Shapovalov, 1929). Following the reports in the 1920s, there were numerous, although sporadic outbreaks of psyllid yellows, including an extremely severe outbreak in 1938 (Daniels, 1934; Jensen, 1938). Most of these early outbreaks were associated with potatoes. Studies by Binkley (1929) indicated that eggs laid by infected psyllids could hatch and the nymphs would feed on plants, but often did not generate disease symptoms. Nymphal feeding, however, when observed in large numbers and for prolonged periods did result in symptom development. Binkley (1929) also suggested that transfer could occur between diseased and healthy tomato and potato plants. Therefore, it was assumed that the psyllid released a toxin during feeding, and that transferable toxin was responsible for the psyllid yellows.

In 1994, a potato defect that afflicted both leaves and fresh tubers of potato plants was detected in Mexico and Central America (Secor and Rivera-Varas, 2004; De Boer et al., 2007; Gudmestad and Secor, 2007). This disorder was given the name "Zebra Chip" and by 2000 it was observed in potato fields in Texas (Munyaneza et al., 2007a). By the mid- to late 2000s, zebra chip was causing substantial losses to potato growers, and had been identified in various other western potato-growing regions (Munyaneza et al., 2007b). In 2004, high densities of *B. cockerelli* were reported on fresh-market tomatoes in both California and Baja, Mexico (Liu and Trumble, 2004; Liu et al., 2006b). This was followed in 2006 by reports of tomato plants with symptoms including: darkened and/or chlorotic veins, curled leaves, and shortened internodes in Arizona greenhouse facilities. Testing of these plants by polymerase chain reaction (PCR) revealed the presence of a species of *Candidatus* Liberibacter Jagoueix et al., but no presence of common tomato viruses. In the same greenhouses, the most common insects identified were individuals of *B. cockerelli* (Crosslin et al., 2010). Also in 2006, *B. cockerelli* were identified from the area around Auckland, New Zealand (Teulon et al., 2009). In 2008, tomatoes from this area were found exhibiting psyllid yellows symptoms (Liefting et al., 2009). The losses from those infestations approached one million New Zealand dollars. By 2009, *B. cockerelli* was identified on both islands of New Zealand (Liefting et al., 2009).

In 2008, tomato plants in California were found to be suffering from infection with a then unknown disease that was associated with also an unknown bacterium. The bacterium was eventually identified as *C.* Liberibacter psyllaurous (née *solanacearum*) (Lso). Lso subsequently was detected in potatoes, tomatoes, cape gooseberry (*Physalis peruviana* L.), and tamarillo (*Solanum betaceum* Cavanilles) (Liefting et al., 2008a,b). Critically, the bacterium was also detected in the psyllid *B. cockerelli* (Hansen et al., 2008). At about the same time that Lso was identified in the United States, it was also detected in infected plants and *B. cockerelli* in New Zealand (Liefting et al., 2008a).

It is now accepted that the psyllid *B. cockerelli* is associated with two different plant disorders. The first is "psyllid yellows," and the second is zebra chip in potatoes and "vein greening disease" in other solanaceous vegetable crops. Lso subsequently was shown to survive in many plants including: eggplant, bell peppers, tamarillo, and various nightshade species (Butler and Trumble, 2012c). Multiple experimental studies have also demonstrated that Lso is transmitted by *B. cockerelli* (Munyaneza et al., 2007a,b; Hansen et al., 2008) which is the only known vector of the bacteria in the United States and New Zealand. Two other haplotypes of Lso also exist in Europe, but are not found in solanaceous plants. They are also transmitted by different species of psyllid (Nelson et al., 2012; Nissinen et al., 2014; Teresani et al., 2014).

Initially, major infestations with *B. cockerelli* in North America were relatively rare (Liu et al., 2006a). However, the occurrence of psyllid yellows and the more recent vein greening disease, in addition to a more permanent presence of the psyllid in many locations, have combined to make *B. cockerelli* a serious pest of tomatoes in the United States, Mexico, and New Zealand. Losses have also been reported in Canada and several Central American countries (Butler and Trumble, 2012c; Munyaneza, 2012).

2. TAXONOMY

B. cockerelli, alternatively referred to as the potato and tomato psyllid, was first described in 1909 from a specimen collected on bell peppers (*Capsicum* sp.) in Colorado (Šulc, 1909). This early description by Šulc (1909) assigned the psyllid the name *Trioza cockerelli*. Šulc's descriptions were followed by the first descriptions of the insect's biology in 1910–11. These included an early description of the psyllid's morphology (Crawford, 1911). Additionally, Crawford moved the psyllid to the new genus *Paratrioza*. Following a synonymization of *Bactericera* and *Triozidae*, the psyllid is now placed in the genus *Bactericera* and the family Triozidae (Burckhardt and Lauterer, 1997; Hodkinson, 2009). Following Crawford (1911), *B. cockerelli*'s morphology was described by Rowe and Knowlton (1935), Tuthill (1945), Burckhardt and Lauterer (1997), Abdullah (2008), and Vargas-Madríz et al. (2013).

3. MORPHOLOGY

Adult *B. cockerelli* (Fig. 7.1A) range in size from 1.3 to 1.9mm (Essig, 1917; Lehman, 1930; Liu and Trumble, 2007). Like many other psyllids and Hemiptera, *B. cockerelli* are often described as "cicada-like" in appearance. Initially, following the final nymphal molt, post-teneral *B. cockerelli* are pale green or light brown. As adults, they are black, but can demonstrate variation in color and some adults are brownish. The first abdominal segment has a white band and the last segment has a white "V-shaped" marking. The psyllids have clear wings that rest on the back of the abdomen. Adult males have six abdominal segments plus an additional segment with the genitalia. Females have five segments and the genitalia segment. Like other insects in the order Hemiptera, *B. cockerelli* has sucking mouthparts with sheathed maxillae and a proboscis. A detailed recent diagnostic description of *B. cockerelli* was published online by Halbert and Munyaneza (2012).

B. cockerelli is hemimetabolous with five wingless nymphal instars (Fig. 7.1B and C). The first instars are light, pale, yellow, or orange while the last instars often turn green (Essig, 1917). Nymphal instars differ significantly in size (antennal length, body width, and body length) and this variation can be used to distinguish between instars (Vargas-Madríz et al., 2013). It is possible to distinguish the sex of nymphs, but this is a complicated process and involves dissection (Carter, 1950).

Eggs of *B. cockerelli* (Fig. 7.1D) are small with an average length of 0.3mm and width of about 0.1mm (Butler and Trumble, 2012c). Eggs are yellowish in color and have an oblong shape often described as "football-like." Typically, eggs are laid on a short (0.2mm) long stalk on the edge (margin) of leaves (Compere, 1915; Lehman, 1930; Pletsch, 1942) and the abaxial side of the leaf; although, this pattern varies among host plants (Prager et al., 2013a, 2014a).

Like all psyllids, *B. cockerelli* is a piercing-sucking insect with specialized mouthparts for feeding on phloem. To date, only a few studies have examined the specific feeding behavior of this psyllid species, and these have mostly been conducted within the context of Lso transmission. Specific studies have examined potatoes (Butler et al., 2012; Pearson et al., 2014) and tomato (Sandanayaka et al., 2014) using Electrical Penetration Graph (EPG) techniques and present similar results. In general, these studies indicate a similar sequence of events in feeding and pathogen transmission which can take as little as an hour, but often much longer (~5h).

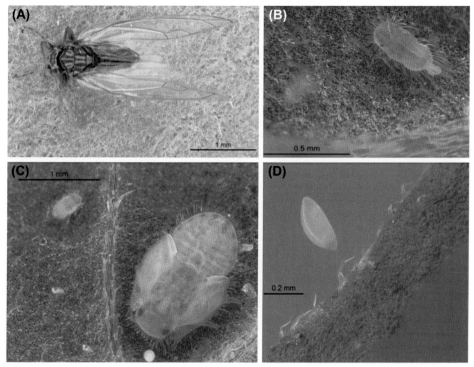

FIGURE 7.1   (A) An adult *Bactericera cockerelli*, (B) Early instar nymph, (C) fifth instar nymph next to a first instar nymph, and (D) egg. *Photos by Ryan Perry.*

4. DISTRIBUTION

The geography and distribution of *B. cockerelli* is a complicated topic that is the focus of substantial current research. In a broad sense, *B. cockerelli* is endemic to North America and probably Central America. However, it has been introduced to, and is invasive in, New Zealand and Canada (Butler and Trumble, 2012c; Munyaneza, 2012).

There are reports of *B. cockerelli* in locations in the western United States and Mexico dating to early 1900s. It generally is believed that the psyllid is native to Mexico, where it has been a documented pest since the 1970s (Garzon et al., 1992; Garzón-Tizanado, 2003; Covarrubias et al., 2006; Lopez et al., 2013). However, the psyllid has increased in range and Cadena-Honojosa and Guzman-Plazola (2003) reported a dramatic increase in midwest Mexico, including the state of Guanajuato where 60% of tomato production was lost in 1990 to the psyllid and Lso. It has also been associated with crop losses of 65% in San Luis Potosi (Diaz et al., 2005), and losses of 35% in northeastern Mexico (Rubio et al., 2011). The psyllid has been documented in many states in the United States including: California, Montana, North Dakota, South Dakota, Nebraska, Kansas, Oklahoma, Texas, and Minnesota (Mills, 1939; Pletsch, 1947). There are records of psyllid in the Canadian provinces of Alberta, British Columbia, and Saskatchewan (Pletsch, 1947; Wallis, 1955). The psyllid is also a recognized pest in Guatemala and Honduras (Tuthill, 1945; Pletsch, 1947; Cranshaw, 1994; Jackson et al., 2009; Crosslin et al., 2010). As noted previously, *B. cockerelli* has been introduced into New Zealand where it is a pest on both greenhouse and field crops of solanaceous plants including pepper, potato, and tomato (Gill, 2006; Martin, 2008; Teulon et al., 2009; Thomas et al., 2011).

While *B. cockerelli* was known in the western United States for decades, it was long assumed that outbreaks in most states were due to migration from southern locations in Mexico (Pletsch, 1947). In part this concept was based on the premise that weather, and in particular, maximum and minimum temperatures, were the primary factors in limiting its endemic geographic range with the psyllid unable to tolerate extremes in temperature. In light of these limitations, many researchers have proposed that the psyllid is migratory, traveling north on the wind currents (Romney, 1939; Pletsch, 1947; Jensen, 1954; Wallis, 1955). Rather than overwinter in the colder environments of the Pacific Northwest and high plains of the United States (Abdullah, 2008; Munyaneza et al., 2009; Swisher et al., 2013a,b), *B. cockerelli* is thought to migrate from "overwintering breeding sites." Initially it was thought that these sites were located in southern Arizona and southern Texas (Romney, 1939). However, Pletsch (1947) noted that the psyllid could be observed much farther south into the east coast area of Mexico. It was thought that those psyllids found in Arizona, New Mexico, and then later more northerly states such as Colorado and Nebraska migrate from these Mexican locations. Recently, the migration concept has been complicated by determination that some *B. cockerelli* overwinter in Washington (United States), and are a genetically distinct haplotype (Swisher et al., 2014).

A similar migration has been proposed for *B. cockerelli* from overwintering locations in western (Baja) Mexico into southern California and throughout the central and northern portions of the state. Supporting this, the northernmost overwintering site for *B. cockerelli* in California was Ventura County (Butler and Trumble, 2012c), while the psyllid was observed occasionally in the summer in the northernmost counties. Thus, while Ventura County may be the northernmost documented overwintering site in California, recent reports of substantial populations in northern California peppers (Trumble, unpublished data) suggest that the northern limit for the Western haplotype is expanding. The potential migration patterns of these psyllids are currently an active research topic for which there is little conclusive information.

The sporadic nature of *B. cockerelli* outbreaks in North America often has been explained by occasional migration of the psyllid into the more northern areas where outbreaks occurred. Associated with this hypothesis was the concept that both maximum and minimum temperature limited the permanent range of the psyllid (Goolsby et al., 2012; Percy, 2012; Tran et al., 2012; Workneh et al., 2013). Consequently, the psyllid was not thought to overwinter in northern states. In an attempt to confirm this hypothesis, many studies of temperature effects on *B. cockerelli* have been conducted.

Some of the initial studies of *B. cockerelli* focused on the role of temperature on development. Those studies determined that both high and low temperatures influenced the rate of development (Lehman, 1930; List, 1939; Wallis, 1955). Specifically, the studies suggested that on potatoes, 26.7°C is optimal for oviposition (List, 1939); this temperature is also suitable for oviposition on tomatoes. Temperatures in excess of about 30°C were found to reduce oviposition, rate of egg hatch, nymphal development and adult survivorship. Temperatures in excess of 35°C were typically fatal (List, 1939). These findings have been subject to subsequent tests under various conditions, and have generally returned similar results (Abdullah, 2008; Yang and Liu, 2009; Yang et al., 2010a,b). Surprisingly, these studies also indicated that *B. cockerelli* is extremely cold-tolerant (Henne et al., 2010; Whipple et al., 2012). Specifically, adults collected in both Washington State and Nebraska can survive temperatures below −10°C for over 14 h while nymphs can tolerate −15°C (Henne et al., 2010; Whipple et al., 2012). Further, reports indicated *B. cockerelli* could survive the winters of the Pacific Northwest of the United States using the nightshade species, *Solanum dulcamara* L., as an overwintering host plant (Sengoda et al., 2013).

When outbreaks of the psyllid occurred in southern California in the early 2000s, researchers began examining the specific geographic origins of the psyllids. Liu et al. (2006b) used a combination of genetic markers to determine that *B. cockerelli* from the western and central United States represented two distinct populations. This work led to further investigations of *B. cockerelli* populations and resulted in a new method for distinguishing the populations (Swisher et al., 2012). Based on the new "high temperature melting" technique, it was determined that there are four distinct *B. cockerelli* haplotypes (Swisher et al., 2012, 2013a,b). Each of these haplotypes was determined to be associated with a specific geographic range and was named accordingly (Swisher et al., 2012, 2013a,b).

5. HAPLOTYPES

As indicated by the naming convention, *B. cockerelli* haplotypes are strongly associated with geographic regions in the United States. The two more broadly distributed haplotypes are the Central and Western. Central haplotype psyllids have been observed throughout Colorado, Idaho, Kansas, Nebraska, New Mexico, North Dakota, Texas, and Wyoming. These are often found in north–central Mexico (Liu et al., 2006b; Swisher et al., 2013a). Western haplotypes have been identified in California, Idaho, Oregon, New Mexico, North Dakota, Washington, and Baja Mexico (Swisher et al., 2012, 2013a,b). Additionally, the source of *B. cockerelli* in New Zealand appears to be the Western haplotype (Swisher et al., 2013a,b). Northwestern haplotype psyllids are rather restricted in range, occurring in Idaho, Oregon, and Washington. Despite these associations, it is notable that some states contain multiple haplotypes of the psyllids. Finally, the Southwestern haplotype is rare and has only been observed in the Rocky mountain regions of Colorado and New Mexico (Swisher et al., 2013a).

Currently, there is a small but increasing body of research on the different haplotypes of *B. cockerelli*, mostly comparing the Western and Central haplotypes. Liu and Trumble (2007) compared the Western and Central haplotypes and found a series of differences with respect to survivorship and fecundity between the haplotypes. They also determined that for the Central haplotype psyllids there were no significant effects of host plant, while host-plant preferences and developmental parameters vary for the Western haplotype. They determined that the Western haplotype psyllids survived better on tomato and developed better than those on peppers. Prager et al. (2014b) further examined this topic and determined that host-plant preferences for oviposition and for settling differ between the Central and Western haplotypes. An important finding was that the Western psyllids had a preference for pepper and tomato versus potatoes. They also found a preference for the natal host plant and that the ability to use the host plants varies with haplotype. It has also been determined that size varies among the haplotypes (Liu and Trumble, 2007; Horton et al., 2014). The Northwest haplotype has larger wings and longer tibia then either the Central or Western haplotypes (Horton et al., 2014). Finally, Swisher et al. (2014) examined *B. cockerelli* samples in Washington State dating back to the late 1990s, and based on this information suggested that the Western haplotype may be relatively new to the region.

6. LIFE CYCLE

Since its appearance as pest in the late 1920s, multiple authors have presented mostly similar descriptions of the *B. cockerelli* life cycle (Lehman, 1930; Knowlton and Janes, 1931; Davis, 1937; List, 1939). *B. cockerelli* is a hemipteran and follows a typical hemimetabolous life cycle that transitions from egg to nymph to adult. The life cycle begins when females lay eggs on the leaves and sometimes stems of host plants, typically following a pre-mating period of 3–5 days and pre-oviposition period of 6–8 days (Abdullah, 2008). Eggs take about 6 days to incubate and hatch after 3–9 days following oviposition. Once the eggs hatch, there are five nymphal instars with nymphal development taking about 20 days. The reported adult life span for *B. cockerelli* in the laboratory varies with factors such as temperature and host plant, but is typically about 41 days (Abdullah, 2008). Notably, the time to development and the number of generations per season, which are associated, are also influenced by a number of factors including: host-plant species, temperature, and humidity (Abdullah, 2008; Yang et al., 2010b; Prager et al., 2014c).

7. HOSTS

A primary complication in management of *B. cockerelli* is its ability to use a large range of host plants. The reported host range exceeds 40 plant species from 20 different families (Crawford, 1911; Essig, 1917; Knowlton and Thomas, 1934; Pletsch, 1942; Wallis, 1955). Within this variation, Wallis (1955) reports a general preference for the families *Solanaceae*, *Convolvulaceae*, and *Lamiaceae*. However, it appears that the psyllid's primary host plants are in the *Solanaceae* with other species serving as alternate or overwintering hosts. Among the reported hosts, the most common are solanaceous vegetable plants including: tomatillo (*Physalis philadelphica* Lam.) bell pepper, chili pepper (*Capsicum* sp.), potato (*S. tuberosum*), eggplant (*Solanum melongena* L.), and tomato (*S. lycopersicum*) (Wallis, 1955; Butler and Trumble, 2012c; Percy, 2012; Prager et al., 2014a,b). Since, in many regions these plants are grown in the same vicinity and during the same season, there are likely many alternate host plants on which the psyllid can develop. This presents the potential for incursion from neighboring untreated fields, weeds in nearby areas, or from urban gardens. A second issue associated with the psyllid's host range is its ability to use numerous other plants as intermediates for migration or overwintering, although many of these species cannot be used for reproduction (Knowlton, 1933a). Other plants reportedly used by psyllid include: water jacket (*Lycium andersonii* A. Gray), desert wolfberry (*Lycium macrodon* A. Gray), matrimony vine/wolfberry (*Lycium barbarum* L.), multiple species of *Datura*, and even species of Pinaceae (Wallis, 1955). Studies also indicate that the psyllid is capable of using species of Solanaceae and Convolvulaceae native to New Zealand including Taewa (*S. tuberosum* ssp. *andigena* and *S. tuberosum*), kumara (*Ipomoea batatas* L. Lam.) and poroporo (*Solanum aviculare* G. Forst).

Within this vast host range, *B. cockerelli* has numerous documented preferences both for plants and varieties, with respect to development, oviposition, and settling. In some instances, these reported patterns might conflict. For example, Knowlton and Thomas (1934) have reported that nymphs may be unable to develop on bell pepper, while other authors report successful growth and reproduction (Liu et al., 2006a; Yang et al., 2010a,b). Similarly, the host plant has been shown to influence fecundity when psyllids are offered a choice of eggplant or bell pepper (Yang and Liu, 2009). Prager et al. (2014b) examined this topic in detail and determined that multiple factors can influence host-plant choices, including the plant on which the psyllid developed, and the psyllid haplotype. In addition to plant species, there is substantial evidence that cultivar (variety) influences attractiveness to potato psyllids in both potato (Diaz-montano et al., 2013; Prager et al., 2014c) and tomato (Liu and Trumble, 2005, 2006).

8. PEST STATUS

B. cockerelli is a pest and economic threat to tomatoes for three different reasons. The most obvious and immediate reason is that the psyllids feed on the phloem of the plants resulting in both weakened stems and substantial accumulation of honeydew. This honeydew accumulation can add weight to the plants thereby stressing stems, lead to fungal infections on the plant, and cause extra cleaning effort or downgrading of harvested fruit.

8.1 Psyllid Yellows

The second more critical problem associated with *B. cockerelli* infestation on tomatoes is a condition known as "psyllid yellows." Psyllid yellows was first described in the early 1920s in association with a series of potato psyllid infestations of potatoes and other solanaceous vegetable crops (Richards, 1928, 1929; Eyer and Crawford, 1933; Daniels, 1934; Eyer, 1937). Attributed to a still unknown toxin (Abernathy, 1991), psyllid yellows is a systemic disorder (Carter, 1939) that is apparently associated with nymphal feeding (Cranshaw, 1994). However, adults have also been shown to produce psyllid yellows symptoms in tomatoes, but only when feeding at much higher densities then nymphs (Daniels, 1934). While responses vary among cultivars, eight nymphs feeding on two-week-old tomato plants will generate these symptoms (Liu et al., 2006a). While feeding appears to be the primary cause of psyllid yellows, it can also be transmitted through grafting (Daniels, 1934; Cranshaw, 1994). Symptoms in tomatoes (Fig. 7.2) can include: spotting, stippling, discoloration of leaves, malformed tissues such as curled or puckered leaves (Chapman, 1985), shortened internodes, enlarged nodes, and stunted fruit, or failure of fruit to set altogether. The most diagnostic symptom is yellowing and cupping of the leaves (Richards, 1933). Finally, an important aspect of psyllid yellows is that in some instances plants may recover with removal of the psyllids. However, this ability is variable with influencing factors including cultivar, age of the plant, density of the infestation, and the duration psyllids were on the plant (Blood et al., 1933; Richards and Blood, 1933; Carter, 1939, 1950).

FIGURE 7.2   (A) A healthy tomato (variety "Yellow Pear") plant and (B–D) leaves exhibiting various psyllid yellows symptoms.

8.2 Vein Greening Disease

Recently, a new *B. cockerelli*–associated threat has arisen in the form of the *C.* Liberibacter solanacearum (Lso) (synonymous with *C.* Liberibacter psyllaurous). Lso is a phloem-limited, Gram-negative, *Alphaproteobacteria* that is a pathogen of many solanaceous plants. The bacterium has been detected in potatoes, bell peppers, eggplant, and tomatoes (Hansen et al., 2008). However, the bacterial titer is higher (>300 times) in tomato than potato (Sengoda et al., 2013). In potato, Lso causes "zebra chip" disease (Fig. 7.3), while in tomatoes infection with Lso leads to a condition known as "vein

FIGURE 7.3   A potato plant (variety "Atlantic") exhibiting characteristic symptoms of zebra chip disease.

greening disease"(Figs. 7.4 and 7.5). Since symptoms first appeared in greenhouse tomatoes in Mexico, vein greening has been identified in numerous locations including tomato fields in California, greenhouses in Arizona, Ontario, and New Zealand.

FIGURE 7.4   Tomato plant in the field exhibiting symptoms of infection with Lso.

Infection with Lso results in an array of symptoms. Among the symptoms in newly infected plants are interveinal chlorosis, curling of leaves, stunting, and vein greening (Figs. 7.4 and 7.5). Infection in older plants also results in bleached leaves in some cultivars, foliar purpling, necrosis, wilting, and eventually death. Most importantly, infection leads to poor-quality fruit with low sugar content, and/or failure of fruit to set.

Symptomatically, vein greening and psyllid yellows are very similar. However, there is increasing research that supports the conclusion that psyllid yellows is not caused by Lso. First, successive grafts of psyllid yellows–infected plants result in a gradual recovery of plants (Daniels, 1954; Cranshaw, 1994). This is not the case with Lso in which grafting results in eventual infection similar to the source plant (Hansen et al., 2008; Crosslin et al., 2010). These factors suggest that a pathogenic microorganism is not involved with psyllid yellows and support the "toxin" hypothesis. The identity of this "toxin"

FIGURE 7.5   (A) Leaf from a healthy tomato plant and (B–C) leaves of tomato plants infected with Lso and exhibiting symptoms.

still remains unknown (Abernathy, 1991). Additionally, psyllid yellows symptoms can be found in plants that test negative for Lso. Thus, feeding by psyllids can lead to psyllid yellows condition, but vein greening results from infection with Lso, which is transmitted by the feeding of the psyllid (Munyaneza et al., 2007a; Hansen et al., 2008; Sengoda et al., 2010).

Currently, detection and positive identification of infection with Lso in either plants or *B. cockerelli* can only be accomplished via PCRs or related molecular techniques. This can be accomplished either using traditional PCR, or through quantitative real-time PCR, which has proven to be more sensitive (Li et al., 2009; Lévy et al., 2013; Wen et al., 2013).

Since Lso was detected, the body of research has increased at a dramatic pace and a particular focus has been the dynamics of transmission by the psyllid vector. The association between Lso and *B. cockerelli* was first described by Munyaneza et al. (2007a). It is now known that Lso can be transmitted by a single psyllid in as little as 2 h (Sandanayaka et al., 2014), while 20 psyllids can result in infection in an hour (Buchman et al., 2011). Given 72 h of exposure, 100% of tomato plants in laboratory studies were infected (Buchman et al., 2011; Sengoda et al., 2013). However, as the number of psyllids feeding on a plant increases, the time required for inoculation decreases (Rashed et al., 2013). EPG studies indicate that while 35 min to an hour is required to reach the phloem (Butler et al., 2012; Sandanayaka et al., 2014), less than a minute is required for inoculation once a psyllid reaches the phloem, and thus inoculation may occur in as little as 35 min (Sandanayaka et al., 2014). Also, nymphs are less efficient than adults at transmitting the virus (Buchman et al., 2011), but they may be more efficient at acquiring Lso (Buchman et al., 2011). Finally, there is no effect of the host plant, on which the psyllid is reared, in their ability to transmit the bacteria to tomatoes (Gao et al., 2009).

The issue of haplotypes is further complicated because not only are there multiple haplotypes of the insect, there also are at least four haplotypes of Lso. Lso haplotype A is found primarily in the western United States (California) but is also detected occasionally in Texas and other central states. Lso recorded from New Zealand is also of the A haplotype. Haplotype B is mostly found in central United States but also occurs in the Pacific Northwest; it has never been found in California. Other haplotypes of Lso have been identified, but these are not found in North America and generally are associated with other crops such as carrots and celery, or are associated with a different species of psyllid (Nelson et al., 2011, 2013; Tahzima et al., 2014; Teresani et al., 2014).

9. MANAGEMENT

9.1 Sampling and Treatment Decisions

One of the most complicated issues in managing *B. cockerelli* is determining when to take management action. Currently, there is no economic threshold for psyllid in any North American crop, including tomatoes. Some thresholds are provided for greenhouse tomato growers in New Zealand, where a psyllid infestation of greater than 2% triggers insecticide treatment of the entire greenhouse. Spot treatments are recommended for 1.5–1.8% infestation (Anonymous, 2008a).

With the lack of a threshold in North America, producers have taken a zero tolerance approach for management of psyllid in potatoes. This approach is indicative of the risk due to Lso. In the absence of Lso, the acceptable level of infestation can be substantially higher as the risk is from psyllid yellows. Since plants may recover from psyllid yellows if timely action is taken, there is more flexibility in management decisions. However, Lso can be transmitted quite rapidly by even a single psyllid. Therefore, when vein greening is a concern, a much lower level of tolerance must be applied. Moreover, both psyllid yellows and Lso infections can cause plant death, but mostly they lead to a reduction in marketable fruit. In a commercial operation, it is typical to experience some loss to disease, therefore growers or pest control advisors (PCAs) may not take action if a small proportion of plants are infected with Lso.

The different haplotypes, numerous potential host plants, and insect's migratory nature complicate the association between sampling for *B. cockerelli* and establishing appropriate thresholds. For example, there is evidence that psyllids do not use all host-plant species or varieties equally. In fact, it was demonstrated that host preferences can vary with cultivar (Liu and Trumble, 2004) and among haplotypes (Prager et al., 2014b). Thus, sampling needs to be conducted throughout the season since it is unclear when psyllids will invade fields in a given location, if they will leave those plants, if there will be multiple infestations, or if the insects will be carrying Lso.

Current recommendations for sampling of *B. cockerelli* are based on potatoes and involve placing yellow sticky cards at the margins of fields near the tops of plants (UCIPM, 2012). These sticky cards are best used as an indicator for the appearance of psyllids and that foliage examination should begin. It is thought that for potatoes, the combination of sticky cards and leaf samples is comprehensive enough to guide decisions (Goolsby et al., 2007). However, there are multiple published objections to the use of sticky cards (Martini et al., 2012), and many authors argue instead for the use of various forms of plant sampling (Butler and Trumble, 2012b; Martini et al., 2012; Yen et al., 2012; Walker et al., 2013; Prager et al., 2012, 2013a, 2014a). In particular, it has been reported that sticky trap results do not correlate to counts on the plants (Prager et al., 2012). Additionally, most of the sampling studies of *B. cockerelli* were conducted on potatoes, and in some pest insects sampling plans and insect distributions have not proven to be consistent among crops or geographic locations (Trumble et al., 1987, 1989). Additionally, the current sampling recommendations for tomato in the United States do not provide guidance for when to take treatment action (UCIPM, 2012). In New Zealand, Walker et al. (2013) recommended potato growers sample 100 middle leaves from 50 randomly selected plants from a row, but again there are no solid recommendations for tomato.

9.2 Within Field and Plant Distribution

Within fields of potato, pepper, and tomato, *B. cockerelli* exhibits an aggregated distribution with noticeable edge effects (Butler and Trumble, 2012b; Martini et al., 2012; Walker et al., 2013). This has been demonstrated in green tomatoes in Mexico in addition to potatoes (Crespo-Herrera et al., 2012). Within tomato plants, *B. cockerelli* exhibits a general preference for the underside of leaves located in the middle-third of plants. However, this changes as plants grow larger (Prager et al., 2014a).

To date, there is only one published sampling protocol for *B. cockerelli* in tomatoes (Prager et al., 2014a). They developed three sequential sampling plans for *B. cockerelli* in California tomatoes based on 27%, 57%, and 70% infestation of fields. The three plans are intended to allow a grower to adjust for their level of risk aversion and perceived threat from Lso. Additionally, the sampling plans indicate that adults are difficult to detect and thus sampling should concentrate on nymphs. Sampling for nymphs also has the advantage that adults may be transient in a field while nymphs require adults to have been present long enough to oviposit. Furthermore, since nymphs are difficult to see in the field and searching plants is time–consuming, they present a binomial plan in which a plant need only be classified as having psyllids present or absent.

Perhaps the biggest complication in sampling and making treatment decisions for psyllids is their ability to use multiple hosts. In particular, psyllids that are reared on other host plants can use tomatoes as hosts and can transmit Lso to tomatoes. This is critically important to sampling and management because psyllids may develop in unmanaged fields or on weedy hosts, and then migrate into tomato fields. This is further complicated by the fact that tomatoes typically are equally or more acceptable to psyllid than either potato or pepper (Prager et al., 2014b).

The threat of *B. cockerelli* moving among plants can be considered in the context of a field study in which the number of adult and juvenile psyllids in adjacent tomato and pepper fields are examined. We conducted such a study in Orange County, California in 2009 and 2010 and found that in both 2009 ($P < .001$, df $= 10$) and 2010 ($P < .001$, df $= 11$) there was a significant difference between the number of adult psyllids in tomato and pepper fields over time (Fig. 7.6). Interestingly, the trend was much more apparent in 2009 where psyllids only appeared in tomato fields on one sampling date. In 2010, psyllids were present in both fields, yet they were typically more abundant in pepper fields (Fig. 7.6). When juvenile psyllids were examined, there was a significant difference between plants ($X^2 = 19.0$, $P < .01$, df $= 1$), years ($X^2 = 28.4$, $P < .01$, df $= 1$), and among weeks ($X^2 = 30.0$, $P < .01$, df $= 11$) (Fig. 7.6). Throughout 2009, this trend was due to the presence of psyllids on pepper plants when there were no (or few) psyllids on tomatoes. However, in 2010 there were more juveniles in

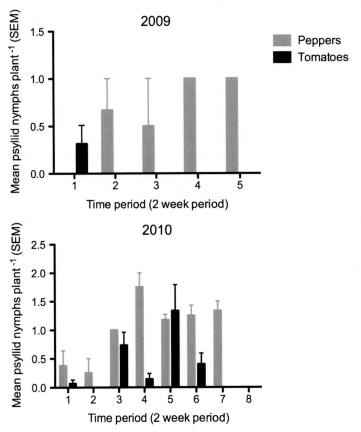

FIGURE 7.6 The mean (SEM) number of nymphs sampled in adjacent fields of tomato and pepper on the same days in 2009–10.

the tomato fields, as opposed to adults that were mostly on peppers. These results indicate that in adjacent fields, sampled with identical methods and on the same days, tomato psyllids are not evenly distributed. Although there were no strong patterns, in 2009 adult psyllids appeared more commonly on pepper than on tomato, a trend which was less distinct but also apparent in 2010. Conversely, juvenile stages were more common in tomato fields than in pepper, even on those dates where adults were prevalent in peppers. These results strongly suggest that sampling must be conducted using methods specific to a given crop, a finding also supported by the differences in within-plant distributions among crops, and the lack of edge effects in peppers relative to tomatoes and potatoes reported by Prager et al. (2013a).

9.3 Biological Control

Biological control is a desirable component of any integrated pest management (IPM) program. Integrating natural enemies requires knowledge of the natural enemy community and their interactions with pest insects. There are numerous reports and surveys of natural enemies of *B. cockerelli* in North America. These include field surveys (Knowlton, 1933b; Romney, 1939) and laboratory assessments (Knowlton, 1933b; Knowlton and Thomas, 1934; Knowlton and Allen, 1936). Many of the field surveys were conducted in potato fields, and reported species of Chrysopidae, Anthocoridae, and Coccinellidae attacking psyllid. Laboratory studies also indicated that Syrphidae larvae, Nabidae, Miridae, and Geocoridae also attack *B. cockerelli* (Knowlton, 1933b; Knowlton and Thomas, 1934; Knowlton and Allen, 1936).

The most comprehensive examination of *B. cockerelli* natural enemies was by Butler and Trumble (2012a) in California. Based on multiple years of surveys in multiple crops, they determined that *Orius tristicolor* (White), *Geocoris pallens* Stål and *Hippodamia convergens* Guérin-Méneville were key natural enemies of *B. cockerelli*. They also suggested that the parasitoid *Tetrastichus triozae* Burks may have some utility as a control agent when pesticides are not used. However, Johnson (1971) examined this species and found limited control of *B. cockerelli*. A primary obstacle to biological control of *B. cockerelli* may be the low rate of parasitism even by psyllid specialists (Romney, 1939; Butler and Trumble, 2012a). In addition, even in those cases where parasitism reached substantial levels, the potential for transfer of the vein green-ing pathogen was high because *T. triozae* attacks third or fourth instar nymphs and feeding by second or third instars can introduce the pathogen into the plant. They suggested that perhaps a greater benefit would occur from psyllid population reductions on adjacent hosts outside the field that would reduce immigration into the crop. Another concern is the occurrence of *Encarsia pergandiella* Howard and *Encarsia peltata* (Cockerell), which hyperparsitize over 5% of the *T. triozae*-parasitized *B. cockerelli* in crop fields in southern California (Butler and Trumble, 2011). The long-term impact of these hyperparasites on any inundative release program of *T. triozae* will require additional investigation.

Butler and Trumble (2012a) also identified many species of spiders preying on *B. cockerelli*. However, they did not identify any particularly effective species. Butler and Trumble (2012a) followed up on these surveys with exclusion cage experiments that demonstrated that natural enemies can play a role in reducing *B. cockerelli* populations. Furthermore, they found that the reduction was due to generalist predators. As mentioned, this is in agreement with some insecticide trials that demonstrate increased populations when less specific insecticides were applied.

Although not as comprehensive as the studies in California, Walker et al. (2011) examined the natural enemies of *B. cockerelli* on potatoes in New Zealand and found that brown lacewings, *Micromus tasmaniae* (Walker) and small hoverflies were most abundant. Additionally, they provided unpublished data from laboratory studies that indicated brown lacewings, 11-spotted ladybird beetles (*Coccinella undecimpunctata* (L.)), large-spotted ladybird beetles (*Harmonia conformis* (Boisduval)), small hoverfly (*Allograpta*) larvae and adults, and damsel bugs were capable of feeding on all life stages of *B. cockerelli*. In New Zealand in 2006–08, surveys were conducted looking for potential control agents for *B. cockerelli*, but they failed to identify any natural enemies that could provide control on tomatoes (Anonymous, 2008a). As of 2013, Plant and Food Research Ltd. (Anderson et al., 2013) has been examining the potential for release of *T. triozae* as a biocontrol strategy.

Al-Jabr (1999) specifically examined the potential of two species of lacewing (*Chrysoperla carnea* (Stephens) and *Chrysoperla rufilabris* (Burmeister)) for control of *B. cockerelli* in tomato greenhouses. Results indicated that both species could develop on *B. cockerelli* but survival in the greenhouse conditions was limited, thus reducing the potential for its use.

Finally, Lacey et al. (2009) tested two species of entomopathogenic fungi for use against the psyllid and found that both *Metarhizium anisopliae* (Metchnikoff) Sorokin and *Isaria fumosorosea* Wize were effective in controlling the psyllid as both adults and nymphs. They further examined commercial formulations of these products and determined that they reduced eggs and nymphs relative to controls while also reducing zebra chip symptoms and plant damage (Lacey et al., 2009, 2011).

9.4 "Green" Compounds and Organically Acceptable Control Methods

To complement or replace the various insecticidal materials applied for control of psyllids, numerous green or biorational materials have been examined. To date, biorational products are more commonly used in greenhouse operations than in the field, especially in New Zealand (Walker et al., 2010, 2011; Jorgensen et al., 2013). Among those materials examined, various levels of efficacy have been reported. For example, laboratory studies on potatoes in New Zealand have demonstrated efficacy of multiple oils including: Sap Sucker Plus, Organic JMS Stylet-Oil (JMS Flower Farms Inc.), Excel Oil (GroChem), and Eco-Oil (Eco Organic Products); none of these materials were found to be repellent or reduce oviposition (Jorgensen et al., 2013). However, there were some associated reductions in nymphs that were likely due to reduced egg hatching (Jorgensen et al., 2013). This finding is somewhat contradictory to the findings of Walker et al. (2010) who found that Eco-Oil was more effective than Excel Oil. However, Eco-Oil's effects are consistent with findings for other products including methylated seed oil and a sprayable wax matrix product (SPLAT, ISCA Technologies, Riverside, California, United States). Tests of those products demonstrated that nymphs and eggs were affected (killed or prevented from hatching) by non-insecticidal materials (Prager and Trumble, unpublished data).

At least two particle film products have been tested against *B. cockerelli* (Yang et al., 2010a,b; Peng et al., 2011). Peng et al. (2011) found that kaolin particle film resulted in reduced egg numbers on tomato plants in the lab, while in the field kaolin reduced numbers of eggs, nymphs, and adults. Similarly, Prager et al. (2013b) examined a limestone particle film on potatoes and found no effect on adult mortality or probing, but a significant reduction in oviposition.

9.5 Chemical Control

Starting with the first major outbreaks of *B. cockererlli* in North America in the 1900s, management has primarily focused on insecticidal approaches. Because of this, there is a long history of research on insecticidal control of *B. cockerelli*. Many of the materials that have been used or tested for control of *B. cockerelli* have been examined on potatoes and not tomatoes.

The earliest control approaches tested were nicotine sulfate and lime sulfur, with the lime sulfur proving effective (List, 1917, 1918). In particular, lime sulfur was repellent to adults and toxic to nymphs. Sulfur had been recommended on many occasions and was demonstrated to increase yields of both potato and tomatoes (List and Daniels, 1934; List, 1935, 1938). Unfortunately, lime sulfur is phytotoxic (List, 1935; Pletsch, 1942). Other materials that have been examined include nicotine, zinc, arsenite sprays, calcium cyanide dusts, and various oils (Knowlton, 1933b; Pletsch, 1942, 1947). In the 1940s growers began using DDT, which was effective but is no longer an option (Hill, 1945, 1948; Pletsch, 1947).

Beginning in the 1960s many different materials were used for control of psyllid. These included the organophosphates phorate, parathion, disulfoton, and demeton (Gerhardt and Turley, 1961; Harding, 1962; Gerhardt, 1966). Some carbamates, such as aldicarb, cloetiocarb, methomyl, and carbaryl have also been used against psyllid, but most were ineffective (Cranshaw, 1985). In the 1980s another series of materials were examined for use in psyllid management. These included diazinon, endosulfan, permethrin, acephate, phorate, disulfoton, and various pyrethroid insecticides (Cranshaw, 1985, 1989). Many of these materials proved effective, however, there is evidence that pyrethroids can lead to increases in psyllid numbers (Prager and Trumble, unpublished data). The use of methomyl and carbamates is known to result in increased pest pressure in potatoes, likely due to the negative effect on natural enemies (Prager and Trumble, unpublished data), but possibly from hormoligosis.

Beginning with the reemergence of psyllid as a pest in the 2000s another round of research on insecticidal management of psyllid has begun. This is largely fueled by the additional threat from Lso and the failure to develop effective or practical non-insecticidal methods of managing psyllid.

Currently, management of *B. cockerelli* is entirely dependent on the application of insecticides (Guenthner et al., 2012; Greenway, 2014). Since the major outbreaks of zebra chip in Texas, numerous insecticidal rotations have been proposed and applied for control of psyllid. These rotations have used materials including: imidacloprid+cyfluthrin, endosulfan, and methamidophos (Goolsby et al., 2007; Guenthner et al., 2012). In potatoes growers are more proactive with weekly insecticide applications than in tomatoes (Guenthner et al., 2012). There are a few specific studies of insecticidal control of *B. cockerelli* in tomatoes. However, these are greenhouse or laboratory-based trials. These studies found that spinosad, acetamiprid, and neem-based compounds are effective in killing *B. cockerelli* adults (Al-Jabr and Cranshaw, 2007). Other work by Liu and Trumble (2004, 2005) examined interactions between tomato cultivars and various insecticides. These studies revealed that imidacloprid, spinosad, pymetrozine, and pyriproxyfen altered psyllid biology. These effects included altered probing behavior, impacts on development and survivorship from egg to adult, and variation in non-feeding behaviors. In New Zealand, where psyllid is an introduced pest on tomato, potato, and bell pepper, multiple materials have been examined in the lab for control of psyllid on bell peppers (Berry et al., 2009). These studies indicated that the contact insecticides dichlorvos, lambda-cyhalothrin, methomyl, tau-fluvalinate, and methamidophos all caused greater than 98% mortality to psyllid. Yet, these materials are not encouraged in modern IPM strategies since they are harmful to humans and beneficial insects.

To date, the most effective material known for knockdown of *B. cockerelli* is abamectin, which has extremely high 24-h mortality (Gharalari et al., 2009; Lacey et al., 2009). Abamectin is currently recommended as part of the University of California IPM strategy for *B. cockerelli* (UCIPM, 2012). Similarly, in New Zealand it is recommended that greenhouse growers spray young plants with abamectin or a similar "IPM compatible" insecticide.

At present, one of the most commonly used and recommended techniques for managing *B. cockerelli* is the soil (in furrow) application of insecticides including phorate or imidacloprid at planting, with imidacloprid being the most commonly used. Imidacloprid is especially useful because it has been shown to interfere with the psyllid's probing behaviors that reduce transmission of Lso (Butler et al., 2011, 2012). Unfortunately, there is substantial evidence that the psyllid may be developing resistance to imidacloprid (Liu and Trumble, 2007; Prager et al., 2013c). This is the case in southern Texas, where psyllids often migrate from Mexico where growers apply imidacloprid many times a growing season (Prager and Trumble, unpublished data). An alternative to imidacloprid is the soil application of thiamethoxam, which is also effective against *B. cockerelli* (Gharalari et al., 2009; Prager et al., 2013c). However, thiamethoxam is highly sensitive to application method and less effective with increased irrigation. Additionally, it is not as effective as imidacloprid for long-term control (Prager et al., 2013c). In New Zealand, greenhouse growers are also advised to use neonicotinoid along with irrigation and apply thiacloprid along with irrigation water (Anonymous, 2008b).

Recently, a new material with a distinct mode of action, cyazapyr (IRAC class 28), has been approved for use in tomatoes in the United States. Preliminary studies indicate that it is effective in controlling *B. cockerelli* and may provide similar results to imidacloprid and thiamethoxam (Trumble and Prager, 2014). Of particular interest is the ability to apply it as a soil treatment in place of neonicotinoid insecticides. This is important for resistance management. In addition to neonicotinoid insecticides, a variety of other selective systemic insecticides are currently used for control of *B. cockerelli*. Among these is pymetrozine, a product that inhibits feeding but is slow acting and has mixed reported efficacy (Liu and Trumble, 2005; Gharalari et al., 2009; Walker et al., 2010). Spiromesifen and spirotetramat have also been shown to be effective against nymphs of psyllid (Liu and Trumble, 2004; Gharalari et al., 2009; Page-Weir et al., 2011), and they are frequently applied as part of insecticidal rotations on potato (Guenthner et al., 2012).

Overall, the lack of robust field data greatly complicates management of psyllid as the growers are often presented with contradictory information, or must rely on the results of studies conducted on other crops. As demonstrated in Table 7.1, even

TABLE 7.1 Reported Efficacy of Insecticidal Materials Evaluated for Control of Potato Psyllids in Potato, Bell Pepper and Tomato Based on Information From Current Pest Management Strategic Plans (PMSP)

Insecticidal Material	Commercial Name	Pepper	Tomato	Potato
Abamectin	AgriMek	-	-	-
Acetamiprid	Asail	Poor	-	Good
Azadirachtin	Neemix	Poor, fair	-	-
Bacillus thurgensis	Bt	Poor	-	-
Beauveria bassiana	Various	Poor	-	-
Carbaryl	Seven	Poor	-	Good, excellent
Carbaryl Bait	Seven	Poor	-	-
Endosulfan	Thionex, Thiodan	-	-	Good, excellent
Esfenvalerate	Asana	-	Fair	Good, fair
Ethoprop	Mocap	-	-	-
Imidacloprid	Provado/Admire/Leverage	Ineffective	-	-
Memthamidophos	Monitor	-	-	Good
Methomyl	Lannate	-	Fair	-
Permethrin	Pounce	-	-	Fair, good
Phorate	Thimet	-	-	Good
Spinosid	Success	Poor	-	-
Sulfur	Sulfur	Good	-	-
Thiamethoxam	Platinum/Actara	-	-	-
Oxyamyl	Vydate	-	Fair	-
Imidacloprid + Cyfluthrin	Leverage	-	-	Good, fair

Hyphen (-) indicates no evaluation included for that pesticide in the PMSP for that crop.
Published online by the Western IPM Center.

the recommendations of institutions such as the Western Regional IPM Center (http://westernipm.org) offers contradictory information. For example, current UCIPM (2012) guidelines for control of psyllid call for the application of abamectin, spiromesifen, and spinosid. Meanwhile, the Western Regional IPM Center's Pest Management Strategic Plans offers little recommendation for control of *B. cockerelli* in tomatoes, but do indicate "fair" results from the use of esfenvalerate, methomyl, and oxyamyl. This obviously contradicts the finding that methomyl leads to increases in psyllid populations in potatoes and peppers (Prager et al., 2016). In the past five or so years, there has been an increase in field evaluations of materials for control of potato psyllids, many of which are described in the various *Proceedings of the Annual Zebra Chip Reporting Sessions* (http://zebrachipscri.tamu.edu). However, few of these results have been integrated into IPM guidelines. Additionally, there is currently only one published scheme for managing potato psyllids in tomatoes in the context of a pest complex (Prager et al., 2016).

10. ECONOMICS OF MANAGEMENT STRATEGIES

The economics of managing *B. cockerelli* in tomatoes can be complicated. In Baja Mexico and southern California, outbreaks of psyllid have resulted in millions of dollars of losses (Liu et al., 2006a). Growers have responded to this threat in two ways. First, they make multiple insecticide applications per week throughout the season. Second, they often grow tomatoes in large screen houses to exclude the insect. These approaches are costly in both labor and materials.

In New Zealand, *B. cockerelli* has been a pest of major concern to greenhouse tomato growers (Ivicevich, 2006; Robertson, 2006a,b; Teulon et al., 2009). This is due to the immediate implications of damaged plants, and also because there are implications for export markets (Robertson, 2008a,b). As a result, growers often have reverted to insecticide use, including the application of materials harmful to beneficial insects (Workman and Pedley, 2007a,b; Teulon et al., 2009). Consequently, there is a substantial risk to progress made toward IPM adoption in these scenarios.

In the United States, California is the largest tomato-growing region where *B. cockererlli* is found, with upward of 80,000 ha grown annually. As in New Zealand, there are well-adopted and successful IPM strategies for tomatoes. However, these are likely threatened by psyllid. In large part this is due to the current reliance on insecticides to control psyllid, combined with the lack of economic thresholds and a tendency to use materials harmful to beneficial insects (see Liu et al., 2012 for details). Additionally, many of the materials used to manage psyllid in other crops that may be effective in tomatoes are still covered under patents and are therefore more costly than "traditional" and older insecticides.

A final problem for managing psyllid in tomatoes is that tomatoes are subject to many plant diseases and viruses. Tomato growers, especially those who grow processing tomatoes, are accustomed to loss from disease. Consequently, they may not specifically target psyllids in their management programs. When an infestation is with Lso-negative psyllids, this may be acceptable because it is sometimes possible to recover from psyllid yellows, and because action may be taken before psyllid populations reach harmful levels. On the other hand, since the Lso bacteria can be transmitted quickly, infected psyllids need to be managed aggressively. However, this is difficult because one cannot determine if a psyllid carries Lso, or a plant is infected with Lso, without complicated molecular techniques. Similarly, the symptoms of Lso can be difficult to distinguish from those of other plant diseases. These factors combine to force growers to evaluate their risk aversion and correspond accordingly, which has economic consequences.

11. FUTURE PROSPECTS

B. cockerelli, the tomato/potato psyllid has been a known pest on tomato in the western United States for nearly a century. For much of this time it was an occasional pest, and most damage was linked to psyllid yellows symptoms. This made management relatively simple, as outbreaks could be treated with insecticide applications. However, over the last decade or so, it has been discovered that *B. cockerelli* is also the vector of *C. Liberibacter solanacearum*, the causal agent of zebra chip and vein greening disease. This makes management of *B. cockerelli* much more difficult. In particular, we are lacking in many of the important tools for optimizing IPM strategies. These include economic and damage thresholds, comprehensive examinations of disease frequency across locations and crops, and explicit tests of insecticides for control of *B. cockerelli* on tomato. We also lack knowledge on the alleged migration of psyllids and are only now starting to understand the biology of the various haplotypes of psyllids and Lso. This is partly due to the complete inability to culture Lso. Finally, we are lacking in IPM schemes that include the psyllid as part of pest complex along with other insects and pathogens. Given that the psyllid is progressively expanding its range and is capable of using numerous hosts, it is likely that both Lso and the psyllid will become an increasing and more consistent problem on tomatoes as it has in bell pepper and potato.

REFERENCES

Abdullah, N.M.M., 2008. Life history of the Potato Psyllid *Bactericera cockerelli* (Homoptera: Psyllidae) in controlled environment agriculture in Arizona. African Journal of Agricultural Research 3 (1), 60–67.

Abernathy, R.L., 1991. Investigation in to the Nature of the Potato Psyllid Toxin (M.S. thesis). Colorado State University, Fort Collins, USA, p. 54.

Al-Jabr, A.M., 1999. Integrated Pest Management of Tomato/Potato Psyllid, *Paratrioza cockerelli* (Sulc) (Homoptera: Psyllidae) with Emphasis on its Importance in Greenhouse Grown Tomatoes (Ph.D. thesis). Colorodo State University, Fort Collins, USA, p. 186.

Al-Jabr, A.M., Cranshaw, W.S., 2007. Trapping tomato psyllid, *Bactericera cockerelli* (Sulc) (Hemiptera: Psyllidae) in greenhouses. Southwestern Entomologist 32 (1), 25–30.

Anderson, S.A., Fullerton, R.A., Ogden, S.C., 2013. TPP and Liberibacter in New Zealand: research programme update and future directions. In: Proceedings of the 13th Annual SCRI Zebra Chip Reporting Session, pp. 28–33.

Anonymous, 2008a. Assessing new biological control agents in greenhouse capsicums and tomatoes. Grower 63 (7), 32–33.

Anonymous, 2008b. Growers Guide to the Management of Tomato/Potato Psyllid in Greenhouse Tomato and Capsicum Crops to Be Read in Association with New Zealand Code of Practice for the Management of the Tomato/Potato Psyllid in Greenhouse Tomato and Capsicum Crops. http://www.freshvegetables.co.nz/assets/PDFs/Psyllid-Growers-Guide-Nov-08.pdf.

Berry, N.A., Walker, M.K., Butler, R.C., 2009. Laboratory studies to determine the efficacy of selected insecticides on tomato/potato psyllid. New Zealand Plant Protection 62, 145–151.

Binkley, A.M., 1929. Transmission studies with the new psyllid-yellows disease of solanaceous plants. Science 70, 615.

Blood, H.L., Richards, B.L., Wann, F., 1933. Studies of psyllid yellows of tomato. Phytopathology 23 (11), 930.

Buchman, J.L., Sengoda, V.G., Munyaneza, J.E., 2011. Vector transmission efficiency of Liberibacter by *Bactericera cockerelli* (Hemiptera: Triozidae) in Zebra Chip potato disease: effects of psyllid life stage and inoculation access period. Journal of Economic Entomology 104 (5), 1486–1495.

Burckhardt, D., Lauterer, P., 1997. A taxonomic reassessment of the triozid genus *Bactericera* (Hemiptera: Psylloidea). Journal of Natural History 31 (1), 99–153.

Butler, C.D., Trumble, J.T., 2011. New records of hyperparasitism of *Tamarixia triozae* (Burks) by Encarsia spp. in California. Pan Pacific Entomologist 87, 130–133.

Butler, C.D., Trumble, J.T., 2012a. Identification and impact of natural enemies of *Bactericera cockerelli* (Hemiptera: Triozidae) in Southern California. Journal of Economic Entomology 105 (5), 1509–1519.

Butler, C.D., Trumble, J.T., 2012b. Spatial dispersion and binomial sequential sampling for the potato psyllid (Hemiptera: Triozidae) on potato. Pest Management Science 142 (6), 247–257.

Butler, C.D., Trumble, J.T., 2012c. The potato psyllid, *Bactericera cockerelli* (Sulc) (Hemiptera: Triozidae): life history, relationship to plant diseases, and management strategies. Terrestrial Arthropod Reviews 5, 87–111.

Butler, C.D., Byrne, F.J., Keremane, M.L., Lee, R.F., Trumble, J.T., 2011. Effects of insecticides on behavior of adult *Bactericera cockerelli* (Hemiptera: Triozidae) and transmission of *Candidatus* Liberibacter psyllaurous. Journal of Economic Entomology 104 (2), 586–594.

Butler, C.D., Walker, G.P., Trumble, J.T., 2012. Feeding disruption of potato psyllid, *Bactericera cockerelli*, by imidacloprid as measured by electrical penetration graphs. Entomologia Experimentalis et Applicata 142 (3), 247–257.

Cadena-Honojosa, M.A., Guzman-Plazola, R., 2003. Distribution, incidence and severity of purple potato rot and abnormal sprouting of potato (*Solanum tuberosum* L.) tubers in the high valleys and mountains of the states of Mexico, Tlaxcala and the Federal District, Mexico. Review Mexico Fitopatholog 21, 248–250.

Carter, W., 1939. Injuries to plants caused by insect toxins. Botanical Review 5, 273–326.

Carter, R.D., 1950. Toxicity of *Paratrioza cockerelli* (Sulc) to Certain Solanaceous Plants (Ph.D. thesis). University of California, USA.

Chapman, R.K., 1985. Insects that poison plants. American Vegetable Grower 31–38.

Compere, H., 1915. *Paratrioza cockerelli* (Sulc). Monthly Bulletin California State Commission Horticulture 4 (12), 574.

Covarrubias, O.A.R., Leon, I.H.A., Moreno, J.I., Salas, J.A.S., Sosa, R.F., Soto, J.T.B., Hinojosa, M.A.C., 2006. Distribution of purple top and *Bactericera cockerelli* Sulc. in the main potato production zones in Mexico. Agricultura Técnica En México 32 (2), 201–211.

Cranshaw, W.S., 1985. Control of potato insects with solid applied systemic insecticides, Greeley CO. Insecticide and Acaricide Tests 10, 133.

Cranshaw, W.S., 1989. Potato insect control. Insecticide and Acaricide Tests 14, 136.

Cranshaw, W.S., 1994. The potato (tomato) psyllid *Paratrioza cockerelli* (Sulc) as a pest of potatoes. In: Zehnder, G.W., Powelson, R.K., Jansson, R.K., Raman, K.V. (Eds.), Advances in Potato Pest Biology. APS Press, St. Paul, MN, pp. 83–98.

Crawford, D.L., 1911. American Psyllidae III (Triozinae). Pamona Journal Entomology 3, 421–453.

Crespo-Herrera, L.A., Vera-Graziano, J., Bravo-Mojica, H., López-Collado, J., Reyna-Robles, R., Peña-Lomelí, A., Garza-García, R., 2012. Spatial distribution of *Bactericera cockerelli* (Sulc) (Hemiptera: Triozidae) on green tomato (*Physalis ixocarpa* (Brot.)). Agrociencia 46 (3), 289–298.

Crosslin, J.M., Munyaneza, J.E., Brown, J.K., Liefting, L.W., 2010. A history in the making: potato zebra chip disease associated with a new psyllid-borne bacterium – a tale of striped potatoes. APSnet Features. http://dx.doi.org/10.1094/APSnetFeature-2010-0110.

Daniels, L.B., 1934. The tomato psyllid and the control of psyllid yellows of potatoes. Colorodo Agricultural Station Bulletin 410, 18.

Daniels, L.B., 1954. The Nature of the Toxicogenic Condition Resulting From the Feeding of the Tomato Psyllid *Paratrioza cockerelli* (Sulc) (Ph.D. thesis). University of Minnesota, USA, p. 119.

Davis, A., 1937. Observations on the life history of the *Paratrioza cockerelli* (Sulc) in Southern California. Journal of Economic Entomology 30 (2), 337–338.

De Boer, S.H., Secor, G., Li, X., Gourley, J., Ross, P., Rivera, V., 2007. Preliminary characterization of the etiologic agent causing zebra chip symptoms in potato. New and old pathogens of potato in changing climate. In: Proceedings of EAPR Pathology Section Seminar, 2–6 July 2007, Hattula, Finland.

Diaz, G.O., Tejada, E.I.M., Avalos, A.L., 2005. Efecto de insecticidas biorracionales y mezclas de hongos sobre *Bactericera cockerelli* (Sulc) (Homoptera: Psyllidae). Entomología Mexicana 5, 539–541.

Diaz-montano, J., Vindiola, B.G., Drew, N., Novy Jr., R.G., Miller, J.C., Trumble, J.T., 2013. Resistance of selected potato genotypes to the potato psyllid (Hemiptera: Triozidae). American Journal of Potato Reserarch 91 (4), 363–367.

Essig, E.O., 1917. The tomato and laurel psyllids. Journal of Economic Entomology 10 (4), 433–444.

Eyer, J.R., 1937. Physiology of psyllid yellows of potatoes. Journal of Economic Entomology 30 (6), 891–898.

Eyer, J.R., Crawford, R.F., 1933. Observations on the feeding habits of the potato psylid (*Paratrioza cockerelli* Sulc.) and the pathological history of the "psyllid yellows" which it produces. Journal of Economic Entomology 26 (4), 846–850.

Gao, F., Jifon, J., Yang, X., Liu, T., 2009. Zebra chip disease incidence on potato is influenced by timing of potato psyllid infestation, but not by the host plants on which they were reared. Insect Science 16 (5), 399–408.

Garzon, T.J.A., Becerra, F.A., Marin, A., Mejia, A.C., Byerly, M.F., 1992. Manejo integrado de la enfermedad "permamente del tomate" en El Bajio. In: Urias, M.C.M., Rodrigues, R., Navarro Bravo, A. (Eds.), Afidos como vectores de virus en Mexico. Contribution a la Ecologia y Control de Afidos en Mexico, pp. 116–128.

Garzón-Tizanado, J.A., 2003. El pulgon Saltador o la *Paratrioza*, una amenaza para la horticultura de Sinaloa. In: Report of the Workshop on *Paratrioza cockerelli* Sulc. As a Pest and Vector of Phytoplasmas in Vegetables. Culiacán, Sinaloa, Mexico, pp. 79–87.

Gerhardt, P.D., 1966. Potato psyllid and green peach aphid control on Kennebec potatoes with Temik and other insecticides. Journal of Economic Entomology 59 (1), 9–11.

Gerhardt, P.D., Turley, D.L., 1961. Control of certain potato insects with soil applications of granulated phorate. Journal of Economic Entomology 54 (6), 1217–1221.

Gharalari, A.H., Nansen, C., Lawson, D.S., Gilley, J., Munyaneza, J.E., Vaughn, K., 2009. Knockdown mortality, repellency, and residual effects of insecticides for control of adult *Bactericera cockerelli* (Hemiptera: Psyllidae). Journal of Economic Entomology 102 (3), 1032–1038.

Gill, G., 2006. Tomato psyllid detected in New Zealand. Biosecurity New Zealand 69, 10–11.

Goolsby, J.A., Adamczyk, J., Bextine, B., Lin, D., Munyaneza, J.E., Bester, G., 2007. Development of an IPM program for management of the potato psyllid to reduce incidence of zebra chip disorder in potatoes. Subtropical Plant Science 59, 85–94.

Goolsby, J.A., Adamczyk, J.J., Crosslin, J.M., Troxclair, N.N., Anciso, J.R., Bester, G.G., Zens, B.A., 2012. Seasonal population dynamics of the potato psyllid (Hemiptera: Triozidae) and its associated pathogen "*Candidatus* Liberibacter solanacearum" in potatoes in the southern great plains of North America. Journal of Economic Entomology 105 (4), 1268–1276.

Greenway, G., 2014. Economic impact of zebra chip control costs on grower returns in seven US States. American Journal of Potato Research 91 (6), 714–719.

Gudmestad, N.C., Secor, G.A., 2007. Zebra chip: a new disease of potato. Nebraska Potato Eyes 19, 1–4.

Guenthner, J., Goolsby, J., Greenway, G., 2012. Use and cost of insecticides to control potato psyllids and zebra chip on potatoes. Southwestern Entomologist 37 (3), 263–270.

Halbert, S.E., Munyaneza, J.E., 2012. Potato Psyllids and Associated Pathogens: A Diagnostic Aid. Florida State Collection of Arthropods, Gainesville, FL. http://www.fscadpi.org/Homoptera_Hemiptera/Potato_psyllids_and_associated_pathogens.pdf.

Hansen, A.H.K., Trumble, J.T., Stouthamer, R., Paine, T.D., 2008. A New Huanglongbing species, *Candidatus* Liberibacter psyllaurous, found to infect tomato and potato, is vectored by the psyllid *Bactericera cockerelli* (Sulc). Applied and Environmental Microbiology 74, 5862–5865.

Harding, J.A., 1962. Tests with systemic insecticides for control of insects and certain diseases on potatoes. Journal of Economic Entomology 55 (1), 62–64.

Henne, D.C., Paetzold, L., Workneh, F., Rush, C.M., 2010. Evaluation of potato psyllid cold tolerance, overwintering survival, sticky trap sampling, and effects of Liberibacter on potato psyllid alternate host Plants. In: Proceedings of the 10th Annual SCRI Zebra Chip Reporting Session, pp. 149–153.

Hill, R.E., 1945. Effects of DDT and other insecticides on several species of potato insects. Nebraska Agricultural Experimental Station Research Bulletin 138, 1–14.

Hill, R.E., 1948. Research on potato insect problems-A review of recent literature. American Potato Journal 25, 107–127.

Hodkinson, I.D., 2009. Life cycle variation and adaptation in jumping plant lice (Insecta: Hemiptera: Psylloidea): a global synthesis. Journal of Natural History 43 (1–2), 65–179.

Horton, D.R., Miliczky, E., Munyaneza, J.E., Swisher, K.D., Jensen, A.S., 2014. Absence of photoperiod effects on mating and ovarian maturation by three haplotypes of potato psyllid, *Bactericera cockerelli* (Hemiptera: Triozidae). Journal of the Entomolgical Society of British Columbia 111, 1–12.

Ivicevich, T., 2006. Another difficult year. Grower 61 (8), 38.

Jackson, B.C., Goolsby, J., Wyzykowski, A., Vitovksy, N., Bextine, B., 2009. Analysis of genetic relationships between potato psyllid (*Bactericera cockerelli*) populations in the United States, Mexico and Guatemala using ITS2 and Inter Simple Sequence Repeat (ISSR) data. Subtropical Plant Science 61, 1–5.

Jensen, J.H., 1938. Psyllid yellows in Nebraska. Plant Disease Reporter 22 (2), 35–36.

Jensen, D.D., 1954. Notes on the potato psyllid, *Paratrioza cockerelli* (Sulc). Pan-Pacific Entomologist 30, 161–165.

Johnson, T.E., 1971. The Effectiveness of *Tetrastichus triozae* Burks (Hymenoptera: Psyllidae) in North Central Colorodo (M.S. thesis). Colorodo State University, USA.

Jorgensen, N., Butler, R.C., Vereijssen, J., 2013. Biorational insecticides for control of the tomato potato psyllid. New Zealand Plant Protection 66, 333–340.

Knowlton, G.F., 1933a. Length of adult life of *Paratrioza cockerelli* (Sulc). Journal of Economic Entomology 26 (3), 730.

Knowlton, G.F., 1933b. Notes on an injurious Utah insects: potato psyllid. In: Proceedings of the Utah Academy of Science, vol. 10, p. 153.

Knowlton, G.F., Allen, M., 1936. Three hemipterous predators of potato psyllid. In: Proceedings of the Utah Academy of Science, vol. 13, pp. 293–294.

Knowlton, G.F., Janes, M.J., 1931. Studies on the biology of *Paratrioza cockerelli* (Sulc). Annals of the Entomological Society of America 24 (2), 283–291.

Knowlton, G.F., Thomas, W.L., 1934. Host plants of the potato psyllid. Journal of Economic Entomology 27 (2), 547.

Lacey, L.A., de la Rosa, F., Horton, D.R., 2009. Insecticidal activity of entomopathogenic fungi (*Hypocreales*) for potato psyllid, *Bactericera cockerelli* (Hemiptera: Triozidae): development of bioassay techniques, effect of fungal species and stage of the psyllid. Biocontrol Science and Technology 19 (9), 957–970.

Lacey, L.A., Liu, T.-X., Buchman, J.L., Munyaneza, J.E., Goolsby, J.A., Horton, D.R., 2011. Entomopathogenic fungi (Hypocreales) for control of potato psyllid, *Bactericera cockerelli* (Šulc) (Hemiptera: Triozidae) in an area endemic for zebra chip disease of potato. Biological Control 56 (3), 271–278.

Lehman, R.S., 1930. Some observations on the life-history of the tomato psyllid. Journal of the New York Entomological Society 38 (3), 307–312.

Lévy, J., Hancock, J., Ravindran, A., Gross, D., Tamborindeguy, C., Pierson, E., 2013. Methods for rapid and effective PCR-based detection of "*Candidatus* Liberibacter solanacearum" from the insect vector *Bactericera cockerelli* streamlining the DNA extraction/purification process. Journal of Economic Entomology 106 (3), 1440–1445.

Li, W., Abad, J.A., French-Monar, R.D., Rascoe, J., Wen, A., Gudmestad, N.C., Levy, L., 2009. Multiplex real-time PCR for detection, identification and quantification of "*Candidatus* Liberibacter solanacearum" in potato plants with zebra chip. Journal of Microbiological Methods 78 (1), 59–65.

Liefting, L.W., Perez-Egusquiza, Z.C., Clover, G.R.C., Anderson, J.A.D., 2008a. A new "*Candidatus* Liberibacter" species in *Solanum tuberosum* in New Zealand. Plant Disease 92, 1474.

Liefting, L.W., Ward, L.I., Shiller, J.B., Clover, G.R.G., 2008b. A new '*Candidatus* Liberibacter' species in *Solanum betacum* (Tamarillo) and *Physalis peruviana* (Cape Gooseberry) in New Zealand. Plant Disease 92 (11), 1588.

Liefting, L.W., Sutherland, P.W., Ward, L.I., Paice, K.L., Weir, B.S., Clover, G.R.G., 2009. A new "*Candidatus* Liberibacter" species associated with diseases of solanaceous crops. Plant Disease 93 (3), 208–214.

List, G.M., 1917. A test of lime-sulphur and nicotine sulfate for the control of the tomato psyllid and effect of these materials upon plant growth. Colorodo Agricultural Station Bulletin 9, 40–41.

List, G.M., 1918. A test of lime sulphur and nicotine sulfate for the control of thetomato psyllid and effect of these materials upon plant growth. Ninth Annual Report State Entomologist Colorado 47, 16.

List, G.M., 1935. Psyllid yellow of tomatoes and control of the psyllid, *Paratrioza cockerelli* (Sulc), by the use of sulfur. Journal of Economic Entomology 28 (2), 431–436.

List, G.M., 1938. Test of certain materials as controls for the tomato psyllid, *Paratrioza cockerelli* (Sulc), and psyllid yellows. Journal of Economic Entomology 31 (4), 491–497.

List, G.M., 1939. The effect of temperature upon egg deposition, egg hatch, and nymphal development of *Paratrioza cockerelli* (Sulc). Journal of Economic Entomology 1, 30–36.

List, G.M., Daniels, L.B., 1934. A promising control for psyllid yellows. Science 79, 2039.

Liu, D., Trumble, J.T., 2004. Tomato psyllid behavioral responses to tomato plant lines and interactions of plant lines with insecticides. Journal of Economic Entomology 97 (3), 1078–1085.

Liu, D., Trumble, J.T., 2005. Interactions of plant resistance and insecticides on the development and survival of *Bactericerca cockerelli* [Sulc] (Homoptera: Psyllidae). Crop Protection 24 (2), 111–117.

Liu, D., Trumble, J.T., 2006. Ovipositional preferences, damage thresholds, and detection of the tomato/potato psyllid (*Bactericera cockerelli* [Sulc]) on selected tomato accessions. Bulletin of Entomological Research 96, 197–204.

Liu, D., Trumble, J.T., 2007. Comparative fitness of invasive and native populations of the potato psyllid (*Bactericera cockerelli*). Entomologia Experimentalis et Applicata 123, 35–42.

Liu, D., Johnson, L., Trumble, J.T., 2006a. Differential responses to feeding by the tomato/potato psyllid between two tomato cultivars and their implications in establishment of injury levels and potential of damaged plant recovery. Insect Science 13 (3), 195–204.

Liu, D., Trumble, J.T., Stouthamer, R., 2006b. Genetic differentiation between eastern populations and recent introductions of potato psyllid (*Bactericera cockerelli*) into western North America. Entomologia Experimentalis et Applicata 118 (3), 177–183.

Liu, T.-X., Zhang, Y.-M., Peng, L.-N., Rojas, P., Trumble, J.T., 2012. Risk assessment of selected insecticides on *Tamarixia triozae* (Hymenoptera: Eulophidae), a parasitoid of *Bactericera cockerelli* (Hemiptera: Trizoidae). Journal of Economic Entomology 105 (2), 490–496.

Lopez, B., Favela, S., Ponce, G., Foroughbakhch, R., Flores, A.E., 2013. Genetic variation in *Bactericera cockerelli* (Hemiptera: Triozidae) from Mexico. Journal of Economic Entomology 106 (2), 1004–1010.

Martin, N.A., 2008. Host plants of the potato/tomato psyllid: a cautionary tale. Weta 16, 12–16.

Martini, X., Seibert, S., Prager, S.M., Nansen, C., 2012. Sampling and interpretation of psyllid nymph counts in potatoes. Entomologia Experimentalis et Applicata 143 (2), 103–110.

Mills, H.B., 1939. Montana insect pests for 1937 and 1938. Bulletin of the Montana Agriculture Experimental Station 366, 32.

Munyaneza, J.E., 2012. Zebra chip disease of potato: biology, epidemiology, and management. American Journal of Potato Research 89 (5), 329–350.

Munyaneza, J.E., Crosslin, J.M., Upton, J.E., 2007a. Association of *Backtericera cockerelli* (Homoptera: Pyllidae) with "Zebra Chip" a new potato disease in southwestern United States and Mexico. Journal of Economic Entomology 100 (3), 656–663.

Munyaneza, J.E., Goolsby, J.M.A., Crosslin, J.M., Upton, J.E., 2007b. Further evidence that Zebra Chip potato disease in the lower Rio Grande Valley of Texas is associated with *Bactericera cockerelli*. Subtropical Plant Science 59, 30–37.

Munyaneza, J.E., Crosslin, J.M., Buchman, J.L., 2009. Seasonal occurrence and abundance of the potato psyllid, *Bactericera cockerelli*, in south central Washington. American Journal of Potato Research 86 (6), 513–518.

Nelson, W.R., Fisher, T.W., Munyaneza, J.E., 2011. Haplotypes of "*Candidatus* Liberibacter solanacearum" suggest long-standing separation. European Journal of Plant Pathology 130 (1), 5–12.

Nelson, W.R., Sengoda, V.G., Alfaro-Fernandez, A.O., Font, M.I., Crosslin, J.M., Munyaneza, J.E., 2012. A new haplotype of "*Candidatus* Liberibacter solanacearum" identified in the Mediterranean region. European Journal of Plant Pathology 135 (4), 633–639.

Nelson, W.R., Munyaneza, J.E., McCue, K.F., Bové, J.M., 2013. The pangaean origin of "*Candidatus* liberibacter" species. Journal of Plant Pathology 95, 455–461.

Nissinen, A.I., Haapalainen, M., Jauhiainen, L., Lindman, M., Pirhonen, M., 2014. Different symptoms in carrots caused by male and female carrot psyllid feeding and infection by '*Candidatus* Liberibacter solanacearum'. Plant Pathology 63, 812–820.

Page-Weir, N.E., Jamieson, L.E., Chhagan, A., Connolly, P.G., Curtis, C., 2011. Efficacy of insecticides against the tomato/potato psyllid (*Bactericera cockerelli*). New Zealand Plant Protection 64, 276–281.

Pearson, C.C., Backus, E.A., Shugart, H.J., Munyaneza, J.E., 2014. Characterization and correlation of EPG waveforms of *Bactericera cockerelli* (Hemiptera: Triozidae): variability in waveform appearance in relation to applied signal. Annals of the Entomological Society of America 107 (3), 650–666.

Peng, L., Trumble, J.T., Munyaneza, E., Liu, T., Munyaneza, J.E., 2011. Repellency of a kaolin particle film to potato psyllid, *Bactericera cockerelli* (Hemiptera: Psyllidae), on tomato under laboratory and field conditions. Pest Management Science 67 (7), 815–824.

Percy, D.M., 2012. An annotated checklist of the psyllids of California (Hemiptera: Psylloidea). Zootaxa 3193, 1–27.

Pletsch, D.J., 1942. The effect of some insecticides on the immature stages of the potato and tomato psyllid, *Paratrioza cockerelli* (Sulc). Journal of Economic Entomology 35 (1), 58–60.

Pletsch, D.J.J., 1947. The potato psyllid *Paratrioza cockerelli* (Sulc), its biology and control. Bulletin of the Montana Agriculture Experimental Station 446, 1–95.

Prager, S.M., Butler, C.D., Trumble, J.T., 2012. Area wide sampling for potato psyllids: comparisons of distributions and scouting strategies on potatoes, tomatoes, and peppers. In: Proceedings of the 12th Annual SCRI Zebra Chip Reporting Session, pp. 48–53.

Prager, S.M., Butler, C.D., Trumble, J.T., 2013a. A sequential binomial sampling plan for potato psyllid (Hemiptera: Triozidae) on bell pepper (*Capsicum annum*). Pest Management Science 69 (10), 1131–1135.

Prager, S.M., Lewis, O.M., Vaughn, K., Nansen, C., 2013b. Oviposition and feeding by *Bactericera cockerelli* (Homoptera: Psyllidae) in response to a solar protectant applied to potato plants. Crop Protection 45, 57–62.

Prager, S.M., Vindiola, B., Kund, G.S., Byrne, F.J., Trumble, J.T., 2013c. Considerations for the use of neonicotinoid pesticides in management of *Bactericera cockerelli* (Šulk) (Hemiptera: Triozidae). Crop Protection 54, 84–91.

Prager, S.M., Butler, C.D., Trumble, J.T., 2014a. A binomial sequential sampling plan for *Bactericera cockerelli* (Hemiptera: Triozidae) in *Solanum lycopersicum* (Solanales: Solanacea). Journal of Economic Entomology 107 (2), 838–845.

Prager, S.M., Esquivel, I., Trumble, J.T., 2014b. Factors influencing host plant choice and larval performance in *Bactericera cockerelli*. PLoS One. 9 (4), e94047. http://dx.doi.org/10.1371/journal.pone.0094047.

Prager, S.M., Lewis, O.M., Michels, J., Nansen, C., 2014c. The influence of maturity and variety of potato plants on oviposition and probing of *Bactericera cockerelli* (Hemiptera: Triozidae). Environmental Entomology 43 (2), 402–409.

Prager, S.M., Kund, G.S., Trumble, J.T., 2016. Low-input, low-cost IPM program helps manage potato psyllid. California Agriculture 70 (2), 89–95.

Rashed, A., Nash, D., Paetzold, L., Workneh, F., Rush, C.M., Nash, T.D., 2013. Transmission efficiency of *Candidatus* Liberibacter solanacearum and potato zebra chip disease progress in relation to pathogen titer, vector numbers and feeding sites. Phytopathology 102 (11), 1–30.

Richards, B., 1928. A new and destructive disease of potato in Utah and its relation to potato psyllid. Phytopathology 18 (1), 140–141.

Richards, B.L., 1929. Psyllid yellows (cause undetermined). Plant Disease Reporter 68, 28.

Richards, B.L., 1933. Notes on injurious Utah insects: potato psyllid. In: Proceedings of the Utah Academy of Science, vol. 10, p. 153.

Richards, B.L., Blood, H.L., 1933. Psyllid yellows of the potato. Journal of Agricultural Research 46, 189–216.

Robertson, K., 2006a. Tomato product group. Potato/tomato psyllid in New Zealand. Grower 61 (9), 44.

Robertson, K., 2006b. Tomato product group. Tomato psyllid. Grower 61 (6), 36.

Robertson, K., 2008a. Fresh tomato group. Industry scan from product group meeting 16 April. Grower 63 (4), 20.

Robertson, K., 2008b. Fresh tomato group. New plant disease – *Candidatus* Liberibacter species found in greenhouse tomato and capsicum crops. Grower 16 (3), 16.

Romney, V.E., 1939. Breeding areas of the tomato psyllid, *Paratrioza cockerelli* (Sulc). Journal of Economic Entomology 32, 150–151.

Rowe, J.A., Knowlton, G.F., 1935. Studies upon the morphology of *Paratrioza cockerelli* (Sulc). Utah Academy Science Proceedings 12, 233–239.

Rubio, C.O., Almeyda, L., Cadena, H., Lobato, S.R., 2011. Relation between *Bactericera cockerelli* and presence of *Candidatus* Liberibacter psyllaurous in commercial fields of potato. Revista Mexicana de Ciencias Agricolas 2, 17–28.

Sandanayaka, W.R.M., Moreno, A., Tooman, L.K., Page-Weir, N.E.M., Fereres, A., 2014. Stylet penetration activities linked to the acquisition and inoculation of *Candidatus* Liberibacter solanacearum by its vector tomato potato psyllid. Entomologia Experimentalis et Applicata 151 (2), 170–181.

Secor, G.A., Rivera-Varas, V., 2004. Emerging diseases of cultivated potato and their impact on Latin America. Revista Latinoamericana de la Papa (Suplemento) 1, 1–8.

Sengoda, V., Munyaneza, J., Crosslin, J., Buchman, J., Pappu, H., 2010. Phenotypic and etiological differences between psyllid yellows and zebra chip diseases of potato. American Journal of Potato Research 87 (1), 41–49.

Sengoda, V.G., Buchman, J.L., Henne, D.C., Pappu, R., Munyaneza, J.E., Pappu, H.R., 2013. "*Candidatus* Liberibacter solanacearum" titer over time in *Bactericera cockerelli* (Hemiptera: Triozidae) after acquisition from infected potato and tomato plants. Journal of Economic Entomology 106 (5), 1964–1972.

Shapovalov, M., 1929. Tuber transmission of psyllid yellows in California. Phytopathology 19 (12), 1140.

Šulc, K., 1909. *Dioza cockerelli* n.sp., a novelty from North America, being also of economic importance. Acta Societatis Zoologicae Bohemoslovenicae Sandanayaka 6 (4), 102–108.

Swisher, K.D., Munyaneza, J.E., Crosslin, J.M., 2012. High resolution melting analysis of the cytochrome oxidase I gene identifies three haplotypes of the potato psyllid in the United States. Environmental Entomology 41 (4), 1019–1028.

Swisher, K.D., Munyaneza, J.E., Crosslin, J.M., 2013a. Temporal and spatial analysis of potato psyllid haplotypes in the United States. Environmental Entomology 42 (2), 381–393.

Swisher, K.D., Sengoda, V.G., Dixon, J., Echegaray, E., Murphy, A.F., Rondon, S.I., Crosslin, J.M., 2013b. Haplotypes of the potato psyllid, *Bactericera cockerelli*, on the wild host plant, *Solanum dulcamara*, in the pacific northwestern United States. American Journal of Potato Research 90 (6), 570–577.

Swisher, K.D., Sengoda, V.G., Dixon, J., Munyaneza, J.E., Murphy, A.F., Rondon, S.I., Crosslin, J.M., 2014. Assessing potato psyllid haplotypes in potato crops in the Pacific Northwestern United States. American Journal of Potato Research 91 (5), 485–491.

Tahzima, R., Maes, M., Unit, P.S., Achbani, E.H., Kaddour, R.H., Swisher, K.D., De Jonghe, K., 2014. First report of "*Candidatus* Liberibacter solanacearum" on carrot in Africa. Plant Disease 98 (10), 1426.

Teresani, G.R., Bertolini, E., Alfaro-Fernandez, A., Martinez, C., Francisco, A.O., Kitajima, E.W., Font, M.I., 2014. Association of "*Candidatus* Liberibacter solanacearum" with a vegetative disorder of celery in Spain and development of a real-time PCR method for its detection. Phytopathology 104 (8), 804–811.

Teulon, D.A.J., Workman, P.J., Thomas, K.L., Nielsen, M.C., 2009. *Bactericera cockerelli*: incursion, dispersal and current distribution on vegetable crops in New Zealand. New Zealand Plant Protection 62, 136–144.

Thomas, K.L., Jones, D.C., Kumarasinghe, L.B., Richmond, J.E., Gill, G.S.C., 2011. Investigation into the entry pathway for tomato potato psyllid *Bactericera cockerelli*. Plant Protection 64, 259–268.

Tran, L.T., Worner, S.P., Hale, R.J., Teulon, D.A., 2012. Estimating development rate and thermal requirements of *Bactericera cockerelli* (Hemiptera: Triozidae) reared on potato and tomato by using linear and nonlinear models. Environmental Entomology 41 (5), 1190–1198.

Trumble, J., Prager, S.M., 2014. Resistance evaluation and insecticide rotation programs for control of potato psyllids. In: Proceedings of the 14th Annual SCRI Zebra Chip Reporting Session, November 9, 2014, Portland, OR, USA.

Trumble, J.T., Edelson, J.V., Story, R.N., 1987. Conformity and incongruity of selected dispersion indices in describing the spatial distribution of *Trichoplusia ni* (Hübner) in geographically separate cabbage plantings. Researches on Population Ecology 29, 155–166.

Trumble, J.T., Brewer, M.J., Shelton, A.M., Nyrop, J., 1989. Transportability of fixed-precision level sampling plans. Researches Population Ecology 31, 325–342.

Tuthill, L.D., 1945. Contributions to the knowledge of the Psyllidae of Mexico (part II). Journal of the Kansas Entomological Society 18 (1), 1–29.

UCIPM, 2012. Emerging Pests in California – Tomato, UCIPM Integrated Pest Management Guidelines. Statewide Integrated Pest Management guidelines University of California, USA. http://ipm.ucanr.edu/PMG/selectnewpest.tomatoes.html.

Vargas-Madríz, H., Bautista-Martínez, N., Vera-Graziano, J., García-Gutiérrez, C., Chavarín-Palacio, C., 2013. Morphometrics of eggs, nymphs, and adults of *Bactericera cockerelli* (Hemiptera: Triozidae), grown on two varieties of tomato under greenhouse conditions. Florida Entomologist 96 (1), 71–79.

Walker, M.K., Butler, R.C., Berry, N.A., 2010. Evaluation of Selected Soft Chemicals as Potential Control Options for Tomato/Potato Psyllid. Plant and Food Research Lincoln, Christchurch, New Zealand, p. 18.

Walker, G.P., Macdonald, F.H., Larsen, N.J., Wallace, A.R., 2011. Monitoring *Bactericera cockerelli* and associated insect populations in potatoes in South Auckland. New Zealand Plant Protection 64, 269–275.

Walker, G.P., Macdonald, F.H., Larsen, N.J., Wright, P.J., Wallace, A.R., 2013. Sub-sampling plants to monitor tomato-potato psyllid (*Bactericera cockerelli*) and associated insect predators in potato crops. New Zealand Plant Protection 66, 341–348.

Wallis, R.L., 1955. Ecological Studies on the Potato Psyllid as a Pest of Potatoes (No. 1107). USDA Technical Bulletin 1107, pp. 1–24.

Wen, A., Johnson, C., Gudmestad, N.C., 2013. Development of a no caps in title PCR assay for the rapid detection and differentiation of "*Candidatus* Liberibacter solanacearum" haplotypes and their spatiotemporal distribution in the United States. American Journal of Potato Research 90 (3), 229–236.

Whipple, S.D., Bradshaw, J.D., Haverson, R.M., 2012. Cold tolerance in potato psyllids. In: Proceedings of the 12th Annual SCRI Zebra Chip Reporting Session, pp. 191–193.

Workman, P., Pedley, R., 2007a. New natural enemies for greenhouse pests. Grower 62 (9), 54–55.

Workman, P., Pedley, R., 2007b. Search for natural control of potato/tomato psyllid. Grower 62 (5), 28.

Workneh, F., Henne, D.C., Goolsby, J.A., Crosslin, J.M., Whipple, S.D., Bradshaw, J.D., Rush, C.M., 2013. Characterization of management and environmental factors associated with regional variations in potato zebra chip occurrence. Phytopathology 103, 1235–1242.

Yang, X.-B., Liu, T.-X., 2009. Life history and life tables of *Bactericera cockerelli* (Homoptera: Psyllidae) on eggplant and bell pepper. Environmental Entomology 38 (6), 1661–1667.

Yang, X.-B.B., Zhang, Y.-M.M., Hua, L., Peng, L.-N.N., Munyaneza, J.E.E., Trumble, J.T., Liu, T.-X.X., 2010a. Repellency of selected biorational insecticides to potato psyllid, *Bactericera cockerelli* (Hemiptera: Psyllidae). Crop Protection 29 (11), 1320–1324.

Yang, X.-B., Zhang, Y.-M., Lei, H., Liu, T.-X., 2010b. Life history and life tables of *Bactericera cockerelli* (Hemiptera: Psyllidae) on potato under laboratory and field conditions in the lower Rio Grande Valley of Texas. Journal of Economic Entomology 103 (5), 1729–1734.

Yen, A.L., Madge, D.G., Berry, N.A., Yen, J.D.L., 2012. Evaluating the effectiveness of five sampling methods for detection of the tomato potato psyllid, *Bactericera cockerelli* (Šulc) (Hemiptera: Psylloidea: Triozidae). Australian Journal of Entomology 52 (2), 168–174.

Chapter 8

Minor Pests

Gerald E. Brust[1], Waqas Wakil[2], Mirza A. Qayyum[2,3]

[1]*CMREC-UMF, University of Maryland, Upper Marlboro, MD, United States;* [2]*University of Agriculture, Faisalabad, Pakistan;* [3]*Muhamamd Nawaz Sharif University of Agriculture, Multan, Pakistan*

1. INTRODUCTION

Damage caused by minor pests of tomatoes is often localized to one area of tomato production. They may also be common in many tomato growing regions, but they cause marketable yield loss sporadically, both in time and space. For many of these minor pests, biological control agents, which are negatively impacted by overusage of pesticides, are important in maintaining these pest populations below economic thresholds. Therefore, it is important to use integrated pest management (IPM) programs in tomato systems to preserve natural enemies while also managing both major and minor pests.

2. STINK BUGS, *EUSCHISTUS CONSPERSUS* UHLER, *THYANTA PALLIDOVIRENS* (STÅL), *NEZARA VIRIDULA* (L.), *CHINAVIA HILARIS* SAY, *EUSCHISTUS SERVUS* (SAY), *HALYOMORPHA HALYS* STÅL, *CHLOROCHROA SAYI* (STÅL), *CHLOROCHROA UHLERI* (STÅL)

The family Pentatomidae is made up of a group of insects commonly called stink bugs that are familiar pests observed during the growing season on a wide variety of plants. While there are many species in this family they all cause very similar damage in tomatoes. The group gets its name because of their ability to release foul smelling odors from scent glands on their thorax located near their middle legs (both in the adults and the immatures) (Aldrich et al., 1995). All stink bugs, adults and immatures have piercing-sucking mouthparts that remove plant fluids. Because there are so many species that are considered to be "stink bug" pests, this section will discuss, in general terms, the various life histories and descriptions of the pests. A few specific species will be discussed in more detail because of their importance as pests in tomato. Although, there are other families of bugs that can attack tomato, Coreidae and Miridae, these will not be discussed in this section.

2.1 Identification

Adult stink bugs are shield-shaped, 1.3–1.8 cm long, 1.2–1.7 cm wide, and tend to be brown, green, or gray (McPherson et al., 1982). Some species have other distinguishing markings such as the brown marmorated stink bug *Halyomorpha halys* Stål that has two white spots on its antennae and alternating dark bands on the thin outer edge of its abdomen (Fig. 8.1). Other species have distinctive red or yellow markings. Some of the more common species of stink bug found as pests of tomato are: Consperse stink bug *Euschistus conspersus* Uhler; Red shouldered stink bug *Thyanta pallidovirens* (Stål) (= *T. acerra*), Southern green stink bug *Nezara viridula* (L.); Green stink bug *Chinavia hilaris* Say (=*Chinavia hilare* Say, *Nezara hilaris* Say, *Acrosternum hilaris* Say, *Pentatoma hilaris* Say); Brown stink bug *Euschistus servus* (Say); Say stink bug *Chlorochroa sayi* Stål; Bagrada bug *Bagrada hilaris* (Burmeister); and Uhler's stink bug *Chlorochroa uhleri* (Stål). Eggs of stink bugs are distinctively barrel-shaped, 1.4 mm long and 1.3 mm wide, laid in tight clustered rows of between 30 and 130 eggs glued together. The eggs tend to be white or green when first laid, later turning pinkish to red or multicolored. Eggs usually are laid on the underside of foliage and hatch in approximately 6–7 days (Miner, 1966; Dietz et al., 1976). Immatures or nymphs are oval-shaped and usually dark colored or black and red when small. As they mature, they transform to adult coloration of green, brown, or gray. Nymphs can have distinctive patterns of spots or other markings that help distinguish one species from another (Dietz et al., 1976). The younger immatures (first to third instars) frequently aggregate while the later instars disperse from the egg-laying site as they mature (Slater and Baranowski, 1978).

Sustainable Management of Arthropod Pests of Tomato. http://dx.doi.org/10.1016/B978-0-12-802441-6.00008-5

FIGURE 8.1 Brown marmorated stink bug.

2.2 Biology

Overwintering stink bugs are in reproductive diapause and often appear reddish-brown (Brennan et al., 1977; Harris et al., 1984). Most stink bug species become active during the first warm days of the growing season. Soon after becoming active they begin to lay eggs. Egg laying activity continues throughout the growing season. The number of generations for stink bugs varies from uni- to multivoltine. In general the further north the stink bug is found, the fewer the number of generations. However, a few species are multivoltine even when found north of 40 degrees latitude (McPherson et al., 1994; Munyaneza and McPherson, 1994). The egg to adult life cycle takes 30–45 days to complete, but varies with temperature. Stink bugs go through an incomplete metamorphosis with five nymphal instars, which takes about 25–30 days to complete (Todd, 1989). Adults live 45–60 days, on average. Adults overwinter in protected areas such as under leaf litter in woods, under crop residue, in cover crops, or on weeds such as wild mustards (Jones and Sullivan, 1981; Javahery, 1990).

2.3 Distribution, Host Range, and Seasonal Occurrence

Stink bugs are found throughout the temperate, neotropical, and tropical crop production areas of the world. Stink bugs feed on a wide variety of food sources from trees such as elderberry, black locust, honey locust, black cherry, dogwoods, basswood, and even pine trees to weeds such as mullein, various grasses, jimson weed, fleabane, wild lettuce, trumpet creeper, sedge, horse nettle, sage, mallow, wild mustards, and shepherd's purse, to many crop plants including apples, beans, blackberry, corn, cotton, cabbage, crucifers, eggplant, peaches, pear, peppers, soybean, and tomato. Over 620 genera of plants are hosts to these pests (McPherson and McPherson, 2000). Stink bugs become active during the first few warm days of the season, and are most active in mid- and late season. They begin to move to overwintering sites as day lengths shorten and temperatures cool.

2.4 Damage, Losses, and Economic Thresholds

The early instar nymphs of stink bugs feed preferentially on the foliage of tomatoes (Simmons and Yeargan, 1988; Hallman et al., 1992). Lye et al. (1988) found that tomatoes would experience a significant reduction in growth with an increase in southern green stink bug density and feeding duration. Older nymphs and adults preferentially feed on green tomato fruit using their mouthparts to remove plant fluids (Lye and Story, 1988). When stink bugs feed they inject enzymes that help breakdown and liquefy tomato fruit cells that can be sucked up (Lye et al., 1988; Jones and Caprio, 1994). This puncture-type of feeding lowers the quality of the fruit often making it unmarketable. The nymphs and adults also may inject yeasts, such as *Eremothecium* spp., that cause fruit rot (Lye et al., 1988; Apriyanto et al., 1989). The brown marmorated stink bug is known to vector *Eremothecium coryli* (Peglion) Kurtzman, which causes extensive fruit rotting (Brust and Rane, 2011).

The feeding wound made by stink bugs also provides an entry point for opportunistic pathogens (Jones and Caprio, 1994; McPherson et al., 1994). The feeding by stink bugs causes tiny punctures surrounded by empty cells that are filled with air that appear as white slightly sunken areas known as "cloudy spot" under the skin of green tomato fruit and as yellow sunken areas in red fruit (Fig. 8.2) (Brust, 2015). Underneath the cloudy spots, the tissue is spongy white (Fig. 8.3).

This damage along with the introduction of yeast and other fruit rotting pathogens increases the number of unmarketable fruit. Tomato fields that undergo heavy stink bug feeding can suffer significant yield (Zalom and Zalom, 1992). Nontreated tomatoes have been shown to have 20% damage from stink bug feeding while insecticide-treated tomatoes had ~4% stink bug damage (Chalfant, 1973). In some cases tomato fruit fed upon by stink bugs can have a bitter taste and mealy texture reducing marketable yield (Callahan et al., 1960).

Developing a threshold for stink bugs has been a challenge for pest management workers. Only a few trapping schemes have been established to monitor stink bugs; these include the consperse and brown marmorated stink bugs (see Section 2.5.1).

FIGURE 8.2   Stink bug feeding damage to tomato (*cloudy spot*).

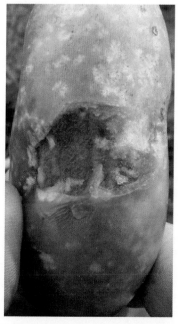

FIGURE 8.3   Stink bug feeding damage to tomato with outer skin removed showing white spongy area.

The adults can be seen at times feeding on plants in the early morning, but as the day becomes hotter they tend to move into the center of the plant (Brust, 2015). Stink bugs also feed extensively at night, which is another reason they are difficult to scout. At times the nymphs can be found in the interior of the tomato plant where they feed on foliage and fruit (Brust, 2015). Often the only thing that alerts one to their presence is the damaged fruit. One threshold for stink bugs is based on the amount of fruit damage in a field. Ten fruits in 40 different locations in a field should be examined, and if 0.5–1% of the fruit has feeding damage treatment is required (LeBoeuf, 2015). Fresh market tomatoes and possibly processing tomatoes sold for solid-pack canning are more sensitive to any feeding damage and need first consideration for treatment (Lye et al., 1988). However, if processing tomatoes are planned for paste then treatment for stink bugs is usually not justified unless the field environment is favorable for yeast or fungal pathogen development. A threshold developed in California (United States) for the processing tomato industry uses one-third to one-half of a stink bug per plant shake (see Section 2.5.1), which on average will result in approximately 5% fruit damage (Anonymous, 2013a).

2.5 Management

2.5.1 Monitoring

Stink bugs are difficult to monitor in tomatoes because they tend to hide in the interior of the plant when temperatures rise, they feed extensively at night, they drop or fly from the plant at slight disturbances, they consistently move in and out of a field to alternate hosts, and their coloration makes them difficult to detect. However, a sequential sampling program was developed for fresh market tomatoes in Louisiana (United States) for *N. viridula* using a visual count to sample populations from horizontal plant strata. A sequential sampling plan using the penultimate fruit cluster as a sample unit was developed to classify populations into low, moderate, and severe damage levels (Lye and Story, 1989). Another sampling program for stink bugs in processing tomatoes in California was developed based on the spatial distribution of the bugs (Zalom et al., 1977). This sampling program uses the plant shake method for detecting stink bugs. This method is performed by beating or shaking the tomato vines and looking for stink bugs on the ground or on a cloth or tray that was placed at the base of the plant (Anonymous, 1998). Stink bug presence often can be detected before any are seen in a field by their watery frass that desiccates into brown spots on leaves and fruit.

Only two pheromones have been developed to monitor stink bugs. One is for the conperse stink bug where a double-cone trap (Anonymous, 2013a) with an aggregation pheromone should be placed along field borders at flowering. The other pheromone that is commercially available is for brown marmorated stink bug, which uses a rocket-type trap that is placed along field edges especially where fields abut woods (Khrimian et al., 2014). The traps are not used to make treatment decisions, but instead are used to begin intense sampling programs in the field for the presence of and damage due to stink bugs.

The University of California Davis (United States) has developed a phenology model for conperse stink bug that helps to predict nymphal emergence (Anonymous, 2013a). The model starts by calculating degree-days (DD) from the first capture of adult stink bugs in pheromone traps using 12°C as the developmental threshold. At 310 DD, intensive scouting in the field for nymphs should begin.

2.5.2 Cultural Control

Stink bugs overwinter in thick foliage in plant material, or under leaf litter. Wooded areas provide prime locations for stink bug diapause, and therefore they are the primary source of stink bugs for invasions into field crops. Stink bugs move out from the wooded areas into nearby vegetation which often consists of weeds or cover crops. Therefore, any weeds utilized by stink bugs around fields scheduled for tomato plantings should be destroyed. Natural enemies have been very important in keeping stink bug numbers low, and their populations should be preserved as much as possible. Calendar-based insecticide applications early in the season leads to reduced natural enemies, potentially resulting in stink bug outbreaks. Natural enemies are shown to increase in tomato fields that border nectar-bearing flowering plants such as alyssum, buckwheat, sunn hemp, and sunflower that were planted early in the growing season (Wang, 2012; Phatak and Diaz-Perez, 2012).

2.5.3 Biological Control

There is a large contingency of natural enemies that prey upon stink bugs and these endemic predators and parasitoids can keep stink bug populations in check. The most vulnerable stage of stink bugs is the egg stage as there are many parasitoids that attack egg masses. In some cases parasitic wasps have been released into tomato production areas. For example, *Trissolcus basalis* (Wollaston) was released in California for control of the southern green stink bug, although there was some disagreement as to the definite efficacy of this parasitoid (Clarke and Walter, 1993; Jones, 1995). Another species of

Trissolcus, Trissolcus japonicus (Ashmead) that parasitizes brown marmorated stink bug (BMSB) eggs has been found in the wild in Maryland (United States) and is being studied in research laboratories for possible release (Anonymous, 2015a). There are 12 families of parasitic Hymenoptera that will parasitize stink bugs, the largest family being Platygastridae (Jones, 1988; Rider, 2015).

There are also other parasitoids such as Tachinids that parasitize stink bug adults and larger nymphs (Jones, 1988; Rider, 2015). Estimates of parasitism rates peak at 80–100% in some locations, but overall rates this high are seldom seen during the season in most locations (Panizzi, 1988; Jones, 1995; Coombs and Khan, 1998; Panizzi and Oliveira, 1999). One common Tachinid genus is *Trichopoda*, and it parasitizes many of the important pest species such as the green and southern green stink bugs (McPherson et al., 1982; Jones, 1988; McLain et al., 1990). These parasitoids are able to use the sex (or aggregation) pheromone released by the male stink bug as a way to locate and parasitize the male (Aldrich et al., 1995). This is probably why male stink bugs of some species such as *N. viridula* are more heavily parasitized than females by the Tachinid *Trichopoda pennipes* (F.) (Mitchell and Mau, 1971; Harris and Todd, 1980).

There are many spiders and insects in the families Carabidae, Staphylinidae, Coccinellidae, Asilidae, Reduviidae, Formicidae, and even other Pentatomidae that feed on stink bugs. These natural enemies can account for a 50–80% reduction in stink bug numbers by the end of a growing season (Ragsdale et al., 1981; Stam et al., 1987).

2.5.4 Chemical Control

Stink bugs are difficult to control even with an efficacious insecticide. However, because large nymphs and adults tend to rest outside the plant canopy in the morning hours an insecticide application at this time would be the most efficacious. One of the problems with managing stink bugs is even if an insecticide treatment works and controls the stink bugs present in the field at the time it is applied, the field is often reinfested within a few days because of the migratory nature of these pests. Studies have shown that there are several insecticide classes such as carbamates and organophosphates (IRAC classification 1) that can reduce most stink bug populations if they contact them. However, these insecticide classes did not control southern green stink bug *N. viridula* populations because their persistence was considered too short to prevent succeeding outbreaks (Waterhouse and Norris, 1987; Martins et al., 1990). Pyrethroid (IRAC classification 3A) and neonicotinoid (IRAC classification 4A) are generally less toxic than older chemicals and their use in the control of stink bugs has increased yields and quality (Catchot et al., 2012; Kuhar et al., 2015). There are also "combination" products available that combine a pyrethroid and a neonicotinoid into one product. These have shown good control of stink bugs (Catchot et al., 2012; Kuhar et al., 2015). While effective, the non-judicious use of pyrethroids can potentially create problems with resistance development and outbreaks of secondary pests if they are over applied; caution should be used with these insecticides.

2.6 Summary

Stink bugs are not a major problem in most tomato production areas of the world because of their sporadic nature as pests. At certain times though, their numbers can increase to damaging levels causing marketable yield reductions. Often times, damage occurs because growers are unaware of the pests until harvesting begins. Even though stink bugs are fairly large insects, they are difficult to monitor without expending time and money. Furthermore, because they are an intermittent pest in both time and space, growers do not spend time scouting for them until it is too late. Stink bugs have a large number of crop and weed hosts available to them throughout a season, thus it is difficult to know from where they may be coming or to where they could be moving as they migrate in and out of tomato fields. Another factor that makes stink bug damage so unpredictable is that natural enemies can play a significant role in keeping their populations below crop-damaging levels. But at certain times in a season, or for certain seasons, the natural enemies either are not effective or stink bug populations increase at a rate at which predators and parasitoids cannot keep up. At this point, growers must rely on pesticides for stink bug control. Fortunately, the pyrethroids and neonicotinoids do a good job of controlling the pest as long as applications are well timed and there is thorough plant coverage.

3. COLORADO POTATO BEETLE, *LEPTINOTARSA DECEMLINEATA* SAY

The Colorado potato beetle (CPB) *Leptinotarsa decemlineata* Say first was identified in the early 1800s on buffalo bur (*Solanum rostratum* Dunal) found in the Rocky Mountain range of the western United States (Casagrande, 1987). In the mid-1800s the beetle began to feed on potato and rapidly became one of the most damaging pests of potatoes and other solanaceous crops in the temperate zone (Ferro et al., 1983; Worner, 1988). Insecticides were able to control the voracious pest in the early years, but it soon developed resistance to most pesticides used (Wilkerson et al., 2005). Pesticide resistance remains a major concern with the management of this pest. This beetle pest is also known as the Colorado beetle, the ten-striped spearman, the ten-lined potato beetle, or the potato bug.

3.1 Identification

CPB adults are pale yellow-orange, oval, and approximately 10 mm long with 10 black longitudinal stripes down the back (Wilkerson et al., 2005). The head has a triangular black spot near its center, and the prothorax has several asymmetrical black markings (Voss, 1989; Capinera, 2001). There are many irregular rows of small indentations (punctuations) on the beetle's back and its legs, and underside of its body is black. Another species, *Leptinotarsa juncta* Germar, the false potato bug, is similar in appearance to CPB, but it is not as common. *L. juncta* is pale yellow with 10 black stripes down its back, but it differs from *L. decemlineata* by having the third and fourth stripe connected, giving the appearance of one wide stripe. In addition, the punctuation on its back is jagged but in straight rows that outline each stripe, and the legs are usually orange. For a more detailed description of the CPB see Hare (1990) and Capinera (2001).

Eggs are 1.7–1.8 mm long and 0.8 mm wide, and yellow-orange in color. Eggs are usually found on the underside of foliage in clusters of 10–90 with 35–50 being the average (Capinera, 2001). Larvae are reddish-orange during the early instars, and they are 1.2–1.5 mm long; the fourth instar larva is 10–14 mm long. Larvae typically have two rows of black spots along their sides, and they are "humped-back" with the abdomen being greatly arched. Larvae have three pairs of thoracic legs, and one proleg at the apex of the abdomen (de Wilde, 1948; Ferro et al., 1985). The false potato beetle larvae are usually pale in color, sometimes appearing white, and have only one row of black spots. Pupae are orange-yellow or brown, and develop from larvae that have dropped from the plant and burrowed 2–5 cm into the soil. The pupal stage lasts 5–10 days, at which time the adults emerge and start the summer cycle again (Walgenbach and Wyman, 1984).

3.2 Biology

Beetles overwinter 5–15 cm below ground as adults. The beetles' flight muscles degenerate while they are in diapause, which lasts approximately 3 months; diapause is broken when soil temperatures rise above 10°C (Logan et al., 1985). Most beetles amass in woody areas next to fields where they had consumed plant material the preceding summer (Ferro et al., 1983; Weber and Ferro, 1993). Once beetles accumulate 75–250 DD (using a 10°C lower threshold) they emerge from the soil and begin short-range searching for food using sight or smell (Ferro, 1985). The emergence of the beetles is timed with the emergence of potatoes. Fields that have not been rotated will be inundated quickly with overwintered adults that either emerged within the field or had to walk only a short distance into the field (Voss and Ferro, 1990; Ferro et al., 1999). In some areas a small percentage (5–15%) of the population will undergo an extended diapause of an additional 1–2 years (Solbreck, 1978; Biever and Chauvin, 1992; Weber and Ferro, 1993). Within days after emergence from diapause the adults have developed their reproductive system and flight muscles (Ferro et al., 1999).

Upon arrival in a field, beetles immediately feed and in 5–6 days begin to lay eggs (Ferro et al., 1985, 1991). Eggs hatch in 5–12 days; females are capable of laying up to 800 eggs per season (Hare, 1990; Weber et al., 1994). There are four larval instars—the first, second, and third instars take 2–3 days to complete, while the fourth instar lasts twice as long (Ferro et al., 1983; Biever and Chauvin, 1992; Ferro, 1993). Under cooler temperatures larvae tend to feed on the upper surface of leaves often near the top of the plant, but as temperatures increase larvae will move into the shade and the interior of the plant (May, 1981; Lactin and Holliday, 1994). Once the fourth instar is full size it passes a number of days as a non-feeding pre-pupa, recognized by its paler color and inactivity. These pre-pupae fall to the ground and burrow into soil to pupate. Depending on several factors such as temperature, day length, and host quality, the adults either emerge in a few weeks, or enter diapause and emerge in the spring. If they emerge in a few weeks they continue reproducing and feeding until short-day length induces them to undergo diapause. Beetles usually fly to their overwintering sites and burrow into the soil (Solbreck, 1978; Voss, 1989). Developmental time from oviposition to adult ranges from 20 to 55 days at optimal temperatures between 26 and 33°C (Walgenbach and Wyman, 1984). There can be 1–3 generations per year depending on location.

3.3 Distribution, Host Range, and Seasonal Occurrence

The CPB is native to Mexico and was found in the early 1800s in Colorado feeding on buffalo bur. After incorporating potato and other solanaceous crops into its diet, it began to expand its range rapidly (Hsiao, 1985; Worner, 1988). The beetle spread north and eastward in a short period of time and by 1859 there was an outbreak of the pest in potato fields west of

Omaha, Nebraska (United States). In just 15 years the CPB had reached the Atlantic coast of the North American continent. Currently its distribution covers North America, (which includes the contiguous United States except for California and Nevada, southern Canada, and Mexico), northern Central America, Europe, and central and western Asia (Jolivet, 1991). CPB could invade much larger areas throughout eastern Asia, northern India, and the temperate zones of the southern hemisphere (Vlasova, 1978; Weber et al., 1994). It is a quarantine pest in the UK, Ireland, and southern parts of Sweden and Finland. The false potato beetle is mostly confined to the eastern United States

While potatoes are the preferred host for the CPB, it also feeds and reproduces on other solanaceous crop plants such as eggplant, pepper, tobacco, and tomato as well as the solanaceous weeds belladonna, common nightshade, buffalo bur, ground cherry, henbane, and horse nettle (Brues, 1940; Gauthier et al., 1981). Very rarely the CPB feeds but not reproduce on other non-solanaceous plants. In choice studies by Hitchner et al. (2008) in Virginia (United States), CPB was more attracted to eggplant than either tomato or pepper. Their study also suggested that the male-produced aggregation pheromone is probably very important in host-plant selection. Additional research has shown that CPBs are able to fine-tune their host choice to locally abundant *Solanum* species (Horton et al., 1988).

CPB emergence and general life cycle is timed with the main crop host, potato. They emerge in the early spring along with the potato crop and continue their life cycle over the summer until shorter day lengths and deteriorating host plants signal the beetles into diapause.

3.4 Damage, Losses, and Economic Thresholds

CPB adult and larvae feed on the foliage and stems of host plants and is the most important defoliator of potato, eggplant, and, at times, tomato throughout North America (Hare, 1990). On an average each CPB larva consumes approximately $40\,cm^2$ of potato leaves per day, while the adult consumes almost $10\,cm^2$ of foliage per day (Ferro et al., 1985; Capinera, 2001). As larvae become larger they consume ever increasing amounts of foliage with first instars consuming 4%, second instars 6%, third instars 16%, and fourth instars consuming 74% of the total foliage (Hare, 1980). Heavy defoliation early in the plant's development can result in greatly reduced yields (Hare and Moore, 1988; Zehnder and Evanylo, 1989). CPB, if left uncontrolled, can significantly reduce yields in tomatoes (Schalk and Stoner, 1979). The larval population resulting from an adult beetle density as small as 0.002 females per plant can cause significant yield losses in indeterminate tomato cultivars (Cantwell et al., 1986). In the midwestern United States, CPB can cause up to 75% crop loss in processing tomatoes. It is estimated that CPB causes hundreds of millions of dollars in crop losses and chemical control expense each year in North America (USDA, 2012).

3.5 Management

3.5.1 Monitoring

There have been some studies using yellow as well as pheromone traps, but the most reliable method of detection is to inspect whole plants for the first signs of feeding damage or for the various life stages of the beetle, which are very evident (Capinera, 2001).

3.5.2 Cultural Control

One of the most effective and oldest management practices for CPB is crop rotation (Hough-Goldstein et al., 1993). Crop rotation was recommended as far back as 1872 (Bethune, 1872). By rotating a solanaceous crop, beetle populations can be reduced by 90% compared with a nonrotated system (Lashomb and Ng, 1984). Grain crops (wheat, oats, and rye) appear to be among the better rotational crops to reduce CPB populations (Wright, 1984).

Other effective cultural controls include altering planting time, changing crop varieties, and using mulches, cover crops, and trap crops. Late planting can disrupt and suppress second generation CPB larvae because adults emerge later in the season on this crop and the shorter day length encourages diapause. Earlier planting can also greatly reduce second generation larvae because the crop is already at or near senescence at the time of their emergence (Weber and Ferro, 1995). Trap crops (using another solanaceous crop that is equally or more preferred as tomato) need to be planted weeks before tomatoes are planted. These plantings are used to keep overwintered beetles from colonizing the main crop in the spring (Ferro et al., 1999). Applying mulch is another possible, albeit management-intensive, practice for CPB control. Mulch, usually straw from crops such as wheat or rye, should cover the field in an 8–10 cm deep layer before CPB emergence. The use of the straw mulch is thought to impact CPB in several ways; by interfering with the adult beetle's ability to locate the field (Ng and Lashomb, 1983; East, 1993) and by increasing predation on eggs and larvae (Brust, 1994; Weber et al., 1994). The use of the mulch will decrease first and second instar larvae (Stoner, 1993,

1997) and decrease potato defoliation by two to five times compared with a non-treated control (Zehnder and Hough-Goldstein, 1990; Brust, 1994). By adding manure to fields instead of synthetic fertilizers it was found that CPB infestations as well as damage were reduced with no reduction in plant vigor (Alyokhin and Atlihan, 2005).

3.5.3 Physical Control

There is one physical control that has had some success for reducing overwintering beetles in a field, but it is very labor intensive. This method comprises digging a trench (1 × 1 m) along the field border that is lined with plastic. Because almost all overwintering beetles walk into the field, these trenches can capture and retain a major proportion of the population trying to colonize the crop in the spring (Misener et al., 1993). Additional physical controls include using propane flamers used to scorch the outer surface of the plant, thereby incinerating or damaging any CPB stage usually located there in the morning hours (Pelletier et al., 1995; Lacasse et al., 1998). Vacuum collectors (large vacuum machines mounted on a tractor) can literally suck the insects off of a plant (Laguë et al., 1999). These two methods are labor intensive and expensive to build and maintain. Another method uses straw mulch in the CPB's overwintering sites. The mulch can be applied to overwintering areas in the fall, and then taken off along with the snow cover in midwinter; this significantly decreases beetle survival by approximately 20% (Milner et al., 1992).

3.5.4 Host-Plant Resistance

Developing genetically modified tomato plants expressing *Bacillus thuringiensis* Berliner delta-endotoxin, which is toxic to the CPB, is a possibility (Grafius and Douches, 2008). However, this technology was presented in 1995, but then quickly discontinued mostly due to consumer concerns about genetically engineered foods. CPB is at a much reduced risk of causing catastrophic and widespread crop losses in tomato compared to potato, and with the public concern about using genetically modified plants it is doubtful that this technology will be adopted in tomato production systems (Grafius and Douches, 2008). More traditional forms of plant breeding for CPB resistance have not been very successful (Flanders et al., 1992; Fisher et al., 2002; Grafius and Douches, 2008; Maharijay and Vosman, 2015).

3.5.5 Biological Control

A number of predatory arthropods attack the CPB (Hough-Goldstein et al., 1993). The most prolific predator is the lady beetle *Coleomegilla maculata* (De Geer), which can devour 35% and 55% of first and second generation CPB eggs, respectively; at times they even feed on small larvae (Hazzard et al., 1991). Ground beetles are a group of predators that can climb potato and tomato plants especially at night and feed on all non-adult stages of the CPB (Brust, 1994). The ground beetle, *Pterostichus chalcites* Say, has been observed feeding on various stages of the CPB in the midwestern and eastern United States (Heimpel and Hough-Goldstein, 1992; Brust, 1994). Several species of *Lebia* in North America have been found to be consistent predators of CPB eggs and small larvae (Logan, 1990; Weber et al., 2006). Other predators, such as stink bugs *Perillus bioculatus* (F.) and *Podisus maculiventris* (Say), have been found to attack beetle larvae, and releases of these predators reduced beetle density and defoliation, which increased potato yields by 65% compared with a non-treated control (Biever and Chauvin, 1992; Hough-Goldstein and McPherson, 1996). Sorokin (1976), Brust (1994), and Drummond et al. (1990) found that several groups of common natural enemies such as ground beetles, lady bugs, predacious stink bugs, phalangids, and spiders could decrease defoliation and increase yields over non-treated controls if their populations were augmented in the field.

Parasitic wasps and flies can at times play a role in the control of CPB. *Edovum* sp. has been shown to parasitize 70–90% of the CPB egg masses on eggplant (Lashomb et al., 1987). But surprisingly the amount of parasitism is only 50% in potatoes and tomatoes (Ruberson et al., 1991; Van Driesche et al., 1991). Tachinid flies (family Tachinidae) can parasitize 25–65% of the second generation of CPB (Lopez et al., 1997).

3.5.6 Biopesticides

The entomogenous fungal pathogen *Beauveria bassiana* (Balsamo-Crivelli) Vuillemin is probably the most widely used biopesticide applied for control of the CPB. One of its major attributes is that it can be applied just like a synthetic pesticide. Studies have shown that *B. bassiana* can reduce beetle populations 35–75% (Cantwell et al., 1986). In another study by Brust (1994) it was found that integrating the use of *B. thuringiensis tenebrionis* with straw mulch (which increased natural enemies) gave results equal to the best pesticide controls. While these studies are encouraging, biopesticides as a whole are typically less effective than synthetic insecticides (Campbell et al., 1985). Biopesticides work best when applied early in an infestation when CPB larvae are small; they do not work very well, if at all, on adults.

3.5.7 Chemical Control

The main method for control of CPB in North America and in most of the world is the use of synthetic insecticides. The reliance on insecticides is often unsuccessful because of the beetle's remarkable ability to quickly develop resistance. Although the beetle has developed resistance to most of the major insecticide classes, not all beetle populations have this ability (Forgash, 1985; Boiteau et al., 1992; Ferro, 1993; Alyokhin et al., 2006; Mota-Sanchez et al., 2006). It is interesting to note that in areas of the world where the beetle is uncommon, insecticides still remain efficacious as the main management tool. In North America more than 30 synthetic chemicals are available for use against this pest and their efficacy varies from one area to another (Alyokhin, 2007).

3.5.8 Resistance Management

The key to resistance management in the CPB is to avoid relying on a single mode of action for its control (Huseth et al., 2014). For instance, over the last 20 years neonicotinoids have been used extensively for CPB control because of their efficacy and ease of use (Mota-Sanchez et al., 2006). However, resistance has developed in some populations of CPB in the eastern United States. New approaches are necessary that will still allow growers to utilize pesticides, but in a manner that will reduce or greatly delay the development of resistance. A very thorough and complete discussion of resistance management can be found in Huseth et al. (2014).

3.6 Summary

CPB is one of the major insect pest of potatoes, but fortunately it is only a sporadic pest in tomato. In most tomato fields the CPB is only a problem in isolated "hot spots" or along field edges. Because of the record of CPB resistance to many insecticides, the judicious use of chemicals along with well-planned chemical rotations is needed (Huseth et al., 2014). Utilizing an integrated approach to CPB management in tomato systems by means of cultural, biological, and other controls in addition to chemicals will result in the most sustainable pest control program.

4. FLEA BEETLES, *EPITRIX* SPP., AND *PHYLLOTRETA* SPP.

There are more than 250 species of flea beetles (Coleoptera, Chrysomelidae, Alticinae) around the world (Kaszab, 1962; Gruev and Tomov, 1986; Brelih et al., 2003) but not all of these are important as plant pests. Important flea beetle pests include the genera *Epitrix* and *Phyllotreta* which feed upon and cause damage to many solanaceous (tomato and potato) and cruciferous (cabbage and broccoli) plants as well as to many other crops (Chenopodiaceae, Cucurbitaceae, Fabaceae) (Mayoori and Mikunthan, 2009; Cuthbertson, 2015). The genus *Epitrix* consists of approximately 180 species globally, with 130 species reported from the neotropics, 12 from North America, and 17 from Europe (Döberl, 2000). For tomato production, the flea beetles are generally considered minor pests with little importance, as compared with other crops such as potatoes, where some species are considered major pests in the United States, Canada, and some European countries (Kabaluk and Vernon, 2000; Antonelli and Davidson, 2001; Ambrosino, 2008).

4.1 Identification and Biology

Many flea beetles are very similar in external appearance, which can make identification rather difficult. Adults in two of the *Epitrix* species (generally known as potato tuber beetles) look very similar in size (1.5–2.0 mm) and color (brown to black). An enlarged femur on the hind leg which is a modified organ for jumping is the characteristic feature that enables flea beetles to hop when disturbed (Döberl, 2000; Anonymous, 2015b). *Phyllotreta* spp. generally known as cabbage beetles are oval shape, metallic black, bluish-green, or bronze in color, and between 1.5 and 3.0 mm in size. In some species, elytra bear longitudinal yellow stripes (Anonymous, 2014). Flea beetle eggs are deposited at the base of plants. Larvae are cylindrical in shape and feed on small roots and root hairs (Anonymous, 2014). The larval stage takes about 1 month to complete. Pupae are yellowish in color (Antonelli and Davidson, 2001).

4.2 Distribution, Host Range, and Seasonal Occurrence

Flea beetles damage several crops as either major or minor pests through the entire European and Mediterranean Plant Protection Organization (EPPO) region. The tomato pest, *Epitrix* spp. can survive in different climate ranges, from a Mediterranean climate of Spain to a temperate climate in the United States and Canada (EPPO, 2015). *Epitrix* species are native to North America with two species having been introduced into Portugal and Spain (Anonymous, 2015a). International trade is the main reason for their introduction to other countries (Jolivet and Hawkeswood, 1995). *Epitrix* utilize different members of the Solanaceae family, particularly tomato and potato, as hosts (Anonymous, 2015a; Cuthbertson, 2015). Adults are strong fliers, and use chemical cues to locate their host plants. Overwintering occurs in the adult stage, in leaves, cracks, and crevices in soil or some other protective structures found in the field. On warm days in midspring, adults start to become active. Usually a second generation occurs in the summer with a third generation being rare (Cranshaw, 2013).

4.3 Damage, Losses, and Economic Thresholds

Damage from flea beetles is caused by adults. It is characterized by small shot holes (small round holes 2–3 mm in size) in the leaves (Mayoori and Mikunthan, 2009; Boavida et al., 2013; Cranshaw, 2013) (Fig. 8.4). Damage begins when the epidermis is peeled away followed by subsequent feeding destroying other leaf tissues giving leaves a sieve-like appearance (Anonymous, 2014). These holes disturb the water conduction and balance in the leaves. This type of damage is more detrimental to seedlings often killing them (Cranshaw, 2013). The pattern of fruit damage is similar to the feeding by the tomato fruitworm *Helicoverpa zea* (Boddie) or tomato pinworm, *Keiferia lycopersicella* (Walsingham) on small tomato fruits (O'Neill, 2009). This type of feeding damage is seen in late-season tomato plantings when leaves undergo senescence due to maturity, water deficiency, or powdery mildew (Anonymous, 2015a). Damage at the 4–5 leaf seedling stage can cause economic loss, especially under hot conditions which results in plant desiccation (Anonymous, 2015a). The foliar feeding of larger plants is generally of little significance to the plant's general health (Anonymous, 2015a). A 10–20% loss of leaf tissue typically does not affect yield (Cuthbertson, 2015), however, damage at later stages can affect fruit quality by causing uneven ripening and thus influencing market sales. The tomato crop should be treated as soon there is 30% (Foster and Flood, 1995).

4.4 Management

Successful management practices are aimed at avoiding insect entry into the field. In fields not previously planted with tomatoes, flea beetle infestations are usually located at borders (O'Neill, 2009). Cultural practices can be useful, particularly when combined with chemical control measures. The impetus in pest management for flea beetles is to select for chemical management, which allows long-term control of flea beetles when beneficial natural enemies are not effective (Antonelli and Davidson, 2001; Ambrosino, 2008).

4.4.1 Trapping (Monitoring/Mass Trapping)

Tomato fields planted after another flea beetle host should be sampled early and thoroughly (O'Neill, 2009). Samplers should look for shot holes in the cotyledons and young leaves. When infestations occur because of nearby alternate hosts, beetles usually remain limited to about 200 ft along tomato field borders (Anonymous, 2013b). Early monitoring should be focused on the first five outside rows and the first 25 ft at the end of rows. A sampling of 30 damaged plants may provide enough information for a population estimate. Spot treatment is recommended when 5–10% of the seedlings are heavily infested (O'Neill, 2009; Anonymous, 2014).

Emergence cages, pyramidal in shape, 1 m² at the base (Dosdall et al., 1996), are reported to be an effective tool for monitoring spring and fall flea beetle populations. The cages are relatively inexpensive, simple to establish and maintain,

FIGURE 8.4 Characteristic flea beetle feeding damage on solanaceous crops.

and the flea beetles can be readily collected and processed from these structures (Ulmer and Dosdall, 2006). A more accurate estimation of flea beetle emergence in the spring uses degree days (Ulmer and Dosdall, 2006). Some plant-derived attractants can also be used to sample flea beetles. Successful capture of flea beetles has been reported by the use of secondary metabolites of plant origin such as isothiocyanates obtained from decomposition of non-volatile glucosinolates (Vincent and Stewart, 1984; Pivnick et al., 1992; Smart and Blight, 2000). The attractant baited crops not only help in monitoring the flight pattern of flea beetles but also play a significant role in the detection of overwintering adults in early spring (Tóth et al., 2007).

4.4.2 Cultural Control

Wider spacing between rows with a low seeding rate often leads to reduced damage by flea beetles (Dosdall et al., 1999). Sometimes, a high seeding rate is recommended to maintain plant numbers when seedlings are attacked by the beetles (Anonymous, 2015b).

4.4.3 Biological Control

Biological control is one of the most significant tools for managing high density pests such as flea beetles. A hymenopterous wasp, *Microctonus vittatae* Muesebeck, serves as a parasitoid of flea beetle and is found throughout the Pacific Northwest (Joyce et al., 2012). A limited level of predation by some generalist predators such as big-eyed bugs (*Geocoris* spp.), lacewing (*Chrysoperla* spp.), and damsel bugs (*Nabis* spp.) has been also reported (Gredler, 2001).

Overwintering flea beetles in the soil allows the use of soil-dwelling entomopathogenic nematodes belonging to Steinernematidae and Heterorhabditidae (Miles et al., 2012). Significant control of *Phyllotreta* spp. under field conditions was reported by Yan et al. (2013) using *Steinernema carpocapsae* Weiser and *Heterorhabditis indica* Poinar, Karunakar & David applied at the rate of 0.75×10^9 IJs/ha. Nematode species *S. carpocapsae* are specialists in suppressing flea beetle larvae in potatoes (Miles et al., 2012). Another widespread parasitic nematode *Howardula phyllotretae* Oldham is reported as a successful biocontrol agent against *Phyllotreta* spp. in the middle and east Black Sea region of Turkey (Yaman, 2002).

The entomopathogenic fungal species *B. bassiana* (Antwi et al., 2007) and *Metarhizium brunneum* Petch (Reddy et al., 2014) have been found highly effective in reducing flea beetle damage and population density. The GHA and ATCC 74040 strains of *B. bassiana* used either as a foliar spray or as a granular application provide control of flea beetle populations (Caldwell et al., 2005).

4.4.4 Chemical Control

In perishable vegetables with a high demand for fresh sales, flea beetles infestations can make the fruits unmarketable (Eyre and Giltrap, 2012; Cuthbertson, 2014). *Epitrix* species can be controlled through the use of insecticides (Döberl, 2000). Liquid and granular insecticides are the best known source of control but rather expensive for season-long management (Smith et al., 1997). When flea beetles threaten crop establishment, one of the viable strategies is seed treatment with a systemic insecticide or foliar application (Soroka et al., 2011). Seed treatment with clothianidin against flea beetles provided better protection to growing seedlings compared with imidacloprid seed treatment (Wilde et al., 2004). In a mortality assay where lambda-cyhalothrin was evaluated, 78.2% mortality was found at a 50% labeled dose, while 94.7% mortality was recorded at a 100% labeled dose (Ekbom and Müller, 2011). Antwi et al. (2007) compared the efficacy of ecorational insecticides including spinosad, *B. bassiana*, azadirachtin, and kaolin clay and the chemical insecticide bifenthrin and seed treatment with thiamethoxam against flea beetles. Flea beetle injury was significantly lower for thiamethoxam, bifenthrin, and spinosad applications. Some insecticides with a narrow spectrum of action were also found safe for some natural enemies. For instance, Yabar (1980) reported deltamethrin had excellent initial and long-lasting control for the management of *Epitrix* spp. and did not affect the populations of predacious carabids. Malathion was found relatively harmless to carabids but did not offer sufficient control of *Epitrix* spp. (Cuthbertson, 2014). Foliar applications of bifenthrin, cyfluthrin, cypermethrin, and deltamethrin were found to be very toxic to *Epitrix* spp. (Vernon and Mackenzie, 1991). Pyrethroid insecticides also offer extended control of flea beetles than the non-pyrethroids insecticides (carbaryl, chlorpyrifos, and methamidophos) (Vernon and Mackenzie, 1991).

4.4.5 Integrated Pest Management

Sustainable management of flea beetles can be carried out through the use of *B. bassiana* and *M. brunneum* and this integrated use was found to be more effective than entomopathogenic nematodes (*S. carpocapsae*), neem, or petroleum spray oils and almost as effective as bifenthrin. Two applications of *B. bassiana* and *M. brunneum* or five applications of bifenthrin were highly effective in reducing flea beetle feeding damage in canola (Reddy et al., 2014). Horticultural oils and some neem-based insecticides exhibit effective repellency against flea beetles (Cranshaw, 2013) which could be helpful in avoiding field infestations.

4.5 Summary

Flea beetles from genera *Epitrix* and *Phyllotreta* serve as minor insect pests causing damage to tomatoes and other solanaceous (potato) and cruciferous (cabbage and broccoli) crops. In tomato production, little attention is paid to flea beetles compared with other crops such as potatoes, where some species are considered major pests in the United States, Canada, and some European countries. Adults are the damaging stage, particularly in spring with the characteristic shot hole feeding damage. These holes disturb the water conduction and balance in the leaves. In fields not previously planted with tomatoes, flea beetle infestations are usually located at the field edges. Spot treatment is recommended when 5–10% of the seedlings are heavily infested. Careful determination of threshold temperature, along with accumulated DDs (requiring 50% emergence) improves the monitoring strategies for flea beetles and helps facilitate the timing of insecticide applications. Foliar applications of bifenthrin, cyfluthrin, cypermethrin, and deltamethrin have been effective against *Epitrix* spp. Pyrethroid insecticides also offer extended control of flea beetles than the non-pyrethroids insecticides (carbaryl, chlorpyrifos, and methamidophos).

5. LEAF-FOOTED BUGS, *LEPTOGLOSSUS PHYLLOPUS* (L.), *PHTHIA PICTA* (DRURY)

Leaf-footed bugs belong to the family Coreidae and occasionally damage fruiting vegetables, fruits, nuts, and ornamental plants (Xiao and Fadamiro, 2010). With piercing-sucking mouthparts, these medium-sized bugs feed on developing plants, particularly fruits and seeds (Mitchell, 2000; Xiao and Fadamiro, 2010; Schuster and Stansly, 2012; Schuster, 2015). The name "leaf-footed bugs" is given due to small leaf-like enlarged portion found on the hind legs of these bugs. Leaf-footed bugs resemble stink bugs (family: Pentatomidae) in their feeding behavior (Ingels and Haviland, 2014).

Generally known species of leaf-footed bugs include *Leptoglossus zonatus* (Dallas), *Leptoglossus clypealis* Heidemann, and *Leptoglossus occidentalis* Heidemann which are native to California and the western United States (Buss et al., 2013). Leaf-footed bugs are polyphagous with many vegetables as preferred hosts, particularly beans, cucumber, pea, pepper, squash, and tomato (Ingels and Haviland, 2014; Schuster, 2015). *Phthia picta* (Drury) is another important species from the family Coreidae that feeds on several important plant species. Both immature and adult bugs can cause damage to tomatoes (Soldi et al., 2012; Schuster, 2015). In addition to the direct damage to leaves and fruits, feeding by these bugs encourage increased attacks by other insects, fungi, and microorganisms. The deterioration hastened by these attacks can cause additional economic losses (Silva and Carvalho, 2001; Silva et al., 2001).

5.1 Identification

Adults of *P. picta* are elongated bugs (Fig. 8.5), with males usually smaller in size (14.2 mm) than females (15.5 mm). They are black in color with yellow or orange pronotal margins and a traverse stripe (Serantes, 1973; Silva et al., 2001). *L. phyllopus* (L.) are usually dark brown with a white band running across the wing covers. The hind tibia is flattened in both genera. Eggs of leaf-footed bugs are commonly metallic in color, ovate in shape, and sometimes flattened laterally (Fig. 8.6). Eggs are deposited along stems in rows or along leaf veins in clusters (Silva et al., 2001; Ingels and Haviland, 2014; Schuster, 2015). In most cases, egg masses contain 10–15 eggs but more than 50 have also been reported. Nymphs tend to aggregate at first and then disperse throughout the plant. There are five nymphal instars that complete development to the adult stage in about 30 days (Silva et al., 2001; Schuster, 2015). Nymphs resemble adults with the exception of being smaller, apterous (without wings), and having a bright orange body. A few morphological differences are found in the species of *Leptoglossus*. For instance, *L. zonatus* bears two yellow spots on the pronotum, *L. clypealis* bears a thorn-like projection on the clypeus, extending from tip of the head, and *L. occidentalis* is without such features. Small leaf-footed bugs sometimes are difficult to identify due to close resemblance with nymphs of the assassin bug (*Zelus renardii* Kolenati) (Ingels and Haviland, 2014).

These bugs overwinter in the adult stage and will aggregate in protected areas like barns, palm fronds, woodpiles, citrus or juniper trees, under peeling bark, or in tree cracks. Severe cold winters may kill most of the adults. Thus severe infestations occur when a high percentage of adults survive a mild winter (Ingels and Haviland, 2014).

FIGURE 8.5   Leaf-footed bug *Leptoglossus* sp.

FIGURE 8.6   Characteristic leaf-footed bug eggs.

5.2 Distribution, Host Range, and Seasonal Occurrence

Leaf-footed bugs are distributed across most of North America including California, Florida, Mexico, Texas, the Caribbean, and South America, including most regions of Brazil (Soldi et al., 2012). Leaf-footed bugs are polyphagous in their feeding behavior, damaging many plant in the Leguminosae and Solanaceae families (Silva and Carvalho, 2001; Silva et al., 2001; Soldi et al., 2012). Preferred vegetables include beans, cucumber, pea, pepper, squash, and tomato. In late spring, thistles and other weeds also are attacked (Ingels and Haviland, 2014; Schuster, 2015). Leaf-footed bugs native to California utilize a wide range of host plants including young citrus fruit, cotton bolls, watermelons, several ornamental trees, and shrubs. *L. zonatus* is considered the most common pest of tomato in California (Ingels and Haviland, 2014). Desert plants including palm trees, Joshua trees, yucca, and many farms, garden, and landscape plants are fed upon by *L. clypealis*. *L. occidentalis* is closely associated with conifer trees. Among garden fruits, leaf-footed bugs feed on almonds, pistachios, and pomegranates. Adults overwinter during the months of September and October until early spring, as the weather becomes warmer during the month of March; they take flight in search of food. The population fluctuates year to year which depends on weather conditions and egg parasitism. The population is greatest after mild winters due to greater adult survival and is directly correlated with rain, availability of food and presence of natural enemies (Ingels and Haviland, 2014).

5.3 Damage, Losses, and Economic Thresholds

Leaf-footed bugs have piercing-sucking mouthparts that are narrow and extend for more than half of the body length. The long proboscis assists in probing and punctures into leaves, shoots, and fruits to suck plant sap (Schuster and Stansly, 2012). Probing depth depends on the body size of the bug as small nymphs generally remain limited to sap at superficial layers while adults can probe deep into fruit (Ingels and Haviland, 2014; Schuster, 2015). After hatching from egg masses deposited under the leaf surface, nymphs start feeding gregariously. Mature nymphs and adults are the most damaging stages. Damage by leaf-footed bugs is usually of little economic concern, but if large numbers attack on few plants, the damage can lead to economic loss (Ingels and Haviland, 2014). In infested tomatoes, fruit punctures are a common symptom; this leads to fruit discoloration upon maturity. Discoloration arises due the presence of fungal yeast, *E. coryli* (formerly known as *Nematospora*) introduced through contaminated mouthparts (Ingels and Haviland, 2014). Such infestations are usually abundant during wet periods. The discoloration affects fruit quality and decreases marketable yield (Ingels and Haviland, 2014; Schuster, 2015). In addition to punctures and discoloration, feeding by leaf-footed bugs can cause fruit abortion and seed malformation (Grimm and Führer, 1998; Xiao and Fadamiro, 2010). The action threshold level is 1 nymph or adult per plant (Schuster, 2015).

5.4 Management

Most of the time leaf-footed bug densities remain low, causing little damage to the tomato crop; no distinct management plan has been developed for the pests (Ingels and Haviland, 2014; Schuster, 2015). Conventionally, the use of broad-spectrum insecticides for other sucking insects has resulted in adequate control of leaf-footed bugs. However, when the pest population exceeds the threshold level, a variety of methods need to be implemented for the effective control of these bugs (Ingels and Haviland, 2014).

5.4.1 Trapping (Monitoring/Mass Trapping)

No special monitoring strategies have been devised for leaf-footed bugs (Haviland, 2007) because of their cryptic nature and spotty distribution. It is uncommon to find the egg masses of these bugs in the field. Most outbreaks of these pests are reported when plants are close to flowering and this is the time to begin sampling (Schuster, 2015). Sampling should be done on the entire plant and slight disturbance of the plant makes it easier to find the bugs. When plants are fruiting, 10 fruits per plant on 6 plants should be examined for damage (Schuster, 2015). Little is known about the pheromone-mediated response of leaf-footed bugs, but Leal et al. (1994) reported behavioral changes of these bugs in response to alarm pheromones. In *P. picta*, the volatile extracts from males elicited responses in females in Y-tube olfactometer bioassays (Soldi et al., 2012). The compound found in the extracts was identified as 5,9,17-trimethylhenicosane (Soldi et al., 2012).

5.4.2 Cultural Control

Adult bugs overwinter in woodpiles, under bark of several trees (eucalyptus, juniper, or cypress), and in outbuildings. Removal of these sites prevents overwintering population buildup and early season infestations. Weeds and senescent crops can be reservoirs for migrating adults (Ingels and Haviland, 2014). Trap crops can be used to attract leaf-footed bugs away from the main crop. The use of a mixed trap cropping system with Peredovik sunflower and NK300 (forage) sorghum worked well in the management of leaf-footed and stink bugs to protect tomatoes (Majumdar and Miles, 2013). In another study, it was reported that leaf-footed bugs prefer soybean and sorghum over tomato (Majumdar, 2013). This trap crop system resulted in a significant reduction of insecticides used in the tomato crop. Foliar insecticide applications were used in a sorghum trap crop when leaf-footed bugs reached peak density, and this provided 78–100% control of the pest without treating the main crop (tomato) (Majumdar et al., 2012).

5.4.3 Biological Control

In countries where leaf-footed bugs are native, several egg and nymphal parasites and parasitoids have been found. *Gryon pennsylvanicum* (Ashmead), *Ooencyrtus johnsoni* (Howard), and *Anastatus pearsalli* Ashmead comprise the egg parasitoid complex of *Leptoglossus* spp. in British Columbia and California (Maltese et al., 2012). Some members of the family Tachinidae, *T. pennipes*, and *Trichopoda plumipes* (F.) are parasitoids of adult leaf-footed bugs (Kuhar et al., 2010). The wasp *G. pennsylvanicum* is an excellent egg parasitoid, and if present in sufficient numbers, it can reduce leaf-footed bug populations (Ingels and Haviland, 2014). *G. pennsylvanicum* also is an important egg parasitoid of *Leptoglossus australis* (F.) and detects potential host-finding cues from male leaf-footed bugs (Yasuda and Tsurumachi, 1995; Yasuda, 1998; Fatouros et al., 2008). Mitchell et al. (1999) monitored the egg parasitism of *L. phyllopus* from 1994 to 1997 in cultivated and wild hosts. The Hymenopteran parasitoids reared from the pest's eggs were *Gryon carinatifrons* (Ashmead), *G. pennsylvanicum*, *Anastatus* spp. (Eupelmidae), and an unidentified encyrtid. Scelionids included about 93% of parasitoids in cowpea (*Vigna unguiculata* (L.) Walp.), wild hosts (*Pyrrhopappus carolinianus* (Walter) DC.), and cultivated blueberry (*Vaccinium corymbosum* L.) plants. The ratio of active parasitoids was 4:1 of *G. carinatifrons*: *G. pennsylvanicum* (Mitchell et al., 1999). A maximum of 72.2% parasitism was found in the wild host, 61.9% in cowpea, and 28.8% in blueberry.

Many generalist predators such as big-eyed bugs (*Geocoris* spp.), damsel bugs (*Nabis* spp.), fire ants (*Solenopsis invicta* Buren), and spiders play a vital role in suppressing field populations of leaf-footed bugs (Ingels and Haviland, 2014). Birds feeding on leaf-footed bugs also contribute in population suppression to some extent (Ingels and Haviland, 2014).

Entomopathogenic fungi, *B. bassiana*, and *Metarhizium anisopliae* (Metchnikoff) Sorokin were found to be pathogenic to adults of leaf-footed bugs in a dipping bioassay. *M. anisopliae* was found as the most efficient fungus with a median lethal concentration (LC_{50}) of 4.34 conidia m/L. During field trials in Nicaragua, when test insecticides were applied against adults of *L. zonatus* and *Pachycoris klugii* Burmeister on physic nut (*Jatropha curcas* L., Euphorbiaceae) plantations using an ultralow volume sprayer, *B. bassiana* was found more effective than malathion or an aqueous extract of ground neem seeds, leading to a 28% increase in total yield (Grimm and Guharay, 1998).

5.4.4 Chemical Control

Season-long management of leaf-footed bugs can be achieved by two applications of insecticides such as zeta-cypermethrin and lambda-cyhalothrin. In field trials conducted on sorghum trap crop, there was 70–90% reduction in the population of leaf-footed bugs without any insecticides needed for the tomato as the main crop (Majumdar and Miles, 2013).

Insecticides from different chemical groups, including pyrthroids (bifenthrin, cypermethrin, cyfluthin, esfenvalerate, lambda-cyhalothrin, and permethrin), carbamates (carbaryl, methomyl andoxamyl), organophosphates (acephate and neonicotinoids acetamiprid, clothianidin, dinotefuran, imidacloprid, and thiamethoxam) have been found to be effective for controlling leaf-footed bug infestations (Kuhar et al., 2010). Neem-based botanicals are potential candidates for the management of leaf-footed bugs, with few negative effects on non-target insects or natural enemies (Ingels and Haviland, 2014).

5.5 Summary

Leaf-footed bugs belong to the family Coreidae. Both immature and adult bugs can damage tomatoes, with small nymphs feeding restricted to epidermal layers while adults can probe deep into fruit. After hatching nymphs feed gregariously, spreading throughout the plant in later instars. Mature nymphs and adults are more damaging stages. In infested tomatoes, fruit punctures are a common symptom of leaf-footed bug attack, and this leads to fruit discoloration during ripening. Damage by leaf-footed bugs is usually of little economic concern, but if large numbers attack a few plants the damage can lead to economic loss. Normally, the use of broad-spectrum insecticides for other sucking insects has resulted in adequate control of leaf-footed bugs. However, when the pest population exceeds the threshold level, a variety of methods need to be implemented for the effective control of these bugs. Insecticides from different chemical groups have been found very effective in controlling leaf-footed bug infestations.

6. LEAFMINERS *LIRIOMYZA SATIVAE* BLANCHARD, *LIRIOMYZA TRIFOLII* (BURGESS), *LIRIOMYZA BRYONIAE* (KALTENBACH), AND *LIRIOMYZA HUIDOBRENSIS* BLANCHARD

Leafminers have gained attention as minor pests in several important crops especially tomatoes. Four well-known species belong to the genus *Liriomyza* (*Liriomyza sativae* Blanchard, *Liriomyza trifolii* (Burgess), *Liriomyza bryoniae* (Kaltenbach), and *Liriomyza huidobrensis* (Blanchard)) feed on 25 different plant families (Spencer, 1990). Members of this genus are polyphagous and have been recorded damaging several vegetable crops in Southeast Asia (Andersen et al., 2002; Winotai and Chattrakul, 2003; Tran et al., 2005, 2007). *Liriomyza* species infestations in tomato can result in crop losses up to 20% due to reduced photosynthetic activity and physical growth (Walker, 2012).

6.1 Identification

All species of *Liriomyza* have similar biological features. They can be difficult to identify; molecular techniques with specific mitochondrial and nuclear genomic sequences (Scheffer, 2000; Scheffer and Lewis, 2001), enzyme-staining techniques, and starch gel electrophoresis (OEPP/EPPO, 1992) are being employed to help differentiate them. Adults are small flies, 1–3 mm in size with the prescutellar area usually dark and yellow, with the scutellum mostly yellow and rarely dark. There is a chitinized ridge present on the hind femora (Spencer, 1990).

Females use their ovipositor to puncture the leaf surface to deposit eggs. The white oval-shaped eggs are about 0.25 mm in length. No differentiation is possible at the egg stage between different species. Larvae are about 0.5 mm in length but may reach 3.0 mm at maturity over three instars. Early instar larvae are translucent at hatching, turn cream colored later with yellow-orange patches appearing in the last instar. The number of spiracles is specific to species with 7–12 pores in *L. bryoniae* and 6–9 in *L. huidobrensis* (Spencer, 1973). Pupae are about 2 mm in size, oval in shape, and slightly flattened ventrally. Several researchers have studied and reported the biology of *Liriomyza* spp. in detail (Spencer, 1973; Waterhouse and Norris, 1987; Murphy and LaSalle, 1999; Kang et al., 2008).

6.2 Distribution, Host Range, and Seasonal Occurrence

L. bryoniae originated in Europe while the three other species (*L. huidobrensis*, *L. sativae*, and *L. trifolii*) are from the New World. *L. huidobrensis* originally was described in 1936 from Argentina (Spencer, 1973). Later this insect rapidly spread to many other regions of the world. In the coastal areas of Peru, it has remained a minor pest where its population is kept in check by a large complex of Hymenopteran parasitoids (Cisneros, 1984). In South East Asia, the exact distribution of different leafminer species is unclear. The earlier invasion of the species was in urban and horticultural areas, which later extended to cropping regions. Rapid migration of *L. huidobrensis* and *L. sativae* was recorded in the early 1990s toward eastern tropical and subtropical regions of Asia, resulting in increased crop losses and management cost (Rauf et al., 2000). In tropical Asia, at higher elevations (>1000 m), *L. huidobrensis* is the dominant species damaging important crops (Sivapragasam and Syed, 1999; Rauf et al., 2000) whereas in lowland areas, *L. sativae* is widely prevalent (Rauf et al., 2000; Andersen et al., 2002). *L. sativae* attacks 40 different hosts belonging to 10 different families in Florida (United States) (Stegmaier, 1966), however, Schuster et al. (1991) observed *L. sativae* on bean, tomato, eggplant, potato, squash pepper and watermelon. *L. trifolii* is recognized one of the major pests of chrysanthemum and celery but in Florida it is reported on 55 different hosts including tomato (Stegmaier, 1966; Schuster et al. 1991). *L. bryoniae* is recognized as a polyphagous insect feeds mainly on tomato, lettuce, cucumber, melon and other vegetables (Spencer, 1964) and *L. huidobrensis* attacks on celery, garden lettuce, broccoli, cauliflower, canation, garlic, onion, pepper tomato and potato (Spencer, 1990). The seasonal activity of the leafminer and its parasitoids can be monitored on potato, scallions, and broccoli by collecting weekly leaf samples (Shepard and Braun, 1998). Cooler temperatures favor *L. huidobrensis* feeding and oviposition (Olivera et al., 1993).

6.3 Damage, Losses, and Economic Thresholds

Leafminers feed on 25 plant families, but tomato is the main crop host; they have been also reported on cucurbits (cucumbers, melons, and watermelon), beans, glasshouse-grown lettuce, and lupines (Spencer, 1989, 1990). Leaf punctures that provide the base for mining within leaves are the early signs of *Liriomyza* infestation. Mines formed by leafminers vary in configuration, and are influenced by the host plant, physical and physiological features of the leaves, and the number of leafminer larvae per leaf. Generally, the shape and pattern of leaf punctures do not give any indication about leafminer species. However, white speckled punctures with a diameter of 0.2 mm on the upper leaf are usually the specific feature of *Liriomyza* species (OEPP/EPPO, 2005). Mining occurs on the upper leaf surface in the palisade tissues and trails of frass appear as dotted, black strips within mines (Fig. 8.7). Plant-mediated responses to mining often appear as discolored mines with dampened black and dried brown areas. Some species-based characteristics of the mines include: *L. trifolii* forming a tightly coiled, almost blotch-like mine; *L. bryoniae* and *L. sativae* forming a more loose, irregular serpentine mine; and *L. huidobrensis* forming irregular serpentine tunnels, restricted by veins within segments of the leaf, and undulating between upper and lower leaf surfaces. For all species, leaf mines increase in size as the larvae mature within the leaf. The mines are ruptured at one end by larvae to exit for pupation. Exit holes usually appear as semicircular slits (Spencer, 1990; OEPP/EPPO, 2005). Leafminer larvae feed internally on leaf tissue and cause damage up to 80% in tomato and its damage mostly prominent against young plants. During oviposition, female can puncture the leaf that ultimately attracts the other adults and also act as entry point for secondary infection (Clark, 2015). One larva per plant at 0-2 leaf stage and one live larva found per 3 leaflets is the threshold of leafminer (Schuster and Smith, 2015).

6.4 Management

The outbreak of leafminers as secondary pests has been often attributed to the elimination of natural enemies (Luckmann and Metcalf, 1994). Studies in the 1940s considered the inclusion of parasitoids as important factors for suppressing the pest population below the economic treatment level (Hills and Taylor, 1951), and outbreaks were recorded only when there

FIGURE 8.7   Leafminer larva *Liriomyza* sp. tunneling in leaf.

were low densities of parasitoids (Oatman and Kennedy, 1976; Ohno et al., 1999). Therefore, narrow spectrum insecticides with the least damage to parasitoid populations should be used (Wene, 1953). Generally, these pest management tactics are preferred by progressive growers that make use of integrated approaches.

6.4.1 Trapping (Monitoring/Mass Trapping)

Placing yellow sticky traps close to the adult flight zone attracts and catches them, allowing rough estimates of field populations. Sticky traps, roller traps, and Taki traps are some of the suggested traps used for *Liriomyza* species in different crops (Anonymous, 2014). Synthetic methyl salicylate (MeSa) also serves as an attractant for *L. bryoniae* and it can be used in pheromone traps. Its effectiveness can be enhanced by choosing a yellow color, which is most preferred by the adults (Būda and Radziute, 2008).

6.4.2 Cultural Control

The entry of *Liriomyza* adults into tomato field can be minimized by planting trap crops such as alfalfa, cotton, and watermelon. Mixed cropping of a trap crop is suggested in heavily leafminer–infested areas to attract the pest to the non-preferred crop. Pest infestations can be reduced at times by efficient use of multiple cropping schemes (Altieri and Letourneau, 1982). Infested residue of the previous crop should be removed or buried immediately after harvest to avoid infestation in the subsequent crop. Larvae are killed by shredding the crop residues before tillage as it lowers the chance of survival of concealed larvae within leaf mines. Weeds should be destroyed in areas around the field as they provide shelter and alternate sites of survival to adult flies (Mau and Kessing, 2007).

6.4.3 Mechanical Control

Along with other IPM tactics, mechanical control measures can be helpful and work as a first line of defense against leaf-mining insects (Nugaliyadde et al., 2000; Meers, 2008). Removal of infected plant parts, uprooting of severely infected plants, and burning them aids in lowering the early infestation of leafminers. Sometimes hand picking and destroying mined leaves is the most useful approach to control the insect pest (Gilberg, 1993; Mahr and Ridgeway, 1993). While labor intensive, this strategy can be a simple and effective control method that preserves natural enemies and does not lead to resistance.

6.4.4 Physical Control

Temperature and humidity–mediated control is possible in mechanically facilitated premises such as greenhouses (Meyer, 2003; Bográn, 2005). Traditional approaches such as manual use of barriers, sticky traps, and fires used to burn crop residues often prove effective in managing many insect pests especially in remote areas (Reddy, 2013). In addition, row covers applied over bedding plants and at field planting can be effective in avoiding leafminer infestation until first bloom.

6.4.5 Host-Plant Resistance

Tomato varieties with low moisture content in their leaves could be used to lower the mining of *Liriomyza* species (Webb and Smith, 1969). A range of differences has been reported regarding the damage by leafminers among tomato cultivars (Wolfenbarger, 1966; Webb and Smith, 1969), breeding material, and *Lycopersicon* spp. (Kelsheimer, 1963; Webb et al., 1971).

6.4.6 Biological Control

Liriomyza leafminers are known to be associated with several Hymenopteran parasitoids (Liu et al., 2009). Most of the reported parasitoids of *L. huidobrensis*, *L. sativae*, or *L. trifolii* are larval and larval-pupal parasitoids. About 40 hymenopterous parasitoids are found in association with the genus *Liriomyza* (Waterhouse and Norris, 1987). In Peru, *Halticoptera arduine* (Walker) (Pteromalidae), *Chrysocharis phytomyzae* (Brethes), *Diglyphus begini* (Ashmead) (Eulophidae), *Diglyphus websteri* (Crawford), and *Ganaspidium* sp. (Eucoilidae) are the dominant members of this complex (Sánchez and de Huiza, 1988; Vignez and de Huiza, 1992). Among these natural enemy fauna, some but not all are oligophagous that respond to the non-native *Liriomyza* spp. (Nicoli, 1997).

Although there is a fairly generalized level of parasitism for leafminers among hymenopterous wasps, there is some evidence of differential parasitism, with the parasitoids preferring some *Liriomyza* species over others (Abe et al., 2005). *L. huidobrensis* is parasitized by over 40 species of natural enemies (Song et al., 2003). A greater density of natural enemies and parasitism rates for *Liriomyza* spp. have been found in many field studies (Root, 1973; Risch et al., 1983). Sometimes the preferred parasitism of a particular species leads to the eradication of that species and replacement with other non-preferred species. For instance, the enhanced parasitism of *L. trifolii* compared with *L. sativae* by *Dacnusa sibirica* Telenga has acted as a major factor in displacing *L. trifolii* with *L. sativae* in Japan (Abe et al., 2005; Abe and Tokumaru, 2008). Natural biological control of *Liriomyza* spp. has been reported with at least 23 species of parasitoids working efficiently to reduce pest populations by over 90% (Waterhouse and Norris, 1987; Greathead and Greathead, 1992; Johnson, 1993). There are some reports of successful release of entomopathogenic nematodes, such as *Steinernema feltiae* (Filipjev) for leafminer control in Europe (Williams and MacDonald, 1995).

For commercial scale use in greenhouse systems, laboratory mass-reared natural enemies are available for *Liriomyza* management. Augmentative biological control utilizing parasitoids has become an emerging field in the United States and other non-European areas for managing *Liriomyza* populations because of the cost-effective applications and compatibility with judicious use of insecticides (Ozawa et al., 2001; Chow and Heinz, 2006).

6.4.7 Autocidal Control

Sterile insect technique has appeared as a vital and innovative idea to suppress *Liriomyza* species that can even work in combination with biological control agents such as *Diglyphus isaea* (Walker). The exposure of late-stage pupae to gamma radiation (160 Gy) induced sterility in both sexes with the fitness of the sterile insects comparable to that of wild-type populations of *Liriomyza* species (Walker, 2012).

6.4.8 Chemical Control

In the 1960s and 1970s, continuous use of chemicals to control leafminer infestations resulted in a lack of emphasis on other management tools. This led to a worldwide leafminer crisis in the mid-1970s (Leibee and Capinera, 1995). Chemical insecticides had been the only tool of *Liriomyza* species management for years (Bartlett and Powell, 1981) which resulted in resistance development against several groups of insecticides (Parrella and Keil, 1985; Mason et al., 1987; MacDonald, 1991). Recently, new insecticides such as abamectin, spinosyns, and cyromazine with novel modes of action are performing well against *Liriomyza* species infestations (Trumble et al., 1997; Reitz et al., 1999). Spinosyn products are also in widespread use against insect pests that cooccur with leafminers, including thrips and lepidopteran pests (Reitz et al., 1999; Demirozer et al., 2012; Reitz and Funderburk, 2012). The use of surfactants can improve surface penetration thus enhancing the efficacy of spinosad against *Liriomyza* larvae (Bueno et al., 2007). Insecticide selection depends on the species of leafminer; in China *L. trifolii* is less susceptible than *L. sativae* to abamectin and cyromazine (Gao et al., 2012).

6.4.9 Integrated Pest Management

Truly IPM programs in tomato can result in efficient and cost-effective control of *Liriomyza* species (Trumble and Alvarado-Rodriguez, 1993; Trumble et al., 1997; Reitz et al., 1999). The integrated use of biological control agents with host-plant resistance and host-specific insecticides has inspired research scientists and progressive growers to realize the full potential of IPM.

6.5 Summary

Tomatoes are attacked by several leafminer species belonging to the genera *Liriomyza*; *L. sativae*, *L. trifolii*, *L. bryoniae*, and *L. huidobrensis*. Damage to tomato crops at times surpasses 20% due to reduced photosynthetic activity which affects plant's physical growth. Farmers often find it difficult to implement management practices due to the concealed activity of the larvae. Major outbreaks are reported in areas where native parasitoids of leafminers have been eliminated by the use of broad-spectrum insecticides. Natural enemies are the important factor needed to suppress leafminer population below the economic treatment level. When insecticides are necessary, those with narrow spectrum of activity and least detrimental impacts on parasitoids should be used.

7. MOLE CRICKETS, *SCAPTERISCUS VICINUS* SCUDDER, *SCAPTERISCUS BORELLII* GIGLIO-TOS, *GRYLLOTALPA GRYLLOTALPA* (L.), *GRYLLOTALPA HEXADACTYLA* PERTY, AND OTHERS

Mole crickets have been reported as insect pests invading field crops, particularly tomatoes, for many years. They damage the basal and underground parts of the plants (Silcox, 2011; Bailey, 2012), feeding exclusively on roots and stems of seedlings; they can girdle the seedling resulting in the complete loss of plants (Villani et al., 2002; Silcox, 2011). The family Gryllotalpidae consists of five genera encompassing field and lawn mole crickets which are distributed throughout the temperate and tropical areas of the world (Otte and Alexander, 1983; Nickle and Castner, 1984; Otte, 1994). Gryllotalpids are polyphagous and can feed on all upland crops (Hudson et al., 2008; Olga et al., 2011). Mole crickets are important pests on many crops such as tomato, potato, vegetables, and turf grasses (Ingrisch et al., 2006). Often Gryllotalpids are macropterous but in rare cases they are brachypterous, micropterous, and apterous (De Villiers, 1985; Frank et al., 1998). *Gryllotalpa gryllotalpa* (L.) (European mole cricket) is the only seven-spurred mole cricket found in Asia (Townsend, 1983). Insects in the genus *Scapteriscus* are two-clawed mole crickets, and they include *Scapteriscus vicinus* Scudder (tawny mole cricket) and *Scapteriscus borellii* Giglio-Tos (southern mole cricket).

7.1 Identification and Biology

Mole crickets generally appear as gray-brown to black covered in ochreous pubescence (Annecke and Moran, 1982; Townsend, 1983; De Villiers, 1985; Cobb, 1998). They are burrowing in behavior and spend most of their lives in vertically and horizontally formed galleries below the soil surface. They possess a prognathous-type head (De Villiers, 1985) with fossorial (digging-type) legs. The forelegs possess two to four strongly sclerotized dactyls (claws) (Nickle and Castner, 1984).

Females lay about 25–60 eggs from early spring to the end of July in underground chambers; eggs hatch in 3–4 weeks (Lake, 2000; Weed, 2003; Doggett, 2015). Hatched nymphs resemble the adults except for the presence of wings in the adult. Nymphs build underground galleries to feed on plant roots (Hudson, 1985). A nymph molts 8–10 times and matures in about 5 months (Walker, 1985). Generally, there are one to two generations per year. Strong flying behavior of mole crickets enables them to cover an area of up to 5 miles. Intraspecific wing length and resultant flight range differs geographically (Semlitsch, 1986; Frank et al., 1998). In males, the hind wing is often vestigial but present in conspecific females (Kavanagh and Young, 1989). Males produce mating sounds within galleries to attract females, after matting females dig nursery chambers to lay eggs (Bennet-Clark, 1989). Mole crickets are nocturnal in feeding and activity is rarely seen in daylight.

7.2 Distribution, Host Range, and Seasonal Occurrence

The Gryllotalpidae is distributed throughout the tropical and temperate regions of the world (Nickle and Castner, 1984; Hill et al., 2002). Generally speaking, mole crickets are cosmopolitan in distribution and have been found across Asia, Africa, Australia, Europe, and North and South America (Hudson and Saw, 1987; Henne and Johnson, 2001).

The plant–host range for mole crickets includes many important field crops (Hudson, 1985; Barbara, 2005; Ingrisch et al., 2006). They have been reported feeding on beans, cereal crops, cash crops, flowers, ornamental plants, grasses, vegetables, weeds, and several other plants (Tindale, 1928; Daramola, 1974; Broadley, 1978; Townsend, 1983; Matsuura et al., 1985; Frank and Parkman, 1999; Kim, 2000; Buss et al., 2002). Oviposition is influenced by soil moisture (Hertl et al., 2001) and to some extent by soil temperature (Brandenburg, 1997). In general, oviposition occurs during spring or early summer (Cobb, 1998; Frank et al., 1998; Potter, 1998; Buss et al., 2002).

7.3 Damage, Losses, and Economic Thresholds

Damage by mole crickets occurs by direct feeding on roots and stems, or by burrowing in the soil around and between roots (Barbara, 2005; Ingrisch et al., 2006). Nymphs are the most active feeding stage and they damage crops by burrowing that leads to plant desiccation due to disturbed soil (Schoeman, 1996). The most susceptible plant stage for damage includes young or newly transplanted seedlings with the main damage to stem, roots, and lower leaves (Frank and Parkman, 1999; Olga et al., 2011). Typical symptoms of damage resembles that of cutworms, with early season feeding causing stand loss by the cutting off of seedling or recently transplanted tomato plants at the soil line and late-season damage consisting of irregular holes chewed into the surface of tomato fruit touching the ground. The damaged areas on the plant also serve as sites for the invasion of plant pathogens (Schuster and Price, 1992).

The presence of mole crickets also increases the activity of several bird and vertebrate pests that dig up and feed on the crickets. This magnifies potential crop losses (Schoeman, 1996; Potter, 1998) and makes the calculation for determining the economic threshold difficult. The estimation of the damage threshold requires it to be divided into three categories: (1) threshold for amount of damage and cost of control; (2) estimation of lowest population level capable of causing economic injury; (3) determining the population at which control should be initiated to avoid EIL (economic injury level) (Dent, 1991). Thresholds could be as low as one adult per gallery per night due to very low tolerance levels (Brandenburg and Williams, 1993; Frank and Parkman, 1999).

7.4 Management

Variable levels of control have been achieved by using different conventional control measures. Mole cricket management recommendations suggest targeting young nymphs to achieve required levels of control (Xia and Brandenburg, 2000; Scharf, 2007). The young nymphal stage is the most susceptible stage to insecticide application (Olga et al., 2011).

7.4.1 Trapping (Monitoring/Mass Trapping)

The general subterranean lifestyle and great mobility of mole crickets make population estimates difficult (Hudson and Saw, 1987); because of this it is difficult to determine a treatment threshold (Hudson, 1985). The generally adopted procedure for population estimation requires the use of linear pitfall traps, soap flushing, and sound traps with male calling songs; the latter is used particularly for *S. borellii* and *S. vicinus* (Lawrence, 1982; Hudson, 1985; Parkman and Frank, 1993; Gadallah et al., 1998).

Most sampling techniques for *Gryllotalpids* under field conditions require the use of liquid solutions consisting of pyrethrins, piperonyl butoxide, dishwashing soap, vinegar, and ammonia to flush the crickets out of the soil (Short and Koehler, 1979; Potter, 1998). In addition, pitfall traps (Lawrence, 1982), estimating surface burrowing (Walker and Nation, 1982; Cobb and Mack, 1989), and soil-core extraction (Williams and Shaw, 1982) can also be used. One cannot rely on a single approach, however, the continuous mapping of infested areas and scouting to record field damage is one of the best methods that may limit the application of insecticides (Weed, 2003).

7.4.2 Cultural Control

Several cultural practices are used to decrease mole cricket populations. For instance, tillage (Denisenko, 1986; Sithole, 1986), flooding (Frank and Parkman, 1999), and burning of dry leaves on infested soil (Denisenko, 1986; Sithole, 1986), particularly in the spring, alter the physical conditions of the soil and disrupt mole cricket galleries desiccating their eggs. Cultural practices that produce a deep, healthy root system create plants that are more tolerant to mole cricket feeding damage (Frank et al., 1998). Generally recommended cultural control practices for pest management include crop rotation (Pimentel, 1982), crop sanitation (Hardison, 1980), mixed cropping (Dempster and Coaker, 1974), tillage (Glass, 1975), fertilizer, and water management practices (Stoll, 1996; Elwell and Mass, 1989).

7.4.3 Mechanical Control

The fact that mole crickets are sluggish during the early morning suggests this to be the best time to employ mechanical practices. Surveys conducted in Bangladesh showed that manual removal of insects was adopted by 72% of the farmers who participated in the study (Dasgupta et al., 2004). This non-chemical method was followed in Pakistan for insect pest management and included manual picking (Ahmed et al., 2004).

7.4.4 Biological Control

Biological control of mole crickets has emerged as a cost-effective approach (Adjei et al., 2000; Buss et al., 2002). In areas with native populations of mole crickets, several parasitoids and predators have been recorded. In addition, introduced populations of biocontrol agents have provided significant control of the mole crickets around the world. For example, *Ormia depleta* (Wiedemann), a tachinid fly was introduced into several counties of Florida (the United States) in 1988. By 1994 it was established and providing a sufficient level of mole cricket control (Frank et al., 1996; Walker et al., 1996). Countries where *O. depleta* established successfully reported lower populations and damage from mole crickets (Frank et al., 1996). *Larra bicolor* (F.) (a digger wasp) is another effective biocontrol organism. The larvae of *L. bicolor* are ectoparasitoids of large *Scapteriscus* nymphs. Studies showed a 70% decline in mole cricket populations in north central Florida after the release of this wasp species (Weed, 2003).

The grubs of *Pheropsophus aequinoctialis* (L.), the South American bombardier beetle, are very efficient predators feeding on the eggs of *Scapteriscus*.

Steinernema scapterisci Nguyen and Smart is an entomopathogenic nematode that was introduced into the United States in 1985 (Frank and Parkman, 1999). This resulted in a significant reduction in nymphal and adult populations of *S. borellii* and *S. vicinus* (Leppla et al., 2007). The only limitation associated to its use is its ineffectiveness against *Scapteriscus abbreviatus* Scudder and small to moderate size *Scapteriscus* mole cricket nymphs (Parkman and Frank, 1993). Pfiffner (1997) tested the efficacy of using the entomopathogenic nematode, *S. carpocapsae* 60 cm deep in soil against mole crickets in tomatoes grown under plastic tunnels and observed an 84% reduction in the mole cricket population.

Other potential biocontrol agents are entomopathogenic fungi, such as *M. anisopliae* and *B. bassiana*, depending upon the virulence of the isolate used (Boucias, 1985). *B. bassiana* provided control of mole crickets similar to that observed under insecticide treatments with imidacloprid, bifenthrin, and deltamethrin (Xia et al., 2000). The finding of the Mole Cricket Biological Control Project (MCBCP) carried out in Florida's agricultural fields showed a reduction in the money spent on mole cricket control in pasture lands by $24/acre annually. Overall a savings of nearly $8 million annually or a cumulative benefit of $260 million was reported. An integrated biological control of mole crickets study was carried out using a Brazilian fly (*Ormia depletaiptera* (Wiedemann)), wasp (*L. bicolor*), and nematode (*S. scapterisci*), which resulted in the reduction of an invasive mole cricket population (*S. vicinus, S. abbreviates, S. borellii*) by up to 95% (Mhina and Thomas, 2013).

7.4.5 Chemical Control

Chemical control has been a conventional practice for the management of mole crickets. This strategy has included the use of food baits in 1900–1940s, use of DDT-based insecticides, and more recently, control with currently used materials. Excessive pesticide use has led to the development of resistance in mole crickets to some pesticides such as chlordane (Kepner, 1985). Mole crickets may have the ability to detect and avoid the insecticides and pathogens in the soil (Brandenburg, 1997; Xia and Brandenburg, 2000). Some pest management tactics are making use of juvenile hormone (JH) analogs that influence the growth and development of the immature mole cricket. Fenoxycarb, a JH analog reduced the egg hatchability and produced developmental deformations in *S. abbreviatus* (Parkman and Frank, 1993), but its application to nymphs did not produce the same results.

Among the new chemicals developed for insect control, imidacloprid has proved to be the most effective mole cricket chemical when applied between the start of egg laying and hatching (Olga et al., 2011). Mole cricket nymphs were also more susceptible to bifenthrin, fipronil, and indoxacarb than adults in laboratory assays (Olga et al., 2011). Fipronil was one of the most potent neurotoxins tested, causing significant neuroexcitation and rapid mortality in *S. vicinus*. The combined use of bifenthrin and imidacloprid increased toxicity and elicited instant knockdown better than either insecticide alone.

7.4.6 Integrated Pest Management

The integrated use of more than one control tactic appears to be the best approach for the management of mole crickets (Silcox, 2011). Sublethal topical application of *B. bassiana* in combination with substrate treatments of diatomaceous earth

(DE) and imidacloprid have been found to suppress populations of the southern mole cricket, *S. borellii*. Thompson and Brandenburg (2006) found that two strains of *B. bassiana* (5977 and 3622) worked synergistically with DE. However, no synergy was found for imidacloprid and DE integration. The integrated use of entomopathogens with other suitable agents such as DE can be used to reduce insecticide applications against mole crickets (Held and Potter, 2012).

7.5 Summary

Mole crickets have been reported as insect pests invading field crops, particularly tomatoes, for many years. They damage the basal and underground parts of the plants. Their burrowing behavior results in them spending most of their lives in vertically and horizontally formed galleries below the soil surface. Gryllotalpidae is distributed throughout the tropical and temperate regions of the world. Damage by mole crickets occurs by direct feeding on roots and stems, or by burrowing in the soil around and between roots. Nymphs are the most active feeding stage and they damage the crop by burrowing that leads to plant/soil disturbance and desiccation. The most susceptible plant stage for damage includes young or newly transplanted seedlings with the main damage to stem, roots, and lower leaves. Typical symptoms of damage resemble that of cutworms. The general subterranean lifestyle and great mobility of mole crickets makes population estimates difficult. The fact that mole crickets are sluggish during early morning hours suggests this as the best time for mechanical practices. Biological control of mole crickets has emerged as a cost-effective approach. Some pest management tactics are making use of JH analogs that influence the growth and development of the immature mole cricket. Among the new chemicals developed for insect control, imidacloprid has proved to be the most effective chemical when applied between the start of egg laying and hatching.

8. ECONOMICS OF MANAGEMENT STRATEGIES

The minor pests discussed above usually are intermittent problems in most tomato fields and infestations usually are confined to a few "hot spots" and along field margins. The best practice is to scout fields when tomato plants are small and when they begin to fruit especially along any field borders with woods, weeds, or a cover crop. Because the damage is usually localized, spot spraying sections of fields where the pest or their damage is found can be effective. Frequent, repeated pesticide applications for any of these pests should be avoided.

Natural enemies are a very important factor in keeping minor pest populations below economic thresholds. The IPM approaches for the major tomato pests discussed in this book are imperative in order to maintain these minor pests below damaging levels. However, if the tomato field is located in an area that consistently produces high densities of any of these minor pests, specific tactics described above may be necessary. These may include using cultural practices, biological control, physical control, and chemical control unique to the minor pest.

9. FUTURE PROSPECTS

One of the more difficult problems with minor pests is knowing how to sample them, which informs when treatments are necessary. Only a few sampling programs or thresholds exist for these pests and most are time-consuming and laborious to implement. What is needed is a quick reliable method of detecting a significant population density of the pests in the field. This may depend on developing effective pheromone traps that can detect the pests as they move into a field before much feeding damage has occurred.

There are two commonly used insecticide classes, pyrethroids and neonicotinoids, that are effective in controlling many of these pests. Overuse of pyrethroids can have negative consequences while neonicotinoids have been implicated as one of the many possible factors in bee decline (Hopwood et al., 2012). Therefore, more selective chemistries are needed that will target the pest but allow the conservation of natural enemies. On the ecological side, a better understanding of how natural enemies interact and respond to environmental conditions and to the pests' population fluctuations is essential for taking advantage of these natural controls. Management programs that integrate different biocontrol agents such as entomophagous nematodes and parasitoids are needed to reduce the dependence on chemical controls.

REFERENCES

Abe, Y., Takeuchi, T., Tokumaru, S., Kamata, J., 2005. Comparison of the suitability of three pest leafminers (Diptera: Agromyzidae) as hosts for the parasitoid *Dacnusa sibirica* (Hymenoptera: Braconidae). European Journal of Entomology 102, 805–807.

Abe, Y., Tokumaru, S., 2008. Displacement in two invasive species of leafminer fly in different localities. Biological Invasions 10, 951–995.

Adjei, M.B., Crow, W.T., Frank, J.H., Leppla, N.C., Smart, G.C.J., 2000. Biological Control of Pasture Mole Crickets with Nematodes. Institute of Food and Agricultural Sciences, University of Florida, Gainesville, FL, USA.

Ahmed, K.U., Rahman, M.M., Alam, M.Z., Ahmed, S.U., 2004. Methods of pest control and direct yield loss assessment of country bean (*Dolichos hyacinth bean*) at farmers field conditions: a survey finding. Pakistan Journal of Biological Sciences 7, 287–291.

Aldrich, J.R., Rosi, M.C., Bin, F., 1995. Behavior correlates for minor volatile compounds from stink bugs (Heteroptera: Pentatomidae). Journal of Chemical Ecology 21, 1907–1920.

Altieri, M.A., Letourneau, D.K., 1982. Vegetation management and biological control in agroecosystems. Crop Protection 1, 405–430.

Alyokhin, A., 2007. Insecticide Resistance in the Colorado Potato Beetle. http://www.potatobeetle.org/resistance/management.html.

Alyokhin, A., Atlihan, R., 2005. Reduced fitness of the Colorado potato beetle (Coleoptera: Chrysomelidae) on potato plants grown in manure-amended soil. Environmental Entomology 34, 963–968.

Alyokhin, A., Dively, G., Patterson, M., Rogers, D., Mahoney, M., Wollam, J., 2006. Susceptibility of imidacloprid-resistant Colorado potato beetles to non-neonicotinoid insecticides in the laboratory and field trials. American Journal of Potato Research 83, 485–494.

Ambrosino, M., 2008. Flea Beetle Pest Management for Organic Potatoes. Oregon State University Extension Service, Carvallis, Oregon, USA. Publication No. EM 8947-E http://extension.oregonstate.edu/catalog/pdf/em/em8947-e.pdf.

Andersen, A., Nordhus, E., Thang, V.T., An, T.T.T., Hung, H.Q., Hofsvang, T., 2002. Polyphagous *Liriomyza* species (Diptera: Agromyzidae) in vegetables in Vietnam. Tropical Agriculture (Trinidad) 79, 241–246.

Annecke, D.P., Moran, V.C., 1982. Insects and Mites of Cultivated Plants in South Africa. Butterworths, Durban, p. 383.

Anonymous, 1998. Insects and related pests. In: Strand, L.L. (Ed.), Integrated Pest Management for Tomatoes, fourth ed. University of California, Davis, California, USA, pp. 36–67. Division of Agriculture and Natural Resources Publication No. 3274.

Anonymous, 2013a. Tomato: Stink Bug. UC IPM Pest Management Guidelines, Statewide Integrated Pest Management Program, Division of Agriculture and Natural Resources. University of California, Davis, California, USA. http://ipm.ucanr.edu/PMG/r783300211.html.

Anonymous, 2013b. Tomato: Flea Beetles. UC IPM Pest Management Guidelines: Tomato, UC IPM Statewide Integrated Pest Management Program. University of California Agriculture and Natural Resources, University of California, Davis, USA. http://ipm.ucanr.edu/PMG/r783301411.html.

Anonymous, 2014. Cabbage Flea Beetles (*Phyllotreta* spp.). http://www.csalomontraps.com/4listbylatinname/pdffajonkentik/phyllotretasppang08.pdf.

Anonymous, 2015a. Asian Wasp, Enemy of Stink Bugs, Found in the United States. National Institute of Food and Agriculture, United State Department of Agriculture, USA. http://www.stopbmsb.org/stink-bug-bulletin/asian-wasp-enemy-of-stink-bugs-found-in-the-united-states/.

Anonymous, 2015b. Potato Flea Beetles *Epitrix* Species. Agri-Food and Biosciences Institute, Newforge Lane, Belfast, Northern Ireland, UK. https://www.afbini.gov.uk/articles/potato-flea-beetles-epitrix-species.

Antonelli, A.L., Davidson, R.M., 2001. Potato Flea Beetles: Biology and Control. Washington State University Cooperative Extension, Pullman, Washington, USA. Extension Bulletin No. 1198E https://research.libraries.wsu.edu/xmlui/bitstream/handle/2376/5251/eb1198.pdf?sequence=1.

Antwi, F.B., Olson, D.L., Knodel, J.J., 2007. Comparative evaluation and economic potential of ecorational versus chemical insecticides for crucifer flea beetle (Coleoptera: Chrysomelidae) management in Canola. Journal of Economic Entomology 100, 710–716.

Apriyanto, D., Sedlacek, J.D., Townsend, L.H., 1989. Feeding activity of *Euschistus servus* and *E. variolarius* (Heteroptera: Pentatomidae) and damage to an early growth stage of corn. Journal of Kansas Entomological Society 62, 392–399.

Bailey, D.L., 2012. Characterization of Biopores Resulting from Mole Crickets (*Scapteriscus* spp.) (M.Sc. thesis). Graduate Faculty of Auburn University, Auburn, Alabama, USA, pp. 1–48.

Barbara, K.A., 2005. Management of Pest Mole Crickets Using the Insect Parasitic Nematode *Steinernema scapterisci* (Ph.D. thesis). Graduate School of the University of Florida, Florida, USA, pp. 1–87.

Bartlett, P.W., Powell, D.F., 1981. Introduction of American serpentine leaf miner, *Liriomyza trifolii*, into England and Wales and its eradication from commercial nurseries, 1977–1981. Plant Pathology 30, 185–193.

Bennet-Clark, H.C., 1989. Songs and physics of sound production. In: Huber, F., Moore, T.E., Loher, W. (Eds.), Cricket Behaviour and Neurobiology. Cornell University Press, Ithaca, New York, USA, pp. 227–261.

Bethune, C.J.S., 1872. Report of the Entomological Society of Ontario for the Year 1871. Hunter, Ross, Toronto, Canada.

Biever, K.D., Chauvin, R.L., 1992. Suppression of the Colorado potato beetle (Coleoptera: Chrysomelidae) with augmentative releases of predaceous stinkbugs (Hemiptera: Pentatomidae). Journal of Economic Entomology 85, 720–726.

Boavida, C., Giltrap, N., Cuthbertson, A.G.S., Northing, P., 2013. *Epitrix similaris* and *Epitrix cucumeris* in Portugal: damage patterns in potato and suitability of potential plants for reproduction. EPPO Bulletin 43, 323–333.

Bográn, C.E., 2005. Biology and Management of *Liriomyza* Leafminers in Greenhouse Ornamental Crops. Texas A&M Agrilife Extension Service Publication No. EEE-00030 The Texas A&M University System, Texas, USA. http://extentopubs.tamu.edu/eee_00030.html.

Boiteau, G., Misener, G., Singh, R.P., Bernard, G., 1992. Evaluation of a vacuum collector for insect pest control in potato. American Potato Journal 69, 157–166.

Boucias, D.G., 1985. Diseases. In: Walker, T.J. (Ed.), Mole Crickets in FloridaFlorida Agricultural Experiment Station Bulletin, vol. 846, pp. 32–35.

Brandenburg, R.L., 1997. Managing mole crickets: developing a strategy for success. Turfgrass Trends 6, 1–8.

Brandenburg, R.L., Williams, E.B., 1993. A Complete Guide to Mole Cricket Management in North Carolina. North Carolina State University, Raleigh, North Carolina, USA.

Brelih, S., Döberl, M., Drovenik, B., Pirnat, A., 2003. *Scopolia*. Gradivo za favno hroscev (Coleptera) Slovenije/Materialien zur Käferfauna (Coleoptera) Slowenien. 1. prispevek/1. Beitrag, Polyphaga: Chrysomeloidea (=Phytophaga): Chrysomelidae: Alticinae, vol. 50, pp. 1–279.

Brennan, B.M., Chang, F., Mitchell, W.C., 1977. Physiological effects on sex pheromone communication in the southern green stink bug *Nezera viridula*. Environmental Entomology 6, 169–173.

Broadley, R.H., July–August 1978. Insect pests of sunflower. Queensland Agricultural Journal 307–314.

Brues, C.T., 1940. Food preferences of the Colorado potato beetle, *Leptinotarsa decemlineata* Say. Psyche 47, 38–43.

Brust, G., 2015. Bad Year for Stink Bugs. College of Agriculture and Natural Resources. University of Maryland, College Park, Maryland, USA. http://extension.umd.edu/learn/bad-year-stink-bugs.

Brust, G., Rane, K., 2011. Transmission of the Yeast *Eremothecium coryli* to Fruits and Vegetables by the Brown Marmorated Stink Bug. University of Maryland Extension, University of Maryland, College Park, Maryland, USA. https://extension.umd.edu/learn/transmission-yeast-eremothecium-coryli-fruits-and-vegetables-brown-marmorated-stink-bug.

Brust, G.E., 1994. Natural enemies in straw-mulch reduce Colorado potato beetle populations and damage in potato. Biological Control 4, 163–169.

Bůda, V., Radziute, S., 2008. Kairomone attractant for the leafmining fly, *Liriomyza bryoniae* (Diptera, Agromyzidae). Zeitschrift für Naturforschung C 63, 615–618.

Bueno, F., Santos, C., Tofoli, R., Pavan, A., Bueno, C., 2007. Reduction of spinosad rate for controlling *Liriomyza huidobrensis* (Diptera: Agromyzidae) in dry beans (*Phaseolus vulgaris* L.) and its impact on *Frankliniella schultzei* (Thysanoptera: Thripidae) and *Diabrotica speciosa* (Coleoptera: Chrysomelidae). BioAssay 2, 3.

Buss, E.A., Capinera, J.L., Leppla, N.C., 2002. Pest Mole Cricket Management. Department of Entomology and Nematology. Florida Cooperative Extension Service, Publication No. ENY-324 University of Florida, Gainesville, Florida, USA. http://polk.ifas.ufl.edu/hort/documents/publications/Pest%20Mole%20Cricket%20Management.pdf.

Buss, L.J., Halbert, S.E., Johnson, S.J., 2013. Pest Alert: *Leptoglossus zonatus* – A New Leaffooted Bug in Florida (Hemiptera: Coreidae). Pest Alert. Florida Department of Agriculture and Consumer Service, Division of Plant Industry, Gainseville, Florida, USA.

Caldwell, B., Rosen, E.B., Sideman, E., Shelton, A.M., Smart, C.D., 2005. Material Fact Sheet *Beauveria bassiana*. Cornell University Organic Resource Guide. http://web.pppmb.cals.cornell.edu/resourceguide/mfs/03beauveria_bassiana.php.

Callahan, P.S., Brown, R., Dearman, A., 1960. Control of tomato insect pests. Louisiana Agriculture 3, 4–5 16.

Campbell, R.K., Anderson, T., Semel, M., Roberts, D., 1985. Management of the Colorado potato beetle using the entomogenous fungus *Beauveria bassiana*. American Potato Journal 62, 29–37.

Cantwell, G.E., Cantelo, W., Schroder, R., 1986. Effect of *Beauveria bassiana* on underground stages of the Colorado potato beetle, *Leptinotarsa decemlineata* (Coleoptera: Chrysomelidae). Great Lakes Entomologist 19, 81–84.

Capinera, J.L., 2001. Handbook of Vegetable Pests. Academic Press, San Diego, California, USA, p. 729.

Casagrande, R.A., 1987. The Colorado potato beetle: 25 years of mismanagement. Bulletin of Entomological Society of America 33, 142–150.

Catchot, A., Musser, F., Cook, D., Gore, J., McPherson, W., Allen, A., 2012. Residual Activity of Selected Insecticides on Green and Brown Stink Bug Adults. Extension Service of Mississippi State University Publication No. 2728 Mississippi State University, USA. https://extension.msstate.edu/sites/default/files/publications/publications/p2728.pdf.

Chalfant, R.B., 1973. Chemical control of southern green stink bug, tomato fruit worm and potato aphid on vining tomatoes in southern Georgia USA. Journal of Georgia Entomological Society 8, 279–283.

Chow, A., Heinz, K.M., 2006. Control of *Liriomyza langei* on chrysanthemum by *Diglyphus isaea* produced with a standard or modified parasitoid rearing technique. Journal of Applied Entomology 130, 113–121.

Cisneros, F.H., June 1984. The need for integrated pest management in developing countries. In: Report of the XXII Planning Conference on Integrated Pest Management, vol. 4, pp. 19–23 Lima, Peru, USA.

Clark, J.K., 2015. Vegetable leaf miner. Fact sheet Statewide IPM program. University of California, California, USA, pp. 1–2.

Clarke, A.R., Walter, G.H., 1993. Biological control of the green vegetable bug *Nezara viridula* (L.) in eastern Australia: current status and perspectives. In: Corey, S.A., Dall, D., Milne, W.M. (Eds.), Pest Control and Sustainable Agriculture. Commonwealth Scientific and Indutrial Research Organisation, Division of Entomology, Canberra, Australia, pp. 223–225.

Cobb, P.P., 1998. Controlling Mole Crickets on Lawns and Turf. Auburn University, Auburn, USA. http://www.aces.eduldepartment/extcomm/publications/anr/anr176/anr176.htm.

Cobb, P.P., Mack, T.P., 1989. A rating system for evaluating tawny mole cricket, *Scapteriscus vicinus* Scudder, damage (Orthoptera: Gryllotalpidae). Journal of Entomological Science 24, 142–144.

Coombs, M., Khan, S.A., 1998. Fecundity and longevity of green vegetable bug, *Nezara viridula*, following parasitism by *Trichopoda giacomellii*. Biological Control 12, 215–222.

Cranshaw, W.S., 2013. Flea Beetles. Insect Series: Home and Garden Factsheet No. 5.592. Colorado State University Extension, Fort Collins, Colorado, USA.

Cuthbertson, A.G.S., 2014. Personal Observation. The Food and Environment Research Agency, Sand Hutton, New York, UK.

Cuthbertson, A.G.S., 2015. Chemical and ecological control methods for *Epitrix* spp. Global Journal of Environmental Science and Management 1, 95–97.

Daramola, A.M., 1974. A review on the pests of Cola species in West Africa. Nigerian Journal of Entomology 1, 21–29.

Dasgupta, S., Meisner, C., Wheeler, D., 2004. Is Environmentally-friendly Agriculture Less Profitable for Farmers? Evidence on Integrated Pest Management in Bangladesh. World Bank Policy Research Working Papers. http://dx.doi.org/10.1596/1813-9450-3417.

De Villiers, W.M., 1985. Orthoptera: Ensifera. In: Scholtz, C.H., Holm, E. (Eds.), Insects of Southern Africa. Butterworths, Durban, South Africa, p. 86.

de Wilde, J., 1948. Developpement embryonnaire et postembryonnaire dy doryphore (*Leptinotarsa decemlineata* Say) en function de la temperature. In: Proceedings of 8th International Congress of Entomology, Stockholm, Sweden, 9–15 August, 1948, pp. 310–321.

Demirozer, O., Tyler-Julian, K., Funderburk, J., Leppla, N., Reitz, S., 2012. *Frankliniella occidentalis* (Pergande) integrated pest management programs for fruiting vegetables in Florida. Pest Management Science 68, 1537–1545.

Dempster, J.P., Coaker, T.H., 1974. Diversification of crop ecosystems as a means of controlling pests. In: Price-Jones, D., Solomon, M.E. (Eds.), Pest and Disease Control. Blackwell, Oxford, UK, pp. 106–114.

Denisenko, M.K., 1986. An effective method of controlling mole crickets. Zaschchita Rastenii 10, 51.

Dent, D., 1991. Insect Pest Management, second ed. CABI Publishing, CAB International, Wallingford, Oxon, UK, p. 424.

Dietz, L.L., Van Duyn, J., Bradley, J., Rabb, R., Brooks, W., Stinner, R.E., 1976. A guide to the identification and biology of soybean arthropods in North Carolina. North Carolina Agricultural Experiment Station Technical Bulletin 238, 1–264.

Döberl, M., 2000. Contribution to the knowledge of the genus *Epitrix* Foudras, 1860 in the Palearctic region (Coleoptera: Chrysomelidae: Alticinae). Mitteilungen des Internationaler Entomologischer Verein 25, 1–23.

Doggett, C., 2015. Mole Crickets. Brochure for Conifer and Hardwood Insects , pp. 153–154. https://rngr.net/publications/forest-nursery-pests/conifer-and-hardwood-insects/mole-crickets/at_download/file.

Dosdall, L.M., Dolinski, M.G., Cowle, N.T., Conway, P.M., 1999. The effect of tillage regime, row spacing, and seeding rate on feeding damage by flea beetles, *Phyllotreta* spp. (Coleoptera: Chrysomelidae), in canola in central Alberta, Canada. Crop Protection 18, 217–224.

Dosdall, L.M., Herbut, M.J., Cowle, N.T., Micklich, T.M., 1996. The effect of tillage regime on emergence of root maggots (*Delia* spp.) (Diptera: Anthomyiidae) from canola. The Canadian Entomologist 128, 1157–1165.

Drummond, F., Suhaya, Y., Groden, E., 1990. Predation on the Colorado potato beetle (Coleoptera: Chrysomelidae) by *Phalangium opilio* (Opiliones: Phalangidae). Journal of Economic Entomology 83, 772–778.

East, D.A., 1993. Colonization of Potato Fields by Post-diapause Colorado Potato Beetles, Leptinotarsa decemlineata (Say) (Ph.D. dissertation). Ohio State University, Columbus, Ohio, USA.

Ekbom, B., Müller, A., 2011. Flea beetle (*Phyllotreta undulata* Kutschera) sensitivity to insecticides used in seed dressings and foliar sprays. Crop Protection 30, 1376–1379.

Elwell, H., Mass, A., 1989. Natural Pest and Disease Control. Zimbabwe Natural Farming Network, Harare, p. 34.

EPPO, 2015. Potato Flea Beetle *Epitrix* spp. Plant Pest Information Note. Department of Agriculture, Fisheries and Food, Backweston Campus, Young's Cross, Celbridge, County Kildare, Dublin, Ireland.

Eyre, D., Giltrap, N., 2012. *Epitrix* flea beetles: new threats to potato production in Europe. Pest Management Science 69, 3–6.

Fatouros, N.E., Dicke, M., Mumm, R., Meiners, T., Hilker, M., 2008. Foraging behavior of egg parasitoids exploiting chemical information. Behavioral Ecology 19, 677–689.

Ferro, D.N., 1985. Pest status and control strategies of the Colorado potato beetle. In: Ferro, D.N., Voss, R.H. (Eds.), Proceedings of the Symposium on the Colorado Potato Beetle, XVII International Congress of Entomology Research Bulletin 704. Massachusetts Agricultural Experiment Station, Amherst, Massachusetts, USA.

Ferro, D.N., 1993. Potential for resistance to *Bacillus thuringiensis*: colorado potato beetle (Coleoptera: Chrysomelidae) – a model system. American Entomologist 39, 38–44.

Ferro, D.N., Alyokhin, A.V., Tobin, D.B., 1999. Reproductive status and flight activity of the overwintered colorado potato beetle. Entomologia Experimentalis et Applicata 91, 443–448.

Ferro, D.N., Logan, L., Voss, R.H., Elkinton, J.S., 1985. Colorado potato beetle (Coleoptera: Chrysomelidae) temperature-dependent growth and feeding rates. Environmental Entomology 14, 343–348.

Ferro, D.N., Morzuch, B.J., Margolies, D., 1983. Crop loss assessment of the colorado potato beetle (Coleoptera: Chrysomelidae) on potatoes in western Massachusetts. Journal of Economic Entomology 76, 349–356.

Ferro, D.N., Tuttle, A.F., Weber, D.C., 1991. Ovipositional and flight behavior of overwintered colorado potato beetle (Coleoptera: Chrysomelidae). Environmental Entomology 20, 1309–1314.

Fisher, D.G., Deahl, K.L., Rainforth, M.V., 2002. Horizontal resistance in *Solanum tuberosum* to colorado potato beetle (*Leptinotarsa decemlineata* Say). American Journal of Potato Research 79, 281–293.

Flanders, K.L., Hawkes, J.G., Radcliffe, E.B., Lauer, F.I., 1992. Insect resistance in potatoes: sources, evolutionary relationships, morphological and chemical defenses, and ecogeographical associations. Euphytica 61, 83–111.

Forgash, A.G., 1985. Insecticide resistance in the Colorado potato beetle. In: Ferro, D.N., Voss, R.H. (Eds.), Proceedings of the Symposium on the Colorado Potato Beetle, XVII International Congress of Entomology. Massachusetts Agricultural Experiment Station, Amherst, Massachusetts, USA. August 20–26, 1984.

Foster, R., Flood, B. (Eds.), 1995. Vegetable insect management: with emphasis on the Midwest. Willoughby, Ohio, USA.

Frank, J.H., Fasulo, T.R., Short, D.E., 1998. Mcricket: Knowledgebase. CD-ROM. Institute of Food and Agricultural Sciences. University of Florida, Gainesville, Florida, USA.

Frank, J.H., Parkman, J.P., 1999. Integrated pest management of pest mole crickets with emphasis on the southeastern USA. Integrated Pest Management Reviews 4, 39–52.

Frank, J.H., Walker, T.J., Parkman, J.P., 1996. The introduction, establishment, and spread of *Ormia depleta* in Florida. Biological Control 6, 368–377.

Gadallah, A.I., Zidan, Z.H., El-Malki, K.G., Amin, A., Eissa, M.A., 1998. Seasonal abundance of *Gryllotalpa africana* adults and efficacy of pitfall traps in reducing its population in qena governorate. Annals of Agricultural Science (Cairo) 3, 911–923.

Gao, Y., Reitz, S.R., Wei, Q., Yu, W., Lei, Z., 2012. Insecticide-mediated apparent displacement between two invasive species of leafminer fly. PLoS One 7, e36622.

Gauthier, N.L., Hofmaster, R.N., Semel, M., 1981. History of Colorado potato beetle control. In: Lashomb, J.H., Casagrande, R. (Eds.), Advances in Potato Pest Management. Hutchinson Ross Publishing Company, Stroudsburg, Pennsylvania, USA, pp. 13–33.

Gilberg, L., 1993. Garden Pests and Diseases. Sunset Books. Sunset Publishing Corporation, California, USA, pp. 67–70.

Glass, E.H., 1975. Integrated Pest Management, Rationale, Potential, Needs and Improvement. Entomological Society of America, p. 141.

Grafius, E.J., Douches, D.S., 2008. The present and future role of insect-resistant genetically modified potato cultivars in IPM. In: Romeis, J., Shelton, A.M., Kennedy, G. (Eds.), Integration of Insect-resistant GM Crops within IPM Programs. Springer, New York, USA, pp. 195–222.

Greathead, D.J., Greathead, A.H., 1992. Biological control of insect pests by insect parasitoids and predators: the BIOCAT database. Biocontrol News and Information 13, 61N–68N.

Gredler, G., 2001. Encouraging Beneficial Insects in your Garden. Pacific Northwest Extension Publication No. PNW550. Washington State University, Pullman, Washington, USA.

Grimm, C., Führer, E., 1998. Population dynamics of true bugs (Heteroptera) in physic nut (*Jatropha curcas*) plantations in Nicaragua. Journal of Applied Entomology 122, 515–521.

Grimm, C., Guharay, F., 1998. Control of leaf-footed bug *Leptoglossus zonatus* and Shield-backed bug *Pachycoris klugii* with entomopathogenic fungi. Biocontrol Science and Technology 8, 365–376.

Gruev, B., Tomov, V., 1986. Coleoptera, Chrysomelidae. Part II. Chrysomelinae, Galerucinae, Alticinae, Hispinae, Cassidinae. In: Josifov, M. (Ed.), Fauna Bulgarica. Academie Scientiarium Bulgaricae, Sofia, Balgaria, p. 388.

Hallman, G.J., Morales, C.G., Duque, M.C., 1992. Biology of *Acrosternum marginatum* (Heteroptera: Pentatomidae) on common beans. Florida Entomologist 75, 190–196.

Hardison, J.R., 1980. Role of fire for disease control in grass seed production. Plant Diseases 64, 641–645.

Hare, J.D., 1980. Impact of defoliation by the Colorado potato beetle on potato yields. Journal of Economic Entomology 73, 369–373.

Hare, J.D., 1990. Ecology and management of the Colorado potato beetle. Annual Review of Entomology 35, 81–100.

Hare, J.D., Moore, R.E., 1988. Impact and management of late-season populations of the colorado potato beetle (Coleoptera: Chrysomelidae) on potato in Connecticut. Journal of Economic Entomology 81, 914–921.

Harris, V.E., Todd, J.W., 1980. Male-mediated aggregation of male, female and fifth-instar southern green stink bugs, *Nezara viridula*, and concomitant attraction of a tachinid parasite, *Trichopoda pennipes*. Entomologia Experimentalis et Applicata 27, 117–126.

Harris, V.E., Todd, J.W., Mullinix, B.G., 1984. Color change as an indicator of adult diapause in the adult southern green stink bug, *Nezera viridula*. Journal of Agricultural Entomology 1, 82–91.

Haviland, D., 2007. Leaf footed bugs in almonds, look before you spray this season. Pacific Nut Producer 24–26.

Hazzard, R., Ferro, D.N., Van Driesche, R.G., Tuttle, A.F., 1991. Mortality of eggs of colorado potato beetle (Coleoptera: Chrysomelidae) from predation by *Coleomegilla maculata* (Coleoptera: Coccinelidae). Environmental Entomology 20, 841–848.

Heimpel, G.E., Hough-Goldstein, J.A., 1992. A survey of arthropod predators of *Leptinotarsa decemlineata* (Say) in Delaware potato fields. Journal of Agricultural Entomology 9, 137–142.

Held, W., Potter, D.A., 2012. Prospects for managing turfgrass pests with reduced chemical inputs. Annual Review of Entomology 57, 329–354.

Henne, D.C., Johnson, S.J., 2001. Seasonal distribution and parasitism of *Scapteriscus* spp. (Orthoptera: Gryllotalpidae) in southeastern Louisiana. Florida Entomologist 84, 209–214.

Hertl, P.T., Brandenburg, R.L., Barbercheck, M.E., 2001. Effect of soil moisture on ovipositional behavior in the southern mole cricket *Scapteriscus borellii* Giglio-Tos (Orthoptera: Gryllotalpidae). Environmental Entomology 30, 466–473.

Hill, P.S.M., Hoffart, C., Buchheim, M., 2002. Tracing phylogenetic relationship in the family Gryllotalpidae. Journal of Orthoptera Research 11, 169–174.

Hills, O.A., Taylor, E.A., 1951. Parasitization of dipterous leafminers in cantaloups and lettuce in the Salt River Valley, Arizona. Journal of Economic Entomology 44, 759–762.

Hitchner, E.M., Kuhar, T.P., Dickens, J.C., Youngman, R.R., Schultz, P.B., Pfeiffer, D.G., 2008. Host plant choice experiments of colorado potato beetle (Coleoptera: Chrysomelidae) in Virginia. Journal of Economic Entomology 101, 859–865.

Hopwood, J., Vaughan, M., Shepherd, M., Biddinger, M., Mader, E., Black, S.H., Mazzacano, C., 2012. Are Neonicotinoids Killing Bees? A Review of Research into the Effects of Neonicotinoid Insecticides on Bees, with Recommendations for Action. The Xerces Society for Invertebrate Conservation Portland, Oregon, USA, p. 44.

Horton, D.R., Capinera, J.L., Chapman, P.A., 1988. Local differences in host use by two populations of the colorado potato beetle. Ecology 69, 823–831.

Hough-Goldstein, J.A., Heimpel, G.E., Bechmann, H.E., Mason, C.E., 1993. Arthropod natural enemies of the Colorado potato beetle. Crop Protection 12, 324–334.

Hough-Goldstein, J.A., McPherson, J., 1996. Comparison of *Perillus bioculatus* and *Podisus maculiventris* (Hemiptera: Pentatomidae) as potential control agents of the colorado potato beetle (Coleoptera: Chrysomelidae). Journal of Economic Entomology 89, 1116–1123.

Hsiao, T.H., 1985. Ecophysiological and genetic aspects of geographic variations of the colorado potato beetle. In: Ferro, D.N., Voss, R.H. (Eds.), Proceedings of the Symposium on the Colorado Potato Beetle, XVII International Congress of Entomology, 20–25 August, 1984, Humberg, Germany.

Hudson, W., Buntin, D., Gardner, W., 2008. Pasture and forage insects. In: Guillebeau, P., Hinkle, N., Roberts, P. (Eds.), Summary of Losses from Insect Damage and Cost of Control in Georgia 2006. University of Georgia College of Agricultural and Environmental Sciences Athens, Georgia, USA, pp. 17–19. Miscellaneous Publications No. 106.

Hudson, W.G., 1985. Ecology of the Tawny Mole Cricket, *Scapteriscus vicinus* (Orthoptera: Gryllotalpidae): Population Estimation, Spatial Distribution, Movement, and Host Relationships (Ph.D. thesis). Entomology and Nematology Department, College of Agricultural and life Sciences, University of Florida, Gainesville, Florida, USA.

Hudson, W.G., Saw, J.G., 1987. Spatial distribution of the tawny mole cricket, *Scapteriscus vicinus*. Entomologia Experimentalis et Applicata 45, 99–104.

Huseth, A.S., Groves, R.L., Chapman, S.A., Alyokhin, A., Kuhar, T.P., Macrae, I.V., Szendrei, Z., Nault, B.A., 2014. Managing Colorado potato beetle insecticide resistance: new tools and strategies for the next decade of pest control in potato. Journal of Integrated Pest Management 5, A1–A18.

Ingels, C., Haviland, D., 2014. Leafooted Bug, Integrated Pest Management for Landscape Professionals and Home Gardeners. Pest Notes Publication No. 74168, UC Statewide Integrated Pest Management Program University of California, Davis, California, USA, pp.1–4. http://ucanr.edu/sites/mgfresno/files/203980.pdf.

Ingrisch, S., Nikouei, P., Hatami, B., 2006. A new species of mole crickets *Gryllotalpa linnaeus*, from Iran (Orthoptera: Gryllotalpidae). Entomologische Zeitschrift Stuttgart 116, 195–202.

Javahery, M., 1990. Biology and ecological adaptation of the green stink bug (Hemiptera: Pentatomidae) in Québec and Ontario. Annals of Entomological Society of America 83, 201–206.

Johnson, M.W., 1993. Biological control of *Liriomyza* leafminers in the Pacific basin. Micronesica Supplement 4, 81–92.

Jolivet, P., 1991. The Colorado beetle menaces Asia (*Leptinotarsa decemlineata* Say) (Coleoptera: Chrysomelidae). L'Entomologiste 47, 29–48.

Jolivet, P., Hawkeswood, T.J., 1995. Host-plants of Chrysomelidae of the World. An Essay about the Relationships between the Leaf-beetles and their Food-Plants. Backhuys Publishers, Leiden, p. 281.

Jones, V.P., 1995. Reassessment of the role of predators and *Trissolcus basalis* in biological control of southern green stink bug (Hemiptera: Pentatomidae) in Hawaii. Biological Control 5, 566–572.

Jones, V.P., Caprio, L.C., 1994. Southern green stink bug (Hemiptera: Pentatomidae) feeding on Hawaiian macadamia nuts: the relative importance of damage occurring in the canopy and on the ground. Journal of Economic Entomology 87, 431–435.

Jones Jr., W.A., 1988. World review of the parasitoids of the southern green stink bug *Nezara viridula* (L.) (Heteroptera: Pentatomidae). Annals of Entomological Society of America 81, 262–273.

Jones Jr., W.A., Sullivan, M.J., 1981. Overwintering habitats, spring emergence patterns and winter mortality of some South Carolina Hemiptera. Environmental Entomology 10, 409–414.

Joyce, P., Miles, C., Murray, T., Snyder, W., 2012. Organic Management of Flea Beetles. A Pacific Northwest Extension Publication No. PNW640. Washington State University, Pullman, Washington, USA.

Kabaluk, J.T., Vernon, R.S., 2000. Effect of crop rotation on populations of *Epitrix tuberis* (Coleoptera: Chrysomelidae) in potato. Journal of Economic Entomology 93, 315–322.

Kang, L., Chen, B., Wei, J.N., Liu, T.X., 2008. Roles of thermal adaptation and chemical ecology in *Liriomyza* distribution and control. Annual Review of Entomology 54, 127–145.

Kaszab, Z., 1962. Leaf beetles – Chrysomelidae. In: Szekessy, V. (Ed.), Magyarország állatvilága. Fauna hungariae. Akadémiai Kiadó, Budapest, Hungary. IX/6, 416.

Kavanagh, W., Young, D., 1989. Bilateral symmetry of sound production in the mole cricket, *Gryllotalpa australis*. Journal of Comparative Physiology A 166, 43–49.

Kelsheimer, E.G., 1963. Tomato varietal resistance to leafminer attack. Proceedings of the Florida State Horticultural Society 76, 134–135.

Kepner, R., 1985. Chemical control of mole crickets. In: Walker, T.J. (Ed.)Walker, T.J. (Ed.), Mole Crickets in Florida. Florida Agricultural Experiment Station Bulletin, vol. 846, pp. 41–48.

Khrimian, A., Zhang, A., Weber, D.C., Ho, H.Y., Aldrich, J.R., Vermillion, K.E., Siegler, M.A., Shirali, S., Guzman, F., Leskey, T.C., 2014. Discovery of the aggregation pheromone of the brown marmorated stink bug (*Halyomorpha halys*) through the Creation of Stereoisomeric Libraries of 1-Bisabolen-3-ols. Journal of Natural Products 77, 1708–1717.

Kim, K.W., 2000. Non-chemical or low-chemical control measures against key insect pests and rats in the ginseng fields. Korean Journal of Applied Entomology 39, 281–286.

Kuhar, T., Jenrette, J., Doughty, H., 2010. Leaf-Footed Bugs. Virginia Cooperative Extension Publication No. VCE 3012/3012-1522 Virginia Polytechnic Institute and State University, Petersburg, Virginia, USA. https://www.pubs.ext.vt.edu/3012/3012-1522/3012-1522.html.

Kuhar, T., Morrison, R., Leskey, T., Aigner, J., Dively, D., Zobel, E., Brust, G., Whalen, J., Cissel, B., Walgenbach, J., Rice, K., Fleischer, S., Rondon, S., 2015. Brown Marmorated Stinkbug in Vegetables: A Synopsis of what Researchers have Learned so far and Management Recommendations. NE IPM Center Publication No.512.

Lacasse, B., Laguë, C., Khelifi, M., Roy, P.-M., 1998. Field evaluation of pneumatic control of Colorado potato beetle. Canadian Agricultural Engineering 40, 273–280.

Lactin, D.J., Holliday, J., 1994. Behavioral responses of Colorado potato beetle larvae to combinations of temperature and insolation, under field conditions. Entomologia Experimentalis et Applicata 72, 255–263.

Laguë, C., Khelifi, M., Gill, J., Lacasse, B., 1999. Pneumatic and thermal control of Colorado potato beetle. Canadian Agricultural Engineering 41, 53–57.

Lake, P.C., 2000. Behaviors of *Pheropsophus Aequinoctialis* (Coleoptera: Carabidae) Affecting its Ability to Locate its Larval Food, Eggs of *Scapteriscus* Spp. (Orthoptera: Gryllotalpidae); and the Effect of Moisture on Oviposition Depth in *Scapteriscus Abbreviatus* (M.S. thesis). University of Florida, Gainesville, Florida, USA, p. 52.

Lashomb, J.H., Ng, Y.S., 1984. Colonization by the colorado potato beetle, *Leptinotarsa decemlineata* (Say) (Coleoptera: Chrysomelidae) in rotated and non-rotated potato fields. Environmental Entomology 13, 1352–1356.

Lashomb, J.H., Ng, Y.S., Jansson, R.K., Bullock, R., 1987. *Edovum puttleri* (Hymenoptera: Eulophidae), an egg parasitoid of colorado potato beetle (Coleoptera: Chrysomelidae)development and parasitism on eggplant. Journal of Economic Entomology 80, 65–68.

Lawrence, K.O., 1982. A linear pitfall trap for mole crickets and other soil arthropods. Florida Entomologist 65, 376–377.

Leal, W.S., Panizzi, A.R., Niva, C.C., 1994. Alarm pheromone system of leaf-footed bug *Leptoglossus zonatus* (Heteroptera: Coreidae). Journal of Chemical Ecology 20, 1209–1216.

LeBoeuf, J., 2015. Stink Bugs in Tomatoes. Ministry of Agriculture, Food and Rural Affairs, Ontario, Canada. http://www.omafra.gov.on.ca/english/crops/hort/news/hortmatt/2015/17hrt15a4.htm.

Leibee, G.L., Capinera, J.L., 1995. Pesticide resistance in Florida insects limits management options. Florida Entomologist 78, 386–399.

Leppla, N.C., Frank, J.H., Adjei, M.B., Vicente, N.E., 2007. Management of pest mole crickets in Florida and Puerto Rico with a nematode and parasitic wasp. Florida Entomologist 90, 229–233.

Liu, T.-X., Kang, L., Heinz, K.M., Trumble, J., 2009. Biological control of *Liriomyza* leafminers: progress and perspective. CAB reviews: perspectives in agriculture, veterinary science. Nutrition and Natural Resources 4, 4.

Logan, P.A., 1990. Summary of Biological Control Activities. 1989–1990. Natural Enemy News 2. University of Rhode Island, p. 10.

Logan, P.A., Casagrande, R.A., Faubert, H.H., Drummond, F.A., 1985. Temperature-dependent development and feeding of immature Colorado potato beetles, *Leptinotarsa decemlineata* Say (Coleoptera: Chrysomelidae). Environmental Entomology 14, 275–283.

Lopez, E., Roth, L.C., Ferro, D.N., Hosmer, D., Mafra-Neto, A., 1997. Behavioral ecology of *Myiopharus doryphorae* (Riley) and *M. aberrans* (Townsend), tachinid parasitoids of the Colorado potato beetle. Journal of Insect Behavior 10, 49–78.

Luckmann, W.H., Metcalf, R.L., 1994. The pest-management concept. In: Metcalf, R.L., Luckmann, W.H. (Eds.), Introduction to Insect Pest Management, third ed. Wiley, New York, USA, pp. 1–34.

Lye, B.H., Story, R.N., 1988. Feeding preferences of the southern green stink bug (Hemiptera: Pentatomidae) on tomato fruit. Journal of Economic Entomology 81, 522–526.

Lye, B.H., Story, R.N., 1989. Spatial dispersion and sequential sampling plan of the southern green stink bug (Hemiptera: Pentatomidae) on fresh market tomatoes. Environmental Entomology 18, 139–144.

Lye, B.H., Story, R.N., Wright, V.L., 1988. Southern green stink bug (Hemiptera: Pentatomidae) damage to fresh market tomatoes. Journal of Economic Entomology 81, 189–194.

MacDonald, O.C., 1991. Responses of the alien leaf miners *Liriomyza trifolii* and *Liriomyza huidobrensis* (Diptera: Agromyzidae) to some pesticides scheduled for their control in the UK. Crop Protection 10, 509–513.

Maharijay, A., Vosman, B., 2015. Managing the colorado potato beetle; the need for resistance breeding. Euphytica 204, 487–501.

Mahr, D.L., Ridgeway, N.M., 1993. The General Approaches to Insect Control: An Overview of Entomology. , vol. 1993http://www.entomology.wisc.edu/mbcn/fea102.html.

Majumdar, 2013. Home Garden Vegetables. Insect Control Recommendations for 2013. Alabama Cooperative Extension System Publication No. IPM-1305. Alabama A&M University and Auburn University, USA.

Majumdar, A., Miles, J., 2013. Vegetable Insecticide Recommendations, Updates 2013. Extension Report of Alabama Cooperative Extension System. Alabama A&M University and Auburn University and Auburn University, Baldwin County, Alabama, USA.

Majumdar, A.Z., Akridge, R., Becker, C., Caylor, A., Pitts, J., Price, M., Reeves, M., 2012. Trap crops for leaffooted bug management in tomatoes. Journal of the NACAA 5, 40–51.

Maltese, M., Caleca, V., Guerrieri, E., Strong, W.B., 2012. Parasitoids of *Leptoglossus occidentalis* Heidemann (Heteroptera: Coreidae) recovered in western North America and first record of its egg parasitoid *Gryon pennsylvanicum* (Ashmead) (Hymenoptera: Platygastridae) in California. Pan-Pacific Entomologist 88, 347–355.

Martins, J.C., Valerio, M.A., Moreira, L.A., 1990. ED formulations for the control of *Nezara viridula* (L., 1758) (Hemiptera: Pentatomidae) on soyabeans. Anais da Sociedade Entomologica do Brasil 19, 51–58.

Mason, G.A., Johnson, M.W., Tabashnik, B.E., 1987. Susceptibility of *Liriomyza sativae* and *Liriomyza trifolii* (Diptera: Agromyzidae) to permethrin and fenvalerate. Journal of Economic Entomology 80, 1262–1266.

Matsuura, H., Oda, H., Ishizak, H., 1985. Damage to Chinese yam by the African mole cricket, *Gryllotalpa africuna* Palisot de Beauvois, and its control by chemicals. Japanese Journal of Applied Entomology and Zoology 29, 36–40.

Mau, R.F.L., Kessing, J.L.M., 2007. *Liriomyza Sativae* (Blanchard). Department of Entomology, University of Hawaii, USA. http://www.extento.hawaii.edu/kbase/crop/type/liriom_s.htm.

May, M.L., 1981. Role of body temperature and thermoregulation in the biology of the Colorado potato beetle. In: Lashomb, J.H., Casagrande, R. (Eds.), Advances in Potato Pest Management. Hutchinson Ross Publishing Company, Stroudsburg, Pennsylvania, USA, pp. 86–104.

Mayoori, K., Mikunthan, G., 2009. Damage pattern of cabbage flea beetle, *Phyllotreta cruciferae* (Goeze) (Coleoptera: Chrysomelidae) and its associated hosts of crops and weeds. American-eurasian Journal of Agricultural and Environmental Sciences 6, 303–307.

McLain, D.K., Marsh, N.B., Lopez, J.R., Drawdy, J.A., 1990. Intravernal changes in the level of parasitation of the southern green stink bug (Hemiptera: Pentatomidae), by the feather-legged fly (Diptera: Tachinidae): host sex, mating status, and body size as correlated factors. Journal of Entomological Science 25, 501–509.

McPherson, J.E., McPherson, R.M., 2000. Major crops attacked. In: McPherson, J.E., McPherson, R.M. (Eds.), Stink Bugs of Economic Importance in America North of Mexico. CRC Press, Washington, DC, USA, pp. 7–36.

McPherson, R.M., Pitts, J.R., Newsom, L.D., Chapin, J.B., Herzog, D.C., 1982. Incidence of tachinid parasitism of several stink bug (Hemiptera: Pentatomidae) species associated with soybean. Journal of Economic Entomology 75, 783–786.

McPherson, R.M., Todd, J.W., Yeargan, K.V., 1994. Stink bugs. In: Higley, L.G., Boethel, D.J. (Eds.), Handbook of Soybean Insect Pests. Entomological Society of American Publishing, Lanham, Maryland, USA, pp. 87–90.

Meers, S., 2008. Physical Control of Pests. Alberta Agriculture and Forestry, Government of Alberta, Canada. http://www1.agric.gov.ab.ca/$department/deptdocs.nsf/all/prm2366.

Meyer, J.R., 2003. Pest Control Tactics. ENT425 Homepage Department of Entomology North Carolina State University, Raleigh, North Carolina, USA.

Mhina, G., Thomas, M., 2013. The Cost Effectiveness of Biological Control: The Case of Invasive Mole Crickets and Florida's Commercial Pastureland. Poster Presentation at the Southern Agricultural Economics Association (SAEA) Annual Meeting, 3–5 February, 2013, Orlando, Florida, USA.

Miles, C., Blethen, C., Gaugler, R., Shapiro-Illan, D., Murray, T., 2012. Using Entomopathogenic Nematodes for Crop Insect Pest Control. Pacific Northwest Extension Publication No. PNW544 Wahington State University, Pullman, Washington, USA. http://cru.cahe.wsu.edu/CEPublications/PNW544/PNW544.pdf.

Milner, M., Kung, K.J.S., Wyman, J.A., Feldman, J., Nordheim, E., 1992. Enhancing overwintering mortality of colorado potato beetle (Coleoptera: Chrysomelidae) by manipulating the temperature of its diapause habitat. Journal of Economic Entomology 85, 1701–1708.

Miner, F.D., 1966. Biology and control of stink bugs on soybean. Arkansas Agriculture Experiment Station Bulletin 708, 1–40.

Misener, G.C., Boiteau, G., McMillan, L.P., 1993. A plastic-lining trenching device for the control of colorado potato beetle: beetle excluder. American Potato Journal 70, 903–908.

Mitchell, P.L., 2000. Leaf-footed bugs (Coreidae). In: Schaefer, C.W., Panizzi, A.C. (Eds.), Heteroptera of Economic Importance. CRC Press, Boca Raton, Florida, USA, p. 828.

Mitchell, P.L., Paysen, E.S., Muckenfuss, A.E., Schaffer, M., Shepard, B.M., 1999. Natural mortality of leaffooted bug (Hemiptera: Heteroptera: Coreidae) eggs in cowpea. Journal of Agricultural and Urban Entomology 16, 25–36.

Mitchell, W.C., Mau, R.F.L., 1971. Response of the female southern green stink bug and its parasite, *Trichopoda pennipes*, to male stink bug pheromones. Journal of Economic Entomology 64, 856–859.

Mota-Sanchez, D., Hollingworth, R., Grafius, E.J., Moyer, D.D., 2006. Resistance and cross-resistance to neonicotinoid insecticides and spinosad in the colorado potato beetle, *Leptinotarsa decemlineata* (Say) (Coleoptera: Chrysomelidae). Pest Management Science 62, 30–37.

Munyaneza, J., McPherson, J.E., 1994. Comparative study of life histories, laboratory rearing and immature stages of *Euschistus servus* and *E. variolarius* (Hemiptera: Pentatomidae). The Great Lakes Entomology 26, 263–274.

Murphy, S.T., LaSalle, J., 1999. Balancing biological control strategies in the IPM of New World invasive *Liriomyza* leafminers in field vegetable crops. Biocontrol News and Information 20, 91N–104N.

Ng, Y.S., Lashomb, J., 1983. Orientation by the Colorado potato beetle (*Leptinotarsa decemlineata* Say). Animal Behavior 31, 617–619.

Nickle, D.A., Castner, J.L., 1984. Introduced species of mole crickets in the United States, Puerto Rico, and the Virgin Islands (Orthoptera: Gryllotalpidae). Annals of the Entomological Society of America 77, 450–465.

Nicoli, G., 1997. Biological control of exotic pests in Italy: recent experiences and perspectives. OEPP Bulletin 27, 69–75.

Nugaliyadde, M.M., Babu, A.G., Karunasena, S., Rathnakumara, A.P., 2000. Management of pests and diseases of potato in Sri Lanka. Proceedings of African Potato Association Conference, Uganda 271–279.

Oatman, E.R., Kennedy, G.G., 1976. Methomyl induced outbreak of *Liriomyza sativae* on tomato. Journal of Economic Entomology 69, 667–668.

OEPP/EPPO, 1992. Quarantine procedures No. 42. Identification of *Liriomyza* spp. OEPP/EPPO Bulletin 22, 235–238.

OEPP/EPPO, 2005. *Liriomyza* spp. OEPP/EPPO Bulletin 35, 335–344.

Ohno, K., Ohmori, T., Takemoto, H., 1999. Effect of insecticide applications and indigenous parasitoids on population trends of *Liriomyza trifolii* in gerbera greenhouses. Japanese Journal of Applied Entomology and Zoology 43, 81–86.

Olga, S.K., Buss, A.E., Scharf, M.E., 2011. Toxicity and neurophysiological effects of selected insecticides on the mole cricket, *Scapteriscus vicinus* (Orthoptera: Gryllotalpidae). Pesticide Biochemistry and Physiology 100, 27–34.

Olivera, C., Bordat, D., Letourmy, P., 1993. Influence of temperature on the laying behaviour of females of the leaf miners *Liriomyza trifolii* and *Liriomyza huidobrensis*. In: Liriomyza Proceedings. Mission de Cooperation Phytosanitaire, pp. 37–48.

O'Neill, M.J., 2009. UC IPM Pest Management Guidelines: Tomato. UC Statewide Integrated Pest Management Program. University of California, Davis, California, USA.

Otte, D., 1994. Orthoptera Species File 1. Crickets (Grylloidea). The Academy of Natural Sciences of Philadelphia, Pennsylvania, USA, pp. 98–100.

Otte, D., Alexander, R.D., 1983. The australian crickets (Orthoptera: Gryllidae). The Academy of Natural Sciences of Philadelphia 17, 448–463.

Ozawa, A., Saito, T., Ota, M., 2001. Biological control of the American Serpentine leafminer, *Liriomyza trifolii* (Burgess), on tomato (*Lycopersicon esculentum*) in greenhouses by parasitoids. II. Evaluation of biological control by *Diglyphus isaea* (Walker) and *Dacnusa sibirica* Telenga in commercial greenhouses. Japanese Journal of Applied Entomology and Zoology 45, 61–74.

Panizzi, A.R., 1988. Parasitism by *Eutrichopodopsis nitens* (Diptera: Tachinidae) of *Nezara viridula* (Hemiptera: Pentatomidae) on different host plants. EMBRAPA Centro Nacional de Pesquisa de Soja 36, 82–83.

Panizzi, A.R., Oliveira, E.D.M., 1999. Seasonal occurrence of tachinid parasitism on stink bugs with different overwintering strategies. Anais da Sociedade Entomológica do Brasil 28, 169–172.

Parkman, J.P., Frank, J.H., 1993. Use of a sound trap to inoculate *Steinernema scapterisci* (Rhabditida: Steinernematidae) into pest mole cricket populations (Orthoptera: Gryllotalpidae). Florida Entomologist 76, 75–82.

Parrella, M.P., Keil, C.B., 1985. Toxicity of methamidophos to four species of Agromyzidae. Journal of Agricultural Entomology 2, 234–237.

Pelletier, Y., McLeod, C.D., Bernard, G., 1995. Description of sublethal injuries caused to the Colorado potato beetle (Coleoptera: Chrysomelidae) by propane flamer treatment. Journal of Economic Entomology 88, 1203–1205.

Pfiffner, L., 1997. Nematoden gegen Maulwurfsgrillen. Tagung Pflanzenschutz im ökologischen Gemüsebau in Heidelberg, 13 February, 1997, Tagungsband. , pp. 20–22.

Phatak, S., Diaz-Perez, J.C., 2012. Managing pests with cover crops profitably, third ed. Sustainable Agriculture Research and Education. http://www.sare.org/Learning-Center/Books/Managing-Cover-Crops-Profitably-3rd-Edition/Text-Version/Managing-Pests.

Pimentel, D., 1982. Perspectives of integrated pest management. Crop Protection 1, 5–26.

Pivnick, K.A., Lamb, R.J., Reed, D., 1992. Response of flea beetles, *Phyllotreta* spp., to mustard oils and nitriles in field trapping experiments. Journal of Chemical Ecology 18, 863–873.

Potter, D.A., 1998. Destructive Turfgrass Pests. Biology, Diagnosis, and Control. Ann Arbor Press, Michigan, USA.

Ragsdale, D.W., Larson, A.D., Newsom, L.D., 1981. Quantitative assessment of the predators of *Nezara viridula* eggs and nymphs within an agroecosystem using ELISA. Environmental Entomology 10, 402–405.

Rauf, A., Shepard, B.M., Johnson, M.W., 2000. Leafminers in vegetables, ornamental plants and weeds in Indonesia: surveys of host crops, species composition and parasitoids. International Journal of Pest Management 46, 257–266.

Reddy, B.S., 2013. Non-pesticidal Management of Pests: Status, Issues and Prospects – A Review. Centre for Economic and Social Studies, Begumpet, Hyderabad-500016.

Reddy, G.V., Tangtrakulwanich, K., Miller, J.H., Ophus, V.L., Prewett, J., 2014. Sustainable management tactics for control of *Phyllotreta cruciferae* (Coleoptera: Chrysomelidae) on canola in Montana. Journal of Economic Entomology 107, 661–666.

Reitz, S.R., Funderburk, J., 2012. Management strategies for western flower thrips and the role of insecticides. In: Perveen, F. (Ed.), Insecticides Pest Engineering. InTech, Rijeka, Croatia, pp. 355–384.

Reitz, S.R., Kund, G.S., Carson, W.G., Phillips, P.A., Trumble, J.T., 1999. Economics of reducing insecticide use on celery through low-input pest management strategies. Agriculture Ecosystems and Environment 73, 185–197.

Rider, D., 2015. Pentatomoids. Department of Entomology, North Dokota State University, Fargo, North Dokota, USA. https://www.ndsu.edu/ndsu/rider/Pentatomoidea/Researchers/researchers.htm.

Risch, S.J., Andow, D.A., Altieri, M.A., 1983. Agroecosystem diversity and pest control: data, tentative conclusion, and new research directions. Environmental Entomology 12, 625–629.

Root, R.B., 1973. Organization of plain arthropod association in simple and diverse habitats: the fauna of collards (*Brassica oleracea*). Ecological Monographs 43, 95–124.

Ruberson, J.R., Tauber, M.J., Tauber, C.A., Gollands, B., 1991. Parasitization by *Edovum puttleri* (Hymenoptera: Eulophidae) in relation to host density in the field. Ecological Entomology 16, 81–89.

Sánchez, G.A., de Huiza, I.R., 1988. *Liriomyza huidobrensis* y sus parasitoides en papa cultivada en Rimac y Cañete, 1986. Revista Peruana de Entomologia 31, 110–112.

Schalk, J.M., Stoner, A.K., 1979. Tomato production in Maryland: effect of different densities of larvae and adults of the colorado potato beetle *Leptinotarsa decemlineata*. Journal of Economic Entomology 72, 826–829.

Scharf, M.E., 2007. Neurological effects of insecticides. In: Pimentel, D. (Ed.). Pimentel, D. (Ed.), Encyclopedia of Pest Management, vol. 2. CRC Press Taylor and Francis Group, Boca Raton, Florida, USA, pp. 395–399.

Scheffer, S.J., 2000. Molecular evidence of cryptic species within the *Liriomyza huidobrensis* (Diptera: Agromyzidae). Journal of Economic Entomology 93, 1146–1151.

Scheffer, S.J., Lewis, M.L., 2001. Two nuclear genes confirm mitochondrial evidence of cryptic species within *Liriomyza huidobrensis* (Diptera: Agromyzidae). Annals of the Entomological Society of America 94, 648–653.

Schoeman, A.S., 1996. Turfgrass insect pests in South Africa. Turf and Landscape Maintenance 7, 15.

Schuster, D.J., 2015. Stink Bugs and Leaf-Footed Bugs. Brochure for IPM Florida. Institute of Food and Agriculture Science Extension, University of Florida, Gainesville, Florida, USA, pp. 65–66.

Schuster, D.J., Gilreath, J.P., Wharton, R.A., Seymour, P.R., 1991. Agromyzidae (Diptera) leafminers and their parasitoids in weeds associated with tomato in Florida. Environmental Entomology 20, 720–723.

Schuster, D.J., Price, J.F., 1992. Seedling feeding damage and preference of *Scapteriscus* spp. mole crickets (Orthoptera: Gryllotalpidae) associated with horticultural crops in west-central Florida. Florida Entomologist 75, 115–119.

Schuster, D.J., Smith, H.A., 2015. Scouting for insects, use of thresholds, and conservation of beneficial insects on tomatoes. Department of Entomology and Nematology, Institute of Food and Agriculture Sciences Extension Publication No. ENY685. University of Florida, Gainesville, Florida, USA.

Schuster, D.J., Stansly, P.A., 2012. Biorational Insecticides for Integrated Pest Management in Tomatoes. Brochure for IPM Florida Publication No. ENY684. Institute of Food and Agriculture Science Extension, University of Florida, Gainesville, Florida, USA, p. 6.

Semlitsch, R.D., 1986. Life history of the northern mole cricket, *Neocurtilla hexadactyla* (Orthoptera: Gryllotalpidae), utilizing Carolina-bay habitats. Annals of the Entomological Society of America 79, 256–261.

Serantes, H.E., 1973. Biologia de *Phthia picta* (Drury) (Hemiptera, Coreidae), vol. 9. Fitotecnica Latinoamericana, Buenos Aries, Argentina, pp. 3–9.

Shepard, B.M., Braun, S.A.R., 1998. Seasonal incidence of *Liriomyza huidobrensis* (Diptera: Agromyzidae) and its parasitoids on vegetables in Indonesia. International Journal of Pest Management 44, 43–47.

Short, D.E., Koehler, P.G., 1979. A sampling technique for mole crickets and other pests in turf grass and pasture. Florida Entomologist 62, 282–283.

Silcox, D.E., 2011. Response of the Tawny Mole Cricket (Orthoptera: Gryllotalpidae) to Synthetic Insecticides and their Residues (M.Sc. thesis). Department of Entomology, North Carolina State University, Raleigh, USA, p. 164.

Silva, R.A.D., Carvalho, G.S., 2001. Aspectos biológicos de *Phthia picta* (Drury, 1770) (Hemiptera: Coreidae) em tomateiro sob condições controladas. Ciência Rural 31, 381–386.

Silva, R.A.D., Flores, P.S., Carvalho, G.S., 2001. Descrição dos estágios imaturos de *Phthia picta* (Drury) (Hemiptera: Coreidae). Neotropical Entomology 30, 253–258.

Simmons, A.M., Yeargan, K.V., 1988. Feeding frequency band feeding duration of insect pests of soybean in response to insecticides and field isolation. Environmental Entomology 6, 501–506.

Sithole, S.Z., 1986. Mole cricket (*Gryllotalpa africana*). Zimbabwe Agricultural Journal 83, 21–22.

Sivapragasam, A., Syed, A.R., 1999. The problem and management of agromyzid leafminers on vegetables in Malaysia. In: Lim, G.S., Soetikno, S.S., Loke, W.H. (Eds.), Proceedings of a Workshop on Leafminers of Vegetables in Southeast Asia. CAB International, Southeast Asia Regional Center (SEARC), Serdang, Selangor, Malaysia, pp. 36–41.

Slater, J.A., Baranowski, R.M., 1978. How to Know the True Bugs. William C. Brown Company Publisher, Dubuque, Lowa, USA.

Smart, L.E., Blight, M.M., 2000. Response of the pollen beetle, *Meligethes aeneus*, to traps baited with volatiles from oilseed rape, *Brassica napus*. Journal of Chemical Ecology 26, 1051–1064.

Smith, I.M., McNamara, D.G., Scott, P.R., Holderness, M., 1997. Epitrix tuberis. In: Quarantine Pests for Europe. second ed. CABI/EPPO, Wallingford, UK, p. 1425.

Solbreck, C., 1978. Migration, diapause, and direct development as alternative life histories in a seed bug, *Neacoryphus bicrucis*. In: Dingle, H. (Ed.), Evolution of Insect Migration and Diapause. Springer, New York, USA, pp. 195–217.

Soldi, R.A., Rodrigues, M.A., Aldrich, J.R., Zarbin, P.H., 2012. The male-produced sex pheromone of the true bug, *Phthia picta*, is an unusual hydrocarbon. Journal of Chemical Ecology 38, 814–824.

Song, L.Q., Xu, Z.F., Gu, D.J., 2003. A review on the parasitoids of *Liriomyza huidobrensis* Blanchard. Natural Enemies of Insects 25, 37–41.

Soroka, J.J., Holowachuk, J.M., Gruber, M.Y., Grenkow, L.F., 2011. Feeding by flea beetles (Coleoptera: Chrysomelidae; *Phyllotreta* spp.) is decreased on canola (*Brassica napus*) seedlings with increased trichome density. Journal of Economic Entomology 104, 125–136.

Sorokin, N.S., 1976. The Colorado potato beetle (*Leptinotarsa decemlineata* Say) and its entomophages in the Rostov Region. Biull. Vses. Nauchn. Issled. Inst. Zashch. Rast 37, 22–27.

Spencer, K.A., 1964. The species-host relationship in the Agromyzidae (Diptera) as an aid to taxonomy. In: Proceedings Xllth International Congress of Entomology London, UK. 8–16 July. pp. 101–102.

Spencer, K.A., 1973. Agromyzidae (Diptera) of Economic Importance. Series Entomologica: Dr. W. Junk, vol. 9, p. 436 The Hague, Netherlands.

Spencer, K.A., 1989. Leaf miners. In: Kahn, R.P. (Ed.). Plant Protection and Quarantine. Selected Pests and Pathogens of Quarantine Significance vol. 2. CRC Press, Boca Raton, Florida, USA, pp. 77–98.

Spencer, K.A., 1990. Host Specialization in the World Agromyzidae (Diptera). Series Entomologica, vol. 45. Kluwer Academic Publishers, Dordrecht, Netherlands, p. 444.

Stam, P.A., Newsom, L.D., Lambremont, E.N., 1987. Predation and food as factors affecting survival of *Nezara viridula* (L.) (Hemiptera: Pentatomidae) in a ecosystem. Environmental Entomology 16, 1211–1216.

Stegmaier, C.E., 1966. Host plants and parasites of *Liriomyza trifolii* in Florida (Diptera: Agromyzidae). Florida Entomologist 49, 75–80.

Stoll, G., 1996. Natural Pest and Disease Control. Magraf Verlag, Germany, p. 56.

Stoner, K.A., 1993. Effects of straw and leaf mulches and sprinkle irrigation on the abundance of colorado potato beetle (Coleoptera: Chrysomelidae) on potato in Connecticut. Journal of Entomological Science 28, 393–403.

Stoner, K.A., 1997. Influence of mulches on the colonization by adults and survival of larvae of the Colorado potato beetle (Coleoptera: Chrysomelidae) in eggplant. Journal of Entomological Science 32, 7–16.

Thompson, S.R., Brandenburg, R.L., 2006. Effect of combining imidacloprid and diatomaceous earth with *Beauveria bassiana* on mole cricket (Orthoptera: Gryllotalpidae) mortality. Journal of Economic Entomology 99, 1948–1954.

Tindale, N.B., 1928. Australasian mole-crickets of the family Gryllotalpidae (Orthoptera). Records of the South Australian Museum 4, 1–42.

Todd, J.W., 1989. Ecology and behavior of *Nezara viridula*. Annual Review of Entomology 34, 273–292.

Tóth, M., Csonka, E., Bakcsa, F., Benedek, P., Szarukan, I., Gomboc, S., Toshova, T., Subchev, M., Ujvary, I., 2007. Species spectrum of flea beetles (*Phyllotreta* spp., Coleoptera, Chrysomelidae) attracted to allyl isothiocyanate-baited traps. Zeitschrift für Naturforschung 62c, 772–778.

Townsend, B.C., 1983. A revision of the Afrotropical mole-crickets (Orthoptera: Gryllotalpidae). Bulletin of the British Museum of Natural History (Entomology) 46, 175–203.

Tran, D.H., Tran, T.T.A., Mai, L.P., Ueno, T., Takagi, M., 2007. Seasonal abundance of *Liriomyza sativae* (Diptera: Agromyzidae) and its parasitoids on vegetables in southern Vietnam. Journal of the Faculty of Agriculture, Kyushu University, Fukuoka, Japan 52, 49–55.

Tran, D.H., Tran, T.T.A., Takagi, M., 2005. Bulletin of the Institute of Tropical Agriculture. Agromyzid Leafminers in Central and Southern Vietnam: Survey of Host Crops, Species Composition and Parasitoids, vol. 28. Kyushu University, pp. 35–41.

Trumble, J.T., Alvarado-Rodriguez, B., 1993. Development and economic evaluation of an IPM program for fresh market tomato production in Mexico. Agriculture Ecosystems and Environment 43, 267–284.

Trumble, J.T., Carson, W.G., Kund, G.S., 1997. Economics and environmental impact of a sustainable integrated pest management program in celery. Journal of Economic Entomology 90, 139–146.

Ulmer, B.J., Dosdall, L.M., 2006. Emergence of overwintered and new generation adults of the crucifer flea beetle, *Phyllotreta cruciferae* (Goeze) (Coleoptera: Chrysomelidae). Crop Protection 25, 23–30.

USDA, 2012. Crop Profile for Omatoes in Michigan. National Institute of Food and Agriculture, United State Department of Agriculture, Washington, USA. http://www.ipmcenters.org/cropprofiles/docs/mitomatoes.pdf.

Van Driesche, R.G., Ferro, D.N., Carey, E., Maher, M., 1991. Assessing augmentative releases of parasitoids using the "recruitment method", with reference to *Edovum puttleri*, a parasitoid of the Colorado potato beetle (Coleoptera. Chrysomelidae). Entomophaga 36, 193–204.

Vernon, R.S., Mackenzie, J.R., 1991. Evaluation of foliar sprays against the tuber flea beetle *Epitrix tuberis* Gentner (Coleoptera: Chrysomelidae), on potato. The Canadian Entomologist 123, 321–331.

Vignez, L.G., de Huiza, I.R., 1992. Niveles de infestacion y parasitismo de *Liriomyza huidobrensis* en papa cultivada sin aplicacion de insecticidas. Revista Peruana de Entomologia 35, 101–106.

Villani, M.G., Allee, L.L., Preston-Wilsey, L., Consolie, N., Xia, Y., Brandenburg, R.L., 2002. Use of radiography and tunnel castings for observing mole cricket (Orthoptera: Gryllotalpidae) behavior in soil. American Entomologist 48, 42–50.

Vincent, C., Stewart, R.K., 1984. Effect of allyl isothiocyanante on field behaviour of crucifer-feeding flea beetles (Coleoptera: Chrysomelidae). Journal of Chemical Ecology 10, 33–40.

Vlasova, V.A., 1978. A prediction of the distribution of Colorado beetle in the Asiatic territory of the USSR. Zaschita Rastenii 6, 44–45.

Voss, R.H., 1989. Population Dynamics of the Colorado Potato Beetle, (*Leptinotarsa decemlineata*) (Say) (Coleoptera: Chrysomelidae), in Western Massachusetts, with Particular Emphasis on Migration and Dispersal Processes (Ph.D. thesis). University of Massachusetts, Amherst, Massachusetts,USA.

Voss, R.H., Ferro, D.N., 1990. Phenology of flight and walking by Colorado potato beetle (Coleoptera: Chrysomelidae) adults in western Massachusetts. Environmental Entomology 19, 117–122.

Walgenbach, J.F., Wyman, J.A., 1984. Colorado potato beetle (Coleoptera: Chrysomelidae) development in relation to temperature in Wisconsin. Annals of the Entomological Society of America 77, 604–609.

Walker, C.S., 2012. The Application of Sterile Insect Technique against the Tomato Leafminer *Liriomyza Bryoniae* (Ph.D. thesis). Department of Life Sciences, Imperial College London, UK.

Walker, T.J., 1985. Systematics and life cycles. In: Walker, T.J. (Ed.), Mole Crickets in Florida. Bulletin of University of Florida, Agricultural Experiment Stations, pp. 3–10.

Walker, T.J., Nation, J.L., 1982. Sperm storage in mole crickets: fall matings fertilize spring eggs in *Scapteriscus aeletus*. Florida Entomologist 65, 283–285.

Walker, T.J., Parkman, J.P., Frank, J.H., Schuster, D.J., 1996. Seasonality of *Ormia depleta* and limits to its spread. Biological Control 6, 378–383.

Wang, K.-H., 2012. Cover Crops as Insectary Plants to Enhance above and below Ground Beneficial Organisms. Hānai' Ai/The Food Provider http://www. ctahr.hawaii.edu/sustainag/news/articles/V11-Wang-insectary-covercrops.pdf.

Waterhouse, D.F., Norris, K.R., 1987. Biological Control. Pacific Prospects. Australian Centre for International Agricultural Research, Inkata Press, Melbourne, p. 454.

Webb, R.E., Smith, F.F., 1969. Effect of temperature on resistance in lima bean, tomato, and chrysanthemum to *Liriomyza munda*. Journal of Economic Entomology 62, 458–462.

Webb, R.E., Stoner, A.K., Gentile, A.G., 1971. Resistance to leaf-miners in *Lycopersicon* accessions. Journal of the American Society for Horticultural Science 96, 65–67.

Weber, D.C., Ferro, D.N., 1993. Distribution of overwintering colorado potato beetle in and near Massachusetts potato fields. Entomologia Experimentalis et Applicata 66, 191–196.

Weber, D.C., Ferro, D.N., 1995. Colorado potato beetle: diverse life history holds keys to management. In: Zehnder, G.W., Jansson, R.K., Powelson, M.L., Raman, K.V. (Eds.), Advances in Potato Pest Biology and Management. American Phytopathological Society Press, Saint. Paul, Minnesota, USA.

Weber, D.C., Ferro, D.N., Buonaccorsi, J., Hazzard, R.V., 1994. Disrupting spring colonization of Colorado potato beetle to nonrotated potato fields. Entomologia Experimentalis et Applicata 73, 39–50.

Weber, D.C., Rowley, D.L., Greenstone, M.H., Athanas, M.M., 2006. Prey preference and host suitability of the predatory and parasitoid carabid beetle, *Lebia grandis*, for several species of *Leptinotarsa* beetles. Journal of Insect Science 6, 1–14.

Weed, A.S., 2003. Reproductive Strategy of *Pheropsophus Aequinoctialis* L.: Fecundity, Fertility, and Oviposition Behavior; and Influence of Mole Cricket Egg Chamber Depth on Larval Survival (M.Sc. thesis). Graduate School, University of Florida, Gainesvella, Florida, USA.

Wene, G.P., 1953. Control of the serpentine leaf miner on peppers. Journal of Economic Entomology 46, 789–793.

Wilde, G., Roozebooom, K., Claassen, M., Janssen, K., Witt, M., 2004. Seed treatment for control of early-season pests of corn and its effect on yield. Journal of Agricultural and Urban Entomology 21, 75–85.

Wilkerson, J.L., Webb, S.E., Capinera, J.L., 2005. Vegetable Pests I: Coleoptera – Diptera – Hymenoptera. UF/IFAS CD-ROM. SW 180. Department of Entomology and Nematology, University of Florida/IFAS Extension, Gainesville, Florida, USA.

Williams, E.C., MacDonald, O.C., 1995. Critical factors required by the nematode *Steinernema feltiae* for the control of the leafminers *Liriomyza huidobrensis*, *Liriomyza bryoniae* and *Chromatomyia syngenesiae*. Annals of Applied Biology 127, 329–341.

Williams, J.J., Shaw, L.N., 1982. A soil corer for sampling mole crickets. Florida Entomologist 65, 192–194.

Winotai, A., Chattrakul, U., 2003. Conservation of natural enemies for biological control of leaf miner fly, *Liriomyza huidobrensis*, in the north of Thailand. In: 6th International Conference on Plant Protection in the Tropics, 11–14 August, 2003, Kuala Lumpur, Malaysia, p. 6.

Wolfenbarger, D.A., 1966. Variations in leaf miner and flea beetle injury in tomato varieties. Journal of Economic Entomology 59, 65–68.

Worner, S.P., 1988. Ecoclimatic assessment of potential establishment of exotic pests. Journal of Economic Entomology 81, 973–983.

Wright, R.J., 1984. Evaluation of crop rotation for control of Colorado potato beetle (Coleoptera: Chrysomelidae) in commercial potato fields on Long Island. Journal of Economic Entomology 77, 1254–1259.

Xia, Y., Brandenburg, R., 2000. Treat young mole crickets for reliable insecticide results. Golf Course Management 63, 49–51.

Xia, Y., Hertl, P.T., Brandenburg, R.L., 2000. Surface and subsurface application of *Beauveria bassiana* for controlling mole crickets (Orthoptera: Gryllotalpidae) in golf courses. Journal of Agriculture and Urban Entomology 17, 177–189.

Xiao, Y.F., Fadamiro, H.Y., 2010. Evaluation of damage to satsuma mandarin (*Citrus unshiu*) by the leaffooted bug, *Leptoglossus zonatus* (Hemiptera: Coreidae). Journal of Applied Entomology 134, 694–703.

Yabar, L.E., 1980. Control of *Epitrix* spp. on potato. Revista Peruana de Entomologia 23, 151–153.

Yaman, M., 2002. Howardula phyllotretae (Tylenchida: Allantonematidae), A nematode parasite of *Phyllotreta undulata* and *P. atra* (Coleoptera: Chrysomelidae) in Turkey. Journal of Asia-Pacific Entomology 5, 233–235.

Yan, X., Han, R., Moens, M., Chen, S., De Clercq, P., 2013. Field evaluation of entomopathogenic nematodes for biological control of striped flea beetle, *Phyllotreta striolata* (Coleoptera: Chrysomelidae). Biocontrol 58, 247–256.

Yasuda, K., 1998. Function of the male pheromone of the leaf-footed plant bug, *Leptoglossus australis* (Fabricius) (Heteroptera: Coreidae) and its kairomonal effect. Japan Agricultural Research Quarterly 32, 161–165.

Yasuda, K., Tsurumachi, M., 1995. Influence of male-adults of the leaf-footed plant bug, *Leptoglossus australis* (Fabricius) (Heteroptera, Coreidae), on host-searching of the egg parasitoid, *Gryon pennsylvanicum* (Ashmead) (Hymenoptera, Scelionidae). Applied Entomology and Zoology 30, 139–144.

Zalom, F.G., Smilanick, J.M., Ehler, L.E., 1977. Spatial pattern and sampling of stink bugs (Hemiptera: Pentatomidae) in processing tomatoes. In: Maciel, G.L., Lopes, G.M.G., Hayward, C., Mariano, R.R.L., Maranhao, E. A. de A. (Eds.), Proceedings of the 1st International Conference on Processing Tomatoes, 18–21 November 1996, Recife, Brazil, pp. 75–79.

Zalom, F.G., Zalom, J.S., 1992. Stink bugs in California tomatoes. California Tomato Grower 35, 8–11.

Zehnder, G.W., Evanylo, G.K., 1989. Influence of extent and timing of Colorado potato beetle (Coleoptera: Chrysomelidae) defoliation on potato tuber production in eastern Virginia. Journal of Economic Entomology 82, 948–953.

Zehnder, G.W., Hough-Goldstein, J., 1990. Colorado potato beetle (Coleoptera: Chrysomelidae) population development and effects on yield of potatoes with and without straw mulch. Journal of Economic Entomology 83, 1982–1987.

Integrated Pest Management of Tomato Pests

Chapter 9

Host-Plant Resistance in Tomato

Michael J. Stout[1,2], Henok Kurabchew[2], Germano Leão Demolin Leite[3]

[1]Louisiana State University Agricultural Center, Baton Rouge, LA, United States; [2]Hawassa University, Hawassa, Ethiopia; [3]Universidade Federal de Minas Gerais, Montes Claros, Brazil

1. INTRODUCTION

Tomato, *Solanum lycopersicum* L. (Solanaceae), is among the world's most widely grown and important vegetable crops. Worldwide production exceeds 140 million metric tons annually, with a value exceeding 50 billion dollars (FAO, 2014). Fresh tomatoes represent approximately 75% of this total. Tomato is rich in minerals, vitamins, and antioxidants that are important to a well-balanced diet. Tomato is also an important dietary component because it contains high levels of lycopene, an antioxidant that reduces the risk associated with several cancers and neurodegenerative diseases (Miller et al., 2002).

In addition to being an important crop, tomato has served as an important model for the study of plant resistance to disease and insect pests. As such, a number of important genetic and molecular tools have been developed in tomato, including a high-quality genome sequence (Menda et al., 2013). A voluminous literature has accumulated on various aspects of plant resistance in tomato, and a comprehensive review of this literature is beyond the scope of this chapter. Rather, this chapter presents an overview of several aspects of host-plant resistance in tomato, including important genetic sources of resistance and mechanisms (bases) of resistance, induced resistance, and the use of resistance in management programs. Emphasis is placed on resistance to arthropod pests, but resistance to pathogenic microorganisms is also discussed, because many of the traits involved in resistance to arthropods are also involved in plant resistance to pathogens and because certain features of tomato resistance to pathogens provide a valuable counterpoint to insect resistance.

Unlike most crop plants, pests and diseases are considered equally important in tomato (Carvalho et al., 2002). Tomatoes are subject to attack by almost 200 species of insects and mites with a diversity of feeding styles across the world. Among the important arthropod pests of tomato plants are piercing/sucking and cell-content feeders such as whiteflies, *Bemisia tabaci* (Gennadius) and *Bemisia argentifolii* Bellows & Perring, thrips in the genus *Frankliniella*, mites in the genus *Tetranychus*, and aphids such as *Myzus persicae* (Sulzer) and *Macrosiphum euphorbiae* Thomas. Some of these piercing/sucking insects are vectors of important viruses, particularly geminiviruses such as tomato yellow leaf curl virus. Among the important chewing insects are the defoliator *Leptinotarsa decemlineata* Say the leaf and fruit borer *Tuta absoluta* (Meyrick) the leafminer *Liriomyza huidobrensis* (Blanchard) and the fruit borers *Neoleucinodes elegantalis* (Guenée) and *Helicoverpa zea* (Boddie) (Gallo et al., 2002). These and other arthropods can cause severe yield losses in tomato; Picanço et al. (2007), for example, found that pests caused, directly or indirectly, a reduction of almost 60% of the total income from tomato culture.

Tomato is also a host to about 200 species of pathogenic microorganisms that affect tomato plants and fruits (Akhtar et al., 2010). These diseases, of bacterial, viral, or fungal origin, can cause severe reductions in yield and fruit quality (Csizinszky et al., 2005; Reis and Lopes, 2006). The most important foliar diseases of tomato in most production areas are the bacterial-speck and bacterial-spot diseases caused by *Pseudomonas syringae* Van Hall pv. *tomato* and *Xanthomonas campestris* Dowson pv. *vesicatoria* (Syn. *Xanthomonas vesicatoria* (Doidge)), respectively (Yu et al., 1995; Blancard, 1997). These two diseases often are found together in mixed infections. Although they are caused by different pathogens, the diseases are marked by similar symptoms and are often confused (Delahaut and Stevenson, 2004; Cuppels et al., 2006). The most important wilt diseases in tomato are fusarium wilt, caused by three races of *Fusarium oxysporum* f. sp. *lycopersici* W.C. Snyder & H.N. Hansen and bacterial wilt, caused by *Ralstonia solanacearum* (Smith) (Yabuuchi et al., 1995). Among other important diseases of tomato are late blight, caused by *Phytophthora infestans* (Mont.) de Bary, gray mold (*Botrytis cinerea* Pers) and several viruses, including tomato spotted wilt virus (vectored by thrips) and tomato yellow leaf curl (transmitted by whiteflies). In Brazil, the main pathogens affecting tomato plants and their fruits are *Erwinia* spp. and *Alternaria solani* Sorauer (Picanço et al., 1998, 2007). Infection of tomato fruit by these pathogens occurs mainly because of injuries during crop management and handling, and damage by insect borers (Bergamin Filho et al., 1995; Zambolim et al., 2000).

In addition to the pathogenic microorganisms, endoparasitic root-knot nematodes (*Meloidogyne* sp.) are considered to be among the most damaging pests of tomato worldwide (Trudgill and Blok, 2001). Disease symptoms caused by root-knot nematodes on susceptible host plants include damaged root systems and the formation of root galls, wilting, chlorosis, and stunted growth resulting in poor yield (Abad et al., 2003).

Currently, repeated applications of synthetic pesticides are used heavily to manage these arthropod and disease pests. The investments required in pesticide-based control can be considerable, and in some cases account for 40% of the total cost of production. There are numerous problems associated with overreliance on pesticides, including the high cost of pesticide applications, exposure of applicators and consumers to pesticides and residues, reductions in populations of beneficial arthropods, selection for insecticide resistance, and other adverse environmental effects (Thomazini et al., 2000; Picanço et al., 1998, 2007). Given the economic, environmental, and societal costs of overreliance on pesticides, it is clear that significant economic and environmental benefits could be gained from the increased use of resistant varieties in tomato pest management.

Plant resistance is variously defined in the literature. For diseases, resistance can be defined as "the ability of an organism to exclude or overcome, completely or to some degree, the effect of a pathogen" (Agrios, 2005). Resistant plants have fewer and/or less severe symptoms, and resistance may result from the inability of a pathogen to colonize, establish, and spread. A distinction is sometimes made between vertical resistance (race specific, highly effective resistance governed by a single gene) and horizontal resistance (non-race-specific, incomplete but broadly effective, and controlled by many genes) (Agrios, 2005). For arthropods, resistance can be defined as the consequence of heritable plant traits that result in a plant being damaged to a lesser degree than plants lacking these traits (Smith, 2005). Arthropod resistance can be manifested as lower preference by arthropod herbivores for resistant plants (antixenosis), by reduced growth, development, or fitness of arthropods on resistant plants (antibiosis), or by lower reductions in quality or yield from a given amount of injury on resistant plants (tolerance). The focus of this chapter is on those traits and mechanisms that limit the amount of injury or degree of infection/infestation caused by pests on resistant hosts, rather than on tolerance.

2. RESISTANCE-RELATED TRAITS IN TOMATO

Plants possess arrays of structural and biochemical traits that together serve as the basis or mechanisms of resistance to pests. These traits can be constitutive (preformed), expressed irrespective of the presence or activity of pests, or inducible, expressed only (or to a greater degree) after attack by a pest. Most resistance-related traits affect arthropods or pathogenic microorganisms directly, but some act indirectly by increasing the effectiveness of natural enemies of the attacking organism. The resistance-related traits that have received the most attention are the secondary compounds, chemical compounds not directly involved in primary metabolic processes that mediate interactions with other organisms (Taiz and Zeiger, 1991). The three main groups of secondary products in plants are terpenes, phenolics, and nitrogen-containing compounds (Taiz and Zeiger, 1991). Cultivated tomato, *S. lycopersicum*, and its wild relatives in this genus contain multiple representatives of these three major groups (Duffey and Stout, 1996; Glas et al., 2012); in addition, tomatoes and wild tomatoes produce methyl ketones and acylsugars, which are present in trichomes and play an important role in resistance to some pests (Fig. 9.1, Table 9.1).

3. MORPHOLOGICAL TRAITS AND SECONDARY METABOLITES AS MECHANISMS OF RESISTANCE IN *SOLANUM*

3.1 Trichomes

Eight different types of trichomes have been distinguished in the genus *Solanum*, four of them glandular and four of them non-glandular (Fig. 9.2) (Glas et al., 2012). Non-glandular trichomes may protect plants from herbivores by interfering with their movements or feeding behaviors (Riddick and Simmons, 2014). The glandular trichomes contribute to the biosynthesis and storage of a large number of secondary plant compounds, such as acylsugars, methyl ketones, and phenolics/oxidative enzymes (Kennedy, 2003; see separate entries below). These compounds may directly toxify pests (Kennedy, 2003). In addition, compounds associated with trichomes are involved in the production of sticky exudates that can entrap small insects such as aphids. In addition to affecting herbivores, trichomes and their associated secondary metabolites can also interfere with the behavior of predators and parasitoids of herbivores (Riddick and Simmons, 2014). Increased densities of glandular trichomes on newly expanding leaves is a component of induced resistance in tomatoes (Boughton et al., 2005).

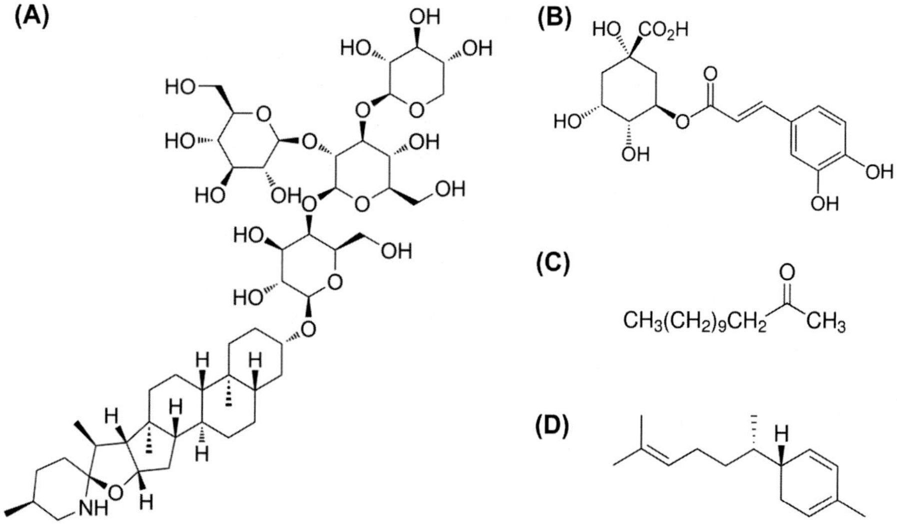

(A)

(B)

(C)

$$CH_3(CH_2)_9CH_2 \overset{O}{\underset{\|}{C}} CH_3$$

(D)

FIGURE 9.1 Examples of secondary chemicals in tomato. (A) Tomatine (https://en.wikipedia.org/wiki/Tomatine). (B) Chlorogenic acid (https://en.wikipedia.org/wiki/Chlorogenic_acid). (C) Tridecanone (http://www.sigmaaldrich.com/catalog/product/aldrich/172839?lang=en&region=US). (D) Zingibrene (https://en.wikipedia.org/wiki/Zingiberene).

TABLE 9.1 Selected Studies Implicating Various Resistance-Related Traits to Arthropod Herbivores in Wild or Cultivated Tomato

Tomato Species or Accession	Resistance-Related Traits Implicated in Resistance	Pests	References and Experimental Approach
Solanum lycopersicum	Type VI trichomes; mono- and sesquiterpenes; flavonoids (rutin)	Flea beetle, *Epitrix cucumeris*; Colorado potato beetle, *Leptinotarsus decemlineata*	Kang et al. (2010) Analysis of mutant line
S. lycopersicum and *Solanum pennellii*	Acylsugars	Spider mites, *Tetranychus urticae*; South American tomato pinworm, *Tuta absoluta*; silverleaf whitefly, *Bemisia argentifolia*	Maluf et al. (2010) Hybridization of low- and high-acylsugar lines/species
S. lycopersicum	Polyphenol oxidase	Common cutworm, *Spodoptera litura*	Mahanil et al. (2008) Transgenic overexpression of enzyme
Solanum habrochaites	Methyl ketones/type VI glandular trichomes	Helicoverpa zea	Farrar and Kennedy(1987) Experimental removal of trichomes
Solanum peruvianum	*Mi-1*gene	Tomato psyllid *Bactericera cockerelli*; potato aphid, *Macrosiphum euphorbiae*	Casteel et al. (2006) and Rossi et al. (1998) Introduction of gene from wild species to cultivated tomato
S. habrochaites	Zingiberene/glandular trichomes	South American tomato pinworm, *Tuta absoluta*	de Azevedo et al. (2003) Correlations among zingiberene levels and resistance in interspecific crosses
S. lycopersicum	Proteinase inhibitors	Tobacco hornworm, *Manduca sexta*	Orozco-Cardenas et al. (1993) Antisense suppression of hormone signaling
Solanum sp.	Tomatine	Tomato fruitworm, *Helicoverpa zea*, beet armyworm, *Spodoptera exigua*	Bloem et al. (1989) Addition of pure compound to artificial diets

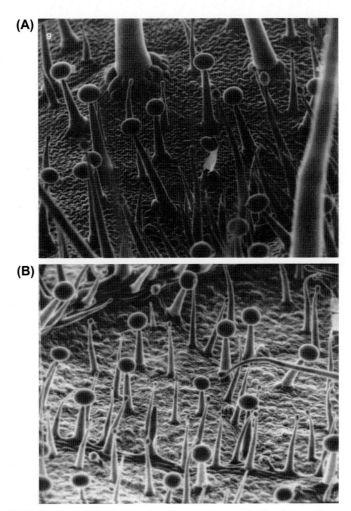

FIGURE 9.2 Photomicrographs (SEM) of tomato wild sp: *Lycopersicon hirsutum* (A, upper leaf surface and B, lower leaf surface). *L. hirsutum* is resistant to TLCV under field conditions, but the resistance is to the vector (*Bemisia tabaci*) due to glandular trichomes, not to TLCV directly. *©2008 Canadian Science Publishing or its licensors. Reproduced with permission Channarayappa, C., Shivashankar, G., Muniyappa, V., Frist, R.H., 1992. Resistance of* Lycopersicon *species to* Bemisia tabaci, *a tomato leaf curl virus vector. Canadian Journal of Botany 70, 2184–2192.*

3.2 Acylsugars

Acylsugars have generated a large amount of interest as resistance traits in *Solanum*. Acylsugars consist of glucose or sucrose backbones to which multiple acyl chains of various lengths are linked via ester bonds. These compounds are synthesized and stored in the glandular trichomes of tomato, particularly type IV trichomes, and are major components of the sticky exudate from these trichomes (Kennedy, 2003). Acylsugars are found in particularly high concentrations and diversity in *Solanum pennellii* Correll and *Solanum habrochaites* S. Knapp & D.M. Spooner (Kim et al., 2012). High levels of acylsugars appear to impart resistance to multiple insect pests. For example, acylsugars inhibit oviposition and feeding by *Liriomyza trifolii* (Burgess) adults and feeding by *M. euphorbiae*, and they negatively affect the growth and survival of *Spodoptera exigua* (Hübner) and *H. zea* (Goffreda et al., 1990; Hawthorne et al., 1992; Juvik et al., 1994).

3.3 Methyl Ketones

Methyl ketones are another type of resistance-related secondary compound typically associated with type VI trichomes in tomato. These compounds, the most important of which are 2-tridecanone (2-TD) (Fig. 9.1C) and 2-undecanone (2-UD), are ketones derived from fatty acids that typically contain 7 to 15 carbons. Both 2-TD and 2-UD have as precursors acetate units, which are condensed to form chains. 2-TD is highly toxic to a range of juvenile insects, increasing larval and pupal mortality, prolonging developmental periods, and causing deformation of pupae in various pests such as *Manduca sexta* (L.), *L. decemlineata*, *T. absoluta*, *Liriomyza sativae* Blanchard, *M. persicae*, *Frankliniella occidentalis*

Pergande, *B. tabaci*, *Tetranychus urticae* Koch and *Aculops lycopersici* (Massee) (Kennedy and Sorenson, 1985; Fery and Kennedy, 1987; Channarayappa et al., 1992; Kumar et al., 1995; Leite et al., 1999a,c,d, 2001). It also deters feeding and oviposition of adult insects. However, 2-TD is not very effective against *H. zea* and *Heliothis virescens* (F.), perhaps because this natural insecticide induces the production of enzymes (monooxygenases) capable of catabolizing this chemical compound by these insects (Rose et al., 1991). 2-UD is also toxic to arthropods, although less so than 2-TD (Dimock et al., 1982). The concentration of 2-UD in the leaves is much lower than the concentration of 2-TD (Lin et al., 1987). However, 2-UD has a synergistic effect with 2-TD (Farrar and Kennedy, 1987). The use of 2-UD as a single resistance factor may have limited use, since this compound only causes deformations in pupae (Farrar and Kennedy, 1988; Farrar et al., 1992).

3.4 Terpenes

Terpenes are lipophyllic compounds synthesized from acetyl-CoA via mevalonic acid and are composed of two or more five-carbon isoprene units. Terpenes are distinguished by the numbers of carbon atoms in their structures: monoterpenes (C_{10}) (e.g., α-pinene, menthol), sesquiterpenes (C_{15}) (e.g., costunolide and gossypol), diterpenes (C_{20}) (e.g., abietic acid and phorbol), triterpenes (C_{30}) (e.g., limonoid, saponins, α-ecdysone, and cardenolide), and polyterpenes ($C_5)_n$ (Taiz and Zeiger, 1991). Tomato sesquiterpenes and derivatives are implicated in direct resistance against herbivores, but may also attract herbivores (Yu et al., 2010; Proffit et al., 2011). For example, tomato lines bred or engineered to express levels of the sesquiterpene 7-epizingiberene comparable to levels found in *S. habrochaites* PI127826 showed enhanced resistance to both spider mites and whiteflies (Bleeker et al., 2012). The resistance was manifested as reduced fecundity and survival of whiteflies and reduced egg laying and population growth of mites. Similarly, in lines resulting from crosses of cultivated tomato with *S. habrochaites*, zingiberene (Fig. 9.1D) content was correlated with resistance to *T. absoluta* (de Azevedo et al., 2003). Terpenoid metabolism in the trichomes of wild and cultivated tomato species differ in important ways and these differences may be responsible for differences in pest resistance (Besser et al., 2009).

As in other plants, volatile terpenes and other compounds emitted following herbivore feeding in tomato can be used as cues by predators and parasitoids to locate prey (indirect resistance; Bruce, 2015). Kant et al. (2004), for example, found that increased volatile emission by tomato plants infested with spider mites (*T. urticae*) coincided with enhanced attraction of predatory mites to infested plants.

3.5 Phenolics

Phenolics are a diverse group of secondary compounds, often derived from shikimate, that include phenylpropanoids, tannins, lignin, and flavonoids. The most important phenolic compounds in tomato are the catecholic phenolics chlorogenic acid (Fig. 9.1B) and rutin. These compounds reduce the growth of noctuids such as *H. zea* and *S. exigua* when added to artificial diet at biologically realistic concentrations (Isman and Duffey, 1982; Duffey and Stout, 1996). A mutant line of cultivated tomato deficient in the production of rutin and terpenes (but not in chlorogenic acid) was more susceptible to flea beetles and the Colorado potato beetle (Kang et al., 2010). The toxic and growth-reducing effects of these phenolics may be related to their participation in the production of reactive oxygen species and/or their conversion to highly reactive quinones by oxidative enzymes such as polyphenol oxidase (Duffey and Stout, 1996).

3.6 Nitrogen-Containing Compounds

Secondary products containing nitrogen are synthesized primarily from amino acids (Taiz and Zeiger, 1991). Tomato, like most of the members of the family Solanaceae, contains alkaloids. It also contains an assortment of resistance-related proteins, and indeed it has been one of the most important models for studying the role of proteins and enzymes in plant resistance since the discovery, by Green and Ryan (1973), that proteinase inhibitors are inducible by wounding in tomato leaves.

Tomatine (Fig. 9.1A) is a steroidal glycoalkaloid saponin present in the foliage and unripe fruits of domesticated tomato and its wild relatives. Alpha-tomatine inhibits the growth and increases the mortality of noctuid larvae such as *H. zea* and *S. exigua* exposed to the compound in artificial diets at levels representative of levels in commercial tomato (Duffey and Stout, 1996). Alpha-tomatine also inhibits the growth of many fungi, although pathogens of tomato often possess the ability to detoxify tomatine (Oka et al., 2006). The deleterious effects of tomatine on insects and pathogens are thought to stem from precipitation of sterols in membranes by tomatine and the resultant destruction of membrane integrity and lysis (Duffey and Stout, 1996). The negative effects of tomatine on insect growth and survival can be alleviated or negated by phytosterols in tomato foliage.

The proteinase inhibitors (PIs) are small polypeptides capable of inhibiting the proteolytic activity of digestive proteases in insect guts. The leaves of *S. lycopersicum* contain two PIs, PI I (an inhibitor of chymotrypsin) and PI II (a trypsin/chymotrypsin inhibitor) (Duffey and Stout, 1996). Several types of evidence support a role for tomato PIs in resistance. Addition of purified PIs to artificial diet reduces the growth of larval lepidopterans and coleopterans. Significant negative correlations between levels of PIs in tomato tissues and insect growth and development have been found; for example, tomato leaves with higher levels of PIs supported reduced growth rates of larval *S. exigua* (Broadway et al., 1986). Transgenic tobacco plants overexpressing PIs from tomato supported markedly reduced growth of *M. sexta* larvae (Johnson et al., 1989), while plants in which expression of PIs was suppressed were more susceptible to insects (Orozco-Cardenas et al., 1993). The mechanism by which PIs lead to reduced growth and development appears to involve both reduction in the digestibility of plant tissues and a toxic effect resulting from a "hypertrophic induction of trypsin synthesis" (Duffey and Stout, 1996; Jongsma and Bolter, 1997).

Other proteins in tomato have been implicated in tomato plant resistance. The oxidative enzyme polyphenol oxidase (PPO), found in both trichomes and leaves of tomatoes, catalyzes the production of quinones from catecholic phenolics such as chlorogenic acid. These quinones are highly reactive and can reduce the nutritive value of leaf tissue by irreversibly derivatizing proteins and rendering them nutritionally unavailable to herbivores (Duffey and Stout, 1996). Oxidation of phenolics by PPO is also involved in the production of sticky exudates from trichomes. As with PIs, a role for PPO in plant resistance is strongly suggested by the inducibility of this enzyme by herbivory and by the strong correlations that exist between resistance and PPO activity. Furthermore, tomato plants engineered to overexpress or underexpress PPO activity show enhanced or reduced resistance to several caterpillar species (Mahanil et al., 2008; Bhonwong et al., 2009). Other oxidative enzymes, such as peroxidase, may contribute to tomato resistance via similar mechanisms (Duffey and Stout, 1996).

The roles of two other inducible enzymes, arginase and threonine deaminase, in tomato resistance have also been studied (Chen et al., 2005). The expression of both enzymes in leaves is induced by jasmonic acid (JA), the major hormone involved in plant responses to feeding by chewing insects, and both enzymes act to lower the nutritive quality of leaf tissues by catabolizing essential amino acids (arginine in the case of arginase and threonine in the case of threonine deaminase). As is the case with both PIs and PPO, overexpression of arginase in transgenic plants led to increase in resistance to a chewing herbivore (*M. sexta*), and reduced growth of *M. sexta* larvae on arginase-overexpressing plants was correlated with reduced levels of arginine in the caterpillar midgut (Chen et al., 2005).

3.7 Interactions of Primary and Secondary Metabolism

In addition to the important role played by resistance-related traits, plant nutrition and primary metabolism are important determinants of resistance in tomato. *T. absoluta* fed on tomatoes subject to nitrogen limitation showed lower survival and pupal weights and prolonged developmental times compared to insects fed on tomatoes grown under optimal N (Han et al., 2014). These results are partly attributable to the importance of nitrogen to insect growth, but tomatoes subjected to N limitation also showed increases in concentrations of chlorogenic acid, rutin, and tomatine (Larbat et al., 2012; Royer et al., 2013), suggesting that increases in plant resistance under nitrogen limitation result from the interaction of plant nutrition and secondary metabolism. Nitrogen fertilization also impacts expression of tridecanone and trichome-mediated resistance in *S. habrochaites* (Barbour et al., 1991).

4. GENETIC RESOURCES FOR PLANT RESISTANCE BREEDING IN *SOLANUM*

4.1 Sources of Resistance: Resistance of Wild and Cultivated Tomato Species to Pests

Tomatoes are native to, and first were domesticated in, South America. In addition to *S. lycopersicum* and its wild form *Solanum lycopersicum* var. *cerasiforme* (Dunal) D.M. Spooner, G.J. Anderson & R.K. Jansen Gray (wild cherry), wild species of tomato include *Solanum pimpinellifolium* L. (currant tomato), *Solanum cheesmaniae* (Riley) Fosburg, *Solanum chmielewskii* (Rick, Kes., Fob. and Holle), *Solanum chilense* (Dunal) Reiche, *Solanum neorickii* D.M. Spooner, G.J. Anderson, & R.K. Jansen, *Solanum peruvianum* (L.), *S. habrochaites* and *Solanum pennellii* Correll (Rick, 1976). In general, wild tomatoes are more resistant to pests than cultivated tomatoes due to elevated production of resistance-related traits. These wild tomatoes thus represent an important source of genetic diversity for the development of pest-resistant varieties (Menda et al., 2013).

Wild relatives of cultivated tomato, particularly *S. chilense*, *S. peruvianum*, *S. habrochaites*, and *S. pimpinellifolium*, have proven to be rich sources of disease-resistance genes (Foolad and Sharma, 2005; Scott and Gardner, 2007). Resistance to over 40 diseases and a few insects have been discovered in wild tomatoes, and at least 20 disease-resistance genes have been bred into tomato cultivars (Ji et al., 2007; Robertson and Labate, 2007). Several resources and approaches have been developed to facilitate the use of this genetic potential in tomato breeding. These resources and approaches include molecular markers to characterize genetic resources (Nuez et al., 2004; Ercolano et al., 2005), estimate genetic relationships (Albrecht et al., 2010), and to manage Genebank accessions (Tanksley and McCouch, 1997), libraries for analyzing diversity in wild species (Eshed and Zamir, 1995; Monforte and Tanksley, 2000), and platforms for developing and screening tomato introgression lines from different wild species (Tripodi et al., 2010). Several cDNA and genomic libraries have been developed from wild tomato species to isolate resistance genes involved in disease responses.

4.1.1 Solanum lycopersicum

The great advantage of using accessions of *S. lycopersicum* in tomato breeding, despite their lower resistance compared to other species, is natural hybridization combined with desirable agronomic characteristics that are not present in wild species. Some varieties of *S. lycopersicum*, such as *cerasiforme*, Giant Orita, and Prince, possess some resistance to pests such as *H. zea*, *T. absoluta*, *M. euphorbiae*, *T. urticae*, and *S. exigua* (Eigenbrode and Trumble, 1993, 1994). This resistance is due, in some cases, to the glandular trichomes and biochemical components in the leaves. In other cases, physical characteristics (thick cuticle) and growth habit (Juvik and Stevens, 1982; Juvik et al., 1982; Leite et al., 2003) influence the resistance to pests. Damage to fruits caused by *T. absoluta* and *H. zea* are negatively correlated with size and total number of fruits (Canerday et al., 1969; Leite et al., 2003). Another factor detected in *cerasiforme* was the presence of alkaloids in fruits, reducing the attack of *S. exigua* (Eigenbrode and Trumble, 1994). Elliger et al. (1981) reported that extracts of *S. lycopersicum* containing glucaric acid (not present in trichomes), when incorporated into artificial diets, reduced the growth of *H. zea* larvae.

The high proportion of non-glandular trichomes (≅95%) generally present in *S. lycopersicum* makes it more susceptible compared to other species of tomato with higher proportions of glandular trichomes. However, those varieties of *S. lycopersicum* that have higher densities of glandular trichomes, especially type VI, also contain rutin, chlorogenic acid, 2-TD, and 2-UD, all of which possess activity against insects and mites. However, there is a catch: the methyl ketones, at low levels, have a positive effect on the biology of these pests (Leite et al., 1999a). Wilkens et al. (1996) reported phenols as chemical groups that confer resistance in *S. lycopersicum* to pests. Silva et al. (1998) found that leaves of *S. lycopersicum* var. Santa Clara had higher total phenolic content than *S. habrochaites* f. *glabratum* PI 134417 and *S. peruvianum* LA 444-1, and that these compounds were positively correlated with the number of *T. absoluta* eggs.

4.1.2 Solanum pennellii

One interesting source, mainly because it is amenable to crosses with *S. lycopersicum*, is *S. pennellii*. High levels of resistance to important pests such as whiteflies and lepidopterans have been found in some genotypes of this species, particularly the LA 716 accession (Kumar et al., 1995). Suinaga et al. (2004b), for example, found resistant accessions of *S. pennellii* harbored a high number of small mines but a low number of large mines of *T. absoluta*. When the bases of resistance have been investigated, evidence points to glandular trichomes (Goffreda et al., 1990), especially acylglucoses such as 2,3,4-tri-*O*-acylated glucose, that are present in type IV trichomes.

The relatively simple inheritance of acylsugars in *S. penellii* and the ease of crossing this species with *S. lycopersicum* suggests it may be possible to use tomato breeding lines derived from interspecific crosses with LA-716 to develop hybrid varieties with favorable agronomic traits and high levels of acylsugar-mediated resistance to pests. Maluf et al. (2010) tested the resistance of hybrids developed from crosses of low and high acylsugar breeding lines to spider mites, silverleaf whitefly, and the South American tomato pinworm (*T. absoluta*). These hybrids had intermediate levels of acylsugars but showed levels of resistance comparable to that of the high acylsugar parent, validating the strategy of crossing low acylsugar lines and high acylsugar lines (with *S. penellii* parentage) for producing arthropod-resistant hybrids.

4.1.3 Solanum pimpinellifolium, Solanum cheesmaniae, Solanum chmielewskii, Solanum galapagense, and Solanum chilense

S. pimpinellifolium is resistant to some pests, such as *T. absoluta*, *S. exigua*, *Liriomyza* spp., and some hemipterans (Lourenção et al., 1984; Eigenbrode and Trumble, 1993). This resistance may be due to chemical compounds (α-tomatine) and physical characteristics (hardness of the fruit cuticle) (Juvik and Stevens, 1982). This species is also a potential source of resistance to *Clavibacter michiganensis* subsp. *michiganensis* Corrig. (Jensen) the causal agent of bacterial canker of tomato (Kabelka et al., 2002).

Solanum cheesmanii Riley is resistant to *Helicoverpa armigera* (Hübner) and *Keiferia lycopersicella* (Walsingham), *S. chmielewskii* is resistant to *Spodoptera littoralis* (Boisduval) and *S. chilense* is resistant to *H. armigera* and *B. tabaci* biotype B and shows moderate antixenosis against *F. occidentalis* (Kumar et al., 1995; Picó et al., 1998). High densities of type III trichomes in hybrids of *S. cheesmanii* f. *minor* and *S. lycopersicon* were correlated with increased resistance to *M. persicae* (Simmons et al., 2005). Kumar et al. (1995) also observed low levels of antixenosis against *F. occidentalis* in the species *S. pimpinellifolium* and *S. chmielewskii*. Lucatti et al. (2013) assessed adult and pre-adult survival and oviposition rate of *B. tabaci* on accessions of *S. cheesmaniae* and *Solanum galapagense* S.C. Darwin & Peralta and found that high levels of resistance in accessions of the latter species were associated with relatively high levels of acyl sugars and type I and IV glandular trichomes.

4.1.4 Solanum peruvianum

Despite the difficulty in crossing *S. peruvianum* with *S. lycopersicum*, which requires use of the embryo rescue technique, *S. peruvianum* has proven resistant to *T. absoluta* (Lourenção et al., 1984), *M. persicae* (Leite et al., 1997), and *B. tabaci* (Channarayappa et al., 1992; Picó et al., 1998). However, the causes of this resistance are still unknown. Leite et al. (1997) concluded that the high density of non-glandular trichomes promoted resistance of *S. peruvianum* LA 444-1 to *M. persicae* and susceptibility to *Frankliniella schulzei* (Trybom). Suinaga et al. (1999) observed a negative correlation between heptadecane and hatchability of *T. absoluta* eggs in leaves of *S. peruvianum* CNPH 101. However, cyclobutanol and hexadecane correlated positively with the number of *T. absoluta* mines in this accession (Suinaga et al., 1999). Suinaga et al. (2004a) attributed resistance of *S. peruvianum* accessions to 4-methyl-2,6-di-t-butylphenol.

S. peruvianum shows resistance to the pathogens *C. michiganensis* subsp. *michiganensis* (Kabelka et al., 2002) and *Oidium neolycopersici* L. Kiss, which causes powdery mildew on tomato (Bai et al., 2004). This species also possesses the *Mi* gene for nematode resistance (Ammiraju et al., 2003).

4.1.5 Solanum habrochaites (syn. Lycopersicum hirsutum Dunal)

The role of methyl ketones in resistance to insect pests has been particularly well studied in *S. habrochaites* f. *glabratum* (Kennedy, 2003). 2-TD is found in the heads of glandular trichomes type VIc (Channarayappa et al., 1992) present in *S. habrochaites* f. *glabratum*, especially in accession PI 134417. In this wild tomato species, 2-tridecanone is present in concentrations up to 5500 μg/g fresh weight, mostly in type VI trichomes. 2-TD also is found in *S. lycopersicum*, but in much lower amounts. 2-UD also occurs in the glandular trichomes in PI 134417 and in other *glabratum* accessions (Farrar and Kennedy, 1987; Weston et al., 1989). Chlorogenic acid and sesquiterpenes (β-caryophyllene and α-humulene) also occur in type VI glandular trichomes of *glabratum*.

Other resistance factors not associated with trichomes occur in *S. habrochaites*. Leite et al. (1998) suggest that the leaf thickness of the PI 134417 could help resist *B. tabaci* biotype B. The high concentration of insoluble fiber and acid detergent cellulose in the leaves of PI 134417 may confer resistance to attack by *T. absoluta*, as levels of these compounds were negatively correlated with the number of eggs of this insect (Silva et al., 1998). Leite et al. (1999b) reported the presence of crystal idioblasts in PI 134417 with resistance to *M. persicae*.

Oliveira et al. (2009) found that resistance to *T. absoluta* was associated with the absence of the compound tricosane; for example, no leaves of accessions HGB-674 and HGB-1497 of *S. habrochaites* were mined by *T. absoluta*, and the compound tricosane was not found in these accessions. Accessions HGB-7236 and HGB-243, with a higher incidence of mined leaves, also contained the highest tetracosane concentrations, suggesting this compound is a susceptibility factor (Oliveira et al., 2009). Tetracosane was also partially implicated in susceptibility to *T. absoluta* by Suinaga et al. (1999). Oliveira et al. (2009) concluded that only two lines, HGB-674 and HGB-1497, appeared promising out of 57 accessions and three varieties of *S. lycopersicum* studied.

Several sesquiterpenes are present in type VI glandular trichomes in *S. habrochaites* f. *typicum*, including α-humulene, α-curcumene, zingiberene, and γ and δ-elemene. Presence of these compounds is correlated with resistance to pests such as *S. exigua* and *T. urticae* (Weston et al., 1989; Eigenbrode and Trumble, 1993; Rahimi and Carter, 1993; Eigenbrode et al., 1994; Freitas et al., 2000a,b). Also, 2,3-dihydrofarnesoic acid, first identified in the leaves of *typicum* by Snyder et al. (1993), has a repellent effect against *T. urticae*. On the other hand, Juvik et al. (1988) showed that the presence of some sesquiterpenes act as an oviposition stimulant for *H. zea*. Ecole (1998) noted that sesquiterpenes farnesene or α-bergamotene, α-humulene or δ-elemene, 2,5,5-trimethyl 1,3,6-heptatriene or (+) or camphene, santalol, and farnesene, and α-bergamotene, β-sinesal or farnesene and farnesol are involved in the resistance of *typicum* LA in 1777 against *T. absoluta*.

Resistant accessions of *S. habrochaites* harbored a high number of small mines but a low number of large *T. absoluta* mines when compared to susceptible *S. lycopersicum* cultivars (Suinaga et al., 2004b). These authors suggested that, when evaluating resistance, the number of small mines should be evaluated together with the number of large mines. Similar results were reported by Ecole et al. (2000, 2001) in the resistant variety LA 1777 of *S. habrochaites* f. *typicum*, compared to susceptible tomato cultivars.

4.2 Considerations in the Breeding of Resistance in Tomato

All species of tomato are diploid ($2n = 2x = 24$) and show similarities in chromosome number and structure. Cultivated tomato shows limited genetic variability, a result of founder events and selection during domestication and development of modern cultivars. Miller and Tanksley (1990) estimated that the genomes of tomato cultivars contain 5% of the genetic variation of their wild relatives (1990). There are tomatoes that are strictly autogamous, such as *S. cheesmaniae* and some *S. pimpinellifolium* populations. The rate of variation within these populations is essentially zero. At the other extreme, total variability is vastly greater in highly outcrossing species such as *S. chilense*, *S. peruvianum*, and *S. pennellii*. An intermediate position with respect to outcrossing is occupied by the species *S. chmielewskii*, *S. pimpinellifolium*, and some wild forms of *S. lycopersicum*. Outcrossing within each group of species (among accessions) is possible. Interspecific crosses occur between groups of *S. habrochaites*, *S. lycopersicum*, *S. pennellii*, *S. cheesmanii*, *S. pimpinellifolium*, *S. neorickii*, and *S. chmielewskii*, and also between groups of *S. peruvianum* and *S. chilense* (Stevens and Rick, 1986). Examples of descendants (F1) obtained by embryo rescue are possible, although difficult, among groups of *S. peruvianum* with *S. lycopersicum* and also with *S. pennellii* and some accessions of *S. peruvianum* (Stevens and Rick, 1986). Outcrossing does not occur between *S. habrochaites* and *S. peruvianum* or *S. chilense* (Stevens and Rick, 1986).

The major goals of tomato breeding focus on increased productivity, tolerance to biotic and abiotic stresses, and improved flavor characteristics and nutritional value of the fruit (Bauchet and Causse, 2012). Tomato breeding aimed at increasing resistance to pests has focused on tomato species in which resistance to pests is known to be present and for which there is some understanding of resistance mechanisms, notably leaf trichomes and associated traits (methyl ketones, phenolics, sesquiterpenes, and ester-sugars), morphological characteristics, and biochemical traits such as α-tomatine.

Genetic improvement of tomato cultivation has followed the process commonly used in autogamous species: introduction, selection, hybridization, and selection from segregating populations. All of these processes are based on the fact that, following every selfing, the percentage of individuals heterozygous for a monogenic character decreases by 50% and, consequently, the percentage of homozygous subjects increases at the same rate (Melo, 1989). The methods of family selection (pedigree), population (bulk), modified family selection (SSD = single-seed descent), retro-crossing (back-cross), and some combinations or modifications of these methods have been used by breeders in the advancement of segregating generations (Melo, 1989).

The genetic bases of some resistance-related traits in tomato have also been studied (Freitas et al., 2002; Fernández-Muñoz et al., 2003). For example, high levels of 2-TD involve at least three independent recessive genes (Fery and Kennedy, 1987), but the genetic control of 2-UD levels is, so far, unknown (Farrar and Kennedy, 1991b). According to Barbosa and Maluf (1994), the heritability of 2-TD in tomato leaves is 60.6%. However, in breeding programs aimed at incorporating this resistance factor into commercial varieties, there has been a gradual diminution in the level of resistance to pests as plants are selected for improved agronomic characteristics (Sorenson et al., 1989; Barbosa and Maluf, 1994).

5. INDUCIBLE RESISTANCE IN TOMATO

5.1 Modes and Systems of Resistance in Tomato

Plant resistance can be constitutive or inducible. Expression of the former type of resistance is not contingent upon the activities of arthropods or pathogens, whereas expression of the latter type is triggered or stimulated by the attack of pathogens or arthropods. The relative importance of these two modes of resistance to overall plant resistance, and the coordination and organization of these various systems of inducible resistance in the plant (see below), are not well understood. Importantly, however, constitutive and inducible resistance appear to rely on the same arrays of resistance traits within a particular plant species.

Studies of plant resistance in tomato have been crucial to demonstrating the importance of inducible resistance as a component of overall plant resistance. Tomato possesses multilayered, complex systems of induced resistance governed by networks of interacting hormones and other signals (Jones and Dangl, 2006; Campos et al., 2014). Both herbivory and pathogen infection trigger extensive transcriptomic reorganizations in tomato, with hundreds of genes upregulated and hundreds downregulated (Reymond et al., 2004; Bosch et al., 2014). Many of the downregulated genes are related to primary metabolism, and induced changes in primary metabolism may contribute to plant tolerance (Gomez et al., 2010).

In contrast, many of the upregulated genes are related to secondary metabolism, and levels and activities of many of the resistance-related traits in tomatoes increase following herbivory or infection (Chen et al., 2005; Thaler et al., 2010; Bosch et al., 2014). As a consequence of these changes in primary and secondary metabolism, the resistance of tomato plants to herbivores and pathogens increases following initial attack.

The systems of induced resistance against pathogenic microorganisms are thought to comprise two tiers or layers (Jones and Dangl, 2006). The basal layer of resistance is induced in plants by highly conserved microbial- or pathogen-associated molecular patterns (PAMPs) and is known as PAMP-triggered immunity (PTI). Suppression or dampening of PTI responses by effectors produced by pathogens can lead to successful colonization of the plant by the pathogen. However, some plants have evolved resistance (R) genes, the products of which directly or indirectly recognize microbial effectors. Recognition of an effector by an R gene leads to an amplified and highly effective version of PTI called effector-triggered immunity (ETI). ETI responses usually involve a hypersensitive response (HR) at the site of attempted infection that prevents further pathogen infection. The recognized effector is called an avirulence (Avr) factor. ETI is more or less synonymous with vertical resistance (Agrios, 2005; Jones and Dangl, 2006).

Plants also possess systems of inducible resistance against arthropod herbivores. Unlike pathogen resistance, however, very few plant–arthropod interactions in tomato or other plants are governed by specific interactions among complementary sets of R genes and Avr genes in the plant and arthropod, and thus arthropod-induced responses appear to have more in common with PTI than with ETI (Campos et al., 2014). Moreover, indirect induced resistance, the most common example of which involves production of volatile compounds that attract natural enemies of a herbivore, appears to be a more important component of induced resistance to arthropods than to pathogens (e.g., Thaler, 1999).

5.2 JA- and SA-Mediated Responses to Arthropods and Pathogens

Pioneering research on the hormonal control of induced resistance has been conducted using tomato (Chen et al., 2006; Ntoukakis et al., 2014). Importantly, it now seems evident that diverse plant responses in tomatoes and other plants to pathogens and herbivores largely, but not entirely, converge on two hormonal pathways, one mediated by salicylic acid (SA) and the other by JA, although it is not yet clear how the various systems of induced resistance against pathogens and herbivores are coordinated or integrated.

Tomato plants respond to chewing insects, mechanical wounding, and some types of necrotrophic pathogens by activating the JA-mediated pathway (Thaler et al., 2010; Campos et al., 2014). An early step in this pathway in tomato involves release of systemin, an 18-amino acid peptide synthesized from a 200-aa precursor protein, prosystemin (McGurl and Ryan, 1992; Bergey et al., 1996; Schilmiller and Howe, 2005; Chen et al., 2006). Systemin binds a plasma membrane-bound receptor kinase (Scheer and Ryan, 2002), resulting in the activation of a phospholipase and the release of linolenic acid. JA and its active derivative, jasmonyl-L-isoleucine (JA-Ile) are synthesized from linolenic acid via a series of well-characterized steps (Bergey et al., 1996; Campos et al., 2014). Eventually, the JA-mediated transcriptional activation of a number of genes leads to increases in secondary compounds (e.g., proteinase inhibitors and polyphenoloxidase) that toxify and reduce the growth of insects (Bergey et al., 1996; Duffey and Stout, 1996). Current evidence implicates JA as the long-distance defense signal within tomato plants (Schilmiller and Howe, 2005).

In contrast to the responses to chewing insects and necrotrophic pathogens mediated by JA, tomato plants respond to biotrophic pathogens and piercing-sucking insects by activating an SA-mediated pathway (Goggin, 2007; Thaler et al., 2010; Campos et al., 2014; Yan and Dong, 2014). SA-mediated responses in general have negative effects on disease development and some sucking insects (Goggin, 2007), but often been shown to have positive effects on the growth and fitness of chewing arthropods (Stout et al., 2006; Goggin, 2007; Thaler et al., 2010). In part, the positive effects of SA on insects stem from inhibition by the SA pathway of JA-mediated responses ("negative crosstalk").

The existence of SA- and JA-mediated responses in tomato plants provides a mechanism whereby pathogenic microorganisms and arthropod herbivores can interact indirectly through induced changes in the shared host plant. Many JA- and SA-mediated changes are long-lasting and systemic, and as a result feeding or infection by an initial attacker can indirectly affect organisms spatially and temporally separated from the initial attacker. These indirect plant-mediated interactions can involve taxonomically disparate organisms (Stout et al., 1998b; Mouttet et al., 2013). There is specificity both in the responses elicited by the initial attacker and in the impact on subsequent attackers (Thaler et al., 2010). These plant-mediated interactions are often asymmetric (i.e., one participant in the interaction is affected more strongly than the other) (Thaler et al., 2010; Mouttet et al., 2013). The form and magnitude of induced plant-mediated interactions in tomato depend on, among other factors, the identity (feeding guild) of both the initial (triggering) and subsequent attackers, the degree of spatial and temporal separation between the interacting organisms, and the severity of attack. Mouttet et al. (2013), for example, documented several symmetrical and asymmetrical plant–mediated interactions among the tomato pests powdery mildew (*O. neolycopersici*), silverleaf whitefly (*B. argentifolii*), and *T. absoluta* on shared tomato hosts. These plant-mediated interactions likely have consequences for community structure and pest management in tomato fields, but these have not been fully elucidated (Thaler et al., 2001).

Tomato plants also exhibit systemic acquired resistance (SAR), a type of SA-dependent induced resistance to pathogens (Gozzo and Faoro, 2013; Yan and Dong, 2014) that intersects and overlaps with many of the phenomena described in the preceding paragraphs. Attempted infection of a resistant plant by an avirulent pathogen results in an HR, in which cells located around the infection site die within a few hours of attempted infection (usually as a part of an ETI response). The HR may directly kill the pathogen, restricting its spread in the plant, and also leads to release of a mobile signal that induces defense mechanisms in uninfected parts of the plant (Hammond-Kosack and Jones, 1997). Cells immediately surrounding the HR develop local acquired resistance (Ross, 1961a), characterized by the reinforcement of cell walls, synthesis of antimicrobial phytoalexins, and expression of pathogenesis-related (PR) genes (Fritig et al., 1998). At sites distant from the HR, SAR is induced (Ross, 1961b). This resistance is correlated with an accumulation of SA, expression of PR proteins, stimulation of other defense-related pathways, and relatively broad-spectrum resistance (Durrant and Dong, 2004). Elicitors of SAR in plants have been discovered and commercialized, and attempts have been made to integrate these elicitors into management programs for diseases in tomato. Obradovic et al. (2005), for example, showed that treatment of greenhouse-grown plants with the SAR elicitor acibenzolar-*S*-methyl (Actigard) reduced the occurrence of typical symptoms of bacterial spot.

Recent studies utilizing tomato mutants and transgenic plants with altered signaling capabilities have confirmed the importance of the JA-SA dichotomy and also have suggested that this dichotomy alone does not provide a framework sufficient to predict interactions among disparate attackers in tomato and other plants (Thaler et al., 2010; Tian et al., 2014). Negative cross talk among the JA and SA pathways, the involvement of other hormones such as ethylene, abscisic acid, or gibberellic acid in signaling networks following attack, and the effects of JA- and SA-induced responses on organisms from the third trophic level all complicate the elucidation of these complex plant-mediated interactions, and as a result the full significance of induced resistance to the management of insect and disease pests of tomato in the field is not yet understood.

5.3 Gene-For-Gene Resistance to Diseases and Breeding for Disease Resistance

Vertical resistance, or ETI, is a particularly important component of tomato resistance to disease-causing microorganisms (Gozzo and Faoro, 2013; Ntoukakis et al., 2014). In contrast to more generalized systems of responses described in the previous section, the bases of ETI responses are specific interactions among sets of corresponding Avr and R genes in the pathogens and plants, respectively (Jones and Dangl, 2006). As described above, the products of Avr genes generated by the pathogen during attempted infection are directly or indirectly recognized by putative receptors encoded by R genes. Although regulation of ETI has a relatively simple genetic basis, the plant responses involved in vertical resistance are highly complex. ETI often involves an HR, activation of the SA pathway, and subsequent expression of SAR.

Vertical resistance is particularly suited for use in practical disease management programs. The complex and concerted responses activated by the recognition of Avr products by R genes are rapid and result in immunity or near-immunity with apparently little cost to the plant. The use of vertical resistance requires no additional efforts from the farmer. Moreover, the introgression of R genes into agronomically acceptable backgrounds, while not trivial, is certainly feasible, and vertical resistance has been used in conventional resistance-breeding programs for decades (Pink, 2002). R genes conferring resistance to fusarium wilt, verticillium wilt, root-knot nematode, alternaria stem canker, gray leaf spot, and some bacterial and viral diseases have been introgressed into commercial tomato varieties. Unfortunately, R gene–mediated resistance is often rapidly circumvented by coevolving pathogens (Pink, 2002). Furthermore, vertical resistance is highly specific to one or a few pathogen strains or races, and thus does not provide broad-spectrum resistance. The structure and organization of resistance loci in the tomato genome have been investigated (Van Ooijen et al., 2007). Several tomato R genes have been isolated by various methods; in fact, the tomato R gene *Pto*, which confers resistance to *P. syringae* pv. *tomato*, was the first resistance (R) gene to be isolated from plants (Tang et al., 1996; Ntoukakis et al., 2014). *Pto* encodes a cytoplasmic serine-threonine protein kinase and confers resistance against strains of *P. syringae* pv *tomato* that express the effector proteins AvrPto or AvrPtoB (Martin et al., 1993). Other R genes isolated from tomato include CF5, CF9, MI1-2, I2, ASC, HERO, VE, BS4, and SW5 (Ori et al., 1997; Parniske et al., 1997; Dixon et al., 1998; Milligan et al., 1998; Brandwagt et al., 2000; Kawchuk et al., 2001; Ernst et al., 2002; Schornack et al., 2004).

As noted above, vertical resistance plays a much smaller role in tomato resistance to arthropods than to pathogens. Of potential note in this respect is the *Mi* gene in tomato. The *Mi* gene confers resistance against several root-knot nematode species (Williamson, 1998). *Mi* encodes a protein with nucleotide-binding and leucine-rich repeat domains and is located near the centromere of the short arm of chromosome 6 (Mehlenbacher, 1995; Milligan et al., 1998). The *Mi* gene was isolated by positional cloning and its identity confirmed by complementation of function (Kaloshian et al., 1998; Milligan et al., 1998). *Mi*-mediated resistance is characterized by a rapid (within 12 h of infection) HR that limits nematode feeding and establishment of a feeding site known as giant cells (Ho et al., 1992; Hwang et al., 2000). This gene was introduced into cultivated tomato from *Lycopersicon peruvianum* (L.) Mill. by embryo rescue. Although nematode infection sometimes occurs on resistant tomato cultivars, the pest generally fails to develop and reproduce to high levels on these resistant tomato genotypes grown under greenhouse or field conditions (Terrell et al., 1983). The gene confers resistance against three major *Meloidogyne* species that infect tomato, *Meloidogyne arenaria* Chitwood, *Meloidogyne incognita* (Kofoid & White) and *Meloidogyne javanica* (Treub) (Martinez de Ilarduya et al., 2001). Interestingly, the *Mi* gene also confers resistance to the potato aphid, *M. euphorbiae* (Rossi et al., 1998), and tomato psyllid, *Bactericera cockerelli* (Šulc) (Casteel et al., 2006), and was the first example of an R gene with activity against such taxonomically disparate organisms (Nombel et al., 2003; Goggin, 2007).

6. PLANT RESISTANCE AS A MANAGEMENT STRATEGY IN TOMATO PRODUCTION

6.1 Current Status

As noted above, R genes for many important diseases have been transferred (introgressed) into commercial varieties, and for these diseases, use of resistant varieties is rightly considered the foundation for management. Fusarium wilt is one example. Isolates of *F. oxysporum* f. sp. *lycopersici* affecting tomato have been grouped into three races according to their ability to infect a set of differential cultivars carrying distinct resistance factors. Three major resistance loci have been genetically characterized in *Solanum* species and all of them have been incorporated into commercial cultivars. Fusarium wilt probably could not be effectively managed without these resistant cultivars. Other diseases for which commercial varieties with resistance are available include the *Verticillium* wilts, Alternaria stem canker, bacterial speck, root-knot nematodes, tobacco and tomato mosaic viruses, and tomato spotted wilt virus (Csizinszky et al., 2005; Jones, 2008). Some varieties have resistance to multiple diseases. Accordingly, resistance to disease pests figures prominently in the recommendations of varieties made by agricultural extension services and in trade and industry publications (e.g., http://ipm.ifas.ufl.edu/resources/success_stories/T&PGuide/pdfs/Chapter2/Tomato_Cultivars.pdf; http://www.ipm.ucdavis.edu/PMG/r783900511.html). New resistant varieties are introduced almost yearly; for example, commercial fresh-market tomato hybrids with vertical resistance to late blight caused by *P. infestans* were recently released by the tomato breeding program at North Carolina State University, United States (Nowicki et al., 2012). In addition, sources of horizontal resistance for several tomato diseases such as early blight, powdery mildew, bacterial canker and bacterial wilt, horizontal resistance have been identified and used to develop varieties with both vertical and horizontal resistance against late blight and powdery mildew (Foolad, 2007).

In contrast, the use of resistant tomato varieties is not a central component of management programs for arthropod pests, as agronomically acceptable varieties with high levels of arthropod resistance are in development but not yet commercially available (Csizinszky et al., 2005). Sources of resistance in the form of genotypes of wild *Solanum* species with high levels of resistance to insects have been identified (see Section 4.1). Some of the plant characters associated with resistance, such as density of glandular trichomes and production of high levels of acylsugars, are under relatively simple genetic control. However, commercialization of resistant varieties has been hindered by poor yield and agronomic characteristics in the donors, and difficulty in introgressing genes for resistance into agronomically acceptable backgrounds (Talekar et al., 2006). As one prominent example, development of tomato varieties resistant to *T. absoluta* has been "intensively pursued" for over two decades, but incorporation of resistance into varieties with acceptable yield and quality has not yet been achieved (Guedes and Picanço, 2012).

Furthermore, despite much interesting work on induced resistance to arthropods in tomato, little progress has been made in developing commercial approaches to using induced resistance. Acibenzolar-*S*-methyl (Actigard) is registered on tomato and is useful against some bacterial diseases such as bacterial speck (McGrath and Smart, 2014). Foliar applications of JA to tomatoes stimulated expression of resistance-related traits, reduced abundance of arthropods in three feeding guilds, and increased rates of parasitism of *S. exigua* caterpillars in field experiments (Thaler et al., 2001). To this point, however, no commercial approaches based on applications of JA to plants have been developed. Recently, Worall et al. (2012) reported that tomato plants grown from seeds treated with JA showed broad-spectrum resistance as seedlings to spider mites, caterpillars, aphids, and the necrotroph *B. cinerea*. Long-lasting protection against pests was not accompanied by reductions in plant growth, suggesting that JA seed treatments may hold potential as a management strategy.

6.2 Complications for the Use of Plant Resistance in Management Programs

One of the limitations of the use of resistant tomato genotypes in pest management is the potential for decreased efficiency of biological control on resistant varieties (Farrar and Kennedy, 1991a, 1993; Barbour et al., 1993; Farrar et al., 1994; Riddick and Simmons, 2014). Sticky exudates produced by glandular trichomes can interfere with the activities of natural enemies (Riddick and Simmons, 2014). Genotypes with high levels of 2-TD and alpha-tomatine adversely affected parasitoids of *S. exigua* and *H. zea* (Duffey and Bloem, 1986; Duffey et al., 1986). Another problem arising from high levels of 2-TD is the induction of resistance to insecticides such as carbaryl by exposure to 2-TD (suggesting a neurotoxic effect of 2-TD). There is evidence that cytochrome P-450s (monooxygenases) are involved in the induction of tolerance to 2-TD following exposure of neonate *H. virescens* to 2-TD or leaves of PI 134417 (Rose et al., 1991), and this detoxication enzyme is also involved in tolerance to insecticides.

The resistance of tomato plants can vary throughout the canopy, with plant stage or age, and with changes in fertilization, temperature, and photoperiod. The density of type VI trichomes and the production of 2-TD exudates are higher in spring than in autumn due to variation of the photoperiod (Weston et al., 1989; Nichoul, 1994). Gianfagna et al. (1992) observed that production of zingiberene in type VI glandular trichomes of leaves of *typicum* is greater under short day conditions (25/20°C temperature, day/night) than under long day conditions (30/25°C temperature, day/night). Levels and inducibilities of PPO and PIs in tomato leaves are influenced by the environment and by plant stage, with highest levels observed in young plants at higher temperatures and light incidence (Green and Ryan, 1973; Stout et al., 1996).

Barbour et al. (1991) observed that increasing rates of 20-20-20 NPK from 1.4 g to 2.8 g/kg soil did not affect the leaf area of PI 134417, but did decrease the density of type VIc trichomes and, consequently, the 2-TD content in leaves. Wilkens et al. (1996) concluded that the concentration of phenols in the leaves of *S. lycopersicum* and its resistance as a function of NK fertilization were consistent with the optimal defense hypothesis that predicted a negative relationship between growth and defense. Induction of PI and PPO activities are not affected by nitrogen fertilization in *S. lycopersicum* (Stout et al., 1998a). With increasing potassium fertilization there is a significant reduction in the density of crystal idioblasts, especially in PI 134417 (Leite et al., 1999b).

Leite et al. (1999b, 2001) observed increases in the contents of 2-TD and crystal idioblasts with increasing plant age and from the base to the apex of the plant canopy in *S. habrochaites* f. *glabratum* PI 134417. This is due to increasing density of trichomes from the base to apex and with increasing age of the plants. Because the apical and middle third of plants are where major attacks of most pests such as *T. absoluta* occur, elevated levels of these resistance factors in these parts of PI 134417 have the potential to reduce tomato susceptibility to pests. The increase in 2-TD and trichomes with age has the advantage of making cultivars bred using PI 134417 as a source of resistance less susceptible to moths when plants are in the reproductive stage. However, these cultivars may show susceptibility to this pest during earlier stages of crop growth. On the other hand, tomato plants produce more PIs when younger (Alarcon and Malone, 1995).

7. CONCLUSION

Tomato is one of the most important vegetables in the world, but it is subject to large losses from both insect pests and pathogenic microorganisms. The development and implementation of tomato varieties with resistance to pests and diseases is a key to reducing losses from these pests and diseases with reduced pesticide inputs. The secondary metabolites (e.g., phenolics, alkaloids, and methyl ketones), resistance-related proteins, morphological traits (e.g., trichomes), and hormonal pathways underlying constitutive and inducible resistance have been more thoroughly studied in tomato than in almost any other plant, although the genetic regulation of these resistance-related traits is insufficiently understood. Furthermore, the availability of tomato genotypes and wild tomato species with high levels of resistance to various pests and diseases represents an outstanding opportunity for the development of resistant varieties, as demonstrated by the commercialization of numerous disease-resistant varieties. Among the wild tomato species, those that appear most promising as sources of resistance are accessions of *S. habrochaites* not only because it is resistant to a large number of insects, but also because it crosses fairly easily with *S. lycopersicum*. Recent research has also focused on *S. pennellii* as another good source for future crosses with commercial species. However, maintaining high levels of resistance and high agronomic quality during the breeding process has proven difficult, particularly for insect-resistant varieties. Moreover, there may be challenges associated with incorporating insect-resistant varieties into management programs.

One important theme emerging from studies of plant resistance in tomato is the complexity of resistance. Multiple resistance-related traits contribute to resistance to any given pest. Suppression of expression of a single resistance-related trait reduces but does not eliminate resistance. For example, suppression of expression of terpenes and rutin in the *odorless-2* mutant reduced but did not eliminate resistance to flea beetles and Colorado potato beetles, and plants were still able to produce proteinase inhibitors (Kang et al., 2010). Interactions among secondary metabolites, such as the inactivation of proteinase inhibitors by polyphenol oxidase (Duffey and Stout, 1996) are important. Plant nutrition (e.g., nitrogen concentration and C:N ratio) strongly influences plant resistance (Han et al., 2014), and primary metabolism and secondary metabolism are intricately intertwined (Barbour et al., 1991; Royer et al., 2013). Attack by arthropods and pathogens increases the resistance of tomatoes to subsequent attackers, and tomato responses to attack involve wholesale transcriptomic and metabolomic changes in the plant (Kant et al., 2015). Different types of attack produce qualitatively different responses in the plant (Kant et al., 2015). Induced responses to arthropods and pathogens are governed by a network of hormones in which JA and SA play prominent roles, but other hormones, and cross talk among hormones, are also important (Thaler et al., 2010; Tian et al., 2014). Expression of resistance-related traits can affect not only individual herbivores but also communities of herbivores. Furthermore, many of the effects of constitutive and inducible resistance on herbivores can be mediated by the activities of natural enemies (Thaler, 1999; Thaler et al., 2001).

Over the past several years, a full range of -omics technologies and genetic approaches (e.g., analysis of mutant lines, generation of transgenic lines) have begun to be applied to tomato to give a more comprehensive view of the location and organization of resistance-related genes and to dissect the responses of tomatoes to pathogens and pests (Table 9.1). The amount of information on various aspects of plant-pest interactions in this crop, coupled with many tools available and the large number of scientists dedicated to research on tomato, creates a synergism that makes important advances likely in the near future. Over the last several years, quantification of the levels of a broad range of metabolites has been documented in tomato–pathogen interactions (López-Gresa et al., 2010). The ability to screen a wide range of metabolites at once will facilitate identification of plant traits involved in resistance and will lead to greater understanding of the metabolic network involved in producing resistant phenotypes (Fernie and Schauer, 2009).

Several recent studies have illustrated the utility of combining data from metabolomics with those from other genomics platforms to provide new insights on both gene annotation (Mintz-Oron et al., 2008) and regulation in complex biological systems (Osorio et al., 2011). These approaches have resulted in the identification of numerous genes likely to be involved in resistance. The integration of genotyping, pheno/morphotyping, and the analysis of the molecular phenotype using metabolomics, proteomics, and transcriptomics will yield novel insights into the interaction of plant genomes with the environment and will also be a powerful strategy for crop improvement. Furthermore, the acceleration in mapping and sequencing techniques and the decreasing costs of next-generation sequencing and metabolomics-based phenotyping will extend the possibilities of gene and marker discovery and genome-wide quantification of gene expression. This will simplify the development of resistant tomato varieties against pests and diseases in the future, either by conventional breeding, marker-assisted selection, or genetic engineering.

ACKNOWLEDGMENTS

This manuscript was approved for publication by the Director of the Louisiana Agricultural Experiment Station manuscript no. 2017-234-31385. We thank Dr. Channarayappa for permission to use the glandular trichome photo. GLDL also thanks "Conselho Nacional de Desenvolvimento Científico e Tecnológico" (CNPq), "Fundação de Amparo 'a Pesquisa do Estado de Minas Gerais" (FAPEMIG) and "Secretaria de Ciência e Tecnologia do Estado de Minas Gerais" for financial support of several research projects. The authors gratefully acknowledge Dr. Waqas Wakil for constructive comments on previous versions of the manuscript and for the invitation to write the chapter. This material is based upon work that is supported by the National Institute of Food and Agriculture, U.S. Department of Agriculture, Hatch accession number 1011556.

REFERENCES

Abad, P., Favery, B., Ross, M.N., Castagnone-Serena, P., 2003. Root-knot nematode parasitism and host response: molecular basis of a sophisticated interaction. Molecular Plant Pathology 4, 217–224.

Agrios, G.N., 2005. Plant Pathology, fifth ed. Academic Press, Elsevier Inc., California, USA, p. 952.

Akhtar, K.P., Saleem, M.Y., Asghar, M., Ahmad, M., Sarwar, N., 2010. Resistance of *Solanum* species to *Cucumber mosaic virus* subgroup IA and its vector *Myzus persicae*. European Journal Plant Pathology 128, 435–450.

Alarcon, J.J., Malone, M., 1995. The influence of plant age on wound induction of proteinase inhibitors in tomato. Physiologia Plantarum 95, 423–427.

Albrecht, E., Escobar, M., Chetelat, R.T., 2010. Genetic diversity and population structure in the tomato-like nightshades *Solanum lycopersicoides* and *S. sitiens*. Annals of Botany 105, 535–554.

Ammiraju, J.S., Veremis, J.C., Huang, X., Roberts, P.A., Kaloshian, I., 2003. The heat-stable root-knot nematode resistance gene *Mi-9* from *Lycopersicon peruvianum* is localized on the short arm of chromosome 6. Theoretical and Applied Genetics 106, 478–484.

Bai, Y., van der Hulst, R., Huang, C.C., Wei, L., Stam, P., Lindhout, P., 2004. Mapping *OI-4*, a gene conferring resistance to *Oidium neolycopersici* and originating from *Lycopersicon peruvianum* LA2172, requires multi-allelic, singlelocus markers. Theoretical and Applied Genetics 109, 1215–1223.

Barbosa, L.V., Maluf, W.R., 1994. Controle genético da resistência de *Lycopersicon* spp. A traça-do-tomateiro [*Scrobipalpuloides absoluta* (Meyrick, 1917) (Lepidoptera – Gelechiidae)]. Horticultura Brasileira 12, 133.

Barbour, J.D., Farrar Jr., R.R., Kennedy, G.G., 1991. Interaction of fertilizer regime with host-plant resistance in tomato. Entomologia Experimentalis et Applicata 60, 289–300.

Barbour, J.D., Farrar Jr., R.R., Kennedy, G.G., 1993. Interaction of *Manduca sexta* resistance in tomato with insect predators of *Helicoverpa zea*. Entomologia Experimentalis et Applicata 68, 143–155.

Bauchet, G., Causse, M., 2012. Genetic diversity in tomato (*Solanum lycopersicum*) and its wild relatives. In: Caliskan, M. (Ed.), Genetic Diversity in Plants. Institut National de la Recherche Agronomique (INRA), Unité de Génétique et Amélioration des Fruits et Légumes (GAFL), France, pp. 133–162.

Bergamin Filho, A., Kimati, H., Amorim, L., 1995. Manual de Fitopatologia: princípios e conceitos (Ceres). Ceres, São Paulo, Brazil.

Bergey, D.R., Hoi, G.A., Ryan, C.A., 1996. Polypeptide signaling for plant defensive genes exhibits analogies to defense signaling in animals. Proceedings of the National Academy of Sciences 93, 12053–12058.

Besser, K., Harper, A., Welsby, N., Schauvinhold, I., Slocombe, S., Li, Y., Dixon, R.A., Broun, P., 2009. Divergent regulation of terpenoid metabolism in the trichomes of wild and cultivated tomato species. Plant Physiology 149, 499–514.

Bhonwong, A., Stout, M.J., Attajarusit, J., Tantasawat, P., 2009. Defensive role of tomato polyphenol oxidases agains cotton bollworm (*Helicoverpa armigera*) and beet armyworm (*Spodoptera exigua*). Journal of Chemical Ecology 35, 28–38.

Blancard, D., 1997. A Colour Atlas of Tomato Diseases: Observations, Identification and Control. John Wiley and Sons, New York, USA, p. 212.

Bleeker, P.M., Mirabella, R., Diergaarde, P.J., VanDoorn, A., Tissier, A., Kant, M.R., Prins, M., de Vos, M., Haring, M.A., Schuurink, R.C., 2012. Improved herbivore resistance in cultivated tomato with the sesquiterpene biosynthetic pathway from a wild relative. Proceedings of the National Academy of Sciences 109, 20124–20129.

Bloem, K.A., Kelley, K.C., Duffey, S.S., 1989. Differential effect of tomatine and its alleviation by cholesterol on larval growth and efficiency of food conversion in *Heliothis zea* and *Spodoptera exigua*. Journal of Chemical Ecology 15, 387–398.

Bosch, M., Berger, S., Schaller, A., Stintzi, A., 2014. Jasmonate-dependent induction of polyphenol oxidase activity in tomato foliage is important for defense against *Spodoptera exigua* but not against *Manduca sexta*. BMC Plant Biology 14, 257–272.

Boughton, A.J., Hoover, K., Felton, G.W., 2005. Methyl jasmonate application induces increased densities of glandular trichomes on tomato, *Lycopersicon esculentum*. Journal of Chemical Ecology 31, 2211–2216.

Brandwagt, B.F., Mesbah, L.A., Takken, F.L., Laurent, P.L., Kneppers, T.J., Hille, J., Nijkamp, H.J., 2000. A longevity assurance gene homolog of tomato mediates resistance to *Alternaria alternata* f. sp. *lycopersici* toxins and fumonisin B1. Proceedings of the National Academy of Sciences 97, 4961–4966.

Broadway, R.M., Duffey, S.S., Pearce, G., Ryan, C.A., 1986. Plant proteinase inhibitors: a defense against herbivorous insects? Entomologia Experimentalis et Applicata 41, 33–38.

Bruce, T.J., 2015. Interplay between insects and plants: dynamic and complex interactions that have coevolved over millions of years but act in milliseconds. Journal of Experimental Botany 66, 455–465.

Campos, M.L., Kang, J.-H., Howe, G.A., 2014. Jasmonate-triggered plant immunity. Journal of Chemical Ecology 40, 657–675.

Canerday, T.D., Todd, J.W., Dilbeck, J.D., 1969. Evaluation of tomatoes for fruitworm resistance. Journal of Georgia Entomological Society 4, 51–54.

Carvalho, G.A., Reis, P.R., Moraes, J.C., Fuini, L.C., Rocha, L.C.D., Goussain, M.M., 2002. Efeitos de alguns inseticidas utilizados na cultura do tomateiro (*Lycopersicon esculentum* Mill.) a *Trichogramma pretiosum* Riley, 1879 (Hymenoptera: Trichogrammatidae). Ciência e Agrotecnologia 26, 1160–1166.

Casteel, C.L., Walling, L.L., Paine, T.D., 2006. Behavior and biology of the tomato psyllid, *Bactericerca cockerelli*, in response to the Mi-1.2 gene. Entomologia Experimentalis et Applicata 121, 67–72.

Channarayappa, C., Shivashankar, G., Muniyappa, V., Frist, R.H., 1992. Resistance of *Lycopersicon* species to *Bemisia tabaci*, a tomato leaf curl virus vector. Canadian Journal of Botany 70, 2184–2192.

Chen, H., Jones, A.D., Howe, G.A., 2006. Constitutive activation of the jasmonate signaling pathway enhances the production of secondary metabolites in tomato. FEBS Letters 580, 2540–2546.

Chen, H., Wilkerson, C.G., Kuchar, J.A., Phinney, B.S., Howe, G.A., 2005. Jasmonate-inducible plant enzymes degrade essential amino acids in the herbivore midgut. Proceedings of the National Academy of Sciences 102, 19237–19242.

Csizinszky, A.A., Schuster, D.J., Jones, J.B., van Lenteren, J.C., 2005. Crop protection. In: Heuvelink, E. (Ed.), Tomatoes. CABI Publishing, Cambridge, MA, USA, pp. 199–235.

Cuppels, D.A., Louws, F.J., Ainsworth, T., 2006. Development and evaluation of PCR- based diagnostic assays for the bacterial – speck and -spot pathogens of tomato. Plant Disease 90, 451–458.

de Azevedo, S.M., Faria, M.V., Maluf, W.R., de Oliveira, A.C.B., de Freitas, J.A., 2003. Zingiberene-mediated resistance to the South American tomato pinworm derived from *Lycopersicon hirsutum* var. *hirsutum*. Euphytica 134, 347–351.

Delahaut, K., Stevenson, W., 2004. Tomato and Pepper Disorders: Bacterial Spot and Speck. Cooperative Extension Publication A 2604, Wisconsin, USA, p. 2.

Dimock, M.B., Kennedy, G.G., Williams, W.G., 1982. Toxicity studies of analogs of 2-tridecanone, a naturally occurring toxicant from a wild tomato. Journal of Chemical Ecology 8, 837–842.

Dixon, M.S., Hatzixanthis, K., Jones, D.A., Harrison, K., Jones, J.D., 1998. The tomato Cf-5 disease resistance gene and six homologs show pronounced allelic variation in leucine-rich repeat copy number. Plant Cell 10, 1915–1925.

Duffey, S.S., Bloem, K.A., 1986. Plant defense-herbivore-parasite interactions and biological control. In: Kogan, M. (Ed.), Ecological Theory and Integrated Pest Management. Wiley, New York, USA, pp. 135–183.

Duffey, S.S., Bloem, K.A., Campbell, B.C., 1986. Consequences of sequestration of plant natural products in plant-insect-parasitoid interactions. In: Boethel, D.J., Eikenbary, R.D. (Eds.), Interactions of Plant Resistance and Predators and Parasitoids of Insects. Wiley, New York, USA, pp. 31–60.

Duffey, S.S., Stout, M.J., 1996. Antinutritive and toxic components of plant resistance against insects. Archives of Insect Biochemistry and Physiology 32, 3–37.

Durrant, W.E., Dong, X., 2004. Systemic acquired resistance. Annual Review of Phytopathology 42, 185–209.

Ecole, C.C., 1998. Resistência do acesso LA 1777 de *Lycopersicon hirsutum* f. typicum a *Tuta absoluta* (Meyrick) (Lepidoptera: Gelechiidae) (Dissertação de Mestrado). Universidade Federal de Viçosa, Brazil, p. 67.

Ecole, C.C., Picanço, M., Moreira, M.D., Magalhães, S.T.V., 2000. Componentes químicos associados à resistência de *Lycopersicon hirsutum* f. *typicum* a *Tuta absoluta* (Meyrick) (Lepidoptera: Gelechiidae). ANAIS: Da Sociedade Entomologica Do Brasil 29, 327–337.

Ecole, C.C., Picanço, M.C., Guedes, R.N.C., Brommonschenkel, S.H., 2001. Effect of cropping season and possible compounds involved in the resistance of *Lycopersicon hirsutum* f. *typicum* to *Tuta absoluta* (Meyrick) (Lep., Gelechiidae). Journal of Applied Entomology 125, 193–200.

Eigenbrode, S.D., Trumble, J.T., 1993. Resistance to beet armyworm, hemipterans, and *Liriomyza* spp. in *Lycopersicon* accessions. Journal America Society Hortitultural Science 118, 525–530.

Eigenbrode, S.D., Trumble, J.T., 1994. Fruit-based tolerance to damage by beet armyworm (Lepidoptera: Noctuidae) in tomato. Environmental Entomology 23, 937–942.

Eigenbrode, S.D., Trumble, J.T., Millar, J.G., White, K.K., 1994. Topical toxicity of tomato sesquiterpenes to the beet armyworm and the role of these compounds in resistance derived from an accession of *Lycopersicon hirsutum* f. *typicum*. Journal of Agricultural and Food Chemistry 42, 807–810.

Elliger, C.A., Wong, Y., Chan, B.G., Waiss Jr., A.C., 1981. Growth inibitors in tomato (*Lycopersicon*) to tomato fruitworm (*Heliothis zea*). Journal of Chemical Ecology 7, 753–758.

Ercolano, M.R., Sebastiano, A., Monti, L., Frusciante, L., Barone, A., 2005. Molecular characterization of *Solanum habrochaites* accessions. Journal of Genetic Breeding 59, 15–20.

Ernst, K., Kumar, A., Kriseleit, D., Kloos, D.U., Phillips, M.S., Ganal, M.W., 2002. The broad-spectrum potato cyst nematode resistance gene (Hero) from tomato is the only member of a large gene family of NBS–LRR genes with an unusual amino acid repeat in the LRR region. The Plant Journal 31, 127–136.

Eshed, Y., Zamir, D., 1995. An introgression line population of *Lycopersicon pennellii* in the cultivated tomato enables the identification and fine mapping of yield-associated QTL. Genetics 141, 1147–1162.

FAO, 2014. FAOSTAT Statistical Yearbook 2014. Statistical Division Food and Agriculture Organization of the United Nations. http://faostat3.fao.org/home/E.

Farrar, R.R., Barbour, J.D., Kennedy, G.G., 1994. Field evaluation of insect resistance in a wild tomato and its effects on insect parasitoids. Entomologia Experimentalis et Applicata 71, 211–226.

Farrar, R.R., Kennedy, G.G., 1987. 2-undecanone, a constituent of the glandular trichomes of *Lycopersicon hirsutum* f. *glabratum*: effects on *Heliothis zea* and *Manduca sexta* growth and survival. Entomologia Experimentalis et Applicata 43, 17–23.

Farrar, R.R., Kennedy, G.G., 1988. 2-undecanone, a pupal mortality factor in *Heliothis zea*: sensitive larval stage and in planta activity in *Lycopersicon hirsutum* f. *glabratum*. Entomologia Experimentalis et Applicata 47, 205–210.

Farrar, R.R., Kennedy, G.G., 1991a. Inhibition of *Telenomus sphingis* an egg parasitoid of *Manduca* spp. by trichome/2-tridecanone-based host plant resistance in tomato. Entomologia Experimentalis et Applicata 60, 157–166.

Farrar, R.R., Kennedy, G.G., 1991b. Relationship of leaf lamellar - based resistance to *Leptinotarsa decemlineata* and *Heliothis zea* in a wild tomato, *Lycopersicon hirsutum* f. *glabratum*, PI 134417. Entomologia Experimentalis et Applicata 58, 61–67.

Farrar, R.R., Kennedy, G.G., 1993. Field cage performance of two tachinid parasitoids of the tomato fruitworm on insect resistant and susceptible tomato lines. Entomologia Experimentalis et Applicata 67, 73–78.

Farrar, R.R., Kennedy, G.G., Roe, R.M., 1992. The protective role of dietary unsaturated fatty acids against 2-undecanone-induced pupal mortality and deformity in *Helicoverpa zea*. Entomologia Experimentalis et Applicata 62, 191–299.

Fernández-Muñoz, R., Salinas, M., Álvarez, M., Cuartero, J., 2003. Inheritance of resistance to two-spotted spider mite and glandular leaf trichomes in wild tomato *Lycopersicon piminellifolium* (Jusl.) Mill. Journal of the American Horticultural Society 128, 188–195.

Fernie, A.R., Schauer, N., 2009. Metabolomics-assisted breeding: a viable option for crop improvement? Trends in Genetics 25, 39–48.

Fery, R.L., Kennedy, G.G., 1987. Genetic analysis of 2-tridecanone concentration, leaf trichome characteristics, and tobacco hornworm resistance in tomato. Journal America Society Horticultural Science 112, 886–891.

Foolad, M., 2007. Genome mapping and molecular breeding of tomato. International Journal of Plant Genomics 1–52.

Foolad, M., Sharma, A., 2005. Molecular markers as selection tools in tomato breeding. Acta Horticulturae 695, 225–240.

Freitas, J.A., Maluf, W.R., Graças Cardosa, M.D., Gomes, L.A.A., Bearzotti, E., 2002. Inheritance of foliar zingibrene contents and their relationship to trichome densities and whitefly resistance in tomatoes. Euphytica 127, 275–287.

Freitas, J.D., Maluf, W.R., Cardoso, M.G., Benites, F.R.G., 2000b. Métodos para quantificação do zingibereno em tomateiro, visando à seleção indireta de plantas resistentes aos artrópodes-praga. Acta Scientiarum 22, 943–949.

Freitas, J.D., Maluf, W.R., Cardoso, M.G., Oliveira, A.C.B., 2000a. Seleção de plantas de tomateiro visando à resistência à artrópodes-praga mediada por zingibereno. Acta Scientiarum 22, 919–923.

Fritig, B., Heitz, T., Legrand, M., 1998. Antimicrobial proteins in induced plant defense. Current Opinion in Immunology 10, 16–22.

Gallo, D., Nakano, O., Silveira Neto, S., Carvalho, R.P.L., Baptista, G.C., Berti Filho, E., Parra, J.R.P., Zucchi, R.A., Alves, S.B., Vendramim, J.D., Marchini, L.C., Lopes, J.R.S., Omoto, C., 2002. Entomologia Agrícola. Fealq, Piracicaba, Brazil.

Gianfagna, T.J., Carter, C.D., Sacalis, J.N., 1992. Temperature and photoperiod influence trichome density and sesquiterpene content of *Lycopersicon hirsutum* f. *hirsutum*. Plant Physiology 100, 1403–1405.

Glas, J.J., Schimmel, B.C.J., Alba, J.M., Escobar-Bravo, R., Schuurink, R.C., Kant, M.R., 2012. Plant glandular trichomes as targets for breeding or engineering of resistance to herbivores. International Journal of Molecular Sciences 13, 17077–17103.

Goffreda, J.C., Steffens, J.C., Mutschler, M.A., 1990. Association of epicuticular sugars with aphid resistance in hybrids with wild tomato. Journal America Society Horticultural Science 115, 161–165.

Goggin, F.L., 2007. Plant – aphid interactions. Molecular and ecological perspectives. Current Opinion in Plant Biology 10, 399–408.

Gomez, S., Ferrieri, R.A., Schueller, M., Orians, C.M., 2010. Methyl Jasmonate elicits rapid changes in carbon and nitrogen dynamics in tomato. New Phytologist 188, 835–844.

Gozzo, F., Faoro, F., 2013. Systemic acquired resistance (50 years after discovery): moving from the lab to the field. Journal of Agricultural and Food Chemistry 61, 12473–12491.

Green, T.R., Ryan, C.A., 1973. Would-induced proteinase inhibitors in plant leaves: a possible defense mechanism against insects. Science 175, 776–777.

Guedes, R.N.C., Picanço, M.C., 2012. The tomato borer *Tuta absoluta* in South America: pest status, management and insecticide resistance. EPPO Bulletin 42, 211–216.

Hammond-Kosack, K.E., Jones, J.D.G., 1997. Plant disease resistance genes. Annual Review of Plant Physiology 48, 575–607.

Han, P., Lavoir, A., Le Bot, J., Amiens-Desneux, E., Desneux, N., 2014. Nitrogen and water availability to tomato plants triggers bottom-up effects on the leafminer *Tuta absoluta*. Scientific Reports 4, 4455.

Hawthorne, D.J., Shapiro, J.A., Tingey, W.M., Mutschler, M.A., 1992. Trichome-borne and artifially applied acylsugars of wild tomato deter feeding and oviposition of the leafminer *Liriomyza trifolii*. Entomologia Experimentalis et Applicata 65, 65–73.

Ho, J.Y., Weide, R., Ma, H.M., Vanwordragen, M.F., Lambert, K.N., Koorneef, M., Zabel, P., Williamson, V.M., 1992. The root-knot nematode resistance gene (Mi) in tomato: construction of a molecular linkage map and identification of dominant cDNA markers in resistant genotypes. The Plant Journal 2, 971–982.

Hwang, C.F., Bhakta, A.V., Truesdell, G.M., Pudlo, W.M., Williamson, V.M., 2000. Evidence for a role of the N terminus and leucine-rich repeat region of the Mi gene product in regulation of localized cell death. The Plant Cell 12, 1319–1329.

Isman, M.B., Duffey, S.S., 1982. Toxicity of tomato phenolic compounds to the fruitworm, *Heliothis zea*. Journal of the American Society for Horticultural Science 107, 67–170.

Ji, Y., Scott, J., Hanson, P., Graham, E., Maxwell, D., 2007. Sources of resistance, inheritance, and location of genetic loci conferring resistance to members of the tomato-infecting begomoviruses. In: Czosnekh, H. (Ed.), Tomato Yellow Leaf Curl Virus Disease. Springer Publishing, Dordrecht, Netherlands, pp. 343–362.

Johnson, R., Narvaez, J., An, G., Ryan, C., 1989. Expression of proteinase inhibitors I and II in transgenic tobacco plants: effects on natural defense against *Manduca sexta* larvae. Proceedings of the National Academy of Sciences 86, 9871–9875.

Jones, J.B., 2008. Tomato Plant Culture in the Field, Greenhouse, and Home Garden, second ed. CRC Press, Boca Raton, Florida, USA.

Jones, J.D.G., Dangl, J.L., 2006. The plant immune system. Nature 444, 323–329.

Jongsma, M.A., Bolter, C., 1997. The adaptation of insects to plant protease inhibitors. Journal of Insect Physiology 43, 885–895.

Juvik, J.A., Babka, B.A., Timmermann, E.A., 1988. Influence of trichome exudates from species of *Lycopersicon* on oviposition behavior of *Heliothis zea* (Boddie). Journal of Chemical Ecology 14, 1261–1278.

Juvik, J.A., Berlinger, M.J., Ben-David, T., Rudich, J., 1982. Resistance among accessions of the genera *Lycopersicon* and *Solanum* to four of the main insect pest of tomato in Israel. Phytoparasitica 10, 145–156.

Juvik, J.A., Shapiro, J.A., Young, T.E., Mutschler, M.A., 1994. Acylglucoses from wild tomatoes alter behavior and reduce growth and survival of *Helicoverpa zea* and *Spodoptera exigua* (Lepidoptera: Noctuidae). Journal of Economic Entomology 87, 482–492.

Juvik, J.A., Stevens, M.A., 1982. Physiological mechanisms of host-plant resistance in the genus *Lycopersicon* to *Heliothis zea* and *Spodoptera exigua*, two insect of the cultivated tomato. Journal of the American Society for Horticultural Science 107, 1065–1069.

Kabelka, E., Franchino, B., Francis, D.M., 2002. Two loci from *Lycopersicon hirsutum* LA 407 confer resistance to strains of *Clavibacter michiganensis* subsp. *michiganensis*. Phytopathology 92, 504–510.

Kaloshian, I., Yaghoobi, J., Liharska, T., Hontelez, J., Hanson, D., Hogan, P., Jesse, T., Wijbrandi, J., Simons, G., Vos, P., 1998. Genetic and physical localization of the root-knot nematode resistance locus *Mi* in tomato. Molecular and General Genetics 257, 376–385.

Kang, J., Liu, G., Shi, F., Jones, A.D., Beaudry, R.M., Howe, G.A., 2010. The Tomato *odorless-2* mutant is defective in trichome-based production of diverse specialized metabolites and broad-spectrum resistance to insect herbivores. Plant Physiology 154, 262–272.

Kant, M.R., Ament, K., Sabelis, M.W., Haring, M.A., Schuurink, R.C., 2004. Differential timing of spider mite-induced direct and indirect defenses in tomato plants. Plant Physiology 135, 483–495.

Kant, M.R., Jonckheere, W., Knegt, B., Lemos, F., Liu, J., Schimmel, B.C.J., Villarroel, C.A., Ataide, L.M.S., Dermauw, W., Glas, J.J., Egas, M., Janssen, A., Van Leeuwen, T., Schuurink, R.C., Sabelis, M.W., Alba, J.M., 2015. Mechanisms and ecological consequences of plant defence induction and suppression in herbivore communities. Annals of Botany 115, 1015–1051.

Kawchuk, L.M., Hachey, J., Lynch, D.R., Kulcsar, F., van Rooijen, G., Waterer, D.R., Robertson, A., Kokko, E., Byers, R., Howard, R.J., Fischer, R., Prufer, D., 2001. Tomato Ve disease resistance genes encode cell surface-like receptors. Proceedings of the National Academy of Sciences 98, 6511–6515.

Kennedy, G.G., 2003. Tomato, pests, parasitoids, and predators: tritrophic interactions involving the genus *Lycopersicon*. Annual Review of Entomology 48, 51–72.

Kennedy, G.G., Sorenson, C.F., 1985. Role of glandular trichomes in the resistance of *Lycopersicon hirsutum* f. *glabratum* to Colorado potato beetle (Coleoptera: Chrysomelidae). Journal of Economic Entomology 78, 547–555.

Kim, J., Kang, K., Gonzales-Virgil, E., Shi, F., Jones, A.D., Barry, C.S., Last, R.L., 2012. Striking natural diversity in glandular trichome acylsugar composition is shaped by variation at the acyltransferase2 locus in the wild tomato *Solanum habrochaites*. Plant Physiology 160, 1854–1870.

Kumar, N.K.K., Ullman, D.E., Cho, J.J., 1995. Resistance among *lycopersicon* species to *Frankliniella occidentalis* (Thysanoptera: Thripidae). Journal of Economic Entomology 88, 1057–1065.

Larbat, R., Olsen, K.M., Slimestad, R., Løvdal, T., Bénard, C., Verheul, M., Bourgaud, F., Robin, C., Lillo, C., 2012. Influence of repeated short-term nitrogen limitations on leaf phenolics metabolism in tomato. Phytochemistry 77, 119–128.

Leite, G.L.D., Costa, C.A., Almeida, C.I.M., Picanco, M., 2003. Efeito da adubação sobre a incidência de traça-do-tomateiro e alternaria em plantas de tomate. Horticultura Brasileira 21, 448–451.

Leite, G.L.D., Picanco, M., Azevedo, A.A., Silva, D.J.H., Gusmao, M.R., 1997. Intensidade de ataque de *Frankliniella schulzei* e *Myzus persicae* em três introduções de *Lycopersicon peruvianum*. Revista Universidade Rural: Série Ciências da Vida 19, 27–35.

Leite, G.L.D., Picanco, M., Bacci, L., Gonring, A.H.R., 1999d. Dose-response regression lines of tridecan-2-one for *Myzus persicae*. Agro-ciencia 15, 135–138.

Leite, G.L.D., Picanco, M., Della Lucia, T.M.C., Moreira, M.D., 1999a. Role of canopy height in the resistance of *Lycopersicon hirsutum f. glabratum* to *Tuta absoluta* (Lep., Gelechiidae). Journal of Applied Entomology 123, 459–463.

Leite, G.L.D., Picanco, M., Guedes, R.N.C., Skowronski, L., 1999b. Effect of fertilization levels, age and canopy height of *Lycopersicon hirsutum* on the resistance to *Myzus persicae*. Entomologia Experimentalis et Applicata 91, 267–273.

Leite, G.L.D., Picanco, M., Guedes, R.N.C., Zanuncio, J.C., 1999c. Influence of canopy height and fertilization levels on the resistance of *Lycopersicon hirsutum* to *Aculops lycopersici* (Acari: Eriophyidae). Experimental & Applied Acarology 23, 633–642.

Leite, G.L.D., Picanco, M., Guedes, R.N.C., Zanuncio, J.C., 2001. Role of plant age in the resistance of *Lycopersicon hirsutum f. glabratum* to the tomato leafminer *Tuta absoluta* (Lepidoptera: Gelechiidae). Scientia Horticulturae 89, 103–113.

Leite, G.L.D., Picanco, M., Zanuncio, J.C., Gonring, A.H.R., 1998. Effect of fertilization levels, age and canopy height of *Lycopersicon* spp. on attack rate of *Bemisia tabaci* (Homoptera: Aleyrodidae). Agronomia Lusitana 46, 53–60.

Lin, S.Y.H., Trumble, J.T., Kumamoto, J., 1987. Activity of volatile compounds in glandular trichomes of *Lycopersicon* species against two insect herbivores. Journal of Chemical Ecolology 13, 837–850.

López-Gresa, M.P., Maltese, F., Belles, J.M., Conejero, V., Kim, H.K., Choi, Y.H., Verpoorte, R., 2010. Metabolic response of tomato leaves upon different plant–pathogen interactions. Phytochemical Analysis 21, 89–94.

Lourenção, A.L., Nagai, H., Zullo, M.A.T., 1984. Fontes de resistência a *Scrobipalpula absoluta* (Meyrick, 1917) em tomateiro. Bragantia 43 (2), 569–577.

Lucatti, A.F., van Heusden, A.W., de Vos, R.C.H., Visser, R.G.F., Vosman, B., 2013. Differences in insect resistance between tomato species epidemic to the Galapagos Islands. Evolutionary Biology 13, 1–12.

Mahanil, S., Attajarusit, J., Stout, M.J., Thipyapong, P., 2008. Overexpression of tomato polyphenol oxidase increases resistance to common cutworm. Plant Science 174, 456–466.

Maluf, W.R., Maciel, G.M., Gomes, L.A.A., Cardoso, M.G., Gonçalves, L.D., da Silva, E.C., Knapp, M., 2010. Broad-spectrum arthropod resistance in hybrids between high- and low-acylsugar tomato lines. Crop Science Society of America 50, 439–450.

Martin, G.B., Brommonschenkel, S.H., Chunwongse, J., Frary, A., Ganal, M.W., Spivy, R., Wu, T., Earle, E.D., Tanksley, S.D., 1993. Map-based cloning of a protein kinase gene confer-ring disease resistance in tomato. Science 262, 1432–1436.

Martinez de Ilarduya, O., Moore, A.E., Kaloshian, I., 2001. The tomato Rme1 locus is required for Mi-1-mediated resistance to root-knot nematodes and the potato aphid. The Plant Journal 27, 417–425.

McGrath, M.T., Smart, C., 2014. Managing bacterial diseases of tomato in the field. Cornell University Vegetable MD Online, Department of Plant Pathology, Ithaca, New York, USA. http://vegetablemdonline.ppath.cornell.edu/NewsArticles/Tom_Bacter_06.html.

McGurl, B., Ryan, C.A., 1992. The organization of the prosystemin gene. Plant Molecular Biology 20, 405–409.

Mehlenbacher, S.A., 1995. Classical and molecular approaches to breeding fruit and nut crops for disease resistance. Hort Science 30, 466–477.

Melo, P.C.T., 1989. Melhoramento genético do tomate (*Lycopersicon esculentum* Mill.). Notas de aula – Curso intensivo de tomaticultura, p. 55 Campinas, SP, Brazil.

Menda, N., Strickler, S.R., Mueller, L.A., 2013. Advances in tomato research in the post-genome era. Plant Biotechnology 30, 243–256.

Miller, E.C., Hadely, C.W., Schwartz, S.J., Erdman, J.W., Boileau, T.M.W., Clinton, S.K., 2002. Lycopene, tomato products, and prostate cancer prevention. Have we established causality? Pure and Applied Chemistry 74, 1435–1441.

Miller, J.C., Tanksley, S.D., 1990. RFLP analysis of phylogenetic relationships and genetic variation in the genus *Lycopersicon*. Theoretical and Applied Genetics 80, 437–448.

Milligan, S., Bodeau, J., Yaghoobi, J., Kaloshian, I., Zabel, P., Williamson, V.M., 1998. The rootknot nematode resistance gene *Mi* from tomato is a member of the leucine zipper, nucleotide binding, leucine-rich repeat family of plant genes. The Plant Cell 10, 1307–1319.

Mintz-Oron, S., Mandel, T., Rogachev, I., Feldberg, L., Lotan, O., Yativ, M., Wang, Z., Jetter, R., Venger, I., Adato, A., Aharoni, A., 2008. Gene expression and metabolism in tomato fruit surface tissues. Plant Physiology 147, 823–851.

Monforte, A.J., Tanksley, S.D., 2000. Development of a set of near isogenic and backcross recombinant inbred lines containing most of the *Lycopersicon hirsutum* genome in a *L. esculentum* genetic background: a tool for gene mapping and gene discovery. Genome 43, 803–813.

Mouttet, R., Kaplan, I., Bearez, P., Amiens-Desneux, E., Desneux, N., 2013. Spatiotemporal patterns of induced resistance and susceptibility linking diverse plant parasites. Oecologia 173, 1379–1386.

Nichoul, P., 1994. Phenology of glandular trichomes related to entrapment of *Phytoseiulus persimilis* A. -H. in the glasshouse tomato. Journal of Horticultural Science 69, 783–789.

Nombel, A.G., Williamson, V.M., Muniz, M., 2003. The root-knot nematode resistance gene Mi-1.2 of tomato is responsible for resistance against the whitefly *Bemisia tabaci*. Molecular Plant-microbe Interactions 16, 645–649.

Nowicki, M., Foolad, M.R., Nowakowska, M., Kozik, E.U., 2012. Potato and tomato late blight caused by *Phytophthora infestans*: an overview of pathology and resistance breeding. Plant Disease 96, 4–17.

Ntoukakis, V., Saur, I.M.L., Conlan, B., Rathjen, J.P., 2014. The changing of the guard: the Pto/Prf receptor complex of tomato and pathogen recognition. Current Opinion in Plant Biology 20, 69–74.

Nuez, F., Prohens, J., Blanca, J., 2004. Relationships, origin, and diversity of Galapagos tomatoes: implications for the conservation of natural populations. American Journal of Botany 91, 86–99.

Obradovic, A., Jones, J.B., Tomol, M.T., Jackson, L.E., Balogh, B., Guven, K., Iriarte, F.B., 2005. Integration of biological control agents and systemic acquired resistance inducers against bacterial spot on tomato. Plant Disease 89, 712–716.

Oka, K., Okubo, A., Kodama, M., Otani, H., 2006. Detoxification of α-tomatine by tomato pathogens *Alternaria alternata* tomato pathotype and *Corynespora cassiicola* and its role in infection. Journal General Plant Pathology 72, 152–158.

Oliveira, F.A., Silva, D.J.H., Leite, G.L.D., Jham, G.N., Picanço, M., 2009. Resistance of 57 greenhouse-grown accessions of *Lycopersicon esculentum* and three cultivars to *Tuta absoluta* (Meyrick) (Lepidoptera: Gelechiidae). Scientia Horticulturae 119, 182–187.

Ori, N., Eshed, Y., Paran, I., Presting, G., Aviv, D., Tanksley, S., Zamir, D., Fluhr, R., 1997. The I2C family from the wilt disease resistance locus I2 belongs to the nucleotide binding, leucine-rich repeat superfamily of plant resistance genes. The Plant Cell 9, 521–532.

Orozco-Cardenas, M., McGurl, B., Ryan, C.A., 1993. Expression of an antisense prosystemin gene in tomato plants reduces resistance toward *Manduca sexta* larvae. Proceedings of the National Academy of Sciences 90, 8273–8276.

Osorio, S., Alba, R., Damasceno, C.M.B., Lopez-Casado, G., Lohse, M., Zanor, M.I., Tohge, T., Usadel, B., Rose, J.K.C., Fei, Z., Giovannoni, J.J., Fernie, A.R., 2011. Systems biology of tomato fruit development: combined transcript, protein, and metabolite analysis of tomato transcription factor (nor, rin) and ethylene receptor (Nr) mutants reveals novel regulatory interactions. Plant Physiology 157, 405–425.

Parniske, M., Hammond-Kosack, K.E., Golstein, C., Thomas, C.M., Jones, D.A., Harrison, K., Wulff, B.B., Jones, J.D., 1997. Novel disease resistance specificities result from sequence exchange between tandemly repeated genes at the Cf-4/9 locus of tomato. Cell 91, 821–832.

Picanço, M., Leite, G.L.D., Guedes, R.N.C., Silva, E.E.A., 1998. Yield loss in trellised tomato affected by insecticidal sprays and plant spacing. Crop Protection 17, 447–452.

Picanço, M.C., Bacci, L., Crespo, A.L.B., Miranda, M.M.M., Martins, J.C., 2007. Effect of integrated pest management practices on tomato production and conservation of natural enemies. Agricultural and Forest Entomology 9, 327–335.

Picó, B., Díez, M.J., Nuez, F., 1998. Evaluation of whitefly-mediated inoculation techniques to screen Lycopersicon esculentum and wild relatives for resistance to tomato yellow leaf curl virus. Euphytica 101, 259–271.

Pink, D.A.C., 2002. Strategies using genes for non-durable resistance. Euphytica 1, 227–236.

Proffit, M., Birgersson, G., Bengtsson, M., Reis Jr., R., Witzgall, P., Lima, E., 2011. Attraction and oviposition of *Tuta absoluta* females in response to tomato leaf volatiles. Journal of Chemical Ecology 37, 565–574.

Rahimi, F.R., Carter, C.D., 1993. Inheritance of zingiberene in *lycopersicon*. Theoretical and Applied Genetics 87, 593–597.

Reis, A., Lopes, C.A., 2006. Tomate em chamas. Revista Cultivar Hortaliças e Frutas. Pelotas 2, 6–8.

Reymond, P., Bodenhausen, N., Van Poecke, R.M.P., Krishnamurthy, V., Dicke, M., Farmer, E.E., 2004. A conserved transcript pattern in response to a specialist and a generalist herbivore. The Plant Cell 16, 3132–3147.

Rick, C.M., 1976. Natural variability in wild species of *Lycopersicon* and its bearing on tomato breeding. Genetica Agraria 30, 249–259.

Riddick, E.W., Simmons, A.M., 2014. Do plant trichomes cause more harm than good to predatory insects. Pest Management Science 70, 1655–1665.

Robertson, L.D., Labate, J.A., 2007. Genetic resources of tomato (*Lycopersicon esculentum*) and wild relatives. In: Razdan, M.K., Mattoo, A.K. (Eds.), Genetic Improvement of Solanaceous Crops. Science Publishers, Enfield, New Hampshire, USA, pp. 25–75.

Rose, R.L., Gould, F., Levi, P.E., 1991. Differences in cytochrome P 450 activities in tobacco budworm larvae as influenced by resistance to host plant allelochemicals an induction. Comparative Biochemistry and Physiology 99, 535–540.

Ross, A.F., 1961a. Localized acquired resistance to plant virus infection in hypersensitive hosts. Virology 14, 329–339.

Ross, A.F., 1961b. Systemic acquired resistance induced by localized virus infections in plants. Virology 14, 340–358.

Rossi, M., Goggin, F.L., Milligan, S.B., Kaloshian, I., Ullman, D.E., Williamson, V.M., 1998. The nematode resistance gene Mi of tomato confers resistance against the potato aphid. Proceedings of the National Academy of Sciences 95, 9750–9754.

Royer, M., Larbat, R., Le Bot, J., Adamowicz, S., Robin, C., 2013. Is the C: N ratio a reliable indicator of C allocation to primary and defence-related metabolisms in tomato? Phytochemistry 88, 25–33.

Scheer, J.M., Ryan, C.A., 2002. The systemin receptor SR160 from *Lycopersicon peruvianum* is a member of the LRR receptor kinase family. Proceedings of the National Academy of Sciences 99, 9585–9590.

Schilmiller, A.L., Howe, G.A., 2005. Systemic signaling in the wound response. Current Opinion in Plant Biology 8, 369–377.

Schornack, S., Ballvora, A., Gurlebeck, D., Peart, J., Baulcombe, D., Ganal, M., Baker, B., Bonas, U., Lahaye, T., 2004. The tomato resistance protein Bs4 is a predicted non-nuclear TIR-NB-LRR protein that mediates defense responses to severely truncated derivatives of AvrBs4 and overexpressed AvrBs3. The Plant Journal 37, 46–60.

Scott, J., Gardner, R., 2007. Breeding for resistance to fungal pathogens. In: Razdan, M.K., Mattoo, A.K. (Eds.). Razdan, M.K., Mattoo, A.K. (Eds.), Genetic Improvement of Solanaceous Crops Tomato, vol. 2. Science Publishers, Edenbridge Limited, British Isles, Enfield, New Hampshire, USA, pp. 421–456.

Silva, C.C.da., Jham, G.N., Picanco, M., Leite, G.L.D., 1998. Comparison of leaf chemical composition and attack patterns of *Tuta absoluta* (Meyrick) (Lepidoptera: Gelechiidae) in three tomato species. Agronomia Lusitana 46, 61–71.

Simmons, A.T., McGrath, D., Gurr, G.M., 2005. Trichome characteristics of F1 *Lycopersicon esculentum* X *L. cheesmanii* f. *minor* and *L. esculentum* X *L.pennellii* and effects on *Myzus persicae*. Euphytica 144, 313–320.

Smith, C.M., 2005. Plant Resistance to Arthropods: Molecular and Conventional Approaches. Springer International Publishing, Dordrecht, The Netherlands, p. 423.

Snyder, J.C., Guo, Z., Thacker, R., Goodman, J.P., Pyrek, J.S.T., 1993. 2,3-dihydrofarnesoic acid, a unique terpene from trichomes of *Lycopersicon hirsutum*, repels spider mites. Journal of Chemical Ecology 19, 2981–2997.

Sorenson, C.E., Fery, R.L., Kennedy, G.G., 1989. Relationship between Colorado potato beetle (Coleoptera: Chrysomelidae) and tobacco hornworm (Lepidoptera: Sphingidae) resistance in *Lycopersicon hirsutum* f. *glabratum*. Journal of Economic Entomology 82, 1743–1748.

Stevens, M.A., Rick, C.M., 1986. Genetics and breeding. In: Atherton, J.G., Rudich, J. (Eds.), The Tomato Crop: A Scientific Basis for Improvement. Chapman and Hall, London, UK, pp. 35–109.

Stout, M.J., Brovont, R.A., Duffey, S.S., 1998a. Effect of nitrogen availability on expression of constitutive and inducible chemical defenses in tomato *Lycopersicon esculentum*. Journal of Chemical Ecology 24, 945–963.

Stout, M.J., Thaler, J.S., Thomma, B.P.H.J., 2006. Plant-mediated interactions between pathogenic microorganisms and arthropod herbivores. Annual Review of Entomology 51, 663–689.

Stout, M.J., Workman, J.S., Workman, K.V., Duffey, S.S., 1996. Temporal and ontogenetic aspects of protein induction in tomato foliage. Biochemical Ecology and Systematics 24, 611–625.

Stout, M.J., Workman, K.V., Bostock, R.M., Duffey, S.S., 1998b. Specificity of induced resistance in foliage of the tomato. Oecologia 113, 74–81.

Suinaga, F.A., Casali, V.W.D., Picanço, M.C., Silva, D.J.H., 2004b. Capacidade combinatória de sete caracteres de resistência de *Lycopersicon* spp. à traça do tomateiro. Horticultura Brasileira 22, 242–248.

Suinaga, F.A., Picanço, M.C., Jham, G.N., Brommonschenkel, S.H., 1999. Causas químicas de resistência de *Lycopersicon peruvianum* (L.) a *Tuta absoluta* (Meyrick) (Lepidoptera: Gelechiidae). Anais da Sociedade Entomológica do Brasil 28, 313–321.

Suinaga, F.A., Picanço, M.C., Moreira, M.D., Semeão, A.A., Magalhães, S.T.V., 2004a. Resistência por antibiose de *Lycopersicon peruvianum* à traça do tomateiro. Horticultura Brasileira 22, 281–285.

Taiz, L., Zeiger, E., 1991. Plant Physiology. The Benjamin/Cummings Publishing Company, Redwood City, California, USA, p. 591.

Talekar, N.S., Opena, R.T., Hanson, P., 2006. *Helicoverpa armigera* management: a review of AVRDC's research on host plant resistance in tomato. Crop Protection 25, 461–467.

Tang, X., Frederick, R.D., Zhou, J., Halterman, D.A., Jia, Y., Martin, G.B., 1996. Initiation of plant disease resistance by physical interaction of AvrPto and Pto kinase. Science 274, 2060–2063.

Tanksley, S., McCouch, S., 1997. Seed banks and molecular maps: unlocking genetic potential from the wild. Science 277, 1063–1066.

Terrell, E.E., Broome, C.R., Reveal, J.L., 1983. Proposal to conserve the name of the tomato as *Lycopersicon esculentum* P. Miller and reject the combination *Lycopersicon lycopersicum* (L.) Karsten (Solanaceae). Taxon 32, 310–314.

Thaler, J.S., 1999. Jasmonate-inducible plant defences cause increased parasitism of herbivores. Nature 399, 686–688.

Thaler, J.S., Agrawal, A.A., Halitschke, R., 2010. Salicylate-mediated interactions between pathogens and herbivores. Ecology 91, 1075–1082.

Thaler, J.S., Stout, M.J., Karban, R., Duffey, S.S., 2001. Jasmonate-mediated induced plant resistance affects a community of herbivores. Ecological Entomology 26, 312–324.

Thomazini, A.P.B.W., Vendramim, J.D., Lopes, M.T.R., 2000. Extratos aquosos de Trichilia pallida e a traça-do-tomateiro. Scientia Agricola 57, 13–17.

Tian, D., Peiffer, M., De Moraes, C.M., Felton, G.W., 2014. Roles of ethylene and jasmonic acid in systemic induced defense in tomato (*Solanum lycopersicum*) against *Helicoverpa zea*. Planta 239, 577–589.

Tripodi, P., Di Dato, F., Maurer, S., Seekh, S.M.B., Van Haaren, M., Frusciante, L., Mahammad, A., Tanksley, S., Zamir, D., Gebhardt, C., Grandillo, S., 2010. A genetic platform of tomato multi-species introgression lines: present and future. In: The 7th Solanaceae Conference, p. 176 Dundee, Scotland.

Trudgill, D.L., Blok, V.C., 2001. Apomictic, polyphagous root-knot nematodes: exceptionally successful and damaging biotrophic root pathogens. Annual Review of Phytopathology 39, 53–77.

Van Ooijen, G., van den Burg, H.A., Cornelissen, B.J., Takken, F.L., 2007. Structure and function of resistance proteins in solanaceous plants. Annual Review of Phytopathology 45, 43–72.

Weston, P.A., Johnson, D.A., Burton, H.T., Snyder, J.C., 1989. Trichome secretion composition, trichome densities, and spider mite resistance of ten accessions of *Lycopersicon hirsutum*. Journal of the American Society for Horticultural Science 114, 492–498.

Wilkens, R.T., Spoerke, J.M., Stamp, N.E., 1996. Differential responses of growth and two soluble phenolics of tomato to resource availability. Ecology 77, 247–258.

Williamson, V.M., 1998. Root-knot nematode resistance genes in tomato and their potential for future use. Annual Review of Phytopathology 36, 277–293.

Worrall, D., Geoff, H.H., Moore, J.P., Glowacz, M., Croft, P., Taylor, J.E., Paul, N.D., Roberts, M.R., 2012. Treating seeds with activators of plant defence generates long-lasting priming of resistance to pests and pathogens. New Phytologist 193, 770–778.

Yabuuchi, E., Kosako, Y., Yano, I., Hotta, H., Nishiuchi, Y., November 1995. Transfer of two *Burkholderia* and an *Alcaligenes* species to *Ralstonia* gen. Microbiology and Immunology 39, 897–904.

Yan, S., Dong, X., 2014. Perception of the plant immune signal salicylic acid. Current Opinion in Plant Biology 20, 64–68.

Yu, G., Nguyen, T.T.H., Guo, Y., Schauvinhold, I., Auldridge, M.E., Bhuiyan, N., Ben-Israel, I., Iijima, Y., Fridman, E., Noel, J.P., Pichersky, E., 2010. Enzymatic functions of wild tomato methylketone synthases 1 and 2. Plant Physiology 154, 67–77.

Yu, Z.H., Wang, J.F., Stall, R.E., Vallejos, C.C., 1995. Genomic localization of tomato genes that control a hypersensitive reaction to *Xanthomonas campestris* pv. *vesicatoria* (Doidge) dye. Genetics 141, 675–682.

Zambolim, L., Vale, F.X.R., Costa, H. (Eds.), 2000. Controle de doenças de plantas hortaliças. Viçosa UFV Editora, p. 879.

Chapter 10

Engineering Insect Resistance in Tomato by Transgenic Approaches

Manchikatla V. Rajam, Sneha Yogindran
University of Delhi South Campus, New Delhi, India

1. INTRODUCTION

Tomato is a major vegetable crop grown in nearly every country of the world. India ranks third in tomato production after China and the United States. Tomatoes are consumed in various ways in many dishes. Aside from being tasty, tomatoes are very healthy as they are a good source of a very powerful antioxidant lycopene, which can prevent the development of many forms of cancer. In addition to lycopene, tomato is a rich source of vitamin A and C and other important nutrients. However, tomatoes can be affected by biotic stresses such as pests and pathogens, which cause massive yield losses. Among the biotic stresses, insect pests are a major threat to yield and if not controlled properly, may cause substantial losses. Hence, insecticides are used extensively for insect control, which has resulted in increased crop production. Yet insect pests have remained a problem when they develop insecticide resistance by means of detoxification mechanisms. Conventional breeding is one of the effective ways to develop resistant plants, although this process is time-consuming and laborious. Therefore, genetic engineering of crop plants for resistance against insect pests is an important goal of agricultural biotechnologists. In the last three decades, genetic engineering has been utilized to generate insect-resistant crop plants, which has led not only to a significant reduction in pesticide use, but also reduced the development of insecticide resistance (Christou et al., 2006). Recently, RNA interference (RNAi) has proven to be a potential alternative tool for raising insect-resistant plants. In this chapter, we review the use of transgenic technology, as well as RNAi, for insect resistance in crop plants, with special emphasis on tomato.

2. CONVENTIONAL METHODS FOR INSECT CONTROL

Many conventional methods have been used to control the devastating effects caused by insects. While these methods have been covered in other chapters of this book, we categorize them under chemical, biological, mechanical, and physical controls. Chemicals are the most widely used method, and the broad nature of insecticides leads to killing pests, inhibiting their feeding, mating, or other essential behaviors required for their development. Commercially available insecticides have been shown to be very effective. Synthetic sex pheromones which affect the mating in insects can also work, especially in large plantings. However, chemical control has many disadvantages. They can affect non-target organisms, including humans, particularly pesticide applicators and other farm workers (Mazid et al., 2011). This has raised the need for alternative control measures. Biological control agents are also widely used for insect control in tomatoes. This includes the use of predators, parasitoids, and microorganisms. Predators directly consume their prey during their whole lifetime, and the larvae of ladybugs are voracious predators of aphids. Parasitoids, on the other hand, lay their eggs on or in the insect host, ultimately killing the pest insect. Wasps are the major insect parasitoids. Pathogenic microorganisms such as bacteria, fungi, and viruses are also used to kill the host insects (Lord, 2005; Van Lenteren, 2012). Biological control is not meant to completely eradicate a pest, but can complement transgenic and other non-conventional approaches. Mechanical control measures can be effective and rapid, but are typically suitable only for pest problems on a small geographic scale due to their labor requirements. For example, handpicking can be used to remove bright colored and/or large insects in small garden plots. Similarly, shaking plants and using a strong spray of water can dislodge some pests from the plants. Physical control measures are also used to keep insect pests from reaching their hosts. These include the use of barriers (window screens on greenhouses) and pheromone traps to draw insects away from the target host plants.

Sustainable Management of Arthropod Pests of Tomato. http://dx.doi.org/10.1016/B978-0-12-802441-6.00010-3

3. NON-CONVENTIONAL APPROACHES FOR INSECT CONTROL: TRANSGENIC STRATEGIES

3.1 *Bacillus thuringiensis* (*Bt*) Toxins

Bacillus thuringiensis Berliner *(Bt)* is a Gram-positive bacterium that produces proteins known as insecticidal crystal proteins (ICPs). These proteins are toxic to a narrow group of insects. The spores produced by these bacteria contain one or more Cry and/or Cyt proteins called δ-endotoxins. These Cry toxins interact with specific receptors on the insect's midgut epithelial cell surface. The host proteases activate these toxins and lead to the formation of an oligomeric pore–like structure that enters into the host membrane. This pore formation leads to ionic leakage in the cells, which kills the insect (Bravo et al., 2007; Crickmore et al., 2010). Hence, *Bt* has turned out to be a novel and potent method for the control of insect pests. *Bt* topical pesticides have been developed and are in extensive use to protect crop plants from insect pests. These *Bt* toxins are non-toxic to vertebrates, and have additional advantages of being safe (eco-friendly), specific, and biodegradable. Usually *Bt* is applied when the insects are in early growth stages as later-stage larvae are more tolerant. Additionally, UV light, weather, and the presence of certain proteases can lead to the degradation of *Bt* toxins. Therefore, topical *Bt* sprays generally have to be sprayed several times for effective control, increasing the cost and the amount of product required (Ranjekar et al., 2003).

These issues related to topical *Bt* application were addressed by introducing *Bt* Cry genes into tobacco and tomato plants (Vaeck et al., 1987; Krattiger, 1996). Tomato plants were transformed with chimeric genes harboring truncated lepidopteran-type *Bt* genes. Tomato plants expressing chimeric *Bt* genes conferred insect resistance to the plants and their subsequent progeny. Although the mRNA accumulation of chimeric transcript was found to be low, the level of protein expressed was enough to kill the larvae of *Manduca sexta* (L.), *Heliothis virescens* (F.), and *Helicoverpa zea* (Boddie) (Fischhoff et al., 1987). Truncated Cry1Aa and Cry1Ab toxin genes have provided resistance to tobacco and tomato against tobacco hornworm larvae (Jouanin et al., 1998).

The expression level of unmodified Cry genes in transgenic plants was low because of the codon bias between pro-karyotic and eukaryotic systems. Therefore, modified Cry genes (Cry1Ab and Cry1Ac) have been introduced for better expression in plant cells (Perlak et al., 1991). Modifications in codon usage, poly A–type signals, and splice sites of ICPs helped in providing resistance against lepidopteran pests of cotton (Cry1Ac) and coleopteran pests of potato (Cry1Aa) (De Maagd et al., 1999). Subsequently, various modified Cry genes were used to transform many crop plants including rice, maize, peanut, soybean, canola, tomato, and cabbage (Sanahuja et al., 2011; Tabashnik et al., 2011). Taking into account the development of tomato transgenics using *Bt* genes, there are many studies which suggest the success of this technology. Transgenic tomato expressing Cry1Ac has shown high levels of protection against larvae of the tomato fruit borer (*Helicoverpa armigera* (Hübner)) (Mandaokar et al., 2000). In another study carried out in Egypt, overexpression of *Bt* (Cry2Ab) gene in transgenic tomato has resulted in the mortality of *H. armigera* and the potato tuber moth *Phthorimaea operculella* (Zeller) when fed on *Bt* tomato (Saker et al., 2011). Recently, a modified truncated *Bt*-CryAb gene was used to produce resistant tomato plants. The transgenic line Ab25E has been selected for further analysis, and showed 100% mortality of second instar *H. armigera* and *Spodoptera litura* (F.) with minimum damage to leaves and fruits (Koul et al., 2014).

Although most pests have been susceptible to *Bt* transgenic plants, resistance was reported as early as 2005 (Tabashnik et al., 2013). For example, the cotton bollworm, *H. zea*, and pink bollworm *Pectinophora gossypiella* (Saunders) showed resistance against Cry1Ac in cotton (Moar et al., 2008; Tabashnik et al., 2008, 2013). Also, in Puerto Rico, fall armyworm, *Spodoptera frugiperda* (Smith) has shown resistance to Cry1F in maize leading to its discontinuation (Matten et al., 2008). To combat resistance, transgenic plants expressing multiple *Bt* toxins have been developed. Transgenic cotton expressing Cry1Ac and Cry2Ab conferred resistance to *H. zea*, *S. frugiperda*, and *Spodoptera exigua* (Hübner) (Zhao et al., 2003).

3.2 Proteinase Inhibitors (PIs)

A wide range of defense proteins are expressed in plants. These include proteinase inhibitors (PIs) and lectins which are induced as a response to insect attack (Ryan, 1990). PIs are ubiquitous and hence they are found in microbes, plants, and animals (Laskowski and Kato, 1980). They regulate the activity of their corresponding proteases and are important for many biological processes. Seeds and tubers have high PI levels suggesting their role as a depot of safe storage forms immune to digestion unless required (Ryan, 1988). PIs were first identified as plant defense proteins in 1972 when they were induced due to wounding and insect herbivory. As a wounding response brought about by insect attack, one of the four classes of proteinases—serine, cytosine, aspartic, and metalloproteinases—is accumulated in plants

(Ryan, 1990). In insects, proteinases release amino acids from dietary proteins and provide required nutrients for larval growth and development. Insect proteinases are inhibited by PIs affecting the digestion which leads to physiological stress on insects as they have to synthesize alternative proteases resulting in growth retardation (Jongsma and Bolter, 1997). Molting, water balance, and enzyme regulation in insects are affected by PIs. Due to their significant inhibitory activity, PIs have been exploited for crop improvement. Cowpea Trypsin inhibitor (CpTi) expression in tobacco showed resistance against *M. sexta* (Hilder et al., 1987). Under laboratory conditions, transgenic tobacco plants expressing CpTi also showed resistance to *S. litura* (Sane et al., 1997). Expression of tomato inhibitor II and potato inhibitor II genes in tobacco resulted in retarded growth of *M. sexta* larvae (Johnson et al., 1989). Different levels of Mustard trypsin inhibitor (MTI-2) expression in transgenic tobacco, *Arabidopsis*, and oilseed rape showed developmental delay and mortality in three lepidopteran pests, *Plutella xylostella* (L.), *Mamestra brassicae* (L.), and *S. litura* (De Leo and Gallerani, 2002). Maize PI gene (MPI) when expressed in rice showed enhanced resistance against striped stem borers (*Chilo suppressalis* (Walker)) (Vila et al., 2005).

The above results clearly suggest that plants PIs are promising antimetabolites that can confer resistance against insect pests. However, insects may change the composition of proteinases to overcome the PI expressed in plants (Jongsma et al., 1995; Broadway, 1995). Studies have revealed that high-level expression of soybean trypsin inhibitor gene in transgenic tobacco failed to confer resistance against *H. armigera*. A number of phytophagous insects have adapted to PIs of their host plants. Their growth and development is not affected by the presence of host PIs in their diet (Jongsma et al., 1995). A possible solution to this problem is to express a combination of PIs for effective resistance. Combined expression of two protease inhibitors has been useful in overcoming the compensatory response shown by insects where there is production of proteases insensitive to these inhibitors. Overexpression of potato PI II and carboxypeptidase inhibitors (PCI) under the control of leaf-specific promoter in tomato resulted in increased resistance to *Heliothis obsoleta* Auctorum and *Liriomyza trifolii* (Burgess) larvae (Abdeen et al., 2005).

Sporamin (trypsin inhibitor) from sweetpotato and CeCPI (phytocystatin) from taro when stacked and transferred to tobacco plants showed delayed growth and development of larvae of *H. armigera* (Senthilkumar et al., 2010). *Nicotiana alata* Link & Otto proteinase inhibitor (NaPI) is found in high levels in female reproductive organs of *N. alata*, an ornamental tobacco. This inhibitor targets the digestive enzymes, trypsin and chymotrypsin, present in the insect midgut. NaPI is a main potato type-II inhibitor (pin II) and has been shown to reduce the growth and development of *Helicoverpa punctigera* Wallengren. But the surviving larvae showed a high level of chymotrypsin activity. To overcome this, *Solanum tuberosum* L. potato type-I inhibitor (StPin1A), having a strong inhibitory activity against NaPI-resistant chymotrypsin activity, was expressed in combination of NaPI. The combination of NaPI and StPin1A in diet and transgenic plants helped in achieving better crop protection as compared to the effect of single inhibitor (Dunse et al., 2010).

A recent study also showed that transgenic rice plants expressing the fused maize proteinase inhibitor (MPI) and the potato carboxypeptidase inhibitor (PCI) into a single ORF (open reading frame) showed resistance to *C. suppressalis*. The MPI-PCI fusion gene was found to be stable for at least three generations without affecting the plant phenotype. Reduction in larval weight of *C. suppressalis* was observed when fed on MPI-PCI expressing rice as compared to control (Quilis et al., 2013).

3.3 α-Amylase Inhibitors (α-AI)

A growing body of research has suggested the use of α-AIs as biotechnological tools for generating transgenic plants with enhanced resistance toward pests. The introduction and expression of bean α-AI gene in pea conferred resistance to bruchid beetles (Shade et al., 1994). The resulting seeds were resistant to larvae of bruchid beetles, *Callosobruchus maculatus* (F.) and pea weevil *Bruchus pisorum* (L.) (Schroeder et al., 1995; Morton et al., 2000). Transgenic pea seeds expressing α-AI under the control of phytohemagglutinin promoter showed high resistance toward pea weevil. The development of weevil was blocked at an early stage even in the T_5 generation. There was minimal damage to the seeds, yet the seed yield was not affected (Schroeder et al., 1995). Extraction of proteinaceous inhibitors from barley and wheat showed its significant inhibitory action on amylase activity of the Sunn pest (*Eurygaster integriceps* Puton) which belongs to a group of the "shield bug" (Scutelleridae) and causes considerable yield losses (Bandani, 2005).

Recently, transgenic tobacco seeds (*Nicotiana tabacum* L.) expressing rye α-AI showed increased mortality in *Anthonomus grandis* Boheman larvae when fed with artificial diet mixed with transgenic seed flour (Dias et al., 2010). Apart from the use in legumes, α-AIs also are used to transform crop plants. Genetic engineering of *Coffea arabica* L. with α-AI1 showed the presence of the inhibitor in seed extracts and also the activity of digestive enzymes of *Hypothenemus hampei* (Ferrari) was significantly affected (Barbosa et al., 2010).

The members of this family have α-amylase inhibitory (α-AIs) activity (Gourinath et al., 2000). α-amylase inhibitors also affect the nutrient utilization by insects and hence function in a similar manner as proteinase inhibitors. α-amylase inhibitors are classified into six different groups based on the structural similarity. They are as follows:

3.3.1 Legume Lectin-Like Inhibitors

These were identified in common beans such as white, red, and black kidney beans. These inhibitors target insect, fungal, and mammalian α-amylases. α-AI1 and α-AI2 belong to this class and contain 240–250 amino acid residues (Marshall and Lauda, 1975; Ho and Whitaker, 1993). Mutation in the primary structures leads to their different specificity toward α-amylases (Grossi de Sa et al., 1997).

3.3.2 Knottin-Like Inhibitors

These are present in the seeds of Amaranth (*Amaranthus hypochondriacus* L.). These inhibitors show strong inhibition of insect α-amylases but do not affect proteases and α-amylases from mammals. They are also known as AAI (α-amylase inhibitor) and are 32 amino acids long (Chagolla-Lopez et al., 1994). AAI is the smallest inhibitor known to date. The three-dimensional structure of AAI shows three antiparallel β strands and is rich in disulfide bonds (Mehrabadi et al., 2012).

3.3.3 Kunitz-Type Inhibitors

This type of α-amylase is found in cereals such as barley, rice, wheat, maize, and cowpea (Micheelsen et al., 2008). They inhibit the α-amylases of insects as well as plants (Alves et al., 2009). BASI (α-amylase/subtilisin inhibitors) have been studied extensively (Mehrabadi et al., 2012) and they are known to have 176–181 amino acid residues (Mundy et al., 1983, 1984). The structure of BASI revealed the presence of two disulfide bonds and 12-stranded β-barrel structure belonging to the family of β-trefoil-fold (Mehrabadi et al., 2012).

3.3.4 γ-purothionin

Isolated from *Sorghum bicolor* (L.) Moench, these inhibitors show strong inhibition against insect and mammalian α-amylases (Bloch and Richardson, 1991). SIα-1, SIα-2, and SIα-3 are the isoinhibitors and comprise 47–48 amino acid residues. These isoforms have eight cysteine residues which forms disulfide bonds (Nitti et al., 1995).

3.3.5 Thaumatin-Like Inhibitors

This class of inhibitors was isolated from maize (*Zea mays* L.) and the proteins are homologous to sweet protein thaumatin present in the fruits of *Thaumatococcus daniellii* (Benn.) Benth., giving them their name (Cornelissen et al., 1986; Hejgaard et al., 1991; Vigers et al., 1991). Zeamatin from maize is a well-studied member of this family and only inhibits insect α-amylase. This has 13 β-strands out of which 11 form a β sandwich at the core of protein (Batalia et al., 1996).

3.3.6 CM Proteins

CM (Chloroform-methanol) proteins were discovered from cereals, and they have 120–160 amino acid residues (Campos and Richardson, 1983). The 0.19 α-AI is the most studied inhibitor of this family and targets α-amylases from insects, bacteria, and mammals (Mehrabadi et al., 2012). Composed of 124 amino acid residues, they act as homodimer (Oda et al., 1997; Franco et al., 2000). These proteins have a typical α-amylase/trypsin double-headed domain, and hence show inhibitory activity against α-amylases (Barber et al., 1986) and trypsin-like enzymes (De Leo and Gallerani, 2002) separately or α-amylase/trypsin-like inhibition at the same time (García-Maroto et al., 1991). X-ray crystallographic analysis showed four major α-helics, a one-turn helix and two short antiparallel β-strands.

Despite these successful reports, α-AI-transgenic crops have not been utilized commercially because of the induction of systemic immunological responses in mice fed with peas expressing the α-AI protein. This is due to the altered post-transcriptional processing of the transgene in the heterologous pea system as bean is the source of the transgene (Prescott et al., 2005).

A recent report showed that peas, chickpeas, and cowpeas expressing bean-αAI as well as non-transgenic beans were all allergenic in BALB/c mice (Lee et al., 2013). An anti-αAI response was seen in mice after consumption of non-transgenic peas lacking αAI due to a cross-reactive response to pea lectin. They also showed that the allergic response of αAI transgenic peas was par with the non-transgenic peas. This has provided an insight into the need of in-depth experiments for such unexpected cross-reactive allergic responses upon consumption of plant products in mice (Lee et al., 2013).

3.4 Lectins

Lectins were discovered by Stillmark in the year 1888 as a compound with hemagglutinating property. Stillmark (1888) extracted this compound from castor beans (*Ricinus communis* L.) and showed its ability to agglutinate red blood cells; it was named ricin. Later on, the first lectin was isolated from *Canavalia ensiformis* (L.) DC. (Jack beans) (Sumner, 1919). For a long period, lectins were called hemagglutinins as they agglutinate erythrocytes and other cells. But, it was found later that all lectins did not show the agglutinating activity. Hence, lectins are presently defined as carbohydrate-binding proteins that bind, with high affinity, to glycoproteins, glycolipids, or polysaccharides through their glycan subunit (Goldstein and Hayes, 1978). Lectins are no longer classified on the basis of agglutination.

Plant lectins are divided into four major groups: merolectins, hololectins, superlectins, and chimerolectins (Van Damme et al., 1998). Merolectins comprise proteins with only one carbohydrate-binding domain. This group does not show the activity of agglutination. Hololectins are proteins having two or more similar carbohydrate-binding domains and hence agglutinate cells. Superlectins, on the other hand, comprise two carbohydrate-binding domains that recognize different carbohydrate structures. Chimerolectins have proteins with one or more carbohydrate-binding domain fused to another domain showing biological activity other than carbohydrate-binding domain. This group of lectins is found abundantly in plants (Van Damme et al., 2008).

3.4.1 Constitutively Expressed and Inducible Lectins

Lectins are distributed extensively in nature and have been isolated from various plant species (Sharon and Lis, 1990; Rüdiger and Gabius, 2001). They are abundantly found in seeds or storage tissues (approximately 0.1–1.0% of the total proteins), hence are considered as storage proteins for growth and development (Van Damme et al., 1998). Constitutively expressed lectins, also known as "classical lectins," accumulate either in vacuoles or in the cell wall and intercellular spaces, or are secreted into the extracellular compartment (Van Damme et al., 2008). Lectins expressing at high concentrations are considered to be involved in plant defense (Etzler, 1986; Guo et al., 2013; Al Atalah et al., 2014; Roy et al., 2014).

Apart from constitutively expressed lectins, there are lectins which are induced and are present in low concentrations (Michiels et al., 2010). These are termed "inducible lectins" and are found in the nucleus and cytoplasm of the plant cell (Van Damme et al., 2003; Lannoo and van Damme, 2010). *Oryza sativa* L. agglutinin (Orysata) was the first inducible lectin to be characterized in rice seedlings (Zhang et al., 2000). Transgenic tobacco expressing orysata showed strong insecticidal activity against *S. exigua*, *Myzus persicae* (Sulzer), and *Acyrthosiphon pisum* Harris (Al Atalah et al., 2014).

Lectin purified from snowdrop bulbs (*Galanthus nivalis* L.) has been well studied. The *G. navalis* agglutinin, or GNA, is a mannose-specific lectin, having four identical 12 kDa subunits (Van Damme et al., 1987). GNA has been reported to show insecticidal activity against the rice brown planthopper (*Nilaparvata lugens* (Stål)) and rice green leafhopper (*Nephotettix cincticeps* Uhler) (Powell et al., 1993). Transgenic tobacco plants expressing GNA under the control of the constitutive promoter, CaMV35S, showed resistance against another family of heteropterans represented by the peach potato aphid (*M. persicae*) (Hilder et al., 1995). Similarly, transgenic potato plants expressing GNA under CaMV35S promoter showed effective control of the tomato moth, *Lacanobia oleracea* (L.), larvae. Larvae caused less damage to the leaves as compared to the control and there was a significant reduction in the insect biomass of moths that fed on GNA plants (Gatehouse et al., 1997). Partial resistance to rice brown planthopper (*N. lugens*) and another hemipteran pest (*Nephotettix virescens* (Distant)) was also obtained by expressing GNA in transgenic rice under tissue-specific (phloem and epidermal layer) or constitutive promoters (Rao et al., 1998; Foissac et al., 2000). Insects fed with transgenic rice showed reduced fecundity and survival rate. Expression of GNA in transgenic potato resulted in significant decrease in the survival and fecundity of the glasshouse potato aphid (*Aulacorthum solani* (Kaltenbach)) (Down et al., 1996). GNA-expressing transgenic plants had reduced damage when fed upon by *Sogatella furcifera* (Horváth) (white backed planthopper), and there was a negative impact on insect survival and fecundity. GNA expression was found to be 0.3% of the total soluble protein, and there was a 90% reduction of nymph survival on the transgenic plants compared to wild-type plants (Nagadhara et al., 2004). Transgenic maize expressing GNA under phloem-specific promoters showed resistance to corn leaf aphids (*Rhopalosiphum maidis* (Fitch)) under greenhouse conditions. The nymph production was considerably reduced by 46.9% on transgenic maize as compared to control (Wang et al., 2005).

Melander et al. (2003) expressed pea lectin in oilseed rape and obtained a significant reduction in pollen beetle (*Meligethes aeneus* (F.)) larval weight and also in larval survival. Mannose-specific lectin from garlic (*Allium sativum* L.) when expressed in transgenic rice resulted in partial resistance to hemipterans, *N. lugens* (brown planthopper), *Nephotettix malayanus* Ishihara & Kawase, and *N. virescens* (green leafhopper), and also showed reduced transmission of rice turgo virus by its insect vector (*N. malayanus* and *N. virescens*) (Saha et al., 2006a,b). *N. tabacum* agglutinin (NICTABA) gene was expressed in *Nicotiana attenuata* Torr. ex S. Watson, which does not carry the gene coding for NICTABA. *Spodoptera littoralis* Boisduval larvae when fed with NICTABA-expressing plants showed up to 70% reduction in weight. Bioassays with *M. sexta* larvae also showed decreased larval weight when fed on plants expressing NICTABA (Vandenborre et al., 2010).

Recently, orysata, a lectin from rice with mannose-specificity, was overexpressed in transgenic tobacco. Detached leaves from the transgenic tobacco lines showed significant mortality, larval weight reduction, and retardation of development in larval stages of *S. exigua* (Al Atalah et al., 2014). Transgenic tobacco plants expressing pea lectin (P-Lec) and the cowpea trypsin inhibitor (CpTI) also were generated to study synergic effects. Higher mortality of tobacco budworm (*H. virescens*) was observed in transgenic plants expressing both proteins in comparison with plant expressing only P-Lec or CpTI (Boulter et al., 1990).

GNA also has been fused with other proteins and evaluated for insect control. When GNA was fused with a spider venom peptide, δ-amaurobitoxin-Pl1a was formed. This was fed to third instar larvae of *M. brassicae*, and 100% larval mortality was observed within 6 days. Pl1a/GNA also caused mortality of the housefly (*Musca domestica* L.). *A. pisum* when fed on Pl1a/GNA-containing diet (1 mg/mL) showed 100% mortality after 3 days of feeding (Yang et al., 2014). This study suggested that using GNA as a carrier with other toxins may prove useful for generating transgenics which may show increased resistance toward insect pests. Similarly, Tajne et al. (2014) came up with a new synthetic gene having *Bt* toxin Cry1Ac and *A. sativum* agglutinin (ASAL). The gene showed significant insecticidal activity for *H. armigera* and *P. gossypiella*.

3.5 Insect Chitinases

Chitinases are hydrolytic enzymes that cause the hydrolysis of chitin. The enzymatic cleavage takes place in a random manner over the entire length of the chitin polymer, and the products are low molecular weight multimers of *N*-acetylglucosamine (GlcNAc) (Kramer and Koga, 1986; Reynolds and Samuels, 1996).

Insect chitinases are involved in cuticle turnover and digestion. These enzymes are found in molting fluid, venom glands, and midguts of insects (Terra and Ferreira, 1994; Krishnan et al., 1994; Terra et al., 1996). Insects periodically shed their exoskeletons and form new ones (Lehane, 1997). Chitinases are involved in this process and are present in the exoskeleton and gut lining of insects (Kramer et al., 1993). Degradation of chitin leads to perforation of the peritrophic matrix and exoskeleton, making insects vulnerable to attack by pathogens. Thus, expression of insect chitinase in plants may affect the growth of insect feeding on these plants owing to an inappropriately timed exposure to chitinase (Kramer and Muthukrishnan, 1997). Expression of cDNA for chitinase obtained from *M. sexta*, in tobacco plants showed partial resistance against *H. virescens*. Control and transformed leaves were fed to first instar larvae of tobacco budworm. Up to 80% reduction in total mass of surviving larvae on transformed leaves as compared to control suggested the potential of insect chitinases as targets (Ding et al., 1998).

3.6 Successes and Limitations of Transgenic Approaches for Insect Control

Transgenic technology has emerged as a potent method to produce insect-resistant plants. Success with this technology could save time and labor which is required for conventional breeding programs. Plants produced by transgenic technology have shown resistance against insects leading to increased crop yield. In combination with other powerful biopesticidal proteins such as proteinase inhibitors, *Bt* could drastically reduce the utilization of chemical pesticides and protect the environment. Plants expressing *Bt* toxin, as discussed above, have provided an effective method for controlling lepidopteran pests (Naranjo, 2011). However, there is a need for the use of pesticides to control the sap-sucking insects, such as aphids, whiteflies, planthoppers, and plant bugs, as there is no *Bt* toxin known with potential insecticidal effects on these pests (Gatehouse and Price, 2011). Proteinase inhibitors and lectins can play a major role in the management of pests which are not susceptible to *Bt*, and they may also play a part in gene pyramiding strategies. At one time, PIs were considered as important components of pest management strategies. However, while there are reports of transgenic plants expressing foreign PI genes, no commercial success has been realized. Understanding the insect defense will play an important role for PIs and other anti-insect components in resistant plants. The availability of plant and insect genome sequences can help us understand the interaction between the host and the insect. Even so, the development of these strategies must also consider the metabolic load required by plants to produce the extra protein used in insect resistance. The genetically modified insect-resistant crops can be used within the integrated pest management (IPM) strategy to obtain better results (Romeis et al., 2008).

4. RNA INTERFERENCE AS A NOVEL ALTERNATIVE TOOL FOR INSECT RESISTANCE

Another technology that offers promise for insect management is RNA interference (RNAi) technology. This technique holds promise for crop protection against insect pests with fewer biosafety issues, since there are no transgene proteins being expressed in transgenic lines (Price and Gatehouse, 2008; Huvenne and Smagghe, 2010; Rajam, 2011, 2012a,b).

4.1 RNAi in Gene Silencing

The discovery of RNA interference (RNAi) proved to be a boon in the area of molecular biology. It first was observed in *Caenorhabditis elegans* (Maupas) where double-stranded RNA (dsRNA) downregulated gene expression by cleaving its counter mRNA (Fire et al., 1998). It is a post-transcriptional gene silencing mechanism leading to cognate mRNA degradation, and it relies on the specificity of the sequence of one strand of the dsRNA and the corresponding complementary transcript. RNAi mainly comprises small-interfering RNAs (siRNAs) and microRNAs (miRNAs) which differ in their biogenesis pathway but bind to their complementary target mRNA and lead to their silencing (Fig. 10.1). Gene knockdown by RNAi has now become a valuable tool to study the function of genes in various organisms. Also, new methods of control of insect pests of important agricultural crops have been developed based on RNAi by genetically engineering the plants to express dsRNA. The advantage of genetically engineered plants expressing dsRNAs is continuous and stable expression of dsRNA in plants without exerting the load of protein expression (Rajam, 2011, 2012a,b).

4.1.1 siRNA-Based Strategies for Insect Control

Small-interfering RNAs (siRNAs) are formed when dsRNA is cleaved by the RNaseIII Dicer. siRNAs are approximately 21 nucleotide duplexes with a 2 bp 3′ overhang. Passenger strands (sense strands) of those duplexes are degraded while retaining the guide strand (antisense strand). The guide strand then is incorporated into a multiprotein complex called RNA-induced silencing complex (RISC). RISC has a catalytic component called Argonaute protein which cleaves single-stranded RNA (ssRNA) molecules having sequence complementarity with a guide strand of the siRNA duplex (Shabalina and Koonin, 2008) (Fig. 10.1). RNA-dependent RNA polymerase (RdRp) then interacts with RISC to generate a fresh lot of dsRNA based on a partially digested target template (Filipowicz, 2005). RdRp activity requires siRNA with 3′-OH annealed to the target mRNA to elongate the dsRNA. The gene coding of the RdRp is found in plants and fungi but not in insects and mammals.

Feeding of dsRNA through transgenic plants expressing the desired hairpin RNA of essential genes of insect has emerged as a potential tool to combat insect attack (Katoch et al., 2013). Transgenic plants expressing dsRNAs against genes of Lepidoptera, Coleoptera, and Hemiptera are common (Mao et al., 2007). Two successful reports of dsRNA-expressing plants came in 2007. Both reports showed that the RNAi pathway could be exploited to control insect pests via in planta expression of dsRNA against well-chosen target genes of insects (Baum et al., 2007; Mao et al., 2007). Transgenic

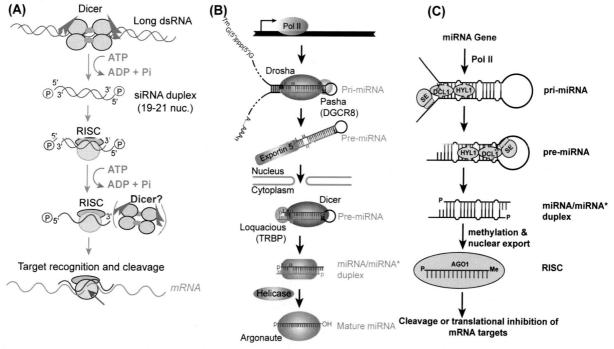

FIGURE 10.1 The biogenesis pathway for (A) siRNAs, (B) miRNAs in animals, and (C) plants. *Adapted from Izquierdo, M., 2005. Short interfering RNAs as a tool for cancer gene therapy, Cancer Gene Therapy 12, 217–227; Du, T., Zamore, P.D., 2005. MicroPrimer: the biogenesis and function of microRNA. Development 132, 4645–4652; and Zhu, J.K., 2008. Reconstituting plant miRNA biogenesis. Proceedings of the National Academy of Sciences USA 105, 9851–9852.*

corn producing dsRNA specific to the gene encoding the A subunit of the *v-ATPaseA* proton pump showed significant mortality of Western corn rootworm larvae (WCR, *Diabrotica virgifera virgifera* LeConte) (Baum et al., 2007). The other report showed that targeting cotton bollworm gut–specific cytochrome P450 gene *CYP6AE14* which gave resistance to gossypol, conferred insect resistance to plants. Cotton bollworm larvae when fed on transgenic tobacco and *Arabidopsis* plants expressing *CYP6AE14*-specific dsRNA showed subsequent sensitivity to gossypol in artificial diet (Mao et al., 2007). The expression of dsRNA of *CYP6AE14* was stable not only in the T1 but also in T2 generation, and showed enhanced resistance to bollworms (Mao et al., 2011).

Huvenne and Smagghe (2010) reviewed different mechanisms by which insects uptake dsRNA. RNAi-mediated gene silencing was successful for the control of the whitefly *Bemisia tabaci* (Gennadius). Actin ortholog, ADP/ATP translocase, α-tubulin, ribosomal protein L9 (RPL9), and V-ATPase A subunit genes were targeted using dsRNAs and siRNAs. High mortality with LC50s of 11.21 and 3.08 μg/mL was observed with RPL9 and V-ATPaseA genes, respectively. The study suggested that the phloem-specific expression of the dsRNAs RPL9 and V-ATPaseA in plants may provide protection against whiteflies (Upadhyay et al., 2011).

Previously, the effect of chemically synthesized siRNAs was evaluated by targeting of the acetylcholinesterase (*AChE*) gene of *H. armigera*. Larvae fed with an artificial diet containing siRNA molecules resulted in mortality, larval growth inhibition, and reduced pupal weight, and it also affected the adult fecundity compared to control larvae (Kumar et al., 2009).

Oral ingestion of dsRNA, expressed in plants triggers target-gene silencing in insects. Tobacco and tomato transgenic plants expressing dsRNA against *H. armigera AChE* when fed to the insects showed larval mortality and developmental delay (Kumar, 2011). The 20-hydroxyecdysone (20E), with its nuclear receptor complex Ecdysone receptor-ultraspiracle (EcR-USP) controls insect molting. Thus larvae feeding on tobacco plants expressing dsRNA of *EcR* showed resistance to pest attack (Zhu et al., 2012).

RNAi also was shown successful for targeting *HaHR3*, a molt-regulating transcription factor gene of *H. armigera*. The coding region of *HaHR3* was targeted by expressing dsRNAs through the L4440 vector in *Escherichia coli* (Migula) and also in a plant transformation vector. *H. armigera* larvae, when fed with the *E. coli* or transgenic plants, considerably decreased the mRNA and protein levels of *HaHR3*, which resulted in developmental delay and larval mortality. The results suggested the potential use of recombinant bacteria for gene silencing studies in insects (Xiong et al., 2013). 3-hydroxy-3-methylglutaryl coenzyme A reductase (HMG-CoA reductase; *HMGR*), a key enzyme in the mevalonate pathway in insects, has also been shown as a potential target for insect control. *HaHMGR* (*H. armigera HMGR*) silencing by RNAi resulted in reduced fecundity, oviposition, and vitellogenin (Vg) mRNA levels. In female moths, ovation rate was significantly decreased by 98% as compared to controls (Wang et al., 2013).

Recently, oral delivery of dsRNA molecules to *S. littoralis* against a gene highly similar to *P102* of *H. virescens*, strongly suppressed the encapsulation and melanization response, while hemocoelic injections did not result in evident phenotypic variations. This suggested that the protein is functionally conserved and plays a role in insect immunity. Hence, based on immunosuppression, dsRNA oral delivery can be exploited to develop novel technologies of pest control (Lelioa et al., 2014). Transgenic tobacco expressing siRNAs against the whitefly *v-ATPaseA* gene led to reduced transcript level of *v-ATPaseA* in whiteflies. The mRNA level was reduced up to 62% leading to mortality (Thakur et al., 2014).

Sitobion avenae (F.) causes serious economic losses to wheat (*Triticum aestivum* L.). Wheat was transformed with RNAi construct against the carboxylesterase (*CbEE4*) gene of *S. avenae*. The resistance gaining of *M. persicae* against organophosphate, carbamate, and pyrethroid pesticides was due to the activity of carboxylesterase. *CbEE4* gene expression was considerably reduced in *S. avenae* fed with transgenic wheat expressing *CbE E4* dsRNA. It was also found that the enzyme CbEE4 from *S. avenae* showed 20–30% hydrolysis within 40min in Phoxim solution after feeding on transgenic wheat, whereas it was 60% in *S. avenae* fed on control plants. This suggested that the target gene was indeed responsible for resistance toward pesticides and this approach may prove useful for reducing pesticide resistance mechanisms in aphids (Xu et al., 2014).

A recent study suggested an interesting way to make plants resistant to insects. Rice and maize roots were irrigated with dsRNA solutions of insect target genes. P450 (*Cyp18A1*) and carboxylesterase (*ces*) genes of planthopper and ACB kunitz-type trypsin inhibitors (KTI) of Asian corn borers were chosen as targets. Increased mortality was observed when planthoppers and Asian corn borers were fed with treated rice and maize, respectively. This provided a clue that crop roots can absorb dsRNAs and trigger the RNAi pathway in both plants and insects. The result clearly demonstrates that root dsRNA soaking can be used as biopesticides during irrigation (Li et al., 2015).

Although RNAi has emerged as a valuable tool for generating resistant plants, it also has some limitations. It has been reported that at least 60bp long dsRNA is required for efficient uptake by the insects (Bolognesi et al., 2012). But when dsRNAs are expressed in plants, they are processed into 21bp long siRNAs by the plant's machinery. Therefore insects take in more of these shorter siRNA sequences than the longer dsRNAs (Vazquez et al., 2010). Chloroplast transformation has provided a solution for this problem as a higher expression of transgenes can be achieved by virtue of the polyploidy nature of its genome (Oey et al., 2009; Ruhlman et al., 2010). Two recent reports have shown the expression of stable, long dsRNAs in the plastid genome which also has the advantage of the absence of transgene silencing in chloroplasts (Verma et al., 2008; Zhang et al., 2015; Jin et al., 2015).

The first report of dsRNA expression in chloroplasts targeted *ACT* and *SHR* genes of Colorado potato beetle (*Leptinotarsa decemlineata* Say; CBP) (Zhang et al., 2015). *ACT* encodes for β-actin and *SHR* encodes for Shrub, an important subunit of the protein complex which works during remodeling of membranes required for vesicle transport. They transformed the plastid genome of potato with *ACT*, *SHR*, and *ACT+SHR* constructs. These three constructs were also used for nuclear transformation so as to compare the level of resistance against insects in nuclear and transplastomic plants. Complete mortality (100%) was seen within 5 days of feeding by CBP on *ACT*-dsRNA-expressing transplastomic plants, whereas mortality was very low with nuclear transgenic lines. While no larvae survived on *ACT*-dsRNA-expressing transplastomic plants, few larvae survived on *SHR*-dsRNA and *ACT+SHR* expressing plants. However, the surviving larvae showed significant growth retardation (Zhang et al., 2015). Similarly, Jin et al. (2015) reported that expression of three dsRNAs targeting Chitinase, Cytochrome P450 (CYPAE14), and V-ATPase genes of *H. armigera* in tobacco chloroplasts led to considerable reduction of these gene transcripts in the insect midgut. They also observed higher larval mortality and lower pupation rate (Jin et al., 2015).

These reports suggest that chloroplast engineering can be exploited as a potential tool for producing large and stable dsRNAs which can control major agricultural pests.

4.1.2 MicroRNA-Based Strategies for Insect Control

MicroRNAs (miRNAs) are small endogenous RNA molecules of 21–24 nucleotides in length that regulate the gene expression at the post-transcriptional level by pairing to the sites within mRNA either by cleavage or translational repression. MicroRNAs were first discovered in *C. elegans* and were shown to be involved in the timed regulation of developmental events (Lee and Ambros, 2001). Since their discovery, computational prediction and small RNA sequencing helped in identifying number of potential miRNAs in various organisms including invertebrates, vertebrates, plants, fungi, and viruses. Understanding the regulatory roles of miRNAs may provide ways for crop improvement. MicroRNA-interference (miR-NAi) technology provides an efficient platform for functional studies and agricultural applications.

miRNAs are generated from their own genes present in intergenic regions and from the protein coding genes, i.e., intragenic regions (Wang, 2009). The miRNA genes are transcribed by RNA pol II as long distinct transcriptional units as well as in clusters of polycistronic units, known as primary miRNAs (pri-miRNAs) (Lagos-Quintana et al., 2001; Kim, 2005). These pri-miRNAs are processed into ~70 nt precursor miRNAs (pre-miRNAs) by the RNase III enzyme, Drosha (Lee et al., 2002). The pre-miRNAs are then transported to the cytoplasm by Exportin-5 and Ran-GTP (Yi et al., 2003), where they are processed further to generate 19–25 nt mature miRNA duplexes miRNA–miRNA* by the RNase III enzyme, Dicer (Grishok et al., 2001). However, in plants the miRNA duplexes are formed in the nucleus from pre-miRNAs by the dicer-like enzyme, Dcl1, which is also involved in the processing of pri-miRNAs to pre-miRNA (Bartel, 2004; Fang and Spector, 2007; Voinnet, 2009). These miRNA duplexes are exported out of the nucleus by Hasty (an Exportin-5 homology protein) (Park et al., 2005). The miRNA duplexes are recruited later into a multiprotein nuclease complex known as RNA-induced silencing complex (RISC), where only the guide strand of miRNA remains with the RISC and the passenger strands (miRNA*) are selectively eliminated in the presence of argonaute proteins (Llave et al., 2002; Tang et al., 2003). This RISC (mi-RISC) then binds to the motifs usually present in 3′UTRs of the target genes by exact or near-exact complementary base pairing, thereby silencing the target gene(s) through cleavage of target mRNA or translational repression, respectively (Carmell et al., 2002; Bartel, 2004; Voinnet, 2009; Wang, 2009) (Fig. 10.1).

Very few studies are available for miRNA-expression profiling after insect attack. A small RNA transcriptome change was observed as a result of herbivory (Pandey et al., 2008). RNA-directed RNA polymerase 1 (RdRp1) is involved in the miRNA biogenesis pathway. Silencing of RdR1 in *N. attenuata* Torr. ex S. Watson resulted in increased susceptibility to insect herbivores. This suggested the possible role of miRNAs in the defense mechanism in plants.

M. persicae (green peach aphid) is a phloem-feeding insect that affects cultivated crops worldwide (Blackman and Eastop, 2000). *M. persicae* produced significantly fewer progeny on *Arabidopsis* microRNA (miRNA) pathway mutants as these mutants responded to aphid infestation with increased induction of phytoalexin deficient3 (PAD3) and production of camalexin. It was also observed that aphids produced less progeny on camalexin containing artificial diet while they were more successful on PAD3 and cyp79b2/cyp79b3 mutants defective in camalexin production. This suggested the role of miRNAs in secondary metabolite regulation with relevance to hemipteran pest resistance (Kettles et al., 2013).

The identification of novel miRNAs of insect pests which are vital for their growth and development, and the discovery of novel miRNAs in host plant in response to pest attack can be utilized for crop protection by two different strategies. First, overexpressing the host plant miRNAs induced in response to pest attack may result in protection. Second, since miRNA genes of the insect are vital for their growth and development, in planta expression of an antisense RNA specific to the miRNA sequences would suppress insect survival.

Alternately, there are other miRNA gene silencing methods (Wang, 2009). For example, endogenous miRNA precursors can be modified in such a way to generate desired small RNAs (sRNAs) and direct gene silencing in either plants or animals (Alvarez et al., 2006; Niu et al., 2006; Schwab et al., 2006). miRNA precursors preferentially produce the miRNA–miRNA* duplex. Naturally occurring miRNA–miRNA* sequences can be replaced by the required duplex sequence without changing structural features such as bulges or mismatches. Artificial miRNAs (amiRNAs) were first generated and used in human cell lines (Zeng et al., 2002), and later in *Arabidopsis* (Parizotto et al., 2004), where they were shown to effectively interfere with reporter gene expression. Later on, it was successfully demonstrated that amiRNAs can efficiently target the endogenous gene similar to that of reporter genes and can be used in other plant species with comparable effectiveness (Alvarez et al., 2006; Schwab et al., 2006).

This technology has been used successfully to generate viral-resistant plants (Ali et al., 2013; Vu et al., 2013), but its use to control insects is limited. Recently, a study showed the use of amiRNA for the control of the aphid, *M. persicae*. amiR-NAs against two different regions of *MpAChE2* gene were expressed in tobacco plants. These transgenic tobacco plants showed better resistance toward aphids attack as compared to plants expressing *MpAChE2*-specific hpRNA. The results suggest that expression of insect-specific amiRNA is a promising and preferable approach to engineer plants resistant to aphids and, possibly, to other plant-infesting insects (Guo et al., 2014).

These novel strategies may prove to be highly potent in protecting our crop plants against important pathogens and pests. However, these strategies have not been explored in plants for insect resistance but have shown to be promising in treating human diseases like cancer and heart disease (Stenvang et al., 2012).

5. CONCLUSION

Insect control in agriculture is feasible in a safe and effective manner. Molecular biology gives us an opportunity to develop techniques which can be used for raising resistant genotypes. *B. thuringiensis* has been used in transgenic plants and has provided effective insect control. RNAi has a great potential for improving the crop yield as it takes the advantage of pathways involved in gene regulation. RNAi-based analyses of gene function have become effortless with the developing genome sequence resources. RNAi can be used to target PI-induced enzymes and genes which help in countering insect adaptation. It is highly specific which makes it a safe method for insect pest management. The success of RNAi mainly depends on the choice of target gene which can effectively kill the insect. Multiple genes can be simultaneously targeted to further strengthen the technology (Rajam, 2011). RNAi has been shown to work efficiently in plants such as tobacco and cotton, and while it has not been demonstrated in tomato, there is promise. Reports suggest that RNAi has potential for pest management, and field trials are required to study the long-term effectiveness of this method. Several aspects have to be taken into account before its practical applicability such as finding suitable target genes, stabilization of dsRNA during and after delivery, and cost-effectiveness for large scale use. The combination of different strategies and application of IPM will improve plant resistance against insects.

ACKNOWLEDGMENTS

We thank University Grants Commission, New Delhi, for Special Assistance Program (DRS-III), and Department of Science and Technology (DST), New Delhi for FIST (Level 2) and DU-DST PURSE (Phase II) program. INSPIRE fellowship to SY by DST is acknowledged.

REFERENCES

Abdeen, A., Virgos, A., Olivella, E., Villanueva, J., Aviles, X., Gabarra, R., Prat, S., 2005. Multiple insect resistance in transgenic tomato plants over-expressing two families of plant proteinase inhibitors. Plant Molecular Biology 57, 189–202.

Al Atalah, B., Vanderschaeghe, D., Bloch, Y., Proost, P., Plas, K., Callewaert, N., Savvides, S.N., Van Damme, E.J.M., 2014. Characterization of a type D1A EUL-related lectin from rice expressed in *Pichia pastoris*. Biological Chemistry 395, 413–424.

Ali, I., Amin, I., Briddon, R.W., Mansoor, S., 2013. Artificial microRNA-mediated resistance against the monopartite begomovirus *Cotton leaf curl Burewala virus*. Virology Journal 10, 231–238.

Alvarez, J.P., Pekker, I., Goldshmidt, A., Blum, E., Amsellem, Z., Eshed, Y., 2006. Endogenous and synthetic microRNAs stimulate simultaneous, efficient, and localized regulation of multiple targets in diverse species. The Plant Cell 18, 1134–1151.

Alves, D.T., Vasconcelos, I.M., Oliveira, J.T.A., Farias, L.R., Dias, S.C., Chiarello, M.D., 2009. Identification of four novel members of kunitz-like α-amylase inhibitors family from *Delonix regia* with activity toward coleopteran insects. Pesticide Biochemistry and Physiology 95, 166–172.

Bandani, A.R., 2005. Effect of plant α-amylase inhibitors on sunn pest, *Eurygaster integriceps* Puton (Hemiptera: Scutelleridae), alpha-amylase activity. Communications in Agricultural and Applied Biological Sciences 70, 869–873.

Barber, D., Sanchez-Monge, R., Méndez, E., Lázaro, A., García-Olmedo, F., Salcedo, G., 1986. New α-amylase and trypsin inhibitors among the CM-proteins of barley (*Hordeum vulgare*). Biochimica et Biophysica Acta 869, 115–118.

Barbosa, A.E., Albuquerque, E.V., Silva, M.C., Souza, D.S., Oliveira-Neto, O.B., Valencia, A., Rocha, T.L., Grossi-de-Sa, M.F., 2010. Alpha-amylase inhibitor-1 gene from *Phaseolus vulgaris* expressed in *Coffea arabica* plants inhibits alpha-amylases from the coffee berry borer pest. BMC Biotechnology 10, 44–51.

Bartel, D.P., 2004. MicroRNAs: genomics, biogenesis, mechanism, and function. Cell 116, 181–197.

Batalia, M.A., Monzingo, A.F., Ernst, S., Roberts, W., Robertus, J.D., 1996. The crystal structure of the antifungal protein zeamatin, a member of the thaumatin-like, PR-5 protein family. Nature Structural Biology 3, 19–23.

Baum, J.A., Bogaert, T., Clinton, W., Heck, G.R., Feldmann, P., 2007. Control of coleopteran insect pests through RNA interference. Nature Biotechnology 25, 1322–1326.

Blackman, R.L., Eastop, V.F., 2000. Aphids on the World's Crops: An Identification and Information Guide Wiley. Chichester, England.

Bloch Jr., C., Richardson, M., 1991. A new family of small (5 kDa) protein inhibitors of insect α-amylases from seeds or Sorghum (*Sorghum bicolor* (L) moench) have sequence homologies with wheat γ-purothionins. FEBS Letters 279, 101–104.

Bolognesi, R., Ramaseshadri, P., Anderson, J., Bachman, P., 2012. Characterizing the mechanism of action of double-stranded RNA activity against western corn rootworm (*Diabrotica virgifera virgifera* LeConte). PLoS One 7, e47534.

Boulter, D., Edwards, G.A., Gatehouse, A.M.R., Gatehouse, J.A., Hilder, V.A., 1990. Additive protective effects of different plant-derived insect resistance genes in transgenic tobacco plants. Crop Protection 9, 351–354.

Bravo, A., Gill, S., Soberón, M., 2007. Mode of action of *Bacillus thuringiensis* Cry and Cyt toxins and their potential for insect control. Toxicon 49, 423–435.

Broadway, R.M., , 1995. Are insects resistant to plant proteinase inhibitors?. Journal of Insect Physiology 41, 107–116.

Campos, F.A.P., Richardson, M., 1983. The complete amino acid sequence of the bifunctional α-amylase/trypsin inhibitor from seeds of ragi (indian finger millet, *Eleusine coracana* gaertn.). FEBS Letters 152, 300–304.

Carmell, M.A., Xuan, Z., Zhang, M.Q., Hannon, G.J., 2002. The Argonaute family: tentacles that reach into RNAi, developmental control, stem cell maintenance, and tumorigenesis. Genes and Development 16, 2733–2742.

Chagolla-Lopez, A., Blanco-Labra, A., Patthy, A., Sánchez, R., Pongor, S., 1994. A novel α-amylase inhibitor from Amaranth (*Amaranthus hypocondriacus*) seeds. Journal of Biological Chemistry 269, 23675–23680.

Christou, P., Capell, T., Kohli, A., Gatehouse, J., Gatehouse, A., 2006. Recent developments and future prospects in insect pest control in transgenic crops. Trends in Plant Science 11, 302–308.

Cornelissen, B.J.C., Hooft van Huijsduijnen, R.A.M., Bol, J.F., 1986. A tobacco mosaic virus-induced tobacco protein is homologous to the sweet-tasting protein thaumatin. Nature 321, 531–532.

Crickmore, N., Zeigler, D.R., Schnepf, E., Van Rie, J., Lereclus, D., Baum, J., Bravo, A., Dean, D.H., 2010. *Bacillus thuringiensis* Toxin Nomenclature. http://www.lifesci.sussex.ac.uk/home/Neil_Crickmore/Bt/.

De Leo, F., Gallerani, R., 2002. The mustard trypsin inhibitor 2 affects the fertility of *Spodoptera littoralis* larvae fed on transgenic plants. Insect Biochemistry and Molecular Biology 32, 489–496.

De Maagd, R.A., Bakker, P.L., Masson, L., Adang, M.J., Sangandala, S., Stiekema, W., Bosch, D., 1999. Domain III of the *Bacillus thuringiensis* delta-endotoxin Cry1Ac is involved in binding to *Manduca sexta* brush border membranes and to its purified amino peptidase N. Molecular Microbiology 31, 463–471.

Dias, S.C., da Silva, M.C.M., Teixeira, F.R., Figueira, E.L.Z., de Oliveira-Neto, O.B., deLima, L.A., 2010. Investigation of insecticidal activity of rye alpha-amylase inhibitor gene expressed in transgenic tobacco (*Nicotiana tabacum*) toward cotton boll weevil (*Anthonomus grandis*). Pesticide Biochemistry and Physiology 98, 39–44.

Ding, X., Gopalakrishnan, B., Johnson, L.B., White, F.F., Wang, X., Morgan, T.D., Kramer, K.J., Muthukrishnan, S., 1998. Insect resistance of transgenic tobacco expressing an insect chitinase gene. Transgenic Research 7, 77–84.

Down, R.E., Gatehouse, A.M.R., Hamilton, W.D.O., Gatehouse, J.A., 1996. Snowdrop lectin inhibits development and decreases fecundity of the glasshouse potato aphid (*Aulacorthum solani*) when administered in vitro and via transgenic plants both in laboratory and glasshouse trials. Journal of Insect Physiology 42, 1035–1045.

Du, T., Zamore, P.D., 2005. MicroPrimer: the biogenesis and function of microRNA. Development 132, 4645–4652.

Dunse, K.M., Stevens, J.A., Lay, F.T., Gaspar, Y.M., Heath, R.L., Anderson, M.A., 2010. Coexpression of potato type I and II proteinase inhibitors gives cotton plants protection against insect damage in the field. Proceedings of National Academy of Sciences USA 107, 15011–15015.

Etzler, M.E., 1986. Distribution and function of plant lectins. In: Liener, I.E., Sharon, N., Goldstein, I.J. (Eds.), The Lectins: Properties, Functions, and Applications in Biology and Medicine. Academic Press, Elsevier Inc., California, USA, pp. 371–435.

Fang, Y., Spector, D.L., 2007. Identification of nuclear dicing bodies containing proteins for microRNA biogenesis in living *Arabidopsis* plants. Current Biology 17, 818–823.

Filipowicz, W., 2005. RNAi: the nuts and bolts of the RISC machine. Cell 122, 17–20.

Fire, A., Xu, S.Q., Montgomery, M.K., Kostas, S.A., Driver, S.E., 1998. Potent and specific genetic interference by double-stranded RNA in *Caenorhabditis elegans*. Nature 391, 806–811.

Fischhoff, D.A., Bowdish, K.S., Perlak, F.J., Marrone, P.G., McCormick, S.M., Niedermeyer, J.G., Dean, D.A., Kusano-Kretzmer, K., Mayer, E.J., Rochester, D.E., Rogers, S.G., Fraley, R.T., 1987. Insect tolerant transgenic tomato plants. Nature Biotechnology 5, 807–813.

Foissac, F., Loc, N.T., Christou, P., Gatehouse, A.M.R., Gatehouse, J.A., 2000. Resistance to green leafhopper (*Nephotettix virescens*) and brown planthopper (*Nilaparvata lugens*) in transgenic rice expressing snowdrop lectin (*Galanthus nivalis* agglutinin; GNA). Journal of Insect Physiology 46, 573–583.

Franco, O.L., Rigden, D.J., Melo, F.R., Bloch Jr., C., Silva, C.P., Grossi De Sá, M.F., 2000. Activity of wheat α-amylase inhibitors towards bruchid α-amylases and structural explanation of observed specificities. European Journal of Biochemistry 267, 2166–2170.

García-Maroto, F., Cabonero, P., Garcia-Olmedo, F., 1991. Site-directed mutagenesis and expression in *Escherichia coli* of WMAI-1, a wheat monomeric inhibitor of insect α-amylase. Plant Molecular Biology 17, 1005–1011.

Gatehouse, A.M.R., Davison, G.M., Newell, C.A., Merryweather, A., Hamilton, W.D.O., Burgess, E.P.J., Gilbert, R.J.C., Gatehouse, J.A., 1997. Transgenic potato plants with enhanced resistance to the tomato moth, *Lacanobia oleracea*: growth room trials. Molecular Breeding 3, 49–63.

Gatehouse, J.A., Price, D.R.G., 2011. Protection of crops against insect pests using RNA interference. Insect Biotechnology 2, 145–168.

Grishok, A., Pasquinelli, A.E., Conte, D., Li, N., Parrish, S., 2001. Genes and mechanisms related to RNA interference regulate expression of the small temporal RNAs that control *C. elegans* developmental timing. Cell 106, 23–34.

Goldstein, I.J., Hayes, C.E., 1978. The lectins: carbohydrate-binding proteins of plants and animals. Advances in Carbohydrates Chemistry and Biochemistry 35, 127–340.

Gourinath, S., Alam, N., Srinivasan, A., Betzel, C., Singh, T.P., 2000. Structure of the bifunctional inhibitor of trypsin and alpha-amylase from ragi seeds at 2.2 Å resolution. Acta Crystallographica Section D 56, 287–293.

Guo, H., Song, X., Wang, G., Yang, K., Wang, Y., Niu, L., Chen, X., Fang, R., 2014. Plant-generated artificial small RNAs mediated aphid resistance. PLoS One 9, e97410.

Guo, P., Wang, Y., Zhou, X., Xie, Y., Wu, H., Gao, X., 2013. Expression of soybean lectin in transgenic tobacco results in enhanced resistance to pathogens and pests. Plant Science 211, 17–22.

Grossi de Sa, M.F., Mirkov, T.E., Ishimoto, M., Colucci, G., Bateman, K.S., Chrispeels, M.J., 1997. Molecular characterization of a bean alpha-amylase inhibitor that inhibits the alpha-amylase of the Mexican bean weevil *Zabrotes subfasciatus*. Planta 203, 295–303.

Hejgaard, J., Jacobsen, S., Svendsen, I., 1991. Two antifungal thaumatin-like proteins from barley grain. Federation of European Biochemical Societies 291, 127–131.

Hilder, V.A., Gatehouse, A.M.R., Sherman, S.E., Barker, R.F., Boulter, D., 1987. A novel mechanism of insect resistance engineered into tobacco. Nature 300, 160–163.

Hilder, V.A., Powell, K.S., Gatehouse, A.M.R., Gatehouse, J.A., Gatehouse, L.N., 1995. Expression of snowdrop lectin in transgenic tobacco plants results in added protection against aphids. Transgenic Research 4, 18–25.

Ho, M.F., Whitaker, J.R., 1993. Purification and partial characterization of white kidney bean (*Phaseolus vulgaris*) alpha-amylase inhibitors from two experimental cultivars. Journal of Food Biochemistry 17, 15–33.

Huvenne, H., Smagghe, G., 2010. Mechanisms of dsRNA uptake in insects and potential of RNAi for pest control: a review. Journal of Insect Physiology 56, 227–235.

Jin, S., Singh, N.D., Li, L., Zhang, X., Daniell, H., 2015. Engineered chloroplast dsRNA silences cytochrome p450 monooxygenase, V-ATPase and chitin synthase genes in the insect gut and disrupts *Helicoverpa armigera* larval development and pupation. Plant Biotechnology Journal 13, 435–446.

Johnson, R., Narraez, J., An, G., Ryan, C.A., 1989. Expression of proteinase inhibitors I and II in transgenic tobacco plants: effects on natural defense against *Manduca sexta* larvae. In: Proceeding of National Academy of Sciences, 86, pp. 9871–9875.

Jongsma, M.A., Bolter, C., 1997. The adaptation of insects to plant protease inhibitors. Journal of Insect Physiology 43, 885–895.

Jongsma, M.A., Bakker, P.L., Peters, J., Bosch, D., Stiekema, W.J., 1995. Adaptation of *Spodoptera exigua* larvae to plant proteinase inhibitors by induction of gut proteinase activity insensitive to inhibition. In: Proceeding of National Academy of Sciences USA, vol. 92, pp. 8041–8045.

Jouanin, L., Bonade-Bottino, M., Girard, C., Morrot, G., Giband, M., 1998. Transgenic plants for insect resistance. Plant Science 131, 1–11.

Katoch, R., Sethi, A., Thakur, N., Murdock, L.L., 2013. RNAi for insect control: current perspective and future challenges. Applied Biochemistry and Biotechnology 171, 847–873.

Kettles, G.J., Drurey, C., Schoonbeek, H., Maule, A.J., Hogenhout, S.A., 2013. Resistance of *Arabidopsis thaliana* to the green peach aphid, *Myzus persicae*, involves camalexin and is regulated by microRNAs. New Phytologist 198, 1178–1190.

Kim, V.N., 2005. MicroRNA biogenesis: coordinated cropping and dicing. Nature Reviews Molecular Cell Biology 6, 376–385.

Koul, B., Srivastava, S., Sanyal, I., Tripathi, B., Sharma, V., Amla, D.V., 2014. Transgenic tomato line expressing modified *Bacillus thuringiensis* cry1Ab gene showing complete resistance to two lepidopteran pests. Springer Plus 3, 84–97.

Kramer, K.J., Corpuz, L.M., Choi, H., Muthukrishnan, S., 1993. Sequence of a cDNA and expression of the gene encoding epidermal and gut chitinases of *Manduca sexta*. Insect Biochemistry and Molecular Biology 23, 691–701.

Kramer, K.J., Koga, D., 1986. Insect chitin: physical state, synthesis, degradation and metabolic regulation. Insect Biochemistry 16, 851–877.

Kramer, K.J., Muthukrishnan, S., 1997. Insect chitinases: molecular biology and potential use as biopesticides. Insect Biochemistry and Molecular Biology 27, 887–900.

Krattiger, A.F., 1996. Insect Resistance in Crops: A Case Study of *Bacillus thuringiensis* (Bt) and Its Transfer to Developing Countries. The International Agricultural Service for the Acquisition of Agribiotech Applications (ISAAA) Ithaca, New York, USA, p. 42.

Krishnan, A., Nair, P.N., Jones, D., 1994. Isolation, cloning and characterization of a new chitinase stored in active form in chitin-lined venom reservoir. Journal of Biological Chemistry 269, 20971–20976.

Kumar, M., 2011. RNAi-mediated Targeting of Acetylcholinesterase Gene of *Helicoverpa armigera* for Insect Resistance in Transgenic Tomato and Tobacco (Ph.D. thesis). University of Delhi, India.

Kumar, M., Gupta, G.P., Rajam, M.V., 2009. Silencing of acetylcholinesterase gene of *Helicoverpa armigera* by siRNA affects larval growth and its life cycle. Journal of Insect Physiology 55, 273–278.

Lagos-Quintana, M., Rauhut, R., Lendecket, W., Tuschl, T., 2001. Identification of novel genes coding for small expressed RNAs. Science 294, 853–858.

Lannoo, N., van Damme, E.J.M., 2010. Nucleocytoplasmic plant lectins. Biochimica et Biophysica Acta 1800, 190–201.

Laskowski Jr., M., Kato, I., 1980. Protein inhibitors of proteinases. Annual Review of Biochemistry 49, 685–693.

Lee, R.C., Ambros, V., 2001. An extensive class of small RNAs in *Caenorhabditis elegans*. Science 294, 862–864.

Lee, R.Y., Daniela Reiner, D., Dekan, G., Moore, A.E., Higgins, T.J.V., Epstein, M.M., 2013. Genetically modified α-Amylase inhibitor peas are not specifically allergenic in mice. PLoS One 8 (1), e52972.

Lee, Y., Jeon, K., Lee, J.T., Kim, S., Kim, V.N., 2002. MicroRNA maturation: stepwise processing and subcellular localization. EMBO Journal 21, 4663–4670.

Lehane, M.J., 1997. Peritrophic matrix structure and function. Annual Review of Entomology 42, 525–550.

Lelioa, I.D., Varricchioa, P., Priscoa, G.D., Marinellia, A., Lascoa, V., 2014. Functional analysis of an immune gene of *Spodoptera littoralis* by RNAi. Journal of Insect Physiology 64, 90–97.

Li, H., Guan, R., Guo, H., Miao, X., 2015. New insights into an RNAi approach for plant defense against piercing-sucking and stem-borer insect pests. Plant Cell and Environment 38, 2277–2285.

Llave, C., Xie, Z., Kasschau, K.D., Carrington, J.C., 2002. Cleavage of Scarecrow-like mRNA targets directed by a class of *Arabidopsis* miRNA. Science 297, 2053–2056.

Lord, J.C., 2005. From Metchnikoff to Monsanto and beyond: the path of microbial control. Journal of Invertebrate Pathology 89, 19–29.

Mandaokar, A.D., Goyal, R.K., Shukla, A., Bisaria, S., Bhalla, R., Reddy, V.S., Chaurasia, A., Sharma, R.P., Altosaar, I., Ananda Kumar, P., 2000. Transgenic tomato plants resistant to fruit borer (*Helicoverpa armigera* Hübner). Crop Protection 19, 307–312.

Mao, Y.B., Cai, W.J., Wang, J.W., Hong, G.J., Tao, X.Y., 2007. Silencing a cotton bollworm P450 monooxygenase gene by plant-mediated RNAi impairs larval tolerance of gossypol. Nature Biotechnology 25, 1307–1313.

Mao, Y.B., Tao, X.Y., Xue, X.Y., Wang, L.J., Chen, X.Y., 2011. Cotton plants expressing CYP6AE14 double-stranded RNA show enhanced resistance to bollworms. Transgenic Research 20, 665–673.

Marshall, J.J., Lauda, C.M., 1975. Purification and properties of Phaseolamin, an inhibitor of α-amylase, from the kidney bean, *Phaseolus vulgaris*. Journal of Biological Chemistry 250, 8030–8037.

Matten, S., Head, G., Quemada, H., 2008. How governmental regulation can help or hinder the integration of Bt crops into IPM programs. In: Romeis, J., Shelton, A.M., Kennedy, G.G. (Eds.), Integration of Insect-Resistant Genetically Modified Crops within IPM Programs. Springer International Publishing AG, Dordrecht, Netherlands, pp. 27–39.

Mazid, S., Kalita, J.C., Rajkhowa, R.C., 2011. A review on the use of biopesticides in insect pest management. International Journal of Advanced Science and Technology 1, 169–178.

Mehrabadi, M., Octavio, L., Franco, Bandani, A.R., 2012. In: Bandani, A.R. (Ed.), Plant Proteinaceous Alpha-Amylase and Proteinase Inhibitors and Their Use in Insect Pest Control, New Perspectives in Plant Protection. InTech. ISBN: 978-953-51-0490-2.

Melander, M., Ahman, I., Kamnert, I., Strömdahl, A.C., 2003. Pea lectin expressed transgenically in oilseed rape reduces growth rate of pollen beetle larvae. Transgenic Research 12, 555–567.

Michiels, K., van Damme, E., Smagghe, G., 2010. Plant-insect interactions: what can we learn from plant lectins? Archives of Insect Biochemistry and Physiology 73, 193–212.

Michelsen, P.O., Vévodová, J., De Maria, L., Ostergaard, P.R., Friis, E.P., Wilson, K., Skjot, M., 2008. Structural and mutational analyses of the interaction between the barley alpha-amylase/subtilisin inhibitor and the subtilisin savinase reveal a novel mode of inhibition. Journal of Molecular Biology 380, 681–690.

Moar, W., Roush, R., Shelton, A., Ferre, J., MacIntosh, S., Leonard, B.R., Abel, C., 2008. Field-evolved resistance to Bt toxins. Nature Biotechnology 26, 1072–1074.

Morton, R.L., Schroeder, H.E., Bate, K.S., Chrispeels, M.J., Armstron, G.E., Higgins, T.J.V., 2000. Bean α-amylase inhibitor 1 in transgenic peas (*Pisum sativum*) provides complete protection from pea weevil (*Bruchus pisorum*) under field conditions. In: Proceeding of National Academy of Sciences USA, vol. 97, pp. 3820–3825.

Mundy, J., Hejgaard, J., Svendsen, I., 1984. Characterization of a bifunctional wheat inhibitor of endogenous α-amylase and subtilisin. FEBS Letters 167, 210–214.

Mundy, J., Svendsen, I.B., Hejgaard, J., 1983. Barley α-amylase/subtilisin inhibitor. In: Isolation and Characterization. vol. 48. Carlsberg Research Communications, pp. 81–90.

Nagadhara, D., Ramesh, S., Pasalu, I., Rao, Y.K., Sarma, N., Reddy, V., Rao, K., 2004. Transgenic rice plants expressing the snowdrop lectin gene (GNA) exhibit high-level resistance to the whitebacked planthopper (*Sogatella furcifera*). Theoretical and Applied Genetics 109, 1399–1405.

Naranjo, S.E., 2011. Impacts of Bt transgenic cotton on integrated pest management. Journal of Agricultural and Food Chemistry 59, 5842–5851.

Nitti, G., Orru, S., Bloch Jr., C., Morhy, L., Marino, G., Pucci, P., 1995. Amino acid sequence and disulphide-bridge pattern of three gamma-thionins from *Sorgum bicolor*. European Journal of Biochemistry 228, 250–256.

Niu, Q.W., Lin, S.S., Reyes, J.L., Chen, K.C., Wu, H.W., Yeh, S.D., Chua, N.H., 2006. Expression of artificial microRNAs in transgenic *Arabidopsis thaliana* confers virus resistance. Nature Biotechnology 24, 1420–1428.

Oda, Y., Matsunga, T., Fukuyama, K., Miyazaki, T., Morimoto, T., 1997. Tertiary and quaternary structures of 0.19 α-amylase inhibitor from wheat kernel determined by X-ray analysis at 2.06 Å resolution. Biochemistry 36, 13503–13511.

Oey, M., Lohse, M., Kreikemeyer, B., Bock, R., 2009. Exhaustion of the chloroplast protein synthesis capacity by massive expression of a highly stable protein antibiotic. Plant Journal 57, 436–445.

Pandey, S.P., Shahi, P., Gase, K., Baldwin, I.T., 2008. Herbivory-induced changes in the small-RNA transcriptome and phytohormone signaling in *Nicotiana attenuata*. In: Proceeding of National Academy of Sciences USA, vol. 105, pp. 4559–4564.

Parizotto, E.A., Dunoyer, P., Rahm, N., Himber, C., Voinnet, O., 2004. In vivo investigation of the transcription, processing, endonucleolytic activity, and functional relevance of the spatial distribution of a plant miRNA. Genes and Development 18, 2237–2242.

Park, M.Y., Wu, G., Gonzalez-Sulser, A., Vaucheret, H., Poethig, R.S., 2005. Nuclear processing and export of microRNAs in *Arabidopsis*. Proceeding of National Academy of Sciences USA 102, 3691—3696.

Perlak, F.J., Fuchs, R.L., Dean, D.A., McPherson, S.L., Fischoff, D.A., 1991. Modification of coding sequence enhances plant expression of insect control protein genes. In: Proceeding of National Academy of Sciences USA, vol. 88, pp. 3324–3328.

Powell, K.S., Gatehouse, A.M.R., Hilder, V.A., Gatehouse, J.A., 1993. Antimetabolic effects of plants lectins and fungal enzymes on the nymphal stages of two important rice pests, *Nilaparvata lugens* and *Nephotettix cinciteps*. Entomologia Experimentalis et Applicata 66, 119–126.

Prescott, V.E., Campbell, P.M., Moore, A., Mattes, J., Rothenberg, M.E., Foster, P.S., Higgins, T.J.V., Hogan, S.P., 2005. Transgenic expression of bean alpha-amylase inhibitor in peas results in altered structure and immunogenicity. Journal of Agricultre and Food Chemistry 53, 9023–9030.

Price, D.R.G., Gatehouse, J.A., 2008. RNAi-mediated crop protection against insects. Trends in Biotechnology 26, 393–400.

Quilis, J., Lopez-Garc, B., Meynard, D., Guiderdoni, E., Segundo, B.S., 2013. Inducible expression of a fusion gene encoding two proteinase inhibitors leads to insect and pathogen resistance in transgenic rice. Plant Biotechnol Journal 1–11.

Rajam, M.V., 2011. RNA interference: a new approach for the control of fungal pathogens and insects. In: Acharya, N.G. (Ed.), Proceedings of the National Symposium on 'Genomics and Crop Improvement: Relevance and Reservations. Ranga Agricultural University, pp. 220–229. February 25–27, 2010.

Rajam, M.V., 2012a. Host induced silencing of fungal pathogen genes: an emerging strategy for disease control in crop plants. Cell and Developmental Biology 1, e118.

Rajam, M.V., 2012b. Micro RNA interference: a new platform for crop protection. Cell and Developmental Biology 1, e115.

Ranjekar, P.K., Patankar, A., Gupta, V., Bhatnagar, R., Bentur, J., Kumar, P.A., 2003. Genetic engineering of crop plants for insect resistance. Current Science 84, 321–329.

Rao, K.V., Rathore, K.S., Hodges, T.K., Fu, X., Stoger, E., Sudhakar, D., Williams, S., Christou, P., Bharathi, M., Bown, D.P., Powell, K.S., Spence, J., Gatehouse, A.M., Gatehouse, J.A., 1998. Expression of snowdrop lectin (GNA) in transgenic rice plants confers resistance to rice brown planthopper. Plant Journal 15, 469–485.

Reynolds, S.E., Samuels, R.I., 1996. Physiology and biochemistry of insect moulting fluid. Advances in Insect Physiology 26, 157–232.

Romeis, J., Shelton, A.M., Kennedy, G.G., 2008. Integration of Insect-Resistant Genetically Modified Crops within IPM Programs. Springer International Publishing AG, Dordrecht, Netherlands, p. 441.

Roy, A., Gupta, S., Hess, D., Das, K.P., Das, S., 2014. Binding of insecticidal lectin *Colocasia esculenta* tuber agglutinin (CEA) to midgut receptors of *Bemisia tabaci* and *Lipaphis erysimi* provides clues to its insecticidal potential. Proteomics 14, 1646–1659.

Rüdiger, H., Gabius, H.J., 2001. Plant lectins: occurrence, biochemistry, functions and applications. Glycoconjugate Journal 18, 589–613.

Ruhlman, T., Verma, D., Samson, N., Daniell, H., 2010. The role of heterologous chloroplast sequence elements in transgene integration and expression. Plant Physiology 152, 2088–2104.

Ryan, C.A., 1988. Proteinase inhibitors gene families: tissue specificity and regulation. In: Goldberg, V. (Ed.), Plant Gene Research. Springer/Verlag Pubs, pp. 223–233.

Ryan, C.A., 1990. Protease inhibitors in plants: genes for improving defenses against insects and pathogens. Annual Review of Phytopathology 28, 425–449.

Saha, P., Dasgupta, I., Das, S., 2006b. A novel approach for developing resistance in rice against phloem limited viruses by antagonizing the phloem feeding hemipteran vectors. Plant Molecular Biology 62, 735–752.

Saha, P., Majumder, P., Dutta, I., Ray, T., Roy, S.C., Das, S., 2006a. Transgenic rice expressing *Allium sativum* leaf lectin with enhanced resistance against sap-sucking insect pests. Planta 223, 329–1343.

Saker, M.M., Salama, H.S., Salama, M., El-Banna, A., Abdel Ghany, N.M., 2011. Production of transgenic tomato plants expressing Cry 2Ab gene for the control of some lepidopterous insects endemic in Egypt. Journal of Genetic Engineering and Biotechnology 9, 149–155.

Sanahuja, G., Banakar, R., Twyman, R.M., Capell, T., Christou, P., 2011. *Bacillus thuringiensis*: a century of research development and commercial applications. Plant Biotechnol Journal 9, 283–300.

Sane, V.A., Nath, P., Aminuddin, N., Sane, P.V., 1997. Development of insect-resistant transgenic plants using plant genes: expression of cowpea trypsin inhibitor in transgenic tobacco plants. Current Science 72, 741–747.

Schroeder, H.E., Gollasch, S., Moore, A., Tabe, L.M., Craig, S., Hardie, D.C., Chrispeels, M.J., Spencer, D., Higgins, T.J.V., 1995. Bean alpha-amylase Inhibitor confers resistance to the pea weevil (*Bruchus pisorum*) in transgenic peas (*Pisum sativum*). Plant Physiology 107, 1233–1239.

Schwab, R., Ossowski, S., Riester, M., Warthmann, N., Weigel, D., 2006. Highly specific gene silencing by artificial microRNAs in *Arabidopsis*. The Plant Cell 18, 1121–1133.

Senthilkumar, R., Cheng, C.-P., Yeh, K.-W., 2010. Genetically pyramiding protease-inhibitor genes for dual broad-spectrum resistance against insect and phytopathogens in transgenic tobacco. Plant Biotechnology Journal 8, 65–75.

Shabalina, S.A., Koonin, E.V., 2008. Origins and evolution of eukaryotic RNAinterference. Trends in Ecology & Evolution 23, 578–587.

Shade, R.E., Schroeder, R.E., Poueyo, J.J., Tabe, L.M., Murdock, L.L., Higgins, T.J.V., Chrispeels, M.J., 1994. Transgenic pea seeds expressing the α-amylase inhibitor of the common bean are resistant to bruchid beetles. Bio/Technology 12, 793–796.

Sharon, N., Lis, H., 1990. Legume lectins – a large family of homologous proteins. Federation of American Societies for Experimental Biology Journal 4, 3198–3208.

Stenvang, J., Petri, A., Lindow, M., Obad, S., Kauppinen, S., 2012. Inhibition of microRNA function by antimiR oligonucleotides. Silence 3, 1.

Stillmark, H., 1888. Uber Ricin, Eines Giftiges Ferment aus den Samen von *Ricinus communis* L. und Anderson Euphorbiacen. Inaugural Dissertation. University of Dorpat, Dorpat, Estonia.

Sumner, J.B., 1919. The globulins of the jack bean, *Canavalia ensiformis*. The Journal of Biological Chemistry 37, 137–142.

Tabashnik, B.E., Brévault, T., Carrière, Y., 2013. Insect resistance to *Bt* crops: lessons from the first billion acres. Nature Biotechnology 31, 510–521.

Tabashnik, B.E., Gassman, A.J., Crowder, D.W., Carriere, Y., 2008. Insect resistance to *Bt* crops: evidence versus theory. Nature Biotechnology 26, 199–202.

Tabashnik, B.E., Huang, F., Ghimire, M.N., Leonard, B.R., Siegfried, B.D., Rangasamy, M., Yang, Y., Wu, Y., Gahan, L.J., Heckel, D.G., Bravo, A., Soberón, M., 2011. Efficacy of genetically modified *Bt* toxins against insects with different mechanism of resistance. Nature Biotechnology 29, 1128–1131.

Tajne, S., Boddupally, D., Sadumpati, V., Vudem, D.R., Khareedu, V.R., 2014. Synthetic fusion-protein containing domains of *Bt* Cry1Ac and *Allium sativum* lectin (ASAL) conferred enhanced insecticidal activity against major lepidopteran pests. Journal of Biotechnology 171, 71–75.

Tang, G., Reinhart, B.J., Bartel, D.P., Zamore, P.D., 2003. A biochemical framework for RNA silencing in plants. Genes and Development 17, 49–63.

Terra, W.T., Ferreira, C., 1994. Insect digestive enzymes: properties, compartmentalization and function. Comparative Biochemistry and Physiology 109B, 1–62.

Terra, W.R., Ferreira, C., Jordao, B.P., Dillon, R.J., 1996. Digestive enzymes. In: Lehane, M.J., Billingsley, P.F. (Eds.), Biology of the Insect Midgut. Chapman and Hall, London, UK, pp. 15–194.

Thakur, N., Upadhyay, S.K., Verma, P.C., Chandrashekar, K., Tuli, R., 2014. Enhanced whitefly resistance in transgenic tobacco plants expressing double stranded RNA of *v-ATPase A* gene. PLoS One 9, e87235.

Upadhyay, S.K., Chandrashekar, K., Thakur, N., Verma, P.C., Borgio, J.F., 2011. RNA interference for the control of whiteflies (*Bemisia tabaci*) by oral route. Journal of Biosciences 36, 153–161.

Vaeck, M., Reynaerts, A., Hoftey, H., Jansens, S., DeBeuckleer, M., Dean, C., Zabeau, M., Van Montagu, M., Leemans, J., 1987. Transgenic plants protected from insect attack. Nature 327, 33–37.

Van Damme, E.J., Allen, A.K., Peumans, W.J., 1987. Isolation and characterization of a lectin with exclusive specificity towards mannose from snowdrop (*Galanthus nivalis*) bulbs. FEBS Letters 215, 140–144.

Van Damme, E.J.M., Peumans, W.J., Barre, A., Rougé, P., 1998. Plant lectins: a composite of several distinct families of structurally and evolutionary related proteins with diverse biological roles. Critical Reviews in Plant Sciences 17, 575–692.

Van Damme, E.J., Lannoo, N., Fouquaert, E., Peumans, W.J., 2003. The identification of inducible cytoplasmic/nuclear carbohydrate-binding proteins urges to develop novel concepts about the role of plant lectins. Glycoconjugate Journal 20, 449–460.

Van Damme, E.J.M., Lannoo, N., Peumans, W.J., 2008. Plant lectins. In: Jean-Claude, K., Michel, D. (Eds.). Advances in Botanical Research. vol. 48. Academic Press, Waltham, MA, USA, pp. 107–209.

Vandenborre, G., Groten, K., Smagghe, G., Lannoo, N., Baldwin, I.T., van Damme, E.J.M., 2010. *Nicotiana tabacum* agglutinin is active against Lepidopteran pest insects. Journal of Experimental Botony 61, 1003–1014.

Van Lenteren, J.C., 2012. The state of commercial augmentative biological control: plenty of natural enemies, but a frustrating lack of uptake. BioControl 57, 1–20.

Vazquez, F., Legrand, S., Windels, D., 2010. The biosynthetic pathways and biological scopes of plant small RNAs. Trends in Plant Science 15, 337–345.

Verma, D., Samson, N.P., Koya, V., Daniell, H., 2008. A protocol for expression of foreign genes in chloroplasts. Nature Protocols 3, 739–758.

Vigers, A.J., Roberts, W.K., Selitrennikoff, C.P., 1991. A new family of plant antifungal proteins. Molecular Plant-Microbe Interaction 4, 315–323.

Vila, L., Quilis, J., Meynard, D., Breitler, J.C., Marfa, V., Murillo, I., Vassal, J.M., Messeguer, J., Guiderdoni, E., San Segundo, B., 2005. Expression of the maize proteinase inhibitor (mpi) gene in rice plants enhances resistance against the striped stem borer (*Chilo suppressalis*): effects on larval growth and insect gut proteinases. Plant Biotechnology Journal 3, 187–202.

Voinnet, O., 2009. Origin, biogenesis, and activity of plant microRNAs. Cell 136, 669–687.

Vu, T.V., Choudhury, N.R., Mukherjee, S.M., 2013. Transgenic tomato plants expressing artificial microRNAs for silencing the pre-coat and coat proteins of a begomovirus, Tomato leaf curl New Delhi virus, show tolerance to virus infection. Virus Research 172, 35–45.

Wang, Z., 2009. Micro RNA Interference Technologies. Springer-Verlag, Berlin.

Wang, Z., Dong, Y., Desneux, N., Niu, C., 2013. RNAi silencing of the HaHMG-CoA reductase gene inhibits oviposition in the *Helicoverpa armigera* cotton bollworm. PLoS One 8, e67732.

Wang, Z., Zhang, K., Sun, X., Tang, K., Zhang, J., 2005. Enhancement of resistance to aphids by introducing the snowdrop lectin gene GNA into maize plants. Journal of Biosciences 30, 627–638.

Xiong, Y., Zeng, H., Zhang, Y., Xu, D., Qiu, D., 2013. Silencing the HaHR3 gene by transgenic plant-mediated RNAi to disrupt *Helicoverpa armigera* development. Integrated Journal of Biological Sciences 9, 370–381.

Xu, L., Duan, X., Lv, Y., Zhang, X., Nie, Z., Xie, C., Ni, Z., Liang, R., 2014. Silencing of an aphid carboxylesterase gene by use of plant-mediated RNAi impairs *Sitobion avenae* tolerance of Phoxim insecticides. Transgenic Research 23, 389–396.

Yang, S., Pyati, P., Fitches, E., Gatehouse, J.A., 2014. A recombinant fusion protein containing a spider toxin specific for the insect voltage-gated sodium ion channel shows oral toxicity towards insects of different orders. Insect Biochemistry and Molecular Biology 47, 1–11.

Yi, R., Qin, Y., Macara, I.G., Cullen, B.R., 2003. Exportin-5 mediates the nuclear export of pre-microRNAs and short hairpin RNAs. Genes Development 17, 3011–3016.

Zeng, Y., Wagner, E.J., Cullen, B.R., 2002. Both natural and designed micro RNAs can inhibit the expression of cognate mRNAs when expressed in human cells. Molecular Cell 9, 1327–1333.

Zhang, J., Khan, S.A., Hasse, C., Ruf, S., Heckel, D.G., Bock, R., 2015. Full crop protection from an insect pest by expression of long double-stranded RNAs in plastids. Science 27, 991–994.

Zhang, W., Peumans, W.J., Barre, A., Houles Astoul, C., Rovira, P., Rougé, P., Proost, P., Truffa-Bachi, P., Jalali, A.A.H., van Damme, E.J.M., 2000. Isolation and characterization of a jacalin-related mannose-binding lectin from salt-stressed rice (*Oryza sativa*) plants. Planta 210, 970–978.

Zhao, J., Cao, J., Li, Y., Collins, H., Roush, R., 2003. Transgenic plants expressing two *Bacillus thuringiensis* toxins delay insect resistance evolution. Nature Biotechnology 21, 1493–1497.

Zhu, J.Q., Liu, S., Ma, Y., Zhang, J.Q., Qi, H.-S., 2012. Improvement of pest resistance in transgenic tobacco plants expressing dsRNA of an insect-associated gene EcR. PLoS One 7, e38572.

Chapter 11

Biological Control in Tomato Production Systems: Theory and Practice

Sriyanka Lahiri, David Orr

North Carolina State University, Raleigh, NC, United States

1. INTRODUCTION

Growers of fresh food products, such as vegetables and fruits, have low tolerance to pest damage due to market expectations and prices (Stern, 1973). Tomato, *Solanum lycopersicum* L. is one such vegetable commodity which originated in the South American Andes (Bai and Lindhout, 2007) and is consumed worldwide. Tomato plants are attacked by a number of pests which cause damage not only to the fruit and stem, but also to the root system of the plant. This is driving scientific investigation toward preventing economic losses incurred by tomato growers worldwide. The use of pesticides often provide adequate pest control, but its misuse or overuse (James et al., 2006; Baimey et al., 2009) can lead to pest resistance, elimination of natural enemy consortia, and accidental exposure to people (Cranham and Helle, 1985; Roberts, 1990; Salazar and Araya, 1997, 2001; Jacobson et al., 1999; Siqueira et al., 2000a,b, 2001; Chandler et al., 2005). As a result of concerns for environmental health, the development of pesticide resistance in pests, and the need for natural enemy conservation, integrated pest management (IPM) approaches for pest control are being researched and implemented. Biological control is an important and promising aspect of IPM (Gabarra et al., 1995). It is defined as "the study, importation, augmentation, and conservation of beneficial organisms to regulate population densities of other organisms" (DeBach, 1964). Biological control is viewed as a beneficial alternative at both pre- and post-harvest stages of tomato (Colak, 2009).

In this chapter, we present the various pests of tomato, their potential biological control agents, and suggested management tactics.

2. BIOLOGICAL CONTROL AGENTS OF TOMATO PESTS

Research based on laboratory, greenhouse, and open-field experiments exploring the efficacy of different groups of potential biocontrol agents has provided the information regarding the mechanisms and extent of possible applications of biocontrol tactics targeting tomato pests. Information, such as source of biocontrol agent, stage of pest most susceptible to biocontrol agent, most effective method of biocontrol agent deployment and biocontrol agents that can be used to complement each other, are discussed.

2.1 Whiteflies and Aphids

Two species of whiteflies, *Trialeurodes vaporariorum* Westwood and *Bemisia tabaci* (Gennadius) attack tomato in greenhouses (Gabarra et al., 1995; Jazzar and Hammad, 2004), while open-field tomato production faces various limiting factors to combat aphid and whitefly-transmitted viruses (Lapidot et al., 2014).

The polyphagous predator *Dicyphus tamaninii* Wagner is a potential biocontrol agent, which controls greenhouse tomato pests such as whiteflies and aphids (Gabarra et al., 1995). The cropping system plays a crucial role in contributing to the efficacy of a natural enemy, as does the prey density. Usually, under high whitefly pressure in greenhouses, *D. tamaninii* preys on greenhouse whitefly, *T. vaporariorum* quite effectively. However, this insect is known to be a facultative predator since it may feed on the fruits when prey is scarce (Gabarra et al., 1988; Alomar, 1994). Greenhouse, studies have shown that this predator does not damage tomato fruits under low prey densities (Gabarra et al., 1995). The efficacy of partially phytophagous predators such as *D. tamaninii* (Dolling, 1991) can be increased by managing natural populations of this predator. For example, conservational biological control programs can be developed by devising decision charts to advise

Sustainable Management of Arthropod Pests of Tomato. http://dx.doi.org/10.1016/B978-0-12-802441-6.00011-5

growers to spray, avoiding whitefly damage and minimizing the risk of injury to the crop from the phytophagous predator (Alomar et al., 1991; Alomar and Albajes, 1996).

The predatory fly, *Coenosia attenuata* Stein, which attacks *B. tabaci* and *T. vaporariorum* in greenhouse tomato (Gerling et al., 2001), is being considered as a biological control agent in Mediterranean greenhouses. Studies have shown an increased predator response during increasing temperatures, more prey flights, and higher conspecific densities (Bonsignore, 2016). The parasitoid *Encarsia formosa* Gahan is a potential biological control agent for *B. tabaci* (Jazzar and Hammad, 2004) as well.

Other insects in the Hemipteran suborder Sternorrhyncha feed on the leaves and stems of tomato, including *Myzus persicae* (Sulzer) and *Macrosiphum euphorbiae* Thomas (Sannino and Espinosa, 2009). Hoverflies (Fig. 11.1) with predatory (or "aphidophagous") larvae (Diptera: Syrphidae) (Fig. 11.2) and lacewings (Neuroptera: Chrysopidae) (Fig. 11.3) are

FIGURE 11.1   Hoverfly. *Photo by Matt Bertone, NCSU.*

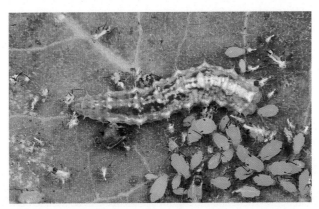

FIGURE 11.2   Hoverfly larvae feeding on aphids. *Photo by Matt Bertone, NCSU.*

FIGURE 11.3   Lacewing larva. *Photo by Matt Bertone, NCSU.*

important natural enemies of aphids (van Rijn et al., 2013). Scarcity of suitable flowering plants in intensified agricultural systems can limit the effectiveness of natural enemies such as the hoverflies (Winkler et al., 2006; Olson and Wäckers, 2007; Tscharntke et al., 2007; Meyer et al., 2009). However, availability of honeydew has been shown to strongly enhance hoverfly survival, even in the absence of floral resources. Nectar and pollen from buckwheat, *Fagopyrum esculentum* Moench, has also been reported to enhance hoverfly adult longevity and egg production (van Rijn et al., 2013). Tian et al. (2016) reported that the entomopathogenic fungus *Isaria fumosorosea* (Wize) caused rapid mortality in *B. tabaci* feeding on tomato and beans. Infection rates on these two crops was higher than in whiteflies that fed on cucumber, *Cucumis sativus* L. and eggplant, *Solanum melongena* L. Therefore, host plants and rates of exposure to fungal conidia affect pathogenicity and virulence of *I. fumosorosea*.

2.2 Caterpillars

The tomato pinworm *Tuta absoluta* (Meyrick) is a devastating pest of tomato. This insect originated in South America, and has spread to several European, Middle Eastern, and North African countries since its discovery in Spain in 2006 (Desneux et al., 2010). Recently, it was reported from India as well (Kumari et al., 2015). *T. absoluta*, being a leafminer damages tomato leaves, terminal buds, flowers, and fruits (Moraes and Normanha Filho, 1982; Haji et al., 1988; Lopes-Filho, 1990; Castelo-Branco, 1992), causing a reduction of crop yield that can be as high as 90% (Apablaza, 1988). Damage is so high because the larvae feed on the mesophyll, thereby affecting photosynthetic capacity of the crop (Urbaneja et al., 2012). Other lepidopteran pests include *Helicoverpa armigera* (Hübner) (Balzan, 2013), *Spodoptera littoralis* Boisduval, and *Lacanobia oleracea* L. Larvae of *S. littoralis* and *L. oleracea* feed on foliage of tomato while *H. armigera* feeds on the fruit.

Fortunately, well-concealed pest stages of the notorious *T. absoluta* have effective natural enemies. Entomopathogenic nematodes (EPNs), *Steinernema carpocapsae* Weiser, *Steinernema feltiae* Filipjev, and *Heterorhabditis bacteriophora* Poinar attack *T. absoluta* larvae, pupae, and adults that remain hidden within galleries in leaves (Batalla-Carrera et al., 2010). The host plant has an important effect on EPN efficiency. For example, the corn earworm, *Helicoverpa zea* (Boddie) feeding on tobacco, has been shown to reduce virulence and reproduction of the EPN *S. riobrave*, when compared to tomato and eggplant diets (Hazir et al., 2016).

Another effective biocontrol agent of the subterranean *T. absoluta* pupa is the entomopathogenic fungus species, *Metarhizium anisopliae* var. *anisopliae* (Metchnikoff) Sorokin that can be applied with irrigation water (Contreras et al., 2014). Certain beneficial fungi benefit from predatory insects, by being spread through long distances as the predator scouts the fields for prey. For example, infective forms of the lepidopteran pathogens *Vairimorpha necatrix* (Kramer) and *L. oleracea* granulovirus (LoGV) are successfully vectored by the spined soldier bug, *Podisus maculiventris* (Say) (Fig. 11.4), via the excreta of the predator after feeding on infected larvae on tomato plants (Down et al., 2004). When five *P. maculiventris* females per plant defecated on the leaves of tomato, the number of newly introduced larvae of *S. littoralis* and *L. oleracea* feeding on the foliage and surviving after 20 days was reduced by 75% and 61%, respectively (Down et al., 2004). In another system, Marti and Hamm (1985) reported that spores of *Vairimorpha* species remained

FIGURE 11.4 Spiny soldier bug. *Photo by Matt Bertone, NCSU.*

viable in the gut of the big-eyed bug, *Geocoris punctipes* (Say) (Fig. 11.5) after feeding on infected larvae of *Spodoptera frugiperda* (J.E. Smith).

However, such associations may prove costly to the predator. For example, a reduction in fecundity and longevity of *P. maculiventris* occurred after feeding on *V. necatrix*-infected prey and a reduction in fecundity was found when feeding on LoGV-infected prey. In spite of the cost, there is demonstrated potential in disseminating infective pathogens to control lepidopteran pests of tomato using *P. maculiventris*. Since these types of interspecies associations of natural enemies may involve a cost for the biocontrol agent, a cost–benefit analysis should be used to decide the plausibility of using such liaisons.

Environmental factors such as nitrogen and water deficit cause a negative bottom-up effect on predators such as *Macrolophus pygmaeus* (Rambur) in tomato crops by depressing predation of hosts such as *Ephestia kuehniella* Zeller eggs and by reducing (up to 30%) predator longevity (Han et al., 2015). This study points to the limitations of mirid predator–based IPM programs in tomato. On the other hand, certain monoterpenes, which are herbivore-induced plant volatiles (HIPVs), can be utilized to manipulate natural enemies such as *M. pygmaeus* to enhance pest management of *T. absoluta* in tomato (De Backer et al., 2015).

T. absoluta is also known to be attacked by an idiobiont Palearctic ectoparasitoid species, a eulophid wasp (*Necremnus artynes* (Walker)) (Fig. 11.6), in Spain and Italy (Desneux et al., 2010; Gabarra and Arnó, 2010; Zappalà et al., 2012). Also, biological control of *T. absoluta* in tomato crops by *Trichogramma* spp. parasitoids has emerged as a promising management tool (Chailleux et al., 2013). The host species in which a parasitoid develops can significantly affect the success of

FIGURE 11.5   Big-eyed bug. *Photo by Matt Bertone, NCSU.*

FIGURE 11.6   Eulophid wasp. *Photo by Matt Bertone, NCSU.*

the parasitoid as a biocontrol agent against other hosts. Chailleux et al. (2013) showed that the mean number of parasitized eggs/female parasitoid ranged from 35.5 to 3.0 when five species of *Trichogramma*, reared on two different hosts, *E. kuehniella* and *T. absoluta* were tested for their potential for biocontrol of *T. absoluta*. Only two of the five parasitoid species reared on *E. kuehniella* and parasitizing *T. absoluta* eggs did slightly better than the ones reared on *T. absoluta*. Overall, however, the lowest parasitism of *T. absoluta* eggs was reported in wasps reared on *T. absoluta*, while the best result was obtained from wasps reared on *E. kuehniella* eggs. Therefore, the authors concluded that parasitism efficiency decreased in the *T. absoluta*–tomato system, and that this pest–plant system appeared unsuitable for the *Trichogramma* species tested, probably due to small size of *T. absoluta* eggs potentially causing morphological malformations. The study also pointed out that releasing both parasitoid and omnivorous predatory mirids may not be effective since the presence of *M. pygmaeus* seemed to deter parasitism in this study. Ultimately, the biological control of *T. absoluta* in tomato by *Trichogramma* spp. should rely on multiple inundative releases, after taking into account the presence of generalist predators (Chailleux et al., 2013). Another factor to consider is the temperature tolerance of the biological control agent. For example, *Trichogramma achaeae* Nagaraja and Nagarkatti, appears to be well adapted to warmer conditions (Kabiri et al., 2010), but not to the low temperatures of the South of France (Voegele et al., 1988). Thus this *Trichogramma* spp. will not be an ideal classical biocontrol agent for this region.

Larval and pupal parasitoids of *T. absoluta*, belonging to 20 different species, were reported from the Spanish Mediterranean Coast, where the most commonly found species were *N.* sp. nr. *artynes*, *Stenomesius* cf. *japonicus* (Ashmead) and *Neochrysocharis formosa* (Westwood) (Gabarra et al., 2014). Other larval parasitoid families found in this study included Braconidae, Ichneumonidae, Pteromalidae and Chalcididae, and all *T. absoluta* egg parasitoids collected from sentinel egg masses on tomato plants belonged to the genus *Trichogramma*.

Dicyphus errans (Wolff) is an indigenous generalist predator species of northwestern Italy, which is used in the IPM of tomato crops. This species has been shown to have potential for controlling *T. absoluta* eggs and first instar larvae (Ingegno et al., 2013). Olfactometer bioassays indicated that adults were attracted to tomato plants either currently or previously infested with *T. absoluta* larvae (Ingegno et al., 2013), showing their ability to track the presence of this pest. Studies showed that female *D. errans* adults were significantly better at consuming *T. absoluta* eggs and preying on first instar larvae compared to males. The authors concluded that further research was needed to evaluate its suitability for mass production.

In an effort to reduce the reliance on chemical treatments for control of *T. absoluta*, laboratory, greenhouse, and open field experiments were conducted using three commercial formulations based on *Bacillus thuringiensis* Berliner (*Bt*) which were Dipel DF [Certis Europe B.V. suc. España (Murcia, Spain)], Turex [Certis Europe B.V. suc. España (Murcia, Spain)], and Costar [Syngenta Agro. S.A. (Madrid, Spain)] (González-Cabrera et al., 2011). The results indicated that spraying *B. thuringiensis*-based formulations were most effective against the highly susceptible first instar larva. However, other instars were also infected, significantly reducing *T. absoluta* numbers compared to non-treated controls. In this study, the treated tomato plants were readily colonized by three species of predatory mirids, which preferentially preyed on *T. absoluta* eggs, while *B. thuringiensis* killed larva. Therefore, the integration of *B. thuringiensis* with biological control efforts using predatory mirids can be a safe strategy to manage *T. absoluta* in tomato.

2.3 Flies

Larvae belonging to unidentified species of the gall midge, *Lasioptera* feed on the tomato stem, thereby indirectly causing fruit loss (Perdikis et al., 2011). Also, tomato fruit fly, *Neoceratitis cyanescens* (Bezzi) has been reported from France as an important pest of the fruit (Brévault and Quilici, 2009).

The gall midge, *Lasioptera donacis* Coutin and Faivre-Amiot, can be infected with the parasitic nematode *Tripius gyraloura* n. sp, which render the infected females sterile (Poinar and Thomas, 2014). Also, a polyphagous parasitoid of tephritid flies, *Fopius arisanus* (Sonan) is reported to attack many fruit flies including *N. cyanescens* (Rousse et al., 2006). *F. arisanus* is an egg/pre-pupal parasitoid of up to 21 tephritid pests. It was introduced in Hawaii in 1947, which led to the successful biocontrol of local *Bactrocera dorsalis* (Hendel) and *Ceratitis capitata* (Wiedemann) populations, becoming the predominant parasitoid of these pests in that region (Haramoto and Bess, 1970). In light of this success, the laboratory rearing of this parasitoid has been developed and research shows that *Bactrocera zonata* (Saunders) can serve as a favorable host for routine colonization of *F. arisanus* for mass production (Rousse et al., 2006). The study also showed that when the parasitoid developed on the tomato fruit fly, *N. cyanescens*, the survival rate was 10–25%. However, since *B. zonata* has been identified as a favorable host of this parasitoid, there is potential for effective augmentation biocontrol using this natural enemy to control tomato fruit flies.

2.4 True Bugs

Nezara viridula (L.) and *Lygus* spp. (Sannino and Espinosa, 2009) are common pests of tomato. In the Sacramento Valley of California, two economically important stink bug species are the consperse *Euschistus conspersus* Uhler and the red shouldered *Thyanta pallidovirens* (Stål) stink bugs (Pease and Zalom, 2010). Both fresh and processing tomatoes suffer from these pests that feed directly on the fruit, causing discoloration upon ripening, development of corky area below the surface, and association with yeast (*Nematospora* spp.) infection of the fruit (Pease and Zalom, 2010).

In California, the exotic *Trissolcus basalis* (Wollaston) has shown 100% parasitization of *N. viridula* egg masses visited by the parasitoid (Ehler, 2002). Other egg parasitoids of *N. viridula* include *Gryon obesum* Masner and *Telenomus podisi* Ashmead and *Ooencyrtus californicus* Girault and *Ooencyrtus johnsoni* (Howard). A potential biocontrol agent also includes the predatory *P. maculiventris* (De Clercq et al., 2002).

An increasing number of studies have reported the effects of non-crop plants on pests and their natural enemies (Gurr et al., 2004). Biocontrol of stink bug pests, with the help of non-crop flowering plants in the agroecosystem, has been examined and found to be a promising avenue. In studies by Bugg and Waddington (1994), Norris and Kogan (2000), and Pease and Zalom (2010), the role of non-crop plants in providing the beneficial insects, such as predators and parasitoids, with favorable habitat was emphasized. Such research, aimed at supporting ecologically based pest management, provides a feasible alternative to chemical control. In support of the benefits of non-crop plants, Pease and Zalom (2010) found that parasitism of sentinel egg masses of consperse stink bug was significantly greater adjacent to sweet alyssum, *Lobularia maritima* (L.) borders compared to no border controls. In addition to research in conservation biocontrol, the possibility of augmentation biocontrol of stink bug has also been examined by assessing various stink bug egg parasitoids for the potential of mass rearing. Due to the unavailability of stink bug eggs in autumn and winter, the mass production of stink bug egg parasitoids for augmentation biocontrol is a challenge (Corrêa-Ferreira and Oliveira, 1998; Doetzer and Foerster, 2013). To solve this problem, Kivan and Kilic (2005) studied the use of pentatomid eggs stored in liquid nitrogen to support sustained parasitoid production throughout the year. Further work reported that the fecundity of *T. basalis* parasitizing eggs of pentatomids, *N. viridula* and *Acrosternum pengue* (Rolston) that were stored in liquid nitrogen, remained unaffected (Doetzer and Foerster, 2013). The same was not the case with *T. podisi*, which showed a reduction in fecundity when parasitizing stored host eggs. Therefore, not all egg parasitoids respond well to stored host eggs but research aimed at identifying the ones that do, can support augmentation biocontrol. Clearly, further research in this field is necessary.

2.5 Psyllids

The tomato–potato psyllid (TPP), *Bactericera cockerelli* Šulc is a major pest of solanaceous crops such as tomato and potato, *Solanum tuberosum* L. in North and Central America (Liu et al., 2006; Munyaneza et al., 2007). It is also widely distributed in the tomato-producing regions of Mexico (Almeyda et al., 2008) and was detected in New Zealand in 2006, rapidly spreading throughout that country since then (Teulon et al., 2009; Thomas et al., 2011). Furthermore, it was reported from Australia's Norfolk Island in April 2014 (Anonymous, 2014). Psyllids cause chlorosis and shortened, swollen internodes of the upper growth of tomato (List, 1925). Perhaps more seriously, they are responsible for the psyllid yellows disease (Munyaneza et al., 2010) and "zebra chip" in potato. The causative agent is an unculturable bacteria, *Candidatus* Liberibacter solanacearum (CLso) Jagoueix, vectored by TPP (Munyaneza et al., 2007). This phytopathogen has been shown to affect tomato volatile organic compounds (VOCs) by downregulating jasmonate, salicylate, and ethylene-regulated defense gene expression (Casteel et al., 2012).

The southern ladybird beetle, *Cleobora mellyi* Mulsant is endemic to Australia (Slipinski, 2007), and it is a voracious predator of TPP in New Zealand (O'Connell et al., 2012). Longevity of the ladybird was shorter on TPP-only diet but could be improved by provisioning the predator with buckwheat, *F. esculentum* (Pugh et al., 2015). Additionally, the "new association" predatory mite, *Amblydromalus limonicus* (Garman and McGregor) in New Zealand is now mass-reared and commercialized, which has great potential as a biological control agent of TPP (Xu and Zhang, 2015). The term "new association" refers to the fact that *A. limonicus* is a local natural enemy in New Zealand (Knapp et al., 2013) and is an effective biological control agent of the invasive TPP.

2.6 Thrips

The western flower thrips, *Frankliniella occidentalis* Pergande are notorious pests of tomato (Gabarra et al., 1995). They are preyed upon by the mirid *D. tamaninii*. Other potential biocontrol agents of these pests include the phytoseiid predatory mites, *Amblyseius swirskii* Athias-Henriot and *Neoseiulus cucumeris* (Oudemans) (Hewitt et al., 2015). Entomopathogenic fungi such as *Beauveria bassiana* (Balsamo-Crivelli) Vuillemin, *M. anisopliae*, *Lecanicillium lecanii* (Zimmerman) Zare and Gams, and *Metarhizium flavoviride* Gams and Rozsypal var *minus* are potential biocontrol agents of tomato thrips, *Thrips tabaci* Lindeman (Hemalatha et al., 2014).

The seasonal variation in the greenhouse climate may affect the efficacy of biocontrol agents. To study the extent to which this is true, small scale and commercial greenhouse studies were conducted on two species of phytoseiid predatory mites, *A. swirskii* and *N. cucumeris*, in controlling western flower thrips on chrysanthemum. The results of this study based on predation and oviposition rates showed that *A. swirskii* performed well in the summer and winter, whereas *N. cucumeris* was recommended as a cost-effective biological control agent in the winter months (Hewitt et al., 2015). Although, this study was conducted on chrysanthemum, it gives a valuable insight into the importance of selecting adaptable biological control agents of pests attacking tomato, such as *F. occidentalis*. Therefore, the ability of the natural enemy to adapt to the seasonal variations should play a crucial role in biocontrol agent selection.

2.7 Mites

A number of mite species such as red tomato spider mite, Tetranychus evansi Baker and Pritchard, two spotted spider mite, *Tetranychus urticae* Koch (Ferrero et al., 2014), carmine spider mite, *Tetranychus cinnabarinus* Boisduval and tomato russet mite, *Aculops lycopersici* (Massee) (Perring and Farrar, 1986; Berlinger et al., 1988) are pests of tomato. A number of phytoseiid predatory mite species are used as biological control agents for pest mites. *Phytoseiulus longipes* Evans successfully develops on *T. evansi* and *T. urticae* (Ferrero et al., 2014). These authors found that the Brazilian and Argentinian populations of *P. longipes* were able to develop and reproduce on *T. evansi* on tomato, but a Chilean population of *P. longipes* was not. In this case, relying exclusively on the Chilean population of predatory mites without prior experimentation would result in a failure of biocontrol efforts. A study conducted by Pappas et al. (2015) showed that feeding of the phytophagous predator, *M. pygmaeus* on tomato plants actually suppressed the performance of subsequently appearing *T. urticae* by reducing egg deposition and adult presence. The reason for the pest suppression is attributed to the induction of plant-defense responses by accumulation of transcripts and the activity of proteinase inhibitors locally and systemically in the tomato plants with prior *M. pygmaeus* exposure.

In addition to predatory mites, entomopathogenic fungi are promising biocontrol agents of *T. urticae*. They can infect *T. urticae* by growing through their cuticle (Chandler et al., 2005). *M. anisopliae, Hirsutella* spp., and *Verticillium lecanii* (Zimmerman) Zare and Gams were found to be most effective in causing mortality in *T. urticae,* attacking glasshouse tomato crops using a single dose (Chandler et al., 2005). In this study, it was also found that three isolates from commercial biopesticides, *B. bassiana* (cultured from "Naturalis-L," Troy Biosciences, Phoenix, TX, US); *Hirsutella thompsonii* Fisher (cultured from "Mycar," Abbott Laboratories, US); and *V. lecanii* (used in "Mycotal" Koppert B.V., The Netherlands) were also pathogenic to *T. urticae*.

2.8 Nematodes

The roots of tomato are attacked by root-knot nematodes, *Meloidogyne* spp. which is usually the most damaging species of nematodes (Affokpon et al., 2011). Free-living soil fungi, *Trichoderma* spp., are regarded as potential biological control agents against root-knot nematodes (Spiegel and Chet, 1998; Sharon et al., 2001). Research targeting the root-knot nematodes was carried out in Benin, West Africa to isolate and identify local parasitic fungal populations in the genus *Trichoderma* that could be utilized as biocontrol agents (Affokpon et al., 2011). Based on field assessment, the authors of this study reported a significant inhibition of nematode reproduction, suppression of root galling, and an increase of tomato yield compared with the non-fungal control treatments.

In addition to this parasitic fungi, three antagonistic microbes including the bacterium *Pseudomonas fluorescens* Migula, fungus *Paecilomyces lilacinus* (Thom) Samson and yeast *Pichia guilliermondii* Wick were found to be effective in controlling root-knot nematode *Meloidogyne incognita* Göldi on tomato under greenhouse conditions. Plants in these treatments experienced enhancing growth parameters such as seed germination, seedling and root-system vigor, and mean shoot length (Hashem and Abo-Elyousr, 2011). However, the study found that the soil-inhabitant cyanobacterium *Calothrix parietina* Nägeli had a negative interaction between either of the biocontrol agents *P. lilacinus* or *P. guilliermondii*, indicating that the presence of *C. parietina* may reduce the efficiency of biocontrol agents in the soil. Such findings are important to pin point the deterrent factors that render a biocontrol agent less effective, which then can follow amelioration efforts to allow the biocontrol agent to reach its full potential.

2.9 Pathogens

Plants may suffer from microbial infections transmitted by various arthropod vectors such as tomato yellow leaf curl virus (transmitted by whiteflies) and potato virus Y (transmitted by aphids) (Berlinger et al., 1988). Additionally, tomato, similar to other nursery-based vegetable crops, are attacked by soil-borne plant pathogens such as fungal and oomycetes pathogens *Pythium ultimum* Trow, *Phoma betae* Frank, *Rhizopus stolonifer* (Ehrenb: Fr.) Vuill. and *Fusarium oxysporum* Schlecht (Shanahan et al., 1992), and other *Pythium* and *Phytophthora* spp. (Dukare et al., 2011). Roots also are affected by other pathogens, such as the soil-borne bacteria *Ralstonia solanacearum* (Smith) which is a Gram-negative bacterium that causes bacterial wilt in tomato (Hayward, 1991; Fujiwara et al., 2011).

Losses of tomato produce can also occur post-harvest. It is estimated that over 20% of fruits and vegetables are lost in developed countries due to post-harvest diseases (Gullino et al., 1991). For example, the fungal pathogen *Botrytis cinerea* Pers. ex Fr. causes rotting of tomato fruits that are damaged during harvesting and packaging (Bello et al., 2008). Severe crown and stem rot in tomatoes is a major problem and is caused by the fungal pathogens, *Rhizoctonia solani* Kühn and *Sclerotium rolfsii* Sacc (De Curtis et al., 2010).

Fortunately, there are many instances of microbial biocontrol agents being examined and employed. Commercial bio-fungicides containing antagonistic bacteria *Bacillus subtilis* Cohn and soil fungi *Trichoderma asperellum Samuels,* Lieckf. and Nirenberg are used to control pathogenic fungi attacking tomato crops (De Curtis et al., 2010). Biological control by microbes is a result of various antagonistic effects such as mycoparasitism, secretion of antibiotics and cell-wall degrading enzymes, competition for space and nutrients, induced systemic resistance (ISR), and rhizobacteria-mediated induced systemic resistance (RISR) (Madi et al., 1997; Pieterse and van Loon, 1999; Pieterse et al., 2002; Harman et al., 2004; Jones and Dangl, 2006; Bakker et al., 2007; Jung et al., 2007; Van der Ent et al., 2008; De Curtis et al., 2010). In search for sources of effective biocontrol microbes, Huang et al. (2013) found that the rhizosphere from diseased plants proved to be a good reservoir of biocontrol bacteria and that the isolates from rhizosphere of diseased tomato plants performed better at reducing disease incidence compared to those isolated from healthy plants. Therefore, it was concluded that the rhizosphere of infected plants is a good source for finding plant growth-promoting rhizobacteria (PGPRs).

Plant growth-promoting rhizobacteria (PGPRs) such as *Bacillus* spp. (Cao et al., 2011; Huang et al., 2011), *Pseudomonas* spp. (Boer et al., 2003; Guo et al., 2004), *Streptomyces* spp. (Sabaratnam and Traquair, 2002; Boukaew et al., 2010) and *P. guilliermondii* (Lahlali et al., 2011) isolated mainly from the rhizosphere of healthy tomato plants are used for biological control of *R. solanacearum* (Vanitha et al., 2009; Wei et al., 2011; Zhao et al., 2011; Huang et al., 2013). Microbial biocontrol agents such as unique lytic phages, φRSL1, have also been shown to control *R. solanacearum* (Fujiwara et al., 2011). It was found that using φRSL1 as a biocontrol agent was far more effective than a phage cocktail consisting of highly virulent phages (Fujiwara et al., 2011). Tomato benefited from treatments with lytic phage φRSL1 at the seedling stage by being less wilted due to limited penetration, growth, and movement of *R. solanacearum* cells.

Fungal and oomycetes pathogens in tomato have been found to be controlled by *P. fluorescens*, a bacterial biocontrol agent. This bacterium produces secondary metabolites such as siderophores, diacetyl-phloroglucinol (DAPG), hydrogen cyanide, and an extracellular protease which it employs to control pathogens (Shanahan et al., 1992; Aarons et al., 2000). Yeast species *Rhodotorula rubra* (Schimon) F.C. Harrison and *Candida pelliculosa* Redaelli were found to be strongly antagonistic to *B. cinerea* in vitro. Further storage experiments indicated that these yeast species could be used as biocontrol agent of gray mold as an environmentally friendly alternative to fungicide treatment to control post-harvest gray mold in tomato (Bello et al., 2008).

Choice of biocontrol agents with the ideal capacity of pest management is crucial. Depending on the adaptability of the beneficial organism, amendments may be required to achieve desired results. For example, the wild strain of *P. fluorescens* has certain limitations such as limited motility and lower competitive colonization ability compared to a triple mutant strain KSW (Barahona et al., 2011). It is therefore suggested by the authors that biocontrol efforts may be supported by utilizing mutated strains of naturally occurring biocontrol agents.

In addition to research involving the determination of the most susceptible pest life stages as well as the possibility of using complimentary biocontrol agents, simple modifications in cultural practices or utilization of mutant strains instead of wild strains were shown to bring about positive results. For example, the application of novel antagonistic bacterial isolates, *Burkholderia cepacia* Yabuuchi et al., and *Pseudomonas* spp. via drip irrigation in Italy turned out to be effective biological control technique. The severity of disease caused by *S. rolfsii* and *R. solani* was reduced by up to 63.8% and 73.2%, respectively (De Curtis et al., 2010). Also, the plant pathogenic fungi *F. oxysporum*, *Pythium debaryanum* Hesse, *Pythium aphanidermatum* (Edson) Fitzp and *R. solani* in tomato, were controlled by compost/compost tea preparations amended by antagonistic cyanobacterial/bacterial cultures, of which *Anabaena oscillarioides* (ex Bornet and Flahault) and *B. subtillis* were most effective (Dukare et al., 2011).

3. BIOLOGICAL CONTROL PRACTICES IN TOMATOES

3.1 Importation Biological Control

Biocontrol of greenhouse whitefly *T. vaporariorum* with the parasitic wasp *E. formosa* is touted as the first successful instance of classical biological control in protected cultivation in Europe, Canada, Australia, and New Zealand with greatest success in the Netherlands and Great Britain (van Lenteren and Woets, 1988; Bale et al., 2008). Other examples are the nematode, *Steinernema scapterisci* Nguyen and Smart Larra, wasp, *Larra bicolor* F. and Brazilian red-eyed fly, *Ormia deplete* (Wiedemann) which are used commercially as biopesticides against mole crickets in Florida tomatoes (Kerr et al., 2014).

3.2 Augmentation Biological Control

Trichoderma spp. currently are used for root-knot nematode management, such as in Kenya (*Trichoderma* spp.-based product, Labuschagne, 2008), Cuba (*Trichoderma harzianum* Rifai-based product, Wabule et al., 2003), Israel (registered as BioNem, Sikora et al., 2000), and United States (available commercially as T-22 Planter Box, Bennett et al., 2009). Also, lacewings have been considered for mass rearing and releasing in tomato crops in California, United States (Wittenborg and Olkowski, 2000). *M. pygmaeus* and *Nesidiocoris tenuis* (Reuter) are endemic natural enemies of *T. absoluta* that commonly appear in tomato crops in the Mediterranean basin (Arnó et al., 2010) and are mass reared and released, resulting in effective control of *T. absoluta* eggs and whitefly. Sometimes these are used along with *B. thuringiensis* that is highly effective in controlling first instar larvae of *T. absoluta* (Urbaneja et al., 2012). An integration of *Bt* treatments with inoculative releases of *N. tenuis* works best in greenhouse tomato (Mollá et al., 2011) and in certain other cropping systems (Urbaneja et al., 2012).

3.3 Conservation Biological Control

Plants with high nectar sources improve the performance of many natural enemies of tomato pests. For example, parasitism of mole crickets in Florida by the *Larra* wasp, has been enhanced by the availability of nectar from the shrubby false buttonweed, *Spermacoce verticillata* L. and patridge pea, *Chamaecrista fasciculata* (Michx.) (Portman et al., 2010). A recent study shows that sowing flower mixture strips among rows of tomato increased abundance and diversity of natural enemy functional groups as well as bees. Interestingly, although damage to plants from lepidopteran pests was highest in plots adjacent to wildflower strips, a higher crop productivity also was recorded from these plots (Balzan et al., 2016). Natural enemies, such as egg parasitoids and the predatory spined stilt bug, *Jalysus wickhami* Van Duzee of pestiferous stink bugs in tomato cultivation in California were found to benefit from sweet alyssum flowers, *L. maritima* bordering the tomato fields (Pease and Zalom, 2010). Conservation of *N. tenuis* and other local natural enemies by significantly reducing the use of insecticides has proved effective in the control of *T. absoluta* in open-field tomatoes in Israel (Harpaz et al., 2011). In the Northeast of Spain, an IPM program in greenhouse and open-field tomato for control of *T. absoluta* based on the predator *M. pygmaeus* takes advantage of large numbers of predators colonizing non-heavily sprayed crops (Gabarra et al., 2004). In addition, insecticides are used selectively and growers use *Bt*, which complement the efficacy *of M. pygmaeus* (Urbaneja et al., 2012). Releasing the predator in tomato nurseries as well as using banker plants (tobacco) also conserves the native predator *Macrolophus caliginosus* Wagner for greenhouse tomato (Arnó et al., 2000; Lenfant et al., 2000). Use of pest management tactics such as host-plant resistance should be examined for compatibility with natural enemy conservation. For example, trichome-based host-plant resistance in *Lycopersicon* species appears to be incompatible with biological control of aphids since the predator, *Mallada signata* Schneider, commonly used as a biocontrol agent of aphids, showed low efficacy due to increased cannibalism and trichome entrapment–related mortality (Simmons and Gurr, 2004).

4. MANAGEMENT STRATEGIES TO SUPPORT BIOLOGICAL CONTROL EFFORTS

There are several crop management practices that have been shown to impact both natural control (control of pests provided by natural processes that may be biotic or abiotic factors, without human intervention) and biological control.

4.1 Adaptability of Natural Enemies

The success of natural enemies depends heavily on their acclimatization to the local climate. For example, isolating and utilizing fungal strains that are locally adapted to the climate and plant-parasitic nematode species seem to be the best management strategy to control root-knot nematodes (Stirling, 1991). Changing the site of natural enemy release can also improve the success of natural enemy. After foliar applications, infective juveniles (IJs) of entomopathogenic nematodes (EPNs) protect themselves from harmful environmental factors such as ultraviolet rays and desiccation by entering galleries made by insects in leaves or hidden foliage (Arthurs et al., 2004). Keeping this behavior of the IJs in mind, efficacy of EPNs to control *T. absoluta* can be increased by spraying the soil to control the last instar larvae as well as adults that emerge from pupae buried in soil (Batalla-Carrera et al., 2010).

4.2 Pesticide Use

Instead of spraying *T. urticae* directly with the commercial biopesticides Naturalis-L and *B. bassiana* isolated from Naturalis-L, greater mortality of the pest can be achieved by exposing the mites to tomato leaflets that have been sprayed previously with conidia suspension (Chandler et al., 2005). Biocontrol of root-knot nematode *Meloidogyne javanica* (Treub) can be achieved by using a combination of nematode trapping fungi *Arthrobotrys oligospora* Fresen and salicylic acid (Mostafanezhad et al., 2014). Other tactics include precisely timed applications of the biopesticide Mycotal, which is a formulation of *V. lecanii* or white halo fungus to control *B. tabaci* without having antagonistic effects on *E. formosa* (Jazzar and Hammad, 2004). The other recommended insecticide that is least harmful to natural enemies, pollinators, and the environment is *B. thuringiensis* (Ahmad et al., 2009; Biondi et al., 2013).

4.3 Habitat Manipulation

Predatory mirids such as *Dicyphus* spp. and *Macrolophus* spp. as well as hymenopteran parasitoids attacking pests of tomato have been known to benefit from access to non-pest resources such as pollen, floral, and extra-floral nectar and insect honeydew (Jervis et al., 1993; Wäckers, 2005; Vandekerkhove and De Clercq, 2010; Wäckers and van Rijn, 2012; Portillo et al., 2012). A strong correlation between abundance of natural enemies such as Coccinellidae, Nabidae, Syrphidae, Chrysopidae (Neuroptera), Thomisidae (Araneae) and availability of wildflower strips was reported in a study conducted in eight organic tomato fields in Italy (Balzan and Moonen, 2014). In this study, the wildflower strips included flowers such as buckwheat, *F. esculentum*, common bean, *Phaseolus vulgaris* L., hairy vetch, *Vicia villosa* Roth, coriander, *Coriandrum sativum* L., fennel, *Foeniculum vulgare* Mill., and borage, *Borago officinalis* L. Therefore, conservation of seminatural vegetation in preexisting field margins and supplementing that with flowering strips is suggested as a valuable management strategy to support arthropod functional diversity in ephemeral crops such as tomato (Balzan and Moonen, 2014). The fecundity and longevity of the parasitoid *N. artynes* were enhanced by the availability of flowering plants such as sweet alyssum, *L. maritima*. In addition, buckwheat, *F. esculentum* enhanced longevity of this wasp (Balzan and Wäckers, 2013).

5. CONCLUSION

Various biocontrol agents have been identified through dedicated research in tomato pest management. Depending on the cropping systems (open-field or greenhouse tomatoes), certain biocontrol agents seem to fair better compared to others. Adaptability of the natural enemies to seasonal variation plays a crucial role in their success as biocontrol agents. It is evident that biological control of pests in tomato is a promising alternative that can be sustained without causing major damage to the environment. Many examples show the implementation of research involving importation, augmentation, and conservation biocontrol. These tactics should be integrated with other IPM approaches. Research also points to the importance of better farm management to conserve natural enemies, utilizing different species of natural enemies that are compatible with each other, and contribute to a holistic IPM program. While the population of the world grows and the need to produce more food increases annually, the conventional agricultural approaches are being modified to accommodate the health of the environment. Biological control is an important tool in facilitating this change in society, which is why the interest for finding potential biocontrol agents for pest management of tomato should be cultivated.

REFERENCES

Aarons, S., Abbas, A., Adams, C., Fenton, A., O'Gara, F., 2000. A regulatory RNA (PrrB RNA) modulates expression of secondary metabolite genes in *Pseudomonas fluorescens* F113. Journal of Bacteriology 182, 3913–3919.

Affokpon, A., Coyne, D.L., Htay, C.C., Agbèdè, R.D., Lawouin, L., Coosemans, J., 2011. Biocontrol potential of native *Trichoderma* isolates against root-knot nematodes in West African vegetable production systems. Soil Biology and Biochemistry 43, 600–608.

Ahmad, M.N., Ali, S.R.A., Masri, M.M.M., Wahid, M.B., 2009. Effect of *Bacillus thuringiensis*, Terakil-1 (R) and Teracon-1 (R) against oil palm pollinator, *Elaeidobius kamerunicus* and beneficial insects associated with *Cassia cobanensis*. Journal of Oil Palm Research 21, 667–674.

Almeyda, L.I.H., Sánchez, S.J.A., Garzón, T.J.A., 2008. Vectores causantes de punta morada de la papa en Coahuila y Nuevo León, México. Agricultura técnica en México 34, 141–150.

Alomar, O., 1994. Els mirids depredadors (Heteroptera: Miridae) en el control integrat de plagues en conreus de tomá quet (Ph.D. thesis). University of Barcelona, Barcelona, Spain.

Alomar, O., Albajes, R., 1996. Greenhouse whitefly (Homoptera: Aleyrodidae) predation and tomato fruit injury by the zoophytophagous predator *Dicyphus tamaninii* (Heteroptera: Miridae). In: Alomar, O., Wiedenmann, R.N. (Eds.), Zoophytophagous Heteroptera: Implications for Life History and Integrated Pest Management. Entomological Society of America, Lanham, Michigan, pp. 155–177.

Alomar, O., Castanäeâ, C., Gabarra, R., Arno, J., Arinäo, J., Albajes, R., 1991. Conservation of native mirid bugs for biological control in protected and outdoor tomato crops. IOBC/WPRS Bulletin 14 (5), 33–42.

Anonymous, 2014. Exotic Pest Alert: Potato-Tomato Psyllid Eradication Plan. The Government of Norfolk Island, South Pacific, p. 2. Media Release 26 June, 2014, Update 5 http://norfolkisland.gov.nf/la/Media%20Releases/2014/2014_06_26%20_%20Exotic%20Pest%20Alert%20%20Potato%20Tomato%20psyllid%20Update5.pdf.

Apablaza, J., 1988. Avances en el control de la polilla del tomate. Evaluacio´n del insecticida cartap. Revista Ciencia Technologia e innovacion 35, 27–30.

Arńo, J., Ariño, J., Español, R., Martí, M., Alomar, O., 2000. Conservation of *Macrolophus caliginosus* Wagner (Het. Miridae) in commercial greenhouses during tomato crop-free periods. IOBC/WPRS Bulletin 23, 241–246.

Arnó, J., Gabarra, R., Liu, T.X., Simmons, A.M., Gerling, D., 2010. Natural Enemies of *Bemisia tabaci*: predators and parasitoids. In: Stansly, P.A., Naranjo, S.E. (Eds.), Bemisia: Bionomics and Management of a Global Pest. Springer International Publishing, Dordrecht, The Netherlands, pp. 385–421.

Arthurs, S., Heinz, K.M., Prasifka, J.R., 2004. An analysis of using entomopathogenic nematodes against above ground pest. Bulletin of Entomological Research 94, 297–306.

Bai, Y., Lindhout, P., 2007. Domestication and breeding of tomatoes: what have we gained and what can we gain in the future? Annals of Botany 100, 1085–1094.

Baimey, H., Coyne, D., Dagnenonbakin, G., James, B., 2009. Plant-parasitic nematodes associated with vegetable crops in Benin: relationship with soil physicochemical properties. Nematologia Mediterranea 37, 225–234.

Bakker, P.A.H.M., Pieterse, C.M.J., Van Loon, L.C., 2007. Induced systemic resistance by fluorescent *Pseudomonas* spp. Phytopathology 97, 239–243.

Bale, J.S., van Lenteren, J.C., Bigler, F., 2008. Biological control and sustainable food production. Philosophical Transactions of the Royal Society B 363, 761–776.

Balzan, M.V., 2013. Conservation of Vegetation and Arthropod Functional Diversity for Multi-pest Suppression and Enhanced Crop Yield in Tomato Cropping Systems. Scuola Superiore di Studi Universitari e di Perfezionamento Sant'Anna, Pisa, Italy.

Balzan, M.V., Bocci, G., Moonen, A.C., 2016. Utilisation of plant functional diversity in wildflower strips for the delivery of multiple agroecosystem services. Entomologia Experimentalis et Applicata 158 (3), 304–319.

Balzan, M.V., Moonen, A., 2014. Field margin vegetation enhances biological control and crop damage suppression from multiple pests in organic tomato fields. The Netherlands Entomological Society Entomologia Experimentalis et Applicata 150, 45–65.

Balzan, M.V., Wäckers, F.L., 2013. Flowers to selectively enhance the fitness of a host-feeding parasitoid: adult feeding by *Tuta absoluta* and its parasitoid *Necremnus artynes*. Biological Control 67, 21–31.

Barahona, E., Navazo, A., Martínez-Granero, F., Zea-Bonilla, T., Pérez-Jiménez, R.M., Martín, M., Rivilla, R., 2011. *Pseudomonas fluorescens* F113 Mutant with enhanced competitive colonization ability and improved biocontrol activity against fungal root pathogens. Applied and Environmental Microbiology 77 (15), 5412–5419.

Batalla-Carrera, L., Morton, A., García-del-Pino, F., 2010. Efficacy of entomopathogenic nematodes against the tomato leafminer *Tuta absoluta* in laboratory and greenhouse conditions. BioControl 55, 523–530.

Bello, G.D., Mónaco, C., Rollan, M.C., Lampugnani, G., Arteta, N., Abramoff, C., Ronco, L., Stocco, M., 2008. Biocontrol of postharvest grey mould on tomato by yeasts. Journal of Phytopathology 156, 257–263.

Bennett, A.J., Mead, A., Whipps, J.M., 2009. Performance of carrot and onion seed primed with beneficial microorganisms in glasshouse and field trials. Biological Control 51, 417–426.

Berlinger, M.J., Dahan, R., Mordechi, S., 1988. Integrated pest management of organically grown greenhouse tomatoes in Israel. Applied Agricultural Research 3 (5), 233–238.

Biondi, A., Zappalà, L., Stark, J.D., Desneux, N., 2013. Do biopesticides affect the demographic traits of a parasitoid wasp and its biocontrol services through sublethal effects? PLoS One 8 (9), e76548.

Boer, M.D., Bom, P., Kindt, F., Keurentjes, J.J.B., Sluis, I.V.D., Loon, L.C.V., Bakker, P.A.H.M., 2003. Control of *Fusarium* wilt of radish by combining *Pseudomonas putida* strains that have different disease-suppressive mechanisms. Phytopathology 93, 626–632.

Bonsignore, C.P., 2016. Environmental factors affecting the behavior of *Coenosia attenuata*, a predator of *Trialeurodes vaporariorum* in tomato greenhouses. Entomologia Experimentalis et Applicata 158 (1), 87–96.

Boukaew, S., Chuenchit, S., Petcharat, V., 2010. Evaluation of *Streptomyces* spp. for biological control of *Sclerotium* root and stem rot and *Ralstonia* wilt of chili pepper. BioControl 56, 365–374.

Brévault, T., Quilici, S., 2009. Oviposition preference in the oligophagous tomato fruit fly, *Neoceratitis cyanescens*. Entomologia Experimentalis et Applicata 133, 165–173.

Bugg, R.L., Waddington, C., 1994. Using cover crops to manage arthropod pests of orchards: a review. Agriculture Ecosystems and Environment 50, 11–28.

Cao, Y., Zhang, Z.H., Ling, N., Yuan, Y.J., Zheng, X.Y., Shen, B., Shen, Q.R., 2011. *Bacillus subtilis* SQR 9 can control *Fusarium* wilt in cucumber by colonizing plant roots. Biology and Fertility of Soils 47, 495–506.

Casteel, C.L., Hansen, A.K., Walling, L.L., Paine, T.D., 2012. Manipulation of plant defense responses by the tomato psyllid (*Bactericera cockerelli*) and its associated endosymbiont *Candidatus* Liberibacter psyllaurous. PLoS One 7 (4), e35191.

Castelo-Branco, M., 1992. Flutuac¸ao populacional da trac¸a-do-tomateiro no Distrito Federal. Horticultura Brasileira 10, 33–34.

Chailleux, A., Biondi, A., Han, P., Tabone, E., Desneux, N., 2013. Suitability of the pest–plant system *Tuta absoluta* (Lepidoptera: Gelechiidae) – Tomato for *Trichogramma* (Hymenoptera: Trichogrammatidae) parasitoids and insights for biological control. Journal of Economic Entomology 106 (6), 2310–2321.

Chandler, D., Davidson, G., Jacobson, R.J., 2005. Laboratory and glasshouse evaluation of entomopathogenic fungi against the two-spotted spider mite, *Tetranychus urticae* (Acari: Tetranychidae), on tomato, *Lycopersicon esculentum*. Biocontrol Science and Technology 15 (1), 37–54.

Colak, A., 2009. Biocontrol products for use against tomato diseases. In: Proceedings of the IInd International Symposium on Tomato Diseases, pp. 413–417 Kusadasi-Turkey, October 8–12, 2007.

Contreras, J., Mendoza, J.E., Martínez-Aguirre, M.R., García-Vidal, L., Izquierdo, J., Bielza, P., 2014. Efficacy of entomopathogenic fungus *Metarhizium anisopliae* against *Tuta absoluta* (Lepidoptera: Gelechiidae). Journal of Economic Entomology 107 (1), 121–124.

Corrêa-Ferreira, B.S., Oliveira, M.C.N., 1998. Viability of *Nezara viridula* (L.) eggs for parasitism by *Trissolcus basalis* (Woll.), under different storage techniques in liquid nitrogen. Annales de la Société Entomologique de Brasil 27, 101–107.

Cranham, J.E., Helle, W., 1985. Pesticide resistance Tetranychidae. In: Helle, W., Sabelis, M.W. (Eds.), Spider Mites, their Biology, Natural Enemies and Control. Elsevier Inc., Amsterdam, Netherlands, pp. 405–421.

De Backer, L., Caparros, M.R., Fauconnier, M.L., Brostaux, Y., Francis, F., Verheggen, F., 2015. *Tuta absoluta*-induced plant volatiles: attractiveness towards the generalist predator *Macrolophus pygmaeus*. Arthropod-Plant Interactions 9 (5), 465–476.

De Clercq, P., Wyckhuys, K., De Oliveira, H.N., Klapwijk, J., 2002. Predation by *Podisus maculiventris* on different life stages of *Nezara viridula*. Florida Entomologist 85 (1), 197–202.

De Curtis, F., Lima, G., Vitullo, D., De Cicco, V., 2010. Biocontrol of *Rhizoctonia solani* and *Sclerotium rolfsii* on tomato by delivering antagonistic bacteria through a drip irrigation system. Crop Protection 29, 663–670.

DeBach, P., 1964. The scope of biological control. In: DeBach, P. (Ed.), Biological Control of Insect Pests and Weeds. Reinhold, New York, USA, pp. 3–20.

Desneux, N., Wajnberg, E., Wyckhuys, K.A.G., Burgio, G., Arpaia, S., Narváez Vasquez, C.A., González-Cabrera, J., Catalán Ruescas, D., Tabone, E., Frandon, J., Pizzol, J., Poncet, C., Cabello, T., Urbaneja, A., 2010. Biological invasion of European tomato crops by *Tuta absoluta*: ecology, geographic expansion and prospects for biological control. Journal of Pest Science 83, 1–19.

Doetzer, A.K., Foerster, L.A., 2013. Storage of pentatomid eggs in liquid nitrogen and dormancy of *Trissolcus basalis* (Wollaston) and *Telenomus podisi* Ashmead (Hymenoptera: Platygastridae) adults as a method of mass production. Neotropical Entomology 42, 534–538.

Dolling, W.R., 1991. The Hemiptera. Natural History Museum Publications, Oxford University Press, Oxford, UK.

Down, R.E., Bell, H.A., Matthews, H.J., Kirkbride-Smith, A.E., Edwards, J.P., 2004. Dissemination of the biocontrol agent *Vairimorpha necatrix* by the spined soldier bug, *Podisus maculiventris*. Entomologia Experimentalis et Applicata 110, 103–114.

Dukare, A.S., Prasanna, R., Dubey, S.C., Nain, L., Chaudhary, V., Singh, R., Saxena, A.K., 2011. Evaluating novel microbe amended composts as biocontrol agents in tomato. Crop Protection 30, 436–442.

Ehler, L.E., 2002. An evaluation of some natural enemies of *Nezara viridula* in northern California. BioControl 47, 309–325.

Ferrero, M., Tixier, M.S., Kreiter, S., 2014. Different feeding behaviors in a single predatory mite species. 1. Comparative life histories of three populations of *Phytoseiulus longipes* (Acari: Phytoseiidae) depending on prey species and plant substrate. Experimental and Applied Acarology 62, 313–324.

Fujiwara, A., Fujisawa, M., Hamasaki, R., Kawasaki, T., Fujie, M., Yamada, T., 2011. Biocontrol of *Ralstonia solanacearum* by treatment with lytic bacteriophages. Applied and Environmental Microbiology 77 (12), 4155–4162.

Gabarra, R., Alomar, O., Castañé, C., Goula, M., Albajes, R., 2004. Movement of greenhouse whitefly and its predators between in- and outside of Mediterranean greenhouses. Agriculture. Ecosystems and Environment 10, 341–348.

Gabarra, R., Arnó, J., 2010. Resultados de las experiencias de control biológico de la polilla del tomate en cultivo de invernadero y aire libre en Cataluña. Phytoma España 217, 66–68.

Gabarra, R., Arnó, J., Lara, L., Verdú, M.J., Ribes, A., Beitia, F., Urbaneja, A., Téllez, M., del, M., Mollá, O., Riudavets, J., 2014. Native parasitoids associated with *Tuta absoluta* in the tomato production areas of the Spanish Mediterranean Coast. BioControl 59, 45–54.

Gabarra, R., Castanä Eâ, C., Bordas, E., Albajes, R., 1988. *Dicyphus tamaninii* Wagner as a beneficial insect and pest of tomato crops in Catalonia. Entomophaga 33, 219–228.

Gabarra, R., Castane, C., Albajes, R., 1995. The mirid bug *Dicyphus tamaninii* as a greenhouse whitefly and western flower thrips predator on cucumber. Biocontrol Science and Technology 5, 475–488.

Gerling, D., Alomar, O., Arno, J., 2001. Biological control of *Bemisia tabaci* using predators and parasitoids. Crop Protection 20, 779–799.

González-Cabrera, J., Mollá, O., Montón, H., Urbaneja, A., 2011. Efficacy of *Bacillus thuringiensis* (Berliner) in controlling the tomato borer, *Tuta absoluta* (Meyrick) (Lepidoptera: Gelechiidae). BioControl 56, 71–80.

Gullino, M.L., Aloi, C., Palitto, M., Benzi, D., Garibaldi, A., 1991. Attempts at biological control of postharvest diseases of apple. Med Fac Landbouww Rijksuniv Gent 56, 195–202.

Guo, J.H., Qi, H.Y., Guo, Y.H., Ge, H.L., Gong, L.Y., Zhang, L.X., Sun, P.H., 2004. Biocontrol of tomato wilt by plant growth-promoting rhizobacteria. Biological Control 29, 66–72.

Gurr, G.M., Scarratt, S.M., Wratten, S.D., Burndt, L., Irvin, N., 2004. Ecological engineering, habitat manipulation and pest management. In: Gurr, G.M., Wratten, S.D., Altieri, M.A. (Eds.), Ecological Engineering for Pest Management. Comstock Publishing Associates, Ithaca, New York, USA, pp. 1–12.

Haji, F.N.P., Olivera, C.A., Amorim, N.M.S., Batista, J.G.S., 1988. Fluctuaçao populational da traça do tomateiro no ub medio Sao Francisco. Pesqui Agropecu Bras 23, 7–17.

Han, P., Bearez, P., Adamowicz, S., Lavoir, A.-V., Amiens-Desneux, E., Desneux, N., 2015. Nitrogen and water limitations in tomato plants trigger negative bottom-up effects on the omnivorous predator *Macrolophus pygmaeus*. Journal of Pest Science 88, 685–691.

Haramoto, F.H., Bess, H.A., 1970. Recent studies on the abundance of the Oriental and Mediterranean fruit flies and the status of their parasites. In: Proceedings of the Hawaiin Entomological SocietyProceedings of the Hawaiin Entomological Society, vol. 20, pp. 551–566.

Harman, G.E., Howell, C.R., Viterbo, A., Chet, I., Lorito, M., 2004. *Trichoderma* species–opportunistic, virulent plant symbionts. Nature Reviews Microbiology 2, 43–56.

Harpaz, L.S., Graph, S., Rika, K., Azolay, L., Rozenberg, T., Yakov, N., Alon, T., Alush, A., Stinberg, S., Gerling, D., November 16–18, 2011. IPM of Tuta Absoluta in Israel. EPPO/IOBC/NEPPO Abstracts' Book of the International Symposium on management of *Tuta absoluta* (tomato borer). , p. 32 Agadir, Morocco.

Hashem, M., Abo-Elyousr, K.A., 2011. Management of the root-knot nematode *Meloidogyne incognita* on tomato with combinations of different biocontrol organisms. Crop Protection 30, 285–292.

Hayward, A.C., 1991. Biology and epidemiology of bacterial wilt caused by *Pseudomonas solanacearum*. Annual Review of Phytopathology 29, 65–87.

Hazir, S., Shapiro-Ilan, D.I., Hazir, C., Leite, L.G., Cakmak, I., Olson, D., 2016. Multifaceted effects of host plants on entomopathogenic nematodes. Journal of Invertebrate Pathology 135, 53–59.

Hemalatha, S., Ramaraju, K., Jeyarani, S., 2014. Evaluation of entomopathogenic fungi against tomato thrips, *Thrips tabaci* Lindeman. Journal of Biopesticides 7 (2), 151–155.

Hewitt, L.C., Shipp, L., Buitenhuis, R., Scott-Dupree, C., 2015. Seasonal climatic variations influence the efficacy of predatory mites used for control of western flower thrips in greenhouse ornamental crops. Experimental and Applied Acarology 65, 435–450.

Huang, J., Wei, Z., Tan, S., Mei, X., Yin, S., Shen, Q., Xu, Y., 2013. The rhizosphere soil of diseased tomato plants as a source for novel microorganisms to control bacterial wilt. Applied Soil Ecology 72, 79–84.

Huang, X.Q., Zhang, N., Yong, X.Y., Yang, X.M., Shen, Q.R., 2011. Biocontrol of *Rhizoctonia solani* damping-off disease in cucumber with *Bacillus pumilus* SQR-N43. Microbiological Research 167, 135–143.

Ingegno, B.L., Ferracini, C., Gallinotti, D., Alma, A., Tavella, L., 2013. Evaluation of the effectiveness of *Dicyphus errans* (Wolff) as predator of *Tuta absoluta* (Meyrick). Biological Control 67, 246–252.

Jacobson, R.J., Croft, P., Fenlon, J., 1999. Resistance to fenbutatin oxide in populations of *Tetranychus urticae* Koch (Acari: Tetranychidae) in UK protected crops. Crop Protection 18, 47–52.

James, B., Godonou, I., Atcha, C., Baimey, H., 2006. Healthy vegetables through participatory IPM in peri-urban areas of Benin. In: Abomey-Calavi (Ed.), Summary of Activities and Achievements 2003–2006. IITA Benin, Africa, p. 134.

Jazzar, C., Hammad, E.A.-F., 2004. Efficacy of multiple biocontrol agents against the sweet potato whitefly *Bemisia tabaci* (Gennadius) (Homoptera: Aleyrodidae) on tomato. Journal of Applied Entomology 128, 188–194.

Jervis, M.A., Kidd, N., Fitton, M., Huddleston, T., Dawah, H., 1993. Flower-visiting by hymenopteran parasitoids. Journal of Natural History 27, 67–105.

Jones, J.D.G., Dangl, J.L., 2006. The plant immune system. Nature 444, 323–329.

Jung, W.J., Mabood, F., Kim, T.H., Smith, D.L., 2007. Induction of pathogenesis-related proteins during biocontrol of *Rhizoctonia solani* with *Pseudomonas aureofaciens* in soybean (*Glycine max* L. Merr.) plants. BioControl 52, 895–904.

Kabiri, F., Vila, E., Cabello, T., 2010. *Trichogramma achaeae*: an excellent biocontrol agent against *Tuta absoluta*. Sting 33, 5–6.

Kerr, C.R., Leppla, N.C., Buss, E.A., Frank, J.H., 2014. Mole Cricket IPM Guide for Florida. UF/IFAS Extension IPM-206 http://edis.ifas.ufl.edu/pdffiles/IN/IN102100.pdf.

Kivan, M., Kilic, N., 2005. Effects of storage at low-temperature of various heteropteran host eggs on the egg parasitoid, *Trissolcus semistriatus*. BioControl 50, 589–600.

Knapp, M., van Houten, Y., Hoogerbrugge, H., Bolckmans, K., 2013. *Amblydromalus limonicus* (Acari: Phytoseiidae) as a biocontrol agent: literature review and new findings. Acarologia 53 (2), 191–202.

Kumari, D.A., Anitha, G., Anitha, V., Lakshmi, B.K.M., Vennila, S., Rao, N.H.P., 2015. New record of leaf miner, *Tuta absoluta* (Meyrich) in tomato. Insect Environment 20 (4), 136–138.

Labuschagne, L., 2008. Is IPM possible for small-scale bean farmers in Africa? Integrated Pest Management, Pesticides News 80, 18–19.

Lahlali, R., Hamadi, Y., Guilli, M.E., Jijakli, M.H., 2011. Efficacy assessment of *Pichia guilliermondii* strain Z1, a new biocontrol agent, against citrus blue mould in Morocco under the influence of temperature and relative humidity. Biological Control 56, 217–224.

Lapidot, M., Legg, J.P., Wintermantel, W.M., Polston, J.E., 2014. Management of whitefly-transmitted viruses in open-field production systems. Advances in Virus Research 90, 147–206.

Lenfant, C., Ridray, G., Schoen, L., 2000. Biopropagation of *Macrolophus caliginosus* (Wagner) for a quicker establishment in southern tomato greenhouses. IOBC/WPRS Bulletin 23, 247–251.

List, G.M., 1925. The tomato psyllid, *Paratrioza cockerelli* Sulc. Colorado State Entomology Circular 47, 16.

Liu, D., Trumble, J.T., Stouthamer, R., 2006. Genetic differentiation between eastern populations and recent introductions of potato psyllid (*Bactericera cockerelli*) into western North America. Entomologia Experimentalis et Applicata 118, 177–183.

Lopes-Filho, F., 1990. Tomate industrial no submédio Sao Fracisco e aspragas que limitamsua produçao. Pesqui Agropecu Bras 25, 183–288.

Madi, L., Katan, T., Katan, J., Henis, Y., 1997. Biological control of *Sclerotium rolfsii* and *Verticillium dahliae* by *Talaromyces flavus* is mediated by different mechanisms. Phytopathology 87, 1054–1060.

Marti Jr., O.G., Hamm, J.J., 1985. Effect of *Vairimorpha* sp. on the survival of *Geocoris punctipes* in the laboratory. Journal of Entomological Science 20, 354–358.

Meyer, B., Jauker, F., Steffan-Dewenter, I., 2009. Contrasting resource-dependent responses of hoverfly richness and density to landscape structure. Basic and Applied Ecology 10, 178–186.

Mollá, O., González-Cabrera, J., Urbaneja, A., 2011. The combined use of *Bacillus thuringiensis* and *Nesidiocoris tenuis* against the tomato borer *Tuta absoluta*. BioControl 56, 883–891.

Moraes, G.J., Normanha Filho, J.A., 1982. Surto de *Scrobipalula absoluta* (Meyrick) em tomateiro no tro´pico semi-a´rido. Pesqui Agropecu Bras 17, 503–504.

Mostafanezhad, H., Sahebani, N., Zarghani, S.N., 2014. Control of root-knot nematode (*Meloidogyne javanica*) with combination of *Arthrobotrys oligospora* and salicylic acid and study of some plant defense responses. Biocontrol Science and Technology 24 (2), 203–215.

Munyaneza, J.E., Crosslin, J., Upton, J.E., 2007. Association of *Bactericera cockerelli* (Homoptera: Psyllidae) with 'Zebra Chip', a new potato disease in southwestern United States and Mexico. Journal of Economic Entomology 100, 656–663.

Munyaneza, J.E., Fisher, T.W., Sengoda, V.G., Garczynski, S.F., Nissinen, A., Lemmetty, A., 2010. Association of '*Candidatus* Liberibacter *solanacearum*' with the psyllid, *Trioza apicalis* (Hemiptera: Triozidae) in Europe. Journal of Economic Entomology 103, 1060–1070.

Norris, R.F., Kogan, M., 2000. Interactions between weeds, arthropod pests, and their natural enemies in managed ecosystems. Weed Science 48, 94–158.

O'Connell, D., Wratten, S., Pugh, A., Barnes, A., 2012. 'New species association' biological control? Two coccinellid species and an invasive psyllid pest in New Zealand. Biological Control 62, 86–92.

Olson, D., Wäckers, F.L., 2007. Management of field margins to maximize multiple ecological services. Journal of Applied Ecology 44, 13–21.

Pappas, M.L., Steppuhn, A., Geuss, D., Topalidou, N., Zografou, A., Sabelis, M.W., Broufas, G.D., 2015. Beyond predation: the zoophytophagous predator *Macrolophus pygmaeus* induces tomato resistance against spider mites. PLoS One 10 (5), e0127251.

Pease, C.G., Zalom, F.G., 2010. Influence of non-crop plants on stink bug (Hemiptera: Pentatomidae) and natural enemy abundance in tomatoes. Journal of Applied Entomology 134, 626–636.

Perdikis, D., Lykouressis, D., Paraskevopoulos, A., Harris, K.M., 2011. A new insect pest, *Lasioptera* sp. (Diptera: Cecidomyiidae), on tomato and cucumber crops in glasshouses in Greece. OEPP/EPPO Bulletin 41, 442–444.

Perring, T.M., Farrar, C.A., 1986. Historical perspective and current world status of the tomato russet mite (Acari: Eriophyidae). Miscellaneous Publications of the Entomological Society of America 63, 1–19.

Pieterse, C.M.J., van Loon, L.C., 1999. Salicylic acid independent plant defence pathways. Trends in Plant Science 4, 52–58.

Pieterse, C.M.J., Van Wees, S.C.M., Ton, J., van Loon, L.C., 2002. Signalling in rhizobacteria-induced systemic resistance in *Arabidopsis thaliana*. Plant Biology 4, 535–544.

Poinar, G.Jr., Thomas, D.B., 2014. *Tripius gyraloura* n. sp. (Aphelenchoidea: Sphaerulariidae) parasitic in the gall midge *Lasioptera donacis* Coutin (Diptera: Cecidomyiidae). Systematic Parasitology 89, 247–252.

Portillo, N., Alomar, O., Wäckers, F., 2012. Nectarivory by the plant-tissue feeding predator *Macrolophus pygmaeus* Rambur (Heteroptera: Miridae): nutritional redundancy or nutritional benefit? Journal of Insect Physiology 58, 397–401.

Portman, S.L., Frank, J.H., McSorley, R., Leppla, N.C., 2010. Nectar-seeking and host-seeking by Larra bicolor, a parasitoid of *Scapteriscus* mole crickets. Environmental Entomology 39, 939–943.

Pugh, A.R., O'Connell, D.M., Wratten, S.D., 2015. Further evaluation of the southern ladybird (*Cleobora mellyi*) as a biological control agent of the invasive tomato–potato psyllid (*Bactericera cockerelli*). Biological Control 90, 157–163.

Roberts, R., 1990. Postharvest biological control of grey mould on apple by *Cryptococcus laurentii*. Phytopathology 80, 526–530.

Rousse, P., Gourdon, F., Quilici, S., 2006. Host specificity of the egg pupal parasitoid *Fopius arisanus* (Hymenoptera: Braconidae) in La Reunion. Biological Control 37, 284–290.

Sabaratnam, S., Traquair, J.A., 2002. Formulation of a *Streptomyces* biocontrol agent for the suppression of *Rhizoctonia* damping-off in tomato transplants. Biological Control 23, 245–253.

Salazar, E.R., Araya, J.E., 1997. Deteccion de resistencia a insecticidas en la polilla del tomate. Simiente 67, 8–22.

Salazar, E.R., Araya, J.E., 2001. Respuesta de la polilla del tomate, *Tuta absoluta* (Meyrick), a insecticidas en Arica. Agricultura Tecnica 61, 429–435.

Sannino, L., Espinosa, B., 2009. I Parassiti Animali Delle Solanacee. Il Sole 24 Ore Edagricole, Milan, Italy.

Shanahan, P., Sullivan, D.J.O., Simpson, P., Glennon, J.D., Gara, F.O., 1992. Isolation of 2, 4-diacetylphloroglucinol from a fluorescent pseudomonad and investigation of physiological parameters influencing its production. Applied and Environmental Microbiology 58, 353–358.

Sharon, E., Bar-Eyal, M., Chet, I., Herrera-Estrella, A., Kleifeld, O., Spiegel, Y., 2001. Biological control of the root-knot nematode *Meloidogyne javanica* by *Trichoderma harzianum*. The American Phytopathological Society 91, 687–693.

Sikora, R.A., Oka, Y., Sharon, E., Hok, C.J., Keren-zur, M., 2000. Achievements and research requirements for the integration of biocontrol into farming systems. Nematology 2, 737–738.

Simmons, A.T., Gurr, G.M., 2004. Trichome-based host plant resistance of *Lycopersicon* species and the biocontrol agent Mallada signata: are they compatible? Entomologia Experimentalis Applicata 113, 95–101.

Siqueira, H.A.A., Guedes, R.N., Picanco, M.C., 2000b. Insecticide resistance in populations of *Tuta absoluta* (Lepidoptera: Gelechiidae). Agricultural and Forest Entomology 2, 147–153.

Siqueira, H.A.A., Guedes, R.N.C., Fragoso, D.B., Magalhaes, L.C., 2001. Abamectin resistance and synergism in Brazilian populations of *Tuta absoluta* (Meyrick) (Lepidoptera: Gelechiidae). International Journal of Pest Management 47, 247–251.

Siqueira, H.A.A., Guedes, R.N.C., Picanco, M.C., 2000a. Cartap resistance and synergism in populations of *Tuta absoluta* (Lep., Gelechiidae). Journal of Applied Entomology 124, 233–238.

Slipinski, A., 2007. Australian Ladybird Beetles (Coleoptera): their Biology and Classification. Australian Biology Resources Study, Canberra, Australia, p. 286.

Spiegel, Y., Chet, I., 1998. Evaluation of *Trichoderma* spp. as a biocontrol agent against soilborne fungi and plant-parasitic nematodes in Israel. Integrated Pest Management Review 3, 1–7.

Stern, V.M., 1973. Economic Thresholds. Annual Review of Entomology 18, 259–280.

Stirling, G.R., 1991. Biological Control of Plant Parasitic Nematodes: Progress, Problems and Prospects. CAB International, Wallingford, UK, p. 282.

Teulon, D., Workman, P., Thomas, K., Nielsen, M., 2009. *Bactericera cockerelli*: incursion, dispersal and current distribution on vegetable crops in New Zealand. New Zealand Plant Protection 62, 136–144.

Thomas, K., Jones, D., Kumarasinghe, L., Richmond, J., Gill, G., Bullians, M., 2011. Investigation into the entry pathway for tomato potato psyllid *Bactericera cockerelli*. New Zealand Plant Protection 64, 259–268.

Tian, J., Hongliang, D., Li, L., Arthurs, S., Hao, C., Mascarin, G.M., Ma, R., 2016. Host plants influence susceptibility of whitefly *Bemisia tabaci* (Hemiptera: Aleyrodidae) to the entomopathogenic fungus *Isaria fumosorosea* (Hypocreales: Cordycipitaceae). Biocontrol Science and Technology 26 (4), 528–538.

Tscharntke, T., Bommarco, R., Clough, Y., Crist, T.O., Kleijn, D., Rand, T.A., Tylianakis, J.M., van Nouhuys, S., Vidal, S., 2007. Conservation biological control and enemy diversity on a landscape scale. Biological Control 43, 294–309.

Urbaneja, A., González-Cabrera, J., Arnó, J., Gabarra, R., 2012. Prospects for the biological control of *Tuta absoluta* in tomatoes of the Mediterranean basin. Pest Management Science 68, 12151222.

Van der Ent, S., Verhagen, B.W.M., Van Doorn, R., Bakker, D., Verlaan, M.G., Pel, M.J.C., Joosten, R.G., Proveniers, M.C.G., van Loon, L.C., Ton, J., Pieterse, C.M.J., 2008. MYB72 is required in early signaling steps of rhizobacteria-induced systemic resistance in *Arabidopsis*. Plant Physiology 146, 1293–1304.

van Lenteren, J.C., Woets, J., 1988. Biological and integrated pest control in greenhouses. Annual Review of Entomology 33, 239–269.

van Rijn, P.C.J., Kooijman, J., Wäckers, F.L., 2013. The contribution of floral resources and honeydew to the performance of predatory hoverflies (Diptera: Syrphidae). Biological Control 67, 32–38.

Vandekerkhove, B., De Clercq, B., 2010. Pollen as an alternative or supplementary food for the mirid predator *Macrolophus pygmaeus*. Biological Control 53, 238–242.

Vanitha, S.C., Niranjana, S.R., Mortensen, C.N., Umesha, S., 2009. Bacterial wilt of tomato in Karnataka and its management by *Pseudomonas fluorescens*. BioControl 54, 685–695.

Voegele, J., Pizzol, J., Babi, A., 1988. The overwintering of some *Trichogramma* species. Les Colloques de l'INRA 43, 275–282.

Wabule, M.N., Ngaruiya, P.N., Kimmins, F.K., Silverside, P.J., 2004. Registration for biocontrol agents in Kenya. In: Proceedings of the Pest Control Products Board Workshop Organized by Kenyan Agricultural Research Institute/Department for International Development Crop Protection Programme, Nakuru, Kenya. May 14–16, 2003, p. 230.

Wäckers, F.L., 2005. Suitability of (extra-) floral nectar, pollen, and honeydew as insect food sources. In: Wäckers, F.L., van Rijn, P.C.J., Bruin, J. (Eds.), Plant-provided Food for Carnivorous Insects: A Protective Mutualism and its Applications. Cambridge University Press, Cambridge, UK, pp. 17–74.

Wäckers, F.L., van Rijn, P.C.J., 2012. Pick and mix: selecting flowering plants to meet the requirements of target biological control insects. In: Gurr, G.M., Wratten, S.D., Snyder, W.E. (Eds.), Biodiversity and Insect Pests: Key Issues for Sustainable Management. John Wiley and Sons, Chichester, UK, pp. 139–165.

Wei, Z., Yang, X.M., Yin, S.X., Shen, Q.R., Ran, W., Xu, Y.C., 2011. Efficacy of *Bacillus* fortified organic fertiliser in controlling bacterial wilt of tomato in the field. Applied Soil Ecology 48, 152–159.

Winkler, K., Wäckers, F.L., Bukovinski-Kiss, G., van Lenteren, J.C., 2006. Nectar resources are vital for *Diadegma semiclausum* fecundity under field conditions. Basic and Applied Ecology 7, 133–140.

Wittenborg, G., Olkowski, W., 2000. Potato aphid monitoring and biocontrol in processing tomatoes. The IPM Practitioner 22 (3), 1–7.

Xu, Y., Zhang, Z.-Q., 2015. *Amblydromalus limonicus*: a "new association" predatory mite against an invasive psyllid (*Bactericera cockerelli*) in New Zealand. Systematic and Applied Acarology 20 (4), 375–382.

Zappalà, L., Bernardo, U., Biondi, A., Cocco, A., Deliperi, S., Delrio, G., Giorgini, M., Pedata, P.P., Rapisarda, C., Tropea Garzia, G., Siscaro, G., 2012. Recruitment of native parasitoids by the exotic pest *Tuta absoluta* in Southern Italy. Bulletin of Insectology 65, 51–61.

Zhao, Q.Y., Dong, C.X., Yang, X.M., Mei, X.L., Ran, W., Shen, Q.R., Xu, Y.C., 2011. Biocontrol of *Fusarium* wilt disease for *Cucumis melo* melon using bio-organic fertilizer. Applied Soil Ecology 47, 67–75.

Chapter 12

Entomopathogenic Nematodes as Biological Control Agents of Tomato Pests

Fernando Garcia-del-Pino[1], Ana Morton[1], David Shapiro-Ilan[2]

[1]Universitat Autònoma de Barcelona, Barcelona, Spain; [2]USDA-ARS, Byron, GA, United States

1. INTRODUCTION

Entomopathogenic nematodes (EPNs) from the families Heterorhabditidae and Steinernematidae are soil-inhabiting organisms that are obligate insect parasites in nature (Kaya and Gaugler, 1993). These nematodes have evolved a mutualistic association with bacteria in the genera *Xenorhabdus* and *Photorhabdus*. *Photorhabdus* is associated with *Heterorhabditis* spp. and is carried in the intestine of infective juveniles (IJs) (Bird and Akhurst, 1983; Silva et al., 2002) (Fig. 12.1).

Xenorhabdus is associated with *Steinernema* spp. and confined to a specific vesicle within the intestine of the IJs. Nematodes locate their potential host by following insect cues (Lewis et al., 2006).

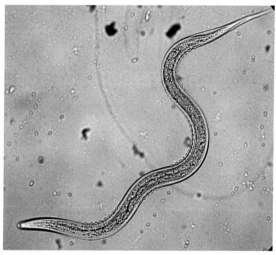

FIGURE 12.1 Infective juvenile of *Heterorhabditis indica* (average length=528 μm and average body diameter=20 μm). *Photo by Juan Morales, USDA-ARS.*

After IJs locate a host, they infect it through an orifice such as the anus, mouth, or spiracles, or by penetrating the cuticle (particularly in *Heterorhabditis* spp.). Once IJs enter the host, they shed their outer cuticle (Sicard et al., 2004) and begin ingesting hemolymph, which triggers the release of symbionts by defecation (in *Steinernema* spp.) or regurgitation (in *Heterorhabditis* spp.) (Martens et al., 2004; Martens and Goodrich-Blair, 2005). The nematode–bacteria complex kills the host within 24–48h through septicemia or toxemia (Dowds and Peters, 2002; Forst and Clarke, 2002). The developing nematodes then consume the bacteria, and liquefied insect tissues metabolized by the bacteria (Kondo and Ishibashi, 1988), mate, and can produce one or more generation before food resources become scarce. Bacteria recolonize the nematodes, which emerge as IJs from the depleted insect cadaver in search of new hosts (Poinar, 1990). These IJs do not feed and can survive in the soil for several months (Fig. 12.2).

Over 100 species of EPNs have been identified worldwide (approximately 80% are steinernematids), and at least 13 of these species have been commercialized (Shapiro-Ilan et al., 2014). In general, the innate virulence against different pest

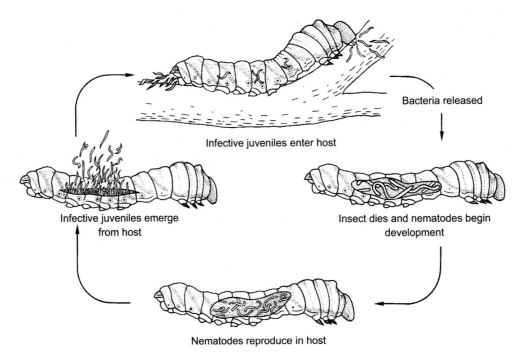

FIGURE 12.2   Generalized life cycle of entomopathogenic nematodes in a Lepidopteran larva.

species varies among EPN species. Additionally, differences among EPN species in host-seeking strategy and tolerance to environmental conditions such as temperature and desiccation can determine the field efficacy of EPNs (Morton and Garcia-del-Pino, 2009). For these reasons, it is necessary to choose the best adapted species or strains of nematode for each pest and for its particular ecological conditions.

EPNs can be applied with most conventional liquid application systems used in horticultural pesticide treatments, including pressurized sprayers, mist blowers, electrostatic sprayers, irrigation systems, or aerial sprays. Handling considerations to apply EPNs are agitation, nozzle type, pressure and recycling time, system environmental conditions, and spray distribution pattern (Grewal, 2002; Shapiro-Ilan et al., 2012).

EPNs have been extensively used in biological control of a variety of economically important pests occupying different habitats (Grewal et al., 2005). Most success has been achieved against soil-dwelling pests or pests in cryptic habitats such as inside galleries in plants where IJs find protection from environmental factors (Begley, 1990; Klein, 1990; Williams and Walters, 2000; Tomalak et al., 2005). The success against foliar pests where IJs are exposed to abiotic factors (UV radiation, desiccation, and extreme temperatures) has been limited in some cases (Begley, 1990). However, EPN formulation to retard desiccation and/or the addition of adjuvants to increase leaf coverage and persistence of the IJs has enhanced the use of EPNs against foliar pests (Williams and Walters, 2000; Arthurs et al., 2004; Head et al., 2004).

Large-scale production and new formulations create marketing opportunities for EPNs in different crops. Horticulture, and particularly vegetables crops, offers a suitable habitat for the use of EPNs (protection from adverse environmental factors, frequent irrigation, soil type, etc.). This chapter will review selected literature on the use of EPNs for control of tomato pests above and below ground and discuss aspects of their efficacy.

2. ROLE OF EPNS IN CONTROL OF PESTS IN TOMATO

2.1 The Tomato Leafminer, *Tuta absoluta* (Meyrick)

The tomato leafminer, *Tuta absoluta* (Meyrick) is an important pest of tomato crops in South America and has recently arrived in Europe affecting tomato plantations. The susceptibility of *T. absoluta* larvae and pupae to three species of entomopathogenic nematodes, *Steinernema carpocapsae* (Weiser), *Steinernema feltiae* (Filipjev), and *Heterorhabditis bacteriophora* Poinar was examined under laboratory conditions by Batalla-Carrera et al. (2010). These authors determined high larval mortality (Fig. 12.3) when a dose of 50 IJs/cm^2 was applied to petri dish experiments: 86.6% mortality with *S. carpocapsae* and 100% with *H. bacteriophora* and *S. feltiae*. In contrast with the larvae, pupae were less susceptible. Percentage of pupae infected by nematodes ranged from 8.3% caused by *S. feltiae*, 5% by *S. carpocapsae*, and 1.7% by *H. bacteriophora*.

FIGURE 12.3 *Tuta absoluta* larva parasitized by entomopathogenic nematodes in soil.

EPNs also are able to find and kill larvae on tomato leaves. In a leaf bioassay conducted in laboratory conditions (Batalla-Carrera et al., 2010), a high level of larval infection (77.1–91.7%) was recorded revealing the nematode's capacity to kill the larvae inside galleries. This is facilitated because tomato leafminer larvae, throughout their development, produce tunnels in the leaves generating entry holes that EPNs can use to penetrate and kill the larvae.

The efficacy of EPNs after foliar application to potted tomato plants was also evaluated under greenhouse conditions (Batalla-Carrera et al., 2010). Tomato plants infected with larvae of *T. absoluta* were sprayed with a 15 mL suspension of 1000 IJs/mL and 0.05% of the oil adjuvant Addit (Koppert). The foliar application of EPNs resulted in an efficacy between 87% for *S. carpocapsae* and 95% for *S. feltiae* and *H. bacteriophora*. One of the major obstacles to EPN efficacy in foliar application is limited persistence. In tomato cropping systems, the period of time that EPNs need to infect the insect is a relevant factor. Median lethal time (LT50) of *S. carpocapsae* applied on tomato leaf surfaces in laboratory conditions is 7.6 h (Garcia-del-Pino et al., unpublished data). Batalla-Carrera et al. (2010) determined that after 1h of nematode exposure mortality rose to 53% and reached up to 86% after 3h. Consequently, a foliar application of EPNs would allow the survival of the IJs long enough to find and infect the larvae on the surface of the leaf by penetrating into the galleries. Once the nematodes penetrate into the galleries they are protected from adverse environmental conditions.

In another series of studies, susceptibility of *T. absoluta* to *Steinernema monticolum* Stock, Choo & Kaya, and *H. bacteriophora* (HP88) was determined in both the laboratory and field (Shamseldean et al., 2014). In leaf bioassays conducted in the laboratory both nematodes showed high levels of virulence (80–100% mortality). Foliar application of the two EPN species in a field trial conducted over 3 years revealed 60–80% *T. absoluta* mortality for *H. bacteriophora* (HP88) and 58–67% mortality when using *S. monticolum*. Thus, the authors concluded that *H. bacteriophora* (HP88) was more virulent than *S. monticolum* against *T. absoluta* in the laboratory and in the field (Shamseldean et al., 2014).

Current biological control strategies using predators, parasitoids, and entomopathogens against *T. absoluta* are focused mainly on controlling adults, eggs, and feeding larvae in the aerial parts of the plant (leaves, stems, or fruits). Thus far, no strategies have been developed to control stages of *T. absoluta* in the soil. The application of EPNs on soil to control last instar larvae, when they drop from the leaves to pupate, as well as emerging adults from the buried pupae could be a complementary strategy to the foliar treatments in the control of *T. absoluta*. The efficacy of soil treatments of three native entomopathogenic nematodes (*S. carpocapsae*, *S. feltiae*, and *H. bacteriophora*) against *T. absoluta* larvae, pupae, and adults was determined under laboratory conditions by Garcia-del-Pino et al. (2013). When larvae dropped into soil to pupate, soil application of nematodes resulted in a high level of larval mortality; 100%, 52.3%, and 96.7% for *S. carpocapsae*, *S. feltiae*, and *H. bacteriophora*, respectively. No mortality of pupae was observed, yet the mortality of adults emerging from soil was 79.1% for *S. carpocapsae* and 0.5% for *S. feltiae*. The susceptibility of emerging adults to *S. carpocapsae* offers a new perspective for the control of this insect. The application of *S. carpocapsae* to the soil surface in tomato plantations could create a nematode barrier that the tomato leafminer adult would have to pass through before reaching the tomato plant. This strategy could increase the effectiveness of EPNs against *T. absoluta*. Although adult mortality caused by nematodes usually occurs within 24–48 h after emergence, during this time adults do not produce progeny because the most prolific oviposition period of the *T. absoluta* females is 7 days after first mating, when they lay 76% of their eggs (Uchôa-Fernandes et al., 1995). Furthermore, as most infected adults will produce nematode progeny, these adult insects may aid nematode dispersal prior to death, although mobility of infected adults prior to succumbing is not yet known.

Garcia-del-Pino et al. (2013) reported on the compatibility of EPNs with some of the common insecticides used against *T. absoluta* larvae such as flubendiamide (Fenos, Bayer), chlorantraniliprole (Altacor, Dupont), and metaflumizone (Alverde, BASF). These insecticides had a negligible effect on nematode survival, infectivity, and reproduction. IJs that survived insecticide exposure were able to infect *T. absoluta* larvae and reproduce inside them. In addition, insect larvae affected by the three insecticides would serve as suitable hosts for the infection and reproduction of the nematodes. These results suggest that larvae of *T. absoluta* that drop from leaves following insecticide application could be suitable hosts for nematodes, thereby increasing their EPN concentration and persistence in the soil.

2.2 The Tomato Budworm, *Helicoverpa armigera* (Hübner) and Armyworms, *Spodoptera* spp.

The tomato budworm, *Helicoverpa armigera* (Hübner) and some species of *Spodoptera*, such as *Spodoptera exigua* (Hübner), *Spodoptera frugiperda* (J.E. Smith), *Spodoptera littoralis* (Boisduval), and *Spodoptera litura* (F.), are extremely polyphagous and constitute some of the most substantial pests of the tomato plant, attacking the leaves and the fruit. The pathogenicity of EPN species of *Steinernema* and *Heterorhabditis* to control these lepidopterans has been widely demonstrated. Various studies showed *H. armigera* to be susceptible to *S. feltiae* (Dutky et al., 1956), *Steinernema glaseri* (Steiner) (Patel and Vyas, 1995), *Steinernema riobrave* Cabanillas, Poinar & Raulston, *S. carpocapsae*, and *Heterorhabditis* sp. (Tahir et al., 1995). EPNs are also pathogenic to armyworms. For example, *S. exigua* was found to be susceptible to *S. feltiae* (Kaya, 1985), *S. litura* was susceptible to different strains of *S. carpocapsae* (Kondo and Ishibashi, 1986; Kondo, 1987), *S. littoralis* was susceptible to *S. carpocapsae* infection (Sikora et al., 1979), and *S. frugiperda* was susceptible to *S. feltiae*, *Steinernema rarum* (Doucet), and *H. bacteriophora* (Doucet et al., 1999).

Evidence suggests the efficacy of EPNs to control each insect species depends on the particular EPN species and strain. Kondo and Ishibashi (1986) compared infectivity of three species of *Steinernema* against larvae of *S. litura* in a petri dish bioassay. Their results showed significant differences in host mortality caused by *S. carpocapsae* (100% larval mortality), *S. feltiae* (70% host mortality), and *S. glaseri* (only 5% larval mortality). Glazer et al. (1991) studied the susceptibility of fourth instar *S. littoralis* to five strains of *S. carpocapsae*, *S. glaseri*, and *H. bacteriophora*. All the strains tested caused 100% mortality with 50 IJs/larva after 96 h, although there were differences in virulence among strains. One strain of *H. bacteriophora* was the most virulent against larvae of *S. littoralis* while *S. glaseri* was the least virulent. In an experiment conducted in pots (Ali et al., 2008), the infection of *H. armigera* was evaluated with five steinernematid species. Results indicated that although the tomato budworm was susceptible to the five EPN species tested, there were differences among species in virulence (with respect to the time to larval death) and the production of IJs per larva was as follows: *Steinernema masoodi* Ali, Shaheen, Pervez and Hussain > *Steinernema seemae* Ali, Shaheen, Pervez and Hussain > *S. carpocapsae* > *S. glaseri* > *Steinernema thermophilum* Ganguly and Singh. Campos-Herrera and Gutierrez (2009) screened virulence of 17 EPN strains from *S. feltiae* and *S. carpocapsae* species against *S. littoralis* larvae, resulting in a high level of mortality of 74–100%. Seenivasan and Sivakumar (2014) evaluated a total of 27 nematode isolates in India belonging to four species, *S. carpocapsae*, *Steinernema siamkayai* Stock, Somsook & Reid, *S. monticolum*, and *H. bacteriophora*, against *H. armigera* and *S. litura*. Although both lepidopterans were susceptible to all of the tested EPN species/strains, the virulence of nematodes varied within each insect species, resulting in mortalities ranging from 41% to 94% in *H. armigera* and from 89% to 93% in *S. litura*. Differences in susceptibility depending on the insect's developmental stage have also been reported. Divya et al. (2010) tested the pathogenicity of *Heterorhabditis indica* Poinar, Karunakar & David against *H. armigera* and *S. litura*. They evaluated the susceptibility of second, third, fourth, and final instar larvae to EPNs in a laboratory experiment. *H. indica* caused 100% mortality in all the instars of both lepidopteran species. However, median lethal time was negatively correlated with increased larval age, revealing that the early instars of both insect species were more susceptible than the older instars. Other laboratory studies reported the susceptibility of pre-pupae and pupae (Fig. 12.4). Ali et al. (2007) observed 43% infectivity of *H. armigera* pre-pupa caused by steinernematids. *S. exigua* pupae showed high susceptibility to *S. carpocapsae*, with 63% pupal mortality (Kaya and Hara, 1981), and to *S. riobrave* (Henneberry et al., 1995).

The susceptibility of eggs of *H. armigera* and *S. litura* to EPN infection recently has been reported with the nematode *S. thermophilum* (Kalia et al., 2014). A dose of 200 IJs of *S. thermophilum* caused ovicidal activity up to 84% when exposing 50 and 100 eggs of *H. armigera* and *S. litura* to the nematode, respectively. The finding of susceptibility in lepidopteran eggs to EPNs is unique and may offer new potential for pest control approaches.

Soil application of EPNs against these lepidopteran pests are commonly used to control the last instars which crawl to the ground and enter crevices or loose soil for pupation, and to prevent adult emergence (Hussain et al., 2014) (Fig. 12.5). Cabanillas and Raulston (1996) reported that the application of nematodes through irrigation systems could be a potential approach for suppressing *Helicoverpa* populations. A microplot study to evaluate the control of *H. armigera* with the nematode *S. masoodi* (Hussain et al., 2014) indicated differences in the level of control relative to the nematode concentration

FIGURE 12.4 *Spodoptera* sp. pupa parasitized by entomopathogenic nematodes.

FIGURE 12.5 Infective juveniles of *Steinernema carpocapsae* on the soil surface after a soil treatment.

applied. The suppression of adult emergence varied from 32% at a nematode concentration of 2×10^9 IJs/ha to 70% at a concentration of 6×10^9 IJs/ha. Control of these pests with foliar EPN application focused to suppression of the early instars has also been tested with success. However, foliar persistence of EPNs is short as the environment exposes the nematodes to unfavorable moisture conditions, high temperatures, and sunlight, which is detrimental to nematode survival (Koppenhöfer, 2000; Georgis et al., 2006). Glazer and Navon (1990) tested *S. feltiae* mixed with two different solutions of antidesiccants, glycerol or folicote, against *H. armigera* and obtained 75% and 95% control of *H. armigera*. Vyas et al. (2002) tested a spray application of a native *Heterorhabditis* sp. to control *H. armigera*. Nematodes (100,000 IJs/m²) were applied alone and together with adjuvants (5% starch mixed with gum Arabic) Both treatments, EPN alone and with adjuvants, reduced the larval population by 32% and 47%, respectively, compared to an untreated control. Navon et al. (2002) developed a calcium-alginate-gel formulation mixed with EPNs to control *S. littoralis* and *H. armigera* larvae, and obtained larval mortality of 89% with *S. feltiae* in the greenhouse. In crops other than tomato, prolonged EPN survival and efficacy in aboveground applications has also been demonstrated using surfactant and polymer formulation (Schroer and Ehlers, 2005) or a sprayable gel (Shapiro-Ilan et al., 2010); these approaches may also be applied to tomato plantings. Thus, in considering the two approaches, the use of foliar application combined with soil application of EPNs could result in better control of these insect pests in the field (Yadav et al., 2008).

The use of EPNs with other biological control agents, such as viruses, fungi, or parasitoids has been developed as a promising alternative. Combined use of EPNs with other biocontrol agents (or other tactics) can result in varying interactions including antagonism, additivity, or synergy. The relationship may vary depending on the specific tactics that are being combined, the rates, the timing of application, and the target host (Shapiro-Ilan et al., 2012). For example, Gothama et al. (1996) reported the additive effect of combining *S. carpocapsae* with a nucleopolyhedrovirus against *S. exigua*. The combination produced higher larval mortality (62%) than either the nematodes (35%) or the virus alone (34%). The effect of *H. bacteriophora* and *S. riobrave* alone and in combination with the fungi *Beauveria bassiana* (Balsamo-Crivelli) Vuillemin and *Metarhizium anisopliae* (Metchnikoff) Sorokin was evaluated by Shairra and Noah (2014) under laboratory conditions against *S. littoralis*. *H. bacteriophora* caused higher mortality in *S. littoralis* larvae than *S. riobrave* (86% and 71% mortality, respectively), and *M. anisopliae* was more effective than *B. bassiana* (80% and 41% mortality, respectively). However, the combined effect of fungi and EPNs resulted in an increased mortality of the host compared to the mortality caused by each agent alone. Atwa et al. (2013) tested *S. carpocapsae* and *H. bacteriophora* in combination with the parasitoid *Microplitis rufiventris* Kokujev against larvae of *S. littoralis*. The interaction between the nematodes and parasitoid was variable. When nematodes were inoculated during the parasitoid egg stage or in the young parasitoid larvae, they were able to reproduce but the parasitoid larvae died. However, when nematodes were inoculated during the late larval instar of *M. rufiventris*, both nematodes and wasps reproduced in the same host and resulted in greater overall mortality on *S. littoralis* larvae than either agent inflicted alone. Before embarking on a combined application program, the nature of interactions (antagonistic, additive, or synergistic) between two tactics should be elucidated (Koppenhöfer and Fuzy, 2002).

2.3 Leafminers, *Liriomyza* spp.

Leafminers of the genus *Liriomyza* number more than 370 species worldwide, of which *Liriomyza trifolii* (Burgess), *Liriomyza bryoniae* (Kaltenbach), *Liriomyza huidobrensis* (Blanchard), and *Liriomyza sativae* Blanchard can inflict significant damage to tomato crops. Damage caused by *Liriomyza* spp. is related to the feeding punctures of adults and the mines produced by the feeding of larvae in the mesophyll tissues of tomato leaves (Spencer, 1973).

EPNs have been proposed as biological control strategies against leafminers (Hara et al., 1993). Even though EPNs cannot enter mines by penetrating the intact leaf cuticle, IJs enter the mine via the oviposition puncture created by the female fly during egg laying, or through any unnatural tear in the mine surface (LeBeck et al., 1993). Temperature and relative humidity (80–90% or higher) are important factors limiting the success of EPNs applied to the foliage for leafminer control. Nevertheless, at 20°C and >90% rh, sufficient IJs of *S. feltiae* were able to enter the leaf tissues within the first 12 h after application to reduce larval survival of *L. bryoniae* (Williams and MacDonald, 1995).

Host-stage suitability of *L. trifolii* to *S. carpocapsae* was reported by LeBeck et al. (1993) showing that all larval stages, the pre-puparium and the early puparium (<1 h after pupation initiation), were susceptible to EPNs. However, mortality was greatest for second-instar larvae (93.3%) and decreased for later developmental stages. *S. carpocapsae* reproduced successfully in all immature stages except for the first-instar and the early puparium. The susceptibility of *L. trifolii* to EPNs also relates differently to the species or strains of nematode tested. Hara et al. (1993) reported mortality rates of *L. trifolii*, ranging from 48% to 98% by 20 strains and/or species of steinernematid and heterorhabditid nematodes. *S. carpocapsae* and *S. feltiae* produced higher mortality (>63%) in *L. trifolii* on bean leaves than *Heterorhabditis* sp. (<33%) when applied at 9×10^9 IJs/ha.

The first results of field trials using *S. carpocapsae* against *L. trifolii* were shown by Harris et al. (1990). Mortality of leafminers on chrysanthemum in a greenhouse after two nematode applications (5×10^8/ha) was equivalent to that obtained by two abamectin applications (0.17 kg[AI]/ha). These results are in agreement with Broadbent and Olthof (1995) who reported that two consecutive applications of *S. carpocapsae* (3 and 4 days after oviposition by leafminer) resulted in >80% mortality of leafminer larvae and 99.6% mortality with abamectin. Williams and Walters (2000) also obtained a mean *L. huidobrensis* mortality of 82 ± 5% after application of *S. feltiae* in a trial on lettuce at a commercial glasshouse, significantly higher than the treatment with the systemic insecticide heptenophos. Similarly, in a trial conducted in a commercial production glasshouse of Chinese brassica infected by *L. huidobrensis* the mean number of pupae was reduced by over 88% after the application of *S. feltiae* (Head and Walters, 2003).

Some studies were conducted on the potential combined use of EPNs with other strategies for leafminer control. Head et al. (2000) determined the compatibility of IJs and chemical insecticides (abamectin, deltamethrin, dimethoate, heptenophos, and trichlorfon) to control larvae of *L. huidobrensis*. Although a limited range of insecticides can be applied simultaneously with *S. feltiae*, high levels of leafminer larval control can be achieved by the application of these nematodes to vegetable foliage previously treated with insecticides. Furthermore, Sher et al. (2000) studied the combined use of *S. carpocapsae* and *Diglyphus begini* (Ashmead) for control of *L. trifolii*. These authors showed that *D. begini* adults were not susceptible to nematode infection. Leafminer larvae parasitized by the wasps were less susceptible to nematode infection, and adult wasps detected and avoided ovipositing on nematode-infected leafminer.

2.4 Hemiptera: Aphididae

Aphids are important pests of cultivated tomato plants and the use of insecticides is still the main control method. Extensive pesticide use may cause the development of resistance in aphids (Needham and Sawicki, 1971; Foster et al., 2002). Alternative control strategies are needed to avoid insecticide resistance and objectionable side effects of chemical insecticides for humans and the environment. Among other biocontrol agents, EPNs have been studied as potential agents to aphid control. Conceivably, aphids could be attacked by steinernematids, but they are generally not deemed to be a favorable host for EPNs (Poinar, 1975). Mráček and Růžička (1990) observed that when *Steinernema* spp. strain *Hylobius* was applied against *Acyrthosiphon pisum* Harris and *Megoura viciae* Buckton in petri dish experiments, only a few adults of *A. pisum* were parasitized and the mortality of *M. viciae* reached 28% after 3 days. Females of this nematode developing in infected aphids grew to pygmy size and mated but did not lay their fertile eggs. More recently, Park et al. (2013) explored the possibility of EPNs as biological control of *Myzus persicae* (Sulzer). In experiments under laboratory conditions, *S. carpocapsae* had low infectivity against nymphs (2% mortality) and adults (6.7% mortality) of *M. persicae*. Only a few IJs invaded the hosts and developed into adult nematodes which grew to about 2 mm long, but did not produce offspring. These authors consider that the low parasitism observed may have been caused by the small body of this aphid.

2.5 Hemiptera: Aleyrodidae

Whiteflies are pests that can adversely affect the success of tomato crops. Damage to commercial tomato plants by these insects can be caused directly through phloem feeding or indirectly by the transmission of plant viruses. Among whiteflies, the sweetpotato whitefly, *Bemisia tabaci* (Gennadius) and the greenhouse whitefly, *Trialeurodes vaporariorum* Westwood are the most problematic for tomato crops.

EPNs have been shown to be effective against *B. tabaci*, but differential susceptibility has been observed among whitefly instars. On tomato plants, when a foliar application of *S. feltiae* (10,000 IJs/mL with 0.02% of the non-ionic wetting agent Agral) at a dose of approximately 160 IJs/cm^2 of leaf area was applied, the first three instars were more susceptible (70–75% mortality) than the fourth (50–55% mortality) (Cuthbertson et al., 2003a). Further studies by Cuthbertson et al. (2007a) demonstrated that the second-instar stage of *B. tabaci* is the most susceptible life stage to *S. feltiae* and *S. carpocapsae*. Nevertheless, Laznik et al. (2011) showed that adults of the greenhouse whitefly *T. vaporariorum* also were susceptible to the attack of *S. feltiae*. In fact, *S. feltiae* can cause adult whitefly mortality similar to levels produced by the chemical insecticide thiamethoxam. In greenhouse experiments, the mean number of whitefly adults per leaf was significantly reduced from 128.5 ± 33.0 on the control leaves to 50.0 ± 12.0 on the leaves treated with EPNs.

Preliminary studies determined that the efficacy of EPNs against immature stages of whiteflies varies among a range of host plants (Head et al., 2004). In these experiments, foliar application of *S. feltiae* (10,000 IJs/mL) against *B. tabaci* recorded mortalities of 32% on tomato, 28% on cucumber, 22% on verbena, 22% on chrysanthemum, and 10% on poinsettia. According to these authors, the differences could be caused by the presence of leaf structures, such as hairs or waxy surfaces that may affect IJ movement or adherence to the leaf surface, thus reducing the efficacy of EPNs. However, further studies by Cuthbertson et al. (2007b) reported that under optimal environmental conditions for nematode activity, no host-plant effects were present and EPNs were equally effective against whitefly larvae on a range of plant foliage types.

Environmental conditions, mainly the ambient temperature and humidity, have been shown to be a critical determinant for nematode survival and efficacy on foliar applications against *B. tabaci* (Cuthbertson and Walters, 2005). The favorable conditions for nematode survival (moderate temperature, high moisture, and limited exposure to sunlight) should be maintained for up to 6–8 h after their application to plant foliage (Williams and MacDonald, 1995). The addition of adjuvants to nematode applications enhance efficacy through their anti-desiccant properties and a reduction in surface tension of suspension droplets (Shapiro-Ilan et al., 2012). The presence of a thin film of water on the leaf could facilitate more effective movement of IJs across the leaf surface when searching for a host. The use of adjuvants in foliar application of *S. feltiae* on tomato plants raised the mortality level of *B. tabaci* from 32% to 63% (Head et al., 2004). Similarly, the combined use of *S. feltiae* (10,000 IJs /mL) and 1% horticultural spraying oil increased the efficacy of EPNs against the second-instar of *B. tabaci* nymphs on collard and hibiscus (Qiu et al., 2008).

Cuthbertson et al. (2007b) provide evidence that the application of *S. feltiae* against second-instar *B. tabaci* on tomato plants under controlled laboratory conditions resulted in 99.2% mortality. A slightly lower level (80%) of *B. tabaci* mortality was obtained in the glasshouse situation. These authors also showed that efficacy was not related to the application rates on tomato plants. Mortality of *B. tabaci* larvae obtained by using 5000 IJs/mL did not differ from the recommended dose rate of 10,000 IJs/mL. Furthermore, there were no differences in mortality between one application of 2500 IJs/mL and two applications of 1000 IJs/mL. The reduction of the application rates of *S. feltiae* on tomato foliage, without reduction in efficacy, offers the potential for more economically viable use of EPNs against whiteflies.

When EPNs are used in integrated pest management (IPM) strategies, it is necessary that nematodes and agrochemical products or other control agents are compatible. The combination of EPNs with chemicals commonly used for the control of whiteflies has been addressed by different studies. Cuthbertson et al. (2003b) showed that better *B. tabaci* larval mortality was achieved when sequential applications of *S. feltiae* along with the chemicals buprofezin, nicotine, and imidacloprid are used. Similar results were obtained by Cuthbertson et al. (2008) following sequential applications of *S. carpocapsae* and thiacloprid, pymetrozine, and imidacloprid, with mortalities up to 94% of *B. tabaci* second instars. Sequential treatments offer a greater flexibility in timing applications against various life stages of the whiteflies (Cuthbertson et al., 2011), further encouraging their use in IPM strategies.

2.6 Thrips, Thysanoptera: Thripidae

The western flower thrips (WFT), *Frankliniella occidentalis* Pergande is an important pest of tomato plants, causing direct and indirect damages. WFT feed on plant tissue by rasping and sucking sap (Childers and Achor, 1995). However, the primary damage caused by WFT is the transmission of Tomato spotted wilt virus (TSWV) to tomato plants. Adults, first, and second-instar larvae feed voraciously on leaves and flowers until they move to the soil for pupation. There are pre-pupae (or pro-pupae) and pupae stages, both of which commonly occur in soil (Arzone et al., 1989).

The high resistance of WFT to chemical insecticides and their cryptic feeding behavior (flowers and leaf axes) make *F. occidentalis* control very difficult (Helyer and Brobyn, 1992). Different biological control agents are presently available and EPNs could provide a potential alternative to chemical insecticides among them (Laznik and Trdan, 2008).

Susceptibility of WFT to different EPN species in laboratory conditions was reported in several studies (Tomalak, 1994; Chyzik et al., 1996; Ebssa et al., 2004; Tomalak et al., 2005). Ebssa et al. (2001) determined the virulence of six different strains of *H. bacteriophora*, *S. feltiae* and *S. carpocapsae* against second-instar larvae and soil-dwelling pre-pupa and pupa of *F. occidentalis*. Under high soil-moisture conditions, *S. feltiae* was highly virulent against late second-instar larvae and pre-pupae of WFT, but less effective against pupae. These authors also showed that rates of 100–200 IJs/cm^2 caused 30–50% mortality. However, high mortality in all soil-dwelling life stages of WFT was obtained only with a high concentration of 400 IJs/cm^2. In similar experimental conditions, Premachandra et al. (2003) determined that *S. feltiae* (Nemaplus strain) and *H. bacteriophora* (HD01 strain) caused 65% and 59% mortalities, respectively, in late second-instar and pupal stages. *S. carpocapsae* (Agriotos strain) and *Steinernema arenarium* (Artyukhovsky) (Anomali strain) caused moderate mortality (40–45%), while *Steinernema* spp. (Morocco strain) and *H. bacteriophora* (Nematop strain) had little effect.

The virulence of EPNs against larvae and adults of WFT on aboveground parts of plants has been studied less. In laboratory conditions, Perme (2005) showed that at high concentration (5000 IJs/mL) the most effective nematode was *H. bacteriophora* (92%), while at low concentration (500–1000 IJs/mL) the most virulent nematodes were *S. carpocapsae* (90%) and *S. feltiae* (82%). Nevertheless, Buitenhuis and Shipp (2005) found that significant mortalities of first (36%) and second instars (28%) were obtained only at very high rates of *S. feltiae* (20,000 and 40,000 IJs/mL), and susceptibility to this nematode in adult WFT was very low (3–6% mortality).

Due to the movement of WFTs from plants to soil over their lifespan, two different approaches for their control with EPNs may be feasible: soil applications and foliar applications. Efficacy of soil applications of EPNs against WFT in compost first was studied by Helyer et al. (1995). These authors showed that *S. carpocapsae* gave 76.6% control of pre-pupal and pupal stages of WFT only when applied at 25×10^4 IJs/L of compost; this high nematode-application rate is uneconomical for glasshouse soil applications. Ebssa et al. (2001) determined that soil applications of *S. feltiae* and *S. carpocapsae* at rates of 100–200 IJs/cm^2 caused 30–50% mortality, but a high concentration of 400 IJs/cm^2 is needed to obtain high mortality (90%) in all soil-dwelling life stages of WFT. Timing and frequency of applications of EPNs also can have a profound effect on WFT control. An early application of EPNs when the WFT density is within the control capacity of nematodes, at a lower IJ concentration (200 IJs/cm^2) but in a repeated manner, can result in higher WFT control than a single application at the higher concentration of 400 IJs/cm^2 (Belay et al., 2005). In IPM strategies, combined releases of different natural enemies lead to a higher WFT control than separate releases of individual biocontrol agents. Combined soil treatments of EPNs and the soil predatory mite *Hypoaspis aculeifer* (Canestrini) are more effective (71–82% reduction of adult emergence) than individual applications of EPNs (46–71%) or the predatory mite alone (46%) (Premachandra et al., 2003). Other studies have shown that simultaneous soil application of EPNs and foliar application of the predatory mites *Amblyseius cucumeris* (Oudemans) produce greater WFT reduction than individual application of each natural enemy (Ebssa et al., 2006).

One of the first greenhouse experiments on foliar application of *S. feltiae* against *F. occidentalis* showed that weekly application of this nematode (125–250×10^6 IJs/100 L/1000 m^2) on chrysanthemum resulted in a WFT adult reduction, especially females, and a minimal crop damage (only occasional flecked petals easily removed at harvest was detected) (Wardlow et al., 2001). Further studies demonstrated that nine foliar applications of *S. feltiae* (2500 IJs/mL) in one cucumber-growing season reduced the damage of WFT on cucumber leaves, never exceeding 10% of the leaf surface (Trdan et al., 2007). The average fruit yield also was positively influenced by EPN applications (37–51% higher in treated vs. non-treated plants). The nine foliar applications of EPNs against WFT resulted in an efficacy similar to three applications of the chemical insecticide abamectin on glasshouse-grown cucumbers (Trdan et al., 2007). The results suggest that, with proper optimization of EPN foliar applications, these biological agents are an effective alternative to the chemical control of WFT.

2.7 Wireworms, Coleoptera: Elateridae

Numerous species of wireworms are severe and widespread agricultural pests causing damage throughout the world to a variety of crops such as potatoes, carrots, onions, tomatoes, aubergines, strawberries, and maize (Parker and Howard, 2001). The difficulty in controlling wireworms with chemical insecticides, and the significant losses caused by this pest in organic crops, has increased research in the use of biological control agents. Therefore, great efforts have been made in finding pathogenic strains of nematodes to control wireworms. Elaterid larvae are generally considered to have poor susceptibility to EPNs. Nevertheless, some work has shown that the susceptibility depends on the nematode species/strains, nematode dose, and the wireworm species.

Danilov (1974) isolated *S. carpocapsae* (agriotos strain) from an *Agriotes lineatus* L. larva, obtaining a 75% control of wireworms with a dose of 1.5×10^6 IJs/m^2. Toba et al. (1983) exposed later instar larvae of the sugar beet wireworm *Limonius californicus* (Mannerheim) to EPNs in a laboratory experiment. *S. feltiae* was more effective than *S. glaseri*, causing mortality of 58% and 7%, respectively. The experiment compared different concentrations of EPNs against *L. californicus* larvae, ranging from 0.4 to 393 IJs/cm^2, and results with *S. feltiae* showed an increase in larval mortality from 13% to 58%, respectively. However, there are other studies that reported poor efficacy of EPNs against wireworms. Morris (1985) reported no efficacy of *S. feltiae* and *H. bacteriophora* against the wireworm *Ctenicera destructor* (Brown) and San-Blas et al. (2012) did not obtain any infection of larvae of *Agriotes* sp. with *S. carpocapsae*.

Field application of EPNs has also been conducted with different species of wireworms. Kovacs et al. (1980) found *S. feltiae* to be effective in reducing corn seedling damage caused by wireworms (species unspecified). Toba et al. (1983) carried out field cage tests to evaluate the efficacy of *S. feltiae* to control *L. californicus*. The mean mortality of these wireworms was 28% for the nematode concentration of 155 IJs/cm^2 of soil, with no increment of efficacy (29%) when concentration was doubled (310 IJs/cm^2). Schalk et al. (1993) reported that *S. carpocapsae* reduced damage caused by the wireworm *Conoderus* spp. (21% control) in a field assay with resistant sweetpotato cultivars, except during a very wet year when nematodes were leached from the rhizosphere. Ester and Huiting (2007) did not find any effect on the protection of potato tubers against wireworm (*Agriotes* spp.) attacks when *S. feltiae* was applied as a furrow treatment at a dose of 545,000 IJs/m^2.

The interest in the development of biological alternatives to control wireworms has led to a search for more virulent EPN strains against wireworms. Toward that end, Ansari et al. (2009) tested five different strains of *H. bacteriophora*, *Heterorhabditis megidis* Poinar, Jackson & Klein, *Heterorhabditis downesi* Stock, Griffin & Burnell, *S. feltiae*, and *S. carpocapsae* against fourth- and fifth-instar larvae of *A. lineatus* in a laboratory assay. The most virulent nematode tested was an *H. bacteriophora* strain which caused 67% mortality. The *S. feltiae* strain did not cause mortality of wireworms, but the other EPN strains provided control ranging between 15% and 50%. In another study, EPN virulence against the dusky wireworm, *Agriotes sordidus* Illiger, was evaluated with 15 native strains belonging to *S. feltiae* and two native strains to an *H. bacteriophora* species in Spain (Campos-Herrera and Gutierrez, 2009). *A. sordidus* was affected only by two *S. feltiae* strains (which caused 7% and 9% larval mortality) and both *H. bacteriophora* strains (which caused 4% and 5% larval mortality). The virulence of different species and strains of EPNs was also tested against the dusky wireworm *Agriotes obscurus* (L.) in the laboratory (Morton and Garcia-del-Pino, 2016). Larvae were exposed to eight nematode strains, six native and two commercial, from the species *S. feltiae*, *S. carpocapsae*, and *H. bacteriophora*. At a dose of 100 IJs/cm^2, a native *S. carpocapsae* was the most virulent strain causing 76% mortality while six other strains produced 9–36% larval mortality. *S. feltiae* was revealed to be ineffective for controlling *A. obscurus*.

The low susceptibility of some wireworms may be due to physical deterrents, such as a dense brush of hairs in the oral cavity, a thick cuticle, valvate spiracles, or muscles closing the anus which protects them from EPN invasion (Eidt and Thurston, 1995). Puza and Mracek (2010) tested the capacity of EPNs to colonize the hosts using *Steinernema affine* (Bovien) and *Steinernema kraussei* (Steiner) in experimental infection of living and freeze-killed wireworms (species unspecified). In their study, living elaterid larvae showed resistance to the nematode infection, and no mortality was observed. However, both nematode species were able to colonize the majority of dead wireworms (up to 70% of individuals colonized by EPNs) and multiply in the cadavers. The authors hypothesized that IJs could enter the cadavers through the anus and/or spiracles, since the constricting muscles of these openings were relaxed in the dead worms. Consistent with these results, Morton and Garcia-del-Pino (2016) reported that when IJs are injected into the hemocoel of larval *A. obscurus*, nematodes are able to reproduce, showing that the routes of entry prevent infection by EPNs.

The increased damage caused by wireworms to agricultural crops, and the reduced efficacy of chemical insecticides to control these pests has led to a greater interest in the potential of EPNs. Current research is focused in two directions; the search for more virulent strains of nematodes to provide a range of promising EPN isolates to test for efficacy against wireworms, and the combining of EPNs with other biological agents to improve the efficacy for controlling this pest.

3. CONCLUSION AND FUTURE PROSPECTS

EPNs have been shown to be effective against a variety of tomato pests, especially when the insects dwell in the soil or live in cryptic habitats such as inside plant galleries. The ability of EPNs to seek and kill insects in these habitats, where chemical insecticides fail, makes these biocontrol agents especially attractive. The proven efficacy and compatibility of EPNs with other biocontrol agents, and with agrochemical products promote their incorporation into IPM strategies used on tomato crops. Furthermore, the resistance to chemical insecticides developed by different pests, European Union restrictions on chemical pesticides in conventional management (Directive 2009/128/EC to achieve the sustainable use of pesticides), and the growth in organic horticultural production all could lead to expanded EPN use in horticulture. Nevertheless, the use of EPNs against aboveground tomato pests is still limited. Factors limiting EPN use against such horticultural pests include efficacies less predictable than are necessary for the successful market penetration of these biocontrol products (Georgis et al., 2006). Suboptimum or unpredictable efficacy may be related to the use of inappropriate nematode species and/or their application under suboptimal conditions such as high temperature, low moisture, and exposure to sunlight. For this reason, to expand the commercial use of EPNs, research should be focused on improving application and formulation technology and the use of appropriate EPN species. Slow-release formulations such as alginate and polyacrylamide gels (Schroer and Ehlers, 2005) or sprayable fire gels (Shapiro-Ilan et al., 2010), wood flour foam (Lacey et al., 2010), and EPN-infected cadavers (Shapiro-Ilan et al., 2003; Del Valle et al., 2008; Dolinski et al., 2015) hold promise for their application in greenhouse-grown and open-field tomatoes. Research should also be focused on discovering new EPN species, more virulent strains, and strains that are best adapted to the tomato crop environment. These beneficial traits of EPNs could be obtained by genetic enhancement via selection, hybridization, or molecular manipulation of currently available populations to improve their biocontrol potential against each pest, and their persistence in the climatic conditions of tomato crops (Shapiro-Ilan et al., 2012, 2014). Once improved strains are obtained it will be important to ensure their biocontrol traits are stabilized, e.g., through the creation of homozygous inbred lines (Bai et al., 2005; Anbesse et al., 2013).

Recently, Ruisheng et al. (2013) suggested there is an added benefit of EPNs for tomatoes. The authors revealed an EPN-induced insect, and disease resistance in tomato plants, when EPNs were applied to the soil surrounding the plants. This is consistent with previous studies which showed a direct antagonistic effect on root-knot/foliar parasitic nematodes and an indirect effect on pests through the activation of defensive mechanisms in the tomato plants themselves (Fallon et al., 2002; Javed et al., 2012). Ruisheng et al. (2013) showed that the application of *S. carpocapsae*-infected wax moth (*Galleria mellonella* (L.)) cadavers to the soil around the roots of tomato plants reduced the rate of armyworm, *S. exigua*, development, reduced egg hatching of *B. tabaci*, and decreased the number of foliar disease spots produced by the bacteria *Pseudomonas syringae* van Hall. These results all support the hypothesis that a soil application of EPNs leads to a lasting benefit in tomato plants by enhancing their defensive capability, and it offers new perspectives for further studies on EPNs in horticultural systems.

Finally, end-user education and marketing support for the use of EPN products could help expand the use of these nematodes, the efficacy of which has already been shown against pests in horticultural production systems.

REFERENCES

Ali, S.S., Pervez, R., Hussain, A.M., Ahmad, R., 2007. Effect of temperature on survival of *Steinernema seemae*, *S. masoodi* and *S. carpocapsae* (Rhabditida: Steinernematidae) and their subsequent infectivity to prepupa of *Helicoverpa armigera* (Hübner). Archives of Phytopathology and Plant Protection 40, 183–187.

Ali, S.S., Pervez, R., Hussain, A.M., Ahmad, R., 2008. Susceptibility of three lepidopteran pests to five entomopathogenic nematodes and *in vivo* mass production of these nematodes. Archives of Phytopathology and Plant Protection 41, 300–304.

Ansari, M.A., Evans, M., Butt, T.M., 2009. Identification of pathogenic strains of entomopathogenic nematodes and fungi for wireworm control. Crop Protection 28, 269–272.

Anbesse, S., Sumaya, N.H., Dörfler, A.V., Strauch, O., Ehlers, R.U., 2013. Selective breeding for desiccation tolerance in liquid culture provides genetically stable in bred lines of the entomopathogenic nematode *Heterorhabditis bacteriophora*. Applied Microbiology and Biotechnology 97, 731–739.

Arthurs, S., Heinz, K.M., Prasifka, J.R., 2004. An analysis of using entomopathogenic nematodes against above-ground pests. Bulletin of Entomological Research 94, 297–306.

Arzone, A., Alma, A., Rapetti, S., 1989. *Frankliniella occidentalis* (Perg.) (Thysanoptera: Thripidae) nuovo fitomizo delle serre in Itali. Information Fitopat 39, 43–48.

Atwa, A.A., Hegazi, E.M., Khafagi, W.E., Abd El-Aziz, G.M., 2013. Interaction of the koinobiont parasitoid *Microplitis rufiventris* of the cotton leafworm, *Spodoptera littoralis*, with two entomopathogenic rhabditids, *Heterorhabditis bacteriophora* and *Steinernema carpocapsae*. Journal of Insect Science 13, 1–14.

Bai, C., Shapiro-Ilan, D.I., Gaugler, R., Hopper, K.R., 2005. Stabilization of beneficial traits in *Heterorhabditis bacteriophora* through creation of inbred lines. Biological Control 32, 220–227.

Batalla-Carrera, L., Morton, A., Garcia-del-Pino, F., 2010. Efficacy of entomopathogenic nematodes against the tomato leafminer *Tuta absoluta* in laboratory and greenhouse conditions. BioControl 55, 523–530.

Begley, J.W., 1990. Efficacy against insects in habitats other than soil. In: Gaugler, R., Kaya, H.K. (Eds.), Entomopathogenic Nematodes in Biological Control. CRC Press, Boca Raton, Florida, USA, pp. 215–232.

Belay, D., Ebssa, L., Borgemeister, C., 2005. Time and frequency of applications of entomopathogenic nematodes and their persistence for control of western flower thrips *Frankliniella occidentalis*. Nematology 7, 611–622.

Bird, A.F., Akhurst, R.J., 1983. The nature of the intestinal vesicle in nematodes of the family Steinernematidae. International Journal of Parasitology 13, 599–606.

Broadbent, A.B., Olthof, T.H.A., 1995. Foliar application of *Steinernema carpocapsae* (Rhabditida, Steinernematidae) to control *Liriomyza trifolii* (Diptera, Agromyzidae) larvae in chrysanthemums. Environmental Entomology 24, 43–435.

Buitenhuis, R., Shipp, J.L., 2005. Efficacy of entomopathogenic nematode *Steinernema feltiae* (Rhabditida: Steinernematidae) as influenced by *Frankliniella occidentalis* (Tysanoptera: Thripidae) developmental stage and host plant stage. Journal of Economic Entomology 98, 1480–1485.

Cabanillas, H.E., Raulston, J.R., 1996. Effects of furrow irrigation on the distribution and infectivity of *Steinernema riobravis* against corn earworm in corn. Fundamental and Applied Nematology 19, 273–281.

Campos-Herrera, R., Gutierrez, C., 2009. Screening Spanish isolates of steinernematid nematodes for use as biological control agents through laboratory and greenhouse microcosm studies. Journal of Invertebrate Pathology 100, 100–105.

Childers, C.C., Achor, D.S., 1995. Thrips feeding and oviposition injuries to economic plants, subsequent damage and host responses to infestation. In: Paker, B.L., Skinner, M., Lewis, T. (Eds.), Thrips Biology and Management. Plenum Press, New York, USA, pp. 31–52.

Chyzik, R., Glazer, J., Klein, M., 1996. Virulence and efficacy of different entomopathogenic nematode species against western flower thrips (*Frankliniella occidentalis*). Phytoparasitica 24, 103–110.

Cuthbertson, A.G.S., Head, J., Walters, K.F.A., Gregory, S.A., 2003a. The efficacy of the entomopathogenic nematode, *Steinernema feltiae*, against the immature stages of *Bemisia tabaci*. Journal of Invertebrate Pathology 83, 267–269.

Cuthbertson, A.G.S., Head, J., Walters, K.F.A., Murray, A.W.A., 2003b. The integrated use of chemical insecticides and the entomopathogenic nematode, *Steinernema feltiae*, for the control of the sweetpotato whitefly, *Bemisia tabaci*. Nematology 5, 713–720.

Cuthbertson, A.G.S., Walters, K.F.A., 2005. Evaluation of exposure time of *Steinernema feltiae* against second instar *Bemisia tabaci*. Tests of Agrochemicals and Cultivars 26, 34–35.

Cuthbertson, A.G.S., Mathers, J.J., Northing, P., Luo, W., Walters, K.F.A., 2007a. The susceptibility of immature stages of *Bemisia tabaci* to infection by the entomopathogenic nematode *Steinernema carpocapsae*. Russian Journal of Nematology 15, 153–156.

Cuthbertson, A.G.S., Walters, K.F.A., Northing, P., Luo, W., 2007b. Efficacy of the entomopathogenic nematode, *Steinernema feltiae*, against sweetpotato whitefly, *Bemisia tabaci*, under laboratory and glasshouse conditions. Bulletin of Entomological Research 97, 9–14.

Cuthbertson, A.G.S., Mathers, J.J., Northing, P., Prickett, A.J., Walters, K.F.A., 2008. The integrated use of chemical insecticides and the entomopathogenic nematode, *Steinernema carpocapsae* (Nematoda: Steinernematidae), for the control of sweetpotato whitefly, *Bemisia tabaci* (Hemiptera: Aleyrodidae). Insect Science 15, 447–453.

Cuthbertson, A.G.S., Blackburn, L.F., Eyre, D.P., Cannon, R.J.C., Millar, J., Northing, P., 2011. *Bemisia tabaci*: the current situation in the UK and the prospect of developing strategies for eradication using entomopathogens. Insect Science 18, 1–10.

Danilov, L.G., 1974. Susceptibility of wireworms to the infestation by the nematode, *Neoaplectana carpocapsae* Weiser, 1955, str. Agriotos. Bulletin of All-Union Research Institute for Plant Protection 30, 54–57.

Del Valle, E.E., Dolinski, C., Barreto, E.L.S., Souza, R.M., Samuels, R.I., 2008. Efficacy of *Heterorhabditis baujardi* LPP7 (Nematoda: Rhabditida) applied in *Galleria mellonella* (Lepidoptera: Pyralidae) insect cadavers to *Conotrachelus psidii*, (Coleoptera: Curculionidae) larvae. Biocontrol Science and Technology 18, 33–41.

Divya, K., Sankar, M., Marulasiddesha, K.N., 2010. Efficacy of entomopathogenic nematode, *Heterorhabditis indica* against three lepidopteran insect pests. Asian Journal of Experimental Biological Sciences 1, 183–188.

Dolinski, C., Shapiro-Ilan, D.I., Lewis, E.E., 2015. Insect cadaver applications: pros and cons. In: Campos-Herrera, R. (Ed.), Nematode Pathogenesis of Insects and Other Pests; Ecology and Applied Technologies for Sustainable Plant and Crop Protection. Springer International Publishing, Switzerland, pp. 207–229.

Doucet, M.M.A., Bertolotti, M.A., Giayetto, A.L., Miranda, M.B., 1999. Host range, specificity, and virulence of *Steinernema feltiae*, *Steinernema rarum* and *Heterorhabditis bacteriophora* (Steinernematidae and Heterorhabditidae) from Argentina. Journal of Invertebrate Pathology 73, 237–242.

Dowds, B.C.A., Peters, A., 2002. Virulence mechanisms. In: Gaugler, R. (Ed.), Entomopathogenic Nematology. CABI Publishing, New York, USA, pp. 79–98.

Dutky, S.R., Thompson, J.V., Hough, W.S., 1956. A Promising Nematode and the Associated Pathogen for Controlling Insect Pests. Entomology Research Branch Circular, USDA, Bethesda, Maryland, USA.

Ebssa, L., Borgemeister, C., Berndt, O., Poehling, H.M., 2001. Efficacy of entomopathogenic nematodes against soil-dwelling life stages of western flower thrips, *Frankliniella occidentalis* (Thysanoptera: Thripidae). Journal of Invertebrate Pathology 78, 119–127.

Ebssa, L., Borgemeister, C., Poehling, H.M., 2004. Effectiveness of different species/strains of entomopathogenic nematodes for control of western flower thrips (*Frankliniella occidentalis*) at various concentrations, host densities, and temperatures. Biological Control 29, 145–154.

Ebssa, L., Borgemeister, C., Poehling, H.M., 2006. Simultaneous application of entomopathogenic nematodes and predatory mites to control western flower thrips *Frankliniella occidentalis*. Biological Control 39, 66–74.

Eidt, D.C., Thurston, G.S., 1995. Physical deterrents to infection by entomopathogenic nematodes in wireworms (Coleoptera: Elateridae) and other soil insect. Canadian Entomologist 127, 423–429.

Ester, A., Huiting, H.F., 2007. Controlling wireworm (*Agriotes* spp.) in a potato crop with biologicals. In: Proceedings of 10th European Meeting "Invertebrate Pathogens in Biological Control: Present and Future", Bari, Italy, pp. 189–196.

Fallon, D.J., Kaya, H.K., Gaugler, R.G., Sipes, B.S., 2002. Effect of entomopathogenic nematodes on *Meloidogyne javanica* on tomato and soyabeans. Journal of Nematology 34, 239–245.

Forst, S., Clarke, D., 2002. Bacteria–nematode symbiosis. In: Gaugler, R. (Ed.), Entomopathogenic Nematology. CABI Publishing, New York, USA, pp. 57–77.

Foster, S.P., Hackett, B., Mason, N., Moores, G.D., Cox, D.M., Campbell, J., Denholm, I., November 18–21, 2002. The BCPC Conference: Pests and Diseases. Resistance to Carbamate, Organophosphate and Pyrethroid Insecticides in the Potato Aphid (Macrosiphum euphorbiae), vols. 1 and 2. Proceedings of an International Conference, Brighton, UK, pp. 811–816.

Garcia-del-Pino, F., Alabern, X., Morton, A., 2013. Efficacy of soil treatments of entomopathogenic nematodes against the larvae, pupae and adults of *Tuta absoluta* and their interaction with the insecticides used against this insect. BioControl 58 (6), 723–731.

Georgis, R., Koppenhöfer, A.M., Lacey, L.A., Bélair, G., Duncan, L.W., Grewal, P.S., Samish, M., Tan, L., Torr, P., van Tol, R.W.H.M., 2006. Successes and failures in the use of parasitic nematodes for pest control. Biological Control 38, 103–123.

Glazer, I., Navon, A., 1990. Activity and persistence of entomoparasitic nematodes tested against *Heliothis armigera* (Lepidoptera: Noctuidae). Journal of Economic Entomology 83, 1795–1800.

Glazer, I., Galper, S., Sharon, E., 1991. Virulence of the nematode (Steinernematids and *Heterorhabditis*)-bacteria (*Xenorhabdus* spp.) complex to the Egyptian cotton leafworm *Spodoptera littoralis* (Lepidoptera: Noctuidae). Journal of Invertebrate Pathology 57, 94–100.

Gothama, A.A.A., Lawrence, G.W., Sikorowski, P.P., 1996. Activity and persistence of *Steinernema carpocapsae* and *Spodoptera exigua* nuclear polyhedrosis virus against *S. exigua* larvae on soybean. Journal of Nematology 28, 68–74.

Grewal, P.S., 2002. Formulation and application technology. In: Gaugler, R. (Ed.), Entomopathogenic Nematology. CABI Publishing, Wallingford, UK, pp. 265–288.

Grewal, P.S., Ehlers, R.U., Shapiro-Ilan, D.I., 2005. Nematodes as Biocontrol Agents. CABI Publishing, Wallingford, UK.

Hara, A.H., Kaya, H.K., Gaugler, R., Lebeck, L.M., Mello, C.L., 1993. Entomopathogenic nematodes for biological control of the leafminer, *Liriomyza trifolii* (Diptera: Agromyzidae). Entomophaga 38, 359–369.

Harris, M.A., Begley, J.W., Warkentin, D.L., 1990. *Liriomyza trifolii* (Diptera: Agromyzidae) suppression with foliar application *Steinernema carpocapsae* (Rhabditida: Steinernematidae) and abamectin. Journal of Economic Entomology 83, 2380–2384.

Head, J., Walters, K.F.A., Langton, S., 2000. The compatibility of the entomopathogenic nematode, *Steinernema feltiae*, and chemical insecticides for the control of the South American leafminer, *Liriomyza huidobrensis*. BioControl 45, 345–353.

Head, J., Walters, K.F.A., 2003. Augmentation biological control using the entomopathogenic nematode, *Steinernema feltiae*, against the South American leafminer, Liriomyza huidobrensis. In: Proceedings of the 1st International Symposium on Biological Control, 13–18 January, 2002, Hawaii, USA. USDA Forest Service, FHTET-03-05, pp. 136–140.

Head, J., Lawrence, A.J., Walters, K.F.A., 2004. Efficacy of the entomopathogenic nematode, *Steinernema feltiae*, against *Bemisia tabaci* in relation to plant species. Journal of Applied Entomology 128, 543–547.

Helyer, N.L., Brobyn, P.J., 1992. Chemical control of western flower thrips (*Frankliniella occidentalis* Pergande). Annals of Applied Biology 121, 219–231.

Helyer, N.L., Brobyn, P.J., Richardson, P.N., Edmonson, R.N., 1995. Control of western flower thrips (*Frankliniella occidentalis* Pergande) pupae in compost. Annals of Applied Biology 127, 405–412.

Henneberry, T.J., Lindegren, J.E., Jech, L.F., Forlow, L., Burke, R.A., 1995. Pink-bollworm (Lepidoptera, Gelechiidae), cabbage-looper, and beet armyworm (Lepidoptera, Noctuidae) pupal susceptibility to steinernematid nematodes (Rhabditida, Steinernematidae). Journal of Economic Entomology 88, 835–839.

Hussain, M.A., Ahmad, R., Ahmad, W., 2014. Evaluation of *Steinernema masoodi* (Rhabditida: Steinernematidae) against soil-dwelling life stage of *Helicoverpa armigera* (Lepidoptera: Noctuidae) in laboratory and microplot study. Canadian Journal of Plant Protection 2, 4–8.

Javed, N., Khan, S.A., Ul-Haq, I., Atiq, M., Kamran, M., 2012. Effect of *Steinernema glaseri* and *Heterorhabditis indica* on the plant vigour and root knot nematodes in tomato roots at different densities and time of applications. Pakistan Journal of Zoology 44, 1165–1170.

Kalia, V., Sharma, G., Shapiro-Ilan, D., Ganguly, S., 2014. Biocontrol potential of *Steinernema thermophilum* and its symbiont *Xenorhabdus indica* against lepidopteran pests: virulence to egg and larval stages. Journal of Nematology 46, 18–26.

Kaya, H.K., 1985. Entomopathogenic nematodes for insect control in IPM system. In: Hass, M.A., Herzog, D.C. (Eds.), Biological Control in Agricultural IPM Systems. Academic Press, Elsevier Inc., USA, pp. 283–302.

Kaya, H.K., Hara, A.H., 1981. Susceptibility of various species of lepidopterous pupa to the entomogenous nematode *Neoaplectana carpocapsae*. Journal of Nematology 13, 291–294.

Kaya, H.K., Gaugler, R., 1993. Entomopathogenic nematodes. Annual Review of Entomology 38, 181–206.

Klein, M.G., 1990. Efficacy against soil-inhabiting insect pests. In: Gaugler, R., Kaya, H.K. (Eds.), Entomopathogenic Nematodes Biological Control. CRC Press, Boca Raton, Florida, USA, pp. 195–214.

Kondo, E., 1987. Size-related susceptibility of *Spodoptera litura* (Lepidoptera: Noctuidae) larvae to the entomogenous nematode *Steinernema feltiae* (Str. DD-136). Applied Entomology and Zoology 22, 560–569.

Kondo, E., Ishibashi, N., 1986. Infectivity and propagation of entomogenous nematodes, *Steinernema* spp. on the common cutworm *Spodoptera litura* (Lepidoptera: Noctuidae). Applied Entomology and Zoology 21, 95–108.

Kondo, E., Ishibashi, N., 1988. Histological and SEM observations on the invasion and succeeding growth of entomogenous nematode, *Steinernema feltiae* (str. DD–136) in *Spodoptera litura* (Lepidoptera: Noctuidae) larvae. Applied Entomology and Zoology 23, 88–96.

Koppenhöfer, A.M., 2000. Nematodes. In: Lacey, L., Kaya, H.K. (Eds.), Field Manual of Techniques in Insect Pathology. Academic Press, San Diego, California, USA, pp. 283–301.

Kovacs, A., Deseo, K.V., Poinar, G., De Leonardis, A., 1980. Prove di lotta contro insetti con applicazione di nematodi entomogeni. In: Atti Giornate Fitopathologiche 1980. Cooperativa Libraria Universitaria Editrice, Bologna, Italy, pp. 449–456.

Koppenhöfer, A.M., Fuzy, E.M., 2002. Comparison of neonicotinoid insecticides as synergists for entomopathogenic nematodes. Biological Control 24, 90–97.

Lacey, L.A., Shapiro-Ilan, D.I., Glenn, G.M., 2010. Post-application of anti-desiccant agents improves efficacy of entomopathogenic nematodes in formulated host cadavers or aqueous suspension against diapausing codling moth larvae (Lepidoptera: Tortricidae). Biocontrol Science and Technology 20, 909–921.

Laznik, Z., Trdan, S., 2008. Entomopathogenic and entomoparasitic nematodes as biological control agents of thrips. Acta Phytopathologica et Entomologica Hungarica 43 (2), 317–322.

Laznik, Z., Znidarcic, D., Trdan, S., 2011. Control of *Trialeurodes vaporariorum* (Westwood) adults on glasshouse-grown cucumber in four different growth substrates: an efficacy comparison of foliar application of *Steinernema feltiae* (Filipjev) and spraying with thiamethoxam. Turkish Journal of Agriculture and Forestry 35, 631–640.

LeBeck, L.M., Gaugler, R., Kaya, H.K., Hara, A.H., Johnson, M.W., 1993. Host stage suitability of the leafminer *Liriomyza trifolii* (Diptera: Agromyzidae) to the entomopathogenic nematode *Steinernema carpocapsae* (Rhabditida: Steinernematidae). Journal of Invertebrate Pathology 62, 58–83.

Lewis, E.E., Campbell, J.C., Griffn, C., Kaya, H.K., Peters, A., 2006. Behavioral ecology of entomopathogenic nematodes. Biological Control 38, 66–79.

Martens, E.C., Goodrich-Blair, H., 2005. The *Steinernema carpocapsae* intestinal vesicle contains a sub-cellular structure with which *Xenorhabdus nematophila* associates during colonization initiation. Cellular Microbiology 7, 1723–1735.

Martens, E.C., Vivas, E.I., Heungens, K., Cowles, C.E., Goodrich-Blair, H., 2004. Investigating mutualism between entomopathogenic bacteria and nematodes. Nematology Monographs and Perspectives: Proceedings of the Fourth International Congress on Nematology 2, 447–462.

Morris, O.N., 1985. Susceptibility of 31 species of agricultural insect pests to the entomogenous nematodes *Steinernema feltiae* and *Heterorhabditis bacteriophora*. Canadian Entomologist 117, 401–407.

Morton, A., Garcia-del-Pino, F., 2009. Ecological characterization of entomopathogenic nematodes isolated in stone fruit orchard soils of Mediterranean areas. Journal of Invertebrate Pathology 102, 203–213.

Morton, A., Garcia-del-Pino, F., 2016. Laboratory and field evaluation of entomopathogenic nematodes for control of *Agriotes obscurus* (L.) (Coleoptera: Elateridae). Journal of Applied Entomology. http://dx.doi.org/10.1111/jen.12343.

Mráček, Z., Růžička, Z., 1990. Infectivity and development of *Steinernema* sp. strain Hylobius (Nematoda, Steinernematidae) in aphids and aphidophagous coccinellids. Journal of Applied Entomology 110, 92–95.

Navon, A., Nagalakshmi, V.K., Shlomit, L., Salame, L., Glazer, I., 2002. Effectiveness of entomopathogenic nematodes in an alginate gel formulation against lepidopterous pests. Biocontrol Science and Technology 12, 737–746.

Needham, P.H., Sawicki, R.M., 1971. Diagnosis of resistance to organophosphorus insecticides in *Myzus persicae* (Sulz.). Nature 230, 125–126.

Park, H.W., Kim, H.H., Cho, M.R., Kang, T.J., Ahn, S.J., Jeon, S.W., Choo, H.Y., 2013. Infectivity of entomopathogenic nematode *Steinernema carpocapsae* Pocheon strain (Nematoda: Steinernematidae) on the green peach aphid *Myzus persicae* (Hemiptera: Aphididae) and its parasitoids. Biocontrol Science and Technology 23 (6), 637–645.

Patel, M.C., Vyas, R.V., 1995. Efficacy of *Steinernema glaseri* against *Helicoverpa armigera* on chickpea in pots. International Chickpea and Pigeonpea Newsletter 2, 39–40.

Parker, W.E., Howard, J.J., 2001. The biology and management of wireworms (*Agriotes* spp.) on potato with particular reference to U.K. Agricultural and Forest Entomology 3, 85–98.

Perme, S., 2005. Testing the Efficacy of Entomopathogenic Nematodes (Rhabditida) Against Foliar Pests of Vegetables (M.Sc. thesis). Department of Agronomy, Biotechnical Faculty, University of Ljubljana, Slovenia, pp. 1–89.

Poinar, G.O., 1975. Entomogenous Nematodes: a Manual and Host List of Insect – Nematode Associations. E.J. Brill, Leiden, The Netherlands, p. 317.

Poinar, G.O., 1990. Biology and taxonomy of Steinernematidae and Heterorhabditidae. In: Gaugler, R., Kaya, H.K. (Eds.), Entomopathogenic Nematodes in Biological Control. CRC Press, Boca Raton Florida, USA, pp. 23–62.

Premachandra, W.T.S.D., Borgemeister, C., Berndt, O., Ehlers, R.U., Poehling, H.M., 2003. Laboratory bioassays of virulence of entomopathogenic nematodes against soil-inhabiting *Frankliniella occidentalis* Pergande (Thysanoptera: Thripidae). Journal of Nematology 5, 539–547.

Puza, V., Mracek, Z., 2010. Does scavenging extend the host range of entomopathogenic nematodes (Nematoda: Steinernematidae)?. Journal of Invertebrate Pathology 104, 1–3.

Qiu, B.L., Mandour, N.S., Xu, C.X., Ren, S.X., 2008. Evaluation of the entomopathogenic nematode *Steinernema feltiae* as a biological control agent of the whitefly, *Bemisia tabaci*. International Journal of Pest Management 54, 247–253.

Ruisheng, A., Orellana, D., Phelan, L., Grewal, P.S., 2013. Effectiveness and durability of the EPN-induced insect and disease resistance in tomato. Journal of Nematology 45, 279–330.

San-Blas, E., Pembroke, B., Gowen, S.R., 2012. Scavenging and infection of different hosts by *Steinernema carpocapsae*. Nematropica 42, 123–130.

Schalk, J.M., Bohac, J.R., Dukes, P.D., 1993. Potential of non-chemical control strategies for reduction of soil insect damage in sweetpotato. Journal of the American Society for Horticultural Science 118, 605–608.

Schroer, S., Ehlers, R.U., 2005. Foliar application of the entomopathogenic nematode *Steinernema carpocapsae* for biological control of diamondback moth larvae (*Plutella xylostella*). Biological Control 33, 81–86.

Seenivasan, S., Sivakumar, M., 2014. Screening for environmental stress-tolerant entomopathogenic nematodes virulent against cotton bollworm. Phytoparasitica 42, 165–177.

Shairra, S.A., Noah, G.M., 2014. Efficacy of entomopathogenic nematodes and fungi as biological control agents against the cotton leaf worm, *Spodoptera littoralis* (Boisd.) (Lepidoptera: Noctuidae). Egyptian Journal of Biological Pest Control 24, 247–253.

Shamseldean, M.S.M., Abd-Elbary, N.A., Shalaby, H., Ibraheem, H.I.H., 2014. Entomopathogenic nematodes as biocontrol agents of the tomato leaf miner *Tuta absoluta* (Meyrick) (Lepidoptera: Gelechiidae) on tomato plants. Egyptian Journal of Biological Pest Control 24, 503–513.

Shapiro-Ilan, D.I., Lewis, E.E., Tedders, W.L., Son, Y., 2003. Superior efficacy observed in entomopathogenic nematodes applied in infected-host cadavers compared with application in aqueous suspension. Journal of Invertebrate Pathology 83, 270–272.

Shapiro-Ilan, D.I., Cottrell, T.E., Mizell, R.F., Horton, D.L., Behle, B., Dunlap, C., 2010. Efficacy of *Steinernema carpocapsae* for control of the lesser peach tree borer, *Synanthedon pictipes*: improved aboveground suppression with a novel gel application. Biological Control 54, 23–28.

Shapiro-Ilan, D., Han, R., Dolinski, C., 2012. Entomopathogenic nematode production and application technology. Journal of Nematology 44, 206–217.

Shapiro-Ilan, D., Han, R., Qiu, X., 2014. Production of entomopathogenic nematodes. In: Morales-Ramos, J., Rojas, G., Shapiro-Ilan, D.I. (Eds.), Mass Production of Beneficial Organisms: Invertebrates and Entomopathogens. Academic Press, Elsevier Inc., San Diego, California, USA, pp. 321–356.

Sher, R.B., Parrella, M.P., Kaya, H.K., 2000. Biological control of the leafminer *Liriomyza trifolii* (Burgess): implications for intraguild predation between *Diglyphus begini* Ashmead and *Steinernema carpocapsae* (Weiser). Biological Control 17, 155–163.

Sicard, M., Brugirard-Ricaud, K., Pages, S., Lanois, A., Boemare, N.E., Brehelin, M., Givaudan, A., 2004. Stages of infection during the tripartite interaction between *Xenorhabdus nematophila*, its nematode vector, and insect hosts. Applied and Environmental Microbiology 70, 6473–6480.

Sikora, R.A., Salem, I.E.M., Klingauf, F., 1979. Susceptibility of *Spodoptera littoralis* to the Entomogenous Nematode *Neoaplectana carpocapsae* and the Importance of Environmental Factors in an Insect Control Program, vol. 44. Medical Faculty Landbouww. University of Ghent, pp. 309–322.

Silva, C.P., Waterfield, N.R., Daborn, P.J., Dean, P., Chilver, T., Au, C.P.Y., Sharma, S., Potter, U., Reynolds, S.E., Ffrench-Constant, R.H., 2002. Bacterial infection of a model insect: *Photorhabdus luminescens* and *Manduca sexta*. Cellular Microbiology 4, 329–339.

Spencer, K.A., 1973. Agromyzidae (Diptera) of economic importance. Series Entomologica 9, 1–418.

Tahir, H.I., Ott, A.A., Hague, N.G.M., 1995. The susceptibility of *Helicoverpa armigera* and *Earias vittella* larvae to entomopathogenic nematode. Afro-Asian Journal of Nematology 5, 161–165.

Toba, H.H., Lindegren, J.E., Turner, J.E., Vail, P.V., 1983. Susceptibility of the Colorado potato beetle and the sugarbeet wireworm to *Steinernema feltiae* and *S. glaseri*. Journal of Nematology 15, 597–601.

Tomalak, M., 1994. Genetic improvement of *Steinernema feltiae* for integrated control of the western flower thrips, Frankliniella occidentalis. IOBC/WPRS Bulletin 17, 17–20.

Tomalak, M., Piggott, S., Jagdale, G.B., 2005. Glasshouse applications. In: Grewal, P.S., Ehlers, R.U., Shapiro-Ilan, D.I. (Eds.), Nematodes as Biological Control Agents. CABI Publishing, Wallingford, UK, pp. 147–166.

Trdan, S., Znidarcic, D., Vidrih, M., 2007. Control of *Frankliniella occidentalis* on glasshouse-grown cucumbers: an efficacy comparison of foliar application of *Steinernema feltiae* and spraying with abamectin. Russian Journal of Nematology 15, 25–34.

Uchôa-Fernandes, M.A., Della-Lucia, T.M.C., Vilela, E.F., 1995. Mating, oviposition and pupation of *Scrobipalpula absoluta* (Meyrick) (Lepidoptera: Gelechiidae). Anais da Sociedade Entomologica do Brasil 24, 159–164.

Vyas, R.V., Patel, N.B., Patel, P., Patel, D.J., 2002. Efficacy of entomopathogenic nematode against *Helicoverpa armigera* on pigeonpea. International Chickpea and Pigeonpea Newsletter 9, 43–44.

Wardlow, L.R., Piggott, S., Goldsworthy, R., 2001. Foliar application of *Steinernema feltiae* for the control of flower thrips. Mededelingen – Faculteit Landbouwkundige en Toegepaste Biologische Wetenschappen 66, 285–291.

Williams, E.C., MacDonald, O.C., 1995. Critical factors required by the nematode *Steinernema feltiae* for the control of the leafminers *Liriomyza huidobrensis*, *Liriomyza bryoniae* and *Chromatomyia syngenesiae*. Annals of Applied Biology 127, 329–341.

Williams, E.C., Walters, K.F.A., 2000. Foliar application of the entomopathogenic nematode *Steinernema feltiae* against leafminers on vegetables. Biocontrol Science and Technology 10, 61–70.

Yadav, Y.S., Pariha, A., Siddiqui, A.U., 2008. Studies on bioefficacy of entomopathogenic nematode against *Helicoverpa armigera*. Journal of Phytological Research 21, 131–134.

Chapter 13

Applications and Trends in Commercial Biological Control for Arthropod Pests of Tomato

Norman C. Leppla[1], Marshall W. Johnson[2], Joyce L. Merritt[1], Frank G. Zalom[3]

[1]University of Florida, Gainesville, FL, United States; [2]University of California, Riverside, CA, United States; [3]University of California, Davis, CA, United States

1. INTRODUCTION

Biological control of insect and mite pests of tomato is feasible if damage by target pest species can be reduced to acceptable levels, and it is possible to manage populations of the remaining pests in a sustainable and nondisruptive manner. Tomato crops are valuable and grown worldwide, therefore the pest complex present at a given location may vary considerably from region to region, as will the community of endemic natural enemies and availability of biological control products. The relative importance of specific pests and applicability of biological control also varies with horticultural practices and type of production, the most important difference being tomatoes grown in protected culture versus open field.

Most tomato production locations have several species of arthropods that are capable of causing economic damage at one or more stages of crop growth, but typically only a subset of these arthropods contains key pests. These pests occur annually and typically require some type of control (Zalom, 2012). The key pests that feed on fruit are considered most damaging due to direct injury and because their feeding is often accompanied by excrement or pathogen-caused decays that render the fruit unmarketable. Foliage feeders are generally less important, although some can severely impact yield and even kill plants outright at high densities (Fig. 13.1). Species that transmit disease-causing pathogens to plants are particularly destructive. Although transplants are commonly used in many types of tomato production, direct seeding is also practiced in field-grown tomatoes in a number of locations worldwide. Seedling pests can reduce plant stands or retard seedling growth when they are numerous, and affected fields may require replanting. Synthetic chemical insecticides have become widely used for managing key tomato pests. Insecticides can prevent these pests from causing economic damage, but are used far too often as cheap "insurance" for anticipated problems and can lead to outbreaks of other arthropods due to disruption of natural enemies (Johnson et al., 1980a). Biological control of the key pests in tomato production, integrated with other compatible tactics including cultural and mechanical methods, represents an alternative approach.

Conservation biological control is a relatively sustainable way of maintaining predators and parasitoids of arthropod pests in tomato production. Natural enemies are conserved by using selective horticultural practices, such as applying insecticides that are minimally disruptive to the natural enemies. There are hundreds of natural enemies of key arthropod pests worldwide that could be conserved to reduce pest populations locally, or enhanced through augmentation of favorable habitat in or around tomato fields. However, this chapter will focus on augmentative biological control (ABC), specifically invertebrate biological control agents (BCAs—arthropod predators and parasitoids) and microbial pesticides that are commercially available or have potential for managing tomato pests.

2. ARTHROPOD PESTS OF TOMATO, INVERTEBRATE BIOLOGICAL CONTROL AGENTS, AND MICROBIAL PESTICIDES

Although BCAs and microbial pesticides are commercially available and applied to control many of the most important arthropod pests of tomatoes worldwide (van Lenteren and Bueno, 2003), there is a great potential for expanding the diversity of products and using them against a wider range of target pests. Some candidates are already identified (Table 13.1) but many other natural enemies could be developed as commercial BCAs. Since the major target pests of tomato are well

Sustainable Management of Arthropod Pests of Tomato. http://dx.doi.org/10.1016/B978-0-12-802441-6.00013-0

FIGURE 13.1 Leafminer *Liriomyza* spp., damage to tomato plant leaves.

known (Finch and Thompson, 1992; Davis et al., 1998; Capinera, 2001; Zalom, 2012; Schuster et al., 2014), this chapter concentrates on species for which commercial BCAs and microbial insecticides are available for use individually or in concert (Table 13.1).

2.1 Lepidoptera

Several lepidopteran species are economically damaging pests of tomato and most are capable of injuring both fruit and foliage. Although some species occur worldwide, it is not uncommon to encounter different but related species occupying similar niches in different geographical regions, e.g., the tomato pinworm, *Keiferia lycopersicella* (Walsingham), and South American tomato moth, *Tuta absoluta* (Meyrick). Although both of these gelechiids are native to the new world, the distribution of *K. lycopersicella* is limited to the southern and western United States and Mexico, while *T. absoluta* is native to Peru and is widespread in all countries in South America. Both species mine foliage of their solanaceous hosts, including tomato, but can also cause extensive damage by feeding on fruit. Following its introduction into Europe, North Africa, and the Middle East, *T. absoluta* has become a serious pest in these regions, causing severe yield loss reaching 100% when left unmanaged. ABC has been considered for this pest (Urbaneja et al., 2012), including hymenopteran and hemipteran predators, and predaceous Acari such as *Amblyseius swirskii* Athias-Henriot (Table 13.1). Predatory mirid bugs have been conserved in greenhouses (Arnó et al., 2000; Albajes and Gabarra, 2003), and two egg predators, *Dicyphus maroccanus* Wagner and *Macrolophus pygmaeus* (Rambur), have been evaluated along with *Bacillus thuringiensis* Berliner to control *T. absoluta* in the Mediterranean region (Zappala et al., 2013). *B. thuringiensis* has also been used to control *K. lycopersicella* in the United States and Mexico in combination with releases of the egg parasitoid *Trichogramma pretiosum* Riley (Trumble and Alvarado-Rodriguez, 1993).

Prior to the emergence of *T. absoluta* as a major tomato pest, fruit-feeding Lepidoptera of the family Noctuidae were considered the most significant insect pests of tomatoes, and many species are still regarded as the key fruit-feeding insects in areas where *T. absoluta* does not yet occur. Noctuids are relatively large, strong flying moths that have a very broad host range and build to very high population levels on both crop and non-crop hosts. They tend to be a greater problem for field-grown tomatoes than those grown in protected culture because noctuids are easy to detect and exclude from structures (Pilkington et al., 2010). Some of the most damaging species are resistant to synthetic chemical insecticides (McCaffery, 1998). Among the noctuids, the old world bollworm, *Helicoverpa* (*Heliothis*) *armigera* (Hübner), and its new world counterpart, the tomato fruit worm, *Helicoverpa* (*Heliothis*) *zea* (Boddie) (Fig. 13.2), are considered the most damaging because their feeding results in deep, watery cavities that are contaminated with feces. Natural enemies, especially *Trichogramma* spp. egg parasitoids (Fig. 13.3) and some generalist predators, can be effective in reducing pest populations. Augmentative releases of *T. pretiosum*, predatory mites, and the minute pirate bug, *Orius insidiosus* (Say), can be used to manage *H. zea*, and *B. thuringiensis* is effective for first or second instar larvae (Table 13.1). Microbial pesticides are currently of limited use, but a commercial formulation of the nuclear polyhedrosis virus can control *H. zea* and other Lepidoptera at a level equal to conventional insecticides (Fig. 13.4). The product was registered by the United States Environmental Protection Agency (EPA) in 1975; however, it is no longer being produced because virus applications cost more than spraying synthetic chemical insecticides.

TABLE 13.1 Important Arthropod Pests of Tomatoes and Their Associated Natural Enemies, Including Commercially Available Species

Pests	Agents	Available Commercially Yes	No	References
Lepidoptera				
Tomato pinworm *Keiferia lycopersicella*	*Apanteles* spp.		X	Schuster (2006)
	Bacillus thuringiensis var *kurstaki*	X		Certis (2016b) and LeBeck and Leppla (2015)
	Trichogramma pretiosum	X		LeBeck and Leppla (2015), Schuster (2006), and Zappala et al. (2013)
South American tomato moth *Tuta absoluta*	*Agathis* spp.		X	Zappala et al. (2013)
	Agathis fuscipennis		X	Zappala et al. (2013)
	Amblyseius swirskii	X		LeBeck and Leppla (2015), van Lenteren (2012), and Zappala et al. (2013)
	Apanteles spp.		X	Zappala et al. (2013)
	B. thuringiensis var *kurstaki*	X		Gonzalez-Cabrera et al. (2011), Jacobson (2004), and LeBeck and Leppla (2015)
	Baryscapus bruchophagi		X	Zappala et al. (2013)
	Brachymeria secundaria		X	Zappala et al. (2013)
	Bracon spp.		X	Zappala et al. (2013)
	Bracon (Habrobracon) didemie		X	Zappala et al. (2013)
	Bracon (Habrobracon) hebetor	X		Biofac Crop Care (2016a), van Lenteren (2000a), and Zappala et al. (2013)
	Bracon (Habrobracon) nigricans (con-colorans, concolor, mongolicus)		X	Biondi et al. (2013) and Zappala et al. (2013)
	Bracon (Habrobracon) sp. near *nigricans*		X	Zappala et al. (2013)
	Bracon (Habrobracon) osculator		X	Zappala et al. (2013)
	Chelonus spp.		X	Zappala et al. (2013)
	Choeras semele		X	Zappala et al. (2013)
	Chrysocharis spp.		X	Zappala et al. (2013)
	Chrysoperla carnea species group	X		LeBeck and Leppla (2015), van Lenteren (2012), and Zappala et al. (2013)
	Cirrospilus spp.		X	Zappala et al. (2013)
	Closterocerus clarus		X	Zappala et al. (2013)
	Cotesia spp.		X	Zappala et al. (2013)
	Diadegma spp.		X	Zappala et al. (2013)
	Diadegma ledicola		X	Zappala et al. (2013)
	Diadegma pulchripes		X	Zappala et al. (2013)
	Dicyphus spp.		X	Zappala et al. (2013)
	Dicyphus errans	X		Ingegno et al. (2013), van Lenteren (2012), and Zappala et al. (2013)
	Dicyphus maroccanus		X	Zappala et al. (2013)
	Dicyphus tamaninii		X	Alomar and Albajes (1996) and Zappala et al. (2013)
	Diglyphus spp.		X	Zappala et al. (2013)
	Diglyphus crassinervis		X	Zappala et al. (2013)
	Diglyphus isaea	X		LeBeck and Leppla (2015), van Lenteren (2012), and Zappala et al. (2013)
	Dineulophus phthorimaeae		X	Luna et al. (2015)
	Diolcogaster spp.		X	Zappala et al. (2013)
	Dolichogenidea litae		X	Zappala et al. (2013)
	Elachertus spp.		X	Zappala et al. (2013)
	Elachertus inunctus species group		X	Zappala et al. (2013)
	Elasmus spp.		X	Zappala et al. (2013)
	Elasmus phthorimaeae		X	Zappala et al. (2013)
	Halticoptera aenea		X	Zappala et al. (2013)
	Hemiptarsenus ornatus		X	Zappala et al. (2013)

Continued

Pests	Agents	Available Commercially Yes	No	References
	Hemiptarsenus zilahisebessi		X	Zappala et al. (2013)
	Hockeria unicolor		X	Zappala et al. (2013)
	Hyposoter didymator		X	Zappala et al. (2013)
	Macrolophus pygmaeus	X		Agrobio (2016a), Bioplanet (2016a), and Koppert (2016a)
	Nabis spp.		X	Desneux et al. (2010) and Zappala et al. (2013)
	Nabis pseudoferus Ibericus	X		Cabello et al. (2009), van Lenteren (2012), and Zappala et al. (2013)
	Necremnus spp.		X	Zappala et al. (2013)
	Necremnus artynes	X		Agrologica (2011) and Zappala et al. (2013)
	Necremnus near *artynes*		X	Zappala et al. (2013)
	Necremnus metalarus		X	Zappala et al. (2013)
	Necremnus tidius		X	Zappala et al. (2013)
	Necremnus near *tidius*		X	Zappala et al. (2013)
	Neochrysocharis spp.		X	Zappala et al. (2013)
	Neochrysocharis formosa (*Closterocerus formosus*)	X		van Lenteren (2012) and Zappala et al. (2013)
	Neoseiulus (*Amblyseius*) *cucumeris*	X		Leng et al. (2011), van Lenteren (2012), and Zappala et al. (2013)
	Nesidiocoris tenuis	X		Agrobio (2016b), Bioplanet (2016b), and Koppert (2016b)
	Orius spp.	X		Syngenta (2016b) and Zappala et al. (2013)
	Orius albidipennis	X		van Lenteren (2012) and Zappala et al. (2013)
	Orius insidiosus	X		Desneux et al. (2010), LeBeck and Leppla (2015), and van Lenteren (2012)
	Pnigalio (*Ratzeburgiola*) *cristatus*		X	Zappala et al. (2013)
	Pnigalio incompletus (*Ratzeburgiola incompleta*)		X	Zappala et al. (2013)
	Pnigalio soemius		X	Zappala et al. (2013)
	Pnigalio spp. *soemius* complex		X	Zappala et al. (2013)
	Pseudapanteles dignus		X	Luna et al. (2015)
	Pteromalus intermedius		X	Zappala et al. (2013)
	Pteromalus semotus		X	Zappala et al. (2013)
	Stenomesius spp.		X	Zappala et al. (2013)
	Stenomesius spp. near *japonicus*		X	Zappala et al. (2013)
	Sympiesis sp. near *flavopicta*		X	Zappala et al. (2013)
	Sympiesis sp. not specified		X	Zappala et al. (2013)
	Tapinoma nigerrimum		X	Zappala et al. (2013)
	Temelucha anatolica		X	Zappala et al. (2013)
	Trichogramma spp.			Zappala et al. (2013)
	Trichogramma achaeae	X		Agrobio (2016c), Biotop (2016), Giraud (2015), and Zappala et al. (2013)
	Trichogramma bourarachae		X	Zappala et al. (2013)
	T. pretiosum	X		LeBeck and Leppla (2015), Oatman and Platner (1971), and van Lenteren (2012)
	Zoophthorus macrops		X	Zappala et al. (2013)
Tomato looper *Chrysodeixis chalcites*	*B. thuringiensis* strains EXP1 and EG 2371		X	Collins et al. (2014) and Vacante et al. (2001)
	B. thuringiensis var. *kurstaki*	X		LeBeck and Leppla (2015) and Plantwise (2016)
	C. carnea	X		LeBeck and Leppla (2015) and van Lenteren (2012)
	Trichogramma achaeae	X		CABI (2016)
	Trichogramma canariensis		X	Pino et al. (2013)
Tomato moth *Lacanobia oleracea*	*B. thuringiensis* var. *kurstaki*	X		LeBeck and Leppla (2015)
	C. carnea	X		LeBeck and Leppla (2015) and van Lenteren (2012)

Continued

TABLE 13.1 Important Arthropod Pests of Tomatoes and Their Associated Natural Enemies, Including Commercially Available Species—cont'd

Pests	Agents	Available Commercially Yes	No	References
	Eulophus pennicornis		X	Morris and Edwards (1995) and Price et al. (2009)
	Steinernema carpocapsae	X		LeBeck and Leppla (2015) and van Lenteren (2012)
	Vairimorpha necatrix		X	Down et al. (2004)
Cabbage moth *Mamestra brassicae*	*B. thuringiensis* var. *kurstaki*	X		LeBeck and Leppla (2015)
	C. carnea	X		Klingen et al. (1996), LeBeck and Leppla (2015), and van Lenteren (2012)
	Microplitis mediator		X	Johansen (1997)
	Podisus maculiventris	X		Filippov (1982) and LeBeck and Leppla (2015)
	Steinernema carpocapsae	X		Beck et al. (2013), LeBeck and Leppla (2015), and van Lenteren (2012)
	Trichograma spp.	X		Johansen (1997)
Beet armyworm *Spodoptera exigua*	*B. thuringiensis aizawai*	X		Certis (2016a)
	B. thuringiensis var. *kurstaki*	X		Certis (2016b)
	Chelonus insularis		X	Capinera (2014c)
	Cotesia marginiventris		X	Alvarado-Rodriguez (1987) and Capinera (2014c)
	Geocoris spp.	X		Biofac Crop Care (2016b) and Capinera (2014c)
	Lespesia archippivora		X	Capinera (2014c)
	Meteorus autographae		X	Capinera (2014c)
	Nabis spp.		X	Capinera (2014c)
	O. insidiosus	X		Capinera (2014c), LeBeck and Leppla (2015), and van Lenteren (2012)
	P. maculiventris	X		Capinera (2014c), LeBeck and Leppla (2015), and van Lenteren (2012)
	Spod-X LC virus	X		Bio-Integral Resource Center (2015) and Certis (2016c)
Egyptian cotton leafworm *Spodoptera littoralis*	*Bacillus subtilis*	X		Ghribi et al. (2012) and LeBeck and Leppla (2015)
	B. thuringiensis var. *kurstaki*	X		Abdelkefi-Mesrati et al. (2011) and LeBeck and Leppla (2015)
Old world bollworm *Helicoverpa (Heliothis) armigera*	*B. thuringiensis* var. *kurstaki*	X		Vijayabharathi et al. (2014)
	Campoletis chlorideae		X	Fatma and Pathak (2011)
	Trichogramma chilonis		X	Fatma and Pathak (2011)
	Various actinomycetes		X	Vijayabharathi et al. (2014)
Tomato fruitworm *Helicoverpa (Heliothis) zea*	*B. thuringiensis* var. *kurstaki*	X		Delannay et al. (1989) and LeBeck and Leppla (2015)
	Baculovirus heliothis (NPV) Gemstar LC, Viron/H		X	Young and Yearian (1974)
	O. insidiosus	X		LeBeck and Leppla (2015) and UCDavis IPM (2013)
	T. pretiosum	X		Oatman and Platner (1971), LeBeck and Leppla (2015), and van Lenteren (2012)
Tomato hornworm *Manduca quinquemaculata*	*B. thuringiensis aizawai*	X		Certis (2016a)
	B. thuringiensis var. *kurstaki*	X		Certis (2016b)
	Cotesia congregata		X	Kessler and Baldwin (2002)
	Geocoris pallens		X	Kessler and Baldwin (2002)
Tobacco hornworm *Manduca sexta*	*Bacillus thuringiensis aizawai*	X		Certis (2016a)
	B. thuringiensis var. *kurstaki*	X		Certis (2016b)
	C. congregata		X	Kessler and Baldwin (2002)
	G. pallens		X	Kessler and Baldwin (2002)
	T. pretiosum	X		LeBeck and Leppla (2015) and Oatman and Platner (1971)
Hemiptera (Whiteflies)				
Glasshouse (greenhouse) whitefly *Trialeurodes vaporariorum*	*Dicyphus hesperus*	X		Lambert et al. (2005), McGregor et al. (2000), and van Lenteren (2012)
	Dicyphus tamaninii		X	Alomar and Albajes (1996)

Continued

TABLE 13.1 Important Arthropod Pests of Tomatoes and Their Associated Natural Enemies, Including Commercially Available Species—cont'd

Pests	Agents	Available Commercially Yes	No	References
	Encarsia formosa	X		Hoddle et al. (1998), LeBeck and Leppla (2015), van Lenteren (2012), and van Lenteren et al. (1996)
	O. insidiosus	X		LeBeck and Leppla (2015) and van Lenteren (2012)
	Macrolophus caliginosus	X		Agrobio (2016a)
Tobacco (sweetpotato) whitefly *Bemisia tabaci* Silverleaf whitefly *B. argentifolii*	Dicyphus hesperus	X		LeBeck and Leppla (2015) and van Lenteren (2012)
	Eretmocerus californicus	X		Rincon Vitova (2017)
	Eretmocerus mundus	X		LeBeck and Leppla (2015) and van Lenteren (2012)
	M. caliginosus	X		Arnó et al. (2000), Gabarra et al. (2006), and van Lenteren (2000a)
	Nesidiocoris tenuis	X		Calvo et al. (2008), Koppert (2016b), and van Lenteren (2012)
	O. insidiosus	X		LeBeck and Leppla (2015) and van Lenteren (2012)
Hemiptera (Aphids)				
Potato aphid *Macrosiphum euphorbiae*	Aphelinus abdominalis	X		LeBeck and Leppla (2015) and van Lenteren (2012)
	Aphidius ervi	X		LeBeck and Leppla (2015) and van Lenteren (2012)
	Chrysoperla spp.	X		LeBeck and Leppla (2015)
	Lecanicillium lecanii	X		Fournier and Brodeur (1999)
	O. insidiosus	X		LeBeck and Leppla (2015) and van Lenteren (2012)
Green peach aphid *Myzus persicae*	Aphidius colemani	X		LeBeck and Leppla (2015) and van Lenteren (2012)
	Aphidius matricariae	X		LeBeck and Leppla (2015) and van Lenteren (2012)
	Aphidoletes aphidimyza	X		Applied Bio-nomics (2016a)
	Chrysoperla spp.	X		LeBeck and Leppla (2015)
	O. insidiosus	X		LeBeck and Leppla (2015) and van Lenteren (2012)
Cotton, melon aphid *Aphis gossypii*	Aphidius colemani	X		LeBeck and Leppla (2015) and van Lenteren (2012)
	Chrysoperla spp.	X		LeBeck and Leppla (2015)
	O. insidiosus	X		LeBeck and Leppla (2015) and van Lenteren (2012)
Hemiptera (Psyllids)				
Tomato (potato) psyllid *Bactericera (Paratrioza) cockerelli*	Beauveria bassiana	X		Lacey et al. (2009) and LeBeck and Leppla (2015)
	Delphastus catalinae	X		Applied Bio-nomics (2016b)
	Metarhizium anisopliae	X		Lacey et al. (2009)
	Tamarixia triozae	X		LeBeck and Leppla (2015)
Hemiptera (Stink Bugs)				
Consperse stink bug *Euschistus conspersus*	Trissolcus spp.		X	Smilanick et al. (1997)
Southern green stink bug *Nezara viridula*	Telenomus podisi		X	Ehler (2002)
	Trichopoda pennipes		X	Ehler (2002)
	Trissolcus basalis		X	Hoffmann et al. (1991)
Thysanoptera (Thrips)				
Western flower thrips *Frankliniella occidentalis*	A. swirskii	X		LeBeck and Leppla (2015) and van Lenteren (2012)
	B. bassiana	X		LeBeck and Leppla (2015)
	N. (Amblyseius) cucumeris	X		LeBeck and Leppla (2015) and van Lenteren (2012)
	Orius spp.	X		LeBeck and Leppla (2015) and Syngenta (2016b)
Tobacco thrips *Frankliniella fusca*	A. swirskii	X		LeBeck and Leppla (2015) and van Lenteren (2012)
	O. insidiosus	X		LeBeck and Leppla (2015) and van Lenteren (2012)
	Thripinema fuscum		X	Carter and Gillert-Kaufman (2015)
	Galendromus occidentalis	X		LeBeck and Leppla (2015)

Continued

TABLE 13.1 Important Arthropod Pests of Tomatoes and Their Associated Natural Enemies, Including Commercially Available Species—cont'd

Pests	Agents	Available Commercially Yes	No	References
Diptera (Leaf Miners)				
Tomato leafminer *Liriomyza bryoniae*	*Chrysocharis oscinidis*		X	CABI-EPPO (2016)
	Dacnusa sibirica	X		CABI-EPPO (2016) and van Lenteren (2012)
	Diglyphus begini		X	van Lenteren (2012)
	D. isaea	X		CABI-EPPO (2016), LeBeck and Leppla (2015), and van Lenteren (2012)
	Opius dimidiatus		X	CABI-EPPO (2016)
	Opius pallipes	X		CABI-EPPO (2016), Rincon Vitova (2001), and van Lenteren (2012)
American serpentine leafminer *Liriomyza trifolii*	*Chrysocharis parksi*		X	Capinera (2014a)
	Diglyphus begini		X	Capinera (2014a) and Tong-Xian et al. (2009)
	Diglyphus intermedius		X	Capinera (2014a)
	Diglyphus pulchripes		X	Capinera (2014a)
Pea leafminer *Liriomyza huidobrensis*	*Dacnusa* spp.		X	Bahlaib et al. (2006)
	D. sibirica	X		LeBeck and Leppla (2015), UCDavis IPM (2010), and van Lenteren (2012)
	Halticoptera circulus		X	Bahlaib et al. (2006)
	Opius spp.		X	Bahlaib et al. (2006)
Vegetable leafminer *Liriomyza sativae*	*Banacuniculus hunteri*[a]		X	Capinera (2014b)
	Chrysonotomyia punctiventris	X	X	Capinera (2014b)
	D. isaea	X		Abd-Raboua (2007), LeBeck and Leppla (2015), and van Lenteren (2012)
	Halicoptera circulis		X	Capinera (2014b)
	Opius dimidiatus		X	Capinera (2014b)
Diptera (Darkwinged Fungus Gnats)				
Bradysia spp.	*B. thuringiensis israelensis* (*Bti*)	X		LeBeck and Leppla (2015) and Pundt (2014)
	Dalotia coriaria (*Atheta coriaria*)	X		LeBeck and Leppla (2015), Pundt (2014), and van Lenteren (2012)
	Hypoaspis miles (*Stratiolaelaps scimitus*)	X		LeBeck and Leppla (2015), Pundt (2014), and van Lenteren (2012)
	Orius spp.	X		LeBeck and Leppla (2015) and Syngenta (2016b)
	Steinernema feltiae	X		LeBeck and Leppla (2015), Pundt (2014), and van Lenteren (2012)
Acari (Mites)				
Two-spotted spider mite *Tetranychus urticae*	*Amblyseius swirski*	X		LeBeck and Leppla (2015) and van Lenteren (2012)
	Neoseiulus (Amblyseius) californicus	X		LeBeck and Leppla (2015) and van Lenteren (2012)
	N. (Amblyseius) cucumeris	X		LeBeck and Leppla (2015) and van Lenteren (2012)
	Phytoseiulus persimilis	X		LeBeck and Leppla (2015) and van Lenteren (2012)
	S. scimitus	X		Applied Bio-nomics (2016d)
Broad mite *Polyphagotarsonemus latus*	*Amblyseius andersoni*	X		LeBeck and Leppla (2015) and van Lenteren (2012)
	Amblyseius (Iphiseius) degenerans	X		LeBeck and Leppla (2015) and van Lenteren (2012)
	N. (Amblyseius) californicus	X		LeBeck and Leppla (2015) and van Lenteren (2012)
Tomato russet mite *Aculops lycopersici*	*Amblyseius andersoni*	X		Syngenta (2016a) and van Lenteren (2012)
	G. occidentalis	X		LeBeck and Leppla (2015)
	Neoseiulus (Amblyseius) fallacis	X		Applied Bio-nomics (2016c)

[a]*Formerly* Ganaspidium hunter.

FIGURE 13.2 Tomato fruitworm, *Heliocoverpa zea*, larva.

FIGURE 13.3 *Trichogramma* sp. wasp parasitizing a tomato fruitworm, *Heliocoverpa zea*, egg. *Photo by Jack Kelly Clark and Frank G. Zalom, University of California Statewide IPM Program, with permission.*

FIGURE 13.4 Beet armyworm, *Spodoptera exigua*, larva killed by *Baculovirus heliothis* (NPV). *Photo by Jack Kelly Clark and Frank G. Zalom, University of California Statewide IPM Program, with permission.*

Noctuids of the genus *Spodoptera*, including the beet armyworm, *Spodoptera exigua* (Hübner), the Egyptian cotton leafworm, *Spodoptera littoralis* (Boisduval) and others, feed on both foliage and fruit, but are not considered as damaging to tomatoes in the open field as is *H. zea*, possibly because they do not feed as deeply into fruit and the wounds may heal. Wounds cause fresh-market tomatoes to be rejected, but there is a wider tolerance for larval holes in tomatoes processed for products such as juice, paste, and ketchup. The microbial pesticide, *B. thuringiensis* subsp. *aizawai,* has enhanced control of *S. exigua* and *Bacillus subtilis* (Ehrenberg) is registered for use against *S. littoralis* (Table 13.1). Hemipteran predators available for ABC of these armyworms include *Geocoris* spp., *O. insidiosus*, and *Podisus maculiventris* (Say). Loopers, also members of the Noctuidae, feed primarily on foliage and the larvae can consume enough leaf area to reduce yields or expose fruit to sunburn when they occur at high densities. They usually can be controlled in tomato crops by natural enemies, particularly hymenopteran egg and larval parasitoids, and applications of *B. thuringiensis* (Delannay et al., 1989). One species, the tomato looper, *Chrysodeixis chalcites* (Esper), which occurs outdoors in Mediterranean tomato-growing areas and in protected culture in northern Europe, is managed with releases of predaceous green lacewings, *Chrysoperla carnea* (Stephens), and *Trichogramma* spp. parasitoids.

Hornworms are very large larvae of the lepidopteran family, Sphingidae, that have the capacity to quickly defoliate plants. The tomato hornworm, *Manduca quinquemaculata* (Haworth), and a related species, the tobacco hornworm, *Manduca sexta* (L.), occur in Mexico north through the United States into Canada and feed only on solanaceous plants. They primarily occur in home gardens and field-grown tomatoes, and are not considered particularly damaging. There are many natural enemies of *M. quinquemaculata* that generally keep it under natural control, one of the most common being an ectoparasitic braconid wasp, *Cotesia congregata* (Say). *B. thuringiensis* also controls it very effectively.

2.2 Hemiptera

Whiteflies damage tomatoes through direct feeding on the foliage, inserting their stylets into leaf veins and extracting nutrients from the phloem. They can occur in very large numbers, reducing plant growth and yield (Johnson et al., 1992). They also excrete honeydew that serves as a substrate for the growth of sooty mold that reduces the quality of harvested fruit. Although they can become damaging in the open field, they are primary pests in glasshouses and other forms of protected culture. Two species are of particular significance, the greenhouse whitefly, *Trialeurodes vaporariorum* Westwood and the tobacco or sweetpotato whitefly, *Bemisia tabaci* (Gennadius) (Trottin-Caudal et al., 2006). Both species transmit plant viruses and biological control may not be possible because very few vectors are required to infect tomato plants (Avilla et al., 2004). Biological control of *T. vaporariorum* and *B. tabaci* is well established in both Europe and North America (Argyriou, 1985; van Lenteren and Woets, 1988; Merino-Pachero, 2007) with the hymenopteran parasitoids *Encarsia formosa* Gahan for *T. vaporariorum* and *Eretmocerus mundus* (Mercet) for *B. tabaci*. Several species of predatory Hemiptera are also commercially produced and distributed for biological control of these pests, e.g., *O. insidiosus, Dicyphus hesperus* Knight, and *Macrolophus caliginosus* Wagner (Table 13.1, whiteflies).

Aphids damage tomatoes through direct feeding on leaves and stems, production of honeydew that is associated with development of sooty mold, and transmission of plant viruses (Blumel, 2004). They are of concern in both open-field production and protected culture. In protected culture, it is preferable to manage aphids through exclusion (Hochmuth and Sprenkel, 2015); however, if they gain entry and spread widely, chemical or biological control is required. The three key species are distributed worldwide: potato aphid, *Macrosiphum euphorbiae* Thomas, green peach aphid, *Myzus persicae* (Sulzer) and cotton or melon aphid, *Aphis gossypii* Glover. Naturally occurring parasitoid guilds that are often habitat-specific and generalist predators, including coccinellids, syrphids, and lacewings, can control these aphids unless they are impacted by insecticides. Commercially available hymenopteran parasitoids and general predators, such as green lacewings, *Chrysoperla* spp. (Fig. 13.5), and the minute pirate bug, *O. insidiosus* (Fig. 13.6), are effectively used in protected culture for controlling aphids (Table 13.1). A microbial insecticide containing the fungus *Lecanicillium lecanii* Zare & Gams is also registered for use on aphids in tomato crops and has shown promise in greenhouse applications (Fournier and Brodeur, 1999).

The tomato psyllid, *Bactericera* (*Paratrioza*) *cockerelli* (Šulc), is a member of the hemipteran family Triozidae. Long considered one of the most destructive potato pests in the western hemisphere, it has become a serious pest of other solanaceous vegetables, including pepper, eggplant, and tomato (Capinera, 2001; Agriculture and Agri-Food Canada, 2011). In recent years, it has gained prominence as a vector of the bacterium *Candidatus* Liberibacter *solanacearum* Jagoueix (syn. *Liberibacter psyllaurous*), commonly called the zebra complex or psyllid yellows. *Bactricera cockerelli* is attacked by a number of generalist natural enemies including lacewing larvae, coccinellids, big-eyed bugs, minute pirate bugs, mirids, nabids, and syrphid larvae, and the parasitoid *Tamarixia triozae* (Burks), which is produced commercially (Table 13.1). Microbial pesticides, including the entomopathogenic fungi *Beauveria bassiana* (Balsamo-Crivelli) Vuillemin and *Metarhizium anisopliae* (Metchnikoff) Sorokin, have produced up to 78% mortality in the field (Lacey et al., 2009).

FIGURE 13.5 *Chrysoperla* sp. larva feeding on a green peach aphid, *Myzus persicae*. *Photo by Jack Kelly Clark and Frank G. Zalom, University of California Statewide IPM Program, with permission.*

FIGURE 13.6 Minute pirate bug, *Orius insidiosus*, feeding on an aphid. *Photo by Jack Kelly Clark and Frank G. Zalom, University of California Statewide IPM Program, with permission.*

Stink bugs are members of the suborder Heteroptera that contains the "true bugs" as opposed to the whiteflies, aphids, and tomato psyllid that are in the suborder Stenorrhyncha. Stink bug species are considered to be primary pests of tomatoes that damage green fruit by inserting their mouthparts into the fruit and injecting salivary secretions. This produces a white corky tissue below the epidermis that remains firm as the fruit ripens and is visible at harvest. Unlike the Stenorrhyncha, stink bugs do not transmit plant viruses, although they may vector the yeast *Nematospora*. Stink bugs usually are only of concern in open-field production because they can be excluded easily from tomatoes produced in protected culture. Stink bugs are relatively large insects, and notoriously difficult to control even with highly toxic insecticides. The most important natural enemies are egg parasitoids in the Hymenoptera family, Scelionidae, and nymphal parasitoids of the Diptera family Tachinidae. The scelionid, *Trissolcus basalis* (Wollaston), has been successfully mass reared and introduced to control the green stink bug, *Nezara viridula* (L.). This pest has spread worldwide from its origin in the Mediterranean area or the African mainland. Smilanick et al. (1997) demonstrated the potential of several native scelionids to control the conspect stink bug, *Euschistus conspersus* Uhler, through ABC in California tomatoes. However, the high cost of rearing the parasitoids on stink bug hosts prevented their adoption by commercial insectaries.

2.3 Thysanoptera

Three species of thrips, the western flower thrips, *Frankliniella occidentalis* Pergande, tobacco thrips, *Frankliniella fusca* (Hinds), and onion thrips, *Thrips tabaci* Lindeman, reportedly damage tomatoes both in protected culture and the open field. They directly injure foliar and floral buds, scar petals and flowers, and silver and crinkle leaves thus reducing yield and fruit quality. They may also cause damage indirectly by transmitting tomato viruses, such as tomato spotted wilt virus, impatiens necrotic spot virus, and tomato chlorotic spot virus. Thrips are tiny and difficult to exclude from protective culture, but elimination of alternate host plants from areas around structures where tomatoes are grown helps to reduce potentially invasive populations (Dissevelt and Ravensberg, 1997; Parrella et al., 1999; van Lenteren, 2000b). Various species of predators attack thrips, especially anthocorids in the genus *Orius*, other generalist hemipteran predators, and spiders. Control with hymenopteran parasitoids has been less effective. Very high levels of parasitism of *F. fusca* by the nematode, *Thripinema fuscum* Tipping & Nguyen, have greatly reduced larval populations in the field, and caused a 50% reduction in the ability of females to transmit tomato spotted wilt virus. Commercial BCAs are available for managing thrips, including *O. insidiosus*, the phytoseiid mites *A. swirskii* and *Neoseiulus* (*Amblyseius*) *cucumeris* (Oudemans) and the fungal pathogen *B. bassiana* (Table 13.1).

2.4 Diptera

Larvae of dipteran leafminers, *Liriomyza* spp., mine leaves (Fig. 13.1) and petioles, reducing the photosynthetic ability of tomato plants and associated yields, and exposing developing fruit to sun scald. The American leafminer, *Liriomyza trifolii* (Burgess), pea leafminer, *Liriomyza huidobrensis* (Blanchard) and vegetable leafminer, *Liriomyza sativae* Blanchard, all originated in the western hemisphere but have greatly expanded their ranges to include protected culture in most tomato-growing regions of the world. They are highly polyphagous, especially *L. trifolii* which has a host range that includes over 400 species in 28 plant families. The tomato leafminer, *Liriomyza bryoniae* (Kaltenbach), has a more limited distribution in Europe and Asia. Hymenopteran parasitoids of dipteran leafminers in Braconidae, Eulophidae, and Pteromalidae families usually control these pests in the absence of insecticides. There are more than 150 species of parasitoids associated with *Liriomyza* spp. (Liu et al., 2011) and some have been mass reared for potential ABC (Liu et al., 2009). Several of these parasitoids are commercially available and used in protected tomato crops (Table 13.1). Predators and entomopathogens are not as effective as parasitoids in controlling dipteran leafminers (Johnson et al., 1980b).

Most species of dark-winged fungus gnats feed on fungi and decaying organic matter and are not considered economic problems. However, larvae of a few species in the genus *Bradysia* attack healthy tissue of tomato plants in greenhouses throughout the world. Fungus gnat problems are often symptomatic of damp conditions and diseased plant roots, and therefore can be mitigated by managing irrigation and promoting root health. ABC targets immature fungus gnats that are found in the upper 2.5 cm of soil. Commercially available predators include a rove beetle, *Dalotia coriaria* Kraatz, and a predatory mite, *Hypoaspis miles* Berlese (*Stratiolaelaps scimitus* (Womersley)) (Table 13.1). The entomopathgenic nematode, *Steinernema feltiae* (Filipjev), and a microbial pesticide, *B. thuringiensis israelensis*, can also be used to control fungus gnats (Table 13.1).

2.5 Acari

Three mites associated with tomato production are regarded as significant pests worldwide. The most common species of spider mite attacking tomato plants in both protected culture and the open field is the two-spotted spider mite, *Tetranychus urticae* Koch. *T. urticae* has the capacity to reproduce very rapidly and build extremely high densities under favorable conditions. Foliar feeding by *T. urticae* causes leaves to desiccate and become chlorotic and necrotic, leading to yield loss and sun scalding of fruit. The broad mite, *Polyphagotarsonemus latus* (Banks), is a worldwide pest on vegetables, including tomatoes, as well as ornamental plants grown in greenhouses. The edges of young leaves usually curl when damaged, become rigid and appear bronzed or scorched. Females of *P. latus* have a phoretic relationship with the whitefly, *B. tabaci*. The tomato russet mite, *Aculops lycopersici* (Massee), has a narrower host range than *T. urticae* or *P. latus*, requiring a perennial alternate host to survive during winter (Rice and Strong, 1962). *A. lycopersici* feeds on tomato foliage, flowers, and young fruit, causing leaf necrosis, bronzing of leaves and stems, flower abortion, and russeting of fruit. If uncontrolled, feeding by *A. lycopersici* kills the infested plant. These mites are controlled in the field by a number of natural enemies, such as predatory thrips, *Orius* spp., and predaceous mites in the families Phytoseiidae and Stigmaeidae. However, ABC of pest mites on tomatoes is limited to several species of Phytoseiidae, and their releases have proven especially successful in protected culture (Table 13.1).

3. BIOLOGICAL CONTROL OF TOMATO PESTS IN PROTECTED CULTURE AND OPEN FIELD PRODUCTION

Biological control of arthropod pests of tomato with predators, parasitoids, and biopesticides depends on the production methods and tomato varieties that are grown, e.g., cherry, roma, grape, pear, cluster, heirloom, beefsteak, and campari. Tomatoes grown under protected culture in glasshouses and structures covered with shade cloth or plastic (e.g., high tunnels) are generally infested by fewer pest species than tomatoes produced in open fields. For various reasons, protected culture has been the primary method of growing tomatoes in Europe for decades and is increasing in North America (Hanafi and Papasolomontos, 1999; Calvin et al., 2012). Almost all commercial tomato production in Canada occurs in technologically advanced glasshouses (Gabarra and Besri, 1999; Cook and Calvin, 2005), while the United States and Mexico are not only expanding protected culture in this type of structure but also in relatively simple shade houses that may not prevent pest entry. The geographical distribution of these tomato production facilities is increasing, potentially exposing the plants to more pest problems in new regions, although the barriers created by protected culture and associated protocols prevent most pests from contacting the tomato plants. In the United States, the hectares of field-grown tomatoes greatly exceeds protected culture, and have declined only slightly as yields have increased. The full range of tomato pests that occur in the field must be managed using cultural practices (e.g., host-free periods), behavioral controls (e.g., mating disruption), applying biopesticides, and conserving natural enemies along with routine pest surveillance.

3.1 Availability of Commercial Natural Enemies for Tomato Pests

Adoption of biological control for tomato pests requires a reliable supply of high-quality natural enemies that are affordable and highly effective (Mhina et al., 2016). The necessary number of products is available, including 219 arthropod natural enemies produced worldwide, and there are tests to determine the quality of the most important species (van Lenteren, 2003, 2012). A current catalog listing natural enemies produced in North America includes 70 species of beneficial nematodes, mites, and insect predators and parasitoids (LeBeck and Leppla, 2015). At least one natural enemy is listed for every arthropod pest of tomato, with many pests having several, along with potentially compatible biopesticides. The catalog identifies 61 suppliers of natural enemies, 35 of which are members of the Association of Natural Biocontrol Producers (ANBP) (http://www.anbp.org/). Adoption of ABC for tomato requires customer support and these companies provide written guidelines, including use and availability of natural enemies, release rates, and general biology of the target pests. The European biological control organization, the International Biocontrol Manufacturer's Association (IBMA), has about 60 active members who produce or supply BCAs. Biorational products can also be located in the 2015 Directory of Least-Toxic Pest Control Products (BIRC, 2015). Therefore, at least in North America and Europe, natural enemies and the know-how to use them are readily available (BIRC, 2015).

3.2 Quality Assurance for Producing Natural Enemies

Producers and suppliers of natural enemies encounter many challenges in delivering high-quality products to tomato growers. Standard problems include developmental and behavioral changes in predators and parasitoids due to laboratory adaptation, facilities, and equipment that malfunction, substandard rearing materials, mistakes made by employees who tend to receive minimal training and supervision, and inadequate packaging, shipping, and handling (Nicoli et al., 1994; van Lenteren and Tommasini, 1999; Leppla, 2013). Many of these problems can be overcome by developing and implementing standard operating procedures (SOPs) that specify every step in producing and delivering the biological control products. SOPs include production control (facilities, equipment, and materials), process control (rearing procedures), and product control (growth stage of the natural enemy delivered) (Leppla, 2013). For product control, standards for BCAs usually describe the following measurable characteristics: species or strain identity and purity, age, size or weight, motility, survival, host location ability, and level of parasitism of the target pest (Smith, 1996). For about 20 of these products, standards and associated tests have been developed by ANBP using a process available from ASTM International (http://www.astm.org/), and some individual companies use the International Organization for Standardization (ISO 9001) program (http://www.iso.org/iso/home.html). ISO International defines a standard as "a document that provides requirements, specifications, guidelines or characteristics that can be used consistently to ensure that materials, products, processes and services are fit for their purpose." The most common approach for testing biological control products is to modify the IOBC international standards for mass-producing insects (van Lenteren et al., 2003, http://users.ugent.be/~padclerc/AMRQC/images/guidelines.pdf). Quality control guidelines are available for about 30 commercially produced BCAs based on these IOBC tests (Nicoli et al., 1994; van Lenteren et al., 2003). Regardless of the approach, at least minimal quality standards are required for all commercial natural enemies purchased by tomato growers to manage arthropod pests. During shipment, however, the natural enemy products are exposed to abnormal circadian rhythms, overcrowding, reduced oxygen concentrations, condensation, and possibly temperature extremes.

3.3 Tomato Grower Evaluation of Natural Enemies

Preservation of the quality of natural enemies shipped by suppliers and received by tomato growers can be assessed by using the "Grower Guide: Quality Assurance of Biocontrol Products" (Buitenhuis, 2014). Tests are described for *Aphelinus abdominalis* (Dalman), *Aphidius* spp., *Aphidoletes aphidimyza* (Rondani), green lacewings (*C. carnea, Chrysoperla rufilabris (Burmeister)*), *Cryptolaemus montrouzieri* Mulsant, *Dacnusa sibirica* Telenga, *D. (Atheta) coriaria, Delphastus catalinae* (Horn), *Dicyphus hesperus* Knight, *Diglyphus isaea* (Walker), *E. formosa, Eretmocerus* spp. *Feltiella acarisuga* (Vallot), *Hippodamia convergens* Guérin-Méneville and other ladybeetles, *Leptomastix dactylopii* Howard, *Orius* spp., phytosiid mite predators (*Amblyseius degenerans* (Berlese), *A. swirskii, Amblyseius andersoni* (Chant), *Neoseiulus californicus* (McGregor), *N. cucumeris, N. fallacis, Phytoseiulus persimilis* Athias-Henriot), *S. feltiae* and other beneficial nematodes, *Stethorus punctillum* (Weise), *S. scimitus* (*H. miles*) and *Trichogramma* spp. Criteria to be checked typically include the product name, company batch number, date received, packaging type and condition, number of organisms and emerging adults in the package, sex ratio, level of movement, and any other pertinent observations on the appearance and performance of the product. The natural enemies must be used as soon as possible and the supplier should be notified immediately if there is a problem.

3.4 The Status of Invertebrate Biological Control Agent Use in Augmentative Biological Control

Tomato growers are adopting ABC more frequently mainly in response to the expansion of pesticide-resistant populations of primary and secondary pests and escalating human health and environmental concerns. Consequently, the number of commercial BCAs available for ABC has increased (van Lenteren, 2012), along with improvements in their rearing (Morales-Ramos et al., 2014). Application methods have also improved continuously (Waage and Greathead, 1988; Parrella et al., 1992; van Lenteren, 2000a; van Driesche and Heinz, 2004), such as the use of sachets and banker plants (Stacey, 1977). Sachets are perforated plastic envelopes of various sizes that are hung in plant canopies. They typically contain both a predator, e.g., predatory mite and a food source (commonly alternate mite prey or pollen). Sachets slowly release the predator (Bale et al., 2008), such as *N. cucumeris* to control western flower thrips on tomatoes (Shipp and Wang, 2003) and cucumbers (Jacobson et al., 2010). Banker plants, or open-rearing systems, are effective when target tomato pest densities are low (Frank, 2010). Crop or noncrop plants are infested with host or prey species that are exposed to BCAs. The plants with the organisms are subsequently moved to the crop infested with target pests. Banker plants were developed initially to manage greenhouse whitefly-infested tomato plants, which were exposed to the aphelinid parasitoid *E. formosa* (Stacey, 1977). The banker plants with parasitized whitefly nymphs were moved to production greenhouses and emerging parasitoids attacked whitefly nymphs on the tomato plants. Although banker plants have been used primarily to manage whiteflies and aphids by incorporating a total of 19 BCAs (Frank, 2010), great potential exists to expand this kind of system to include more pests and natural enemies on a variety of greenhouse and field crops. Moreover, ABC could be integrated with other compatible management tactics, such as sterile insect technique, use of kairomones, and habitat diversification using various plant species (Sivinski, 2013). Improvements in BCAs and methods for their use will increase their availability and adoption (Collier and Van Steenwyk, 2004, 2006; van Lenteren, 2006). However, expansion in the development and use of both new and well-established BCAs continues to be complicated by governmental regulations. Delays and costs associated with assessing the risk of importing and releasing BCAs impedes the worldwide sharing of highly effective products (Bolckmans, 1999; Cock et al., 2010; van Lenteren, 2012).

3.5 Microbial Pesticides Available for Tomato Pests

Biopesticides have been broadly defined to include biochemical pesticides (pheromones and other attractants), microbial pesticides, and pesticides resulting from genetic manipulations (Glare et al., 2012). Numerous biopesticides are commercially available for managing arthropods in field and greenhouse tomato crops (Lipa and Smits, 1999; IR-4, 2010; BIRC, 2015). Although biopesticides constitute only about 5% of the global pesticide market (Seiber et al., 2014), their market share has grown significantly over the past few years as the use of certain synthetic chemical pesticides has declined (Thakore, 2006; Glare et al., 2012) and because of a greater emphasis on pesticide resistance management. Microbial pesticides are preparations containing species of bacteria, fungi, viruses, or protozoa as the main ingredient. Currently, microbial pesticides account for approximately 75% of the global biopesticide market with the bacterial agent, *B. thuringiensis kurstaki*, and other subspecies being the most widely used (Hassan and Gökçe, 2014; Olson, 2015). The 67 registered *B. thuringiensis* products that are marketed in more than 450 formulations alone generate annual sales exceeding $90 million (Jindal et al., 2013), and they are also the most common biopesticides used in tomato production. Other registered microbial pesticides used on tomato crops include the fungi *B. bassiana*, *M. anisopliae* and *L. lecanii*. These infect taxa that are not controlled by *Bacillus* spp., including the western flower thrips (Ansari et al., 2007), potato psyllid (Lacey et al., 2009), and greenhouse whitefly (Ravensberg et al., 1990). A number of insect viruses infect Lepidoptera larvae but only a few, such as nuclear polyhedrosis viruses specific to the beet armyworm, *S. exigua*, and tomato fruitworm, *H. zea*, have been registered for use against tomato pests.

3.6 Advantages of Using Microbial Pesticides

Microbial pesticides have many advantages relative to synthetic chemical pesticides, including lower cost of development and registration, fewer side effects, and the possibility of less selection for resistance in the target pests. A novel chemical pesticide routinely requires $250 million and 9 years to reach the marketplace; whereas, development of a microbial product may be limited to only $10 million over 4 years (Olson, 2015). Most microbial pesticides are not toxic to humans and other non-target vertebrates (Leng et al., 2011); therefore, they are exempt from residue limits on fresh and processed foods (Olson, 2015). This is particularly important for export crops that are shipped globally and subject to international maximum residue levels (Marrone, 2007). Moreover, since 2003 many chemical pesticides (>300) have been officially banned from the European Union (EU) to restrict residue levels, especially products deemed to be human endocrine disrupters (Leng et al., 2011; Lacey et al., 2015). When microbial pesticides are applied either alone or in combination with lower rates of chemical pesticides, consumer exposure to regulated pesticide residues is reduced while crop yields remain high (Regnault-Roger, 2012; Hassan and Gökçe, 2014). Food contamination by microbial pesticides is also unlikely because of their brief persistence (Hassan and Gökçe, 2014). Low-to-no reentry intervals and minimal post-harvest intervals also favor adoption of microbial pesticides instead of chemical pesticides (Olson, 2015). Because of their complex mode of action, it is unlikely that microbial pesticide use will lead to resistance in the target pests (Marrone, 2007; Copping, 2013). Also, these pesticides are generally composed of few volatile organic compounds, so they can be used to replace chemicals that cause air pollution (Marrone, 2007).

3.7 Challenges in Using Microbial Pesticides

Major challenges to increasing the acceptance and utilization of microbial pesticides are marketplace competition, standardization of production, and regulatory requirements. Competing with established, highly effective chemical pesticide distributors can be overwhelming for microbial pesticide suppliers who must convince growers to convert from their time-tested methods and use relatively novel, unproven products (Olson, 2015). While production of chemical pesticides is dominated by fewer than 10 massive global corporations, microbial pesticide companies number in the hundreds, with more than 50 controlling no more than 60% of the total market (Leng et al., 2011; Olson, 2015). These companies produce microbial pesticides by fermentation using raw materials readily available from agriculture with production wastes often recycled to farms as fertilizer (Marrone, 2007). Nevertheless, production costs are relatively high and current microbial pesticide products are more expensive than most chemical pesticides, a major limiting factor in many potential markets (Lacey et al., 2015). Substandard production, handling, and distribution of microbial products can lead to loss of effectiveness and also reduce sales (Fravel et al., 1999; Mishra et al., 2015). Mass-production technologies for microbial pesticides must be improved and standardized to minimize problems associated with contamination, formulation potency, reduced effectiveness, and limited shelf life (Jindal et al., 2013). Registration of a new microbial pesticides discovered by screening candidate microorganisms can be delayed because regulatory agencies use a framework designed for synthetic chemical pesticides (Lacey et al., 2015). Thus, the normally cost-competitive registration of a microbial pesticide can become a lengthy and expensive procedure (Jindal et al., 2013), and there is a need for greater harmonization of registration requirements across international boundaries, including acceptance of generic safety data (Regnault-Roger, 2012; Lacey et al., 2015). Given these constraints, research and development for microbial pesticides is more attractive to start-ups and small companies with limited budgets than to large global corporations (Olson, 2015).

Another major difficulty in adopting microbial pesticides is grower education (Leng et al., 2011), especially demonstrations that illustrate how microbial pesticides fit into IPM programs and what should be expected from these products (Jacobson, 2004; Marrone, 2007). Microbial pesticides are sometimes regarded as inferior to chemical pesticides in several ways, e.g., amounts required, speed-of-kill, and high life stage specificity (Leng et al., 2011). Although rapid degradation of most microbial pesticides can be considered advantageous when compared with chemical pesticides, frequent applications are required for optimal effectiveness. Many growers are confused and skeptical about microbial pesticides, brand-name recognition is lacking, and expectations are high for "silver bullets" as experienced with many chemical pesticides (Marrone, 2007; Mishra et al., 2015). The relatively long periods required by microbial pesticides to kill their hosts have also hampered their effectiveness and acceptance by potential users (Jindal et al., 2013). Moreover, growers often lack knowledge about how to handle and apply microbial pesticides. The added labor and expense often is a deterrent that inhibits the growth of the microbial pesticide market (Olson, 2015).

4. TRENDS IN COMMERCIAL BIOLOGICAL CONTROL

Microbial pesticides could create a new era of sustainable agriculture by providing alternatives to some of the most problematic chemical pesticides (Dufour, 2015; Mishra et al., 2015). In an effort to expand markets for microbial pesticides and BCAs, the historic first meeting of the fledgling "International Federation of Biocontrol/Biopesticide Associations" (now named "BioProtection Global") was held in 2015. Founding members of the federation were the Biopesticide Industry Alliance (BPIA), International Biocontrol Manufacturers Association (IBMA), the South African Bioproducts Organization (SABO), Associacao Brasileira das Empresas de Controle Biologica (ABC Bio), and ANBP. Objectives of the federation include: (1) advancing biocontrol globally; (2) international harmonization of regulations; (3) access and benefit sharing of genetic resources; and (4) building relationships with allied industries and institutions, e.g., organic and sustainable agriculture, and relevant intergovernmental organizations (IGOs) and non-governmental organizations (NGOs). An impetus for formation of the global federation was the continued expansion of "biorational" products (biopesticides, biostimulents, and BCAs) and the recent purchase of smaller biorational product companies by large multinational chemical companies, e.g., BASF, Monsanto, Dow, DuPont, Syngenta, and Bayer.

Efforts to integrate microbial pesticides and BCAs have been ongoing for many years. Oatman et al. (1983) demonstrated that major lepidopteran fruit pests (e.g., *H. zea*) of tomato could be managed successfully using sprays of *B. thuringiensis* var. *kurstaki* (Dipel) combined with inundative releases of the egg parasitoid *T. pretiosum*. The level of control was not significantly different between ABC and weekly treatments with the broad spectrum pesticide methomyl. However, methomyl sprays led to increases in *Liriomyza* leafminers because of reductions in several species of hymenopteran parasitoids that normally kept them from becoming abundant (Johnson et al., 1980b). Due to detrimental effects on predators and parasitoids, along with many other problems, efforts are being made to dramatically reduce the use of synthetic chemical pesticides (Marrone, 2007), creating opportunities to employ safer microbial pesticides in combination with effective BCAs. However, each microbial pesticide must be evaluated individually with respect to its impact on BCAs. Johnson and Krugner (2004) reviewed 29 studies on the impacts of commercial organic agriculture products (four insect pathogens, Spinosad, and neem/azadirachtin) on 49 natural enemy species among 23 arthropod families, including insects, mites, and spiders. Many of the toxins used in legally compliant organic pesticides either killed or debilitated, via sublethal effects, many parasitoids and predators commonly found in agricultural crops, such as tomato.

Effective BCAs and microbial pesticides are available for managing many agricultural pests, but more have been discovered that should be commercialized. Due to the development of niche markets, e.g., glasshouse crops, organic production, and certain field crops, BCA use has been growing at a rate of about 15–20% per year (Bolckmans, 1999; van Lenteren, 2012), and probably has exceeded $200 million (Bolckmans, 2008). The sustained development of microbial pesticides is essential for increasing the quality and safety of agricultural production and enhancing the economic value of agricultural products, such as tomatoes and other vegetables (Leng et al., 2011). Progress in bringing new microbial pesticides to the marketplace will require additional investments in research to accelerate the discovery of potentially useful microorganisms and to enhance the efficacy, delivery, and persistence of the resulting products. Regulatory assistance is needed to specifically address living microorganisms, minimize expensive non-target testing, and decrease the time required for registration of microbial products (Glare et al., 2012). Currently, annual global sales of biopesticides are at least $3 billion depending on which organisms are included (Olson, 2015). If the global pesticide market is about $60 billion, biopesticides account for 5%, with an annual growth rate of 12% (Koivunen et al., 2013). Additional investments in accelerating development of these safer biological control products will yield significant economic and environmental returns (Leppla and King, 1996).

5. ECONOMICS OF ADOPTING AUGMENTATIVE BIOLOGICAL CONTROL

In addition to being effective, BCAs and biopesticides must be affordable, and growers and pest management advisors must know how to use them appropriately in tomato cropping systems. Complete biological control of greenhouse tomato pests is routinely accomplished with BCAs, but can be complicated and is usually site-specific. Accordingly, most major suppliers require customers to contact them for guidance on how to manage their pest problems. It is necessary to accurately define key variables, such as the pest species and their abundance, possible use of chemical pesticides, and a range of complex cultural practices. This information is required to determine the species of natural enemies needed, the timing, release rates, and associated cost, and even to decide if biological control is feasible. The cost of biological control products varies widely and is not advertised usually because suppliers work with customers to design and price successful pest management programs. As a generalized example for large-scale greenhouse tomato production, a major supplier of *E. formosa* recommends the following weekly regimen: (1) At planting—in a clean greenhouse, apply at a rate of 0.25 parasitoids/m^2 if no whitefly is detected but, if whitefly is detected, increase the rate to 0.5/m^2; (2) Main season— if the weekly count is over 2 whitefly per sampling card, double the rate to 1 parasitoid/m^2; and (3) Late season—*E. formosa* should be maintained until the last 1–2 weeks of production (http://www.appliedbio-nomics.com/). If it would take 16 weeks to produce a tomato crop, the weekly release rates are 0.25 parasitoids/m^2 for the first 2 weeks (500 parasitoids/ha), 0.5/m^2 for the next 4 weeks (2000 parasitoids/ha), and 1/m^2 for the last 10 weeks (10,000 parasitoids/ha). The grower would need 12,500 parasitoids per ha and, at an approximate price of US$8 per 1000, the cost would be $100/ha ($0.10/m^2). Due to this economy of scale, a large well-managed tomato greenhouse in Canada had an annual cost of $140/ha ($0.14/m^2) for whitefly control with *E. formosa* (B. Spencer, Applied Bio-nomics). However, the cost of biological control would escalate rapidly if the grower needed to manage spider mites concurrently with *S. scimitus* (*H. miles*) or *P. persimilis* at 50 to 100/m^2, russet mites with *A. fallacis* at 2 mites/m^2, aphids with *A. aphidimyza* at 0.2–0.6/m^2, thrips with *A. cucumeris* at 100 mites per plant, and potato psyllid with *D. catalinae* at 1/m^2. A natural enemy producer listed the following prices for these natural enemies: *S. scimitus* (10,000 at $27.00), *P. persimilis* (1000 at $32.50), *A. fallacis* (1000 at $32.50), *A. aphidimyza* (250 at $27.00), *A. cucumeris* (1000 at $80), and *D. catalinae* (100 at $53.00) (http://www.rinconvitova.com/index.htm). For comparison, a fictitious 16-week pest management strategy using two chemical pesticides, Acramite (bifenazate) for mites ($160/851 g/ha) and Pyganic (pyrethrins) for thrips, whiteflies, aphids, and other pests ($60/1.1 L/ha) would cost about $220/ha ($0.22/m^2). Thus, although there would be an additional cost for mite biological control, use of BCAs probably would be less expensive for tomato production in certain situations, such as protected culture, and preferred to eliminate ineffective or highly toxic insecticide use, protect pollinator bumblebees, avoid worker reentry or pre-harvest intervals (Olson, 2015), preclude insecticide application costs (materials, equipment, training, record keeping, etc.), meet organic certification requirements, comply with government regulations, and satisfy consumer demands. Consequently, commercial biological control for arthropod pests of tomato will continue to increase globally as new BCAs and microbial pesticides are discovered and become commercialized.

REFERENCES

Abdelkefi-Mesrati, L., Boukedi, H., Dammak-Karray, M., Sellami-Boudawara, T., Jaoua, S., Tounsi, S., 2011. Study of the *Bacillus thuringiensis* Vip3Aa16 histopathological effects and determination of its putative binding proteins in the midgut of *Spodoptera littoralis*. Journal of Invertebrate Pathology 106, 250–254.

Abd-Raboua, S., 2007. Biological control of the leafminer, *Liriomyza trifolii* by introduction, releasing, evaluation of the parasitoids *Diglyphus isaea* and *Dacnusa sibirica* on vegetables crops in greenhouses in Egypt. Archives of Phytopathology and Plant Protection 39, 439–443.

Agriculture and Agri-Food Canada, 2011. Crop Profile for Greenhouse Tomato in Canada. http://publications.gc.ca/site/eng/9.697480/publication.html.

Agrobio, 2016a. MACRO Control. http://www.agrobio.es/products/pest-control/macrocontrol-macrolophus-caliginosus-thrip-caterpillar-whitefly-control/?lang=en.

Agrobio, 2016b. NESIDIO Control. http://www.agrobio.es/products/pest-control/nesidiocontrol-nesidiocoris-tenuis-thrip-caterpillar-whitefly-control/?lang=en.

Agrobio, 2016c. TRICHO Control. http://www.agrobio.es/products/pest-control/trichocontrol-trichogramma-achaeae-caterpillar-control/?lang=en.

Agrologica, 2011. Información sobre *Necremnus artynes*. http://www.agrologica.es/informacion-plaga/avispa-parasitoide-necremnus-artynes/.

Albajes, R., Gabarra, R., 2003. Conservation of mirid bugs for biological control in greenhouse tomatoes: history, successes and constraints. In: Roche, L., Edin, M., Mathieu, V., Laurens, F. (Eds.), Colloque International Tomate Sous Abri, Protection Integree-Agriculture Biologique, pp. 17–18 Avignon, France.

Alomar, Ò., Albajes, R., 1996. Greenhouse whitefly (Homoptera: Aleyrodidae) predation and tomato fruit injury by the zoophytophagous predator *Dicyphus tamaninii* (Heteroptera: Miridae). In: Alomar, Ò., Wiedenmann, R.N. (Eds.), Zoophytophagus Heteroptera: Implications for Life History and Integrated Pest Management, Entomological Society of America, pp. 155–177 Lanham, Maryland.

Alvarado-Rodriguez, B., 1987. Parasites and disease associated with larvae of beet armyworm *Spodoptera exigua* (Lepidoptera: Noctuidae) infesting processing tomatoes in Sinaloa, Mexico. Florida Entomologist 70, 444–448.

Ansari, M.A., Shah, F.A., Whittaker, M., Prasad, M., Butt, T.M., 2007. Control of western flower thrips (*Frankliniella occidentalis*) pupae with *Metarhizium anisopliae* in peat and peat alternative growing media. Biological Control 40, 293–297.

Applied Bio-nomics, 2016a. Aphidoletes, *Aphidoletes Aphidimyza*. http://www.appliedbio-nomics.com/products/aphidoletes/.

Applied Bio-nomics, 2016b. Delphastus, *Delphastus catalinae*. http://www.appliedbio-nomics.com/products/delphastus/.

Applied Bio-nomics, 2016c. *Neoseiulus (Amblyseius) fallacis*. http://www.appliedbio-nomics.com/products/fallacis/.

Applied Bio-nomics, 2016d. Stratiolaelaps, *Stratiolaelaps scimitus*. http://www.appliedbio-nomics.com/products/stratiolaelaps/.

Argyriou, L.C., 1985. Some data on integrated control of greenhouse pests in Greece. In: Cavalloro, R. (Ed.), Integrated and Biological Control in Protected Crops, Proceedings of a Meeting of the EC Experts' Group/Heraklion. Balkema, A.A, Rotterdam, Netherlands, pp. 64–73.

Arnó, J., Ariño, J., Español, R., Martí, M., Alomar, Ò., 2000. Conservation of *Macrolophus caliginosus* Wagner (Hemiptera: Miridae) in commercial greenhouses during tomato crop-free periods. Bulletin OILB/SROP 23, 241–246.

Association of Natural Biocontrol Producers (ANBP). http://www.anbp.org.

Avilla, J., Albajes, R., Alomar, Ò., Castañé, C., Gabarra, R., 2004. Biological control of whiteflies on vegetable crops. In: Heinz, K.M., Van Driesche, R.G., Parrella, M.P. (Eds.), Biocontrol in Protected Culture. Ball Publishing, Batavia, Netherlands, pp. 171–184.

Bahlaib, C.A., Goodfellow, S.A., Stanley-Horn, D.E., Halletta, R.H., 2006. Endoparasitoid assemblage of the pea leafminer, *Liriomyza huidobrensis* (Diptera: Agromyzidae), in Southern Ontario. Environmental Entomology 35, 351–357.

Bale, J.S., van Lenteren, J.C., Bigler, F., 2008. Biological control and sustainable food production. Philosophical Transactions of the Royal Society of London B: Biological Sciences 363, 761–776.

Beck, B., Brusselman, E., Nuyttens, D., Moens, M., Temmerman, F., Pollet, S., van Weyenberg, S., Spanoghe, P., 2013. Improving the biocontrol potential of entomopathogenic nematodes against *Manestra brassicae*: effect of spray application technique, adjuvants and an attractant. Pest Management Science 70, 103–112.

Biofac Crop Care, 2016a. Bracon hebetor. http://www.biofac.com/storegrainusgmrl/USDA/Stored_Grain_Pests/stored_grain_pests.html.

Biofac Crop Care, 2016b. *Geocoris* sp. http://www.biofac.com/Biological_Agents/Biological_Control/biological_control.html.

Bio-Integral Resource Center (BIRC), 2015. Directory of least-toxic pest control products. The IPM Practitioner. 34, 1–48. http://www.birc.org/Final2015Directory.pdf.

Biondi, A., Desneux, N., Amiens-Desneux, E., Siscaro, G., Zappala, L., 2013. Biology and developmental strategies of the Palaearctic parasitoid *Bracon nigricans* (Hymenoptera: Braconidae) on the neotropical moth *Tuta absoluta* (Lepidoptera: Gelechiidae). Journal of Economic Entomology 106 (4), 1638–1647.

Bioplanet, 2016a. Macrolophus pygmaeus. http://bioplanet.it/en/macrolophus-pygmaeus-2/.

Bioplanet, 2016b. Nesidiocoris tenuis. http://bioplanet.it/en/nesidiocoris-tenuis-2/.

Biotop, 2016. *Trichogramma achaeae* BIOTOP Product TRICHOTOP TA. http://www.biotop-solutions.com/agriculture/professional/market-gardening-horticulture.html.

Blumel, S., 2004. Biological control of aphids on vegetable crops. In: Heinz, K.M., van Driesche, R.G., Parrella, M.P. (Eds.), Biocontrol in Protected Culture. Ball Publishing, Batavia, Netherlands, pp. 297–312.

Bolckmans, K.J.F., 1999. Commercial aspects of biological pest control in greenhouses. In: Albajes, R., Gullino, M.L., van Lenteren, J.C., Elad, Y. (Eds.), Integrated Pest, Disease Management in Greenhouse Crops. Kluwer Publishers, Dordrecht, Netherlands, pp. 310–318.

Bolckmans, K.J.F., 2008. De insectenfabriek. In: Osse, J., Schoonhoven, L., Dicke, M., Buiter, R. (Eds.), Natuur als bondgenoot: biologische bestrijding van ziekten en plagen. Bio-Wetenschappen en Maatschappij, Den Haag, Netherlands, pp. 51–52.

Buitenhuis, R., 2014. Grower Guide: Quality Assurance of Biocontrol Products. Vineland Research and Innovation Centre, Ontario, Canada. http://www.vinelandresearch.com/sites/default/files/grower_guide_pdf_final.pdf.

Cabello, T., Gallego, J.R., Fernandez-Maldonado, F.J., Soler, A., Beltran, D., Parra, A., Vila, F., 2009. The damsel bug *Nabis pseudoferus* (Hemiptera: Nabidae) as a new biological control agent of the South American Tomato Pinworm, *Tuta absoluta* (Lep.:Gelechiidae), in tomato crops in Spain. IOBC WPRS Bulletin 49, 219–223.

CABI, 2016. *Trichogramma achaeae* Natural Enemy of Tomato Looper, *Chrysodeixis chalcites* (Golden Twin-Spot Moth). Crop Protection Compendium DataSheet http://www.cabi.org/isc/datasheet/13243.

CABI-EPPO, 2016. *Liriomyza bryoniae* Data Sheet. Centre for Agriculture and Biosciences International, European and Mediterranean Plant Protection Organization. http://www.eppo.int/QUARANTINE/data_sheets/insects/LIRIBO_ds.pdf.

Calvin, L., Thornsbury, S., Cook, R.L., June 28 , 2012. Recent Trends in the Fresh Tomato Market. Vegetables and Pulses Outlook, Economic Research Service. USDA, pp. 26–35.

Calvo, J., Bolckmans, K., Stansly, S., Urbaneja, A., 2008. Predation by *Nesidiocoris tenuis* on *Bemisia tabaci* and injury to tomato. BioControl 54, 237–246.

Capinera, J.L., 2001. Handbook of Vegetable Pests. Academic Press, New York, USA.

Capinera, J.L., 2014a. American Serpentine Leafminer *Liriomyza Trifolii* (Burgess) (Insecta: Diptera: Agromyzidae). Featured Creatures Entomology and Nematology Department, University of Florida. http://entnemdept.ufl.edu/creatures/veg/leaf/a_serpentine_leafminer.htm.

Capinera, J.L., 2014b. Vegetable Leafminer *Liriomyza sativae* Blanchard (Insecta: Diptera: Agromyzidae). Featured Creatures Entomology and Nematology Department, University of Florida, http://entnemdept.ufl.edu/creatures/veg/leaf/vegetable_leafminer.htm.

Capinera, J.L., 2014c. Beet Armyworm *Spodoptera exigua* (Hübner) (Insecta: Lepidoptera: Noctuidae). Featured Creatures Entomology and Nematology Department, University of Florida. http://entnemdept.ufl.edu/creatures/veg/leaf/beet_armyworm.htm.

Carter, E., Gillett-Kaufman, J., 2015. Tobacco Thrips *Frankliniella fusca* (Hinds) (Insecta: Thysanoptera: Thripidae). Featured Creatures Entomology and Nematology Department, University of Florida. http://entnemdept.ufl.edu/creatures/VEG/THRIPS/Frankliniella_fusca.htm.

Certis, 2016a. Javelin Biological Insecticide. http://www.certisusa.com/pdf-labels/Javelin_WG_label.pdf.

Certis, 2016b. Costar Biological Insecticide Label. http://www.certisusa.com/pdf-labels/costar-label.pdf.

Certis, 2016c. Spodx, Virus Insecticide Specific for the Beet Armyworm. http://www.certisusa.com/pest_management_products/insecticidal_viruses/spod-x_biological_insecticide.htm.

Cock, M.J.W., van Lenteren, J.C., Brodeur, J., Barratt, B.I.P., Bigler, F., Bolckmans, K., Cônsoli, F.L., Haas, F., Mason, P.G., Parra, J.R.P., 2010. Do new access and benefit sharing procedures under the convention on biological diversity threaten the future of biological control? BioControl 55, 199–218.

Collier, T., Van Steenwyk, R., 2004. A critical evaluation of augmentative biological control. Biological Control 31, 245–256.

Collier, T., Van Steenwyk, R., 2006. How to make a convincing case for augmentative biological control. Biological Control 39, 119–120.

Collins, L., Korycinska, A., Baker, R., 2014. Rapid pest risk analysis for *Chrysodeixis chalcites*. The Food and Environment Research Agency.

Cook, R., Calvin, L., April 2005. Greenhouse Tomatoes Change the Dynamics of the North American Fresh Tomato Industry. ERR-2 USDA, Economics Research Service. http://www.ers.usda.gov/publications/err2/err2.pdf.

Copping, L., 2013. Biopesticide market opportunities–strategic knowledge sharing and networking event. Outlooks on Pest Management 24, 136–138.

Davis, R.M., Hartz, T.K., Lanini, W.T., Marsh, R.E., Zalom, F.G., (technical coordinators), 1998. Integrated Pest Management for Tomatoes, fourth ed. University of California Division of Agriculture and Natural Resources Publication 3274, p. 118.

Delannay, X., LaVallee, B.J., Proksch, R.K., Fuchs, R.L., Sims, S.R., Greenplate, J.T., Marrone, P.G., Dodson, R.B., Augustine, J.J., Layton, J.G., Fischhoff, D.A., 1989. Field performance of transgenic tomato plants expressing the *Bacillus thuringiensis* var. Kurstaki insect control protein. Nature Biotechnology 7, 1265–1269.

Desneux, N., Wajnberg, E., Wyckhuys, K.A.G., Burgio, G., Arpaia, S., Narvaez-Vasquez, C.A., Gonzalez-Cabrera, J., Ruescas, D.C., Tabone, E., Frandon, J., Pizzol, J., Poncet, C., Cabello, T., Urbaneja, A., 2010. Biological invasion of European tomato crops by *Tuta absoluta*: ecology, geographic expansion and prospects for biological control. Journal of Pest Science 83, 197–215.

Dissevelt, M., Ravensberg, W.J., 1997. The present status and future outlook of the use of natural enemies and pollinators in protected crops. In: Goto, E., Kurata, K., Hayashi, M., Sase, S. (Eds.), Plant Production in Closed Ecosystems: The International Symposium on Plant Production in Closed Ecosystems, Narita, Japan, August 26–29, 1996. Springer, Dordrecht, Netherlands.

Down, R.E., Bell, H.A., Kirkbride-Smith, A.E., Edwards, J.P., 2004. The pathogenicity of *Vairimorpha necatrix* (Microspora: Microsporidia) against the tomato moth, *Lacanobia oleracea* (Lepidoptera: Noctuidae) and its potential use for the control of lepidopteran glasshouse pests. Pest Management Science 60, 755–764.

Dufour, R., 2015. Biorationals: Ecological Pest Management Database. National Center for Appropriate Technology, ATTRA. https://attra.ncat.org/attra-pub/biorationals/.

Ehler, L.E., 2002. An evaluation of some natural enemies of *Nezara viridula* in Northern California. BioControl 47, 309–325.

Fatma, Z., Pathak, P.H., 2011. Food plants of *Helicoverpa armigera* (Hübner) and extent of parasitism by its parasitoids *Trichogramma Chilonis* Ishii and *Campoletis chlorideae* Uchida – a field study. International Journal of Entomology 2, 31–39.

Filippov, V., 1982. Integrated control of pests of vegetable crops grown in the open in Moldavia. Acta Entomologica Fennica 40, 6–9.

Finch, S., Thompson, A.R., 1992. Pests of cruciferous crops. In: McKinlay, R.G. (Ed.), Vegetable Crop Pests. Macmillan, Basingstoke, UK, pp. 87–138.

Fournier, V., Brodeur, J., 1999. Biological control of lettuce aphids with the entomopathogenic fungus *Verticillium lecanii* in greenhouses. IOBC/WPRS Bulletin 22, 77–80.

Frank, S.D., 2010. Biological control of arthropod pests using banker plant systems: past progress and future directions. Biological Control 52, 8–16.

Fravel, D.R., Thodes, D.J., Larkin, R.P., 1999. Production and commercialization of biocontrol products. In: Albajes, R., Gullino, M.L., van Lanteren, J.C., Elad, Y. (Eds.), Integrated Pest, Disease Managemnet in Greenhouse Crops. Kluwer Publishers, Dordrecht, The Netherlands, pp. 365–376.

Gabarra, R., Besri, M., 1999. Tomatoes. In: Albajes, R., Gullino, M.L., van Lenteren, J.C., Elad, Y. (Eds.), Integrated Pest, Disease Management in Greenhouse Crops. Kluwer Publishers, Dordrecht, The Netherlands, pp. 420–434.

Gabarra, R., Zapata, R., Castañé, C., Riudavets, J., Arnó, J., 2006. Releases of *Eretmocerus mundus* and *Macrolophus caliginosus* for controlling *Bemisia tabaci* on spring and autumn greenhouse tomato crops. IOBC/WPRS Bulletin 29, 71–76.

Ghribi, D., Abdelkefi-Mesrati, L., Boukedi, H., Elleuch, M., Ellouze-Chaabouni, S., Tounsi, S., 2012. The impact of the *Bacillus subtilis* SPB1 biosurfactant on the midgut histology of *Spodoptera littoralis* (Lepidoptera: Noctuidae) and determination of its putative receptor. Journal of Invertebrate Pathology 109, 183–186.

Giraud, M., 2015. *Trichogramma achaeae* as an IPM Tool in Tomato Greenhouses. In Vivo Agro Solutions, France. https://www.ior.poznan.pl/plik,2056, giraud-marion-pdf.pdf.

Glare, T., Caradus, J., Gelernter, W., Jackson, T., Keyhani, N., Köhl, J., Marrone, P., Morin, L., Stewart, A., 2012. Have biopesticides come of age? Trends in Biotechnology 30, 250–258.

Gonzalez-Cabrera, J., Molla, O., Monton, H., Urbaneja, A., 2011. Efficacy of *Bacillus thuringiensis* (Berliner) in controlling the tomato borer, *Tuta absoluta* (Meyrick) (Lepidoptera: Gelechiidae). BioControl 56, 71–80.

Hanafi, A., Papasolomontos, A., 1999. Integrated production and protection under protected cultivation in the Mediterranean region. Biotechnology Advances 17, 183–203.

Hassan, E., Gökçe, A., 2014. Production and consumption of biopesticides. In: Singh, D. (Ed.), Advances in Plant Biopesticides. Springer, India, pp. 361–379.

Hochmuth, R.C., Sprenkel, R.K., 2015. Exclusion Methods for Managing Greenhouse Vegetable Pests. UF/IFAS. Extension Digital Information Source (EDIS), ENY-846 https://edis.ifas.ufl.edu/in730.

Hoddle, M.S., van Driesche, R.G., Sanderson, J.P., 1998. Biology and use of the whitefly parasitoid *Encarsia formosa*. Annual Review of Entomology 43, 645–669.

Hoffmann, M.P., Davidson, N.A., Wilson, L.T., Ehler, L.E., Jones, W.A., Zalom, F.G., 1991. Imported wasp helps control southern green stink bug. California Agriculture 45, 20–22.

Ingegno, B.L., Ferracini, C., Gallinotti, D., Alma, A., Tavella, L., 2013. Evaluation of the effectiveness of *Dicyphus errans* (Wolff) as predator of *Tuta absoluta* (Meyrick). Biological Control 67, 246–252.

International Biocontrol Manufacturers Association (IBMA). http://www.ibma.ch.

IR-4, 2010. Biopesticide and Organic Database for Integrated Pest Management. http://www.ir4.rutgers.edu/Biopesticides/Labeldatabase/index.cfm.

Jacobson, R.J., 2004. IPM program for tomato. In: Heinz, K.M., van Driesche, R.G., Parrella, M.P. (Eds.), Biocontrol in Protected Culture. Ball Publishing, Batavia, Netherlands, pp. 457–471.

Jacobson, R.J., Chandler, D., Fenlon, J., Russell, K.M., 2010. Compatibility of *Beauveria bassiana* (Balsamo) Vuillemin with *Amblyseius cucumeris* Oudemans (Acarina: Phytoseiidae) to control *Frankliniella occidentalis* Pergande (Thysanoptera: Thripidae) on cucumber plants. Biocontrol Science and Technology 11, 391–400.

Jindal, V., Dhaliwal, G.S., Koul, O., 2013. Pest Management in 21st century: roadmap for future. Biopesticides International 9, 1–22.

Johansen, N.S., 1997. Mortality of eggs, larvae and pupae and larval dispersal of the cabbage moth, *Mamestra brassicae*, in white cabbage in south-eastern Norway. Entomologia Experimentalis et Applicata 83, 347–360.

Johnson, M.W., Caprio, L.C., Coughlin, J.A., Tabashnik, B.E., Rosenheim, J.A., Welter, S.C., 1992. Impact of greenhouse whitefly, *Trialeurodes vaporariorum* (Westwood), (Homoptera: Aleyrodidae) on yield of fresh market tomatoes. Journal of Economic Entomology 85, 2370–2376.

Johnson, M.W., Krugner, R., July 13–15, 2004. Impact of legally compliant organic pesticides on natural enemies. In: Hoddle, M. (Ed.), Proceedings, California Conference on Biological Control IV, Berkeley, California, pp. 69–75.

Johnson, M.W., Oatman, E.R., Wyman, J.A., 1980a. Effects of insecticides on populations of the vegetable leafminer and associated parasites on summer pole tomatoes. Journal of Economic Entomology 73, 61–66.

Johnson, M.W., Oatman, E.R., Wyman, J.A., 1980b. Natural control of *Liriomyza sativae* on pole tomatoes in southern California. Entomophaga 25, 193–198.

Kessler, A., Baldwin, I.T., 2002. *Manduca quinquemaculata's* optimization of intra-plant oviposition to predation, food quality, and thermal constraints. Ecology 83, 2346–2354.

Klingen, I., Johansen, N.S., Hofsvang, T., 1996. The predation of *Chrysoperla carnea* (Neurop., Chrysopidae) on eggs and larvae of *Mamestra brassicae* (Lepidoptera: Noctuidae). Journal of Applied Entomology 120, 363–367.

Koivunen, M., Duke, S.O., Coats, J.C., Beck, J.J., 2013. Pest management with natural products. In: Beck, J.J., Coats, J.R., Duke, S.O., Koivunen (Eds.), Pest Management with Natural Products, ACS Symposium Series. vol. 1141. American Chemical Society, Washington, DC, USA, pp. 1–4.

Koppert, 2016a. Mirical *Macrolophus pygmaeus* (formerly known as *Macrolophus caliginosus*). http://www.koppert.com/pests/tuta-absoluta/product-against/mirical/.

Koppert, 2016b. NESIBUG *Nesidiocoris Tenuis*. http://www.koppert.com/pests/tuta-absoluta/product-against/nesibug/.

Lacey, L.A., De La Rosa, F., Horton, D.R., 2009. Insecticidal activity of entomopathogenic fungi (Hypocreales) for potato psyllid, *Bactericera cockerelli* (Hemiptera: Triozidae): development of bioassay techniques, effect of fungal species and stage of the psyllid. Biocontrol Science and Technology 19, 957–970.

Lacey, L.A., Grzywacz, D., Shapiro-Ilan, D.I., Frutos, R., Brownbridge, M., Goettel, M.S., 2015. Insect pathogens as biological control agents: back to the future. Journal of Invertebrate Pathology 132, 1–41.

Lambert, L., Chouffot, T., Tureotte, G., Lemieux, M., Moreau, J., 2005. Biological control of greenhouse whitefly (*Trialeurodes vaporariorum*) on interplanted tomato crops with and without supplemental lighting using *Dicyphus hesperus*. Bulletin OILB/SROP 28.

LeBeck, L.M., Leppla, N.C., 2015. Guidelines for Purchasing and Using Commercial Natural Enemies and Biopesticides in North America. UF/IFAS Electronic Data Information Source (EDIS) IPM-146, UN-849.

Leng, P., Zhang, Z., Pan, G., Zhao, M., 2011. Applications and developmental trends in biopesticides. African Journal of Biotechnology 10, 19864–19873.

Leppla, N.C., 2013. Concepts and methods of quality assurance for parasitoids and predators. In: Morales-Ramos, J.A., Shapiro-Ilan, D. (Eds.), Mass Production of Beneficial Organisms. Elsevier, pp. 277–317.

Leppla, N.C., King, E.G., 1996. The role of parasitoid and predator production in technology transfer of field crop biological control. Entomophaga 41, 343–360.

Lipa, J.J., Smits, P.H., 1999. Microbial control of pests in greenhouses. In: Albajes, R., Gullino, M.L., van Lenteren, J.C., Elad, Y. (Eds.), Integrated Pest, Disease Management in Greenhouse Crops. Kluwer Publishers, Dordrecht, The Netherlands, pp. 295–309.

Liu, T.X., Kang, L., Heinz, K.M., Trumble, J., 2009. Biological control of *Liriomyza* leafminers: progress and perspective. CAB reviews: perspectives in agriculture, veterinary science. Nutrition and Natural Resources 4 (4), 1–16.

Liu, T.X., Kang, L., Lei, Z., Hernandez, R., 2011. Hymenopteran parasitoids and their role in biological control of vegetable *Liriomyza leafminers*. In: Liu, T., Kang, L. (Eds.), Recent Advances in Entomological Research: From Molecular Biology to Pest Management. Springer, Beijing, China, pp. 376–403.

Luna, M.G., Pereyra, P.C., Coviella, C.E., Nieves, E., Savino, V., Salas Gervassio, N.G., Luft, E., Virla, E., Sanchez, N.E., 2015. Potential of biological control agents against *Tuta absoluta* (Lepidoptera: Gelechiidae): current knowledge in Argentina. Florida Entomologist 98, 489–494.

Marrone, P.G., 2007. Barriers to adoption of biological control agents and biological pesticides. CAB Reviews: Perspectives in Agriculture, Veterinary Science, Nutrition and Natural Resources 2, 51.

McCaffery, A.R., 1998. Resistance to insecticides in heliothine Lepidoptera: a global view. Philosophical Transactions of the Royal Society of London B: Biological Sciences 1376 (353), 1735–1750.

McGregor, R.R., Gillespie, D.R., Park, C.G., Quiring, D.M.J., Foisy, M.R.J., 2000. Leaves or fruit? The potential for damage to tomato fruits by the omnivorous predator, *Dicyphus hesperus*. Entomologia Experimentalis et Applicata 95, 325–328.

Merino-Pachero, M., 2007. Almeria finally forced to turn green. Fruit and Vegetable Technology 7, 23–25.

Mhina, G.J., Leppla, N.C., Thomas, M.H., Solis, D., 2016. Cost effectiveness of biological control of invasive mole crickets in Florida pastures. Biological Control 100, 108–115.

Mishra, J., Tewari, S., Singh, S., Arora, N.K., 2015. Biopesticides: where we stand. In: Arora, N.K. (Ed.), Plant Microbes Symbiosis: Applied Facets. Springer, India, pp. 37–75.

Morales-Ramos, J.A., Guadalupe Rojas, M., Shapiro-Ilan, D.I., 2014. Mass Production of Beneficial Organisms. Elsevier, Amsterdam, Netherlands.

Morris, C.G., Edwards, J.P., 1995. The biology of the ectoparasitoid wasp *Eulophus pennicornis* (Hymenoptera: Eulophidae) on host larvae of the tomato moth, *Lacanobia oleracea* (Lepidoptera: Noctuidae). Bulletin of Entomological Research 85, 507–513.

Nicoli, G., Benuzzi, M., Leppla, N.C. (Eds.), 1994. Proceedings Seventh Workshop of the IOBC Global Working Group, Quality Control of Mass Reared Arthropods, Rimini, Italy. September 13–16, 1993.

Oatman, E.R., Platner, G.R., 1971. Biological control of the tomato fruitworm, cabbage looper, and hornworms on processing tomatoes in Southern California, using mass releases of *Trichogramma pretiosum*. Journal of Economic Entomology 64 (2), 501–506.

Oatman, E.R., Wyman, J.A., Van Steenwyk, R.A., Johnson, M.W., 1983. Integrated control of the tomato fruitworm (Lepidoptera: Noctuidae) and other lepidopterous pests on fresh market tomatoes in Southern California. Journal of Economic Entomology 76, 440–445.

Olson, S., 2015. An analysis of the biopesticide market now and where it is going. Outlooks on Pest Management 26, 203–206.

Parrella, M.P., Hansen, L.S., van Lenteren, J., 1999. Glasshouse environments. In: Enhanced Biological Control through Pesticide Selectivity, Handbook of Biological Control. Academic Press, Elsevier Inc., USA, pp. 819–839.

Parrella, M.P., Heinz, K.M., Nunney, L., 1992. Biological control through augmentative releases of natural enemies: a strategy whose time has come. American Entomologist 38, 172–179.

Pilkington, L.J., Messelink, G., van Lenteren, J.C., Le Mottee, K., 2010. Protected biological control, biological pest management in the greenhouse industry. Biological Control 52, 216–220.

Pino, M.Del, Hernandez-Suarez, E., Cabello, T., Rugman-Jones, P., Stouthamer, R., Polaszek, A., 2013. *Trichogramma canariensis* (Insecta: Hymenoptera: Trichogrammatidae) a parasitoid of eggs of the twinspot moth *Chrysodeixis chalcites* (Lepidoptera: Noctuidae) in the Canary Islands. Arthropod Systematics and Phylogeny 71, 169–179.

Plantwise, 2016. Technical Fact Sheet on Golden Twin-Spot Moth *Chrysodeixis chalcites*. *B. thuringiensis* var. *kurstaki* control of *Chrysodeixis chalcites* Plantwise Knowledge Bank http://www.plantwise.org/KnowledgeBank/Datasheet.aspx?dsid=13243.

Price, R.G., Bell, H.A., Hinchliffe, G., Fitches, E., Weaver, R., Gatehouse, J.A., 2009. A venom metalloproteinase from the parasitic wasp *Eulophus pennicornis* is toxic towards its host, tomato moth (*Lacanobia oleracae*). Insect Molecular Biology 18, 195–202.

Pundt, L., 2014. Biological Control of Fungus Gnats. University of Connecticut. 2007, Revised 2014 http://ipm.uconn.edu/documents/raw2/html/666.php?aid=666.

Ravensberg, W.I., Malais, M., van der Schaff, D.A., 1990. *Verticillium lecanii* as a microbial insecticide against glasshouse whitefly. In: Brighton Crop Protection Conference, Pests and Diseasesvol. 1, pp. 265–268.

Regnault-Roger, C., 2012. Trends for Commercialization of Biocontrol Agent (Biopesticide) Products. In: Mérillon, J.M., Ramawat, K.G. (Eds.), Plant Defence: Biological Control. Springer, Netherlands, pp. 139–160.

Rice, R.E., Strong, F.E., 1962. Bionomics of the tomato russet mite, *Vasates lycopersici* (Massee). Annals of the Entomological Society of America 55, 431–435.

Rincon Vitova, 2017. Catalog of Beneficials. Rincon-Vitova Insectaries Inc., Ventura, California, USA. http://www.rinconvitova.com/catalog-beneficials.htm.

Schuster, D., 2006. Tomato pinworm: *Keiferia lycopersicella*. In: Gillett, J.L., HansPetersen, H.N., Leppla, N.C., Thomas, D.D. (Eds.), Grower's IPM Guide for Florida Tomato and Pepper Production. pp. 77–78. http://ipm.ifas.ufl.edu/pdfs/Tomato_Pinworm.pdf.

Schuster, D.J., Zalom, F.G., Gilreain, D., 2014. Arthropod pests. In: Jones, J.B., Zitter, T.A., Momol, T.M., Miller, S.A. (Eds.), Compendium of Tomato Diseases and Pests, second ed. American Phytopathological Society Press, Saint Paul, Minnesota, USA, pp. 120–127.

Seiber, J.N., Coats, J., Duke, S.O., Gross, A.D., 2014. Biopesticides: state of the art and future opportunities. Journal of Agricultural and Food Chemistry 62, 11613–11619.

Shipp, J.L., Wang, K., 2003. Evaluation of *Amblyseius cucumeris* (Acari: Phytoseiidae) and *Orius insidiosus* (Hemiptera: Anthocoridae) for control of *Frankliniella occidentalis* (Thysanoptera: Thripidae) on greenhouse tomatoes. Biological Control 28, 271–281.

Sivinski, J., 2013. Augmentative biological control: research and methods to help make it work. CAB Reviews 8, 26.

Smilanick, J.M., Zalom, F.G., Ehler, L.E., 1997. Evaluation of scelionid egg parasitoids for stink bug control in California tomato fields. In: Hanafi, A., Achouri, M., Baudoin, W.O. (Eds.), Production and Protection Integree (PPI), pp. 45–49Agadir, Morocco.

Smith, S.M., 1996. Biological control with *Trichogramma*: advances, successes, and potential of their use. Annual Review of Entomology 41, 375–406.

Stacey, D.L., 1977. Banker plant production of *Encarsia formosa* Gahan and its use in control of glasshouse whitefly on tomatoes. Plant Pathology 26, 63–66.

Syngenta, 2016a. Anderline aa (*Amblyseius andersoni*) for Tomato Russet/Rust Mite Control. http://www.bfgsupply.com/order-now/product/157/3551/syngenta-bioline-anderline-aa-amblyseius-andersoni-spider-mite-rust-russet-mite-control–125000-units-5-L-bag-brandvermiculite.

Syngenta, 2016b. Oriline for Thrips Control. http://www.bfgsupply.com/order-now/product/157/3575/syngenta-bioline-oriline-i-orius-insidiosus-thrip-control–5-x-50-million-unites-gel-trays.

Thakore, Y., 2006. The biopesticide market for global agricultural use. Industrial Biotechnology 2, 194–208.

Tong-Xian, L., Kang, L., Heinzand, K.M., Trumble, J., 2009. Biological Control of *Liriomyza leafminers*: Progress and Perspective. Centre for Agriculture and Biosciences International. http://www.cabi.org/bni/FullTextPDF/2009/20093049924.pdf.

Trottin-Caudal, Y., Chabrière, C., Fournier, C., Leyre, J.M., 2006. Current situation of *Bemisia tabaci* in protected vegetable crops in the South of France. Bulletin OILB/SROP 29, 53–58.

Trumble, J.T., Alvarado-Rodriguez, B., 1993. Development and economic evaluation of an IPM program for fresh market tomato production in Mexico. Agriculture, Ecosystems & Environment 43 (3-4), 267–284.

UCDavis IPM, 2010. Pea Leaf Miner *Liriomyza huidobrensis*, *Dacnusa Sibirica*. http://www.ipm.ucdavis.edu/PMG/r280300911.html.

UCDavis IPM, 2013. Tomato Fruitworm. UC Pest Management Guidelines for Tomato Fruitworm, *Helicoverpa* (*Heliothis*) *zea* http://www.ipm.ucdavis.edu/PMG/r783300111.html.

Urbaneja, A., Gonzalez-Cabrera, J., Arno, J., Gabarra, R., 2012. Prospects for the biological control of *Tuta absoluta* in tomatoes of the Mediterranean basin. Pest Management Science 68, 1215–1322.

Vacante, V., Benuzzi, M., Palmeri, V., Brafa, G., 2001. Experimental trials of microbiological control of the Turkey moth (*Chrysodeixis chalcites* (Esper)) in Sicilian greenhouse crops. Informatore Fitopatologico 51, 73–76.

van Driesche, R.G., Heinz, K.M., 2004. An overview of biological control in protected culture. In: Heinz, K.M., van Driesche, R.G., Parrella, M.P. (Eds.), Biocontrol in Protected Culture. Ball Publishing, Batavia, Netherlands, pp. 1–24.

van Lenteren, J.C., 2000a. Success in biological control of arthropods by augmentation of natural enemies. In: Gurr, G., Wratten, S.D. (Eds.), Biological Control: Measures of Success. Kluwer Academic Publishers, Netherlands, pp. 77–103.

van Lenteren, J.C., 2000b. A greenhouse without pesticides: fact or fantasy? Crop Protection 19, 375–384.

van Lenteren, J.C., 2003. Commercial availability of biological control agents. In: van Lenteren, J.C. (Ed.), Quality Control and Production of Biological Control Agents, Theory and Testing Procedures. CABI Publishing, Cambridge, Massachusetts, pp. 167–179.

van Lenteren, J.C., 2006. How not to evaluate augmentative biological control. Biological Control 39, 115–118.

van Lenteren, J.C., 2012. The state of commercial augmentative biological control: plenty of natural enemies, but a frustrating lack of uptake. BioControl 57, 1–20.

van Lenteren, J.C., Bueno, V.H.P., 2003. Augmentative biological control of arthropods in Latin America. BioControl 48, 123–139.

van Lenteren, J.C., Hale, A., Klapwijk, J.N., van Schelt, J., Steinberg, S., 2003. Guidelines for quality control of commercially produced natural enemies. In: van Lenteren, J.C. (Ed.), Quality Control and Production of Biological Control Agents, Theory and Testing Procedures. CABI Publishing, Cambridge, Massachusetts, pp. 265–303.

van Lenteren, J.C., Tommasini, M.G., 1999. Mass production, storage, shipment and quality control of natural enemies. In: Albajes, R., Gullino, M.L., van Lenteren, J.C., Elad, Y. (Eds.), Integrated Pest, Disease Management in Greenhouse Crops. Kluwer Publishers, Dordrecht, Netherlands, pp. 276–294.

van Lenteren, J.C., van Roermund, H.J.W., Sutterlin, S., 1996. Biological control of greenhouse whitefly (*Trialeurodes vaporariorum*) with the parasitoid *Encarsia formosa*: how does it work? Biological Control 6, 1–10.

van Lenteren, J.C., Woets, J.V., 1988. Biological and integrated pest control in greenhouses. Annual Review of Entomology 33, 239–269.

Vijayabharathi, R., Kumari, B.R., Gopalakrishnan, S., 2014. Microbial agents against *Helicoverpa armigera*: where are we and where do we need to go? African Journal of Biotechnology 13, 1835–1844.

Waage, J.K., Greathead, D.J., 1988. Biological control: challenges and opportunities. Philosophical Transactions of the Royal Society, London, UK. B. 318, 111–128.

Young, S.Y., Yearian, W.C., 1974. Persistence of *Heliothis NPV* on foliage of cotton, soybean, and tomato. Environmental Entomology 3, 253–255.

Zalom, F.G., 2012. Management of arthropod pests. In: Davis, R.R., Pernezny, K., Broome, J.C. (Eds.), Tomato Health Management. American Phytopathological Society Press, Saint Paul, Minnesota, USA, pp. 49–64.

Zappala, L., Biondi, A., Alma, A., Al-Jboory, I.J., Arno, J., Bayram, A., Chailleux, A., El-Arnaouty, A., Gerling, D., Guenaoui, Y., Shaltiel-Harpaz, L., Siscaro, G., Stavrinides, M., Tavella, L., Vercher Aznar, R., Urbaneja, A., Desneux, N., 2013. Natural enemies of the South American moth, *Tuta absoluta*, in Europe, North Africa Middle East and their potential use in pest control strategies. Journal of Pest Science 86, 635–647.

Chapter 14

Protection of Tomatoes Using Bagging Technology and Its Role in IPM of Arthropod Pests

Germano Leão Demolin Leite, Amanda Fialho
Universidade Federal de Minas Gerais, Montes Claros, Brazil

1. INTRODUCTION

Solanum lycopersicum L. (Solanaecae) fruit production is globally important with an annual production of 170 million tons (FAO, 2014). Tomato production requires significant investment with pest control accounting for 40% of the total cost. In Brazil, pest control requires up to 40 spray treatments/crop in the rainy season (Rodrigues Filho et al., 2003). Continued pesticide use may result in resistant insects, which can reduce fruit quality, and pesticides can also contaminate the environment, the applicator, and the consumer if not applied correctly (Picanço et al., 1998, 2007). Organic cultivation of tomatoes and other vegetable crops has been increasing approximately 9% a year throughout the world (Paull, 2011). Tomato plants that have been well nourished by organic matter are said to produce healthy fruit with higher dry matter content, flavonoid, and ascorbic acid content compared to fruit from conventional production (Premuzic et al., 1998; Stertz et al., 2005; Mitchell et al., 2007). One method of reducing damage due to insects and diseases of tomato fruits is the use of bags. The use of waxed-paper, translucent plastic bags, and tissue non-tissue (TNT) fabric to protect fruit, starting when they are small (in general, after its formation), is one of the oldest and most effective control practices. These techniques have been found to be effective in preventing attacks by the fruit borers *Annona crassiflora* Mart. (Leite et al., 2012), *Malus domestica* Borkh. (Teixeira et al., 2011), *Psidium guajava* L. (Bilck et al., 2011), *Pouteria caimito* Radlk (Nascimento et al., 2011), and *Mangifera indica* L. (Graaf, 2010). In addition, bagging fruit with glassine paper resulted in a 67% reduction in infestation rates of *Neoleucinodes elegantalis* (Guenée) compared with non-bagged tomatoes (Rodrigues Filho et al., 2003). Clearly, bagging can prevent damage by insects and reduce pesticide use without interfering with fruit formation and color development (Lebedenco, 2006; Leite et al., 2014).

2. TOMATO PRODUCTION IN BRAZIL AND IPM

Integrated pest management (IPM) can reduce the number of spray treatments needed to control pests and diseases when producing tomatoes (Picanço et al., 2007). Treatments may include selective use of insecticides (Leite et al., 1998), mineral oil as an insecticide (Picanço et al., 1998), more effective and selective compounds (i.e., new insecticides) (Silvério et al., 2009), and plant extracts (Barbosa et al., 2011). Other techniques used to control pests and diseases include vertical staking of plants (Picanço et al., 1996, 1998), greater spacing between plants (Picanço et al., 1998), adequate fertilization (Leite et al., 2003), polyculture (Picanço et al., 1996), crop rotation (Leite et al., 2011), and mating disruption (Welter et al., 2005). Moreover, use of resistance to pests derived from wild tomato species and rustic accessions of *S. lycopersicum* (Leite et al., 1999, 2001; Oliveira et al., 2009), disease resistant varieties (Paula and Oliveira, 2003), use of organic substances to attract or repel pests (Oliveira et al., 2009; Leite et al., 2011), natural and applied biological control (Pratissoli et al., 2005; Barbosa et al., 2011; Picanço et al., 2011), fruit bagging (Lebedenco, 2006; Leite et al., 2014), and organic farming (Mitchell et al., 2007) are the techniques used to manage pests and diseases.

3. IMPACT OF FRUIT BAGGING ON *TUTA ABSOLUTA* (MEIRYCK), *NEOLEUCINODES ELEGANTALIS* (GUENÉE), AND *HELICOVERPA ZEA* (BODDIE)

Tomato fruit borers cause substantial damage in tomatoes in Brazil. These borers include *Tuta absoluta* (Meiryck) (Fig. 14.1A and B), *N. elegantalis*, and *Helicoverpa zea* (Boddie) (Picanço et al., 1998).

Only 0.3% of fruits bagged with organza fabric or TNT (Fig. 14.2A and B) were damaged by the insect borers (*T. absoluta*, *N. elegantalis*, and *H. zea*), while 23% of fruit was damaged in non-bagged clusters (Leite et al., 2014). Clusters bagged with paper (Fig. 14.2C) had 7.1% fewer fruits injured by *H. zea* versus fruit not bagged. Overall non-bagged fruits (Fig. 14.2D) had a higher proportion of fruit damaged by insect borers: 7.5% by *T. absoluta*, 32.9% by *N. elegantalis*, and 28.7% by *H. zea* than bagged fruit (Leite et al., 2014). Serious damage by borers was reduced and greater fruit quality was observed in clusters bagged with organza fabric and TNT compared with non-bagged fruit. This finding agrees with the results of Rodrigues Filho et al. (2003) who found reduced damage caused by *N. elegantalis* on clusters of tomatoes bagged with glassine paper compared to insecticide treatment. Efficient control resulted from greater protection by the bags, which were completely sealed to prevent the penetration of these insects (i.e., oviposition on the fruits). Jordão and Nakano (2002) observed that a protective paper cone open at the bottom reduced damage by *N. elegantalis* by 70%, and by 40% for *H. zea*, but this type of protection was not effective against *T. absoluta*. The 4% damaged fruit on clusters bagged with paper was greater than the threshold of 1% of the fruit being injured by fruit borer (Alvarenga, 2004; Leite et al., 2014). This level of damage may be a result of the weakness of the paper (Fig. 14.2E), especially after rainfall, even when the bags were replaced periodically (Leite et al., 2014). Similar results were observed by Leite et al. (2012) with fruits of "araticum" *A. crassiflora*, bagged to protect them against fruit borers. These authors found that paper bags were destroyed by rain thus allowing a high level of insect damage.

4. IMPACT OF FRUIT BAGGING ON *ERWINIA* SPP. AND *ALTERNARIA SOLANI* SORAUER

The main pathogens affecting tomato plants and their fruits in Brazil are *Erwinia* spp. and *Alternaria solani* Sorauer (Fig. 14.1C and D) (Picanço et al., 1998, 2007). Infection of tomato fruit by these pathogens occurs mainly because of injuries during crop management and handling and damage by insect borers (Bergamin Filho et al., 1995). Bagging tomato fruit can

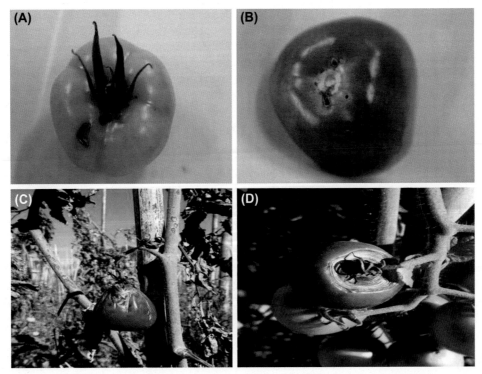

FIGURE 14.1 (A) and (B) Tomato fruits attacked by *Tuta absoluta*, (C) tomato fruit attacked by *Erwinia* spp., and (D) tomato fruit attacked by *Alternaria solani*. *Photos by Leite, G.L.D. (A), Rocha, F.S. (B), Catão, H.C.R.M. (C and D), with permission.*

prevent damage by these diseases (Lebedenco, 2006). Bagging tomato clusters with TNT showed a greater disease reduction (93.3%) of fruit damage by *A. solani*, followed by tissue organza with an 80.9% reduction compared with fruit bagged with paper, microperforated plastic bags, or non-bagged clusters (Leite et al., 2014). The reduction of disease by *Erwinia* spp. was greater on fruit bagged with TNT and organza fabric, at 86.3% and 78.4%, respectively, and 47.5% lower with microperforated plastic bags (Leite et al., 2014). These authors observed that TNT and organza fabric bags were more effective in preventing diseases by *A. solani* and *Erwinia* spp. because these coverings provided (1) greater protection against the spread of *A. solani* spores, which were mainly spread by wind as well as by splashing with rain and irrigation water (Zambolim et al., 2000); (2) higher gas exchange, avoiding the accumulation of moisture inside bagged clusters; and (3) increased protection against damage by insects, which also reduced the spread of *Erwinia* spp. (Bergamin Filho et al., 1995). Also, the lower effectiveness of bagging clusters with microperforated plastic against *Erwinia* spp. and *A. solani* shows that this material is inappropriate for organic tomato production (Fig. 14.2F). This may be a result of the reduced gas exchange and increased moisture accumulation, which favors the formation of a suitable environment for these pathogens (Leite et al., 2014).

FIGURE 14.2   (A) Torganza bag, (B) TNT bag, (C) paper bag, (D) control, (E) damaged paper bag, and (F) plastic bag.

5. IMPACT OF FRUIT BAGGING ON THRIPS AND MITES

Frankliniella schultzei (Trybom) attacks tomato plants and is most numerous in tomato flowers. Other thrips pests include *Frankliniella occidentalis* Pergande, which occasionally causes goldfleck damage to tomato fruit, and *Thrips palmi* Karny (Ghidiu et al., 2006; Kakkar et al., 2012). *Tetranychus urticae* Koch is an important mite pest on tomatoes that can feed directly on the fruit decreasing fruit marketability (Meck et al., 2013). *Aculops lycopersici* (Massee) is another important mite pest affecting tomato fruit yield and quality (Duso et al., 2010; Houten et al., 2013). Leite et al. (2014) did not observe any thrips or mite damage in tomato fruits that were bagged. However, no other research was found about the use of bags against these pests. We believe that the use of TNT and organza bags is able to reduce the damage due to these pests in tomato fruits, but field experiments are needed.

6. IMPACT OF FRUIT BAGGING ON FRUIT PRODUCTION, QUALITY, AND PRODUCTION COST

Fruit bagging can impact fruit production, its quality, and production cost. Leite et al. (2014) tested four kinds of fruit bags against fruit borers and diseases in tomato plants: bagging with organza fabric, TNT covering, a microperforated plastic, brown paper bags, and no bagging (control). The percentage of flowers aborted (average of 0.36%) and fresh fruit mass (average of 91.87 g) were statistically similar among all of the treatments (Leite et al., 2014). Total number of fruits of "extra" size/cluster/plant were similar for clusters bagged with organza fabric (3.88 fruits in four clusters) or TNT (3.80 fruits in four clusters) compared with the control (0.45 fruit in four clusters) (Leite et al., 2014). Fruits bagged with TNT (0.04%) or organza fabric (0.02%) had significantly fewer severe defects than the other treatments (0.12%), while the incidence of fruit with slight damage (0.04%) was similar among treatments (Leite et al., 2014). Bagged and non-bagged fruit clusters had similar percentages of aborted flowers and similar formation of fresh mass, which suggest that the material of these bags did not affect tomato fruit formation (Leite et al., 2014). Lebedenco (2006) measured similar production rates of fresh mass for bagged and non-bagged tomato fruit. The average fresh weight/fruit of 91.87 g resulted in a yield of 29.44 tons/ha.

Fruit size differed only for those with medium diameters, with the highest percentage (28.4%) for clusters bagged with microperforated plastic (Leite et al., 2014). They observed that clusters bagged with TNT or organza fabric had a higher percentage of "extra" fruit types with large (55.4% and 53.1%, respectively) and medium (19.8% and 17.7%, respectively) diameter compared to non-bagged clusters. On the other hand, clusters bagged with microperforated plastic bags or non-bagged showed 22.9% and 17.7% large fruits and 11.1% and 6.3% medium fruits, respectively (Leite et al., 2014). The percentage of "extra" fruit types with small diameters (4.14%) was similar among treatments (Leite et al., 2014). Bagging with either organza or TNT provided more than twice the fruit yield of the control. Tomato fruit classified as "extra" have higher quality and market value (CEAGESP, 2000).

Leite et al. (2014) reported that skin color did not differ among treatments, with 92% of the fruit having the desired color pattern established by MAPA (Ministry of Agriculture, Livestock and Food Supply). The total soluble solids content (Brix) differed among treatments, with similar values for organza fabric (3.52 °Brix), TNT (3.67 °Brix) and control treatments (3.83 °Brix), which were greater than other treatments (2.77 °Brix) (Leite et al., 2014). The development of fruit bagged either with TNT (41.03 days) or organza (40.72 days) to first harvest exhibited about a 3.86 degree days (DD) compared to the control (37.01 days) (Leite et al., 2014). They suggested that similar skin color (normal production of lycopene), similar Brix degrees, and the roughly 3 DD in starting the tomato harvest either with organza fabric or TNT bagging were important because of the need to maintain the quality of the fruit (i.e., skin color and Brix degrees). The pattern of ripe tomato fruits and crude soluble solids that ranged from 3.5 to 6.0 °Brix showed that bagging fruit either with organza fabric or TNT maintained the proper temperature conditions for producing high-quality fruits (Alvarenga, 2004).

Leite et al. (2014) noted that the economic gain for bagging with organza fabric and TNT was significantly greater than for other treatments (Table 14.1). Bagging with TNT or organza yielded increases in profit over the control treatment of 373.7% and 331.4%, respectively (Leite et al., 2014). The authors recommended the reuse of organza fabric bags for up to five crops to provide an economic gain similar to bagging with TNT (Table 14.2).

The >300% increase in production of tomatoes bagged either with organza fabric or with TNT compared to non-bagged clusters offsets the higher production cost by lowering losses to insect borers and to diseases (Leite et al., 2014). The value of organic tomatoes ranged from US$2.54 to US$7.15/kg and averaged US$5.15/kg compared to an average of US$1.00/kg for conventionally grown tomatoes (Martins et al., 2006) in the consumer markets of São Paulo State (Brazil), the largest

TABLE 14.1 Profits From Bagging for the Production (US$) of 1 ha of Organic Tomatoes by Bagging Four Clusters of Fruits per Tomato Plant

Bagging Type	Profit in US$/ha
TNT	119,958a
Organza	106,365a
Paper	70,050b
Plastic	46,996b
No bagging	32,096c

Values followed by same letter (a, b, c) do not differ by contrast analysis at 5% probability.
Adapted from *Leite, G.L.D., Fialho, A., Zanuncio, J.C., Reis Junior, R., Costa, C.A., 2014. Bagging tomato fruits: a viable and economical method of preventing diseases and insect damage in organic production. Florida Entomologist 97, 50–60.*

TABLE 14.2 Total Cost (US$) of Inputs and Services (Total Cost) and Profit per Hectare From Bagging With Organza Estimated on the Basis of Reusing the Organza Bags for Five Cropping Seasons

	Total US$
Total cost of production except bagging	17,690
Total cost of production	27,386
Profit	114,708

Adapted from *Leite, G.L.D., Fialho, A., Zanuncio, J.C., Reis Junior, R., Costa, C.A., 2014. Bagging tomato fruits: a viable and economical method of preventing diseases and insect damage in organic production. Florida Entomologist 97, 50–60.*

producer of organic vegetables in Brazil (Valarini et al., 2007). The profitability of organic tomatoes in protected cultivation is 59.9% (summer) and 113.6% (winter) with a production cost 17.2% lower than conventional tomatoes (Luz et al., 2007). The lower yield of organic tomatoes (30 tons/ha) compared to conventional production (56 tons/ha) can be offset by the higher market value of organic fruit, which can be 304% higher than tomatoes grown using conventional techniques. This higher value of organic tomatoes can more than compensate for the difficulties encountered when producing organic tomatoes (Martins et al., 2006). Growing a tomato crop is a high-risk enterprise with the fruit quality and market determining the prices.

The demand for foods with less risk to human health (Diniz et al., 2006) extends to organically grown tomatoes that may have higher concentrations of flavonoids with antioxidant properties, which can be effective against cardiovascular diseases and some types of cancer (Mitchell et al., 2007).

7. FUTURE PROSPECTS

We believe that the use of bagging fruit technique is a great option for tomato production (organic and conventional production), and it is necessary to disclose this technique to farmers and test it in different world regions for other pests that attack fruit. The organza fabric can be reused numerous times, but should be cleaned in water with 1% sodium hypochlorite (Machado et al., 2001). Bagging of tomato fruits with the organza material reused for five consecutive seasons can provide an economic gain similar to that of bagging with TNT. In addition to economic profitability, bagging tomato fruits has major environmental benefits. The bagging of tomato fruit clusters with TNT or organza fabric were more effective in reducing damage by insect borers and diseases than other bagging treatments and control. These two treatments produced a greater number of "extra" grade fruit, without negative effects on the color of the skin, fresh weight, Brix degrees, or flower abortion, thus resulting in greater economic gains.

REFERENCES

Alvarenga, M.A.R., 2004. Tomate: produção em campo, em casa-de-vegetação e em hidroponia. UFLA, Lavras.

Barbosa, F.S., Leite, G.L.D., Alves, S.M., Nascimento, A.F., D'ávila, V.A., Costa, C.A., 2011. Insecticide effects of *Ruta graveolens*, *Copaifera langsdorffii* and *Chenopodium ambrosioides* against pests and natural enemies in commercial tomato plantation. Acta Scientiarum Agronomy 33, 37–43.

Bergamin Filho, A., Kimati, H., Amorim, L., 1995. Manual de Fitopatologia: princípios e conceitos (Ceres). São Paulo, Ceres.

Bilck, A.P., Roberto, S.R., Grossmann, M.V.E., Yamashita, F., 2011. Efficacy of some biodegradable films as pre-harvest covering material for guava. Scientia Horticulturae 130, 341–343.

CEAGESP, 2000. Programa Brasileiro para Modernização da Horticultura. Normas de Classificação do Tomate. CQH/CEAGESP, São Paulo.

Diniz, L.P., Maffia, L.A., Dhingra, O.D., Casali, V.W.D., Santos, R.S.H., Mizubuti, E.S.G., 2006. Avaliação de produtos alternativos para o controle da requeima do tomateiro. Fitopatol Brasileira 31, 171–179.

Duso, C., Castagnoli, M., Simoni, S., Angeli, G., 2010. The impact of eriophyoids on crops: recent issues on *Aculus schlechtendali*, *Calepitrimerus vitis* and *Aculops lycopersici*. Experimental & Applied Acarology 51, 151–168.

FAO (Food and Agriculture Organization of the United Nations), 2014. Production Yearbook. Rome, Italy www.fao.org/faostat/en/#data/QC.

Graaf, D., 2010. Developing a systems approach for *Sternochetus mangiferae* (Coleoptera: Curculionidae) in South Africa. Journal of Economic Entomology 103, 1577–1585.

Ghidiu, G.M., Hitchner, E.M., Funderburk, J.E., 2006. Goldfleck damage to tomato fruit caused by feeding of *Frankliniella occidentalis* (Thysanoptera: Thripidae). Florida Entomologist 89, 279–281.

Houten, Y.M., Glas, J.J., Hoogerbrugge, H., Rothe, J., Bolckmans, K.J.F., Simoni, S., Arkel, J., Alba, J.M., Kant, M.R., Sabelis, M.W., 2013. Herbivory-associated degradation of tomato trichomes and its impact on biological control of *Aculops lycopersici*. Experimental & Applied Acarology 60, 127–138.

Jordão, A.L., Nakano, O., 2002. Ensacamento de frutos do tomateiro visando ao controle de pragas e à redução de defensivos. Scientia Agricola 59, 281–289.

Kakkar, G., Seal, D., Stansly, P.A., Liburd, O.E., Kumar, V., 2012. Abundance of *Frankliniella schultzei* (Thysanoptera: Thripidae) in flowers on major vegetable crops of South Florida. Florida Entomologist 95, 468–475.

Lebedenco, A., 2006. Eficiência de métodos de controle de pragas do tomateiro (Lycopersicom esculentun Mill) na região de Presidente Prudente. 51 f. Dissertação (Mestrado em Agronomia). Universidade do Oeste Paulista-UNOESTE, São Paulo.

Leite, G.L.D., Picanço, M., Guedes, R.N.C., Gusmão, M.R., 1998. Selectivity of insecticides with and without mineral oil to *Brachygastra lecheguana* (Hymenoptera: Vespidae), a predator of *Tuta absoluta* (Lepidoptera: Gelechiidae). Ceiba 39, 3–6.

Leite, G.L.D., Picanço, M., Della Lucia, T.M., Moreira, M.D., 1999. Role of canopy height in the resistance of *Lycopersicon hirsutum* f. *glabratum* to *Tuta absoluta* (Lep., Gelechiidae). Journal of Applied Entomology 123, 459–463.

Leite, G.L.D., Picanço, M., Guedes, R.N.C., Zanuncio, J.C., 2001. Role of plant age in the resistance of *Lycopersicon hirsutum* f. *glabratum* to the tomato leafminer *Tuta absoluta* (Lepidoptera: Gelechiidae). Scientia Horticulturae 89, 103–113.

Leite, G.L.D., Costa, C.A., Almeida, C.I.M., Picanço, M., 2003. Efeito da adubação sobre a incidência de traça-do-tomateiro e alternaria em plantas de tomate. Horticultura Brasileira 21, 448–451.

Leite, G.L.D., Picanço, M., Zanuncio, J.C., Moreira, M.D., Jham, G.N., 2011. Hosting capacity of horticulture plants for insect pests in Brazil. Chilean Journal of Agricultural Research 71, 383–389.

Leite, G.L.D., Souza, M.F., Souza, P.N.S., Fonseca, M.M., Zanuncio, J.C., 2012. The bagging of *Annona crassiflora* fruits to control fruit borers. Acta Scientiarum Agronomy 34, 253–257.

Leite, G.L.D., Fialho, A., Zanuncio, J.C., Reis Junior, R., Costa, C.A., 2014. Bagging tomato fruits: a viable and economical method of preventing diseases and insect damage in organic production. Florida Entomologists 97, 50–60.

Luz, J.M.Q., Shinzato, A.V., Silva, M.A.D., 2007. Comparação dos sistemas de produção de tomate convencional e orgânico em cultivo protegido. Bioscience Journal 23, 7–15.

Machado, J.C., Oliveira, J.A., Vieira, M.G.G.C., Alves, M.C., 2001. Inoculação artificial de sementes de soja por fungos utilizando solução de manitol. Revista Brasileira Sementes 23, 95–101.

Martins, V.A., Waldemar Filho, P.C., Bueno, C.R.F., 2006. Preços de frutas e hortaliças da agricultura orgânica no mercado varejista da cidade de São Paulo. Information Economics and Policy 36, 42–51.

Meck, E.D., Kennedy, G.G., Walgenbach, J.F., 2013. Effect of *Tetranychus urticae* (Acari: Tetranychidae) on yield, quality, and economics of tomato production. Crop Protection 52, 84–90.

Mitchell, A.E., Hong, Y.J., Koh, E., Barrett, D.M., Bryant, D.E., Denison, R.F., Kafka, S., 2007. Ten-year comparison of the influence of organic and conventional crop management practices on the content flavonoids in tomatoes. Journal of Agricultural and Food Chemistry 55, 6154–6159.

Nascimento, W.M.O., Muller, C.H., Araújo, C.S., Flores, B.C., 2011. Ensacamento de frutos de abiu visando à proteção contra o ataque da mosca-das-frutas. Revista Brasileira Fruticultura 33, 48–52.

Oliveira, F.A., Silva, D.J.H., Leite, G.L.D., Jham, G.N., Picanço, M., 2009. Resistance of 57 greenhouse-grown accessions of *Lycopersicon esculentum* and three cultivars to *Tuta absoluta* (Meyrick) (Lepidoptera: Gelechiidae). Scientia Horticulturae 119, 182–187.

Paula, R.S., Oliveira, W.F., 2003. Resistência do tomateiro *Lycopersicon esculentum* ao patógeno *Altenaria solani*. Pesquisa Agropecuaria Tropical 33, 89–95.

Paull, J., 2011. The uptake of organic agriculture: a decade of worldwide development. Journal of Social and Development Sciences 2, 111–120.

Picanço, M., Leite, G.L.D., Madeira, N.R., Silva, D.J.H., Myamoto, N.A., 1996. Efeito do tutoramento do tomateiro e seu policultivo com o milho no ataque de *Scrobipalpuloides absoluta* (Meyrick) e *Helicoverpa zea* (Bod). Neotropical Entomology 25, 175–180.

Picanço, M., Leite, G.L.D., Guedes, R.N.C., Silva, E.E.A., 1998. Yield loss in trellised tomato affected by insecticidal sprays and plant spacing. Crop Protection 17, 447–452.

Picanço, M.C., Bacci, L., Crespo, A.L.B., Miranda, M.M.M., Martins, J.C., 2007. Effect of integrated pest management practices on tomato production and conservation of natural enemies. Agriculture and Forest Entomology 9, 327–335.

Picanço, M.C., Bacci, L., Queiroz, R.B., Silva, G.A., Miranda, M.M.M., Leite, G.L.D., Suinaga, F.A., 2011. Social wasp predators of *Tuta absoluta*. Sociobiology 58, 1–13.

Pratissoli, D., Thuler, R.T., Andrade, G.S., Zanottie, L.C.M., Faria Da Silva, A., 2005. Estimativa de *Trichogramma pretiosum* para controle de *Tuta absoluta* (Lepidoptera: Pyralidae) em tomateiro estaqueado. Pesquisa Agropecuaria Brasileira 40, 715–718.

Premuzic, A., Bargiela, M., Garcia, A., Rondina, A., Na Lorio, A., 1998. Calcium, iron, potassium, phosphorus and vitamin c content of organic and hydroponic tomatoes. Hortscience 33, 255–257.

Rodrigues Filho, I.L., Marchior, L.C., Silva, L.V., 2003. Análise da oviposição de *Neoleucinodes elegantalis* (Guén, 1854) (Lep.: Crambidae) para subsidiar estratégia de manejo. Agronomia 37, 23–26.

Silvério, F.O., Alvarenga, E.S., Moreno, S.C., Picanco, M.C., 2009. Synthesis and insecticidal activity of new pyrethroids. Pest Management Science 65, 900–905.

Stertz, S.C., Santo, A.P.E.E., Bona, C., Freitas, R.J.S., 2005. Comparative morphological analysis of cherry tomato fruits from three cropping systems. Scientia Agricola 62, 296–298.

Teixeira, R., Amarante, C.V.T., Boff, M.I.C., Ribeiro, L.G., 2011. Controle de pragas e doenças, maturação e qualidade de maçãs "imperial gala" submetidas ao ensacamento. Revista Brasileira Fruticultura 33, 394–401.

Valarini, P.J., Frighetto, R.T.S., Schiavinato, R.J.C.C., Sena, M.M., Balbinot, L.J.P.R., 2007. Análise integrada de sistemas de produção de tomateiro com base em indicadores edafobióticos. Horticultura Brasileira 25, 60–67.

Welter, S.C., Pickel, C., Millar, J.G., Cave, F., Van Steenwyk, R.A., Dunley, J., 2005. Pheromone mating disruption offers selective management options for key pests. California Agriculture 59, 16–22.

Zambolim, L., Vale, F.X.R., Costa, H., 2000. Controle de doenças de plantas hortaliças. Viçosa Editora Universidade Federal de Viçosa, Brazil.

Chapter 15

Integrated Pest Management Strategies for Tomato Under Protected Structures

Srinivasan Ramasamy[1], Manickam Ravishankar[2]

[1]AVRDC – The World Vegetable Center, Tainan, Taiwan; [2]World Vegetable Center, Ranchi, India

1. INTRODUCTION

Tomato (*Solanum lycopersicum* L.) is one of the most widely grown vegetables in the world. It is grown on more than 4.76 million ha with a production of nearly 165 million tons. Asia and Africa account for about 80% of the global tomato area, with about 72% of world output (FAO, 2013). China accounts for more than one-fifth of the world's tomato production area; United States and India together account for another one-fifth. Although tomato requires a relatively cool, dry climate for high yield and quality (Nicola et al., 2009), it is adapted to a wide range of climatic conditions from temperate to hot and humid tropics (Naika et al., 2005). In the tropics, tomato production is severely constrained by insect, mite, and nematode pests. The major pests include fruit borer (*Helicoverpa armigera* (Hübner)), common armyworm (*Spodoptera litura* (F.)), beet armyworm (*Spodoptera exigua* (Hübner)), whitefly (*Bemisia tabaci* (Gennadius)), leafminers (*Liriomyza* spp.), spider mites (*Tetranychus urticae* Koch, *Tetranychus cinnabarinus* Boisduval, *Tetranychus evansi* Baker & Pritchard), and root-knot nematodes (*Meloidogyne* spp.) (Srinivasan, 2010). Growers rely heavily on chemical pesticides to protect their tomato crops. For example, farmers in southern India spray chemical pesticides more than 50 times during a cropping season in tomato (Nagaraju et al., 2002). Pesticide misuse has adverse effects on the environment and human health, and also increases the cost of production. For instance, the share of the cost of pesticides to total material input cost was 31% for tomato in the Philippines (Orden et al., 1994).

One option to reduce pests on tomato and thus reduce the overuse of chemical pesticides is to grow tomato under protective structures. Cultivation of vegetables under protective structures such as net houses has become popular in recent years in many countries. Growing tomatoes under protective structures not only keeps pests away, but also enables farmers to produce this crop during off-season, when tomato usually sells for higher prices in the market. For instance, off-season tomato production in net houses increases the period of fruit availability from the last week of January to the first week of June in Punjab, India (Cheema et al., 2004). Tomato production under protected conditions is increasing due to high productivity and returns per unit area during the off-season (Tahir and Altaf, 2013). However, there is substantial cost in construction and operations of protective agriculture (Sethi et al., 2009). In addition, damage from pests inside net houses can be as serious as open-field production, causing net-house growers to use pesticides indiscriminately (Kaur et al., 2010). This chapter aims to compile the key pests and effective and affordable control technologies for their management on tomato under protective cultivation.

2. BIOECOLOGY AND DAMAGE POTENTIAL OF MAJOR PESTS

2.1 Aphids, *Aphis gossypii* Glover

Aphids are important insect pests found during early crop stages on tomato under protective cultivation (Kaur et al., 2010). Since high relative humidity favors aphid population growth (Chakraborty, 2011), protective structures offer a conducive environment for the multiplication of aphids. Aphid infestations occur mostly during the cool dry season. Aphids reproduce through parthenogenesis (virgin birth) and are viviparous (give birth to live young). The adult color is highly variable, from light green to greenish brown. Both wingless and winged forms occur. The wingless forms are most abundant, but winged forms (Fig. 15.1) are produced under high aphid densities or when host-plant quality is inferior. Aphids possess a pair of black cornicles on the dorsal side of the abdomen which are used to release alarm pheromones. Aphids are mostly found in groups, and each female produces about 20 nymphs a day; the nymphs mature to adults in a week (Srinivasan, 2009). Both

Sustainable Management of Arthropod Pests of Tomato. http://dx.doi.org/10.1016/B978-0-12-802441-6.00015-2

FIGURE 15.1 Winged aphids on tomato.

the nymphs and adults possess piercing/sucking mouthparts. They occur in large numbers on the tender shoots and lower leaf surfaces, and suck the plant sap. Slightly infested leaves exhibit yellowing, while severe aphid infestations cause young leaves to curl and become deformed, stunt plant growth, and reduce the number and quality of flowers and fruits. Aphids also produce honeydew, which leads to the development of sooty mold on leaf surfaces that reduces the photosynthetic efficiency of the plants.

2.2 Thrips *Thrips palmi* Karny and *Scirtothrips dorsalis* Hood

Thrips palmi Karny is widely distributed in South and Southeast Asia and Oceania. *T. palmi* is polyphagous and is known to feed on tomato, potato, hot pepper, watermelon, muskmelon, bottle gourd, cucumber, pumpkin, squash, among others. This species is commonly known as "melon thrips" because of its preferential feeding on cucurbits. *Scirtothrips dorsalis* Hood predominantly an Asian species but also present in Africa and Greater Caribbean (Kumar et al., 2014), prefers to feed on hot pepper. However, it is also reported to damage tomato (Meena et al., 2005; Kaur et al., 2010). Thrips attack tomatoes mostly during the dry season.

Each thrips female lays about 200 eggs, which take from 3 days to 2 weeks to hatch depending on the temperature. Nymphs have two active feeding stages, and the nymphal stage varies from 4 days to 2 weeks depending on the temperature. The pupal stage consists of the pre-pupa and pupa. The pre-pupal stage lasts for one to 2 days, and the pupal stage lasts for another one to 3 days. *T. palmi* adults are yellow in color, and there are three red ocelli on the top of their heads in a triangular form. There is also a pair of setae near the ocellar triangle, and the abdominal pleurotergites are without setae (Mound, 1996). The head and legs of *S. dorsalis* are pale. The head is wider than longer, and has an ocellar triangle with closely spaced transverse lines. The pronotum has transverse lines and possesses four pairs of posterior marginal setae (Mound, 2007). Adults live from 2 weeks to 2 months.

Thrips adults and nymphs prefer to feed on foliage and flowers, sometimes on young fruit too. Slightly infested leaves exhibit silvery feeding scars on the lower leaf surfaces, especially along the midrib and veins. In severe infestations, the leaves turn yellow or brown and dry on the lower leaf surfaces. Infested fruit is scarred and deformed. Thrips also transmit tospoviruses. Tomato spotted wilt virus (TSWV) in the United States, Spain, Hawaii, Taiwan, and Argentina, and Peanut bud necrosis virus (PBNV) in India are two important tospoviruses that infect tomato. Depending on the stage of the crop and season, diseases caused by these pathogens result in yield losses up to 80–100% (Ramana et al., 2011).

2.3 Whitefly, *Bemisia tabaci* (Gennadius)

Whiteflies are widely distributed in tropical and subtropical regions, and in greenhouses in temperate regions. *B. tabaci* is highly polyphagous, and is known to feed on several vegetables including tomato, eggplant, okra, other field crops, and weeds. Hot, dry conditions favor the whitefly, and heavy rain showers drastically reduce population buildup. This insect prefers the lower leaf surfaces. The whitefly adult is soft-bodied and light yellow in color. The body and wings are covered with powdery wax and the wings are held over the body like a tent (Fig. 15.2). Adult males are slightly smaller in size than the females. Adults live one to 3 weeks.

FIGURE 15.2   Adult whitefly (*Bemisia tabaci*).

The females lay eggs on the underside of tomato leaves, more on hairy leaf surfaces. Each female can produce as many as 300 eggs in its lifetime. The egg stage lasts 3–5 days during summer and 5–33 days in winter (David, 2001). The first instar nymph has antennae, eyes, and three pairs of well-developed legs. The nymphs are flattened, oval-shaped, and greenish yellow. The legs and antennae are atrophied during the next three instars and they are immobile during the remaining nymphal stages. The last nymphal stage has red eyes, and is sometimes incorrectly called a pupa because it is non-feeding. This last nymphal instar lasts about 9–14 days during the summer and 17–73 days in winter (David, 2001). Adults emerge from the last nymphal instar through a T-shaped slit, leaving an empty exuvia.

Both the adults and nymphs suck the plant sap and reduce the vigor of the plant. In severe infestations, the leaves turn yellow and drop off. They also secrete large quantities of honeydew, which favors the growth of sooty mold on leaf surfaces. *B. tabaci* acts as a vector for several viral pathogens including Tomato yellow leaf-curl virus (TYLCV). Plants infected by TYLCV show stunted growth with erect shoots. Leaflets curl upward and inward, are reduced in size, and have yellowing along the margins and interveinal chlorosis. The flowers wither and droop, and fruit set is reduced or nil.

2.4 Leafminers, *Liriomyza* spp.

Leafminers in the genus *Liriomyza* can cause severe infestations in protected tomato production. *Liriomyza bryoniae* (Kaltenbach) is known to occur in China, India, Japan, Korea, Taiwan, and Vietnam in Asia, and Egypt and Morocco in Africa. However, *Liromyza sativae* Blanchard, *Liriomyza trifolii* (Burgess), and *Liriomyza huidobrensis* (Blanchard) may also cause damage in tomato. *L. huidobrensis* occurs mostly in high elevations, whereas *L. sativae* occurs in low elevations (Srinivasan, 2010). *L. bryoniae* is a polyphagous insect and is recorded as a damaging pest on several plant species in at least 16 families (Spencer, 1990), although it prefers to feed on Cucurbitaceae. The major host plants are tomato, melon, watermelon, cucumber, cabbage, and lettuce.

L. byroniae will be used to describe the general biology of the leafminer flies causing problems in tomato; *L. sativae* and *L. huidobrensis* are similar. The adults are small, gray flies; adult males are slightly smaller in size than the females. The mesonotum is shiny black, femur is yellow, and subsequent leg segments are brownish. The longevity of adult females is 3–12 days (Cheng, 1994) and males live for a shorter time than females (Parrella, 1987). Each female can lay as many as 184 eggs in its lifetime (Lee et al., 1990). Eggs hatch in about three to 7 days. Larvae have four instars feeding inside the leaf tissues, and the larval stage lasts one to 2 weeks. The last instar larvae cuts a hole in the leaf and drops to the soil, where the insects pupate. Pupae are oval-shaped and yellow to brown in color. The pupal period lasts 8–11 days depending on temperature (Parrella, 1987).

The adult females create a leaf puncture and eggs are deposited in these punctures. After every leaf puncture, the females feed from them; hence, the leaf punctures can also be considered feeding punctures. Males are unable to create their own punctures, and feed from punctures created by females. Leaf puncturing can reduce photosynthesis and may kill young plants in severe infestations (Parrella, 1987). The larvae feed on the leaf mesophyll and cause irregular mines on leaf surfaces. In severe cases, several mines are formed on the same leaf (Fig. 15.3), which drastically reduce photosynthesis and thus yield.

FIGURE 15.3 Mines caused by leafminer (*Liriomyza* spp.).

2.5 Spider Mites, *Tetranychus urticae* Koch, *Tetranychus cinnabarinus* Boisduval, *Tetranychus evansi* Baker & Pritchard

Spider mites are serious pests of vegetable crops including tomato. Low relative humidity favors the multiplication of mites and precipitation is an important abiotic factor that restricts spider mite populations (Tehri et al., 2014). *T. urticae* is commonly known as the two-spotted spider mite or the red spider mite. It is minute in size and varies in color, from green to greenish yellow, to brown or orange red with two dark spots on the body. Eggs are round, and the egg stage lasts 2–4 days. It has a 6-legged larval stage and two 8-legged nymphal stages (protonymph and deutonymph). The life cycle is completed in 1–2 weeks, depending on temperature, and there are multiple overlapping generations per year (Malinoski et al., 2006). *T. cinnabarinus* is commonly known as carmine spider mite. This mite is similar in size and biology to the two-spotted spider mite, but it is carmine in color. *T. evansi* is known as the tomato red spider mite (Navajas et al., 2013), and has a similar biology to the two-spotted spider mite. This is the predominant species in several countries in Africa on tomato and other solanaceous vegetables (Wekesa et al., 2005). In Asia, it has been reported in Taiwan (Ho, 2011).

Spider mite damage leads to the formation of several white or yellow speckles on the leaves. In severe infestations, leaves will become completely desiccated and drop-off. The mites also produce webbing on the leaf surfaces (Fig. 15.4); under severe conditions, the whole plant becomes wrapped in webbing. Under high population densities, the mites move to the tip of the leaf or top of the plant and congregate to form a ball-like mass (Fig. 15.4). The mites in these masses are carried to newer leaves or plants by wind (Brandenburg and Kennedy, 1982).

FIGURE 15.4 Webbing and ballooning of red spider mite.

2.6 Oriental Leafworm Moth, *Spodoptera litura* (F.)

S. litura is the predominant species on tomato in tropical South Asia and Southeast Asia. It is a polyphagous and highly mobile insect and a pest of economic importance on many agricultural and horticultural crops. *S. litura* is a major lepidopteran pest on tomato in net houses; the adult females lay eggs on the roof or side walls, and the neonate larvae can easily enter the structure (Talekar et al., 2003).

The adult is a stout-bodied moth with a wingspan of about 40 mm. The adults are usually brown and the forewings have numerous crisscross streaks on a cream or brown background. The hind wings are white with a brown patch along the border. The eggs are laid in groups of 200–300, and covered with brown hairs from the body of the female (Muniappan et al., 2012). The eggs take about 3–5 days to hatch. The neonate larvae are gregarious, and young larvae feed in groups (Fig. 15.5). However, they disperse when they become older to feed individually. The larvae are nocturnal and feed actively during the night hours. During the day, the larvae hide in the soil, in cracks and crevices, or in plant debris in the field. The larval stage lasts 15–30 days, depending on temperature, and pupation takes place in the soil. The pupae are shiny reddish-brown, and this stage lasts from 1 to 3 weeks (Muniappan et al., 2012).

The neonate larvae feed on leaf surfaces and skeletonize the leaf while mature larvae cause general defoliation, leaving only the main veins of the leaves. Rarely, *S. litura* larvae feed on the immature stages of tomato fruit, and sometimes they will girdle the young seedlings at the soil surface resulting in lodged plants.

2.7 Nematodes, *Meloidogyne incognita* (Kofoid & White), *Meloidogyne javanica* (Treub), and *Meloidogyne arenaria* Chitwood

Plant parasitic nematodes are a major constraint in protected cultivation of tomato due to crop susceptibility, a favorable environment (especially temperature, moisture, and relative humidity), lack of awareness of nematode problems among growers, and nematode interactions with other microbes. The ideal conditions provided by protected cultivation and continuous year-round availability of the host plant often results in high population buildup of parasitic nematodes. Root-knot nematodes (*Meloidogyne* spp.) are a concern in most of the tomato-growing areas of the world and the most severe damage is caused by three species of root-knot nematode: *Meloidogyne incognita* (Kofoid & White), *Meloidogyne javanica* (Treub), and *Meloidogyne arenaria* Chitwood (Sikora and Fernández, 2005; Anwar and McKenry, 2010). Tomato yield losses of up to 80% in heavily infected soils have been recorded (Kaskavalci, 2007). Severe infestations can kill a tomato plant outright (Kamran et al., 2010). Nematode infestations can increase a plant's vulnerability to various diseases such as *Fusarium*, *Ralstonia solanacearum* (Smith) and *Pythium* (Rivera and Aballay, 2008).

Nematode feeding results in root rot, lesions, discoloration, and deformities such as galls. The uptake of water and nutrients is greatly reduced in infected plants. Above ground symptoms of root-knot nematode infestation are stunting, chlorosis or yellowing, and often wilting of plants or foliage in warm weather though the soil may have an adequate supply of moisture. The number, size, and quality of fruits are reduced. Below ground symptoms are typical root knots or galls produced on roots that eventually lead to reduced water uptake. Several such galls give a clubbed, distorted, and cracked appearance to roots. Plants may be killed where nematode damage is severe (Mitkowski and Abawi, 2003).

Meloidogyne species are sedentary endoparasites. The infective second stage juveniles hatch from eggs in moist soil, move freely in the soil, and penetrate into the roots just behind the root tip. Once inside the cortical tissue, the juveniles

FIGURE 15.5   Early larvae of *Spodoptera litura*.

establish feeding sites where several root cells around the pests' heads become enlarged, forming giant cells that constitute a nutrient source from which the nematodes take nutrients. Juveniles enlarge and swell as they develop into adult females. Within the galls, pear-shaped females lay eggs in a gelatinous matrix known as an egg mass. The life cycle is completed in 4–6 weeks depending on soil temperature. Nematodes are active in warm and moist soils that support the growth of host plants, and can complete several generations in one season, increasing their population densities by several fold (Ciancio and Mukerji, 2009).

Nematodes rarely move more than a few inches in a thin film of water within the soil. However, indirectly they can be moved with infected transplants and infested soil. Inside the net houses, nematodes are spread by implements used for plowing and cultivation. Nematodes can also spread from one area to another through irrigation or drainage water. They thrive in hot climates but can survive in temperate conditions as well (Strajnar et al., 2011).

3. INTEGRATED PEST MANAGEMENT STRATEGIES IN PROTECTED TOMATO PRODUCTION

3.1 Cultural Control

Since most tomatoes grown in protected culture are of similar varieties, crop rotation can be effective in reducing pests on these crops. Some insect pests, such as *S. litura* have a wide host range of vegetable crops, so if growers plant tomato after tomato, or other host plants such as hot pepper, brassicas, legumes, the damage will be higher from emerging *S. litura* that pupated in the soil during a previous crop cycle. In this situation, the tomato crop should be rotated with a non-host cereal crop or flowers. Crop rotation with non-host crops such as cereals or grasses is also effective in suppressing nematode populations. Of course, growers should select the best rotational crop that will maximize their profits.

Growing tomato in the vicinity of other host plants should also be avoided, because *S. litura* adults can easily migrate to the newer tomato crop. Eggs of *S. litura* are laid in masses, and hence the young larvae that remain on a plant can be identified based on the feeding damage. Groups of these larvae should be hand picked and destroyed (Ghosh, 1989). Similar to *S. litura* many other pests of tomato are polyphagous; they have multiple host plants for feeding and survival ranging from cultivated crops to weeds. The selected site for tomato seedling production should be clean and located away from these host plants or weeds. If seedlings happen to be grown in those areas, the seedling trays should be covered with insect-proof (50–64 mesh) nylon net to prevent the entry of aphids, thrips, and whiteflies. If the finer mesh size is not available, coarser (30 or 40) mesh size can be used and neem-based pesticides should be sprayed on the net surfaces. Ultraviolet (UV)-reflective plastic or straw mulches can reduce the number of whiteflies and thrips landing on tomato crops (Stavisky et al., 2002; Rajasri et al., 2011). UV-absorbing plastic films are a very promising tool to protect greenhouse tomato from thrips (López-Marin et al., 2011). These mulches offer promising control of several insect vectors, such as aphids, thrips, and whiteflies (Narayanasamy, 2013).

Covering moistened soil with a transparent polythene sheet for 1–2 months during the warmer months is called soil solarization. This method offers major advantages, including simultaneous control of pests, soil-borne pathogens, nematodes, and weeds, and also improves soil physical properties (Stapleton, 2000). Diseases caused by *Meloidogyne* spp. have been successfully controlled by soil solarization (Grinstein et al., 1995; Rao and Krishnappa, 1995). The main limitation is the dependence on warmer months with sufficiently high temperatures and sunshine to kill pests, pathogens, and weed. Heating the soil up to 70°C by means of aerated steam can be used for soil sterilization between cropping cycles in facilities that have heating systems. Steam can be useful in nursery facilities for treating potting media and propagation beds. The limitation of this method is the high cost of power and water.

As previously stated, continuous cropping increases soil-borne diseases and nematodes. Grafting has become a useful technique for the production of repeated crops of tomato. Grafting tomato scions on resistant root stocks of eggplant and *Solanum torvum* Sw. was shown to be an effective approach against soil-borne diseases and nematodes (Gisbert et al., 2011). In many European and other Mediterranean countries, grafting susceptible plants onto nematode-resistant root stocks is being used in protected culture systems for a number of annual greenhouse crops including tomato, eggplant, and various cucurbits (Giotis et al., 2012).

The addition of organic amendments to the soil can be used to reduce nematodes on crop plants. For instance, mustard oil cake can be used in managing *M. incognita* in tomato under net-house conditions (Kaur and Srinivasan, 2013). Amendments including peat, manure, and compost are useful for increasing the water- and nutrient-holding capacity of the soil, especially in sandy soils. Because nematodes more readily damage water-stressed plants, increasing the soil's capacity to hold water can lessen the effects of nematode injury. Organic amendments enhance the activity of natural enemies and improve soil fertility and structure.

3.2 Colored Sticky Traps

Yellow sticky traps are a common method for monitoring many pests, including aphids, whiteflies, and leafminer adults. Use of yellow sticky traps in seedling production areas at the rate of 1–2 traps/50–100 m^2 can trap significant numbers of whiteflies. Lu et al. (2012) showed that yellow sticky traps significantly suppressed the population increase of adult and immature whiteflies in greenhouses. Similarly, thrips are highly attracted to bright blue color; blue sticky traps can be set up every 2–3 m^2 in a greenhouse for effective control (Murai, 2002).

3.3 Pheromone Traps

Insect sex pheromones have long been used as monitoring and mass-trapping tools in IPM strategies, especially for lepidopteran insects. Sex pheromones of *S. litura* are commercially available in many countries, and can be used for monitoring as well as mass trapping (Srinivasan et al., 2015). Sex pheromone lures for *S. litura* should be installed in traps placed 45–60 cm above the canopy level for effective attraction (Srinivasan, 2010). The lures should be replaced once every 2–3 weeks, depending upon the prevailing weather conditions.

3.4 Biological Control

Predators and parasitoids offer promising control of pests in protective structures. Leafminers have several parasitoids. For instance, *Gronotoma micromorpha* (Perkins) (larval-pupal parasitoid), *Chrysocharis pentheus* (Walker), *Neochrysocharis formosa* (Westwood) and *Diglyphus isaea* (Walker) (larval parasitoids), *Halticoptera circulus* (Walker) and *Opius phaseoli* Fischer (pupal parasitoids) are known to occur in Asia, including Japan, Malaysia, Sri Lanka, and Taiwan (Lee et al., 1990; Sivapragasam and Syed, 1999; Niranjana et al., 2005; Abe, 2006). Predators such as green lacewings (*Chrysoperla* sp.), predatory mites (e.g., *Neoseiulus cucumeris* (Oudemans)), and predatory thrips (e.g., *Haplothrips* spp.) may feed on thrips (McDougall and Tesoriero, 2011). The ladybird beetles (*Menochilus* sp. and *Coccinella* sp.) and green lacewings are efficient predators of aphids (Hazarika et al., 2001). Inundative release of ladybird beetles (*Coccinella septempunctata* (L.)) at 5000 beetles per hectare also can suppress aphid populations (Yadav and Singh, 2015).

Several predators of spider mites occur in many countries. For instance, *Stethorus* spp., *Oligota* spp., *Anthrocnodax occidentalis* Felt, and *Feltiella minuta* Felt are known to occur in Taiwan (Ho, 2000). Predatory mites such as *Phytoseiulus persimilis* Athias-Henriot and several species of *Amblyseius*, especially *Amblyseius womersleyi* (Schicha) and *Amblyseius fallacies* (Garman), can be used to control spider mites. They are more effective in protective structures and in high humidity conditions. Green lacewing [*Mallada basalis* (Walker) and *Chrysoperla carnea* (Stephens)] are also effective generalist predators of spider mites. Third-instar *C. carnea* can consume 25–30 spider mite adults per day; however, it needs supplemental food for long-term survival (Hazarika et al., 2001).

Natural enemies help keep pest populations on tomato in check under protected cultivation. However, broad-spectrum insecticides are harmful to most of these natural enemies. Therefore the use of natural enemies and pesticides that are safer to natural enemies should be considered in an overall pest management program.

3.5 Biopesticides

Commercially available biopesticides based on *Bacillus thuringiensis* Berliner (*Bt*), *S. litura* nucleopolyhedrovirus (SlNPV), and neem (*Azadirachta indica* A. Juss.) can be used against *S. litura* (Nathan and Kalaivani, 2006) and other sucking pests. Proper rotation should be adopted while applying *Bt* formulations to avoid development of resistance. For instance, *B. thuringiensis* subsp. *kurstaki* formulations may be rotated with *B. thuringiensis* subsp. *aizawai* formulations against *S. litura*. An organic salt (Lastraw) and an oil-based formulation of *Beauveria bassiana* (Balsamo-Crivelli) Vuillemin (Myco-Jaal) were also found to reduce aphids, thrips, whiteflies, and mites (Tashpulatova, 2010; Kaur and Srinivasan, 2014).

Currently, biofumigation is of great interest as a technology for control of root-knot nematodes. The incorporation of cruciferous plants and those that release the nematicidal compound, isothiocyanate, can contribute to insect control (Kirkegaard et al., 1993; Stirling and Stirling, 2003). However, it has been stated repeatedly that the amount of nutrient-rich organic matter incorporated into the soil also stimulates microorganisms that produce ammonia, which is also highly nematicidal (Bello et al., 1998). In addition, research on agents that work against root-knot nematodes and do not have a detrimental impact on the environment is becoming increasingly important due to bans on chemical pesticides. Recent advances have produced quite a number of commercially available biocontrol products. Some of the well-accepted commercial products contain the bacteria *Bacillus firmus* Bredemann and Werner, *Pasteuria penetrans* (Sayre and Starr) and fungus *Purpureocillium lilacinus* (Thom) Luangsa-ard. According to a study by Wilson and

Jackson (2013), the key products at the moment are VOTiVO (*B. firmus*), DiTera (*Myrothecium verrucaria* (Alb. and Schwein) Ditmar), and BioAct (*P. lilacinus*). Thus, various microbial pesticides, botanical pesticides, and organic salts can be used to control major insect, mite, and nematode pests on tomato under protected cultivation.

4. FUTURE PROSPECTS

There are two types of protective structures: permanent and temporary. These structures vary in design across countries. Simple net-house and plastic-house structures are widely used in South and Southeast Asia to grow tomatoes, especially during the off-season when tomatoes are not grown in the field. Although these structures can prevent entry of damaging insects if appropriate maintenance procedures are adopted, most of the time the pest pressure becomes high inside the structures because of operational issues or structural flaws. Once the pests get into the structures, they rapidly proliferate and cause enormous damage because of the absence of natural enemies. Pest damage and yield losses can be extremely high, which prompts growers to rely on chemical pesticides. However, recent developments in improving protective structures have demonstrated that they can be effective and sustainable for vegetable production (Bhatnagar and Narayan, 2011).

Recent research has shown that alternate technologies can significantly reduce pesticide use in tomato under protective structures. For example, biopesticide products, particularly botanicals, provide good levels of invertebrate pest control in UK tomatoes in glasshouses (George et al., 2015). Grafting of tomatoes on eggplant significantly enhanced the off-season tomato production under protective structures in the Philippines (Boncato, 2008). It is encouraging to note that the availability and use of IPM component technologies including microbial and botanical pesticides have increased in recent years in response to changes in policy. For instance, use of botanical and biological products increased 56-fold in Malaysia from 2007 to 2010. About a threefold increase in the use of mineral oils in pest management was reported during the same period in Thailand (FAO, 2011). There is still a long way to go to realize the impact of these IPM technologies. More adaptive research is required, as well as commercial production and/or availability of IPM technologies. These technologies, combined with high quality protective structures can significantly increase tomato production and economic returns to vegetable growers in the near future.

REFERENCES

Abe, Y., 2006. Exploitation of the serpentine leafminer *Liriomyza trifolii* and tomato leafminer, *L. bryoniae* (Diptera: Agromyzidae) by the parasitoid *Gronotoma micromorpha* (Hymenoptera: Eucoilidae). European Journal of Entomology 103, 55–59.

Anwar, S.A., McKenry, M.V., 2010. Incidence and reproduction of *Meloidogyne incognita* on vegetable crop genotypes. Pakistan Journal of Zoology 42, 135–141.

Bello, A., Gonzalez, J.A., Arias, M., Rodríguez-Kabana, R. (Eds.), 1998. Alternatives to Methyl Bromide for the Southern European Countries. Phytoma-españa, DG XI EU. CSIC, Valencia, Spain.

Bhatnagar, P.R., Narayan, M., 2011. Annual Report 2010–2011, AICRP on Application of Plastics in Agriculture. Central Institute of Post-Harvest Engineering and Technology, Punjab Agricultural University, Ludhiana, India, p. 46. http://www.ciphet.in/upload/file/FOR%20PDF.pdf.

Boncato, T.A., 2008. Growth and yield performance of grafted tomato varieties under protective structures during off-season production. Benguet State University Research Journal 16, 53.

Brandenburg, R.L., Kennedy, G.G., 1982. Intercrop relationships and spider mite dispersal in a corn/peanut agro-ecosystem. Entomologia Experimentalis et Applicata 32, 269–276.

Chakraborty, K., 2011. Incidence of aphid, *Aphis gossypii* Glover (Hemiptera: Aphididae) on tomato crop in the agro climatic conditions of the northern parts of West Bengal, India. World Journal of Zoology 6 (2), 187–191.

Cheema, D.S., Kaur, P., Kaur, S., 2004. Offseason cultivation of tomato under nethouse conditions. Acta Horticulturae 659, 177–181.

Cheng, C.H., 1994. Bionomics of the leafminer, *Liriomyza bryoniae* Kalt. (Diptera: Agromyzidae) on muskmelon. Chinese Journal of Entomology 14, 65–81.

Ciancio, A., Mukerji, K.G., 2009. Integrated Management of Fruit Crops and Forest Nematodes. Springer, Netherlands, p. 346.

David, B.V., 2001. Elements of Economic Entomology (Revised and Enlarged Edition). Popular Book Depot, Chennai, India, p. 590.

Food and Agriculture Organization (FAO), 2011. FAOSTAT. http://faostat.fao.org.

Food and Agriculture Organization (FAO), 2013. FAOSTAT. http://faostat3.fao.org/browse/Q/QC/E.

George, D.R., Banfield-Zanin, J.A., Collier, R., Cross, J., Birch, A.N.E., Gwynn, R., O'Neill, T., 2015. Identification of novel pesticides for use against glasshouse invertebrate pests in UK tomatoes and peppers. Insects 6 (2), 464–477.

Ghosh, M.R., 1989. Concepts of Insect Control. New Age International Limited Publishers, New Delhi, India, p. 282.

Giotis, C., Theodoropoulou, A., Cooper, J., Hodgson, R., Shotton, P., Shiel, R., Eyre, M., Wilcockson, S., Markellou, E., Liopa-Tsakalidis, A., Volakakis, N., Leifert, C., 2012. Effect of variety choice, resistant rootstocks and chitin soil amendments on soil-borne diseases in soil-based, protected tomato production systems. European Journal of Plant Pathology 134 (3), 605–617.

Gisbert, C., Prohens, J., Raigón, M.D., Stommel, J.R., Nuez, F., 2011. Eggplant relatives as sources of variation for developing new rootstocks: effects of grafting on eggplant yield and fruit apparent quality and composition. Scientia Horticulturae 128, 14–22.

Grinstein, A., Kritzman, G., Hetzroni, A., Gamliel, A., Mor, M., Katan, J., 1995. The border effect of soil solarization. Crop Protection 14, 315–320.

Hazarika, L.K., Puzari, K.C., Wahab, S., 2001. Biological control of tea pests. In: Upadhyay, R.K., Mukerji, K.G., Chamola, B.P. (Eds.), Biocontrol Potential and its Exploitation in Sustainable Agriculture: Insect Pests. Springer, USA, pp. 159–180.

Ho, C.C., 2000. Spider-mite problems and control in Taiwan. Experimental and Applied Acarology 24, 453–462.

Ho, C.C., 2011. Monitoring on Two Exotic Spider Mites in Taiwan, pp. 1–9. http://www.agnet.org/htmlarea_file/activities/20110826121346/paper-246339602.pdf.

Kamran, M., Anwar, S.A., Javed, N., Khan, S.A., Sahi, G.M., 2010. Incidence of root knot nematodes on tomato in Sargodha, Punjab. Pakistan Journal of Zoology 28, 253–262.

Kaskavalci, G., 2007. Effect of soil solarization and organic amendment treatments for controlling *Meloidogyne incognita* in tomato cultivars in Western Anatolia. Turkish Journal of Agriculture 31, 159–167.

Kaur, S., Kaur, S., Srinivasan, R., Cheema, D.S., Lal, T., Ghai, T.R., Chadha, M.L., 2010. Monitoring of major pests on cucumber, sweet pepper and tomato under net-house conditions in Punjab, India. Pest Management in Horticultural Ecosystems 16 (2), 148–155.

Kaur, S., Srinivasan, R., 2013. Evaluation of organic soil amendments against root-knot nematode, *Meloidogyne incognita,* in eggplant under nethouse conditions. Green Farming 4(2), 190–193.

Kaur, S., Srinivasan, R., 2014. Evaluation of an organic salt and entomopathogenic fungal formulation against insect and mite pests on sweet pepper under net-house conditions in Punjab, India. Pest Management in Horticultural Ecosystems 20, 141–147.

Kirkegaard, J.A., Gardner, P.A., Desmarchelier, J.M., Angus, J.F., 1993. Biofumigation – using Brassica species to control pests and diseases in horticulture and agriculture. In: Wratten, N., Mailer, R.J. (Eds.), Proceedings of the 9th Australian Research Assembly on Brassicas. Agricultural Research Institute, Wagga Wagga, New South Wales, Australia, pp. 77–82.

Kumar, V., Seal, D.R., Kakkar, G., 2014. Chilli Thrips *Scirtothrips dorsalis* Hood (Insecta: Thysanoptera: Thripidae). EENY-463 Entomology and Nematology Department, UF/IFAS Extension, p. 9. http://edis.ifas.ufl.edu/pdffiles/in/in83300.pdf.

Lee, H.S., Lu, F.M., Wen, H.C., 1990. Effects of temperature on the development of leafminer, *Liriomyza bryoniae* (Kaltenbach) (Diptera: Agromyzidae) in Taiwan. Chinese Journal of Entomology 10, 143–150.

López-Marin, J., Rodríguez, M., González, A., 2011. Effect of ultraviolet-blocking plastic films on insect vectors of virus diseases infesting tomato (*Lycopersicon esculentum*) in Greenhouse. Acta Horticulturae 914, 175–179.

Lu, Y., Bei, Y., Zhang, J., 2012. Are yellow sticky traps an effective method for control of sweetpotato whitefly, *Bemisia tabaci*, in the greenhouse or field? Journal of Insect Science. 12, 113. http://www.insectscience.org/12.113.

Malinoski, M.K., Davidson, J., Raupp, M., 2006. Spider Mites, p. 2. https://extension.umd.edu/sites/default/files/_images/programs/hgic/Publications/HG13%20Spider%20Mites.pdf.

McDougall, S., Tesoriero, L., 2011. Western Flower Thrips and Tomato Spotted Wilt Virus, p. 7. www.industry.nsw.gov.au/publications.

Meena, R.L., Ramasubramanian, T., Venkatesan, S., Mohankumar, S., 2005. Molecular characterization of *Tospovirus* transmitting thrips populations from India. American Journal of Biochemistry and Biotechnology 1, 167–172.

Mitkowski, N.A., Abawi, G.S., 2003. Root-knot nematodes. The Plant Health Instructor. . http://dx.doi.org/10.1094/PHI-I-2003-0917-01. http://www.apsnet.org/edcenter/intropp/lessons/nematodes/pages/rootknotnematode.aspx.

Mound, L.A., 1996. The Thysanoptera vector species of tospoviruses. Acta Horticulturae 431, 298–309.

Mound, L.A., 2007. Oriental tea Thrips (*Scirtothrips dorsalis*). Pest and Diseases Image Library. http://www.padil.gov.au/pests-and-diseases/pEST?MAIN/136432.

Muniappan, R., Shepard, B.M., Carner, G.R., Ooi, P.A.-C., 2012. Arthropod Pests of Horticultural Crops in Tropical Asia. CAB International, Oxfordshire, UK, p. 168.

Murai, T., 2002. The pest and vector from the East: *Thrips palmi*. In: Marullo, R., Mound, L. (Eds.), Proceedings of the 7th International Symposium on Thysanoptera. , pp. 19–32. http://www.ento.csiro.au/thysanoptera/Symposium/Section1/2-Murai.pdf.

Nagaraju, N., Venkatesh, H.M., Warburton, H., Muniyappa, V., Chancellor, T.C.B., Colvin, J., 2002. Farmers' perceptions and practices for managing tomato leaf curl virus disease in southern India. International Journal Pest Management 48, 333–338.

Naika, S., Van Lidt de Jeude, J., de Goffau, M., Hilmi, M., Van Dam, B., 2005. Cultivation of tomato. In: Van Dam, B. (Ed.), Production, Processing and Marketing. Digigrafi, Wageningen, The Netherlands.

Narayanasamy, P., 2013. Biological Management of Diseases of Crops. Integration of Biological Control Strategies with Crop Disease Management Systems, vol. 2. Springer Science and Business Media, p. 364.

Nathan, S.S., Kalaivani, K., 2006. Combined effects of azadirachtin and nucleopolyhedrovirus (SpltNPV) on *Spodoptera litura* Fabricius (Lepidoptera: Noctuidae) larvae. Biological Control 39, 96–104.

Navajas, M., de Moraes, G.J., Auger, P., Migeon, A., 2013. Review of the invasion of *Tetranychus evansi*: biology, colonization pathways, potential expansion and prospects for biological control. Experimental and Applied Acarololgy 59, 43–65.

Nicola, S., Tibaldi, G., Fontana, E., 2009. Tomato production systems and their application to the tropics. Acta Horticulturae 821, 27–33.

Niranjana, R.F., Wijeyagunesekara, H.N.P., Raveendranath, S., 2005. Parasitoids of *Liriomyza sativae* in farmer fields in the Batticaloa district. Tropical Agricultural Research 17, 214–220.

Orden, M.E.M., Patricio, M.G., Canoy, V.V., 1994. Extent of Pesticide Use in Vegetable Production in Nueva Ecija: Empirical Evidence and Policy Implications. Research and Development Highlights 1994. Central Luzon State University, Republic of the Philippines, pp. 196–213.

Parrella, M.P., 1987. Biology of *Liriomyza*. Annual Review of Entomology 32, 201–224.

Rajasri, M., Vijaya lakshmi, K., Prasada Rao, R.D.V.J., Loka Reddy, K., 2011. Effect of different mulches on the incidence of *Tomato leaf curl virus* and its vector whitefly *Bemisia tabaci* in tomato. Acta Horticulturae 914, 215–221.

Ramana, C.V., Venkata Rao, P., Prasada Rao, R.D.V.J., Kumar, S.S., Reddy, I.P., Reddy, Y.N., 2011. Genetic analysis for *Peanut bud necrosis virus* (PBNV) resistance in tomato (*Lycopersicon esculentum* Mill.). Acta Horticulturae 914, 459–463.

Rao, V.K., Krishnappa, K., 1995. Soil solarization for the control of soil borne pathogen complexes with special reference to *M. incognita* and *F. oxysporum* f. sp. *ciceri*. Indian Phytopathology 48, 300–303.

Rivera, L., Aballay, E., 2008. Nematicide effect of various organic soil amendments on *Meloidogyne ethiopica* whitehead on potted vine plants. Chilean Journal of Agricultural Research 68, 290–296.

Sethi, V.P., Dubey, R.K., Dhath, A.S., 2009. Design and evaluation of modified screen nethouse for off-season vegetable raising in composite climate. Energy Conversion and Management 50 (12), 3112–3128.

Sikora, R.A., Fernández, E., 2005. Nematodes parasites of vegetables. In: Luc, M., Sikora, R.A., Bridge, J. (Eds.), Plant Parasitic Nematodes in Subtropical and Tropical Agriculture, second ed. CABI International, UK, pp. 319–392.

Sivapragasam, A., Syed, A.R., 1999. The problem and management of agromyzid leafminers on vegetables in Malaysia. In: Lim, G.S., Soetikno, S.S., Loke, W.H. (Eds.), Proceedings of a Workshop on Leafminers of Vegetables in Southeast Asia, Serdang, Malaysia. CAB International, Southeast Asia Regional Centre, pp. 36–41.

Spencer, K.A., 1990. Host Specialization in the World Agromyzidae (Diptera). Series Entomologica 45. Kluwer Academic Publishers, Dordrecht, The Netherlands.

AVRDC Publication No. 10-740. In: Srinivasan, R. (Ed.), 2010. Safer Tomato Production Methods: A Field Guide for Soil Fertility and Pest Management. AVRDC – The World Vegetable Center, Shanhua, Taiwan, p. 97.

Srinivasan, R., 2009. Insect and Mite Pests on Eggplant: A Field Guide for Identification and Management. AVRDC Publication No. 09-729. AVRDC – The World Vegetable Center, Shanhua, Taiwan, p. 64.

Srinivasan, R., Lin, M.-Y., Su, F.-C., Sopana Yule, Chuanpit Khumsuwan, Thanh Hien, Vu Manh Hai, Lê Đức Khánh, Bhanu, K.R.M., 2015. Use of insect pheromones in vegetable pest management: successes and struggles. In: Chakravarthy, A.K. (Ed.), New Horizons in Insect Science: Towards Sustainable Pest Management. Springer India, Bangalore, India, pp. 231–237.

Stapleton, J.J., 2000. Soil solarization in various agricultural production systems. Crop Protection 19, 837–841.

Stavisky, J., Funderburk, J., Brodbeck, B.V., Olson, S.M., Andersen, P.C., 2002. Population dynamics of *Frankliniella* spp. and tomato spotted wilt incidence as influenced by cultural management tactics in tomato. Journal of Economic Entomology 95, 1216–1221.

Stirling, G.R., Stirling, A.M., 2003. The potential of Brassica green manure crops for controlling root-knot nematode (*Meloidogyne javanica*) on horticultural crops in a subtropical environment. Australian Journal of Experimental Agriculture 43, 623–630.

Strajnar, P., Sirca, S., Knapic, M., Urek, G., 2011. Effect of Slovenian climatic conditions on the development and survival of the root-knot nematode *Meloidogyne ethiopica*. European Journal of Plant Pathology 129, 81–88.

Tahir, A., Altaf, Z., 2013. Determinants of income from vegetables production: a comparative study of normal and off-season vegetables in Abbottabad. Pakistan Journal of Agricultural Research 26 (1), 24–31.

Talekar, N.S., Su, F.C., Lin, M.Y., 2003. How to Produce Safer Leafy Vegetables in Nethouses and Net Tunnels. Asian Vegetable Research and Development Center, Shanhua, Tainan, Taiwan, p. 18.

Tashpulatova, B., 2010. Control of thrips in cucumber greenhouse using Lastraw® preparation. CAC News. 43, 9–10. http://cac-program.org/cac_news/en/cac43e.pdf.

Tehri, K., Gulati, R., Gero, M., 2014. Host plant responses, biotic stress and management strategies for the control of *Tetranychus urticae* Koch (Acarina: Tetranychidae). Agricultural Reviews 35 (4), 250–260.

Wekesa, V.W., Maniania, N.K., Knapp, M., Boga, H.I., 2005. Pathogenicity of *beauveria bassiana* and *Metarhizium anisopliae* to the tobacco spider mite, *Tetranychus evansi*. Experimental and Applied Acarology 36, 41–50.

Wilson, M.J., Jackson, T.A., 2013. Progress in the commercialization of bionematicides. BioControl 58, 715–722.

Yadav, S., Singh, S.P., 2015. Bio-intensive integrated management strategy for mustard aphid *Lipaphis erysimi* Kalt. (Homoptera: Aphididae). Journal of Applied and Natural Science 7 (1), 192–196.

Chapter 16

Integrated Pest Management Strategies for Field-Grown Tomatoes

James F. Walgenbach
North Carolina State University, Mills River, NC, United States

1. INTRODUCTION

The concepts and definitions of integrated pest management (IPM) have been extensively reviewed and debated (Bajwa and Kogan, 1996; Kogan, 1998; Prokopy and Kogan, 2003; Ehler, 2006), with at least 67 definitions proposed by a diversity of pest management stakeholders (Bajwa and Kogan, 1996). There is a general agreement, however, that the basic tenets of IPM include a decision-based process that relies on the coordinated use of multiple tactics to manage pest populations that is acceptable from an economic, environmental, and societal perspective. Implied in the definition is that decisions are based on knowledge of the ecology of pest(s) in relevant agroecosystems, pest density estimates are obtained through monitoring programs, and treatment thresholds are used to trigger control actions. In its most elementary form, IPM consists of the use of monitoring programs and economic thresholds to dictate insecticide applications. However, more advanced programs, referred to by some as ecologically based IPM (NRC, 1996), consist of the coordinated use of multiple tactics which are biologically intense and largely preventive in nature, to suppress or control a complex of pests. Regardless of the level of IPM used, the goal is to minimize pesticide use and the risks they pose to the environment and human health.

Despite the numerous research advances in developing new pest management tactics and technology during the last half century (Kennedy and Sutton, 1999; Jepson and Kogan, 2007), some propose that there is little evidence of progress in implementing IPM at the farm level (Barfield and Swisher, 1994; Benbrook and Groth, 1996; Ehler and Bottrell, 2000; Zalucki et al., 2009), citing agriculture's continued reliance on pesticides as a key management tool (Ehler, 2006). However, there are economic and logistic issues that serve as formidable and legitimate barriers to implementing more biologically intense IPM programs on high-value crops such as tomatoes. Nonetheless, considerable progress has been made in understanding the ecology of tomato insect pests in agroecosystems throughout the world, the development of sampling and threshold levels, and the development of non-chemical management approaches. This chapter will review some of the major achievements in tomato IPM as well as constraints to progress.

2. TOMATO PRODUCTION FACTORS AFFECTING IPM PROGRAMS

Numerous factors affect the options available for managing pest populations, including the diversity of the pest complex, value of the target crop(s), environmental issues, and regulatory actions that often reflect societal concerns over agriculture's impact on the environment and human health. All of these factors operate at the local level, and thus there is diversity in the level of IPM practiced. Higher levels of IPM can be complex and information intense, and the programs are therefore affected by the level of public and private support available to growers.

Tomatoes are grown throughout the world, with the top five producers as of 2011 being China with approximately 30% of world production, followed by India (10.5%), United States (7.9%), Turkey (6.9%), and Egypt (5.1%) (http://www.agribenchmark.org/horticulture/sector-country-farm-information0/tomato.html). Despite the diversity of production regions, the arthropod pest complex is remarkably similar among different production areas. Although no worldwide list of arthropod tomato pests has been compiled, Lange and Bronson (1981) estimated that there are probably >100 different species. However, only about 30–35 species are potential pests (Zalom, 2012). Those that have been most intensively researched during the past 10–20 years are shown in Fig. 16.1.

From a production system and economic perspective, tomatoes are essentially grown as three different crops; field-grown processing, field-grown fresh market, and protected fresh market (i.e., greenhouse or tunnel production). The value of fresh

Sustainable Management of Arthropod Pests of Tomato. http://dx.doi.org/10.1016/B978-0-12-802441-6.00016-4

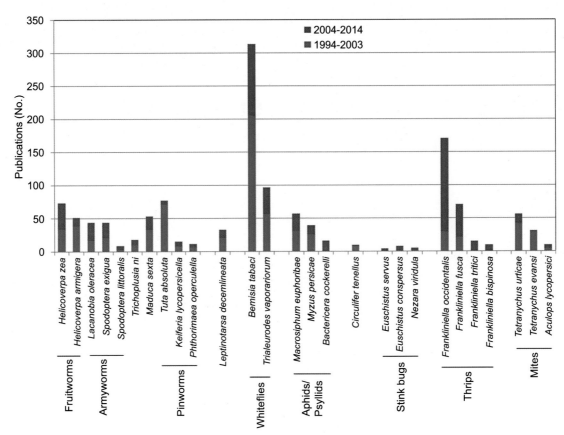

FIGURE 16.1 Number of studies during the past 20 years reported for various pests of tomato. Values represent the number generated from an AGRICOLA search inserting the scientific name of the insect and the world tomato.

market tomatoes is generally 9–12 times higher than that of processing crops (USDA-NASS, https://quickstats.nass.usda.gov/), which affects the monetary inputs afforded to produce the crop and tolerance levels for damage. In addition, processing tomato production is a highly mechanized operation, while fresh market production requires high labor inputs for pruning, staking, stringing, harvesting, and farmworkers' safety concerns can differentially affect the pesticide options on the two crops. Processing crops usually are contracted for a predetermined price and a set tolerance level for damage, while fresh market tomatoes are sold on a fluctuating market with a zero tolerance level for damage. The protected-culture tomato industry is rapidly growing throughout the world, and these production systems offer expanded opportunities for managing pests using physical and biological control options.

3. THRESHOLDS AND SAMPLING PLANS

The use of sampling plans and economic thresholds is the foundation of IPM, because it provides a means for assessing pest population densities and an economic basis for applying control measures—usually insecticides. The economic injury level (EIL) concept was first proposed by Stern et al. (1959) as the "lowest pest population density that will cause economic damage," and was later quantified by Pedigo et al. (1986) with the expression of $EIL = C/V \times I \times D \times K$, where C = cost of management, V = market value of the crop, I = injury units per pest, D = damage per unit injury, and K = proportionate reduction in the insect population. Stern et al. (1959) defined the economic threshold level (ETL) as "the population density at which control measures should be initiated to prevent an increasing pest population from reaching the economic injury level." ETLs do not have a mathematical definition and are often set using subjective methods. Yet this is the pest density at which action is taken and the terms "action threshold" and "treatment threshold" are used interchangeably.

A diversity of sampling plans and economic threshold levels exist for many of the key arthropod pests of tomatoes (Table 16.1). Two general approaches have been taken for establishing thresholds for the Heliothine complex on tomato, including *Helicoverpa zea* (Boddie), *Helicoverpa armigera* (Hübner), and *Heliothis virescens (F.)*. There is

TABLE 16.1 Resources for Sampling Programs and Threshold Levels for Some Key Arthropod Pests of Tomato

Insect	Type of Crop, Location	Description	References
Heliothine			
Helicovera zea	Processing, GA	Sequential sampling plan based on eggs, with critical density of 0.02 eggs/leaf	Nilakhe et al. (1982)
H. zea	Processing, CA	Sequential sampling plan based on eggs. Threshold adjusted based on level of egg parasitism	Hoffmann et al. (1990, 1991)
Helicoverpa armigera	Processing, New Zealand	Larval-based economic threshold adjusted from 1 to 8.3 larva/plant-based on level of larval parasitism	Cameron et al. (2001); Walker et al. (2010)
H. zea	Fresh, VA	Egg-based threshold to initial sprays (≥10% plants with eggs, with subsequent sprays based on fruit damage (≥3%)	Kuhar et al. (2006)
Helicoverpa spp.	Fresh, Australia	Egg-based sequential sampling plant, with a threshold of 0.5 eggs/plant	Dawson et al. (2006)
H. zea	Fresh, AL	Egg-based threshold (presence of any healthy *H. zea* eggs) to dictate insecticide	Zehnder et al. (1995)
H. armigera	Processing, Spain	In-season fruit damage threshold of 3% to prevent end-of-season damage of >3%	Torres-Vila et al. (2003)
Pinworm			
Keiferia lycopersicella	Fresh, CA	Pheromone trap threshold of 10 moths/trap/night over 1.5 weeks to initiate weekly insecticide sprays	van Steenwyk and Oatman (1983)
K. lycopersicella	Fresh, CA	Economic threshold of 3–4 larvae/3 m of row (0.5 larvae/plant)	Wiesenborn et al. (1990)
Tuta absoluta	Greenhouse, Spain	Binomial sampling plan based on proportion of midplant leaves with larval infestations. Plan uses thresholds of 1% and 3% infested fruit	Cocco et al. (2015)
Leafminers			
Liriomyza spp.	Fresh, FL	Sequential sampling plan using a lower and upper threshold 10% and 40% infested leaves, respectively	Wolfenbarger and Wolfenbarger (1966)
Liriomyza spp.	Fresh, CA	Sequential sampling plan based on yellow sticky traps for adults and pupal trays for pupal densities	Zhender and Trumble (1985)
Liriomyza trifolii	Greenhouse, Korea	Sequential binomial sampling plan using fluctuating threshold of 3 (early) and 7 (late season) mines per leaf	Lee et al. (2007)
Stink Bugs			
Nezara viridula	Fresh, LA	Sequential sampling plan based on sampling penultimate fruit clusters for and using bug-days to classify populations as low, moderate, and high	Lye et al. (1989); Lye and Story (1988)
Aphids			
Macrosiphum euphoribae	Processing, OH	Sequential sampling plan based on upper or upper plus middle leaves	Walker et al. (1984b)
M. euphoribae	Fresh, NC	Binomial sampling plan based on third most recently expanded leaf. Threshold level of 25% infested leaves	Walgenbach (1994); Walgenbach (1997)

Continued

TABLE 16.1 Resources for Sampling Programs and Threshold Levels for Some Key Arthropod Pests of Tomato—cont'd

Insect	Type of Crop, Location	Description	References
M. euphoribae	Processing, CA	Binomial sampling plan using upper leaves, and a threshold of 50–60% infested leaves	Hummel et al. (2004)
Thrips			
Frankliniella spp.	Fresh, NC	Fixed-precision level sequential sampling plan based on 10-flower sample unit	Cho et al. (1995)
Frankliniella spp.	Fresh, FL	Adult and nymphal thrips exhibited differential reference for upper and lower plant strata, with adults preferring upper flowers, and nymphos lower plant foliage	Reitz (2002)
Frankliniella spp.	Fresh, GA	Sampling of thrips during pre-flowering periods was most precise and efficient using a styrofoam beat cup method	Joost and Riley (2004)
Whiteflies			
Bemisia tabaci	Fresh, FL	Within plant distribution of different life stages with optimum sample sizes	Schuster (1998)
B. tabaci	Fresh, FL	Action threshold of 5 nymphs per 10 leaflets to minimize irregular ripening disorder	Schuster (2002)
B. tabaci	Fresh, Brazil	Sequential sampling plan based on sampling nymphs per leaf and adults with beat tray. Economic threshold of 4 nymphs/leaf and 1 adult/tray	Gusmão et al. (2006)
Trialeurodes vaporariorum	Greenhouse, Korea	Fixed-precision sequential sampling plan based on yellow sticky cards spaced >12.5 m to monitor adults	Kim et al. (2001)
Multiple species	Greenhouse	Review of use of progress in using yellow sticky cards factors for monitoring whiteflies, and identification of gaps in their use as decision-making tools	Pinto-Zevallos and Vänninen (2013)
Spider Mites			
Tetranychus urticae	Fresh, Kenya	Sequential binomial sampling plan with threshold based on predictive exponential model of population increase. Threshold of 27.9 mites/leaf, or 83% infested leaves, 5 weeks after transplanting	Knapp et al. (2006)
T. urticae	Fresh, NC	Action threshold of 8 mite per leaflet on upper plant leaf	Meck et al. (2013)

usually a minimum tolerance level between 2% and 5% damaged fruit in processing tomatoes, therefore in-season economic threshold levels are based on pest densities that will prevent damage from exceeding tolerance levels at harvest (Hoffmann et al., 1991; Cameron et al., 2001; Torres-Vila et al., 2003). In California, the *H. zea* threshold is based on an adjustable egg density, which varies with the level of parasitized eggs (Hoffmann et al., 1991). In New Zealand processing tomatoes, the *H. armigera* threshold level is based on larval sampling, and the threshold is adjusted based on larval parasitism (Walker et al., 2010). Because of the higher value and lower tolerance for damage of fresh market compared to processing tomatoes, growers by default seek to minimize the damage to the greatest extent possible. Consequently, damage tolerance levels are very low and thresholds are often based on the detection of very low numbers of eggs (Zehnder et al., 1995; Kuhar et al., 2006). The use of thresholds in processing tomatoes has been highly successful in reducing pesticide use, with reductions of up to 40% in California (Grieshop et al., 1988) and 95% in New Zealand (Walker et al., 2010).

A diversity of sampling plans and thresholds exist for several other common pests of tomato, including direct counts of individuals on plants, and the use of pheromone traps and yellow sticky cards, the latter being common in greenhouse production (Pinto-Zevallos and Vänninen, 2013). From the IPM practitioner's perspective, the number of samples required to reliably estimate a population often exceeds what is practical (Cullen et al., 2000), therefore sampling for pest management decisions is usually a compromise between reliability and cost. The development of binomial or presence/absence sampling is one approach to reduce sampling time without sacrificing precision. Binomial sampling is particularly useful for arthropods that are difficult to count and increase to high numbers, such as aphids, whiteflies, and spider mites (Table 16.1).

4. CULTURAL CONTROL

Cultural pest control refers to the manipulation of the crop production system or cultural practices to reduce or eliminate pest populations. Cultural control practices can range from simple concepts such as adjusting planting dates to avoid pest infestations to more complex farmscaping approaches that may include adjusting the spatial and temporal arrangement of an agroecosystem. Cultural practices can be complex to implement, particularly those that rely on interactions of multiple plant species and arthropods. For this reason, their use is more common in small-scale diversified farming systems than large-scale monoculture tomatoes.

Perhaps the most common and easily implemented cultural pest management practice is the concept of avoidance. Practices such as adjusted planting dates and crop rotations have long been recognized at the local or farm level to avoid pest infestations. Avoidance measures can be particularly effective when implemented on an area-wide scale. For example, host-free periods have been successfully used to manage several insect-vectored viruses of tomato. Implementation of an area-wide rouging program of old, overwintered tomato crops in a 10 km² area of Cyprus was used to eliminate primary inoculum sources of tomato yellow leaf curl virus (TYLCV) before emergence of its vector, the silverleaf whitefly *Bemisia tabaci* (Gennadius). This practice contributed to a reduction of TLCV-infected plants from 40% to 50% before the program began, and to <5% once the rouging program was initiated (Ioannou, 1987). Implementation of crop-free periods have also been highly successful in reducing whitefly-vectored tomato viruses in Israel (Ucko et al., 1998) and the Dominican Republic (Hilje et al., 2001).

Reflective mulches have long been used as a cultural tactic to manage the spread of insect-vectored plant viruses on several vegetable crops including tomato. Mulches, usually aluminum-based, reflect ultraviolet light and repel incoming insects and prevent them from alighting on plants. This system was initially developed based on the knowledge of the flight behavior of aphids (Kring, 1972), but is also relevant to thrips (Scott et al., 1989) and whiteflies (Kring and Schuster, 1992). It is most often used to manage non- or semi-persistently transmitted viruses in which the vector transmits the virus quickly, before insecticides can kill the insect. Reflective mulches have been used successfully to reduce the incidence of viruses transmitted by *B. tabaci*, including tomato mottle virus (Csizinszky et al., 1999) and tomato leaf curl virus (Rajasri et al., 2011), aphids and transmission of cucumber mosaic virus and potato virus Y (Yoltas et al., 2003; Sikora and Murphy, 2005; Fanigliulo et al., 2009), and thrips transmission of tomato-spotted wilt virus (Greenough et al., 1990; Stavisky et al., 2002; Riley and Pappu, 2004). In addition to reflective mulches, rice husk (Rajasri et al., 2011) and living mulches consisting of either peanut or coriander reduced *B. tabaci* populations on tomatoes and delayed the incidence of begomoviruses. The relatively high cost and concerns about disposal of reflective mulches have been the obstacles to more widespread use of this strategy.

Intercropping consists of planting multiple crops in close proximity to promote beneficial interactions and to use growth resources more efficiently, such as light, water, and nutrients. Intercropping is a common practice where cultivated land is a limited resource. This diversification strategy can also aid pest management by reducing pest colonization and increasing natural enemy populations by creating refuges (Andow, 1991). Intercropping tomato and maize in Nigeria resulted in reduced aphid densities and increased predatory spiders in tomatoes (Umeh and Onukwu, 2005). Intercropping tomato with coriander (*Coriandrum sativum* L.) and Gallant soldier (*Galinsoga parviflora* Cav.) resulted in lower abundance and higher diversity of arthropod pests, and higher abundance and diversity of predaceous arthropods compared with tomato monocultures, which led to lower numbers of adults and eggs of the key lepidopteran pest *Tuta absoluta* (Meyrick) (Medeiros et al., 2009). When used as a living mulch with tomato, coriander was also observed to contribute to reductions of *B. tabaci* populations and the incidence of begomovirus (Hilje and Stansly, 2008). Conversely, certain living mulches have sometimes lead to increased pest problems on tomatoes, as was the case with red clover and stink bugs (Hummel et al., 2002). Knowledge of how individual pests and specific combinations of crops and intercrops interact is critical to employing this strategy.

Trap cropping is a form of habitat manipulation in which stands of plants are grown to attract insect pests away from the target crop, or concentrating pests in a certain part of the field where they can be destroyed (Hokkanen, 1991). One of the first examples of trap crops in tomatoes was use of a tall variety of African marigold to suppress damage by *H. armigera* in India (Srinivasan et al., 1994). Planting rows of marigold at intervals ranging from every 10 to 20 rows of tomato reduced

H. armigera damage to tomato fruit by 9.3 and 3.9 times, respectively. The Colorado potato beetle (CPB) is a serious pest of solanaceous crops (including potato, tomato, and eggplant) in North America and Europe. Cultivated potatoes are considered a more attractive and suitable host than tomato for CPB, and planting potatoes several weeks before setting tomato plants can attract and retain beetles and prevent them from infesting tomatoes (Hunt and Whitefield, 1996). *B. tabaci* is a serious pest of vegetables in Florida that also vectors tomato TYLCV to tomato. Taking advantage of the greater attractiveness of whiteflies to cucurbits versus tomatoes, Schuster (2004) showed that planting a squash crop on the perimeter of tomatoes resulted in reduced whitefly populations and incidence of TYLC in tomato as compared to a tomato crop surrounded by other tomatoes.

In addition to non-crop plants, tillage systems can also affect arthropod populations. For instance, a no-till system consisting of dead winter rye residue reduced initial colonization of tomato plants by overwintered Colorado potato beetle compared to a conventional till system, which subsequently led to lower CPB populations and fewer insecticide applications (Zehnder and Linduska, 1987; Hunt, 1998).

Knowledge of the impact of different plant species on pest and beneficial organisms can be used in more large-scale farmscape approaches to manage pests. For instance, weedy borders surrounding processing tomato fields in California were found to be a source of consperse stink bugs, *Euschistus conspersus* Uhler, infesting tomato fields. Managing the border with tillage and herbicides reduced the densities of this pest in adjacent tomatoes (Pease and Zalom, 2010). In this same study, sweet alyssum (*Lobularia maritima* (L.)), a long-flowering, high nectar-producing brassica that is an excellent resource for several hymenopteran parasitoids (Johanowicz and Mitchell, 2000), was evaluated as a companion crop to enhance stink bug natural enemy population. This cultural practice did increase parasitism of stink bug eggs in tomatoes, but only later in the season when parasitoids did not affect damage levels. Nonetheless, these types of studies provide the groundwork to better understand trophic interactions at the field level in tomato agroecosystems.

5. HOST-PLANT RESISTANCE

Host-plant resistance is an attractive pest management option that can reduce pesticide use and provide relatively durable, long-term management of pest populations. Remarkable progress has been made in identifying disease-resistant genes in wild tomato varieties and incorporating these traits into commercial cultivars of *Solanum lycopersicon* L. (Gardner and Scott, 2012). In contrast, the development of arthropod-resistant cultivars is lagging far behind that of diseases. One of the few insects for which resistance does exist in commercial tomato cultivars is that of the potato aphid, *Macrosiphum euphorbiae* Thomas. This resistance was attributed to a novel gene (*Meu-1*) that was thought to be closely linked to the nematode resistant gene *Mi* (Kaloshian et al., 1995), both of which were obtained from the wild cultivar *Lycopersicon peruvianum* (L.) Mill (Medina-Filho and Tanksley, 1983). It is now known that the *Mi* gene confers resistance to nematodes and potato aphid (Milligan et al., 1998; Rossi et al., 1998) as well as several biotypes of the silverleaf whitefly, *B. tabaci* (Nombela et al., 2003).

While insect resistance is currently lacking in commercial tomato cultivars, there is extensive knowledge on arthropod-resistant traits in wild *Solanum* species, some of which have been incorporated into breeding lines. This resistance is largely associated with allelochemicals in type IV and VI glandular trichomes of accessions of *Solanum* (=*Lycopersicum*) *hirsutum* (=*Solanum habrochaites* S. Knapp & D.M. Spooner) f. *glabratum*, *S. habrochaites* f. *glabratum*, *Solanum pennellii* (Correll) D'Arcy, and *Solanum pimpinellifolium* L., and confers resistance to a wide range of tomato arthropod pests (Farrar and Kennedy, 1991). Type IV glandular trichomes of *S. pennellii* (Blauth et al., 1998) and *S. pimpinellifolium* (Alba et al., 2009) contain high concentrations of acyl sugars that confer resistance to a broad spectrum of tomato pests, including aphids, whiteflies, tomato fruitworm, armyworm, leafminers, and spider mites. Additional allelochemicals associated with resistance include the sesquiterpenes zingiberene in type VI trichomes of accessions of *S. habrochaites*, and 2-tridecanone and 2-undecanone in type VI trichomes of *Solanum hirsutum*, f. *glabratum* Dunal. Incorporating trichome-based resistance into commercial cultivars is challenging, because inheritance is polygenic and linked to undesirable traits in wild donor plants. These trichome-mediated defense traits also have generally negative effects on parasites and predators (Kennedy, 2003), and this can have implications that need to be considered when deploying resistant plants in the field.

Genetically modified (GM) crops with insect-resistant traits are a key component of modern agriculture that can increase sustainability by reducing pesticide and fuel use, increase yields, and enhance effects on biological diversity (NRC, 2010). Despite the high adoption rates and benefits realized to producers of GM crops throughout the world, primarily those of cotton and maize (James, 2013), this technology remains largely unavailable to producers of tomatoes and other fruits and vegetables. Public concern about the safety of GM vegetables has prevented the adoption of this technology by the fruit and vegetable industry (Nap et al., 2003; Anderson et al., 2007). Other than China (James, 2012), there is no known world production of GM tomatoes. Nonetheless, considerable progress has been made in engineering tomatoes for pest resistance and improved market qualities, including the first commercially produced GM crop in the United States—the delayed-ripening

Flavr Savr tomato that was available from 1994 to 1997 (commercialized in 1994 and ceased 1997) (Bruening and Lyons, 2000). Numerous transgenic tomatoes with resistance to lepidopteran pests based on expression of *Bacillus thuringiensis* Berliner (*Bt*) proteins (see Mandoakar et al., 2000; Saker et al., 2011) and proteinase inhibitors (Abdeen et al., 2005) have been developed, as are tomatoes with resistance to viruses transmitted by whiteflies (Sengoda et al., 2012), aphids (Zhang et al., 2012), and thrips (Fedorowicz et al., 2005). Despite the absence of commercially grown GM tomatoes anywhere in the world other than China, only corn, soybeans, cotton, and potatoes exceed tomatoes in the number of USDA-APHIS issued R&D field releases of GM varieties in 2013 (Fernandez-Cornejo et al., 2014).

6. BIOLOGICAL CONTROL

There is extensive information on biological control programs for tomatoes in protected environments (see Chapter 15). The closed systems of greenhouses are ideally suited to biological control, because it is feasible to limit dispersal of natural enemies, avoid invasion of pests from outside sources, and create an environment more favorable to natural enemy than pest populations (Perdikis et al., 2008). Biological control plays a less prominent role in field-grown tomatoes, although there are several examples of programs that have used natural enemy populations to reduce insecticide use.

6.1 Aphids

The green peach aphid, *Myzus persicae* (Sulzer), is a common pest of tomato in both field and protected environments. Aphids have many predators and parasites in field-grown tomatoes, but natural enemies are generally ineffective in maintaining populations below damaging levels (Walker et al., 1984a; Perring et al., 1988; Walgenbach, 1994). Parasitism of potato aphid by *Aphidius* spp. is hampered by high levels of hyperparasitism, and occurrence of the most effective predator, the syrphid *Aphidoletes aphidimyza* (Rondani), is not predictable (Farrar et al., 1986). Rates of parasitism are inversely density dependent, and predation by *A. aphidimyza* has been most effective in the early season when aphid densities are low. Thus the use of practices that suppress aphids, such as resistant cultivars, would be expected to enhance the efficacy of biological control. Natural enemies commonly used in greenhouse tomato crops include the parasitoids *Aphidius ervi* Haliday, *Aphidius colemani* Viereck, *Aphelinus abdominalis* (Dalman), and the syrphid *A. aphidimyza* (Perdikis et al., 2008).

6.2 Mites

There are three potential mite pests of tomato, including the two-spotted spider mite (*Tetranychus urticae* Koch), probably the most common worldwide pest of tomato, the tomato russet mite (*Aculops lycopersici* (Massee)), and the tomato red spider mite (*Tetranychus evansi* Baker and Pritchard), the latter being an invasive species native to South American, but now occurring in most subtropical areas (Navajas et al., 2013). In contrast to the success of augmentative releases of predatory phytoseiid mites in greenhouse tomatoes with *Phytoseiulus persimilis* Athias-Henriot (van Lenteren, 2000), biological control in the field has been less successful. Augmentative releases have not been shown to be practical in field-grown crops and natural populations of phytoseiid predators are sporadic because of variable climatic conditions. In addition, there can be a delay in the response of some phytoseiids to phytophagous mites while the predators adapt to the glandular trichome defense of tomatoes (Drukker et al., 1997; Sato et al., 2011; van Houten et al., 2013). Entomopathogenic fungi, particularly *Neozygites floridana* (Weiser and Muma), is widespread in distribution and can decimate spider mite populations under high humidity conditions (Duarte et al., 2009). Unfortunately, this fungus is highly sensitive to several fungicides used for managing diseases of tomato in many areas of the world (Klingen and Westrum, 2007; Weskesa et al., 2008).

Following the introduction of *T. evansi* into Africa in the 1990s, there was an effort to identify and import natural enemies from South America, the native range of *T. evansi*. This work is summarized by Navajas et al. (2013), but two candidate natural enemies that appeared to play an important role in maintaining *T. evansi* at low densities in much of Brazil were the phytoseiid *Phytoseiulus longipes* Evans (Furtado et al., 2006) and the fungal pathogen *N. floridana* (Weskesa et al., 2007). Introductions and assessments of these natural enemies in Africa have not yet been completed.

6.3 South American Pinworm

Despite a rich natural enemy complex of *T. absoluta* in its native regions of South America (Desneux et al., 2010; Luna et al., 2012), management in this region relies extensively on chemical control (Guedes and Picanco, 2012). Following the invasion of *T. absoluta* in Europe, several surveys revealed that a complex of native mirid predators and hymenopteran parasites were potentially effective biological control agents (Luna et al., 2011; Urbaneja et al., 2012; Zappala et al., 2013).

Natural enemies were suspected of helping to suppress populations within 2–3 years of *T. absoluta*'s arrival in Spain (Urbaneja et al., 2012). To date, the greatest progress has been the one made in protected crops with generalist mirid predators *Macrolophus pygmaeus* (Rambur) and *Nesidiocoris tenuis* (Reuter) and the egg parasitoid *Trichogramm achaeae* Nagaraja and Nagarkattii, while programs in field-grown crops are focusing on conservation methods.

6.4 Fruitworms

Worldwide, tomato fruitworms, including *H. zea*, *H. armigera*, and *H. virescens*, are among the most important pests of field-grown tomatoes. Damage threshold levels are extremely low, especially on high-value fresh market crops, and preventive control practices are often necessary to minimize the damage. *Trichogramma* spp. egg parasitoids are well suited for preventive control, because they kill the egg and prevent the larvae from boring into fruit. Despite high rates of parasitism that often exceed 50% by naturally occurring *Trichogramma* populations (Oatman et al., 1983; Roltsch and Mayse, 1983; Hoffmann et al., 1990), biological methods alone do not provide the necessary level of control. Oatman and Platner (1971, 1978) originally demonstrated that mass releases of *Trichogramma pretiosum* Riley were effective in reducing damage to processing tomatoes. This concept of mass releases of *Trichogramma* for fruitworm control was evaluated as a component of a tomato IPM program on fresh market tomatoes in Mexico (Trumble and Alvarado-Rodriguez, 1993). The mass-release component consisted of weekly releases of 100,000 *T. pretiosum* per hectare over a 5–9 week period. While this was effective in reducing fruitworm and armyworm damage, it did not improve control over the standard insecticide treatment. Due to the low cost of insecticides and high crop value, *Trichogramma* releases are not commonly used (Mills and Daane, 2005). However, conservation of naturally occurring populations is possible with selective insecticide use (Campbell et al., 1991; Hewa-Kapuge et al., 2003; Momanyi et al., 2012).

Processing tomato crops offer greater opportunities for incorporating biological control into IPM programs, because there are established minimum tolerance levels for fruit damage. For example, in California processing tomatoes where *H. zea* is a key late-season pest, the benefits of parasitism of eggs by *Trichogramma* spp. (Hoffmann et al., 1990) in reducing the need for insecticides was achieved by developing a sequential sampling plan with an adjustable threshold level based on rates of egg parasitism (Hoffmann et al., 1991). A similar program exists for managing *H. armigera* on processing tomatoes in New Zealand, except that the program relies on larval rather than egg parasitoids. The braconids *Cotesia kazak* (Telenga) and *Microplitis croceipes* (Cresson) were introduced into New Zealand in 1977 and 1986, respectively, and quickly became established and led to increased levels of *H. armigera* larval parasitism in several crops (Walker and Cameron, 1989). In tomatoes, parasitism levels increased from <1% before introduction to 60–80% by 1992 (Cameron et al., 2006). *C. kazak*, which parasitizes small larvae before they enter fruit, is the dominate parasite. Larval threshold levels were adjusted to account for the level of parasitism—threshold levels increase with increasing levels of parasitism—while maintaining fruit damage levels below 5% tolerance level (Walker et al., 2010). Incorporation of these parasitoids into tomato IPM programs, along with the use of compatible insecticides, led to a 95% reduction in insecticide use in New Zealand tomatoes (Cameron et al., 2009).

6.5 Leafminers

There are several species of leaf-mining *Liriomyza* that can be important pests of tomatoes, including the serpentine leafminer, *Liriomyza trifolii* (Burgess), vegetable leafminer, *Liriomyza sativae* Blanchard, and tomato leafminer, *Liriomyza bryoniae* (Kalt.). Because leafminers are indirect pests and the threshold for damage is relatively high (Kotze and Dennill, 1996), naturally occurring parasites often keep leafminer populations at non-damaging levels when conserved with selective insecticide use. The most common hymenopteran parasites of *Liriomyza* spp. are Eulophids in the genus *Diglyphus*, and Brachonids in the genus *Opius* and *Dacnusa* (Liu et al., 2009). Classic study by Oatman and Kennedy (1976) demonstrated that populations of *L. sativae* could be induced with applications of methomyl due to its adverse effects on parasites. This work remains relevant today, and selective insecticide use has helped to minimize the importance of leafminers as pests of field-grown tomatoes.

7. SEMIOCHEMICAL CONTROL

Pheromone-mediated mating disruption has been evaluated as a management strategy for the tomato pinworm, *Keiferia lycopersicella* (Walsingham), and South American pinworm *T. absoluta*. Both of these insects are candidates for mating disruption because they have a restricted host range that includes tomato as a primary host, and neither of them has a strong migratory behavior. van Steenwyk and Oatman (1983) obtained high levels of suppression of phero-mone trap captures, but damage suppression was not considered sufficient to provide economic control. Immigration of female moths from adjacent non-disrupted fields was cited as a concern. Mating disruption of *K. lycopersicella* was a component of an IPM program compared to conventional chemical control in Mexican tomatoes, and while the IPM program was more cost-efficient, it was not possible to attribute lower levels of pinworm fruit damage solely to mating disruption (Trumble and Alvarado-Rodriguez, 1993). More recent studies with *T. absoluta* in Spain showed that mating disruption was efficacious in tomatoes grown in high containment, protected crops, but not in open-field grown tomatoes (Vacas et al., 2011). However, the high cost of *T. absoluta* pheromone is an obstacle to implementation. Finally, mass trapping of *T. absoluta* in greenhouse crops in Argentina resulted in reduced larval infestations on a consistent basis (Lobos et al., 2013).

8. CHEMICAL CONTROL

Synthetic insecticides have been the primary tool used to manage arthropods in field-grown tomatoes for decades, and will likely remain a critical element in the near future. While some lament the lack of progress in moving away from pesticides in IPM programs (Ehler, 2006), the ease and efficiency of using pesticides are formidable obstacles to overcome on crops with stringent quality standards such as tomatoes. Numerous studies from a diversity of locations have demonstrated the value of insecticides in enhancing the profitability of tomato production, particularly when used in the context of an IPM program (Wiesenborn et al., 1990; Walgenbach and Estes, 1991; Trumble and Alvarado-Rodriguez, 1993; Nault and Speese, 2002; Yardim and Edwards, 2003; Engindeniz, 2006). It is noted, however, that production economic analyses do not capture costs associated with pesticide impacts on the environment or human health (Rushing et al., 1995; Ngowi et al., 2007; Grovermann et al., 2013).

8.1 Change in the Type of Pesticides

The type and amount of pesticides used on tomatoes have changed dramatically during the past 20-plus years. These changes have largely been the result of regulatory decisions and pesticide resistance; the former being a trend in agriculture in general (Osteen and Fernandez-Cornejo, 2012), and the latter having impacts more at the local level. Introduction of synthetic pyrethroids in the late 1970s and 1980s resulted in a dramatic reduction in insecticide active ingredients applied—insecticide use on vegetables declined by almost 50% between the early 1970s and mid 1980s—because of the lower use rates of pyrethroids compared to many of the organophosphate and carbamate insecticides that they replaced. Public concerns about the effect of pesticide residues in the United States food supply led to a revamping of pesticide laws with the Food Quality Protection Act of 1996 (FQPA). The FQPA led to more stringent regulatory standards, and the crop protection industry responded by developing new products with unique modes of action and more friendly human health and environmental profiles. Among those new groups commonly used in tomato production are the neonicotinoids, oxadiazines, diamides, spinosyns, tetonic acids, and several different types of insect-growth regulators. In contrast to older products that had a broad spectrum of pest activity, this new generation of insecticides have more narrow ranges of pest activity, which offers opportunities for enhancing biological control programs. It is noteworthy that while many pest management scientists have lauded the appearance of narrow-spectrum insecticides, the number of premix products containing one or more active ingredients with different modes of action is the evidence that broad-spectrum control remains a desirable goal for agricultural community.

The availability of several different classes of insecticides with systemic activity is creating opportunities for delivery methods that can minimize negative impacts of insecticide use on human health and the environment. The vast majority of insecticides are applied to tomatoes as foliar applications with high-pressure sprayers. Negative impacts on human health and the environment can arise from the inevitable drift of pesticides to non-target sites, erosion of soil which releases pesticide-contaminated particles into water resources, and exposure of farm workers to pesticides that accumulate on plant surfaces. The latter is particularly relevant to fresh market tomatoes that require considerable hand labor to stake, string, and harvest. Many of these negative consequences can be mitigated with the use of systemic insecticides delivered to the root system via drip irrigation or drench applications. Neonicotinoids have been available for this delivery method for several decades, but control was limited to whiteflies, aphids, flea beetles, and certain species of thrips; foliar insecticides were still required for lepidopteran pests. The recent registration of several systemic diamides with lepiodopteran activity, including chlorantraniliprole and cyantraniliprole, broadens the spectrum of pests that can be managed with soil applied insecticides (Kuhar et al., 2010).

8.2 Insecticide Resistance

There are several key arthropod pests of tomato that historically have been problematic to control because of insecticide resistance. Resistance among highly mobile, polyphagus lepidopteran pests, including the Heliothine complex of *H. zea*, *H. armigera*, and *H. virescens*, as well as the beet armyworm, *Spodoptera exigua* (Hübner), is usually not associated with insecticide use practices on tomatoes. Rather, resistance in these pests is more common in higher acreage crops such as cotton and corn that are responsible for insecticide exposure to large portions of populations. The widespread use of pyrethroids that provided effective and inexpensive control of the Heliothine complex on multiple crops beginning in the 1980s eventually led to resistant populations, particularly in cotton production regions around the world (Ahmad, 2007; Hopkins and Pietrantonio, 2010; Wilson et al., 2013). The migratory behavior of these species plays an important role in the insecticide-resistance dynamics, affecting insecticide performance on a diversity of crops at considerable distances from the source of resistance (Pietrantonio et al., 2007; Avilla and González-Zamora, 2010). Concern about the development of *Bt* resistance in Heliothine populations due to widespread plantings of cotton and corn modified with *Bt Cry* proteins has largely not materialized. However, populations of the cabbage looper, *Trichoplusia ni* (Hübner), in Canadian vegetable greenhouses have developed resistance to repeated applications of *Bt* var. *kurstaki* (Janmaat, 2007), which is a concern in an environment where selective insecticide use is important in the maintenance of biological control programs.

In contrast to the Heliothines, *T. absoluta*, has a more restricted host range that includes the family Solanaceae, with tomato being the primary host. In its native South America, *T. absoluta* has developed resistance to virtually all classes of insecticides used against this pest (see Lietti et al., 2005; Silva et al., 2011; Campos et al., 2014), including organophosphates, pyrethroids, abamectin, cartap, indoxacarb, chitin synthesis inhibitors, and spinosad. Analysis of spatial patterns of resistance suggests that *T. absoluta* resistance is a local phenomenon related to the frequency of insecticide use and weather (Silva et al., 2011). A high proportion of *T. absoluta* populations in the Mediterranean Basin, which originated from an accidental introduction in Europe in 2006, carry pyrethroid-resistant alleles (Haddi et al., 2012). Indeed, high levels of resistance were detected in Greece, where potential resistance to chlorpyrifos and metaflumizone were also suspected (Roditakis et al., 2012). Current resistance management focuses on rotation of insecticides with different modes of action, along with the development of nonchemical tactics.

Another group of pests that have developed resistance to a wide range of insecticides are whiteflies, including *Trialeurodes vaporariorum* (Westwood) and *B. tabaci*, particularly among the B and Q biotypes of *B. tabaci* (Stansly and Naranjo, 2010). The greatest concern has been resistance development among the neonicotinoids, which have played a key role in managing whiteflies since the registration of imidacloprid in the early to mid-1990s. The dynamics of resistance is quite different in protected versus open-field tomatoes. For example, in open-field production in Florida where tomatoes are just one of several crops that serve as transient hosts on a local level, surveys of *B. tabaci* from multiple farms from 2000 to 2010 exhibited a trend of increasing resistance to neonicotinoids (Schuster et al., 2010; Caballero et al., 2013). In southeast Spain where tomatoes are grown predominately in greenhouses, resistance was related to management strategies used by individual operators. Operations that relied largely on chemical control had significant resistance to multiple insecticides, while those populations from houses that relied on biological control had the most susceptible populations (Fernández et al., 2009).

9. FUTURE PROSPECTS

Tomato pest management programs will be impacted by many factors in the coming years, some of the most prominent being invasive pests, climate change, pesticide regulatory issues, and the availability of GM tomato cultivars. It is anticipated that pest problems will increase in many areas of the world due to the spread of invasive pests and climate change. The challenge will be to develop pest management strategies that are sustainable, preserve biodiversity, and are compatible with a growing population in which the buffer between agriculture and residential areas continues to diminish.

The introduction of invasive pests is a problem of rising importance on a global scale. International trade has been implicated as a primary cause of the spread of pest species to new areas (Westphal et al., 2008), and this spread is expected to continue for the foreseeable future. While managing invasive pests is not fundamentally different from that of native pests, it complicates IPM programs by the need to manage more pests. Additionally, more pests limit the range of tactics available and affect the scale at which management must operate (Venette and Koch, 2009). The absence of effective natural control agents, at least during the initial stages of spread, can lead to large densities on a regional scale, and chemical control is often the only viable option under these circumstances. Increased insecticide use can disrupt biological and cultural control programs for existing pests from which the system will not recover easily. Some of the more important invasive pests of tomato in recent years include *T. absoluta*, which was introduced from South America into Europe and the North African

Mediterranean Basin (Desneux et al., 2010), the brown marmorated stink bug (*Halyomorpha halys* Stål) introduced from China to North America (Leskey et al., 2012) and Europe (Gariepy et al., 2014), the tomato red spider mite (*T. evansi*) introduced from South America to sub-Saharan Africa, the Mediterranean region, Hawaii, and Taiwan (Navajas et al., 2013), and *Liriomyza* spp., which after 50 years continues to spread to new areas of the world (e.g., Rauf et al., 2000; Andersen et al., 2002; Nyambo et al., 2011). Continued movement of these pests to new areas is probable.

The most recent assessment in 2013 by the Intergovernmental Panel on Climate Change (IPCC) predicts mean global surface temperature to rise by 1–3.7°C by 2100 (Stocker et al., 2013). Combined with rising levels of atmospheric carbon dioxide, changes of this magnitude are expected to significantly affect global agriculture and the population dynamics and status of arthropod pests (Sharma and Prabhakar, 2014). Foremost among these effects is the altered vulnerability of agriculture to invasive species (Ziska et al., 2011), e.g., regions previously not suitable for establishment of certain pests will become suitable. Impacts on native pests at the local level also will occur, including altered overwintering ranges, population growth and phenology, and interactions with natural enemies and host plants (Sun et al., 2011; Sharma and Prabhakar, 2014). There are many gaps in our ability to respond to these changes, but modeling efforts are helping to predict climate change effects on the expansion of pests (Ulrichs and Hopper, 2008), and the tritophic interactions among pests, host plants, and natural enemies (Gutierrez et al., 2008).

The suite of pesticides now available to growers represents a considerable improvement over those of 15–20 years ago in terms of safety to humans, the environment, and natural enemy populations. They primarily include new chemistry products with low use rates and pest specificity, and biopesticides with even greater specificity and safety. This trend will likely continue in view of the general public's concern and perception about environmental and food safety issues. Multinational agrochemical companies have the ability to quickly incorporate new research technology into their search for new active ingredients (Lamberth et al., 2013), which will result in a continued flow of new products with unique modes of action that meet regulatory and market demands. While biopesticides account for <5% of the global pesticide market (Bailey et al., 2010), this will likely increase with several large agrochemical companies now investing in this technology.

Host-plant resistance arguably offers the greatest opportunity for advances in tomato pest management. It is easy to use, compatible with other pest management tactics, and can greatly reduce pesticide use. However, these benefits are unlikely to occur until consumer skepticism of the safety of genetically modified fresh produce is overcome. Consequently, progress is being hamstrung by the need to rely on traditional breeding techniques to move insect-resistant genes into commercial cultivars. Although, *Solanum* spp. is the primary source of insect resistance, linkage drag associated with polygenic traits is slowing the progress. Recent studies suggest that there is a willingness among consumers to buy genetically modified vegetables, assuming that genes are not from other species (Lusk and Rozon, 2006; Lim et al., 2014).

As previously noted, there are numerous genetically modified tomatoes with insect-resistant traits that have been developed. Advances in recombinant DNA technology and RNAi technology (Katoch et al., 2013) greatly expand the sources of resistant traits to incorporate into tomatoes. Based on the success of other GM crops (Romeis et al., 2008), the eventual availability of GM tomato crops offers great promise to develop more IPM programs. The ultimate success in implementing these programs at the field level, however, will depend on pest management scientists integrating this technology to programs that avoid pest-resistance development, determine compatibility with biological control programs, and understand shifts in the importance of various pests (Kennedy, 2008). This will require research at the applied ecology level, and development of educational programs to facilitate implementation of new technology.

REFERENCES

Abdeen, A., Virgos, A., Olivella, E., Villanueva, J., Aviles, X., Gabarra, R., Prat, S., 2005. Multiple insect resistance in transgenic tomato plants over-expressing two families of plant proteinase inhibitors. Plant Molecular Biology 57, 189–202.

Ahmad, M., 2007. Insecticide resistance mechanisms and their management in the *Helicoverpa armigera* (Hübner). Journal of Agricultural Research 45, 319–335.

Alba, J.M., Montserrat, M., Fernandez-Munoz, R., 2009. Resistance to the two-spotted spider mite (*Tetranychus urticae*) by acylsucroses of wild tomato (*Solanum pimpinellifolium*) trichomes studied in a recombinant inbred line population. Experimental and Applied Acarology 47, 35–47.

Andersen, A., Nordhus, E., Thang, V.T., An, T.T.T., Hung, H.Q., Hofsvang, T., 2002. Polyphagous *Liriomyza* species (Diptera: Agromyzidae) in vegetables in Vietnam. Tropical Agriculture 779, 241–246.

Anderson, J.C., Wachenheim, C.J., Lesch, W.C., 2007. Perceptions of genetically modified and organic foods and processes. AgBioForum 9, 180–194.

Andow, D., 1991. Vegetational diversity and arthropod population response. Annual Review of Entomology 36, 561–586.

Avila, C., González-Zamora, J.E., 2010. Monitoring resistance of *Helicoverpa armigera* to different insecticides used in cotton in Spain. Crop Protection 29, 100–103.

Bailey, K.L., Boyetchko, S.M., Langle, T., 2010. Social and economic drivers shaping the future of biological control: a Canadian perspective on the factors affecting the development and use of microbial biopesticides. Biological Control 52, 221–229.

Bajwa, W.I., Kogan, M., 1996. Compendium of IPM Definitions (CID) – What Is IPM and How Is It Defined in the Worldwide Literature? IPPC Publication No. 998 Integrated Plant Protection Center (IPCC), Oregon State University, Corvallis, OR 97331, USA. http://www.ipmnet.org/ipmdefinitions/index.pdf.

Barfield, C.S., Swisher, M.E., 1994. Integrated pest management: ready for export? Historical context and internationalization of IPM. Food Reviews International 10, 215–267.

Benbrook, C., Groth, E., 1996. Pest Management at the Crossroads. Consumers Union, Yonkers, New York, USA.

Blauth, S.L., Churchill, G.A., Mutschler, M.A., 1998. Identification of quantitative trait loci associated with acylsugar accumulation using interspecific populations of the wild tomato, *Lycopersicon pennellii*. Theoretical and Applied Genetics 96, 458–467.

Bruening, G., Lyons, J.M., 2000. The case of the FLAVR SAVR tomato. California Agriculture 45, 6–7.

Caballero, R., Cyman, S., Schuster, D.J., 2013. Monitoring insecticide resistance in biotype B of *Bemisia tabaci* (Hemiptera: Aleyrodidae) in Florida. Florida Entomologist 96, 1243–1256.

Cameron, P.J., Walker, G.P., Herman, T.J.B., Wallace, A.R., 2001. Development of economic thresholds and monitoring systems for *Helicoverpa armigera* (Lepidoptera: Noctuidae) in tomatoes. Journal of Economic Entomology 94, 1104–1112.

Cameron, P.J., Walker, G.P., Herman, T.J.B., Wallace, A.R., 2006. Incidence of the introduced parasitoids *Cotesia kazak* and *Micoplitis croceipes* (Hymenoptera: Braconidae) from *Helicoverpa armigera* (Lepidoptera: Noctuidae) in tomatoes, sweet corn, and lucerne in New Zealand. Biological Control 39, 375–384.

Cameron, P.J., Walker, G.P., Hodson, A.J., Kale, A.J., Herman, T.J.B., 2009. Trends in IPM and insecticide use in processing tomatoes in New Zealand. Crop Protection 28, 421–427.

Campbell, C.D., Walgenbach, J.F., Kennedy, G.G., 1991. Effect of parasitoids on lepidopterous pests in insecticide-treated and untreated tomatoes in western North Carolina. Journal of Economic Entomology 84, 1662–1667.

Campos, M.R., Rita, A., Rodrigues, S., Silva, W.M., Barbosa, T., Silva, M., Silva, V.R.F., Guedes, R.N.C., Siqueira, H.A.A., 2014. Spinosad and the tomato borer *Tuta absoluta*: a bioinsecticide, an invasive pest threat, and high insecticide resistance. PLoS One 9 (8), e103235.

Cho, K.J., Eckel, C.S., Walgenbach, J.F., Kennedy, G.G., 1995. Spatial distribution and sampling procedures for *Frankliniella* spp. (Thysanoptera: Thripidae) in staked tomato. Journal of Economic Entomology 88, 1657–1665.

Cocco, A., Serra, G., Lentini, A., Deliperi, S., Delrio, G., 2015. Spatial distribution and sequential sampling plans for *Tuta absoluta* (Lepidoptera: Gelechiidae) in greenhouse tomato crops. Pest Management Science 71, 1311–1323.

Csizinszky, A.A., Schuster, D.J., Polston, J.E., 1999. Effect of ultra-violet reflective mulches on tomato yields and on the silverleaf whitefly. HortScience 34, 91–914.

Cullen, E.M., Zalom, F.G., Flint, M.L., Zilbert, E.E., 2000. Quantifying trade-offs between pest sampling time and precision in commercial IPM sampling programs. Agricultural Systems 66, 99–113.

Dawson, J., Hamilton, A.J., Mansfield, C., 2006. Dispersion statistics and a sampling plan for *Helicoverpa* (Lepidoptera: Noctuidae) on fresh market tomatoes (*Lycopersicon esculentum*). Australian Journal of Entomology 45, 91–95.

Desneux, N., Wajnberg, E., Wyckhuys, K.A.G., Burgio, G., Arpaia, S., Narvaez-Vasquez, C.A., Gonzalez-Carbera, J., Catalan, D., Tabone, E., Frandon, J., Pizzol, J., Poncet, C., Cabello, T., Urbaneja, A., 2010. Biological invasion of European tomato crops by *Tuta absoluta*: ecology, geographic expansion and prospects for biological control. Journal of Pest Science 83, 197–215.

Drukker, B., Janssen, A., Ravensberg, W., Sabelis, M.W., 1997. Improved control capacity of the mite predator *Phytoseiulus persimilis* (Acari: Phytoseiidae) on tomato. Experimental and Applied Acarology 21, 507–518.

Duarte, V.S., Silva, R.A., Wekesa, V.W., Rizzato, F.B., Dias, C.T.S., Delalibera Jr., I., 2009. Impact of natural epizootics of the fungal pathogen *Neozygites floridana* (Zygomycetes: Entomophthorales) on population dynamics of *Tetranychus evansi* (Acari: Tetranychidae) in tomato and nightshade. Biological Control 51, 81–90.

Ehler, L.E., 2006. Integrated pest management (IPM): definition, historical development, and implementation, and the other IPM. Pest Management Science 62, 787–789.

Ehler, L.E., Bottrell, D.G., 2000. The illusion of integrated pest management. Issues in Science and Technology 25, 61–64.

Engindeniz, S., 2006. Economic analysis of pesticide use on processing tomato growing: a case study for Turkey. Crop Protection 25, 534–541.

Fanigliulo, A., Comes, S., Crescenzi, A., Momol, M.T., Olson, S.M., Sacchetti, M., Ferrara, L., Caliguri, G., 2009. Integrated management of viral diseases in field-grown tomatoes in Southern Italy. Acta Horticulturae 808, 387–391.

Farrar, C.A., Perring, T.M., Toscano, N.C., 1986. A midge predator of potato aphids on tomatoes. California Agriculture 40, 9–10.

Farrar Jr., R.R., Kennedy, G.G., 1991. Inhibition of *Telenomous sphingis* (Ashmead) (Hymenoptera: Scelionidae), an egg parasitoid of *Manduca* spp. (Lepidoptera: Sphingidae), by trichome/2-tridecanone-based host plant resistance in tomato. Entomologia Experimentalis et Applicata 60, 157–166.

Fedorowicz, O., Bartozewski, G., Kamińska, M., Stoeva, P., Niemirowicz-Szczytt, K., 2005. Pathogen-derived resistance to tomato spotted wilt virus in transgenic tomato and tobacco plants. Journal of the American Horticultural Society of America 130, 218–224.

Fernández, E., Gravalos, C., Haro, P.J., Cifuentes, D., Bielza, P., 2009. Insecticide resistance status of *Bemisia tabaci* Q-biotype in South-Eastern Spain. Pest Management Science 65, 885–891.

Fernandez-Cornejo, J., Wechsler, S., Livingston, M., Mitchell, L., February 01, 2014. Genetically Engineered Crops in the United States. USDA-Economic Research Service Report 162 USDA, USA. http://ssrn.com/abstract=2503388.

Furtado, I.P., de Moraes, G.J., Kreiter, S., Knapp, M., 2006. Search for effective natural enemies of *Tetranychus evansi* in south and southeast Brazil. Experimental and Applied Acarology 40, 157–174.

Gardner, R.G., Scott, J.W., 2012. Tomato breeding and cultivar selection. In: Davis, R.M., Pernezny, K., Broome, J.C. (Eds.), Tomato Health Management. APS Press, Saint Paul, Minnesota, USA, pp. 31–36.

Gariepy, T.D., Haye, T., Fraser, H., Zhang, J., 2014. Occurrence, genetic diversity, and potential pathways of entry of *Halyomorpha halys*, in newly invaded areas of Canada and Switzerland. Journal of Pest Science 87, 17–28.

Greenough, D.R., Black, L.L., Bond, W.P., 1990. Aluminum-surfaced mulch: an approach to the control of tomato spotted wilt virus in solanaceous crops. Plant Disease 74, 805–808.

Grieshop, J.I., Zalom, F.G., Miyao, G., 1988. Adoption and diffusion of itegrated pest management innovations in agriculture. Entomological Society of America 75 (2), 72–78.

Grovermann, C., Schreinemachers, P., Berger, T., 2013. Quantifying pesticide overuse from farmer and societal points of view: an application to Thailand. Crop Protection 53, 161–168.

Guedes, R.N.C., Picanco, M.C., 2012. The tomato borer *Tuta absoluta* in South America: pest status, management and insecticide resistance. Bulletin of the European and Mediterranean Plant Protection Organization 42, 211–216.

Gusmão, M.R., Picanco, M.C., Guedes, R.N.C., Galvan, T.L., Pereira, E.J.G., 2006. Economic injury level and sequential sampling plan for *Bemisia tabaci* in outdoor tomato. Journal of Applied Entomology 130, 160–166.

Gutierrez, A.P., Ellis, C.K., d'Oultremont, T., Luigi, P., 2008. Climate change effects on poikilotherm tritrophic interactions. Climate Change 87, 167–192.

Haddi, K., Berger, M., Bielza, P., Cifuentes, D., Field, L.M., Gorman, K., Rapisarda, C., Williamson, M.S., Bass, C., 2012. Identification of mutations associated with pyrethroid resistance in the voltage-gated sodium channel of the tomato leaf miner (*Tuta absoluta*). Insect Biochemistry and Molecular Biology 42, 506–513.

Hewa-Kapuge, S., McDougall, S., Hoffmann, A.A., 2003. Effects of methoxyfenozide, indoxacarb, and other insecticides on the beneficial egg parasitoid *Trichogramma* nr. *brassicae* (Hymenoptera: Trichogrammatidae) under laboratory and field conditions. Journal of Economic Entomology 96, 1083–1090.

Hilje, L., Costa, H.S., Stansly, P.A., 2001. Cultural practices for managing *Bemisia tabaci* and associated viral diseases. Crop Protection 20, 801–812.

Hilje, L., Stansly, P.A., 2008. Living ground covers for management of *Bemisia tabaci* (Gennadius) (Homoptera: Aleyrodidae) and tomato yellow mottle virus (ToYMoV) in Costa Rica. Crop Protection 27, 10–16.

Hoffmann, M.P., Wilson, L.T., Zalom, F.G., Hilton, R.J., 1990. Parasitism of *Heliothis zea* (Lepidoptera: Noctuidae) eggs: effect on pest management decision rules for processing tomatoes in the Sacramento Valley of California. Environmental Entomology 19, 753–763.

Hoffmann, M.P., Wilson, L.T., Zalom, F.G., Hilton, R.J., 1991. Dynamic sequential sampling plan for *Helicoverpa zea* (Lepidoptera: Noctuidae) eggs in processing tomatoes: parasitism and temporal patterns. Environmental Entomology 20, 1005–1012.

Hokkanen, H.M.T., 1991. Trap cropping in pest management. Annual Review of Entomology 36, 119–138.

Hopkins, B.W., Pietrantonio, P.V., 2010. The *Helicoverpa zea* (Boddie) (Lepidoptera: Noctuidae) voltage gated sodium channel and mutations associated with pyrethroid resistance in field-collected adult males. Insect Biochemistry and Molecular Biology 40, 385–393.

Hummel, N.A., Zalom, F.G., Miyo, G.M., Underwood, N.C., Villalobos, A., 2004. Potato aphid, *Macrosiphum euphorbiae* (Thomas) in tomatoes: plant canopy distribution and binomial sampling on processing tomatoes in California. Journal of Economic Entomology 97, 490–495.

Hummel, R.L., Walgenbach, J.F., Hoyt, G.D., Kennedy, G.G., 2002. Effects of production system on vegetable arthropods and their natural enemies. Agriculture, Ecosystems and Environment 93, 165–176.

Hunt, D.W.A., 1998. Reduced tillage practices for managing the Colorado potato beetle in processing tomato production. HortScience 33, 279–282.

Hunt, D.W.A., Whitefield, G., 1996. Potato trap crops for control of Colorado potato beetle (Coleoptera: Chrysomelidae) in tomatoes. Canadian Entomologist 128, 407–412.

Ioannou, N., 1987. Cultural management of tomato yellow leaf curl disease in Cyprus. Plant Pathology 36, 367–373.

James, C., 2012. Global status of commercialized biotech/GM crops: 2012. In: International Service for the Acquisition of Agri-Biotech Applications Brief No. 44. Ithaca, New York, USA.

James, C., 2013. Global status of commercialized biotech/GM crops: 2013. In: International Service for the Acquisition of Agribiotech Applications Brief No. 46. Ithaca, New York. USA.

Janmaat, A.F., 2007. Development of resistance to the biopesticide *Bacillus thuringiensis kurstaki*. In: Vincent, C., Goettel, M.S., Lazarovits, G. (Eds.), Biological Control: A Global Perspective: Case Studies From Around the World. Center for Agriculture and Bioscience International Publishing, Wallingford, UK, pp. 179–184.

Jepson, P.C., Kogan, M., 2007. Perspectives in Ecological Theory and Integrated Pest Management. Cambridge University Press, Cambridge, England.

Johanowicz, D., Mitchell, E.R., 2000. Effects of sweet alyssum flowers on the longevity of theparasitoid wasps *Cotesia marginiventris* (Hymenoptera: Braconidae) and *Diadegma insulare* (Hymenoptera: Ichneumonidae). Florida Entomologist 83, 41–47.

Joost, P.H., Riley, D.G., 2004. Sampling techniques for thrips (Thysanoptera: Thripidae) in preflowering tomato. Journal of Economic Entomology 97, 1450–1454.

Kaloshian, I., Lange, W.H., Williamson, V.M., 1995. An aphid-resistant locus is tightly linked to the nematode-resistance gene, *Mi*, in tomato. Proceedings of the National Academy of Sciences 92, 622–625.

Katoch, R., Sethi, A., Thakur, N., Murdock, L.L., 2013. RNAi for insect control: current perspective and future challenges. Applied Biochemistry and Biotechnology 171, 847–873.

Kennedy, G.G., 2003. Tomato, pests, parasitoids, and predators: tritrophic interactions involving the genus *Lycopersicon*. Annual Review of Entomology 48, 51–72.

Kennedy, G.G., 2008. Integration of insect-resistant genetically modified crops within IPM programs. In: Romeis, J., Shelton, A.M., Kennedy, G.G. (Eds.), Integration of Insect-Resistant Genetically Modified Crops within IPM Programs. Springer Publishing International, Dordrecht, Netherlands, pp. 1–26.

Kennedy, G.G., Sutton, T.B., 1999. Emerging Technologies for Integrated Pest Management: Concepts, Research, and Implementation. American Pathological Society Press, Saint Paul, Minnesota, USA.

Kim, J.K., Park, J.J., Park, H., Cho, K., 2001. Unbiased estimation of greenhouse whitefly, *Trialeurodes vaporariorum*, mean density using yellow sticky trap in cherry tomato greenhouses. Entomologia Experimentalis et Applicata 100, 235–243.

Klingen, I., Westrum, K., 2007. The effect of pesticides used in strawberries on the phytophagous mite *Tetranychus urticae* (Acari: Tetranychidae) and its fungal natural enemy *Neozygites floridana* (Zygomycetes: Entomophthorales). Biological Control 43, 222–230.

Knapp, M., Sarr, I., Gilioli, G., Baumgärtner, J., 2006. Population models for threshold-based control of *Tetranychus urticae* in small-scale Kenyan tomato fields and for evaluating weather and host plant species effects. Experimental and Applied Acarology 39, 195–212.

Kogan, M., 1998. Integrated pest management: historical perspectives and contemporary developments. Annual Review of Entomology 43, 243–270.

Kotze, D.J., Dennill, G.R., 1996. The effect of *Liriomyza trifolii* (Burgess) (Diptera: Agromyzidae) on fruit production and growth of tomato, *Lycopersicon esculentum* (Mill) (Solanaceae). Journal of Applied Entomology 120, 231–235.

Kring, J.B., 1972. Flight behavior of aphids. Annual Review of Entomology 17, 461–492.

Kring, J.B., Schuster, D.J., 1992. Management of insects on pepper and tomato with UV-reflective mulches. Florida Entomologist 75, 119–129.

Kuhar, T.P., Nault, B.A., Hitchner, E.M., Speese, J., 2006. Evaluation of action threshold-based insecticide spray programs for tomato fruitworm management in fresh-market tomatoes in Virginia. Crop Protection 25, 604–614.

Kuhar, T.P., Walgenbach, J.F., Doughty, H.B., 2010. Control of *Helicoverpa Zea* in Tomatoes With Chlorantraniliprole Applied Through Drip Chemigation. Plant Health Progress. http://dx.doi.org/10.1094/PHP-2009-0407-01-RS. https://www.plantmanagementnetwork.org/pub/php/research/2010/tomato/.

Lamberth, C., Jeanmart, S., Luksch, T., Plant, A., 2013. Current challenges and trends in the discovery of agrochemicals. Science 341, 742–746.

Lange, W.H., Bronson, L., 1981. Insect pests of tomatoes. Annual Review of Entomology 26, 345–371.

Lee, D.H., Park, J.J., Lee, J.H., Shin, K.I., Cho, K.J., 2007. Evaluation of binomial sequential classification sampling plan for leafmine of *Liromyza trifolii* (Diptera: Agromyzidae) in greenhouse tomatoes. International Journal of Pest Management 53, 59–67.

Leskey, T.C., Hamilton, G.C., Nielsen, A.L., Polk, D.C., Rodriguez-Saona, C., Bergh, J.C., Herbert, D.A., Kuhar, T.P., Pfeiffer, D., Dively, G.P., Hooks, C.R.R., Raupp, M.J., Shrewsbury, P.M., Krawczyk, G., Shearer, P.W., Whalen, J., Koplinka-Loehr, C., Myers, E., Inkley, D., Hoelmer, K.A., Lee, D.H., Wright, S.E., 2012. Pest status of the brown marmorated stink bug, *Halyomorpha halys* in the USA. Outlooks on Pest Management 23, 218–226.

Lietti, M.M.M., Botto, E., Alzogaray, R.A., 2005. Insecticide resistance in Argentine populations of *Tuta absoluta* (Meyrick) (Lepidoptera: Gelechiidae). Neotropical Entomology 34, 113–119.

Lim, W., Miller, R., Park, J., Park, S., 2014. Consumer sensory analysis of high flavonoid transgenic tomatoes. Journal of Food Science 79, 1212–1217.

Liu, T.X., Kang, L., Heinz, K., Trumble, J., 2009. Biological control of *Liriomyza* leafminers: progress and perspective. CAB Reviews: Perspectives in Agriculture, Veterinary Science, Nutrition, and Natural Resources 4 (4), 1–16.

Lobos, E., Occhionero, M., Werenitzky, D., Fernandez, J., Gonzalez, L.M., Rodriguez, C., Calvo, C., Lopez, G., Oehlschlager, A.C., 2013. Optimization of a trap for *Tuta absoluta* Meyrick (Lepidoptera: Gelechiidae) and trials to determine the effectiveness of mass trapping. Neotropical Entomology 42, 448–457.

Luna, M.G., Sanchez, N.E., Pereyra, P.C., Nieves, E., Savino, V., Luft, E., Virla, E., Speranza, S., 2012. Biological control of *Tuta absoluta* in Argentina and Italy: evaluation of indigenous insects as natural enemies. Bulletin of the European and Mediterranean Plant Protection Organization 42, 260–267.

Luna, M.G., Wada, V.I., La Salle, J., Sánchez, N.E., 2011. *Neochrysocharis formosa* (Westwood (Hymenoptera: Eulophidae), a newly recorded parasitoid of the tomato moth, *Tuta absoluta* (Meyrick) (Lepidoptera: Gelechiidae), in Argentina. Neotropical Entomology 40, 412–414.

Lusk, J.L., Rozon, A., 2006. Consumer acceptance of ingenic foods. Biotechnology Journal 1, 1433–1434.

Lye, B.H., Story, R.N., 1988. Spatial dispersion and sequential sampling plan of the southern green stink bug (Hemiptera: Pentatomidae) on fresh market tomatoes. Environmental Entomology 18, 139–144.

Lye, B.H., Story, R.N., Wright, V.L., 1989. Damage threshold of the southern green stink bug, *Nezara viridula*, (Hemiptera: Pentatomidae) on fresh market tomatoes. Journal of Entomological Science 23, 366–373.

Mandoakar, A.D., Goyal, R.K., Shukla, A., Bisaria, S., Bhalla, R., Reddy, V.S., Chaurasia, A., Sharma, R.P., Altosaar, I., Kumar, P.A., 2000. Transgenic tomato plants resistant to fruit borer (*Helicoverpa armigera* Hubner). Crop Protection 19, 307–312.

Meck, E.J., Kennedy, G.G., Walgenbach, J.F., 2013. Effects of *Tetranychus urticae* (Acari: Tetranychidae) on yield, quality and economics of tomato production. Crop Protection 52, 84–90.

Medeiros, M.A., Sujii, E.R., Morais, H.C., 2009. Effect of plant diversification on abundance of South American tomato pinworm and predators in two cropping systems. Horticultura Brasileira 27, 300–306.

Medina-Filho, H.P., Tanksley, S.D., 1983. Breeding for nematode resistance. In: Evans, D.A., Sharp, W.R., Ammirato, P.V., Yamada, Y. (Eds.), Handbook of Plant Cell Culture, vol. 1. MacMillan, New York, USA, pp. 904–923.

Milligan, S.B., Bodeau, J., Yaghoobi, J., Kaloshian, I., Zabel, P., Williamson, V.M., 1998. The root knot nematode-resistance gene *Mi* from tomato is a member of the leucine zipper, nucleotide binding, leucinerich repeat family of plant genes. The Plant Cell 10, 1307–1319.

Mills, N.J., Daane, K.M., 2005. Nonpesticide alternatives can suppress crop pests. California Agriculture 59, 23–28.

Momanyi, G., Maranga, R., Sithanantham, S., Agong, S., Matoka, C.M., Hassan, S.A., 2012. Evaluation of persistence and relative toxicity of some pest control products to adults of two native Trichogrammatid species in Kenya. Biocontrol 57, 591–601.

Nap, J.P., Escaler, P.L.J., Coner, A.J., 2003. The release of genetically modified crops into the environment. Plant Journal 33, 1–18.

Nault, B.A., Speese III, J., 2002. Major insect pests and economics of fresh market tomato in eastern Virginia. Crop Protection 21, 359–366.

Navajas, M., de Moraes, G.J., Auger, P., Migeon, A., 2013. Review of the invasion of *Tetranychus evansi*: biology, colonization pathways, potential expansion and prospects for biological control. Experimental and Applied Acarology 59, 43–65.

Ngowi, A.V.F., Mbise, T.J., Ijani, A.S.M., London, L., Ajayi, O.C., 2007. Smallholder vegetable farmers in North Tanzania: pesticides use practices, perceptions, cost and health effects. Crop Protection 26, 1617–1624.

Nilakhe, S.S., Chalfant, R.B., Phatak, S.C., Mullinix, B., 1982. Tomato fruitworm: development of sequential sampling and comparison with conventional sampling in tomatoes. Journal of Economic Entomology 75, 416–421.

Nombela, G., Williamson, V.M., Muñiz, M., 2003. The root-knot nematode resistance gene *Mi-1.2* of tomato is responsible for resistance against the whitefly *Bemisia tabaci*. Molecular Plant-Microbes Interactions Journal 16, 645–649.

NRC, 1996. Ecologically Based Pest Management: New Solutions for a New Century. National Research Council, National Academies Press, Washington, DC, USA.

NRC, 2010. The Impact of Genetically Engineered Crops on Farm Sustainability in the United States. National Research Council, National Academies Press, Washington, DC, USA, p. 250.

Nyambo, B., Ekesi, S., Chabi-Olaye, A., Sevgan, S., 2011. Management of alien invasive insect pest species and diseases of fruits and vegetables: experiences from East-Africa. Acta Horticulturae 911, 215–333.

Oatman, E.R., Kennedy, G.G., 1976. Methomyl induced outbreak of *Liriomyza sativae* on tomato. Journal of Economic Entomology 69, 667–668.

Oatman, E.R., Platner, G.R., 1971. Biological control of the tomato fruitworm, cabbage looper, and hornworms on processing tomatoes in southern California using mass releases of *Trichogramma pretiosum*. Journal of Economic Entomology 64, 501–506.

Oatman, E.R., Platner, G.R., 1978. Effect of mass releases of *Trichogramma pretiosum* against lepidopterous pests on processing tomatoes in southern California, with notes on host egg population trends. Journal of Economic Entomology 71, 896–900.

Oatman, E.R., Platner, G.R., Wyman, J.A., Van Steenwyk, R.A., Johnson, M.W., Browning, H.W., 1983. Parasitization of lepidopterous pests on fresh-market tomatoes in southern California. Journal of Economic Entomology 76, 452–455.

Osteen, C.D., Fernandez-Cornejo, J., 2012. Economic and policy issues of US agricultural pesticide use trends. Pest Management Science 69, 1001–1025.

Pease, C.G., Zalom, F.G., 2010. Influence of non-crop plants on stink bug (Hemiptera: Pentatomidae) and natural enemy abundance in tomatoes. Journal of Applied Entomology 134, 626–636.

Pedigo, L.P., Hutchins, S.H., Higley, L.G., 1986. Economic injury levels in theory and practice. Annual Review of Entomology 31, 341–368.

Perdikis, D., Kapaxidi, E., Papadoulis, G., 2008. Biological control of insect and mite pests in greenhouse solanaceous crops. European Journal of Plant Science and Biotechnology 2 (Special Issue 1), 125–144.

Perring, T.M., Farrar, C.A., Toscano, N.C., 1988. Relationships among tomato planting date, potato aphids (Homoptera: Aphididae), and natural enemies. Journal of Economic Entomology 81, 1107–1112.

Pietrantonio, P.V., Junek, T.A., Parker, R., Mott, D., Siders, K., Troxclari, N., Vargas-Camplis, J., Westbrook, J.K., Vassiliou, V.A., 2007. Detection and evolution of resistance to the pyrethroid cypermethrin and *Helicoverpa zea* (Lepidoptera: Noctuidae) populations in Texas. Environmental Entomology 36, 1174–1188.

Pinto-Zevallos, D.M., Vänninen, I., 2013. Yellow sticky traps for decision-making in whitefly management: what has been achieved? Crop Protection 47, 74–84.

Prokopy, R.J., Kogan, M., 2003. Integrated pest management. In: Resh, V.H., Cardé, R.T. (Eds.), Encyclopedia of Insects. Academic Press, Elsevier Inc., California, USA, pp. 4–9.

Rajasri, M., Lakshmi, K.V., Rao, R.D.V., Reddy, K.L., 2011. Effect of different mulches on the incidence of tomato leaf curl virus and its vector whitefly *Bemisia tabaci* in tomato. Acta Horticulturae 914, 215–221.

Rauf, A., Shepard, M., Johnson, M.W., 2000. Leafminers in vegetables, ornamental plants and weeds in Indonesia: surveys of host crops, species composition and parasitoids. International Journal of Pest Management 46, 257–266.

Reitz, S.R., 2002. Seasonal and within plant distribution of *Frankliniella* thrips (Thysanoptera: Thripidae) in North Florida tomatoes. Florida Entomologist 85, 431–439.

Riley, D.G., Pappu, H.R., 2004. Tactics for management of thrips (Thysanoptera: Thripidae) and tomato spotted wilt virus in tomato. Journal of Economic Entomology 97, 1648–1658.

Roditakis, E., Skarmoutsou, C., Staurakaki, M., 2012. Toxicity of insecticides to populations of tomato borer *Tuta absoluta* (Meyrick) from Greece. Pest Management Science 69, 834–840.

Roltsch, W.J., Mayse, M.A., 1983. Parasitic insects associated with Lepidoptera on fresh-market tomato in Southeast Arkansas. Environmental Entomology 12, 1708–1713.

Romeis, J., Shelton, A.M., Kennedy, G.G., 2008. Integration of Insect-Resistant Genetically Modified Crops Within IPM Programs. Springer, Dordrecht, Netherlands.

Rossi, M., Goggin, F.L., Milligan, S.B., Kaloshian, I., Ullman, D.E., Williamson, V.M., 1998. The nematode resistance gene *Mi* of tomato confers resistance against the potato aphid. Proceedings of the National Academies of Sciences 95, 9750–9754.

Rushing, J.W., Spell, L., Cook, W.P., 1995. Accumulation of pesticides in tomato packinghouse wastewater and the influence of integrated pest management on reducing residues. HortTechnology 5, 243–247.

Saker, M.M., Salsma, H.S., Salama, M., El-Banna, A., Abel Ghany, N.M., 2011. Production of transgenic tomato plants expressing *Cry* 2Ab gene for the control of some lepidopterous insects endemic in Egypt. Journal of Genetic Engineering and Biotechnology 9, 149–155.

Sato, M.M., de Moraes, G.J., Haddad, M.L., Wekesa, V.W., 2011. Effect of trichomes on the predation of *Tetranychus urticae* (Acari: Tetranychidae) by *Phytoseiulus macropilis* (Acari: Phytoseiidae) on tomato, and the interference of webbing. Experimental and Applied Acarology 54, 21–32.

Schuster, D.J., 1998. Intraplant distribution of immature lifestages of *Bemisia argentifolii* (Homoptera: Aleyrodidae) on tomato. Environmental Entomology 27, 1–9.

Schuster, D.J., 2002. Action threshold for applying insect growth regulators to tomato for management of irregular ripening caused by *Bemisia argentifolii* (Homoptera: Aleyrodidae). Journal of Economic Entomology 95, 372–376.

Schuster, D.J., 2004. Squash as a trap crop to protect tomato from whitefly-vectored tomato yellow leaf curl. International Journal of Pest Management 50, 281–284.

Schuster, D.J., Mann, R.S., Toapanta, M., Cordero, R., Thompson, S., Cyman, S., Shurtleff, A., Morris, R.F., 2010. Monitoring neonicotinoid resistance in biotype B of *Bemisia tabaci*, in Florida. Pest Management Science 66, 186–195.

Scott, S.J., McLeod, P.J., Montgomery, F.W., Handler, C.A., 1989. Influence of reflective mulch on incidence of thrips (Thysanoptera: Thripidae: Phlaeothripidae) in staked tomatoes. Journal of Entomological Science 24, 422–427.

Sengoda, V.G., Tsai, W., De La Pena, R.C., Green, S.K., Kenyon, L., Huges, J., 2012. Expression of the full-length coat protein gene of *tomato leaf curl Taiwan virus* is not necessary for recovery phenotype in transgenic tomato. Journal of Phytopathology 160, 213–219.

Sharma, H.C., Prabhakar, C.S., 2014. Impact of climate change on pest management and food supply. Integrated Pest Management. In: Aborl, D.P. (Ed.), Integrated Pest Management: Current Concepts and Ecologial Perspective. Elsevier, Amsterdam, Netherlands, pp. 23–35.

Sikora, E.J., Murphy, J.F., 2005. Identification and management of cucumber mosaic virus in Alabama. Acta Horticulturae 695, 191–194.

Silva, G.A., Picano, M.C., Bacci, L., Crespo, A.L.B., Rosado, J.F., Guedes, R.N., 2011. Control failure likelihood and spatial dependence of insecticide resistance in the tomato pinworm, *Tuta absoluta*. Pest Management Science 67, 913–920.

Srinivasan, K., Krishna Moorthy, P.N., Raviprasad, T.N., 1994. African marigold as a trap crop for the management of the fruit borer *Helicoverpa armigera* on tomato. International Journal of Pest Management 40, 56–63.

Stansly, P.A., Naranjo, S.E., 2010. *Bemisia* Bionomics and Management of a Global Pest. Springer, Dordecht, Netherlands.

Stavisky, J., Funderburk, J., Brodbeck, B.V., Olson, S.M., Andersen, P.C., 2002. Population dynamics of *Frankliniella* spp. and tomato spotted wilt incidence as influenced by cultural management tactics in tomato. Journal of Economic Entomology 95, 1216–1221.

Stern, V.M., Smith, R.F., van den Bosch, R., Hagen, K.S., 1959. The integrated control concept. Hilgardia 29, 81–101.

Stocker, T.F., Qin, D., Plattner, G.K., Tignor, M., Allen, S.K., Boschung, J., Nauels, A., Xia, Y., Bex, V., Midgley, P.M., 2013. Climate change 2013: the physical science basis. In: Working Group I Contribution to the Fifth Assessment Report of the Intergovernmental Panel on Climate Change. Cambridge University Press, Cambridge, UK, p. 1535.

Sun, Y.C., Yin, J., Chen, F.J., Wu, G., Ge, F., 2011. How does atmospheric elevated CO_2 affect crop pests and their natural enemies? Case histories from China. Insect Science 18, 393–400.

Torres-Vila, L.M., Rodríguez-Molina, M.C., Lacasa-Plasencia, A., 2003. Impact of *Helicoverpa armigera* larval density and crop phenology on yield and quality losses in processing tomato: developing fruit count-based damage thresholds for IPM decision-making. Crop Protection 22, 521–532.

Trumble, J.T., Alvarado-Rodriguez, B., 1993. Development and economic evaluation of an IPM program for fresh market tomato production in Mexico. Agriculture, Ecosystems and Environment 43, 267–284.

Ucko, O., Cohen, S., Ben-Joseph, R., 1998. Prevention of virus epidemics by a crop-free period in the Arava Region of Israel. Phytoparasitica 26, 313–321.

Ulrichs, C., Hopper, K.R., 2008. Predicting insect distributions from climate and habitat data. Biocontrol 53, 881–894.

Umeh, V.C., Onukwu, D., 2005. Development of environmentally-friendly tomato insect pest control options under tropical conditions. Journal of Vegetation Science 11, 73–84.

Urbaneja, A., Gonzalez-Cabrera, J., Arno, J., Gabarra, R., 2012. Prospects for the biological control of *Tuta absoluta* in tomatoes of the Mediterranean basin. Pest Management Science 68, 1215–1222.

Vacas, S., Alfaro, C., Primo, J., Navarro-Llopis, V., 2011. Studies on the development of a mating disruption system to control the tomato leafminer *Tuta absoluta* Povolny (Lepidoptera: Gelechiidae). Pest Management Science 67, 1473–1480.

van Houten, Y.M., Glas, J.J., Hoogerbrugge, H., Rothe, J., Blockmans, K.J.F., Simoni, S., van Arkel, J., Alba, J.M., Kant, M.R., Sabelis, M.W., 2013. Herbivory-associated degradation of tomato trichomes and its impact on biological control of *Aculops lycopersici*. Experimental and Applied Acarology 60, 127–138.

van Lenteren, J.C., 2000. A greenhouse without pesticides: fact or fiction? Crop Protection 19, 375–384.

van Steenwyk, R.A., Oatman, E.R., 1983. Mating disruption of tomato pinworm (Lepidoptera: Gelechiidae) as measured by pheromone trap, foliage, and fruit infestation. Journal of Economic Entomology 76, 80–84.

Venette, R.C., Koch, R.L., 2009. IPM for invasive species. In: Radcliffe, E.B., Hutchison, W.D., Cancelado, R.E. (Eds.), Integrated Pest Management: Concepts, Tactics, Strategies and Case Studies. Cambridge University Press, Cambridge, England, pp. 424–436.

Walgenbach, J.F., 1994. Distribution of parasitized and nonparasitized potato aphid (Homoptera: Aphididae) on staked tomato. Environmental Entomology 23, 795–804.

Walgenbach, J.F., 1997. Effect of potato aphid (Homoptera: Aphididae) on yield, quality and economics of staked tomato production. Journal of Economic Entomology 90, 996–1004.

Walgenbach, J.F., Estes, E.A., 1991. Economics of insecticide use on staked tomatoes in western North Carolina. Journal of Economic Entomology 85, 888–894.

Walker, G.P., Cameron, P.J., 1989. Status of introduced larval parasitoids of tomato fruitworm. New Zealand Plant Protection 42, 229–232.

Walker, G.P., Herman, T.J.B., Kale, A.J., Wallace, A.R., 2010. An adjustable action threshold using larval parasitism of *Helicoverpa armigera* (Lepidoptera: Noctuidae) in IPM for processing tomatoes. Biological Control 52, 30–36.

Walker, G.P., Lowell, R.N., Simonet, D.E., 1984a. Natural mortality factors acting on potato aphid (*Macrosiphum euphorbiae*) populations in processing-tomato fields in Ohio. Environmental Entomology 13, 724–732.

Walker, G.P., Madden, L.V., Simonet, D.E., 1984b. Spatial dispersion and sequential sampling of the potato aphid, *Macrosiphum euphorbiae* (Homoptera: Aphididae), on processing tomatoes in Ohio. Canadian Entomologist 116, 1069–1075.

Weskesa, V.W., de Moraes, G.F., Knapp, M., Delalibera, I.J., 2007. Interactions of two natural enemies of *Tetranychus evansi*, the fungal pathogen *Neozygites floridana* (Zygomycetes: Entomophthorales) and the predatory mite, *Phytoseiulus longipes* (Acari: Phytoseiidae). Biological Control 41, 408–414.

Weskesa, V.W., Knapp, M., Delalibera Jr., I., 2008. Side-effects of pesticides on the life cycle of the mite pathogenic fungus *Neozygites floridana*. Experimental and Applied Acarology 46, 287–297.

Westphal, M.I., Browne, M., MacKinon, K., Noble, I., 2008. The link between internal trade and the global distribution of invasive alien species. Biological Invasions 10, 391–398.

Wiesenborn, W.D., Trumble, J.T., Oatman, E.R., 1990. Economic comparison of insecticide treatment programs for managing tomato pinworm (Lepidoptera: Gelechiidae) on fall tomatoes. Journal of Economic Entomology 83, 212–216.

Wilson, L., Downes, S., Khan, M., Whitehouse, M., Baker, G., Grundy, P., Maas, S., 2013. IPM in the transgenic era: a review of the challenges from emerging pests in Australian cotton systems. Crop and Pasture Science 64, 737–749.

Wolfenbarger, D.A., Wolfenbarger, D.O., 1966. Tomato yields and leafminer infestations and a sequential sampling plan for determining need for control treatments. Journal of Economic Entomology 59, 279–283.

Yardim, E.N., Edwards, C.A., 2003. An economic comparison of pesticide application regimes for processing tomato. Phytoparasitica 31, 51–60.

Yoltas, T., Baspinar, H., Aydin, A.C., Yildirim, E.M., 2003. The effect of reflective and black mulches on yield, quality and aphid populations on processing tomato. Acta Horticulturae 613, 267–270.

Zalom, F.G., 2012. Management of arthropod pests. In: Davis, R.M., Pernezny, K., Broome, J.C. (Eds.), Tomato Health Management. The American Phytopathological Society, Saint Paul, Minnesota, USA, pp. 49–64.

Zalucki, M.P., Furlong, M.J., Adamson, D., 2009. The future of IPM: whither or wither? Australian Journal of Entomology 48, 85–96.

Zappala, L., Biondi, A., Alma, A., Al-Jboory, I.J., Arno, J., Bayram, A., Chailleux, A., El-Arnaouty, A., Gerling, D., Guenaoui, Y., Shaltiel-Harpaz, L., Siscaro, G., Stavrinides, M., Tavella, L., Aznar, R.V., Urbaneja, A., Desneux, N., 2013. Natural enemies of the South America moth, *Tuta absoluta*, in Europe, North Africa, and Middle East, and their potential use in pest control strategies. Journal of Pest Science 86, 635–647.

Zehnder, G.W., Linduska, J.J., 1987. Influence of conservation tillage practices on populations of Colorado potato beetle (Coleoptera: Chrysomelidae) in rotated and non-rotated tomato fields. Environmental Entomology 16, 135–139.

Zehnder, G.W., Sikora, E.J., Goodman, W.R., 1995. Treatment decisions based on egg scouting for tomato fruitworm, *Helicoverpa zea* (Boddie), reduce insecticide use in tomato. Crop Protection 14, 683–687.

Zhang, N., Liu, C., Yang, F., Dong, S., Han, Z., 2012. Resistance mechanisms to chlorpyrifos and F392W mutation frequencies in the acetylcholine esterase ace1 allele of field populations of the tobacco whitefly, *Bemisia tabaci*, in China. Journal of Insect Science 12, 41.

Zhender, G.W., Trumble, J.T., 1985. Sequential sampling plans with fixed levels of precision for *Liriomyza* species (Diptera: Agromyzidae) in fresh market tomatoes. Journal of Economic Entomology 78, 138–142.

Ziska, L.H., Blumenthal, D.M., Runion, G.B., Diaz-Soltero, H., 2011. Invasive species and climate change: an agronomic perspective. Climate Change 105, 13–42.

Section IV

Registration and Regulation

Chapter 17

Agricultural Pesticide Registration in the United States

Keith Dorschner, Daniel Kunkel, Michael Braverman

Rutgers University, Princeton, NJ, United States

1. LEGISLATIVE HISTORY AND BACKGROUND OF PESTICIDE REGULATION IN THE UNITED STATES

Pesticide use and registration in the United States is currently regulated by the United States Environmental Protection Agency (EPA), which was established in 1970 in the wake of increasing concerns regarding environmental pollution. The new Agency was cobbled together from various programs from within the Department of Health, Education and Welfare, Department of the Interior, Department of Agriculture, and the Atomic Energy Commission. This diverse collection of governmental units provided the EPA with the appropriate range of skills required to regulate pesticides within a single agency. EPA's regulatory authority has evolved over time as a result of a number of important statutes, keeping pace with the changing needs and demands of society.

Federal pesticide regulation began in the United States with the passage of the Federal Insecticide Act (FIA) in 1910, the purpose of which was to protect farmers and consumers from fraudulent or adulterated products. By the World War II era, chemistry had advanced to the point where numerous new compounds with agricultural value had significantly expanded the pesticide market. The FIA had ensured quality pesticide products but failed to anticipate the massive increases in pesticide use and the potential for environmental damage and human health risk concerns that followed. In response, the United States Congress passed the Federal Insecticide, Fungicide, and Rodenticide Act (FIFRA) of 1947.

As originally passed, FIFRA charged the United States Department of Agriculture (USDA) with the responsibility for pesticide regulation. FIFRA underwent a major revision in 1972 which transferred pesticide regulation authority to the EPA, with a greater emphasis toward protection of the environment and public health. FIFRA was then amended in 1988 (known as FIFRA 88) to allow many pesticide products registered prior to 1984 to be reevaluated and reregistered using modern standards. It was amended again in 1996 with the passage of the Food Quality Protection Act (FQPA). The FQPA further modernized the safety standards to account for cumulative and aggregate risk associated with pesticides. Aggregate risks consider all routes of exposure from a given pesticide, including food, water, occupational, companion animal, and household use. Cumulative exposure seeks to address the combined risk of pesticides with a common mode of toxicity. FQPA also implemented a 15-year chemical review cycle for every registered pesticide product. This ensures that pesticides will continue to meet the United States government's evolving safety standards.

The stricter safety standards under FQPA resulted in a large number of pesticides being either removed from the market or having use patterns modified to reduce exposure and risk. All crops, including major agronomic crops and small acreage specialty crops, were impacted. These actions by the EPA further encouraged pesticide registrants to develop new products that are deemed to have lower risk compared to older products. Almost all of the new pesticides registered since 1996 have superior human health and environmental risk profiles compared to their predecessors. This has allowed growers the opportunity to transition away from more hazardous pest management technologies while still retaining effective pest control. Another benefit of many of the new products is that they are more compatible with integrated pest management (IPM) principles.

The most recent amendments to FIFRA have been made in an effort to streamline and provide a more predictable EPA registration process. The Pesticide Registration Improvement Act (PRIA) was passed by Congress in 2003 and amended in 2007 and 2012. The PRIA basically provides a fee structure for EPA service to review pesticide registration requests. Registration actions in three divisions of EPA's Office of Pesticide Programs (OPPs) are subject to PRIA fees: Antimicrobials, Biopesticides and Pollution Prevention, and Registration. The category of action (work effort by EPA), the amount of the

Sustainable Management of Arthropod Pests of Tomato. http://dx.doi.org/10.1016/B978-0-12-802441-6.00017-6

pesticide registration service fee, and the corresponding decision review periods are all prescribed in the statute (http://www2.epa.gov/pria-fees/fy-201415-fee-schedule-registration-applications). The goal of this amendment is to create a more predictable evaluation process for affected pesticide decisions, and couple the collection of individual fees with specific decision review periods, or timelines, which provide resources to the three divisions noted. The legislation also promotes shorter decision review periods for reduced-risk pesticide applications. The shorter review period is meant to encourage the submission of pesticides that offer less risk to man and the environment than currently registered alternative products.

The Federal Food, Drug, and Cosmetic Act (FFDCA, originally passed in 1938) is another significant law regulating pesticides in the United States. This law authorizes the setting of tolerances (otherwise known internationally as Maximum Residue Limits, or MRLs) for pesticide residues in foods, usually expressed in parts per million. A tolerance level is the legally allowed level of pesticide residues in or on a food or animal feedstuff after use, according to Good Agricultural Practices. In practice, a pesticide use which follows EPA-registered label directions will not result in residues of that pesticide on food or feed that exceed the tolerance level. The tolerance is specific for commodities or groups of commodities. In addition to being part of the EPA's safety risk assessment, the tolerance level is also used for enforcement purposes, helping ensure and monitor legal usage. In the absence of a tolerance level (or exemption from the requirement of a tolerance) any pesticide residues in food or feed are considered illegal, as are pesticide residues that exceed their established tolerance. Such foods are subject to seizure by the United States or state governments, whether domestically grown or imported.

In most instances in the United States, biopesticides do not have specific numerical tolerances, but are exempt from tolerance based on the lack of a toxic endpoint, and quite often, their innate presence in the environment. In some cases when a biopesticide product naturally occurs in the environment it is not possible to distinguish residues from an application of the biopesticide from that which can be naturally present on the commodity.

To make a registration decision, EPA examines the ingredients of a pesticide along with the site or crop on which the product will be used, as well as other parameters such as the amount, frequency, timing of application, and storage and disposal practices. EPA evaluates the pesticide to ensure that it will not have unreasonable adverse effects on humans, environment, and non-target species. This evaluation process requires the product registrant to provide certain data to EPA for review in order to make a registration decision, commonly known as a safety finding. A pesticide cannot be legally used in the United States until it has been reviewed and registered with EPA's OPPs.

2. ORGANIZATIONAL STRUCTURE OF EPA'S OFFICE OF PESTICIDE PROGRAMS

The EPA OPPs has the primary responsibility of accepting pesticide product registration applications, reviewing those applications, and issuing final product registrations. The OPP has nine divisions, plus the Office of the Director (http://www.epa.gov/pesticides/contacts/index.htm#office).

Antimicrobial pesticides are regulated through EPA's Antimicrobial Division (AD). The antimicrobial pesticides regulated by AD are not intended for use in foods and include hard surface disinfectants, sanitizers, and preserving agents. These products can be extremely important considering their use in public health to control human pathogens and vectors of human pathogens. This division handles all product registrations, amendments, and reregistrations of such pesticide products. AD reviews all data within the division, does risk assessments, and issues registration decisions directly.

Known as a science division, the Biological and Economic Analysis Division (BEAD) supports pesticide regulation by providing use-related information and economic analysis. This division provides details on how pesticides are actually used which is valuable in exposure assessments, considering pesticide needs, and impacts of regulatory decisions. BEAD has laboratories which validate analytical methods used to measure pesticide residues and test public health antimicrobial products to ensure they work as intended. The laboratories serve as technical experts to states and support compliance monitoring by a number of federal entities, including FDA.

Biological pesticides are regulated by the Biopesticide and Pollution Prevention Division (BPPD). Biopesticides include microbials, plant incorporated protectants, and other biotechnology-related products. Biochemical biopesticides are also regulated in this division, which include plant extracts, pheromones, minerals, and other naturally occurring materials of limited toxicological concern. Such products routinely require a preclassification process to determine the proper EPA division within which the pesticide will be evaluated and regulated.

The classification of biochemical biopesticides is distinguished from conventional pesticides which are reviewed in the Registration Division (RD). BPPD also seeks to limit pesticide impacts by promoting IPM activities, coordinating a Pesticide Environment Stewardship Program, and providing grants. Similar to AD, BPPD reviews all data within the division and can do its own risk assessment and make its own registration decisions. In general, the overarching types of health and environmental data for biopesticides and conventional pesticides are similar. The primary difference is that biopesticide requirements are set up in a tiered approach, with Tier 1 data being acute studies and then higher tier studies being subchronic and chronic, etc. In a vast majority of cases, registrations are mostly completed with data at the Tier 1 level.

Frequently, these requirements can be fulfilled through information in the public scientific literature since many biopesticide active ingredients are commonly present in the environment with significant ongoing human exposure.

There are two other significant science divisions at EPA. The Environmental Fate and Effects Division (EFED) is responsible for reviewing and assessing data submitted to EPA related to environmental properties and effects of conventional pesticides. The other important science division is Health Effects Division (HED). HED evaluates the data submitted to EPA concerning exposure and impacts of pesticide registration on humans and domestic animals.

All conventional chemical pesticides are regulated by the RD. This division is responsible for registration approvals, registration amendments, setting pesticide tolerances, approving experimental use permits (EUPs), and emergency exemptions (in the case of emergency pest problems that cannot be addressed with products already registered). RD relies almost exclusively on the scientific reviews generated by BEAD, EFED, and HED in their regulatory decision process.

By law, all pesticide registrations must be periodically reviewed to ensure safety in light of the evolving knowledge of pesticide impacts. These reviews are conducted by the Pesticide Re-Evaluation Division (PRD). PRD has evolved out of the EPA response to the FIFRA 88 registration review process. It is a relatively new division in which EPA has integrated reregistration and tolerance reassessment activities to more effectively accomplish the goals outlined in FQPA.

The remaining EPA divisions are the Field and External Affairs Division (FEAD), providing links to many of the United States regional EPA offices, and the Information Technology and Resources Management Division (ITRMD). FEAD is responsible for Congressional interactions with EPA, coordination with and assistance to states and tribal authorities, international field activities, and other EPA communication and outreach programs. ITRMD provides information support, records management, pesticide incident and adverse impacts monitoring, and computer technical support. OPP's personnel and budget are also managed by this division.

3. TYPES OF PESTICIDE REGISTRATIONS UNDER FIFRA

There are several types of registration actions through which pesticides can be used legally in the United States:

3.1 Federal Registration Actions

Under Section 3 of FIFRA, EPA can register pesticides for use throughout the United States, although some pesticides are registered by EPA for more limited use in only certain states. In addition, states, tribes, and territories can place further restrictions on EPA-registered pesticides used or sold within their own jurisdictions. Therefore, the final label, although initially approved by EPA, is granted by these authorities for local use. There may be a significant lag time between label approval by EPA and final registration in a state and in some cases, a state may decline to register the use altogether.

3.2 Experimental Use Permits

Under Section 5 of FIFRA, EPA can allow manufacturers to field test pesticides under development. Manufacturers of conventional chemical pesticides are required to obtain Experimental Use Permits (EUPs) before testing new pesticides or new uses of registered pesticides, if they conduct experimental field tests on 10 acres or more of land or one acre or more of water. Experimental use of biopesticides also requires EUPs when used in these settings.

3.3 Emergency Exemptions

Under Section 18 of FIFRA, EPA can allow state and federal agencies to permit the unregistered use of a pesticide in a specific geographic area for a limited time if emergency pest conditions exist. Usually, this arises when growers and others encounter a pest problem on a site for which there is either no effective registered pesticide available, or for which there is a registered pesticide that would be effective but is not yet approved for use on that particular site or crop. These exemptions can also be approved for public health and quarantine reasons.

3.4 State-Specific Registrations

Under Section 24(c) of FIFRA, states can register a new pesticide product for any use, or a federally registered product for an additional use (or change in use pattern), as long as there is both a demonstrated "special local need," a tolerance for the pesticide (or exemption from a tolerance), or other clearance under FFDCA (that would not change the safety finding of the product or use). However, EPA must review the 24(c) label and can disapprove a State's special local need registration if they determine that it is not appropriate.

4. REGISTRATION PROCESS FOR A CONVENTIONAL (I.E., MAN-MADE) PESTICIDE IN REGISTRATION DIVISION

The Registration Process has five main steps: Front-End Processing, Science Reviews, Risk Assessment and Peer Review, Risk Management and Regulatory Decision, and finally the Federal Register publication (Fig. 17.1).

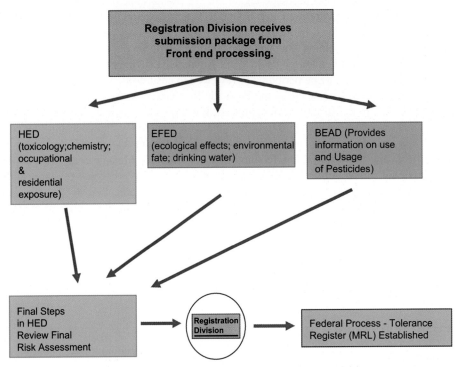

FIGURE 17.1 Pesticide review process in the registration division.

The front-end process takes place in the ITRMD where the registration submission (commonly referred to as "package") is reviewed for completeness as outlined in EPA's Pesticide Registration Notice 2011–3 (https://www.epa.gov/sites/production/files/2014-04/documents/pr2011-3.pdf) and is billed according to the PRIA. The package is then provided to RD for further processing. There is a recommended report format that EPA strongly encourages that can be found at: https://www.epa.gov/pesticide-registration/prn-2011-3-standard-format-data-submitted-under-fifra-and-certain-provisions. The use of these templates provides ease of review, consistency, quality, efficiency, and transparency. These templates provide applicants with important information regarding the specific data elements for each study and identifies key components used by OPP to evaluate the suitability of a study for regulatory purposes. Submitters are further encouraged to submit the reports electronically.

5. SCIENCE REVIEWS

Once the package is received and logged in by RD, it moves to the Science Divisions for review. The Science Review process is based on Standard Operating Procedures and internal EPA policies and guidance. All data submitted to EPA must be from studies conducted according to Good Laboratory Practice standards (http://www.ecfr.gov/cgi-bin/text-idx?SID=e8e8c217a719a588f03e927b0c618fba&mc=true&node=pt40.26.160&rgn=div5) or a clear indication of why data may not meet these requirements. HED does a primary review involving animal toxicity data to identify toxicological endpoints for humans. These include short-term (or acute), intermediate-term (subchronic), and long-term exposure (chronic) studies. The routes of exposure tested include not only oral and dermal but also inhalation exposure. The EPA evaluates the data for toxic effects with regard to cancer, organ toxicity, reproductive effects, birth defects, and endocrine effects.

For the typical pesticide there is also a series of environmental fate studies required for the primary review by EFED. These lab and field studies characterize persistence, mobility, and bioaccumulation of the product. Water resources data are used to help with modeling and monitoring data to estimate the potential exposure of these pesticides and to help refine estimates of exposure to humans. Other ecological studies are needed to estimate risk to terrestrial and aquatic organisms;

therefore, studies are provided to determine acute and chronic toxicity. The data are initially reviewed by OPP scientists or contractors, and peer reviews are performed by EPA senior scientists. Depending on results from primary studies, additional, more specific studies may be required. Data reviews along with toxic endpoint recommendations are forwarded to the appropriate EPA risk assessor. For a complete listing of the testing guidelines please see: https://www.epa.gov/test-guidelines-pesticides-and-toxic-substances/final-test-guidelines-pesticides-and-toxic.

BEAD contributes to the process by providing science reviews concerning what sites are being treated and for which pests. They examine how the pesticide is being applied, including equipment, timing, safety precautions, and limitations. Also provided are assessments of the quantity of pesticide used (such as rates of application, area treated, and total pesticide amounts to be applied). The information supplied by BEAD can help determine any additional data requirements and estimates of human and environmental exposure for risk analysis.

6. RISK ASSESSMENT

The universal equation for risk is: *Risk = Hazard × Exposure*. The types of risk assessment vary depending on use pattern and type of compound being registered. EPA assesses the human risks through a variety of sources, including diet (food and water), residential (oral, dermal, and inhalation), and occupational (dermal and inhalation). All of these consider short-term, intermediate-term, and long-term (chronic) carcinogenic effects. There may also be cumulative risk assessments for chemicals that have a common mechanism of toxicity (for example, all organophosphate insecticides share a common mode of toxicity).

The first aspect of risk assessments are informed by the science reviews. These data reviews identify and quantify potential health effects that may occur from different types of pesticide exposure and are used to set the toxicological endpoints, LOEL (the lowest observable effect level) which is the lowest data point at which a toxic or adverse effect is observed, and NOEL (the no observed effect level) which is the highest data point at which there is no observed toxic or adverse effect. These are considered in the context of the reference dose (RfD), which is the daily oral exposure of a pesticide, uncertainty factors and sensitive population subgroups included, that is unlikely to have any deleterious effects during a lifetime.

EPA considers the full spectrum of a pesticide's potential health effects. These studies also examine dose responses that provide dose levels at which adverse effects are observed in test animals. These dose levels then can be used to calculate an equivalent dose in humans. The potential hazard to humans is estimated by extrapolating from animal toxicity studies. EPA then applies safety or uncertainty factors to further protect humans and to account for any variability that occurs when extrapolating data from animals to humans. The safety factors include a 10× factor for interspecies variation as well as a 10× factor for intraspecies (individual) variation (a total 100-fold protection factor). The FQPA also requires an additional 10× safety factor for special sensitivity to infants or children. However, this safety factor may be lowered to 3× or eliminated if data are available to show that there is not an increased risk to these sensitive population subgroups.

Human pesticide exposure is determined by data and modeling, water exposure is determined by using models and monitoring data. Spray drift models and data are used as well as dietary exposure and occupational exposure models. Occupational and residential exposure evaluates dermal and inhalation exposures both during and post-application for situations such as agricultural workers who mix and load chemicals, or for home owners who use products on their lawns and home gardens. Since the passage of FQPA, the EPA essentially considers all possible routes of exposure (known as aggregate exposure).

The EPA also conducts ecological (terrestrial and aquatic) assessments that consider acute and chronic exposures. This risk assessment is intentionally conservative to account for the wide ranges in pesticide sensitivity observed in the environment. EPA currently has an increasing focus on protecting endangered species (https://www.epa.gov/endangered-species) and pollinators. While the human risk assessment is conducted within the HED, the ecological risk assessment is conducted in the EFED. EFED evaluates risk to non-target organisms (i.e., wildlife, both aquatic and terrestrial organisms) and determines whether additional analysis is needed to assess risk to endangered species. They also assess the impacts to surface water and groundwater quality.

The EPA must assess pesticide risks to endangered species. The EPA and its federal partners, the USDA and the United States Fish and Wildlife Service (FWS) must consult to refine the risk estimate for endangered species. They are currently using an interim approach provided by recommendations of National Academy of Sciences' National Research Council report, entitled, "Assessing Risks to Endangered and Threatened Species from Pesticides" (http://www.nap.edu/catalog/18344/assessing-risks-to-endangered-and-threatened-species-from-pesticides). This new approach provides species protections while also streamlining the consultation process and allowing ample opportunity for stakeholder engagement as early as possible. It is a difficult balance and can be resource-intensive at times, especially when considering the specialty fruit and vegetable crops which may be grown in close proximity to areas protected for endangered-species habitat.

In assessing environmental risks for pesticides in water, including risks to people and to aquatic species, EPA uses a multistep process that includes computer models together with product-specific and species-specific data (https://www.epa.gov/water-research/methods-models-tools-and-databases-water-research#models). Estimating the deposition of pesticide constituents on a water surface following an application is frequently a first step, and the accepted models for this are the United States Forest Service's AGDISP and the closely related AgDRIFT; for drift beyond the computational limits of these programs, a linked Gaussian extension model has also been developed. Estimating the concentrations of pesticide materials in water requires a suite of models that take into account diffusion, chemical transformations, binding into sediment, and other chemical and physical processes. GENEEC (GENeric Estimated Environmental Concentration) is the Tier 1 screening model for assessing exposure of pesticides to aquatic organisms and the environment, and FIRST (FQPA Index Reservoir Screening Tool) provides a Tier 1 evaluation of pesticide levels in drinking water. For more refined evaluations, EPA uses the linked field-scale models PRZM-3 and EXAMS II (collectively known as PRZM-EXAMS); for aquatic assessments, PRZM-EXAMS uses a standard pond scenario, while drinking water assessments are based on an Index Reservoir and on Percent Crop Area exposed. Finally, SCI-GROW is used by EPA to model pesticide concentrations in groundwater. Regardless of the model(s) used, estimated concentrations of pesticide constituents are compared with toxicology data to move from estimates of exposure to estimates of risk.

Throughout the science review and risk assessment process, there are a series of technical reviews and peer reviews that take place. Peer review panels comprise senior scientists with specialized expertise in topics being assessed. These scientists comprise the Science Assessment Review Committees (SARCs) that are used extensively at OPP. They ensure consistency on decisions and methodology, make sure there are no gaps in data or information, conduct quality review checks of studies, and consider updated guidance documents. The SARCs will oversee decisions on a variety of issues in OPP such as toxic effects including toxicological endpoint selection and cancer considerations, FQPA Safety Factor, pesticide metabolism, and routes of pesticide exposure.

In certain cases, the EPA will consider having a Scientific Advisory Panel (SAP) that consist of experts outside of the EPA to resolve certain issues, or to comment on new or existing policies. For example, the OPPs used a SAP to develop policy for the use of FQPA 10× Safety Factor for children, statistical methods for use of composite data in acute dietary exposure assessment, and use of watershed-derived percent crop areas as refinement tool in FQPA drinking water exposure assessments for tolerance reassessment. This SAP comprised experts from areas such as the National Institute of Health, medical universities. The recommendations from this panel were responsible for providing much of the information used to develop many of EPA's current procedures.

7. RISK REFINEMENT

After the passage of the FQPA, the reference dose for a given pesticide has been compared to that of a "risk cup." The risk cup can be thought of as containing the total extent of all exposures to a pesticide. Potential exposures that are added include water, residential, occupational, and food. When the proverbial cup is full, the maximum RfD has been reached and there are no more exposures permissible. During the initial review of a new product, the EPA will assume that residues will occur at the level of the registrant's proposed tolerance, that all the crops to be registered have the maximal residues, and that 100% of the crop acreage will be treated. This is a very broad, overestimate of the actual exposure, but if EPA finds the risk associated with such high exposure acceptable, there will be no need to further refine the estimate. On the other hand, if the risk cup "overflows" using this scenario, then EPA has the ability to do refinements that may still allow for new or additional uses to be registered.

EPA RD considers all of the "Risks and Benefits" of the product and develops methods to manage these risks. This process often requires a very close working relationship with BEAD. In many cases it requires a decision-making process that may also require additional data. These data may include actual treated acre estimates or actual data from water monitoring studies. The pesticide product risk manager weighs results of the risk assessment against predetermined levels of concern. Every risk category (e.g., occupational, ecological, aggregate) has a predetermined level of concern and if triggered will cause the agency to further analyze that risk or use. EPA continues to determine whether additional data could reduce the risk estimate. They also include an economic assessment based upon crop loses, projected use, and product performance. If the risk estimates are as refined as possible, then the EPA product manager seeks risk mitigation measures, which may include reduced application rates, spray drift control measures, use of additional protective clothing, reentry restrictions, or other label restrictions.

Each phase of refinement is called a "tier" and there are three basic levels starting with no refinements at "Tier I," to the risk assessment being highly refined at "Tier III" (Fig. 17.2). If, through the refinement and mitigation process, risks are no longer of concern, then the registration process will likely proceed at that point and eventually be granted.

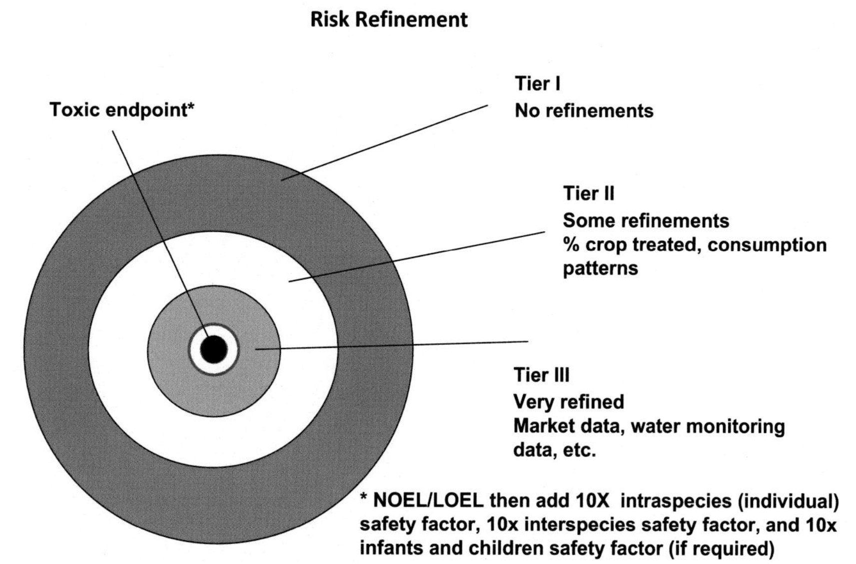

Risk Refinement

Toxic endpoint*

Tier I
No refinements

Tier II
Some refinements
% crop treated, consumption
patterns

Tier III
Very refined
Market data, water monitoring
data, etc.

* NOEL/LOEL then add 10X intraspecies (individual)
safety factor, 10x interspecies safety factor, and 10x
infants and children safety factor (if required)

FIGURE 17.2   Risk refinement example.

8. FEDERAL REGISTER PUBLICATION PROCESS

The Federal Register publication process is required for two phases of the registration process. The first phase is the notification process which usually occurs shortly after the registration package has been submitted to EPA. This Federal Register notice announces the receipt of an application to register pesticide products either containing new active ingredients or new uses for already registered products that have not been included in any previously registered products under the provisions of section 3(c)(4) of the FIFRA, as amended (covered under 40 *Code of Federal Regulations* part 180). This posting allows for public comment and provides a brief description of the studies that have been submitted and what the submitter is proposing for a pesticide tolerance (MRL) level. The second phase of the Federal Register publication process occurs once a decision has been made, and approved by EPA management. This announcement is made public through a Federal Register Final Rule.

The final process requires a number of review and comment periods among the EPA staff at RD (along with consultation with HED and EFED) with the office of General Council. Once RD has completed, prepared, and signed the final document, it is then prepared and presented to the Federal Register staff for publication as the Final Rule. The final signature for new active ingredients is provided by the OPPs Director. The final registration (label) is provided at the same time the Federal Register Final Rule is published and the process is complete.

9. RECENT ADVANCES IN THE REGISTRATION PROCESS

9.1 Global Joint Reviews

Several years ago, the EPA entered into a cooperative relationship with a number of other national regulatory agencies to begin conducting joint reviews for new pesticide submissions. A global joint review is an evaluation of a pesticide submission through work sharing between two or more countries. The participating regulatory authorities establish a schedule for the review and determine the work split for the various scientific disciplines. The assigned countries do the primary study reviews, and then a secondary review (also known as a peer review) is completed by each country. After the reviews are agreed upon, each country uses the data reviews to develop individual risk assessments for the active ingredient, taking into account the specific uses being registered in their jurisdiction.

The goals of global joint reviews are to save resources, harmonize review requirements to the extent possible, and harmonize toxicological end points and MRLs (i.e., tolerances) to the greatest extent possible. Since the initiation of global joint reviews, there has been significant progress in harmonizing MRLs and reducing trade irritants associated with pesticide usage.

A recent example of a global joint review is noted in Table 17.1, where the insecticide chlorantraniliprole (from E. I. du Pont de Nemours and Company) was jointly reviewed by five countries. The various countries split the data review process, then shared the conclusions, and registered this new, very low-risk insecticide in less than 2 years (submitted in 2007

TABLE 17.1 Global Joint Review Assignments for Chlorantraniliprole

Chemical	Toxicology	Residue Chemistry	Ecotoxicology	Environmental Fate	Product Chemistry
Chlorantraniliprole	United States	Australia	United Kingdom	Ireland	Canada

and registered in 2009). As a result of this stunning achievement, the MRLs established globally were nearly harmonized (i.e., identical in value for individual crops). Growers could use this new safer product with confidence that residues on exported agricultural commodities would not be illegal in the importing country. The product was reviewed in Codex (Joint meeting of Pesticide Residues) shortly thereafter and Codex MRLs were set, as well as registrations in other countries. Chlorantraniliprole is currently registered in over 100 countries around the globe.

9.2 Crop Grouping

Residue extrapolation is the process by which the residue levels on representative commodities are utilized to estimate residue levels on related commodities in the same commodity group or subgroup. This is also known as crop grouping. Representative commodities are chosen based on their commercial importance, and the similarity of their morphology and residue characteristics to other related commodities in the grouping. Ideally, representative commodities are the most economically important commodities in production and/or consumption and have a greater dietary burden than other crops belong to the grouping.

Residue extrapolation is a common consideration utilized by regulators internationally. The practice ensures that data requirements are only at a level that is scientifically justified in conducting risk assessment and to ensure the regulatory process does not become unnecessarily burdensome, especially for minor crops. (EPA considers crops grown under 300,000 acres to be minor crops.) Crop grouping not only reduces the burden for regulators but also for data generators. Minor crop producers have benefited most from this process because lower registration costs encourage more registrations on minor crops and the regulators have fewer reports to review. See http://www2.epa.gov/pesticide-registration/pesticide-tolerance-crop-grouping-revisions for an update on the EPA process to expand the existing crop group regulations.

10. SUMMARY

Pesticides are a key component of modern day pest management practices. Utilized in a vast number of agricultural commodities, it is critical that these substances are not only effective against the target pest, but safe for applicators, consumers, and the environment. With over a century of pesticide regulation in the United States, the processes of insuring this safety are well developed and presented above. The laws governing pesticide regulation have evolved to ensure that the highest standards are utilized. While the system in the United States is highly sophisticated with rigorous risk assessments, it is also transparent and open to external observation and public scrutiny. At this same time, the global community has developed a high level of sophistication to provide scrutiny not only at the national level within each country, but also at regional (NAFTA) and global (Codex and OECD) scales. As the liberalization of trade and interdependence among countries increases in an effort to provide stable food supplies across the world, even greater cooperation and harmonization among countries will be needed. There will be greater effort placed on newer products and biopesticides. To that end, the cooperative efforts in pesticide regulation will serve to harmonize standards resulting in sustainable, lower risk products that will support global food security.

Index

Printed in the United States
By Bookmasters